Development of Engineering and Innovation in the Scenarios of the 4th Industrial Revolution (EXPOTECNOLOGIA 2020)

IOP Conference Series: Materials Science and Engineering
Volume 1154

Cartagena, Colombia
4 - 6 November 2020

Editors:

Cesar Gomez Sierra
Yesid Tarriba Lezama

ISBN: 978-1-7138-4622-2
ISSN: 1757-8981

Printed from e-media with permission by:

Curran Associates, Inc.
57 Morehouse Lane
Red Hook, NY 12571

Some format issues inherent in the e-media version may also appear in this print version.

Institute of Physics
Dirac House, Temple Back
Bristol BS1 6BE UK

Phone: 44 1 17 929 7481
Fax: 44 1 17 920 0979

techtracking@iop.org

Additional copies of this publication are available from:

Curran Associates, Inc.
57 Morehouse Lane
Red Hook, NY 12571 USA
Phone: 845-758-0400
Fax: 845-758-2633
Email: curran@proceedings.com
Web: www.proceedings.com

TABLE OF CONTENTS

IOP Conf. Series: Materials Science and Engineering 1154 (2021) 011001 doi:10.1088/1757-899X/1154/1/011001

Expotecnología 2020

Organized by

UNITECNAR

FACULTAD DE CIENCIAS DE LA INGENIERÍA UNITECNAR

Sponsors:

ACIEM Capítulo Bolívar · COPNIA · ACOFI Asociación Colombiana de Facultades de Ingeniería · LACCEI · confedi

Robotic People · LABTECH Knowledge Engineering www.labtech.org · Impointer · CVR Electronica · MAPEI

SIEMENS · PANDUIT · STD

IOP Conf. Series: Materials Science and Engineering 1154 (2021) 011001 doi:10.1088/1757-899X/1154/1/011001

Expotecnología 2020 was an event that took place in the city of Cartagena de Indias, from November 4 to 6, 2020, with the slogan "Development of Engineering and Innovation in the scenarios of the 4th Industrial Revolution." This eighteenth version was a meeting that brought together the scientific and academic community in spaces for the dissemination and transfer of knowledge, where innovation was the essence and a fundamental part of the event. Expotecnología was created in the Faculty of Engineering Sciences - FACI, of the Antonio de Arévalo University Foundation - UNITECNAR and in this version it maintained as mission, the development of the interaction between science, technology and society, through establishing of knowledge networks and academic and professional cooperation. The main objective was to address the challenges that emerge from the everyday life of the engineering community, the problems immersed in the development of projects and programs of Higher Education Institutions (IES), organizations and research centers at the national and international level. In the development of the day of presentations in a virtual way, it was possible to promote the dissemination of the research work of professionals, teachers and students from all areas of engineering, framed in four (4) thematic axes, as follows:

- New technologies: Big data, Artificial Intelligence, Business Intelligence, Software Engineering, Education and new technologies and Mobile Applications.
- Petrochemicals and process control: Materials Science, Instrumentation and Control Systems, Robotics and Automation and Industrial Maintenance.
- Electricity, electronics and telecommunications: IoT, Fiber optics, Digital Signal Processing, Wireless Networks, Mobile Communications, Power Systems, Energy Storage, Smart Grids, Renewable Energies, Operation and Distribution, Energy Dispatch, Energy Markets
- Construction Summit: Modern Building Systems and Computational Modeling.

On this occasion, the event was attended by 483 people, representatives from 25 national and 12 international universities. 97 virtual presentations were made, of which 66 articles were selected for this volume of the IOP Conference Cycle: Materials Science and Engineering and 19 articles for the book Engineering, Innovation and the fourth industrial revolution., After a process of Peer Review. The plenary conferences were in charge of 2 national and 8 international speakers.

IOP Conf. Series: Materials Science and Engineering 1154 (2021) 011001 doi:10.1088/1757-899X/1154/1/011001

Invited speakers:

PhD. Javier Rosero García, National University of Colombia

PhD. Jesús Camacho Yamayo, National University of Colombia

PhD. Adriana Páez Pino, president of GEDC Latam

PhD. Aryanne Quintal, Organization of American States - OAS

PhD. Carlos Peralta Delgado, General Manager Musuxchay Consultores at Prospectiva S.A.C

PhD. Laura Romero Robles, Director of Nanotechnology Engineering and Chemical Sciences at ITESM Campus Monterrey

PhD. Liliana Estela Rathmann, Dean of the Faculty of Engineering - Universidad Atlántida Argentina

PhD. María Larrondo Petrie, Executive Director of LACCEI.

PhD. Roberto Giordano Lerena, Dean of the Faculty of Engineering - FASTA University

PhD. Wanessa Carla Gazoni, University of Campinas - Brazil.

Organizing Committee

President

Msc. Libis Valdez Cervantes, Antonio de Arévalo University Foundation

Editorial committee

Msc. César Gómez Sierra, Antonio de Arévalo University Foundation,
Msc. Yesid Tarriba Lezama, Antonio de Arévalo University Foundation

IOP Conf. Series: Materials Science and Engineering 1154 (2021) 011001 doi:10.1088/1757-899X/1154/1/011001

Technical committee

Msc. César Gómez Sierra, Antonio de Arévalo University Foundation,
Msc. Yesid Tarriba Lezama, Antonio de Arévalo University Foundation
Msc. Libis Valdez Cervantes, Antonio de Arévalo University Foundation

Scientific Committee

President

Msc. César Gómez Sierra, Antonio de Arévalo University Foundation,

Coordinator

Msc. Yesid Tarriba Lezama, Antonio de Arévalo University Foundation

Special thanks

We greatly thank all the authors for their valuable participation in Expotecnología 2020 and all the peer reviewers for their commitment and dedication in reviewing the papers to be published in IOP and Book chapters.

Msc. César Gómez Sierra - Msc. Yesid Tarriba Lezama
Editors of the Proceedings of EXPOTECNOLOGIA 2020

IOP Conf. Series: Materials Science and Engineering 1154 (2021) 011002 doi:10.1088/1757-899X/1154/1/011002

Peer review declaration

All papers published in this volume of *IOP Conference Series: Materials Science and Engineering* have been peer reviewed through processes administered by the Editors. Reviews were conducted by expert referees to the professional and scientific standards expected of a proceedings journal published by IOP Publishing.

- **Type of peer review:**

  Double-blind

  For the review of the papers, a call was made by email voluntarily requesting some pairs for this purpose. After having the peer bank, the assignment was made on the EasyChair platform according to their profiles and the congress to which they were applying the different articles. After having all the opinions of the peers, we proceeded to send the authors for their review and correction.

- **Conference submission management system:**

  All the articles that were going to be submitted, began to be received from the authors, through the EasyChair platform, likewise they were assigned to a temporary agenda in which the date and time of the conference were scheduled within the framework of the event. As it was carried out virtually, we used the Zoom platform for all the presentations during the three days that the event lasted.

- **Number of submissions received:** 106
- **Number of submissions sent for review:** 106
- **Number of submissions accepted:** 66
- **Acceptance Rate (Number of Submissions Accepted / Number of Submissions Received X 100):** 62.26%
- **Average number of reviews per paper:** 2
- **Total number of reviewers involved:** 40
- **Any additional info on review process:** The review was carried out under the guidelines, policies and IOP criteria described in the document: Proceedings peer

review policy. It should be noted that through the EasyChair platform, the reviewers had access to some questions in a questionnaire where they submitted their comments and the authors could read and update them.

Evaluation criteria.

* Valid in time.
* Conflict of interests
* Pertinence according to the event.
* Confidentiality and legitimacy

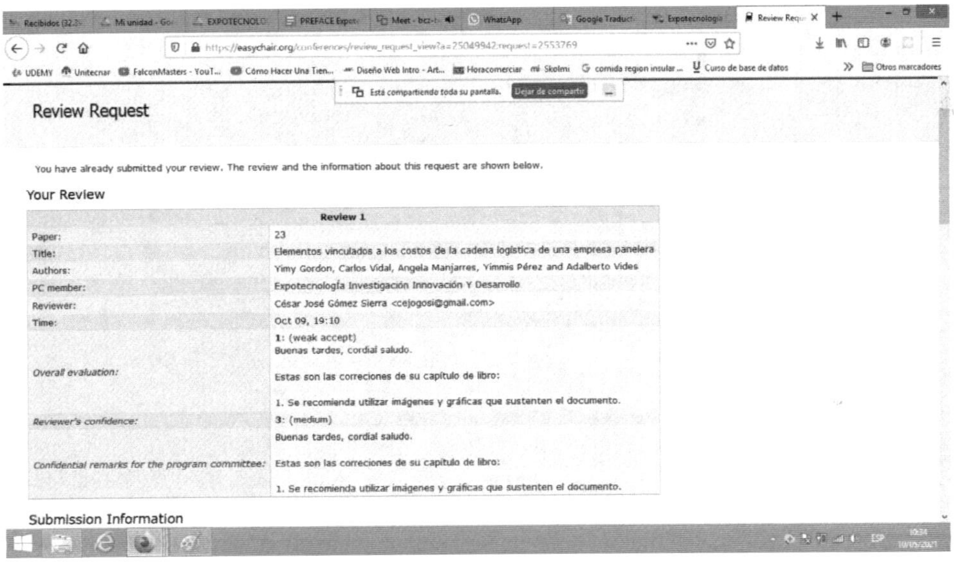

- **Contact person for queries:** Cesar Gomez Sierra y Yesid Tarriba Lezama
- **Contact email:** expotecnologia@unitecnar.edu.co

IOP Conf. Series: Materials Science and Engineering 1154 (2021) 012001 doi:10.1088/1757-899X/1154/1/012001

Robotics in Shared Service Centers in Emerging Economies

James Pérez-Morón and Lina Marrugo-Salas.
School of Business, Universidad Tecnológica de Bolivar, Campus Tecnológico Km 1. Vía Turbaco, Cartagena de Indias, Colombia. jperez@utb.edu.co y lmarrugo@utb.edu.co

Corresponding author´s email address: jperez@utb.edu.co

Abstract. The purpose of this paper provides a better understanding of Robotic Process Automation- RPA phenomena on the accounting / management activities of Shared Service Centers-SSCs in emerging economies specifically in Latin America - LATAM. Due to the rawness of RPA concept in this attractive region for SSCs, this study aims to explore how automation will affect the future of accounting work in SSCs. This research uses a qualitative approach; it has chosen a single case study meth-od based on purposive sampling as well as semi-structured interviews. A literature review was conducted to analyze the relevant research papers related to RPA within the Scopus Web of Science databasesranging from year 2006 until pre-sent. An analysis has been outlined following the Preferred Reporting Items for Systematic Review and Meta-Analyses (PRISMA) statement. This study provides an understanding of organizational changes on accounting /management processes. 33% of all positions from the SSC were eliminated, which its 40% were part of the accounting staff. The existing accounting positions after RPA were associated to decision makings and information analysis. The case demonstrates that RPA generates cost savings, is not perfect and re-quires initial human interaction and monitoring. To the best of author´s knowledge, there is no article on RPA in SSC´s based in LATAM. The authors help to close this gap, particularly in equipping SCSs in LATAM with knowledge in accounting and management areas that may be automated.

Introduction. Robotic revolution or fourth industrial revolution [1], [2], [3] refers to the progressive automation of repetitive works and processes, mundane tasks [4], [5], even those that require certain cognitive abilities [6] have been replaced by machines, technologies or robots, which has modified the nature ofthe employment and economies and the nature of the work in the future is expected to change fundamentally as well [7], [8], [9], [10]. Robots and Robotic Process Automation (RPA) are already impacting the work done by qualified humans, and it´s generating structural changes in the labor scene where jobs are created, transformed and destroyed. RPA refers to a computer software configuration [11] that automates manual processes [12] This is due to the scientific and technological advances that have taken place in recent decades, plus the need for companies to generate greater competitiveness in the market, which has been increasingautomation of processes at work [13], [14], [3], although this obviously depends on the business sector,the country and the financial capacity of the organizations - it is clear that not all of them are moving atthe same pace. Bhatnagar (2019) [1] and Comar et al (2018) [14] show concrete examples of the development of new technologies, such as artificial intelligence, big data, cloud computing, machine learning and robotics, by stating that they "do not exert their influence individually but, combined, generate synergiesthat multiply their effect and make it, in some way, unpredictable.". Issa, Sun and Vasarhelyi (2016) [15] also identified further advanced technologies like Scanning and OCR, Electronic records and plentiful computing, Blockchain, Smart contracts and Large data stores. Nowadays there are more complex humans-machine interactions, with a high-paid / low-paid increase in jobs as well as a reduction of medium-paid jobs in secondary sector [15]. Additionally, previous researches have also demonstrated that technology will replace and complement jobs. Tedious tasks and earnings tend to disappear but upgrading them for analytical moreproductive jobs [17], [18], [19].

Table 1. Summarizes the WOS review
Source: Author´s own elaboration

Authors	Journal	Title	H Index	Description
Balsmeier, B., and Woerter, M. (2019).	Research Policy	Is this time different? How digitalization influences job creation and destruction.	206	Increased investment in digitalization is associated with increased employment of high-skilled workers and reduced employment of low-skilled workers, with a slightly positive net effect
Nomaler, Ö., and Verspagen, B. (2019).	Economics of Innovation and New Technology	Perpetual growth, the labor share, and robots	28	New technologies may lead to so-called perpetual growth. factor-eliminating technological progress reduces the role of labor in the production process and that this leads to a rising wage rate but ever-declining share of wage income
Piva, M., and Vivarelli, M. (2018).	Eurasian Business Review	Technological change and employment: Is Europe ready for the challenge?	11	An economy based on R&D, knowledge and innovation—points in the right direction in terms of job creation, specifically in high-tech sectors
Buyst, E., Goos, M. and Salomons, A. (2018)	Oxford Review of Economic Policy	Job polarization: an historical perspective	75	Uses the the hypothesis of Routine-Biased Technological Change (RBTC) to illustrate the employment impacts due to technology
Hershbein and Kahn (2018)	American Economic Review	Do Recessions Accelerate Routine- Biased Technological Change? Evidence from Vacancy Postings	-253	Their results show that recessions accelerate the job routines related to technology, also how skilled workers become key nowadays.
Castro Silva and Lima (2017)	Research Policy	Technology, employment and skills: A look into job duration	206	Technological intensity reinforces the positive relationship between skills and job duration. Authors also showed the strong relationship between human capital and technological jobs/firms
Woods (2017)	Journal of the Knowledge Economy	The Effect of Technological Change on the Task Structure of Jobs and the Capital-Labor Trade-Off in US Production	20	Authors used the Cobb-Douglas model transformed into the Solow growth model to illustrate the effect of technology on the task structure of jobs, capital/labor ratio, productivity and skills workers
Cirillo (2017)	Economics of Innovation and New Technology	Technology, employment and skills	28	This paper investigates the relationships between technological change and structural determinants of employment changes
Martynovich and Lundquist. (2015).	Regional Studies	Technological Change and Geographical Reallocation of Labour: On the Role of Leading Industries.	105	Authors analyzed the effects of technology over regions to attract, rellocate and retain workers

Graetz and Michaels (2015) demonstrated that the use of industrial robots in a specific business sector increased labor productivity and wages of workers. However, for Bendel (2018) [20], all these changes could also be seen in a negative way; co-robots can support workers but can also displace them which involve both opportunities and risks. This study has the following research question: RQ1: How will automation affect the future of accounting work, specifically, in a Shared Service Centers SSCs? SSCs refer to an organizational model very popular in LATAM in which a company consolidates routine business functions performed by multiple operational entities into a single one that provides services for the entire company [19], [20]. SSCs are very useful for organizing accounting processes [21], [22], [23], [24], [25] and for companies looking forward cost savings [26], [22]. SSCs combine thebest practices of a centralized model and a decentralized model [27], [28], [29]. SSCs have a complex social and organizational arrangement [30] performed in a highly technologicalenvironment [31], [23]. And most businesses rather move business support functions to service centers where activities are processed as efficiently as possible [32], [27], [26]. SSCs abound in LATAM due to its geostrategic position and offers a wide variety such as lower costs, high quality service, tighter risk controls and data transparence [33].

1. What is Robotic Process Automation? For Bhatnagar (2019) [1] Robotic process automation (or RPA) is an emerging form of business processautomation technology based on the notion of software robots or artificial intelligence (AI) workers. The Institute of Electrical and Electronics Engineers- IEEE Standards Association (2017) [36] defines RPA as: ‘‘A preconfigured software instance that uses business rules and predefined activity choreography to complete the autonomous execution of a combination of processes, activities, transactions, and tasks in one or more unrelated software systems to deliver a result or service with human exception management’’. Robotics is a technology to reshape processes and encompasses three different terms: Robotic Process Automation software -RPA-, Cognitive Technology, and Artificial Intelligence. This research only addresses the impact of the first term, in this respect RPA is defined by Deloitte1 as "fast, non-invasive software with a return on investment of less than a year (...) a great alternative toreduce or eliminate workload of people in bulky back-office processes (repetitive processes of finance, accounting , human resources, supply chain management, customer service, et al) and integrates - in a non-invasive way with the IT environments of companies - with practically any existing application in a process”. The importance of RPA is increasing every day more. Some of the reasons are explained by Moffittet al., (2018) [4] as follows: “The business process improvement literature and professional auditing literature suggest that RPA can result in improved processes and economies of scale when the steps to perform a rules-based task are repetitive and manual. Conversely, RPA is less appropriate for those tasks that require elements of human judgment, that have uncertain outcomes, or that occur infrequently. When implementing RPA for the first time, organizations should look for easy wins; hence, complex and subjective tasks should be avoided”. Flak (2020) [36] includes the application of the automatic -methods and procedures for the substitution of the human person (operator) - to the control of industrial processes. The industries related to automation are mainly manufacturing and processes. According to Bendel (2018) [20] to implement the use of robots, the first thing to consider is the characteristic of the task -which is dirty, boring, dangerous or difficult- in reference to the 4Ds of robotics, since those tasks are more likely to be automated and robotized.

1.1 Impacts of Robotics Process Automation RPA. Walvei (2016) [39] and Flak (2020) [37] warned that "smart" automation can affect highly qualified jobs, including, lawyers, doctors, and accountants. Arntz, Gregory, and Zierahn (2016) [7], showed that people working in the areas of accounting and auditing face a potential for automation of the 98%, which is a first alert about the impact on the accounting profession. Now, some countries are already betting on total automation or automation in manufacturing, in which an industries integrate machinery tangible physical and digital processes, looking for such machines are capable of deciding and cooperating with others as well as with the human population, according to Atzori, Lera and Morabito (2010) [40] this is possible due to the term internet of things " Concept that refers to the digital interconnection of everyday objects and machines with internet" ; in the end, it is creating autonomous processes, intelligent factory, and headquarters [34]. It can be said that machines are becoming more intelligent, that is, they are increasingly able to detecttheir own performance and health, act on this information for themselves and communicate it explicitlyto operators and other machines [35] which definitely affects the way as one relationship with people and their traditional use, with negative and positive consequences that these leads-.Although, in order to generate automation process - aimed at organizations that think about incorporating it or have already generated it - certain considerations must be taken into account since it depends on (i) the organization and characteristics of the workplace; (ii) the previous existence and degree of investments in automation technologies, and (iii) the education of workers (training and assimilation skills). Chui et al., (2017) [6] identified other considerations when determining the implications for companies, namely, what kind of automation is possible with current technology and what is likely as technology evolves; what factors besides technical feasibility should be considered when making decisions about automation, where and how much to automate to better capture the value of long-term automation. In the end, all is part of reinventing and preparing to a new environment that has no return, where the accountants must enhance their creativity, their ability to solve problems and constantly update if they do not want to be replaced. In this regard, Comar et al., (2018) [14] argue that "social renewal, advance, and progress, go through the reinvention of professionals, who come out of their" comfort zone"establish new relationships" with themselves and with their interest groups. In fact, just Osan and Somers (2017) [42] propose how to leverage the fourth revolution since they require new human capacities, the processes necessary to store and manage the data, the professionals with the capacity to generate useful information from these and the functions of continuous improvement that turn knowledge into sustainable yield changes. It is also not a matter of being a victim or extremist, in fact, in the study that was previously cited, Arntz, Gregory, and Zierahn (2016) [7] concluded that it "clearly points to the need to focus more in the possible inequalities and requirements of (re) formation derived from technological change instead of the general threat of unemployment that technological progress could or could not cause.". That said, robotics, updating new technologies are necessary steps for economic development, but the impacts generated in employment should be guaranteed and carefully studied as a macroeconomic indicator. Accountants, of course, should not be oblivious to these discussions that are generated from different areas.

2. Methodology. This research reviews the literature using papers published between 2006 and 2020. The literature search was done independently by two researchers who searched for papers in various databases (Scopus and Web of Science) using the following keywords "robotic process automation, automation, shared servicecenter, back office, accounting, technology, labor changes". These search terms were initially broad in scope with the intention to cover a wide breadth of knowledge in answering the research question. Then, key word synonyms were identified and combinedwith key words to allow a more comprehensive search. The search used both 'free terms' and 'index terms' funneled using the Bolean operators (OR, AND) and truncations. This review was conducted byfollowing the reporting checklist of the Preferred Reporting Items for Systematic Reviews and Meta- Analyses [36]. Figure 1 presents the results of PRISMA analysis:

Figure 1. Results of PRISMA analysis.
Source: Author´s own elaboration

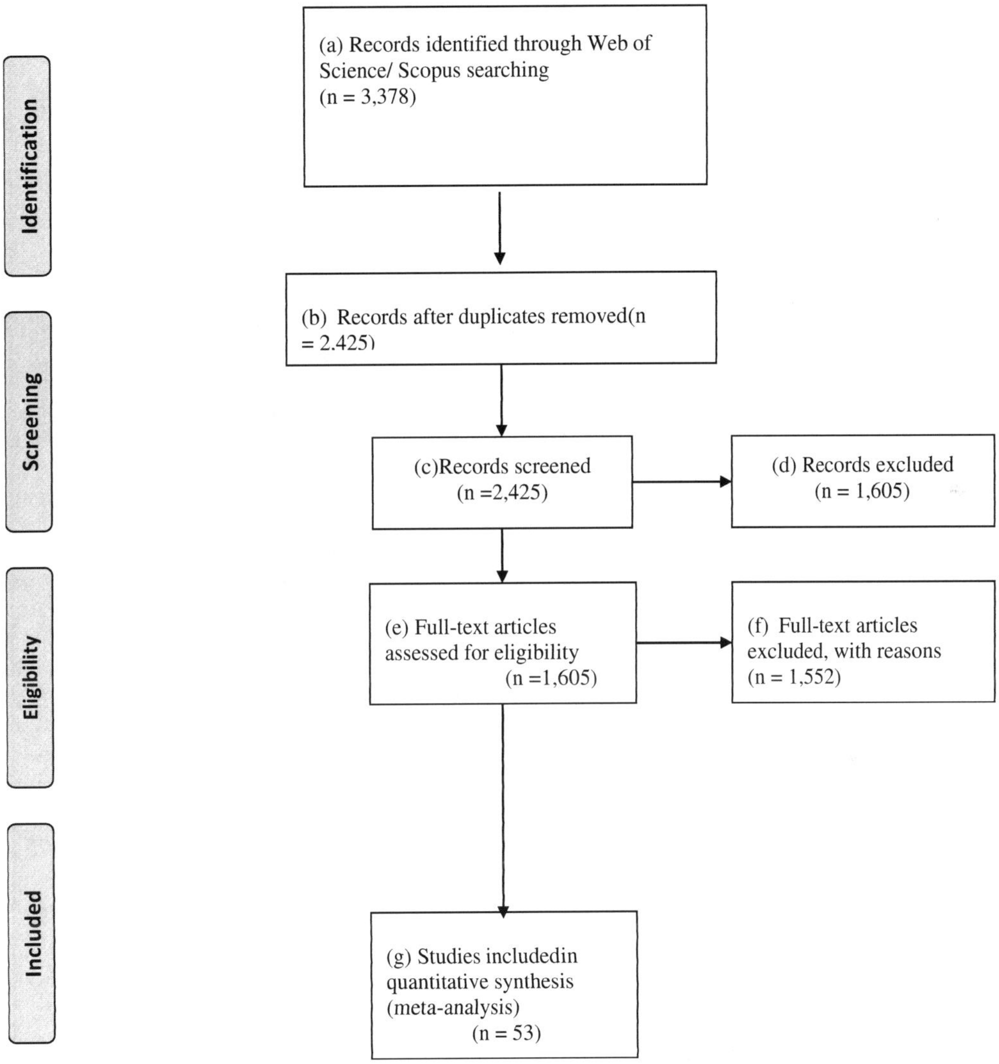

The review conducted in this study is guided by the main question: How will automation affect the future of accounting work in Shared Service Centers SSCs? More specifically, other guiding sub-questions are listed as follows:

I. Q1: What is the current state of RPA from the accounting perspective?
II. Q2: How has RPA been implemented in the field of SSCs?

This research excludes the following papers indexed in WOS and Scopus: papers published in letters, congresses, editorials, note, incorrect affiliations, inaccessible abstracts, short surveys, and wrong titles. Finally, only 53 relevant papers closely or partially related were found within the five keywords selected. This paper follows the qualitative research methodology in response to calls for investigations, suchas in the case of Herbert (2014) [44] and Kreiner (2006) [45], and the experiences of accountants in the context of shared service work described in Lepisto et al., (2018) [32]. This study subscribes to exploratory and descriptive case-based research, selected based on purposive sampling [37]. It is a single case with the holistic design of the instrumental type to be developed on a single unit of analysis given its revealing and character [38].

3.The Shared Service Center ´´GLEco´´. The study was conducted in LATAM. The organization being studied, referred to as "GLEco" is a company created in 2001 in North America by a group of investors who established as its goal to designinnovative products of high - tech water sports. Its corporate headquarters are in the United States, it hasdistribution centers in Asia and the Shared Services Center of the Caribbean and Latin America in Colombia, South America was founded in 2001. The company has won numerous worldwide awards and recognitions. This paper aims to observe the implementation and impacts of RPA in some accounting processes and in the organizational structure of the company. The situation experienced by the company is shown chronologically, from the moment in which it is decided to implement RPA and the evaluation after the decision. Initially, permission is requested from the main headquarter in the USA to start collecting data on randomly selected managers and employees. A few weeks later and after several sessions where consensus was not reached, finally approval is received from senior management to start documenting the case, but not before signing a new NDA (Non-Disclosure Agreement) or Confidentiality Agreementwhere the first numeral states that the generic information could be shared and the company's patented information should be kept in absolute confidentiality such as: " idea, discovery, research, data,formulations, specification, process, technique, algorithm, architecture, know-how, invention, plan, drawing, outline, outline of products, document, manual, report, market studies, photography, sample, prototype, customer lists, price list, detailed product description, business plan, marketing plan, financial information or work in progress " as well as the identity of the officials who are part of the study. Subsequently, the interviews to the accounting, computer experts, clients and logistics staff of the officein Colombia began via telephone and email. The company went through a series of events which have put to the test the management capacity ofits directives and which led to the search for initiatives to reduce costs and seek greater efficiency in the processes, which led to the emergence of RPA as a response to such concerns.

3.1 *Seasonal Products.* The main market was the North American one and during the winter season (December to March approximately) the sales of nautical products fell drastically what showed that the items were put under a strong seasonality that the sales underwent strong oscillations depending on the time of the year. The sustained fall in revenues led to the search for initiatives to reduce costs. In this regard, Ma et al., (2018) [48] point out those seasonal products give rise to a new media scenario and pose both challenges and opportunities for companies. However, and as established by Lobos Andrade and Muñoz Ibáñez (2005) [49], seasonality is an external factor that cannot be controlled by producers, which impacts not only production processes; it also establishes the seasonal component asthe most relevant of the time series data used to analyse price stability and marketing margins.

3.2 Asia and Intellectual Property. The new future products began to be manufactured in a non-owned factory in Asia in 2014, with industrial designs, prototypes, and patents. The new product line was a success in the first months. However, during the same period, the decrease in the frequency of purchases by some distributors and representatives was identified and when inquiring about the reasons, it was detected that the Asian company had taken intellectual property, modified designs, and prototypes and was directly marketing the "new" products at a lower price, even with regular customers. The violation of intellectual property rights in country X has always been considered a major concern for foreign companies. Since 2001 country X has made great strides in protecting the intellectual property. However, country X still requires special attention to international standards regarding intellectual property, cooperation with foreign peers and recognition of the need for new forms of public-private partnerships [39]. Although they have several international patent applications that surpasseven those of countries like Germany and South Korea and like Japan and the USA, their innovations do not demonstrate dominant influence in the rest of the world, they have a very low commercial value, additionally China is in the first position in the list of countries with "priority surveillance" for not providing adequate and effective protection of intellectual property according to the United States. Pending orders had to be produced in alternate factories, at less competitive prices. The transition took a while, several customers decided to wait for their orders, if they receive a compensation(additional accessories, credits for future orders or international freight payment), while others decidedto cancel their orders and requested the return of the deposits paid. All of this hit even more intensely the company´s finance and it became even more important for top management to reduce costs. The legal process against those who stole the intellectual property began and is still ongoing, pending the decision of the Court.

3.3 Main Strategic Decisions. To counteract the seasonality of the products, a marketing and penetration strategy was initiated towards new markets such as the European, Australian and Asian Pacific Islands. The research anddevelopment department began to work on expanding the portfolio of products that met the brand'sown innovative character, which could be sold in a sustained manner independent of the season and complement the existing line. Creation of a manufacturing and distribution center in the USA. Bank loans are acquired to materialize this project since with the background of intellectual property it was difficult to return sensitive information to any other factory. This decision was jointly supported by offering the possibilityof being able to regain available inventory in the main market. The opening of the Shared Services Center (SCS) in Colombia, South America for the Caribbean and Latin America aimed to improve service levels, continue to reduce qualitative and quantitative costsin personnel, processes, technology, and property and consolidate administrative functions such as Logistics and International Transportation, Human Resources, Finance, Purchasing, and Accounting.

3.4 Implementation of RPA in SCS in LATAM. The SCS employs 15 people detailed as follows: 1 General Manager, 1 Human Resources Manager, 1Accountant, 2 Accounting Assistants, 2 Logistics and Transportation Assistants, 1 Production Coordinator, 2 System Engineers, and 5 Salesmen. when senior management began to consider the RPA, it began by reviewing all process maps in conjunction with all the administrative staff to determine which were the most manual and repetitive tasks that were likely to be automated, without sharing any activity with the company's IT staff. The selected RPA Tool was BdPath (name changed), and the accounting software was QuickBooks Online Plus (QBO), independently of the volume of transactions, which focuses on organization, productivity and insights, key characteristics for this SSC. QBO can sync with hundreds of different apps which automate multiple tasks to, below it is QBO´s apps used and their main features in the SSC:

3.4.A Payroll Automation (this feature takes care of global employee's net payment, employment taxes, reports, worked hours, among others), which uses the app "QuickBooks Online Payroll" powered by KeyPay

3.4.B Invoice/Inventory Automation (Invoicing and Inventory have been streamlined with "Shopify for QuickBooks Online" app, create templates, shipping orders, invoices automatically and export all sales to QBO, using any gadget)

3.4.C Inventories, sales through multiple distribution channels (in physical, social networks), handling of transport documents and production orders are handled in a single site ("Shopify for QuickBooks Online" app dropships from a third-party warehouse)

3.4.D Payment Automation (with "Shopify for Quickbooks Online" app the company can accept online payments from different currencies via credit card or bank transfer. This app is integrated with around 100 worldwide gateways and offers a merchant account).

The main types of automation processes identified were:

1. Specific Processes: Transactional processes that are part of a larger function within the company, are simple and repetitive. Eg: Inventory management, Receipt, and processing ofinvoices within

accounts payable, accounts receivable, reports, marketing campaigns

2. Multifunctional Processes: Processes that are executed through multiple functions in an organization. Eg: Bank reconciliations, Payroll
3. End-to-end processes: These are complete processes carried out through multiple tasks. They usually involve several officials generating reprocesses and possible bottlenecks. Eg: Payment to suppliers, purchase orders, import and export and international cargo transportation

By automating the above processes, the main results were the following:o Cost reduction: 60-70%

- o Accuracy: 100%
- o Employees in the new processes: 1
- o Saving time: 2800 hours

3.5 Example of an accounts receivable process, end-to-end (before and after). To highlight the benefits of RPA this paper will show some data regarding the application of RPA forthe process of accounts receivable in the company. Under standard conditions, the duration times of the processes and people involved in the process do not vary, also considering that there are no errorsthat delay the process. See Figure 2:

Figure 2. Accounts Receivable Workflow. After
Source: Author´s own elaboration

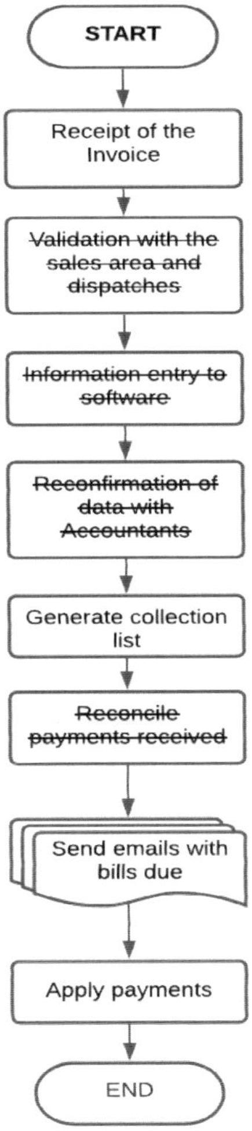

IOP Conf. Series: Materials Science and Engineering 1154 (2021) 012001 doi:10.1088/1757-899X/1154/1/012001

3.6 Process of Accounts Receivable of Invoice not paid without Automating (See Table 2).
Estimated time of completion: 20 minutesPeople involved in the process: 3;Accuracy: 80%

Table 2. Wage level in US Dollars.
Source: Author´s own elaboration

Position	Before RPA	After RPA	Reduction	%
General Manager	$ 2,500	$ 2,500	0	0
H.R. Manager	$ 2,000	$ 2,000	0	0
Accountant	$ 1,500	$ 1,500	0	0
Accounting Assistant (2)*$480 each	$ 960	0	$ 960	100%
Logistics and Transport Assistants (2) *$480 each	$ 960	0	$ 960	100%
Production Manager	$ 1,000	0	$ 1,000	100%
Systems Engineer (2)	$ 960	$ 960	0	0
TOTAL	**$ 9,880**	**$ 6,960**	**$ 2,920**	**29%**

Automated Accounts Receivable Process of Invoice not paid (eliminated processes have been crossed out). Estimated time of completion: 3 minutesPeople involved in the process: 1 Accuracy: 100%

4. Discussion -The organizational structure changed after RPA. The implementation of RPA not only affected the accounting part but also had an impact in logistics and transport and production. At first the intention of the company was to retain human talent;however, it was impossible because as they have conceived the charges most of the activities, they performed were routines that had been automated or were in the process of being. It is concluded that the impacts were from the labor point of view, with a reduction in the jobs. Bendel(2018) [20] stated that one of the concerns that comes with automation process is "the impact on a weakened labor markets marked by job insecurity and high unemployment"; for this case, although therewas no significant impact because it was a small office, attention should be paid to the multiplier effect,since if these practices are extended, the impacts can be much greater. At the same time, greater process efficiency was achieved by reducing waiting time by customers; this no longer must wait days for their documentation (commercial invoice, packing list and certificate of origin as well as other trade documents) to proceed with the importation or logistics processes.

Figure 3. Organizational pyramid before and after RPA

Source: Author´s own elaboration

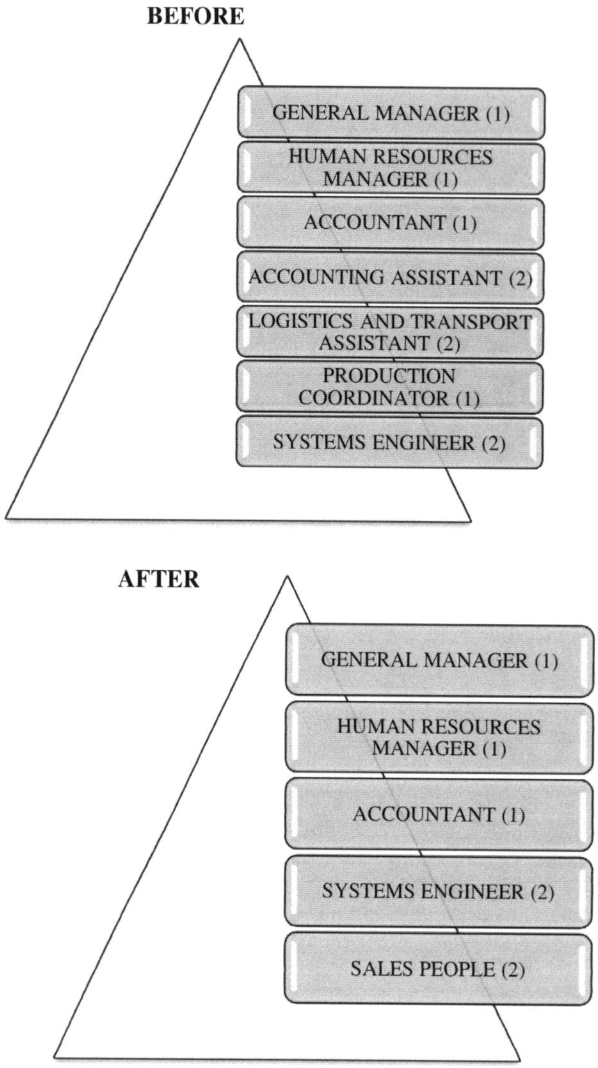

The implementation of RPA accomplished the original goal of cost reduction, a total of $2,920 USD permonth in reduction equivalent to a 29% reduction of the payroll. Five of fifteen positions of the SSC inLATAM (33%) were eliminated, two of five positions (40%) were part of the accounting staff (one of the positions was the AR Specialist and the other one was the Inventory Analyst, see their functions in figure 3). Essentially, the tedious / routinized jobs disappeared meanwhile the others still exist due to the skills their jobs require and that cannot be automated (at least yet), like creativity, critical thinking, emotional intelligence and ability to "read" people. This last finding is in aligned with Cortes, Jaimovich and Siu (2017) [51] and the World Economic Forum (2016) when stated that the changing nature of work, and skills instability–the rapid change in the skills requirements of all existing jobs–are nowadays among the most important drivers of change. It is also aligned with the European Commission (2016) when confirmed that the impact of concurrent disruptions in technological, demographic and socioeconomic level will transform the employment landscape and skills requirements. The existing accounting positions after RPA were associated to decision makings and information analysis. The case demonstrates that RPA generates cost savings, is not perfect and requires initial human interaction and monitoring.

Figure 4. Accounting positions removed and their functions

Source: Author´s own elaboration

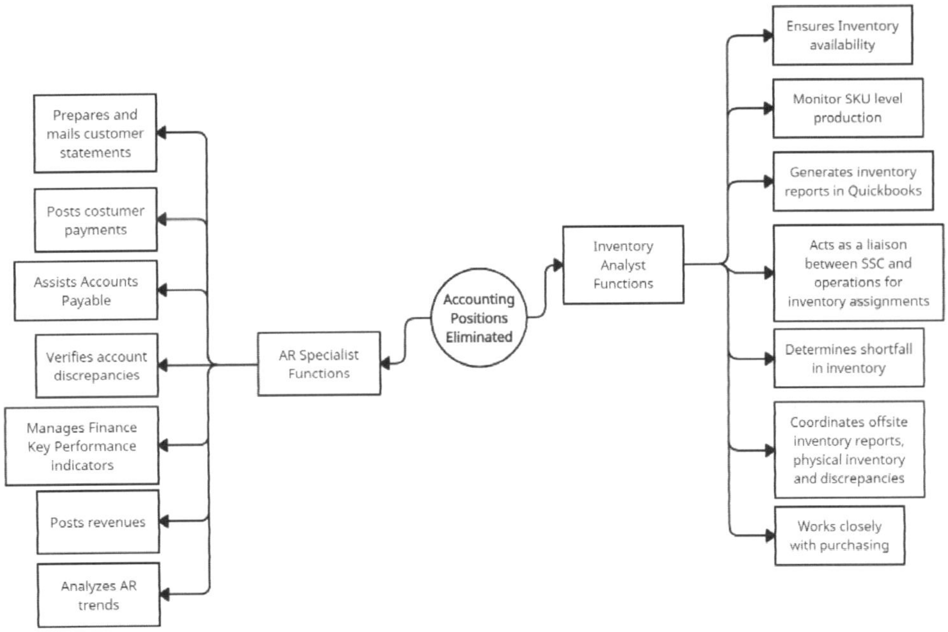

IOP Conf. Series: Materials Science and Engineering 1154 (2021) 012001 doi:10.1088/1757-899X/1154/1/012001

The key person contacted in the SSC was the controller. After several attempts to set an appointment emails and calls, the controller finally agreed to meet the interviewers to know further the purpose of this research. The meetings were in the SSC´s office and it was necessary to meet several times to clarify the intentions of the researchers and the potential benefits for the SSC. The controller helped to contact upper management in corporate headquarter until final approval to share the information was provided. The first round of semi-structured interviews started with two guide questions:

Q1: What is RPA?

Q2: What is your experience with an RPA Tool called BdPath (robot software)?

During the interviews, the respondents provided in-depth details of the functioning of their areas andthe main bottlenecks identified, which explained why upper management considered RPA as a solutionto overcome their difficulties. The cost of implementation an RPA Tool was a big concern (the respondents thought the cost was high). All the information gathered served as the primary guide for this research´s writing. This study collected data through semi-structured interviews (lasted from one up to two hours). In total, 37 hours of interviews were undertaken from October to December 2018 with nine respondents including the chief executive officer, chief information officer, chief technical officer (in USA), the operational manager, the controller, the logistics manager, the production manager, the human resourcemanager, accountants (in SSC in Latin America). All interviews have been recorded and then the transcription process completed accordingly. These results are in Table 3.

Table 3. Details of Interviews.
Source: Author´s own elaboration

Item	Name and Position	Interview Session	
		Frequency	Duration (hours)
1	Chief Executive Officer	1	3
2	Chief Information Officer	2	6
3	Chief Technical Officer	2	6,5
4	Operational Manager	2	5
5	Controller	2	4
6	Logistics and Transport Manager	1	2
7	Human Resource Manager	1	3
8	Accountants	2	5,5
9	Production Manager	1	2
	TOTAL		**37**

Surprisingly, some collaborators had already heard and read about it, which was not something imposed: "When I first heard about RPA was a few years ago during college. I´m really glad we´re consideringRPA for our company....... Actually, I think it´ll improve the company performance across the board. It will be a bit awkward to have a robot as a new co-worker [laugh])" (Richard, Accountant, 38 years old). Others felt skeptical when heard about RPA: "A robot may do a great job but I´m not quite sure how well it can do when it comes to numbers or figures...Upper management will always want a human being to be responsible for counting the beans of the company. These tasks are still too sensitive to handle for a machine, I guess". (Marcus, FinancialAnalyst, 55 years old).

Finally, the initial motivation to reduce costs was always clear, however, there are also other benefitsthat derive from the use of RPA. Two employees spoke about their experience after implementation: "As one of the architects behind the implementation of RPA, to "see people go" was not easy. Hightalented, smart and committed employees, and more important, personal friends in some cases, lost theirjobs. Having a robot as a co-worker was something that all employees/professionals aren´t ready for. On the other hand, the productivity and accuracy increased dramatically, key processes in several areas were streamlined and senior management considered that RPA was worth it. From a personal perspective, my role as a controller changed positively, I had to develop new powerful skills and improveother set of abilities. Now, I was more analytical oriented, and began to spend more time making strategic/educated decisions and with higher quality information." (Joseph, Controller, 38 years old). "Information is way more reliable now and is handy 24/7. The person in charge of that task prior RPA was good, however cannot beat the current software in terms of efficiency and accuracy." (Richard,Accountant, 41 years old). One of the exclusive distributors reflected on the experience: "Multiple times in the past I was given the shipping documents and accounting/financial informationa bit late. That caused me some fees with customs. Now, that bottle neck is gone. Once I place an order,triggers are pulled immediately into the new tool you guys implemented and I get all the documents I require right away, no waiting time"... So I appreciate these types of improvement and now we can focus on closing more sales and expand". (Mary, Distributor, 60 years old).

5.Directions for future research. This study introduces the concept of RPA in SSCs in Latin America. The findings will help other SSCsto incorporate RPA in their operations. Future research is required in how to use other new technologies(Cognitive Technology, Blockchain, Big Data, Machine Learning, Internet of Things and Artificial Intelligence) in SSCs in Latin America as well as other areas that can be automated (logistics, legal, compliance, HR, purchasing, finance, payroll) following prior research studies' methods [52]. Specifically, future research could focus on the following questions:

1.How will these new technologies create and help managing jobs, organizational problems and business relations in LATAM?
2. How do the business environments in the SSC host country influence the success or failure of RPA implementation in LATAM?
3.What are the cost impacts on SSCs from different organizational contexts? Will those results vary/how much once multiple in-depth case studies and cross-country data be used?
4.What are the opportunities and challenges faced by SSCs in emerging countries while implementing RPA?
5.What new technologies will SSCs implement in each country of LATAM? In what areas of theSSC? Can the same technology be applied in all SSCs with presence in countries of LATAM across the board?
6.How will accountants and all workers in general be motivated in the machine economy?

7. What will be the consequences and how will we address the income inequality?
8. What is the bias in new technologies? What are their sources and main impacts?

6.Conclusions. It is a fact that the incorporation of new technologies and the fourth industrial revolution in the businesssector is here to stay and is constantly evolving. Neither accounting nor SSC´s should be left out of global trends. It is an excellent opportunity for the empowerment of the profession and defense for a more analytical and less technical role. The accounting information system (AIS) course require updates.Accountants of 21st century with IT and data analytics knowledge make a perfect combination and gather the new skills an accountant should have information creation and maintenance, data analysis, and IT infrastructure [53], [15].

LATAM is a valley for SSCs who need to remain competitive by adapting these new technologies in addition to other factors that the region offers (labor arbitrage, LATAM environment, geographic proximity to North America, et al). Part of the new SSCS approach is to become strategic decision centers and to be attractive for different company´s sizes. Competitiveness that will allow SSCs to enhance their operations, control and transparency [35].

It is clear that jobs will be lost, tasks and positions in administrative management will be created, modified and disappeared, however, one aspect that must be considered is a strong social preference forthe provision of certain tasks and services by humans as opposed to machines, there is a social value associated with humans who perform certain tasks that tend to preserve their comparative advantage. [7], social, interpersonal and analytical skills are still important, despite the specialization of work, as accountants constantly negotiate with clients, analyses their processes and respond to open-ended queries [27].

Companies should anticipate this reality and should carry out an exhaustive inventory of the activitiesof the organization and create a heat map where the automation potential is high, in this way, all the associated risks will be reduced and planned in advance. This requires universities to update themselves in these trends, interdisciplinary work groups are created, and company-academy partnerships are established to strengthen these labor competencies [54].

RPA can improve productivity, accuracy, and compliance, and allows professionals to focus on morecreative and more valuable work. It is a mistake to believe that it takes a lot of money and investments in RPA to see big changes in operational efficiency. The implementation of these systems reduces costs,improves the accuracy and speed of a work process, reducing human negligence, but can also affect psychosocial aspects such as motivation and commitment of employees, due to staff reductions. These technologies speed up the collection and processing of data; although it can also generate the rejection of technology because of the concern to lose their jobs [55]. The current supply of vendors capable of automating processes in the company is very varied and offers competitive market prices, research estimates that it will be further reduced with the passage of time and increases their degree of reliability.

The results show that due to its low cost, development and efficiency increase, it is expected that theuse of the RPA gradually grows in specific processes (accounts payable, inventories), multifunctional (reconciliations of invoices and banking) and cutting-edge processes (payroll and purchases). Other accounting processes that will benefit from RPA are billing, payroll, preparation of financial reports andother financial activities of the company, such as cash flow, investment, debt, financial planning and decision making [55].

The Accounting Big Four are also part of this wave, implementing and patenting key tools in their daily operations: document review platforms, asset inspection, automate auditing processes, drones to monitor inventory, Natural Language Processing, among other. They have also partnered with high-techplayers: IBM Watson, Ayasdi, Narrative Science, Kira, H2O.ai. [56].

Finally, it should be noted that the accountants will not disappear; they should simply adjust their role pointing to consulting, analysis and value-added services. Here the strengthening of soft skills andgreater knowledge about RPA and ICTs becomes vital for universities, which should incorporate the RPA approach in updating the accounting curricula to foster new skills in IT audit, systems design, anddata analytics to accountants [53] so they can respond to nowadays challenges and requirements from employers [57], [58], [59].

7. References

[1] N. Bhatnagar, Handbook of Experimental Pharmacology, 2019.
[2] A. Hirschi, The fourth industrial revolution: issues and implications for career research and practice, 2018, pp. 192-204.
[3] T. Nam, Citizen attitudes about job replacement by robotic automation, vol. 109, 2019, pp. 39- 49.
[4] K. Moffit, A. Rozario y M. Vasarhelyi, Robotic process automation for auditing, 2018.
[5] K. Schawb, The fourth insutrial revolution. New York: Crown Business, 2016.
[6] M. Chui, J. Manyika, K. George y M. Miremadi, «McKinsey & Company,» 2017. [En línea]. Available: https://www.mckinsey.com/business-functions/operations/our-insights/human-plus- machine-a-new-era-of-automation-in-manufacturing.
[7] M. Arntz, T. Gregory y U. Zierahn, The risk of automation for jobs in OECD countries: a comparative analysis, 2016, pp. 1-34.
[8] E. Brynjolfsson y A. McAfee, The second machine age: work progress and prosperity in a time of brilliant technologies, 2014.
[9] E. Ford, Rise of the robots technology and the threat of a jobless future, 2015.
[10] C. Frey y M. Osborne, the future of employment: how susceptible are jobs to computerisation? 2017, pp. 254-280.
[11] L. Willcocks, M. Lacity y A. Craig, «The IT Function and Robotic Process Automation,» The Outsourcing Unit Working Research Paper Series, 2015.
[12] C. Bataller, A. Jacquot y T. S.R, Robotic Process Automation, 2017.
[13] A. S, M. P, C. B y N. D, A novel automated Finalcial Transaction System Using Natural Language Proccesing, 2020.
[14] C. Comar, G. Duncan y K. Miller, FlipTube technology promotes clean manipulation of forensic samples on automated robotic workstations, 2017.
[15] H. Issa, T. Sun y M. Vasarhelyi, «Research Ideas for Artificial Intelligence in Auditing: The Formalization of Audit and Workforce Supplementation,» Journal of Emerging Technologies in Accounting, vol. 13, 2016.
[16] E. Buyst, M. Goos y A. Salomons, «Job Polarization: An Historical Perspective».
[17] D. Acemoglu y H. Autor D, «Skills, Tasks and Technologies: Implications for employment and earnings,» de Handbook of Labor Economics .
[18] D. H, «The Task Approach to Labour Markets: An Overview».
[19] H. Autor D y M. J. Handel, «Putting Tasks to the Test: Human Capital, Job Tasks, and Wages,»
Journal of Labor Economics , 2013.
[20] O. Bendel, Co-robots from an ethical perspective, vol. 141, 2018, pp. 275 - 288.
[21] N. Su, R. Akkiraju, N. Nayak y R. Goodwin, Shared Services Transformation: Conceptualization and Valuation from the Perspective of Real Options, 2009.
[22] F. Sum, I. Paula, G. Tortorella, A. Pontes y F. R.T, «Analysis of the Implementation of a Lean Service in a Shared Service Center: A Study of Stability and Capacity,» IEEE

Transactions on Engineering Management, pp. 1-13, 2019.
[23] A. Bhimani y L. Willcocks, Digitization ´Big Data´ and the transformation of accounting information, vol. 44.
[24] D. Harritz, «Role of Management Devices in Enacting Strategy - Case study of shared service centre,» Journal of Accounting and Organizational Change, vol. 12, pp. 504-521, 2016.
[25] J. Lindvall y E. Iveroth, «Creating a Global Network of Shared Service Centres for Accounting,»
Journal of Accounting and Organizational Change, vol. 7, nº 3, pp. 278-305, 2011.
[26] J. Oliveira y S. Clegg, «Paradoxical Puzzles of Control and Circuits of Power,» de Qualitative Research in Accounting and Management, vol. 12, 2015, pp. 425-451.
[27] W. Seal y I. Herbert, «Shared Service Centers and The Role of the Finance Function: Advancing the Iron Cage?» Journal of Accounting and Organizational Change, vol. 9, nº 2, pp. 188-205, 2013.
[28] R. Raudla y K. Tammel, «Creating Shared Service Centers for Public Sector Accounting,»
Accounting, Auditing and Accountability Journal, vol. 28, pp. 158-179, 2015.
[29] D. Howcroft y H. Richardson, «The Back Office Goal: Exploring connections and contradictions in shared service centres,» Work, Employment and Society, vol. 26, pp. 111-127, 2012.
[30] B. Quinn, R. Cooke y A. Kris, Shared Services: Mining for Corporate Gold, Englewood Cliffs, NJ , 2000.
[31] D. Schulman, M. Harmer y J. Dunleavy, Shared Services - Adding Value to the Business Units, Hoboken, NJ, 1999.
[32] L. Lepisto, J. Dobroszek, S. Moilanen y E. Zarzycka, «Being a Management Accountant in a Shared Service Centre,» Journal of Accounting and Organizational Change, 2018.
[33] I. Herbert y W. Seal, «Shared Services as a new organisational form: some implications for management accounting,» The British Accounting Review, vol. 44, nº 2, pp. 83-97, 2012.
[34] S. Bridelli, S. Wesneck y F. Martins, Shared Services Center: A Definitive Solution for The Administrative Processes? vol. 1, 2005, pp. 1-8.
[35] Deloitte, Shared Services Centers in Latin America, 2015.
[36] I. C. A. Group, IEEE Guide for Terms and Concepts in Intelligent Process Automation, New York, NY, 2017.
[37] O. Flak, System of organizational terms as methodological concept in replacing human managers with robots, vol. 70, 2020, pp. 479-500.
[38] Deloitte, ¿Qué es Robotics Process Automation? 2018.
[39] U. Walvei, «Digitalization and Structural Labour Market Problems: The Case of Germany,» ILO Research Paper , 2016.
[40] L. Atzori, A. Lera y G. Morabito, The Internet of Things: A Survey, 2010, pp. 2787 - 2805.
[41] A. Frank, L. Dalenogare y N. Ayala, «Industry 4.0 technologies: implementation patterns in manufacturing companies,» International Journal of Production Economics, 2019.
[42] A. Osan y K. Somers, Optimizing production in the age of the machine, 2017.
[43] A. Liberati, D. Altman, J. Tetzlaff, C. Mulrow, P. Gotzsche, J. Loannidis y D. Moher, The PRISMA Statement for Reporting Systematic Reviews and Meta-analyses of Studies that Evaluate Healthcare Interventions: Explanation and Elaboration.
[44] I. Herbert y W. Seal, «A Knowledge Management Perspective to Shared Service Centers: A Case Study of a Finance SSC,» Shared Service as a New Organizational Form, pp. 133-151, 2014.

[45] G. Kreiner, B. Ashforth y S. D.M, «Identity Dynamics in Ocuppational Dirty Work: Integrating Social Identity and System Justification Perspective,» Organization Science, vol. 17, nº 5, pp. 619-636, 2006.
[46] M. Patton, Qualitative Research and Evaluation Methods, 3 ed., 2002.
[47] R. Yin, Case Study Research and Methods, Thousand Oaks, 1994.
[48] J. Ma, D. Huang, D. Markovitch y B. Ratchford, «High or Low Season,» European Journal of Marketing, vol. 52, pp. 1956-1980, 2018.
[49] G. Lobos y T. Muñoz, «Indices de estacionalidad de los precios recibidos por los productores de manzanas chilenas,» Pesquiza Agropecuaria Brasileira, vol. 40, nº 11, pp. 1051-1057, 2005.
[50] X. Y y R. Suttmeier, «China´s Post-WTO Technology Policy: Standards, Software and The Changing Nature of Techno-Nationalism,» The National Bureau of Asian Research, 2004.
[51] G. Cortes, N. Jaimovich y H. Siu, «Disappearing routine jobs: who, how and why,» Journal of Monetary Economics, vol. 91.
[52] J. Paul y A. Rosado-Serrano, «Gradual Internationalization vs Born-Global/International New Venture Models,» International Marketing Review, 2019.
[53] J. Coyne, E. Coyne y K. Walker, «A Model to Update Accounting Curricula for Emerging Technologies,» Journal of Emerging Technologies in Accounting, vol. 13, pp. 161-169, 2016.
[54] D. Drum y A. Plvermacher, «Accounting Automation and Insight at the Speed of Thought,» Journal of Emerging in Accounting, vol. 13, 2016.
[55] D. Fernandez y A. Aman, «Impacts of Robotic Process Automation on Global Accounting Services,» Asian Journal of Accounting and Governance, vol. 9, 2018.
[56] D. Faggaella, «AI in the Accounting Big Four - Comparing Deloitte, PwC, KPMG, and EY,» 2019.
[57] Association to Advance Collegiate Schools of Business AACSB, 2014.
[58] Deloitte, Tech Trends 2013: Elements of Postdigital, 2013.
[59] PricewaterhouseCoopers, «Data Driven: What Students Need to Succeed in a Rapidly Changing Business World?» 2015.
[60] H. Castro Silva y F. Lima, Technology Employment and Skills: A look into job duration, 2017.
[61] V. Cirillo, Technology, Employment and Skills.
[62] Deloitte, The Robots are ready, are you? 2018.
[63] B. Hershbein y L. Kahn, «Do Recessions Accelerate Routine-Biased Technological Change?» American Economic Review, pp. 1737-1772, 2018.
[64] M. Lacity y J. Willcocks, «A New Approach to Automating Services,» Massachusetts Institute of Technology Sloan Management Review, 2016.
[65] Lighthizer, «Office of The United States Trade Representative,» 2018.
[66] M. Martynovich, Lundquist y J. M, «Technological Change and Geographical Reallocation of Labour: On the Role of Leading Industries,» de Regional Studies, 2015, pp. 1633-1647.
[67] O. Nomaler y B. Verspagen, «Perpetual Growth, The Labor Share and Robots,» de Economics of Innovation and New Technology, 2019, pp. 1-19.
[68] M. Piva y M. Vivarelli, «Technological change and employment: is Europe ready for the challenge?» Eurasian Business Review, 2018.
[69] F. Silva, M. Ostos y O. Gonzalez, Automatización Robótica de Procesos (RPA), 2017.
[70] K. Tammel, «Shared Services and Cost Reduction Motive in The Public Sector,» International Journal of Public Administration, vol. 40, nº 9, pp. 792-804, 2017.
[71] M. Williams, «An Old Model of Social Class? Job Characteristics and the NS-SEC

Schema,» de Work, Employment and Society, 2016, pp. 153-165.
[72] J. Woods, «The Effect of Technological Change on the Task Structure of Jobs and the Capital- Labor Trade-Off in US Production,» Journal of the Knowledge Economy, pp. 739-757, 2015.

Analysis of the Economic and Environmental Feasibility of a Home Automation System

G. Pico[1], Y. Llosas[1], P. Maques[2], E. Mendoza[3], R. Nuñez[4], N. Castro[5], J. Nuñez[5*]

[1]Department of Electricity. Universidad Técnica de Manabí, Ecuador
[2]Department of Energy and Environment. Instituto Tecnológico de Leira, Portugal
[3]Automation Department. Universidad Católica de Santiago de Guayaquil, Ecuador
[4]Faculty of Engineering Student. Universidad del Norte, Colombia
[5]Department of Energy, Universidad de la Costa, Colombia

[*]Corresponding author: jnunez22@cuc.edu.co

Abstract. The current research was carried out to analyze the economic feasibility and contribution to homes and buildings automation in the field of environmental protection. The objective that led this study was to explore and investigate the increase in comfort and safety, as well as, the reduction in the electrical energy consumption, so that, it is possible to achieve greater energy efficiency in the No 3 building at the Technical University of Manabí (UTM) in Ecuador. In order to, investigate deeply the economic feasibility; it is relevant to be aware of the electric power and automation equipment cost that will be integrated into the Z-wave protocol. Moreover, the electric bill study contracted by the UTM this was developed initially with the monetary value of the electric power, which was regulated by the Electricity Control Agency. On regards to the environmental analysis, it was shown that the projection uses of the domotic system in the teaching building No. 3, not only reflected some economic and energy savings but also, it presented a considerable contribution to the environment, due to, the emissions reduction in CO_2 to the atmosphere. To accomplish this environmental analysis, the CO_2 Emission Factor was considered, regulated by the Ministry of the Environment, which is 0.7079 Kg CO_2/kWh. It is important to mention that this value is applicable to all systems that are connected to the national electrical system. To validate the fulfillment of the outlined objective, it was considered two very important scenarios. The first one reflected the reduction of polluting emissions to the atmosphere, because of the use of the photovoltaic located in this build. The second one referred to the theoretical energy saving that will be obtained as the implementation of the home automation system.

Keywords. Energy management, smart grid, home automation, smart buildings, renewable energy sources.

1. Introduction

Currently, various systems that use so-called clean energies for electricity production are being studied and implemented, which must be efficient and comply with the safety protocols to achieve reliability in these systems [1]. One of the most used are Photovoltaic Solar Systems (SFV) which can be used autonomously, connected to the network or a combination of both [2]. In [3] a methodology is developed that allows monitoring the performance of photovoltaic systems by detecting faults. The research uses practical data from a 7.5 kWp Grid-Connected Photovoltaic Solar System (SFVCR), concluding that the methodology used was able to detect and identify 100% of the simulated faults. In [4], a hybrid renewable energy generation plant is proposed that could supply a percentage of the total energy demand and reduce the environmental impact of conventional energy generation in the Caribbean region of Colombia.

The hybrid plant works with a photovoltaic (PV) system and wind turbine systems, connected in parallel with the grid to supply a fraction of the total energy demand. The proposed system features 441 photovoltaic panels and 3 wind turbines, reaching a net current cost (NPC) of $ 11.8 million and a low CO_2 production of 244.1 tons per year.

In recent years, different home automation control systems are being studied and developed to increase efficiency and comfort, in addition to reducing the costs of energy carriers. These systems are generally implemented and performed with various communication protocols to achieve the expected results [6], [7]. In [8] a wireless sensor network oriented to home automation applications is designed and implemented. In the experimental tests that were performed, a 12-meter line-of-sight range was obtained between sensor nodes, guaranteeing full coverage of the area of a single-family house.

In our research, a home automation system is developed using the Z-Wave protocol, which consists of a mesh-shaped network that uses radio waves to communicate different smart devices with each other, also allowing interaction with different WEB pages. The platform used, "Vera Home Automation", allows to easily program our intelligent system giving the user the possibility of managing it in a simple, efficient and centralized way [9, 10, 11, 12, 13].

The description of the configuration of the implemented protocol is made from the use of the VERA platform [9] and the economic feasibility analysis and the contribution that the use of the home automation system has to the environment by reducing electricity consumption are made [14, 15, 16, 17, 18, 19, 20] and thus obtain greater energy efficiency in teacher building No. 3 of the Technical University of Manabí (UTM) in Portoviejo, Ecuador [9].

For the economic feasibility analysis, the cost of electrical energy and the smart equipment that will be connected to each other using the Z-Wave protocol must be known [21, 22, 23, 24, 25]. In Ecuador, the energy cost is regulated by the Electricity Regulation and Control Agency and an analysis is made of the contracted electricity bill, which for the UTM is medium voltage with regulated hourly demand. In addition, an analysis of the environmental impact of the use of this technology in teaching building No. 3 is carried out, which not only reflects savings on the economic and energy side, but also on the considerable contribution to the environment [26, 27, 28, 29, 30] by reducing CO_2 emissions to the atmosphere. To carry out this study, a CO_2 Emission Factor of 0.7079 Kg CO_2 eq/kWh was considered, data obtained from the Ministry of the Environment and which is applicable to all those connected to the National Electricity System (SEN). Due to the foregoing, two scenarios are considered: the first reflects the reduction of polluting emissions from the use of the photovoltaic micro-station located in this building, and the second refers to the economic and energy savings obtained by implementing the home automation system.

2. Methodology

The development of the research work has several stages, first, the scenes are added, in which the name must be located for each of the scenes that you want to add. In these scenes, the devices to be used in the home automation project can be selected, Figure 1, and they can be programmed the times in which they will carry out specific tasks [8].

Figure 1. Selection of events to perform

We can view the different types of installed devices by clicking on "Select a device". If some of them are not properly installed, you must go to the Devices option and click on the "Add devices" button where the additional Device options are displayed. In Figure 2 you can see all the devices that are compatible with this protocol.

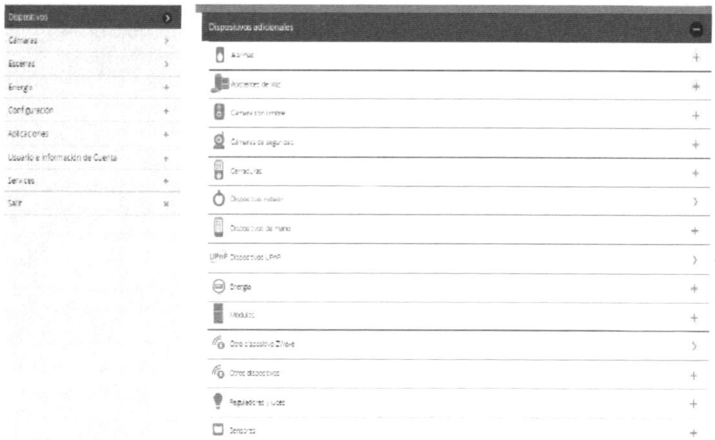

Figure 2. Devices that work with the Z-Waze protocol

Once the device that you want to add in the project has been identified, the procedure that the interface of the Vera Home page indicates must be followed. For example, if we want to add the "6 in 1 Multisensor" device, the technical specifications must be found in the manual of the device, and when clicking on it the following page is displayed, see Figure 3.

Figure 3. Registration window of the equipment to be installed

Figure 4 shows the registration of all the devices that will make up the system.

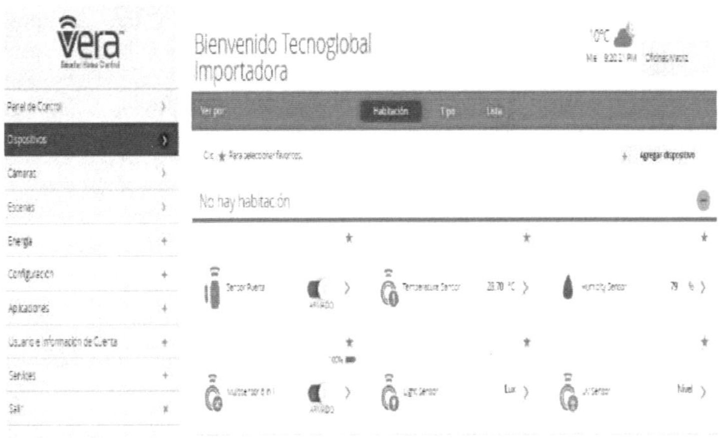

Figure 4. Devices added in the system

Once the device has been added, the scenes are created, see Figure 5. To access it, place the cursor on the word “Scene” shown on the left side of the screen and then click on the word "Add a scene" located on the right side of the window.

Figure 5. Introduction to creating a home automation scene

Then the device to which the scene will be created is selected with the conditions required by the user, see Figure 6. In this case the 6-in-1 Multisensor is used.

Figure 6. Scene creation window

Subsequently, the option that best suits the function to be performed is chosen. This device is responsible for detecting movement as long as the 6-in-1 Multisensor is armed or disarmed; finally, the order is validated, see Figure 7.

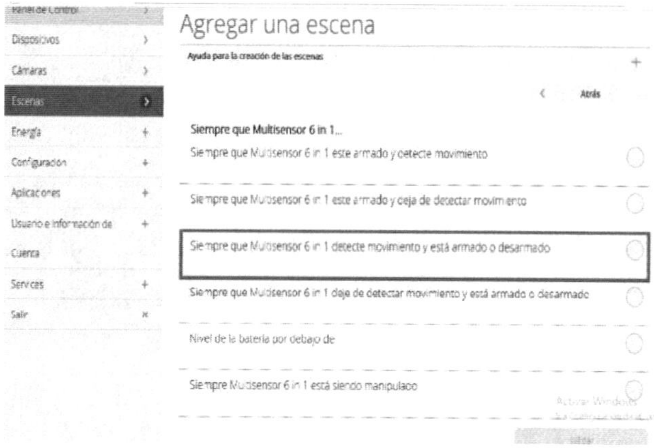

Figure 7. Device selection window for a scene

To continue with the configuration process, click on “Next Step”, as shown in Figure 8.

Figure 8. Programming options of the selected device

Figure 9 indicates the action that the device needs to perform, clicking on "What do you want to happen?", Provided that the 6-in-1 Multisensor is armed or disarmed.

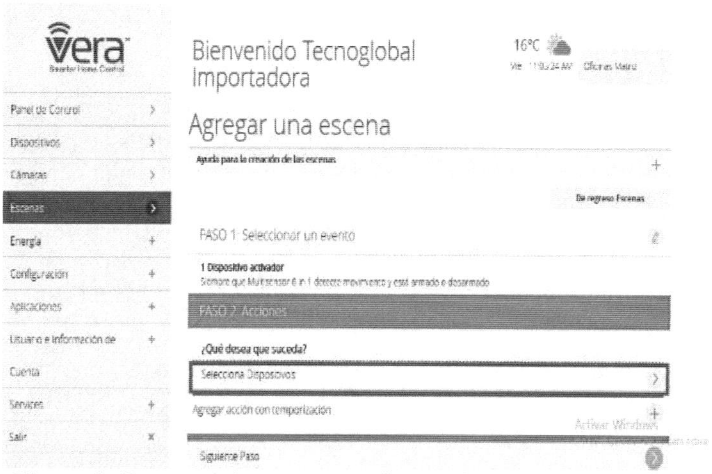

Figure 9. Programming corresponding to the action to be executed.

Figure 10 shows how the place where the home automation element is to be located within the room is selected. In this case it is going to be located in one of the outlets that are located in the teachers' office of Teacher Building No. 3 of the "Electricity Career".

Figure 10. Selection of devices to use

The window that shows the state of the switch, on or off, when the outlet is armed or disarmed, is shown in Figure 11. In this case, the ignition is chosen, and the order is validated.

Figure 11. Selection of the command to be executed by the device

To finalize the scene configuration, the "Step 3" option must be pointed out, which defines the scene's execution time, its working mode, as well as the user to whom it must be notified. It should be noted that to receive the notification you must download an application called VeraMobie available in the Play Store and that is compatible for Android and IOs operating systems.

Figure 12 shows how you define how you want to run the scene. There are four options available: vacation, home, night, and absent; from which you can select the ones you consider necessary, having the possibility to mark them all. In our case, all are marked and then the word "Done" is clicked.

Figure 12. Scene execution modes

Once the configuration of the work scene is completed, the economic feasibility analysis of the proposed project is carried out. In the investigation, it became necessary to know the cost of electrical energy, and the cost of home automation equipment that works with the Z-Wave protocol.

In the economic analysis, it was necessary to know the rate contracted by the UTM, which is a medium voltage rate with hourly demand. This rate is regulated by the Electricity Regulation and Control Agency, which also regulates existing penalties due to a low power factor (fp). Said agency establishes that when the average value of the power factor is less than 0.6, the distribution company will make a notification to inform the user of the suspension of the electric power service. This service will be restored when the value of the f.p > 0.6 (Electricity Regulation and Control Agency, 2018).

3. Discussion and Results

Table 1 shows the total monetary value that the UTM has to pay. The value to be paid is $ 822.29 and takes into account the cost of the electrical energy consumed plus the penalty for low power factor.

Table 1. Electricity billing in a trial month

Monthly energy consumption without the application of the Z-Wave home automation protocol				
	Energy Cost	Consumption	Unit	Total value
Power 7:00 am-22:00pm	$ 0.095	4288.25	kWh	$ 407.38
Energy 22:00 pm-7:00am	$ 0.077	287.74	kWh	$ 22.16
Monthly demand	$ 4.576	29	kW	$ 132.70
			Amount to be paid	$ 562.24
Average power factor	0,63			
Low power factor penalty (fp)	0,46		Low penalty fee payment	$ 258.63
			Commercialization	$ 1.41
			Total value to pay	$ 822.29

By implementing the home automation system using the Z-Wave communication protocol, savings of over 30% of total electricity consumption can be obtained. The economic analysis also showed that up to 60% can be saved on the building's lighting system. Finally, it is considered that by varying the temperature in the air conditioning system by one degree, savings of up to 5% will be obtained; for this reason, the proposed proposal assumes a savings of 35%.

Table 2. Monthly billing of the monthly electrical energy consumed by applying the Z-Wave home automation.

Energy consumption of the building applying the Z-Wave home automation protocol							
	Energy Cost	Consumption	% saving	Consumption with Z-Wave	(C.E-C.A.D)	Unit	Total Values
7:00am 22:00pm	$ 0.095	4288.25	0.35	1500.89	2787.36	kWh	$ 264.80
22:00pm 7:00am	$ 0.077	287.74	0.65	187.03	100.71	kWh	$ 7.75
Monthly Demand	$ 4.576	29	0.35	10.15	18.85	kW	$ 86.26
						Amount to be paid	$ 358.81
Power factor (fp)			0.63				
Low fp penalty			0.46			Low fp penalty	$ 165.05
						Commercialization	$ 1.41
						Total value to pay	$ 525.27

Table 2 shows the electric energy billing which considers the possible estimated monthly savings. In addition, it is observed that the total value to be paid is $ 525.27, which demonstrates the energy savings that can be obtained.

In Figure 13 it is observed that the theoretical saving is $ 297.19 per month, so there will be an annual saving of $ 356.27. If it is considered, in the near future, to generalize the results obtained to other university buildings, it would considerably increase energy savings, improving the return time of the investment made. In addition to contributing to energy savings, the implementation of the home automation system project will contribute to the improvement of the power factor in said building.

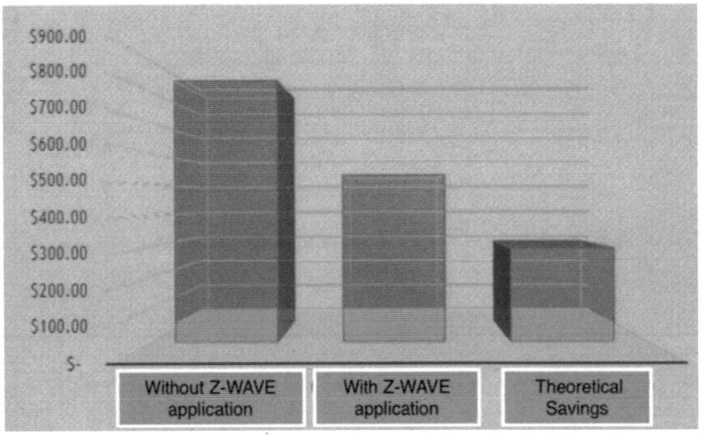

Figure 13. Theoretical savings by applying the Z-Wave Protocol

The data for the analysis of the feasibility of the project are shown in Table 3, which shows the approximate values of the cost of implementation of the home automation devices of the system, the annual maintenance, the discount rate and the useful life of the installation.

Table 3. Preliminary data for feasibility analysis

Elements	Costs	Rate off	10%
Home Automation Devices	$ 17 357.84		
Annual maintenance	$ 100.06	Useful life (years)	20
Total value	$ 17 457.84		

Table 4 allows visualizing the final results of the feasibility analysis, where it is shown that the implementation of the home automation project is feasible, considering only building No. 3, since the payback time of the investment is 8 years.

Table 4. Results of the feasibility analysis

VAN	$ 12152.49
TIR	19.4 %
ROI	1.70
PAY-BACK (years)	8

In Figure 14 you can see the possible behavior of the feasibility analysis.

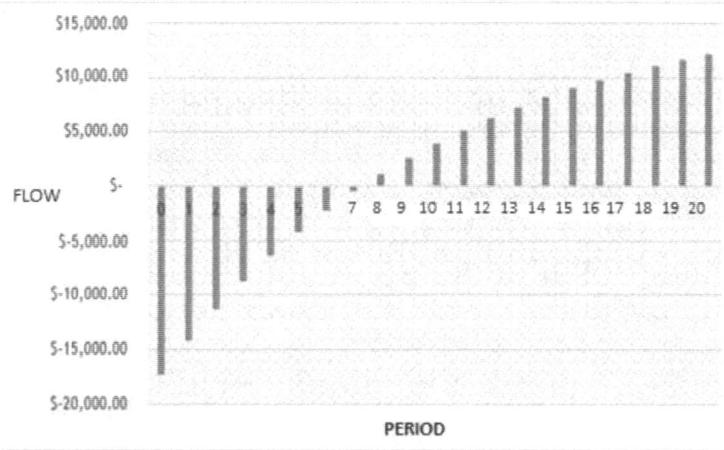

Figure 14. Behavior of the economic flow of the home automation project installation

Table 5 shows the environmental impact produced using the mini-PV plant. In it you can see the total emissions generated into the atmosphere by analyzing different consumption; it also refers to the amount of tons of CO_2 that would cease to be emitted into the atmosphere thanks to the implementation of the photovoltaic micro-plant, which value is 2.55 tons of CO_2 equivalent. Total emissions to the atmosphere are 36.32 tons of CO_2 equivalent, which allows determining the annual consumption of the building, which is 51304.28 kWh.

Table 5. Environmental impact from the use of the mini-photovoltaic plant

	Annual consumption	Unit	Gas emission	Unit
Power consumed from the grid	54911.89	kWh	38.87	ton CO_2
Annual energy produced by the mini-plant	3607.66	kWh	2.55	ton CO_2
Total equivalent energy	51304.28	kWh	36.32	ton CO_2

Knowing the decrease in CO_2 emissions to the environment by the implementation of the photovoltaic micro-plant, the study aims to further reduce these emissions using the Z-Wave protocol. Table 6 collects the results obtained from this analysis.

Table 6. Environmental impact of the implementation of the home automation system

	Annual consumption	Unit	Gas emission	Unit
Total energy using the Z-Wave protocol	34656.86	kWh	24.53	ton CO_2
Energy consumed from 10:00 pm to 7:00 am	1208.52	kWh	0.85	ton CO_2
Total energy consumed	33448.34	kWh	23.68	ton CO_2

4. Conclusions

For the implementation of the proposed home automation system, an economic and environmental analysis was_performed, resulting in the feasibility of its installation. This home automation system, in addition to contributing to energy efficiency, comfort and safety, will have the capacity to contribute to the reduction and correction of different electrical anomalies that exist in this building.

The result obtained in terms of projected savings is 35%, which corresponds to a value of $ 297.19 per month, $ 3566.27 per year. These values demonstrate the feasibility of this proposal by reducing 15.19 tons of CO_2 equivalent in emissions of polluting gases into the atmosphere.

In order to achieve a greater reduction in CO_2 emissions into the atmosphere, the consumption of electrical energy from the network must be reduced, and for this, the implementation of the home automation system has been considered, so in future research, the design of a photovoltaic plant will be carried out. Have the capacity to cover all the energy consumption of the UTM, and with it the implementation of an energy storage system. In addition, to obtain a better optimization of the electrical energy from this renewable energy source, it must be connected to the home automation system, which is intended to be done through the use of intelligent electrical networks based on the creation of an artificial neural network.

References

[1] Valencia, G., Núñez, J., & Acevedo, C. (2019). Research Evolution on Renewable Energies Resources from 2007 to 2017: A Comparative Study on Solar, Geothermal, Wind and Biomass Energy. International *Journal of Energy Economics and Policy*. 9 (6). Pp 242-253. DOI: https://doi.org/10.32479/ijeep.8051

[2] Lenz, J. M., H. C. Sartori., & J. R. Pinheiro. (2018). Defining Photovoltaic Mission Profile for the Pre-Design of Static Converters. *IEEE Latin America Transactions*, Vol. 16, No. 5, pp. 1402-1409. DOI: 10.1109/TLA.2018.8408434.

[3] Núñez, J. R., Benítez, I., Proenza, R., Vázquez, L., & Díaz, D. (2020). Methodology of fault diagnosis for grid-connected photovoltaic systems of network connection. *Rev. Iberoam. Automática e Informática Ind.*, vol. 17, no. 1, p. 94, Jan. 2020, doi: 10.4995/riai.2019.11449.

[4] Barrozo, F., Valencia, G., Obregón, L., Arango-Manrique, A., & Núñez, J. (2020). Energy, Economic, and Environmental Evaluation of a Proposed Solar-Wind Power On-Grid System Using HOMEPro®: A Case Study in Colombia. *Energies*, vol. 13, no. 7, pp. 1662. https://doi.org/10.3390/en13071662.

[5] Cedeño Núñez, V. E., & Ruiz Vasco, J. C. (2013). Diseño e implementación de un módulo de control domótico de arquitectura centralizada y distribuida basada en Lonworks. Departamento de Eléctrica y Electrónica. Carrera de Ingeniería en Electrónica e Instrumentación. ESPE. Extensión Latacunga. http://repositorio.espe.edu.ec/handle/21000/7198.

[6] Cuenca, L. (2016). Aplicación de las Redes Inteligentes para el aumento de la Eficiencia Energética en las Redes Eléctricas. *Revista RIEMAT*, vol 1, pp. 49–53.

[7] Nuñez, J., Benítez, I., Rodríguez, A., Díaz, D., & Oliveira, D. (2019). Tools for the Implementation of a SCADA System in a Desalination Process. *IEEE Latin America Transactions*, vol. 17, no. 11, pp. 1858–1864, 2019. 10.1109/TLA.2019.8986424.

[8] Mendoza, E. et al. (2020). Red de sensores inalámbricos multisalto para sistemas domóticos de bajo costo y área extendida. *Revista Iberoamericana de Automática e Informática industrial*, [S.l.]. v. 17, n. 4, p. 412-423. doi:https://doi.org/10.4995/riai.2020.12301.

[9] Morganti, G., Perdon, A. M., Conte, G., & Scaradozzi, D. (2009). Multi-agent system theory for modelling a home automation system. *Lecture Notes in Computer Science*. vol. 5517, PART 1, pp. 585-593. DOI: 10.1007/978-3-642-02478-8_74
[10] Intriago, M., Llosas, Y., Franco, P., Marques, Pico, G., & Nuñez, J. R. (2019). Sizing of a networked self-consumption system at the Technical University of Manabí-Ecuador, in CISCI 2019 - Decima Octava Conferencia Iberoamericana en Sistemas, Cibernetica e Informatica. Memorias, vol. 1, pp. 6–11.
[11] Lapiedra, R., Devece, C., A., & Guiral, J. (2011). Introducción a la gestión de sistemas de información en la empresa. Editorial Universitat Jaume I. 1st Edition. pp 72. ISBN: 978-84-693-9894-4
[12] Calzada, L., & Abreu, J. L. (2009). El impacto de las herramientas de inteligencia de negocios en la toma de decisiones de los ejecutivos. *International Journal of Good Conscience*, 4(2). pp. 17-52.
[13] López, N. M., Vela, J. P., & Mondejar, J. C. (2010). Diseño y explotación de almacenes de datos: Conceptos básicos de modelado multidimensional. E-book, v.1.0. Editorial Club Universitario. pp 103. ISBN: 978-84-9948-546-1.
[14] Chung, I. (2013). Distributed Intelligent Microgrid Control Using Multi-Agent Systems. *Engineering*, 5(1), 1–6. https://doi.org/10.4236/eng.2013.51B001.
[15] Milanes, C., Planas, J., Pelot, R., & Núñez, J. R. (2020). A new methodology incorporating public participation within Cuba's ICZM program. *Ocean & Coastal Management*, vol. 186, 105101. https://doi.org/10.1016/j.ocecoaman.2020.105101.
[16] Oketunji, T., & Omodara, O. (2011). Design of Data Warehouse and Business Intelligence System: A case study of Retail Industry. pp. 12-13, 21-22.
[17] Barrio, J., Abreu, H., Pina, I., & Alvarez, J. (2018). SmartGrid proposal in communities of Guamá Municipality of Santiago de Cuba Province. *Journal of Engineering and Technology for Industry Applications*, 4(14), 66-74. https://doi.org/10.5935/2447-0228.201831.
[18] Aguilar-Meléndez, A., et al. (2017). Development and Validation of Software CRISIS to Perform Probabilistic Seismic Hazard Assessment with Emphasis on the Recent CRISIS2015. *Computación y Sistemas*, 21(1), 67-90.
[19] Nuñez-Alvarez, J. R., Benítez, I., & Llosas, Y. (2020). Communications in Flexible Supervisor for Laboratory Research in Renewable Energy. *IOP Conf. Ser.: Mater. Sci. Eng.* 844 012016. DOI:10.1088/1757-899X/844/1/012016
[20] Valencia, G., Nuñez, J., & Vanegas, M. (2019). Data set on wind speed, wind direction and wind probability distributions in Puerto Bolivar – Colombia. *Data in Brief.* Vol 27, 104753. https://doi.org/10.1016/j.dib.2019.104753
[21] Nuñez-Alvarez, J. R., et al. (2020). Design of a Fuzzy Controller for a Hybrid Generation System. *IOP Conf. Ser.: Mater. Sci. Eng.* 844 012017. DOI: 10.1088/1757-899X/844/1/012017
[22] Milanés-Batista, C., Tamayo-Yero, H., De Oliveira, D., & Nuñez-Alvarez, J. R. (2020). Application of Business Intelligence in studies management of Hazard, Vulnerability and Risk in Cuba. *IOP Conf. Ser.: Mater. Sci. Eng.* vol. 844 012033. doi:10.1088/1757-899X/844/1/012033
[23] Ardila-Gamboa, C., & Ballesteros-Riveros, F. (2028). Data Envelopment Analysis to measure relative performance based on key indicators from a supply network with reverse logistics. *INGE CUC*, vol. 14, no. 2, pp.137-146, 2018. DOI: http://doi.org/10.17981/ingecuc.14.2.2018.13
[24] Acosta-Vega, R. K., Ospino-Ayala, O. J., & Valencia-Espejo, V. E. (2017). Design of an Enterprise Resource Planning system (ERP) for a microenterprise. *INGE CUC*, 13(1), 84-100.

[25] Intriago, M. G., et al. (2020). Use of the Sunny Design Web software with a view to be dimensioning the network-connected system at the Technical University of Manabí. *Conference on Complexity, Informatics and Cybernetics*, CICIC 2020, Volume 1, Pages 11-15, Code 159951.
[26] Palomino, K., Reyes, F., Núñez, J., Valencia, G., & Herrera, R. (2020). Wind Speed Prediction Based on Univariate ARIMA and OLS on the Colombian Caribbean Coast. *Journal of Engineering Science and Technology Review.* 13 (3), pp 200 – 205. doi:10.25103/jestr.133.22
[27] Botero, C. M., Arrizabalaga, M., Milanés, C. & Vivas, O. (2017). Governance indicators for coastal risk management in Colombia. *Revista Luna Azul*, 45, 227-251. DOI: 10.17151/luaz.2017.45.12. http://200.21.104.25/lunazul/downloads/Lunazul45_12.pdf
[28] Grimaldo Guerrero, J. W., et al. (2017). Forecast electricity demand model using predicted values of sectorial gross domestic product: Case of Colombia. *Revista Espacios*. 38(22), pp. 3.
[29] Merchán, E. V., et al. (2020). An analysis of electricity generation with renewable resources in Germany. *International Journal of Energy Economics and Policy*, 10(5), pp. 361–367, 2020. DOI: https://doi.org/10.32479/ijeep.9369
[30] Grimaldo Guerrero, J. W., et al. (2018). Analysis of the Energetic and Productive Effects Derived by the Installation of a Conveyor Belt in the Metal-mechanic Industry. *International Journal of Energy Economics and Policy*, 18(6), pp. 196-201, DOI: https://doi.org/10.32479/ijeep.7066

IOP Conf. Series: Materials Science and Engineering 1154 (2021) 012003 doi:10.1088/1757-899X/1154/1/012003

Recognition of Colombian Alphabeth in Sign Language Using Deep Learning Techniques

E Arrieta-Rodríguez, R E Monterroza-Barrios, P L Torres-Alvarez,G E Castro-Lozano

Universidad del Sinú Elías Bechara Zainum Cartagena

E-mail: investigacionsistemas@unisinucartagena.edu.co

Abstract. This work presents the development of an algorithm for sign language alphabet recognition Colombian in images, using the deep learning technique. For this, a data set consisting of each letter of the alphabet from A up to Z and a total of 4043 images with a resolution of 150x150 pixels of 3 channels, a red neural was built convolutional with 256 layers, with an accuracy of 75%.

1. Introduction

Sign language is a language used throughout the world by audio-impaired people, but with notable differences between countries, specifically in this work it is covered in Colombia [8]. People with hearing disabilities present some problems at a personal, social, and work level, which make it difficult for them to fully include society, for this reason this work proposes the construction of a tool that facilitates interaction between audio-impaired people and the rest of the society.

2. Related word

Numerous investigations have been carried out on the incorporation of sign language in computer resources in the educational field, both in our country and in the world, this section mainly focuses on national and Latin American studies and experiences, due to the importance of in this dissertation he awards the social context that characterizes the educational field where Information Technology -IT- is incorporated.

One project titled "Recognition of sign language using image processing" developed by Ananya Roy and Sandhya Arora, January 2018. It proposes the system recognizes the alphabet of sign language by calculating the distance metric of Bhattacharyya between the histograms of the captured image, processed and the stored image and outputting that alphabet whose image histogram has a smaller Bhattacharyya distance with the histogram of the captured image [1].

Also the project "Automatic voice recognition method for the classification of vowels to Colombian sign language." developed by Andres Arias and David Rubiano, 2018. It proposes the system recognizes the vowels of the Colombian alphabet by means of audio in order to recognize the letter by audio and return the vowel by Colombian sign language [1].

Additionally, we found that in Mexico City, D.F., at the *Instituto Técnico Nacional* in the computing research center, the editor Fausto Pável Priego Pérez, carried out a project with the name of *Recognition of images of Mexican sign language.* This work proposes a system of two main modules which consist of learning and recognition. Use a video camera device (Kinect device) to detect the pattern or sign of each word; Two methods are used for the recognition of the sign, in the first one requires some type of specific gloves to recognize the sign. In the second, the Kinect device is used for recognition [6].

In 2012 at the Córdoba Faculty of the Technological University - Argentina, he presented a research project based on the Kinect system of the Xbox game console, an interpreter and translator software was produced for both sign language and natural language, so allows users to interact in different ways, it has a sign for each word and also for each letter in case you want to spell a non-existent word [4].

Financially, several works are identified, such as [2],[7], [5] which are related to sign language, with the application of different techniques from computer science.

3. Methods and Materials

This section describes the process of collecting, organizing and finally building the database of the sign language alphabet, through photographic registration. It also describes the process of modeling, design, normalization, division of the data and the diagram of the data.

3.1. Data collection

Taking into account that a database of images with the sign alphabet was not available in a local context, it was decided to build it manually. For this, the taking of photos was done, using cell phone cameras, this was necessary to meet the minimum amount required for the training of the model. The two smartphones that were used were the ZenFone Max (M2) and the Neffos y5s. The collection was made by the two researchers, and the participants who served as models to take photos of their hands indicating each letter of sign language. 4043 photos were taken for training and testing of the model.

3.2. Database Construction

Following this, the segmentation process by password was carried out; This process is also called the labeling process, where the names of each folder represent a letter of the Colombian alphabet in sign language. 4043 photos were taken of the 26 letters of the alphabet in sign language, in the data set there are two types of images due to the difference in dimensions that was caused by the use of the two devices for taking the images. To solve this problem, the data dimensions are standardized in the model.

3.3. Modeling

Figure 1 presents the design of the algorithm for the detection of sign language, this raises the segmented development in two phases, the first is the creation of the data set and the second correspondence to the construction stage of the neural network model convolutional [3].

3.4. Data Normalization

It is useful to normalize the data to make learning easier for machine learning. By normalizing we mean putting all the data on the same scale. It was necessary to take the data matrix and perform a normalization by applying a division of the matrix containing the training data between 255 (training data / 255 the maximum value of a pixel), Keras was used to process the images at the scale of data pixels 1/255, this normalization allowed the matrix operations that were carried out in training to be much faster and more efficient both for the multiplication that is carried out with the weights that are initialized and for the update of weights with respect to the level of error to be decreased.

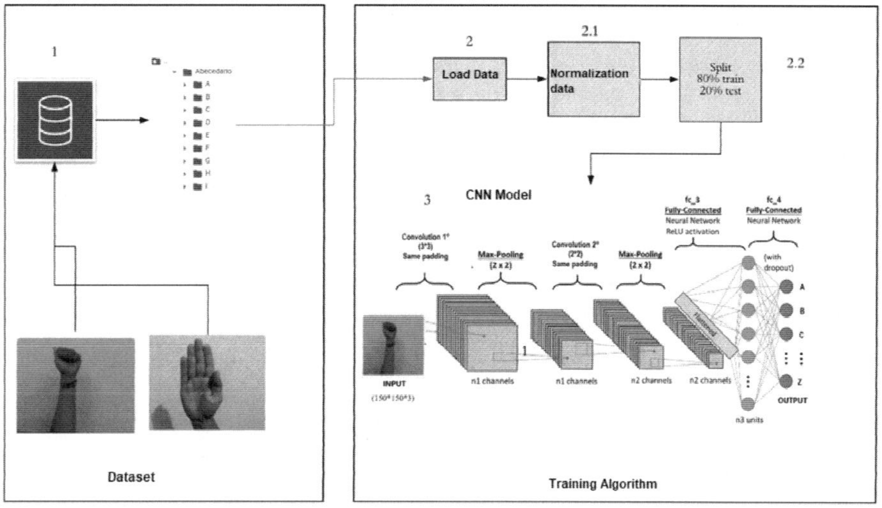

Figure 1. Design of the sign language prediction algorithm using CNN

3.5. Split Data

The Split technique is applied, which operates with a division of the total data in a distribution of 80% for training data and 20% for evaluation, the objective of having these divided data was relevant to us to corroborate with the evaluation data, if our generalized models with unknown data (show Figure 2).

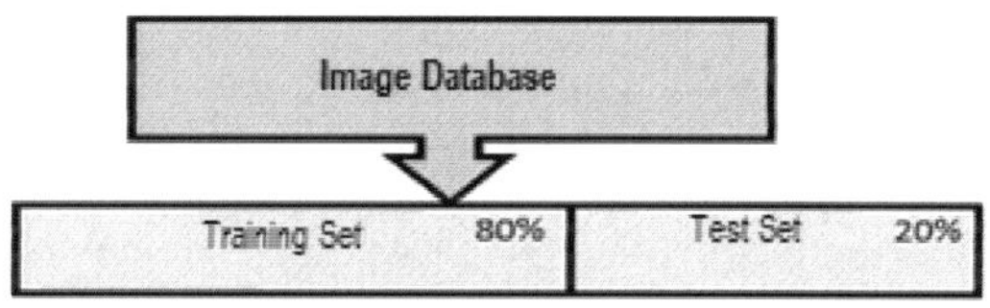

Figure 2. Design of the sign language prediction algorithm using CNN

3.6. Load Data

With the data set already built and saved in the Google drive tool, using the unique Google Colaboratory libraries to load the data into it. In it, the Python code was executed with the TensorFlow and Keras libraries for the construction and evaluation of the CNN (Convolutional Neural Network) model.

4. Architecture

In the architecture of the model the functions that helped to build the convolutional neural network are defined, then it will be briefly explained how these functions(Show Figure 3).

(i) Sequential(): This function is a stack of linear layers, which are created by passing an instance list of layers to the constructor.

(ii) add(convolution2D()): This layer creates a convolution core that is interlaced with the input layers to produce an output tensor. When using this layer as the first layer in a model, certain keywords are provided which are filters, kernel size, strides, padding, input shape and activation.

- Filters: is the dimension of the convolution output.
- Kernel size: specifies the dimensions of the input image.
- Strides: convolution strides far and wide.
- Padding: same results in padding the input so that the output is the same length as the original input.
- Input shape: indicates that the expected input will be batches of vectors of number of dimensions.
- Activation: can be used through one activation layer or it can be supported by all layers.

(iii) MaxPooling2D: Maximum grouping operation for spatial data. MaxPooling2D is a 3D shape tensor (weight, height, channel) the height and weight dimensions tend to decrease as we enter the hidden layers of the neural network.Where Units means Pool size: tuple of 2 integers that reduces the scale (vertical, horizontal). (2, 2) will halve the input in both spatial dimensions. If only an integer is specified, the same window length will be used for both dimensions.

(iv) Flatten: Converts the elements of the input image array to a flat array.

(v) Dense: With this instruction we add a hidden layer of the neural network. As Units means the dimensionality of the output space.

(vi) Dropout: Dropout is randomly setting a rate fraction of the input units to 0 or 1 on each update during the training time, which helps avoid overfitting.

(vii) Compile: Set up the model for training.

- Optimizer: instance of the optimizer.
- Loss: instance loss function, if the model has multiple outputs, you can use a different loss in each output passing a dictionary or a list of losses.
- Metrics: List of metrics that the model will evaluate during training and testing.

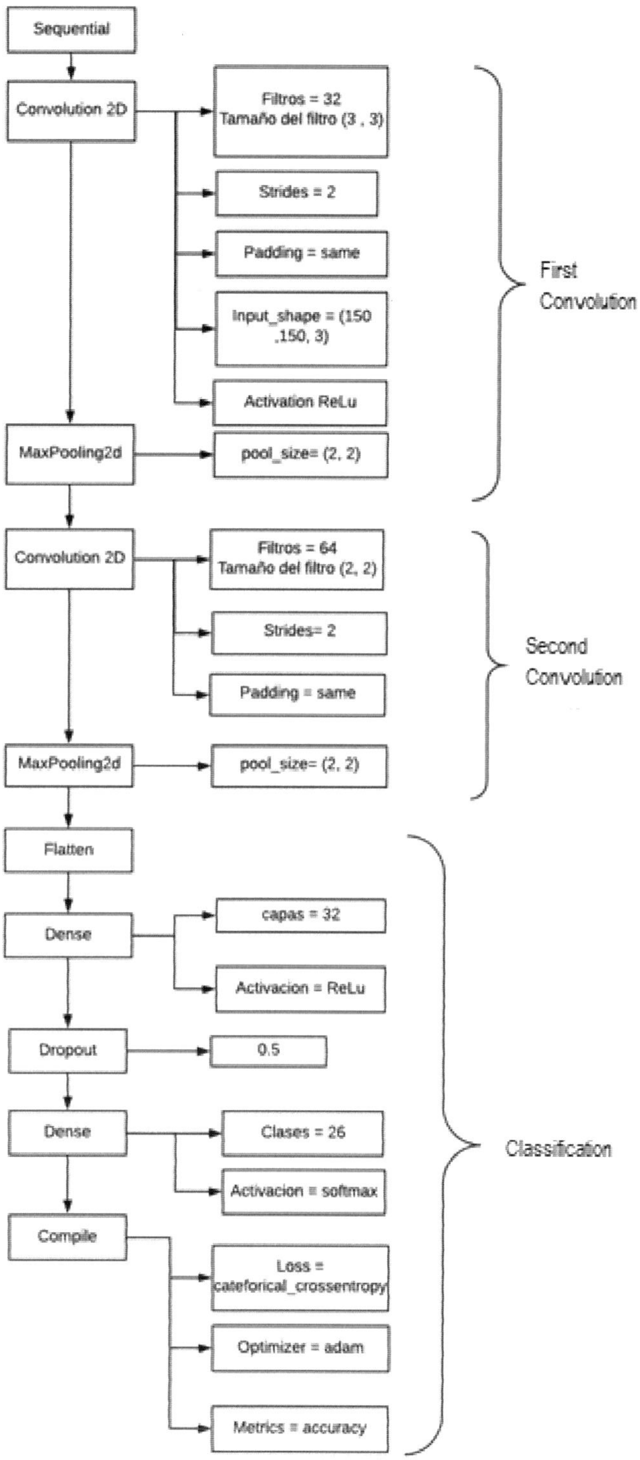

Figure 3. Model Diagram

IOP Conf. Series: Materials Science and Engineering 1154 (2021) 012003 doi:10.1088/1757-899X/1154/1/012003

5. Trainig and Test Model

Once the suitable data set has been obtained and also the machine learning technique that will be used to develop the letter classifier for sign language. The convolutional neural network technique was used, the model training was carried out with 80% of the data set and 20% of the set was used for the validation of the model. In the training it was done in 15 epochs (epoch) that are the times that the model will be trained, with 100 steps per epoch (sample batches) it is the one that says when one epoch ends and the other begins.

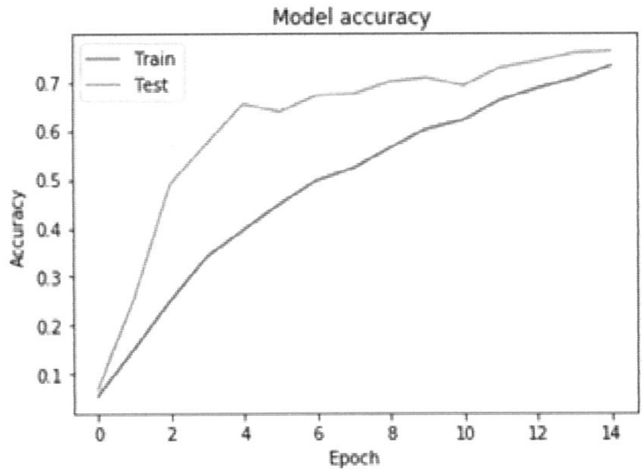

Figure 4. Accuracy

Previous in Figure 4 we can see the progress of the model's accuracy as the epochs are iterated, discriminated between the training (train) and the validation (test) of the model.

Next in Figure 5 shows the loss of the model in the passage of each epoch, both in the training process and in tests. As a result, a close loss in training and testing is obtained, and finally at time 12 the model reaches a very similar loss.

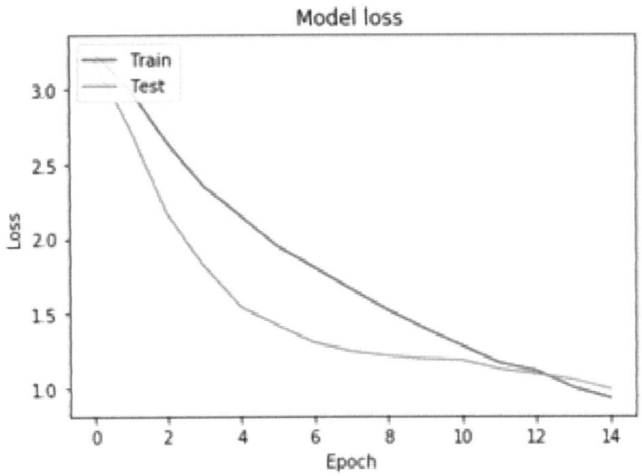

Figure 5. Loss

6. Results

After training the machine learning model, it is necessary to evaluate the performance of the classifier to determine its viability; one of the most used metrics when evaluating an algorithm is

IOP Conf. Series: Materials Science and Engineering 1154 (2021) 012003 doi:10.1088/1757-899X/1154/1/012003

the confusion matrix. This gives us relevant information about the performance of the classifier, since from it other metrics are extracted that give true information about it.

The confusion matrix allows the visualization of the performance of an algorithm and helps to validate the successes in the prediction and the frequency of errors. The variables H, N and Z (letters of the alphabet) have low correct answers, since in sign language these images have similarities with others (show Figure 6).

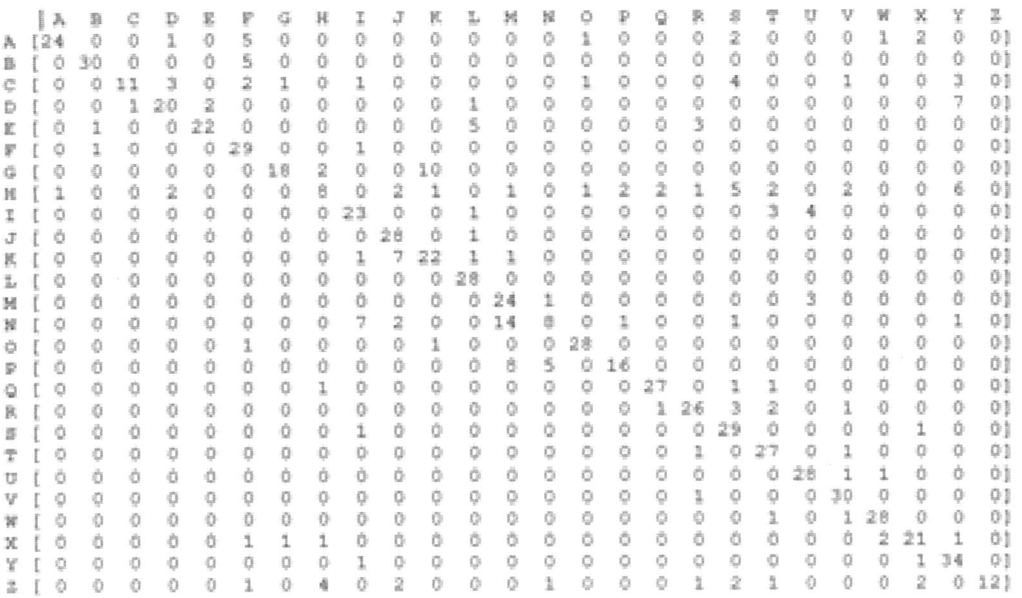

	A	B	C	D	E	F	G	H	I	J	K	L	M	N	O	P	Q	R	S	T	U	V	W	X	Y	Z
A	24	0	0	1	0	5	0	0	0	0	0	0	0	0	1	0	0	0	2	0	0	0	1	2	0	0
B	0	30	0	0	0	5	0	0	0	0	0	0	0	0	0	0	0	0	0	0	0	0	0	0	0	0
C	0	0	11	3	0	2	1	0	1	0	0	0	0	0	1	0	0	0	4	0	0	1	0	0	3	0
D	0	0	1	20	2	0	0	0	0	0	0	1	0	0	0	0	0	0	0	0	0	0	0	0	7	0
E	0	1	0	0	22	0	0	0	0	0	0	5	0	0	0	0	0	3	0	0	0	0	0	0	0	0
F	0	1	0	0	0	29	0	0	1	0	0	0	0	0	0	0	0	0	0	0	0	0	0	0	0	0
G	0	0	0	0	0	0	18	2	0	0	10	0	0	0	0	0	0	0	0	0	0	0	0	0	0	0
H	1	0	0	2	0	0	0	8	0	2	1	0	1	0	1	2	2	1	5	2	0	2	0	0	6	0
I	0	0	0	0	0	0	0	0	23	0	0	1	0	0	0	0	0	0	0	3	4	0	0	0	0	0
J	0	0	0	0	0	0	0	0	0	28	0	1	0	0	0	0	0	0	0	0	0	0	0	0	0	0
K	0	0	0	0	0	0	0	0	1	7	22	1	1	0	0	0	0	0	0	0	0	0	0	0	0	0
L	0	0	0	0	0	0	0	0	0	0	0	28	0	0	0	0	0	0	0	0	0	0	0	0	0	0
M	0	0	0	0	0	0	0	0	0	0	0	0	24	1	0	0	0	0	0	0	3	0	0	0	0	0
N	0	0	0	0	0	0	0	0	7	2	0	0	14	8	0	1	0	0	1	0	0	0	0	0	1	0
O	0	0	0	0	0	1	0	0	0	0	1	0	0	0	28	0	0	0	0	0	0	0	0	0	0	0
P	0	0	0	0	0	0	0	0	0	0	0	0	8	5	0	16	0	0	0	0	0	0	0	0	0	0
Q	0	0	0	0	0	0	0	1	0	0	0	0	0	0	0	0	27	0	1	1	0	0	0	0	0	0
R	0	0	0	0	0	0	0	0	0	0	0	0	0	0	0	0	1	26	3	2	0	1	0	0	0	0
S	0	0	0	0	0	0	0	0	1	0	0	0	0	0	0	0	0	0	29	0	0	0	0	1	0	0
T	0	0	0	0	0	0	0	0	0	0	0	0	0	0	0	0	0	1	0	27	0	1	0	0	0	0
U	0	0	0	0	0	0	0	0	0	0	0	0	0	0	0	0	0	0	0	0	28	1	1	0	0	0
V	0	0	0	0	0	0	0	0	0	0	0	0	0	0	0	0	0	1	0	0	0	30	0	0	0	0
W	0	0	0	0	0	0	0	0	0	0	0	0	0	0	0	0	0	0	0	1	0	1	28	0	0	0
X	0	0	0	0	0	1	1	1	0	0	0	0	0	0	0	0	0	0	0	0	0	0	2	21	1	0
Y	0	0	0	0	0	0	0	0	1	0	0	0	0	0	0	0	0	0	0	0	0	0	0	1	34	0
Z	0	0	0	0	0	1	0	4	0	2	0	0	0	1	0	0	0	1	2	1	0	0	0	2	0	12

Figure 6. Confusion matrix

Figure 7 shows precision, recall, f1-score, support in detail. The precision provides information about their performance of how many they have captured, recall gives information from a classifier of how many have failed, f1-score is the weighted average of precision and recovery, and support were the number of images that were used for each letter to make matrix.

The accuracy of the model is 0.75 percent precision, failing more where the f1-score has the lowest value. This implies that the model has to be trained with a greater number of images to obtain a higher accuracy result.

7. Conclusions

At the end of the algorithm tests, we conclude that the code works with a margin of error of approximately 25

- Initially, a manual data set was built which was used to train the model.
- The neural network was then trained with the training data set, which then led to the evaluation of the model to wait for the results given in a classification report and a confusion matrix.
- Finally, the validation stage is reached. It was a work in conjunction with the test stage, since after observing the test results it was possible to analyze results to determine the best behavior of the training parameters.

Classification Report

	precision	recall	f1-score	support
A	0.96	0.67	0.79	36
B	0.94	0.86	0.90	35
C	0.92	0.41	0.56	27
D	0.77	0.65	0.70	31
E	0.92	0.71	0.80	31
F	0.66	0.94	0.77	31
G	0.90	0.60	0.72	30
H	0.50	0.22	0.31	36
I	0.66	0.74	0.70	31
J	0.68	0.97	0.80	29
K	0.65	0.69	0.67	32
L	0.76	1.00	0.86	28
M	0.50	0.86	0.63	28
N	0.53	0.24	0.33	34
O	0.90	0.93	0.92	30
P	0.84	0.55	0.67	29
Q	0.90	0.90	0.90	30
R	0.79	0.79	0.79	33
S	0.62	0.94	0.74	31
T	0.73	0.93	0.82	29
U	0.80	0.93	0.86	30
V	0.81	0.97	0.88	31
W	0.88	0.93	0.90	30
X	0.78	0.78	0.78	27
Y	0.65	0.94	0.77	36
Z	1.00	0.46	0.63	26
accuracy			0.75	801
macro avg	0.77	0.75	0.74	801
weighted avg	0.77	0.75	0.74	801

Figure 7. Metrics

This work is presented as an initial approach to the use of Deep Learning techniques applied in a real problem, such as communication in sign language, in search of achieving an inclusive society for this population.

References

[1] Arora, S., Roy, A.: Recognition of sign language using image processing. International Journal of Business Intelligence and Data Mining **13**(1-3), 163–176 (2018)
[2] Castro Estévez Luis Alberto, C.G.K.K.: Herramienta de apoyo para la interpretación de lenguaje de señas mexicano (hilsem) (2015)
[3] Crisvill: Avances en redes neuronales. medium (2016), `https://medium.com/espanol/avances-en-redes-neuronales-705c2efe5`
[4] Matías Reggi, Fabián Bertetto, E.R.: Inventaron software que interpreta lenguaje de señas. buendiario (2015), `https://www.buendiario.com/inventaron-software-interpreta-lenguaje-senas`
[5] de Oliveira Andrade, R.: Software convierte voz en lenguaje de señas para sordos. scidev.net (2014), `https://www.scidev.net/america-latina/comunicacion/noticias/software-convierte-voz-en-lenguaje-de-se-as-par`
[6] Pérez, F.P.: Reconocimiento de imágenes del lenguaje de señas mexicano. México, DF: Instituto Politécnico Nacional (2012)
[7] Saavedra, Y.: Microsoft planea usar kinect para interpretar el lenguaje de señas. hipertextual (2013), `https://hipertextual.com/2013/07/kinect-lenguaje-senas`
[8] Salazar, M.C., Molina, D.B.: Instituto nacional para sordos–insor-observatorio social

IOP Conf. Series: Materials Science and Engineering 1154 (2021) 012004 doi:10.1088/1757-899X/1154/1/012004

Tools for the Implementation of an Inmotic System in the Imperial Hotel in Santiago de Cuba, Cuba

J. R. Nuñez[1a], B. Torres[2], I. Benítez[3], C.B. Milanés[4], E. Noriega[5]

[1,5] Department of Energy, Faculty of Engineering, Universidad de la Costa, Calle 58 No 55-66, 080002, Barranquilla, Colombia.
[2] Department of Development and Automation of Advanced Systems, CEDAI, Ave Las Américas, s / n, 90300, Santiago de Cuba, Cuba.
[3] Department of Automation, Universidad de Oriente, Ave Las Américas, s / n, 90300, Santiago de Cuba, Cuba.
[4] Civil and Environmental Department. Universidad de la Costa, Colombia. ORCID: 0000-0003-2560-8859

[a]Corresponding author: jnunez22@cuc.edu.co

Abstract. Given the inefficiency of the control systems present in many hotels in Cuba and the need to automate and intelligently manage the electrical energy of the Imperial Hotel in Santiago de Cuba, the design and implementation of an imotic control using Tac technology is carried out View of Schneider Electric. The implemented system has a decentralized, scalable, modular, sustainable, and upgradeable architecture, which allows migrating to other similar or superior technological variants if necessary. In the design and implementation of the system, the different field sensors, the programmable robots, and the specific purpose controllers were selected, which communicate through Lonworks and TCP / IP protocols with the SCADA created using the Tac Vista 5.1.9 software. As a result of the investigation, the supervision and control of the lights, the bathroom odor extractor, and the temperature in the rooms are achieved. In addition, the automatic control of the lights in the corridors and exterior areas of the hotel is achieved, and finally, the sequence of automatic switching on or off of the air conditioning units in the lobby and restaurant respectively is programmed. All this leads to greater comfort, energy savings, and quality of the services provided, which at the same time becomes a tool for the management of the hotel by the people in charge.

Keyword. Automatic, sensors, control systems, controllers.

1. Introduction

Because in Cuba the main source of income to the country's economy is tourism, it has become necessary to build new hotels and modernize existing ones [1, 2]. The main tourist areas of the country are present in cities such as Havana, Varadero, Holguín, and the northern keys of the center of the territory, being served by national and foreign companies such as Gaviota, Cubanacan, Gran Caribe, Meliá, among others [3, 4].

In [2] the design of an inmotic system capable of achieving automation of the residential blocks of the 5-star Hotel plus Santiago de Cuba is carried out, where the networks used, their architecture, and topologies based on a distributed system where structured cabling is described Horizontal is achieved with UTP twisted pair and vertical cabling with optical fiber, as control elements programmable logic controllers and different field elements are used for the measurement and execution of control actions where supervision is carried out through a SCADA. In [5] a prototype of an automated system to open or close blinds or curtains that allow natural light to enter the rooms of a building is developed using a logical and electromechanical architecture for its operation, comparing the advantages of this method with the traditional.

In [6] a study of energy saving in different spaces using the Unified Modeling Language (UML) and the simulation of multiple agent systems is presented. The work demonstrates the benefits of using artificial intelligence and modern information and communication technologies for energy management.

Furthermore, in [7] an architecture is proposed that uses the Internet of Things (IoT) to control buildings using free technologies.

In [8] an automation system is designed and implemented to control the processes within a solar conversion unit using a programmable logic controller from Schneider Electric. Finally, the startup, operation, and shutdown sequences could be solved in safe conditions, as well as monitoring the working parameters of a Stirling motor together with a solar concentrator to increase the energy efficiency of the installation.

The Imperial Hotel is a majestic building that is located in the central Enramadas street in the center of the city of Santiago de Cuba and is considered one of the jewels of the city's architecture, figure 1. This hotel has just been subjected to a capital maintenance process, which took into account the current state of the hotel situation in the province and the country, and which also met international standards in terms of beauty and comfort. The Hotel is made up of a lobby bar, three living floors with a total of 42 rooms, including 3 suites and one room for the disabled, it has a Snack-Bar-Grill and a Roof Garden.

Figure 1. Vista del Hotel Imperial. *(Source: https://images.app.goo.gl/ouUHuNkNeW42Td379)*

In the Imperial Hotel, although all the luminaires installed in the common areas use Led technology, they are switched on and off utilizing on / off switches, so these luminaires are switched on in the afternoon and switched off in the morning. The same happens with the luminaires inside the rooms, which will remain on if the guest does not decide to turn them off.

The air conditioning systems installed in the 42 rooms are multi-split, that is, with a single compressor, more than one space can be air-conditioned. These Split have their temperature regulation using a thermostat that sends a signal to the compressor when it reaches the temperature selected by the host. This system works correctly, but it is not very efficient, since the guest can leave the room and leave the air conditioning on, spending energy unnecessarily.

A common in this Hotel, as in most Cuban hotels, the odor extractor systems in the bathrooms are directly connected to the switch that controls the lighting on and off, so its operation is conditioned to the luminaire remaining On, and once it is turned off, the extractor stops, even if it has not evacuated all odors. Practice is that the host leaves the luminaire on so that the extractor continues to operate, which causes unnecessary expense and excessive use of the equipment [4].

Based on the particularities of the Hotel and the knowledge of the working modes of its systems, strategies were designed to ensure that the proposed inmotic system fulfilled the task of automatically turning the lights on and off, depending on the presence of people in the room, that conditions the operation of the air conditioning to the presence of people and that, in addition, the control of the bathroom odor extractor does not depend on the switch that controls the lighting on and off. On the other hand, it is necessary that the system achieves the automatic on and off of the luminaires in the interior corridors depending on their use, as well as achieving control of the lighting of the exterior façade lamps depending on the level of existing lighting. or the schedule of the day. Finally, and of great importance for the saving of energy carriers, the sequential on and off of the air conditioners is required automatically.

2. Materials and methods

When analyzing the main technology manufacturers associated with the world of automation and control, which, in addition, had a presence in our country and complied with the requirements of Cuban standards for the construction and automation of hotel facilities, the technological solution was chosen. that best suits the specific needs of the hotel [4], in this case, the TAC Vista of Schneider Electric.

The TAC Vista is a building management system that allows you to monitor and control different systems, such as lighting, access control, security, among others. Based on the Lonworks open protocol, it is among the most adaptable systems in the industrial sector and with simple computer technologies. Its open architecture enables seamless integration with third-party products and offers full interoperability.

The TAC Vista supervision and control system controls and manages the hotel's automation system, facilitating the functions of the graphic display of processes, automatic alarm distribution, and a wide range of different data analysis functions [9, 10, 11, 12]. It is a distributed system composed of 3 levels of hierarchy, figure 2.

- Supervision Level (1): Central Supervision and Control System supported on the SCADA "TAC Vista 5 with Terminal application Server, Workstation and Webstation, communicating with Router CEA-709 / IP-852, communication gateway with LON protocols and Modbus over Ethernet TCP / IP, which allows the integration of Xenta controllers in the management system.
- Control Level (2): Xenta 121 FC parameterizable controllers and Modicon M241 programmable controllers that carry out distributed control of the hotel.
- Field instrumentation level (3): Group of sensors and final action elements that act directly with the processes to be controlled and supervised.

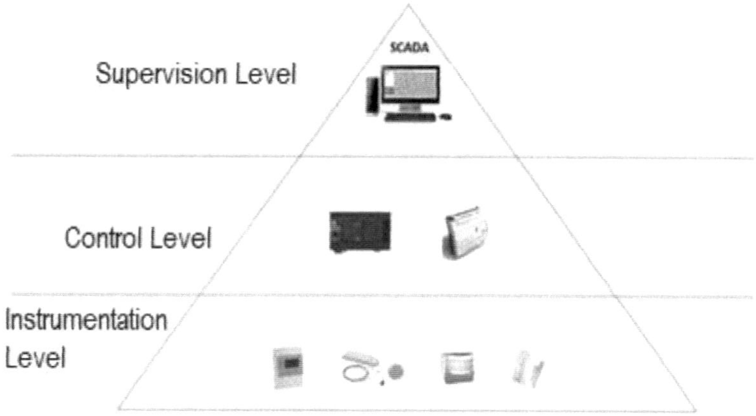

Figure 2. Automation pyramid at the Hotel Imperial

2.1. Protocols and Topologies

The communication configuration between the different hierarchy levels is made up as follows: a central computer with the TAC Vista 5 software as supervisor and rector of all the systems; all rooms use the TAC Xenta 121 as a controller, which is a parameterizable specific-purpose controller, where the presence sensors, door and window contacts are connected, as well as the STR150 temperature sensor [13, 14, 15, 16, 17, 18].

All these room controllers are connected via a Lonworks network and reach a RouterCEA-709 / IP-852 placed on each floor, which performs the function of a gateway and transmits via Ethernet to SCADA [19, 20, 21, 22, 23]. The other systems are governed by an M241 PLC that communicates via Ethernet with SCADA [24, 25, 26, 27, 28, 29, 39]. Figure 3 shows the communication network implemented in the hotel.

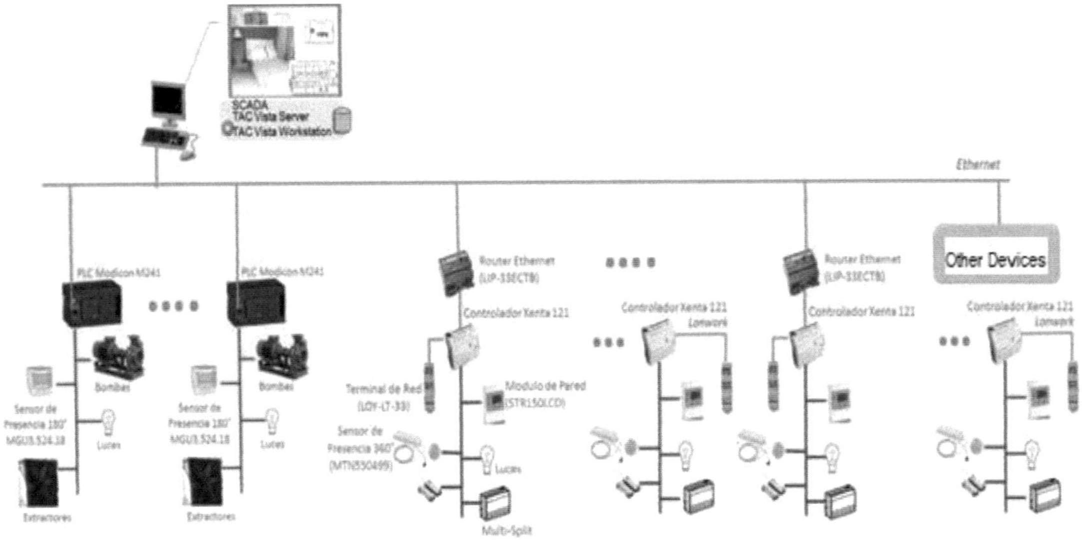

Figure 3. General communication scheme

2.2. Control of systems in rooms

The objective of the automatic system in the rooms is to control the lights, the bathroom exhaust fan and turn the air conditioning on or off once the desired temperature is reached, with the premise of achieving and maintaining comfort parameters, as well as controlling o decrease the energy consumption rates of the installation to optimize operating costs and extend the life of the equipment. In the rooms, the temperature and the presence of people in it are taken as variables to measure, in addition, the state of opening of doors and windows is taken into account. The devices on which the controller operates are the lights in the room, the power in the outlets, and the switching on or off of the air conditioning compressor, see figure 4.

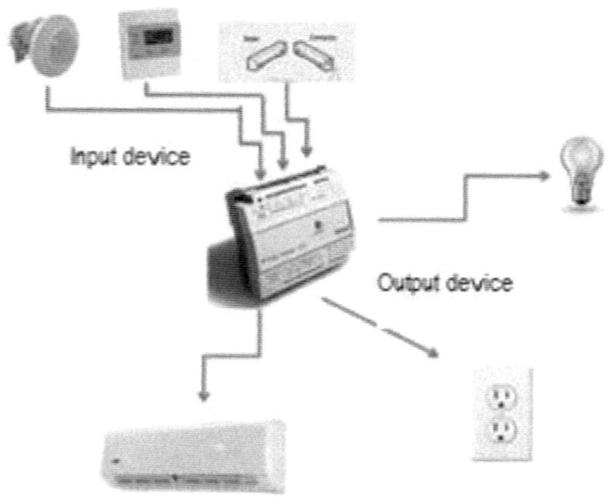

Figure 4. Input and output diagrams to the Xenta 121 FC controller

The rooms have the Xenta 121 specific purpose parameterizable controller, to which the signal of the Argus 360o presence system is connected. Bticino magnetic contacts detect the opening of doors and windows and also send the signal to the Xenta 121 FC controller. For the control of the bathroom extractor, it has an iRTC time relay A9E16067, parameterizable from 0.1 seconds to 100 hours.

The room temperature is measured and indicated through the temperature sensor with display STR 150, this sensor is the one that sends the SCADA the temperature values that are displayed in the mimics. Figure 5 specifies the location of the sensors within the rooms.

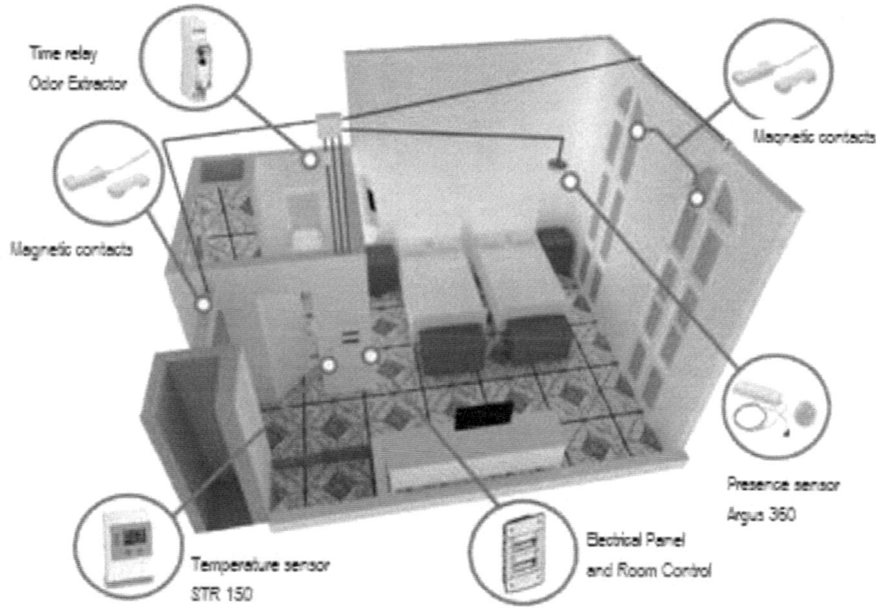

Figure 5. Location of the sensors inside the rooms

2.3. Xenta 121 FC Controller Operating Modes

The Xenta 121 FC Specific Purpose Controller has three basic modes of operation and a set of programmable exceptions:

1. Busy: When the presence sensor detects that there are people, the controller automatically goes into busy mode. In this state, it is not operated on any control element, it will be the guest who

decides at will if he wants to turn the lights on or off. The control of the air conditioning is carried out through the same multi-split equipment, where the guest can choose the temperature they want, this temperature is displayed on the SRT 150 wall module and sent to SCADA.

2. Unoccupied: This mode is activated once the time set in the controller has elapsed without the presence sensor having detected people in the room. In this state, it is actuated on a control contactor that is located in the electrical distribution box in the room, deactivating the main circuit that controls the bedroom lights and outlets. The room temperature setpoint value is also changed to 26 °C. To exit this state, it is necessary to activate the presence detector or to open the door.
3. Standby: This mode starts when the controller has been in idle mode for the time programmed in the controller. In this mode, the temperature setpoint value is changed again up to 28 °C. The controller exits this mode once the presence sensor that detects the presence or the magnetic contacts is activated.

Exceptions are special cases of equipment functionality and are configured to support basic operating modes and maintain comfort parameters. Depending on one or more conditions, they act to activate or deactivate Xenta 121 outputs, send parameters to SCADA, among others. The block diagram corresponding to the operation of the sample controller in Figure 6.

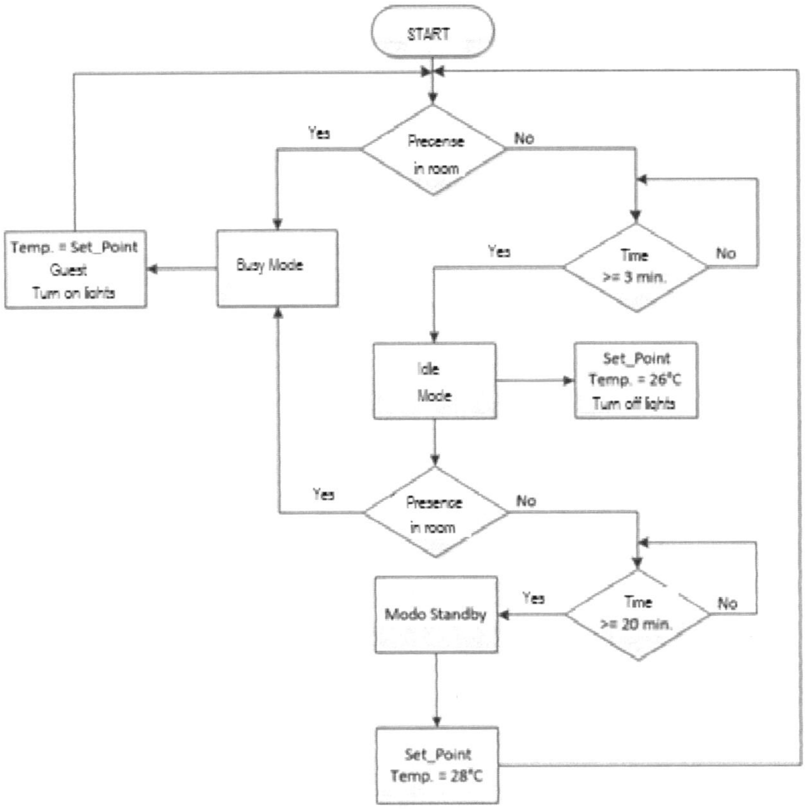

Figure 6. Block diagram of the operation of the Xenta 121

2.4. Lighting control in corridors

The control system for turning the luminaires on and off in the corridors uses 4 presence sensors of the type MGU3.524.18 and is designed to satisfy visual needs, creating healthy, safe, and comfortable environments. The system works in such a way that if it detects presence, the lighting circuit is activated, and then, after a time of 1 minute without detecting any presence, the circuit in question is turned off. This time can be modified in SCADA. The block diagram is shown in Figure 7.

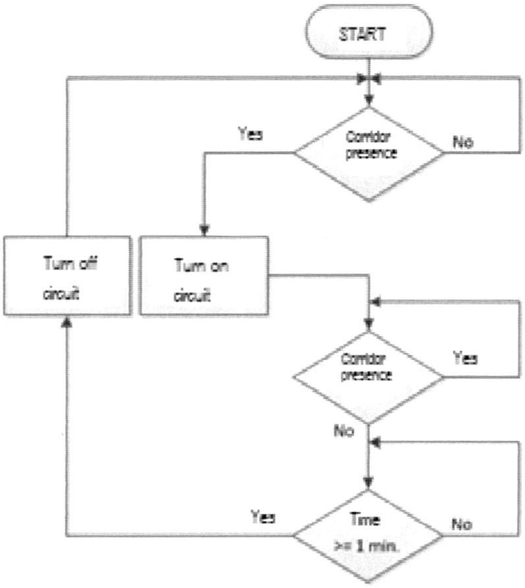

Figure 7. Block diagram of hallway light control

2.5. Control of external lights

For the control of the external lighting, a redundant control system has been designed, taking as a measurement the signal of two SLO300 luminosity sensors that are located in the East and West parts of the hotel, respectively, and taking into account the schedule in which said lighting is carried out. measurement.

Depending on the signal from these sensors, the system will be able to turn on the outdoor lights at night, and on days when the level of natural light is not optimal, generally on very cloudy or rainy days. Figure 8 shows the block diagram of the programming logic used to control the external lighting.

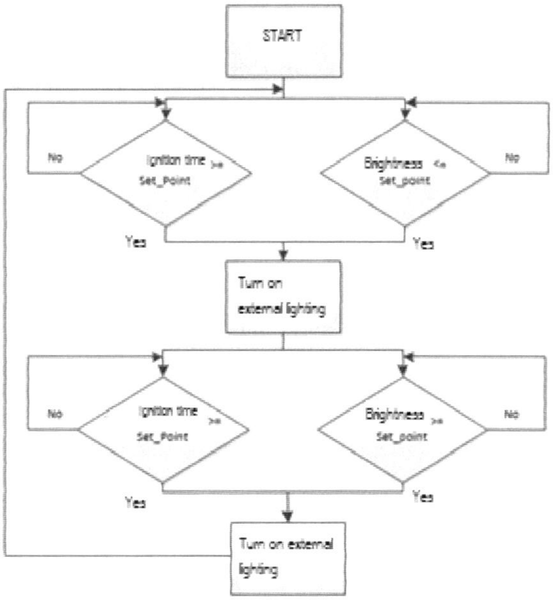

Figure 8. Block diagram of the exterior lighting control

2.6. Control and monitoring of air conditioners

Large fan coils are called air conditioners or air handling units (UTA). This equipment controls the temperature of a room and is used in large spaces such as meeting rooms, hotel lobby, restaurants, among others.

The Imperial Hotel has two air conditioners used in the Lobby and the Restaurant. Each of them has its temperature control so it was not necessary to control this variable, however, this logic was brought to the PLC to automatically turn it on and off, waiting for the command signal sent from the control board, as shown in Figure 9. The operating states of the compressors and fans of both air conditioners are represented in the SCADA, allowing it to change the on/off status.

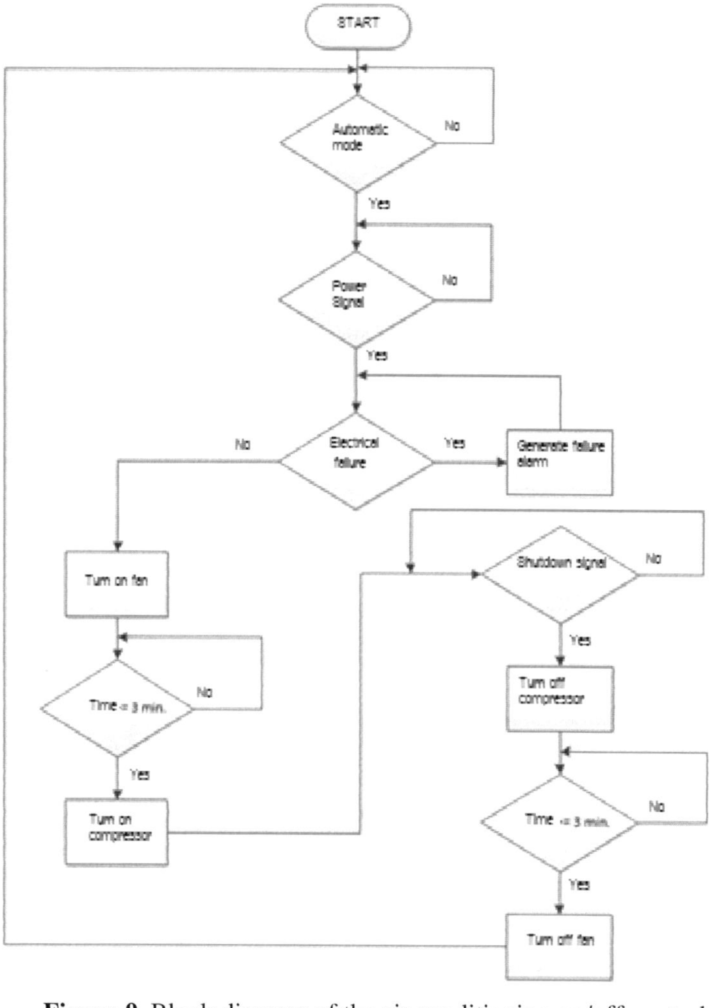

Figure 9. Block diagram of the air conditioning on/off control

3. Results and Discussion

The hotel's SCADA communicates through an Ethernet network, created specifically for the implemented imotic system and which is physically separated from the hotel's work network, thus accessing the different PLCs installed on the different levels of the building. The Xenta 121 FC controllers communicate through a gateway that converts the Lonwork network into an Ethernet network, in this case, they are assigned an IP address to achieve communication between them, Figure 10.

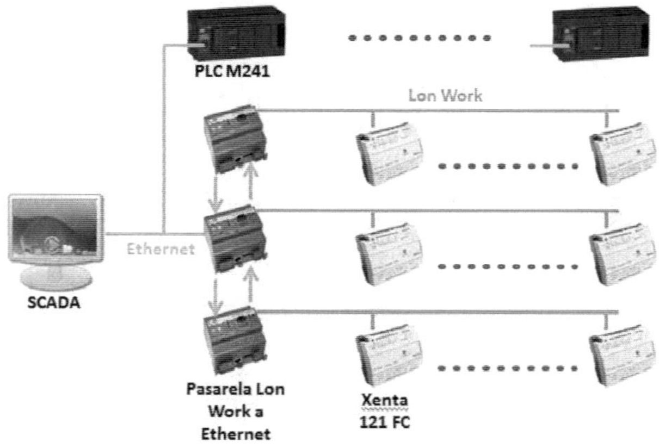

Figure 10. SCADA communication with the different devices

3.1. Supervision and Control System

The SCADA created for the supervision and control of the system is made up of a group of screens that represent the different systems of the hotel, which allow remote management from a control room of the different variables. The SCADA is in charge of managing a total of 546 variables distributed in the PLCs and Xenta 121 controllers. Table 1 shows the distribution of the variables by equipment and hotel levels.

Table 1. Distribution of variables by teams and levels

Floors	PLC variables	Variables Xenta 121 FC	Total Variables
Lobby	45	0	45
First	33	126	159
Second	33	126	159
Third	33	126	159
Roof Garden	24	0	24
Total	**123**	**378**	**546**

When starting the application, user authentication is necessary because different levels of the hierarchy have been established and only authorized users can change parameters. It is accessed through the screen shown in Figure 11.

Figure 11. TAC Vista login window

IOP Conf. Series: Materials Science and Engineering 1154 (2021) 012004 doi:10.1088/1757-899X/1154/1/012004

After accessing the hotel supervision and control system, the home screen shown in Figure 12 is displayed, which allows the user to navigate through the different links and also provides information on the status of the lighting, indicating in green that it is switched on.

Figure 12. Home screen

One of the main tasks of the research was to have a system capable of determining and displaying the electrical energy consumed in the hotel to keep a historical record and make comparisons with energy consumption before the implementation of the system. Figure 13 shows the screen that collects the value of the energy consumed per hour, day, month, and year.

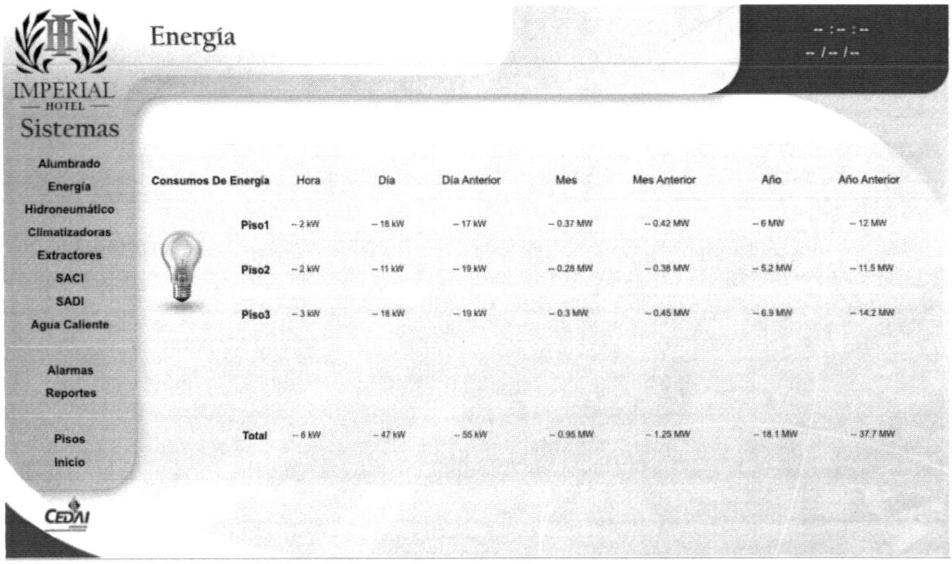

Figure 13. Energy consumed window

4. Conclusions

The solution provided has been designed so that it works continuously and efficiently, in addition, with the decentralized topology, each system has its independent controller, guaranteeing that, in the event of a fault in one of them, the rest of the systems continue to work uninterruptedly.

All the devices and elements used to belong to the same Tac Vista family, allowing full compatibility between them. On the other hand, the controllers used in the rooms have the same programming, this guarantees their replacement and reprogramming in a short time.

The supervisory system allows us to know about what is happening in the hotel in real-time and to make timely decisions if required. The technical staff has software that allows you to navigate through the different systems of the hotel and that shows their current states. From SCADA you can control and change different parameters of the system, activate the circuits that control the corridor lights, set the waiting times, and know if any of them are in electrical failure, among others. Furthermore, the lighting levels necessary for activating exterior lights and time controls are easily configurable from the supervisory system screens, without the need to reprogram any controller.

References

[1] Acosta Corzo, A., et al., 2015. Predictive controller experiences in two Havana hotels. *Cuban Magazine of Engineering*, vol. 6, no. 2, pp. 17-24. DOI: https://doi.org/10.1234/rci.v6i2.348

[2] Nuñez-Alvarez, J. R, Benítez-Pina, I., & Carbonell-Mestre D., 2018. Tools for the design of an inmotic system in the residential block of a five star plus hotel. *Journal of Engineering and Technology for Industrial Applications*, vol. 4, no. 13, pp 166-175. DOI: https://dx.doi.org/10.5935/2447-0228.20180024

[3] Costa, Adriana V. et al., 2011. Model for the Energy Prediction of a Hotel Installation. *Iberoamerican Journal of Automation and Industrial Computing*, vol. 8, no. 4, pp. 309-322. DOI: https://doi.org/10.1016/j.riai.2011.09.001

[4] Salinas Chávez, E., & La O Osorio, J. A., 2006. Tourism and sustainability: from theory to practice in Cuba. *Tourism Notebooks*, vol. 17, pp. 201-221. https://revistas.um.es/turismo/article/view/17931

[5] Gutiérrez, S., Álvarez, J., Velázquez, R., & Lay, A., 2017. Use of automated blinds in smart buildings for energy savings: A mexican case, *IEEE 37th Central America and Panama Convention*, Managua, pp. 1-5, DOI: 10.1109/CONCAPAN.2017.8278509

[6] Saba, D., Degha, H. E., Berbaoui, B., Laallam, F. Z., & Maouedj, R., 2017. Contribution to the modeling and simulation of multiagent systems for energy saving in the habitat, *International Conference on Mathematics and Information Technology*, pp. 204-208, DOI: 10.1109/MATHIT.2017.8259718

[7] Rojas, C. A., Chanchí, G. E., & Villalba, K. M., 2020. Proposal of an IoT architecture for the domotic and inmotic control of buildings. *Revista Iberica de Sistemas e Tecnologias de Informacao*, vol. 2020, Issue E27, pp 28-40.

[8] Mocanu, D. A., et al., 2020. PLC Automation and Control Strategy in a Stirling Solar Power System. *Energies*, vol. 13, pp. 1917. https://doi.org/10.3390/en13081917

[9] Aste, N., Manfren, M., & Marenzi, G., 2017. Building Automation and Control Systems and performance optimization: A framework for analysis", *Renewable and Sustainable Energy Reviews*, vol. 75, pp. 313–330, 2017.

[10] Buitrago, O., Palacio, O., Britto, R., & Adarme Jaimes, W., 2016. Methodological proposal for the selection of the configuration of imotic distribution centers using data envelopment analysis. I'll engineer. *Chilean engineering magazine*, vol. 24, no. 3, pp. 480-492. https://dx.doi.org/10.4067/S0718-33052016000300012

[11] Strauch-Gómez, F., et al., 2017. Inmotics: sustainability and comfort. *Revista Facultad de Ingeniería*, vol. 26, no. 46, pp. 131-139. https://doi.org/10.19053/01211129.v26.n46.2017.7325

[12] Querol, O., 2016. Saving and energy efficiency with Domotics and Inmotics. *Real Estate Magazine*, vol. 159, pp. 38-43.

[13] Pérez, E., 2015. Los sistemas SCADA en la automatización industrial. *Tecnología en Marcha*, vol. 28, no. 4, pp. 3-14. DOI: 10.18845/tm.v28i4.2438

[14] Barzaga, L., Mompie, R. C., & Valdes, B., 2016. SCADA systems for the automation of CIGB production processes. *Electronic, Automatic and Communications Engineering. EAC*, vol. 37, no.1, pp. 20-37.

[15] Núñez, A., J. R., et al., 2020. Methodology of fault diagnosis forgrid-connected photovoltaic systems of network connection. *Revista Iberoamericana de Automática e Informática industrial*, vol. 17, no. 1, pp. 94-105, DOI: https://doi.org/10.4995/riai.2019.11449

[16] Barrozo, F., Valencia, G., Obregon, L., Arango-Manrique, A., & Núñez-Álvarez. J. R., 2020. Energy, Economic, and Environmental Evaluation of a Proposed Solar-Wind Power On-Grid System Using HOMER Pro®: A Case Study in Colombia. *Energies*, vol. 13, no. 7, pp. 1662.

[17] Valencia, G., Núñez, J., & Vanegas, M., 2019. Data set on wind speed, wind direction and wind probability distributions in Puerto Bolivar – Colombia. Data in Brief, vol. 27, pp. 104753. https://doi.org/10.1016/j.dib.2019.104753

[18] Palomino, K., Reyes F., Núñez J., Valencia, G., & Herrera, R., 2020. Wind Speed Prediction Based on Univariate ARIMA and OLS on the Colombian Caribbean Coast. *Journal of Engineering Science and Technology Review*, vol. 13, no. 3, pp 200-205.

[19] Nuñez, J. R., et al., 2019. Tools for the Implementation of a SCADA System in a Desalination Process. *IEEE Latin America Transactions*, vol. 17, no. 11, pp. 1858-1864, DOI: 10.1109/TLA.2019.8986424

[20] Intriago, M.G., et al., 2019. Sizing of a networked self-consumption system at the Technical University of Manabí, Ecuador. *CISCI 2019 - Decima Octava Conferencia Iberoamericana en Sistemas, Cibernetica e Informatica*, vol. 1, pp. 6-11.

[21] Nuñez-Alvarez, J. R., Benítez, I., & Llosas, Y., 2020. Communications in Flexible Supervisor for Laboratory Research in Renewable Energy. *IOP Conf. Ser.: Mater. Sci. Eng*, vol. 844, 012016. DOI:10.1088/1757-899X/844/1/012016

[22] Nuñez-Alvarez, J. R., et al., 2020. Design of a Fuzzy Controller for a Hybrid Generation System. *IOP Conf. Ser.: Mater. Sci. Eng*, vol. 844, 012017. DOI: 10.1088/1757-899X/844/1/012017

[23] García-Guarín, P. J., Cantor-López, J., Cortés-Guerrero, C., Guzmán-Pardo, M. A., & Rivera-Rodríguez, S., 2019. Implementation of the VNS-DEEPSO algorithm for the energy dispatch in smart distributed grid. *INGE CUC*, vol. 15, pp. 142-154.

[24] Vélez, V. E., Henao, C. A., & Calvo A. F., 2018. Design and Implementation of a Wired Intercommunication Prototype for Hospital Care. *INGE CUC*, vol. 14, no. 1, pp. 101-112. DOI: http://doi.org/10.17981/ingecuc.14.1.2018.9

[25] Milanés, C., Planas, J. A., Pelot, R., & Nuñez, J. R. (2020). New Methodology with Public Participation for Environmental Energy Planning within ICZM in Cuba. *Ocean and Coastal Management*, vol. 186, no. 15, pp. 105101. DOI: 10.1016/j.ocecoaman.2020.105101

[26] Milanés-Batista, C. et al. (2020). Application of Business Intelligence in studies management of Hazard, Vulnerability and Risk in Cuba. *IOP Conf. Ser.: Mater. Sci. Eng*, vol, 844. 012033.

[27] Nuñez Alvarez, J. R, Zamora, Y., Pina, I. B., & Angarita, E. N. Demilitarized network to secure the data stored in industrial networks, *International Journal of Electrical and Computer Engineering (IJECE)*, vol. 11, no. 1, pp. 611, 2021.

[28] Andramuño, J., et al., 2021. Intelligent distributed module for local control of lighting and electricaloutlets in a home, *J. Phys.: Conf. Ser.*, vol. 1730, 012001.

[29] Mendoza, E., et al., 2021. Deliberative architecture for smart sensors in the filtering operation of a water purification plant, *J. Phys.: Conf. Ser.*, vol. 1730, 012088.

[30] Merchán, E. V., et al. (2020). An analysis of electricity generation with renewable resources in Germany. *International Journal of Energy Economics and Policy*, 10(5), pp. 361–367, 2020. DOI: https://doi.org/10.32479/ijeep.9369

Development of a mobile application for Early Warning Systems and risk management in Cuba

Celene B. Milanés[1a], Ángel R. Batista[2], Ricardo A. Núñez[3], Humberto T. Yero[4]

[1] Civil and Environmental Department. Universidad de la Costa. ORCID: 0000- 0003-2560-8859
[2] Automatic Control Department. Universidad de Oriente.
[3] Faculty of Engineering. Universidad del Norte
[4] Faculty of Engineering, Telecommunications, Informatics and Biomedical. Universidad de Oriente

[a] Corresponding author: cmilanes1@cuc.edu.co

Abstract. Information and communication technologies (ICT) play a determining role in tackling climate change. ICTs are increasingly used in surveillance, prediction and support actions in adapting to these environmental risks. The archipelago of Cuba is one of the most vulnerable due to the negative effects of climate change. The island nation adopts the benefits offered by ICTs, in order to promote risk reduction due to the impact of natural and anthropogenic threats. The article presents the design of a computer tool called SATEMovil, which was conceived as a mobile phone application for receiving early alerts in the face of extreme threats. The research exposes the operation of the Early Warning Systems and mobile applications in Cuba and the process of design and creation of the SATEMovil App. The application contributes to adequate integrated risk management in Cuba, by favoring the identification and continuous monitoring of natural hazards, through textual reports and dynamic graphics. Its use optimizes the existing relationship, between social vulnerability indices and the level of uncertainty in the population, in the six identified threats. The result obtained enhances the country's preparedness to reduce and confront risk.

Keywords. *Threats, risk reduction, climate change, adaptation, mobile telephony, App.*

1. Introduction

The Republic of Cuba is geographically located in the Caribbean Sea, very close to the coasts of the United States and Mexico. It constitutes an archipelago made up of the largest island in the Antilles, the Isla de la Juventud, and more than 4,195 cays and islets [1]. In Cuba, natural threats such as hurricanes and earthquakes have been analyzed as the most extreme in the country [2]. The central and western regions of the Cuban archipelago are always the most impacted by hydrometeorological events, not so the eastern region, reported with the lowest historically reported impacts. In the eastern zone, high seismic activity is the most dangerous [1,3], (See Figure 1).

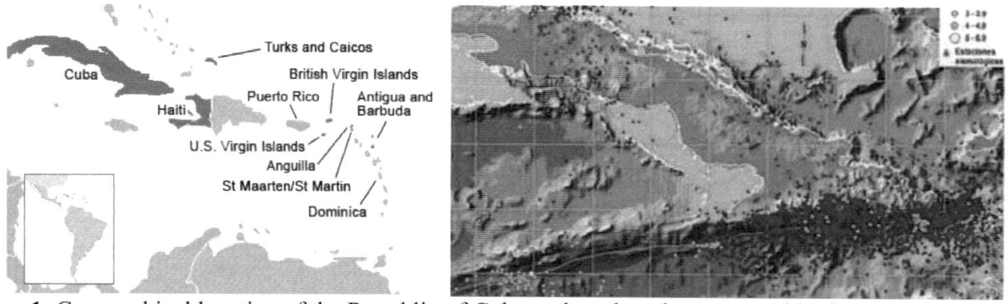

Figure 1. Geographical location of the Republic of Cuba and earthquakes occurred in the southeast region of the country. (Source: www.unicef.org and www.cubasi.cu).

One of the ways used to guide communities in case of extreme events is the Early Warning System (EWS). These warning systems are in place in the country to address various types of threats [4, 5]. The EWSs is based on maintaining a sustained exchange of information and data analysis, between the center or system that intervenes in the detection and monitoring of threats with the relevant authorities, who finally make the appropriate decisions and guide the management of the country and the population on the various types of measures to be implemented.

In Cuba, for the information to be processed, the participation of a group of institutions involved in the EWS is required. These are: the National Institute of Hydraulic Resources-INRH (www.hidro.gob.cu/es), Cuban Institute of Radio and Television-ICRT (www.icrt.gob.cu), Institute of Meteorology-INSMET (www.insmet.cu) and the Civil Defense- DC (www.cubadefensa.cu), as well as the highest government bodies. Sometimes this information arrives late to the directors of each institution, due to failures in the electrical network or in the email servers through which these alerts are sent, this may result in a delay in transmitting the information to the decision-makers and these in turn to the population, who transmit the information using radio or television or by other means of broadcasting. In the country, there is a poor use of information technologies for risk management and therefore information is not always received in a sagacious way.

In most of the provinces of Cuba, different types of EWS have been carried out, as is the case of the Local System of Environmental Early Warning [1]. Another example is the EWS of the Bahía de Santiago de Cuba ecosystem, created in 2008 [6]. This EWS requests that the population report the various contingencies found in the land, air, and marine media in the aforementioned bay. This EWS required back then, that the citizen have a nearby landline to make the incident report. This was rarely done, either because of the absence of this transmission equipment, or because of the high costs of cellular telephony at that time. The aforementioned limitations show how little access there is to information by mobile phone.

Currently in Cuba, obtaining credits has been achieved with rapid growth in the application of mobile telephony [7, 8]. This constitutes an opportunity that should be seized today by Cuban communities. The gradual decrease in the prices of calls and the free reception of messaging, as well as the increase in the use of smartphones that can manage mobile applications for access to information in an easy way and wherever the user is. find, constitute other advantages [9]. However, it is of utmost importance to continue looking for options that allow the population to access this information using these new communication alternatives, without the need for internet access by mobile telephony, which to date has not been implemented on the island. Due to this situation, the research question arises: *How to design a mobile application for the EWS, which allows information from alerts to natural threats to reach the population in a faster and more efficient way?*

2. Materials and methods

Since an integrated approach is required to understand the early warning systems (EWS), it was necessary to analyze aspects relative to this term in the Cuban context. This section describes the methodological framework carried out to analyze the factors in relationship with EWS and the App SATEMóvil.

2.1. *Creation of SATEmovil*

The Early Warning System in Mobile Telephony SATEMovil is a tool that contributes to adequate integrated risk management, which favors the identification, dynamics, and monitoring of different threats, which allows monitoring forecasts on the probabilities and magnitudes of impacts, reducing the uncertainties in the population about the identified threats and promote less social vulnerability, by enhancing the population's preparation for their confrontation [10]. In the SATEmovil system, 3 types of ways of accessing the information of the alerts are defined, they are: SMS (Short Message Service); MMS (Multimedia Messaging System) and Correo Nauta. The latter is part of a strategy that tries to seek convergence in the use of ETECSA. This company offers the possibility of checking the email inbox from cell phones.

Among the conditions established for the use of SATEMovil are: a) having access to data transmission services; b) have enabled, the mail access service called NAUTA on the mobile network (activation that is only carried out by personal request in ETECSA's business units); c) have configured the NAUTA mail, attending to the parameters that guarantee the correct operation of this email application. In this case it is important to clarify that when it is used as a modem it is not necessary to make the configuration. Users can receive the information from any of these variants, always as a choice and that suits their needs. But it is worth noting the use of SMS at present because to date it is the one with the fewest errors in our country, in addition to the fact that reception is free at no cost, unlike MMS, and Correo Nauta, which generates expenses on the balance of the user when accessing the GPRS.

2.2. *SATEMovil requirements and functionalities*

One of the main elements for the definition of the SATEMovil functionalities and the achievement of the proposed objective, was the analysis of the information needs of the clients, translated into requirements, both in characteristics and functions, which is known within of software engineering as the lifting of functional and non-functional demands of the application. The following were considered as functional requirements or requirements (RF):

RF1: Show Introduction in the Application: this requirement when the user opens the application for the first time will be shown a short introduction about the functionalities that the application will have.

RF2: Sign up for Alerts: It will allow users to sign up for the arrival of alert information via SMS, MMS, or Email.

RF3 Unsubscribe from an Alert: The user can at any time unsubscribe from any alert to which he is already registered, this will allow him to no longer receive information on that particular alert.

RF4: Show Alerts Information: This will allow displaying the information of each type of alert, for this the application will automatically access each access path of information that the user has registered, that is, if the user registered to an SMS alert, the application will access your SMS mailbox to obtain the information.

RF5: Show Geographic Location of the Alert: Depending on the type of alert, the user will be able to see on a geographic map the location of the event that generated the alert.

RF6: Reports: Here the user can contribute to the entities that manage the alerts, sending them some type of information that provides these institutions with important information for their investigations. In this case a survey.

RF7: Locate on the Map: The user will be able to locate the map of the application the sites that he frequents the most, which will allow him to send this information in the reports that are generated.

RF8: Configuration: In this requirement, the user will be able to configure the application if a change in the information reception infrastructure is necessary. For this you can configure the mobile number where the alerts will arrive, among other important values for the correct operation of the application.

As part of the non-functional requirements of SATEMovil are the following:

1. Appearance or external interface: It will have a simple and friendly interface with colors that are pleasing to the user with different shades of reds and whites. It will make use of discreet icons allowing easy navigation between the different views in a comfortable way for the user.

2. Usability: The application can be used by users who do not need advanced computer skills, therefore, the application must be flexible and easy to use. It will be possible to visualize in the different types of existing screen sizes, in addition, containers and filtering of the content types are used.

3. Performance: The application must implement several transactions per second to give a quick response to the user and avoid unnecessary delays, also depending on the availability of the infrastructure created. The response times and speed of information processing will be faster, as well as how the application is running in the background will not affect the performance of the mobile device.

4. Reliability: The information stored, processed and read by the application will be reliable, since it will have managers who will allow the manipulated information to be reliable, it must also have mechanisms that allow rapid recovery from any failure that has occurred.

5. Security: The information handled by the application must be carefully protected against corruption and states of inconsistency and the personal information of users within their devices must be found safe from any leak, which is why the application only access the information for which it was created.

6. Support: Possible semi-annual updates.
7. Software: Android 4.4 or higher.
8. Legal: The application must be adjusted and governed by the stipulated laws to comply with the related processes. In this case, the application will be patented to avoid plagiarism.
9. Efficiency: It must ensure quick responses, as well as allow access to information effectively, efficiently and with great dynamism in order to capture customer attention.

2.3 Development methodology of the SATEMovil App

For the creation of the SATEMovil App, the agile methodology of application development Extreme Programing (XP) was used [18]. Among the practices that this methodology proposes, the following were adopted: 1) Development of adaptable solutions instead of foresight, in the face of constantly changing customer requirements, 2) Simplicity in design, allowing speedy construction of parts of the application and the ease of its maintenance and 3) High level of feedback to improve the functional capabilities of the software, by enhancing customer-project member communication in the various phases of its development.

Pressman (2010) states that "*Software engineering includes processes, methodological approaches, and tools that allow complex and computer-based systems to be developed on time and with quality. The software process incorporates five structural activities: communication, planning, modeling, construction, and deployment that are applicable to all software projects* "[11]. The SATEMovil App offers support for 5 types of alerts, which are: 1) Severe Local Storms; 2) Droughts; 3) Threat to Livestock; 4) Hurricanes or Tropical Cyclones and 5) Earthquakes. Its development does not yet contemplate the threats derived from the anthropogenic origin.

This methodology of software development or, software engineering, as it is also called, is supported by diagrams. The sequence diagram defined in the Unified Modeling Language (UML), is one of the most used methods to identify the behavior of a system [19]. Represents the objects on the stage and the sequence of messages exchanged between them, to carry out the functionality described by a transaction system. Furthermore, it is frequently used to validate use cases and to appreciate design logic dynamically [20, 21]. Figures 2-A and B show the sequence and case use diagrams of SATEMovil.

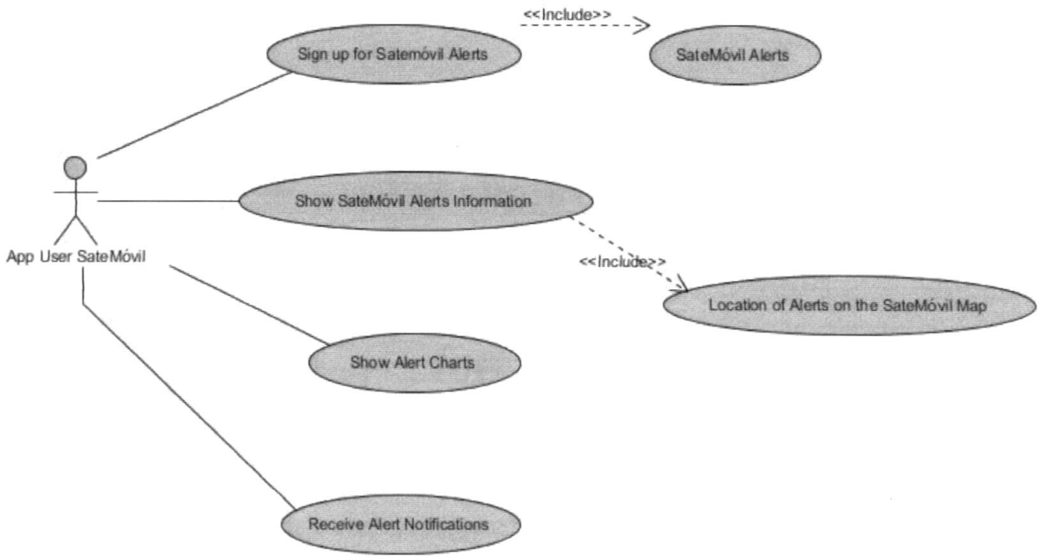

Figure 2-A. Methodological diagram of sequence for App SATEMovil.

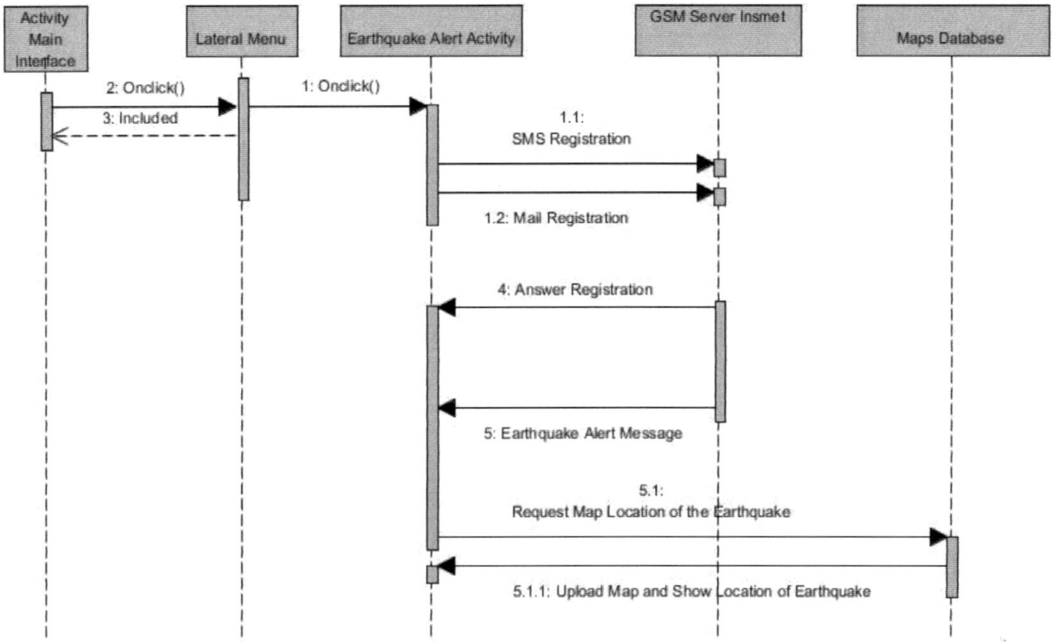

Figure 2-B. Methodological diagram of use case for App SATEMovil

Figure 2A shows the graphic representation that is made between the different messages that are exchanged within the SATEMovil application, which the end-user does not perceive. Everything is executed correctly for the optimal operation of the project. Figure 2-B shows the sequence diagram of an activity when earthquake alerts are recorded. Similarly, other sequence diagrams can be made for the remaining alerts. Each use case tells a stylized story of how an end-user (who has a number of possible roles) interacts with the system and in specific circumstances. The story can be a narrative text, a guideline for tasks or interactions, a description based on a format, or a diagrammatic representation. Regardless of form, a use case illustrates the software or system from the end-user point of view [19]. Therefore, the diagram of use cases of the EWSEMóvil App symbolizes in a general way the different actions that the user will be able to carry out on the application allowing them to understand its operation.

2.4 Data sources and selection of tools and variables for the implementation of the SATEMovil.

The data sources that support the mobile application were collected through surveys of managers and support staff of the institutions that work on early warning systems, including Civil Defense and the Provincial Meteorological Center, as entities fundamental in the generation of Hydrometeorological alerts. The National Seismological Research Center (CENAIS), in charge of issuing seismic alerts among other entities with jurisdiction in Cuban territory, was also consulted.

For the development of the application, an analysis was made of all the existing technologies, paying special attention to those that are free. Considering the number of applications, the two leading operating systems or mobile platforms in the market, both by revenue and by the number of applications, are the Apple is the system and Google's Android [12].

For the creation of SATEMóvil, the Android platform is selected for its versatility for the design of applications and for being the most prevalent on mobile devices within Cuba. Likewise, it provides greater transparency in its programming, since it is open-source, with countless examples, tutorials, available resources, and a wide and growing community of developers.

The Java language that is used in the development of mobile applications for the previously exposed platform is used, as well as the use of XML Markup Language that allows modeling the different user interfaces that the application will have.

The persistence of the data is managed using the own storage mechanisms of the information within the mobile devices such as the use of the SQLITE database and the use of the SharedPreferences class in the Java language to store the configurations that are executed within the application. In addition, our own algorithms for the treatment of character strings were implemented, which allowed us to interpret the formats of incoming messages and display them to the user in readable text. For the representation of the alerts, geographically, the geographic data obtained from the OpenStreetMap project and the MapsForge programming library were used for the use of offline and free maps.

3. Results and discussion

3.1 Early Warning Systems (EWS) in mobile applications

With the rapid growth of the world population, mobile phones and networks derived from this technology are being used more to communicate alert and coordinate preparedness and response actions [10]. At the same time, alerts through electronic messaging are widely used in the dissemination of mass messages. In the Google Play application stores for devices with Android platform, as well as the App Store for devices with iOS platform, there are countless free applications that implement early warning systems for different types of alerts, among them the following can be mentioned :

Earthquake Alert: The application is capable of reporting earthquakes worldwide, thanks to the fact that it gathers the data network of seismology entities and you can share the information on social networks.

Seismic Detector: It is an application that reports on temblor movements and can give warnings seconds before it is perceived. According to the description of the application, *"the Seismic Detector research project is aimed at developing an open collaboration early warning system, based on smartphones, to detect earthquakes in real-time and generate alerts for the population" [12].*

MyShake: This new application that turns a smartphone into a mobile seismometer is being tested by scientists in California [13].

SkyAlert: This multiplatform application emits the alarm signal with a time frame of 60 seconds before any event occurs [14,15].

All these applications, regardless of whether they show and perform the alerts EWSisfactorily, have the limitation that they need continuous access to the Internet or EWSellite use to access the information. In developed countries, this is not a problem, but in the case of Cuba, the infrastructure created to access the Network of Networks from mobile phones does not yet exist, so these applications do not work correctly in our country. In response to this problem, it is vitally important to have a mobile application to obtain alerts in the Cuban nation, using available technologies.

3.2. *Deployment of SATEMovil*

SATEMovil is made up of several elements that ensure its correct operation [15], the different elements and the details of their interaction are shown in Figure 3.

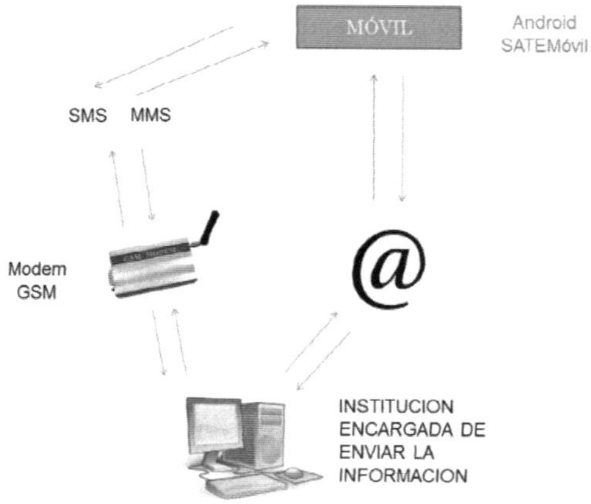

Figure 3. Deployment Diagram

Among the elements of the system is a computer where the computer system that processes the information and sends it manually or automatically will be found. To issue and send the information, the GSM mobile network is used. The General Packet Radio Service (GPRS) or general packet service via radio is also used, which is an extension of this system and allows the transmission of non-switched (or packet) data.

This technology is also used in Cuba for the MMS multimedia messaging service and the @nauta service [16]. This is a new service that has been in use for several years. Each one with a configuration defined by the supplier of the aforementioned Cuban company ETECSA S.A. According to the current price list for the different services, the cost of using the GPRS is 1.00 CUC x 50 MB of data transferred for the NAUTA mail [17, 18]. Furthermore, it should be noted that not all mobile devices in our country have the ability to use email [22-24].

All the institutions involved in sending the information have this system deployed and therefore the information will be successfully distributed with the SATEMovil application. It mainly opts for the use of SMS since its reception is free for the Cuban population.

3.2.1. SATEMovil Design and Functionalities and information to process

For this, the first time the user accesses the EWSEMóvil mobile application, the application will display a loading screen with the application's logo where it will quickly initialize itself for use (See Figure 2). Likewise, as the application is initialized, the different resources available on mobile devices to obtain or send the information is shown, in this case, SMS, MMS or Email [25-27], as well as the map report of the occurrence of the event in real-time (see Figure 4 AC).[1]

Figure 4-A **Figure: 4-B** **Figure: 4-C**

Figure 4. Different environments of the EWSEMóvil App. 4-A. Presentation of the App; 4-B. Data collection format: 4-C. Location on the map of the application of some events. (*Source: Modification from Ravelo and Milanés 2018 [15]*).

When the application runs, the user, by selecting the icon with three horizontal lines, displays a menu where the different types of alerts to which he can subscribe and begin to receive notifications of these are displayed (see Figure 5). By selecting an alert type, the user will be able to subscribe and receive the information through different channels that they need: SMS, MMS or Email, the first being the most recommended for users (See Figure 6).

[1] *The figures and screenshots of the App are placed in Spanish because it is the official language in which the tool was designed.*

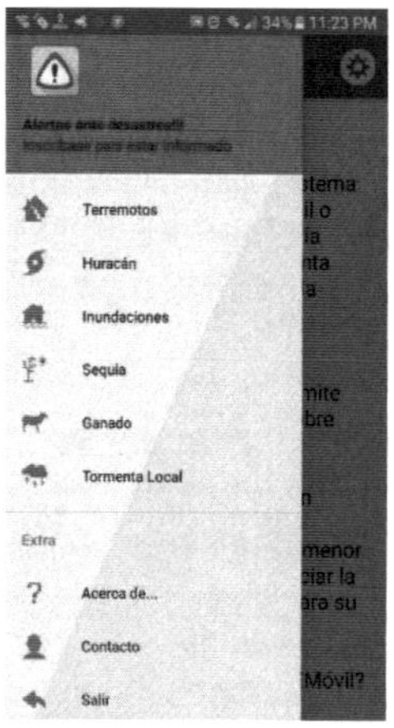

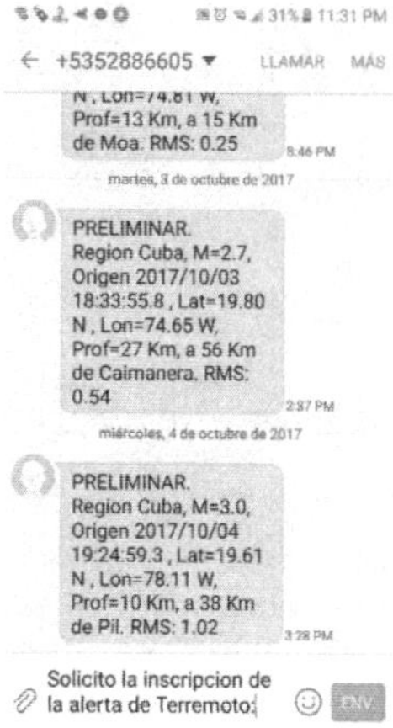

Figure 5. Alerts menu. **Figure 6.** Alert Registration (Earthquake)

3.2.2. Alerts in SATMóvil for Earthquakes

In the case of earthquakes, a message is received every time an earthquake is generated, it is sent to a list of users that are defined in CENAIS, mainly for the researchers of the highest hierarchy in the center and some decision-makers within the provincial government and municipalities. The message that is sent arrives in a reduced format in acronyms, so that the application is able to process it and display it in a more understandable format (see Figure 7). With this information, the user can access the location of the earthquake that was generated (See Figure 8).

Figure 7. Earthquake Alerts **Figure 8.** Geographical Location of Earthquake

3.2.3. *SATEMovil Alerts for Hurricanes*

For hurricane alerts, INSMET is proposed to use a specific format to be sent in messages, which allows it to send a single message and not several in one, thus avoiding the over-EWSuration of the communication channel and that the information reaches quickly to the various recipients. For this, the proposed example format is: *Hurricane. Nom: Irma, Cat: 5, Lat: 18.23N, Lon: 75.05 W, Vmax: 650 km / h, Pat: 1004 hPa, Date: 09/27/2017, Time: 05:00 [20].*

From here the application gets the category, being able, in addition, in case the category is not sent, to be able to know it by the speed of the winds, as well as if it is a hurricane or a tropical storm. The information is analyzed by SATEMovil and is shown as in Figure 9 to later be located on the map (see Figure 10).

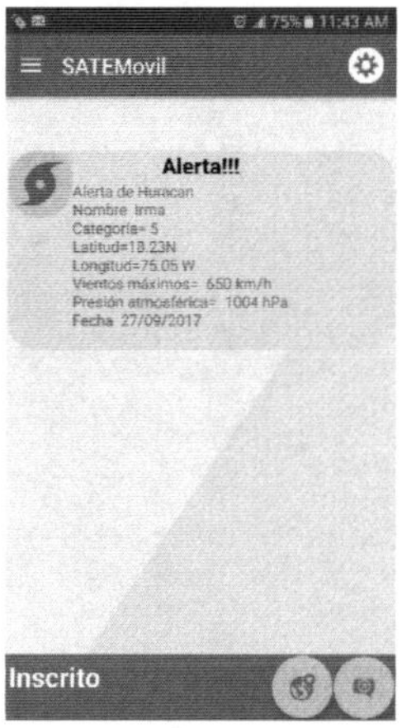

Figure 9. Hurricane Warning (INSMET) Figure **10.** Hurricane Geographic Location

3.2.4. Alerts on SATEMovil for Droughts

The Institute of Meteorology of Cuba, by its acronym INSMET (www.insmet.cu), prepares a monthly, Climate Surveillance Bulletin that reports on the behavior of the climate. To do this, different variables extracted from meteorological stations located in strategic locations are analyzed. These variables are: the oceanic and atmospheric conditions, the behavior of the main meteorological variables of the month that ends where the different temperature ranges are analyzed (average, minimum average, maximum average) as well as the relative humidity, rainfall, the thermal sensation between others. The proposed format to send by would be: Sequia.CM: -10, -1.6; GM: -20,1.7; JAM: -10, -1.7; PS: -15, -1.9; SL: 13.4; STGO: 23 , 1.4; SF: 7.2.3; SLM: -18, -1.9; TF: 8, -0.3; PRO: Drought expected; Date: 09/27/2017; Time: 05:00

In this processed message (see Figure 11), the first value displayed is defined for each province as a value such as the Standardized Precipitation Index (ISP), to show the behavior of drought in each region of the province. In addition, the amount of rainfall made per cubic millimeters will be received for each municipality. The user clicking on the received message EWSEMóvil gives the possibility to graph this information for a better understanding of it in two different ways (see Figure 12).

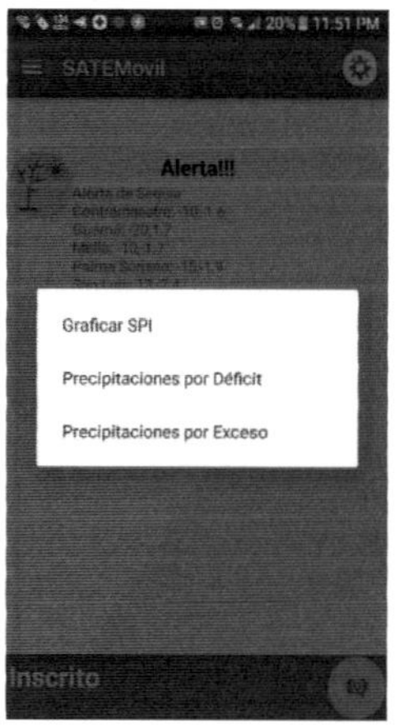

Figure 11. Drought Alert **Figure 12. Graphs available to represent the Drought.**

A pie chart for the ISP is shown are calculated in general percentages of each province of the country (see Figure 13A-C), which can be Extreme, Severe, Moderate or Weak depends on its values and bar graphs for rainfall for each municipality in each province that may be Deficit or Excess (see Figure 13 BC). It is also possible to save such graphics within the mobile device as an image for later use by the user.

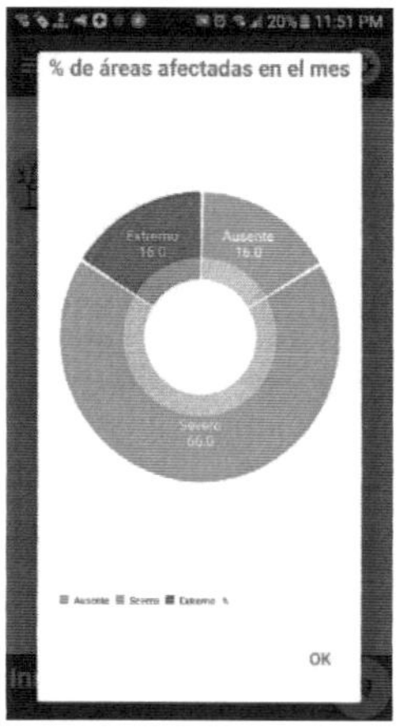

Figure 13-A

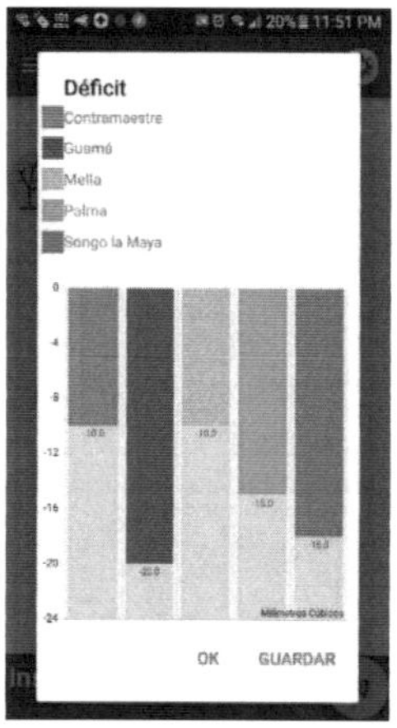

Figure 13-B

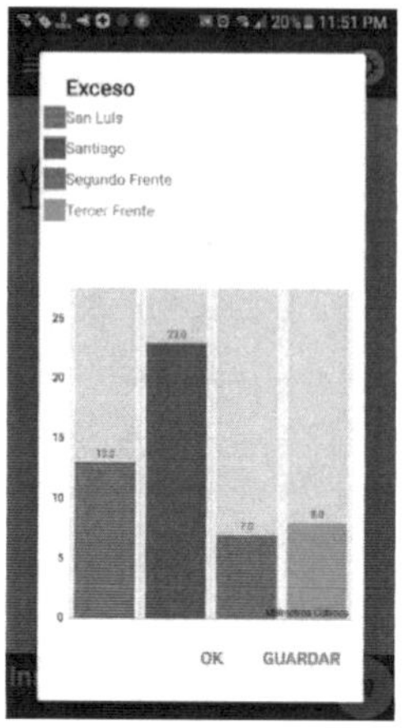

Figure 13-C

Figure 13. Different environments of the *SATEMovil* App. 13-A: Pie Chart for the Precipitation Index; 13-B: Bar Graph Precipitation by Deficit by provinces; 13-C: Bar Graph Precipitation by Excess by provinces. (Source: Modification from Ravelo and Milanés 2018 [15]).

3.2.5. Alerts in SATEMovil to threaten comfort for livestock

For this type of alert, the INSMET generates an Agrometeorological Bulletin that is, every 10 days for each month in progress, addressed to the main producers in the province. Here are different information such as: the phases of the moon in the behavior of vegetable crops and the influence it has on animals, monitoring agricultural drought, vegetation conditions, behavior of forest fires, behavior of pests and diseases in crops, among others. Special attention to Cattle Comfort is also featured within this app. This type of comfort is defined as the area with a temperature range, within which the animal can be without it being necessary to activate its thermal self-regulation mechanism

It is then when the center specialist calculates the Humidity Temperature Index, which is used to know the degree of caloric stress present in cattle [28, 29]). This information needs to be sent to the producers, as well as some advice on what to do in case of stress or not. This is why the message format is proposed: *Threat to livestock. Descen: 1, DS: GL: NE, GC: EL, GP: EM, ES: GL: NE, GC: EL, GP: EM, Month: September, Time: 05: 00 PM.* With this information, the producer or user within SATEMovil receives it better processed (see Figure 14) and is informed of the recommendations that must be taken into account to achieve greater livestock production (Figure 15).

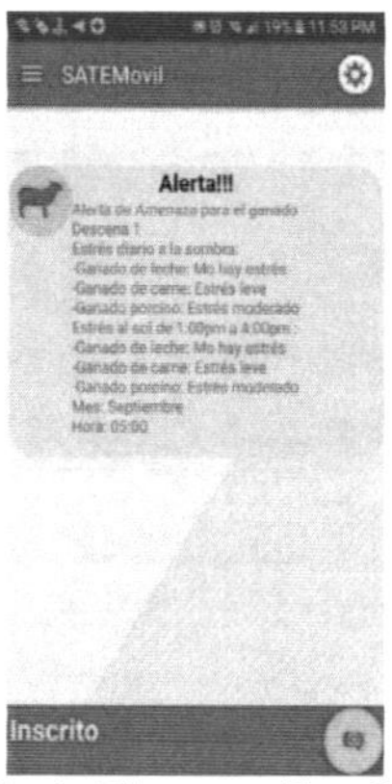

Figure14. Cattle Threat or Comfort Alert

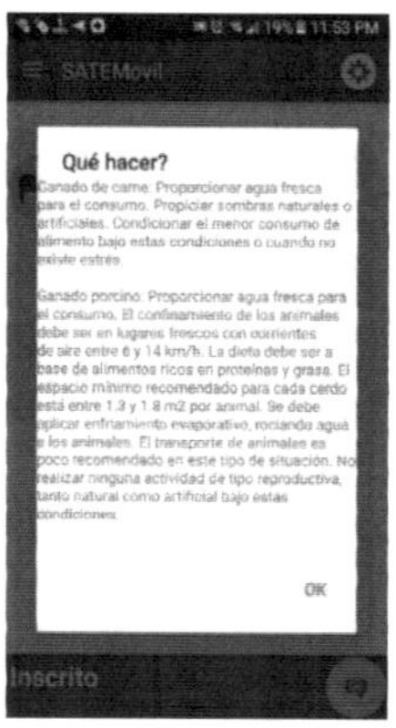

Figure15. Livestock Tips.

IOP Conf. Series: Materials Science and Engineering 1154 (2021) 012005 doi:10.1088/1757-899X/1154/1/012005

3.2.6. Alerts on SATEMovil for Floods and Local Storms

For floods, the INSMET accesses various resources such as radar tables and identifies the most extensive and significant floods due to intense rains, located mainly in municipalities with surfaces of wide plains, low land inclination, and poorly permeable soils.

The type of message to be proposed is in the following format: *Flood. Place: Guamá, Lat: 19.97N, Lon: 76.40 W, Date: 09/27/2017, Time: 05:00.* This is processed (see Figure 16) and will allow you to see the location of the event on the application map [30, 31] (Figure 17).

Severe local storms usually occur at any time of the year. Generally, these can occur in the months of March to September with greater frequency, being May the month with the highest probability of occurrence. These local storms are characterized by the appearance of electrical discharges and heavy rainfall in the form of showers[32]. The validity period of a storm cloud is short, lasting around one or two hours. In this topic, the proposed message is: *TLS. Type: Hail, Reg: Guamá, Lat: 19.96N, Lon: 76.32W, Date: 09/27/2017, Time: 05:00,* in the same way as with floods, information is handled by EWSEMóvil (see Figure 18) and can be located on the map as well (Figure 19).

Figure 16. Flood Alert

Figure 17. Geographic Location of Flood

Figure18. Geographic Location of Flood

Figure 19 Geographic Location of Flood

3.3. Management of information security from SATEMovil

All the information accessed by the SATEMovil application of the device is confidential, so the user can have the guarantee that their information is safe at all times. The objective of the application will always be to obtain, through the different routes used in Cuba, the information about the early alerts and represent it to the user in the appropriate format. The application will always be running, so if the user is taking any other action on their device and an alert is received, they will be notified of this alert. If the user wants to access it for more details, it will automatically redirect them to the application.

4. Conclusions

The Republic of Cuba has achieved great development in the implementation of Early Warning Systems. These EWSs allow surveillance and monitoring of various threats of natural origin that represent a danger to the population and the economy. Despite this development achieved in the EWS of conventional origin, in the country, there is some evidence of the use of tools linked to the use of mobile telephony. The development of a mobile application in Cuba like SATEMovil, allows the reception of early alerts on six different types of extreme natural threats. The success in its development will depend on the level at which this App is socialized with the entire population. SATEMovil favors timely decision-making in the short, medium, and long term. The system is nourished by the gaps in scientific knowledge that, in the Cuban case, the implementation of the EWS in the country today has.

SATEMovil becomes an innovative software for Cuban society on the subject of computerization on disaster risk management. This application is of great importance as it will ensure that the population and decision-makers can be informed before, during, or after the occurrence of any threat that may affect civil society.

The limitation of the use of EWSEMol only focuses on those people who use another type of mobile phone with a platform that is not Android, say iOS, Symbian, or another system. Identifying this weakness allows, over time, to consider creating a new version for each of these platforms or operating systems. To guarantee this, it is necessary to think about incorporating new functionalities that make it a primary application in all Cuban mobile devices.

Acknowledgments
The authors would like to thank the great contribution received from the National Project named "National Project about Studies of Danger, Vulnerability, and Risks caused by sea flooding in Santiago de Cuba province." In the same way, we like to thank the Index Project grant numbers INV.1106-01-002-15] called "Cultural practices and environmental certification of beaches: a contribution for the sustainable development in the insular states". This Project provided the necessary funds for the Interchange Program between a Professor and Researcher from the Research Groups for Environmental Management and Sustainability (GESSA) and Marine and Coastal Management (GeMarc) at the Universidad de la Costa, with one student from Universidad del Norte, and Cuban professors at the Universidad de Oriente.

References
[1]. Botero, C. y Milanés Celene. 2015. (Edtrs). Contributions to marine-coastal governance. Risk management, governance and coastal districts. 554 p. Research book. Publications Fund of the Sergio Arboleda University, Bogotá, Colombia. ISBN: 978-958-8866-67-3. (Versión rústica). ISBN: 978-958-8866-68-0. (Version pdf). http://repository.usergioarboleda.edu.co/handle/11232/934 DOI: https://doi.org/10.22518/9789588866673 ; URI: http://hdl.handle.net/11232/934
[2]. Milanés B, Celene, Botero Saltarén C., Arenas Granados. P., Cabrera. J.A., 2014. "Integrated Coastal Management in Cuba and Colombia: A Comparative Analysis". *Ocean Yearbook*. Volume 28. Leiden/Boston: Martinus Nijhoff, 672-697 pp. http://www.brill.com/products/book/ocean-yearbook
[3]. Milanés B. Celene y Pérez M. Ofelia. 2012. "An Inquiry into Land-Use Planning and Integrated Coastal Zone Management: The Cuban Experience". *Ocean Yearbook* 26 Magazine. Dalhousie University. Canadá. Martinus NIJHOFF Publishers. Boston. 509-532 pp. Available in http://www.brill.com/ocean-yearbook-26. DOI:10.1163/22116001-90000081
[4]. Valencia, G., Nuñez, J., Vanegas, M. 2019. Data set on wind speed, wind direction and wind probability distributions in Puerto Bolivar – Colombia. *Data in Brief.* Vol 27, 104753. https://doi.org/10.1016/j.dib.2019.104753
[5]. Milanés B., Celene; Galbán .R. L., Olaya Coronado ,N. J. 2017. *Amenazas, riesgos y desastres: Visión teórico-metodológico y experiencias reales*. Libro de investigación. 306 p. ISBN: 987-958-8921-44-0 (Digital). Avaible in https://repositorio.cuc.edu.co/handle/11323/1156
[6]. Botero C.M., Pereira C.I.; Milanes C.B.; Pranzinid E. (2020). Dataset of human interventions as anthropogenic perturbations on the Caribbean coast of Colombia. Data in Brief. Volume 31, August 2020, 105847 https://doi.org/10.1016/j.dib.2020.105847
[7]. Milanés B. Celene. 2012 Coastal environmental units for management in Santiago de Cuba: delimitation and priorities for action. Architecture and Urbanism Magazine. Volume XXXIII No. 3. p. 83-97. Consult at http://revistascientificas.cujae.edu.cu/Revistas/Arquitectura/Vol-XXXIII/3-2012/08_au_03_2012.pdf
[8]. Milanés. C., Lastra, R. y Sierra-Correa P. 2019. Case studies on integrated management of coastal zones in Ibero-America: management, risk and good practices (1 ed.). Barranquilla: University of the Coast Corporation.Available in https://repositorio.cuc.edu.co/handle/11323/5502
[9]. Barrozo, A., Valencia, G., Obregon, L. G., Arango-Manrique, A., Núñez Álvarez, J. R. 2020. Energy, Economic, and Environmental Evaluation of a Proposed Solar-Wind Power On-grid System Using HOMER Pro®: A Case Study in Colombia. *Energies*, 13(7), 1662.
[10]. Milanés B, Celene y Pacheco M. Alicia., 2011 Coastal settlements in the bay of Santiago de Cuba: study of its urban vulnerability. Architecture and Urbanism Magazine. Vol. XXXII, No 3. 18-26 pp. Available in http://es.scribd.com/doc/86054921/Revista-Arquitectura-y-Urbanismo-3-2011
[11]. R. S. Pressman. 2010. Software Engineering: A Practical Approach, 7th ed. México D. F., México: McGraw-Hill.
[12].Finazzi Francesco. 2019. Earthquake Detector. Real-time alerts. Avalilable in https://apps.apple.com/do/app/sismo-detector/id1449893235 Chcked 13th, may 2019.
[13]. Amos Jonathan. 2016. The App that turns your smartphone into an earthquake detector. Available in www.bbc.com cheked 18th february on 2016

[14]. Milanés Batista Celene. 2015. The experience of the Southeast region of Cuba in confronting climate change. *Architecture and Urbanism Magazine* vol. XXXVI, no 2, 2015, ISSN 1815-5898. 120-127 pp.
[15]. Ravelo Angel & Milanés B C. 2018. Contribution of mobile telephony to Early Warning Systems to manage risk in Cuba. Chapter 3 p. 54-85 in Milanes B Celene and Szlafsztein Claudio Fabian. 2018. Methodological experiences for risk management. ISBN: 978-958-8921-69-3 (Digital). Editorial Universitaria de la Costa, EDUCOSTA S.A.S. 208 p. Available in http://repositorio.cuc.edu.co/xmlui/handle/11323/1686
[16]. Milanés B. Celene. 2011. Comparative methodological analysis of land use planning under integrated coastal management approaches in Cuba: proposal of parameters and variables to consider. 1-18 pp. Electronic magazine Science on your PC, No.3. Available in http://www.redalyc.org/articulo.oa?id=181322267002
[17]. Botero, C. M.; Arrizabalaga, M.; Milanés, C. & Vivas, O. 2017. Governance indicators for coastal risk management in Colombia. *Revista Luna Azul*, 45, 227-251. DOI: 10.17151/luaz.2017.45.12 Available in: http://200.21.104.25/lunazul/downloads/Lunazul45_12.pdf
[18]. Tamayo Yero Humberto Alejandro y Milanés Batista Celene. 2018. Educational software for training in risk issues and integrated management of coastal zones. *Architecture and Urbanism*,vol. XXXIX, no 3, septiembre- diciembre 2018, pp. 113-123, ISSN 1815-5898. Available in http://rau.cujae.edu.cu/index.php/revistaau/article/view/512/484
[19]. Pressman, R. 2005. Software Engineering: a practitioner's approach, 6th ed. McGraw-Hill, NewYork
[20]. Ambler, S. 2005. The elements of UML 2.0 style. Cambridge University Press, New York
[21]. Fowler, M. and Scott, K. 1997. UML distilled: applying the standard object modeling language. Addison-Wesley Series Editors, Essex.
[22]. Nuñez, J., Benítez, I., Rodríguez, A., Díaz, D., Oliveira, D. 2019. Tools for the Implementation of a SCADA System in a Desalination Process. *IEEE Lat. Am. Trans.*, vol. 17, no. 11, pp. 1858–1864, 2019. 10.1109/TLA.2019.8986424.
[23]. Planas, F.J.A., Milanés, B.C., Fanning, L.M. and Botero, C.M. 2016. Validating Governance Performance Indicators for Integrated Coastal and Ocean Management in the Southeast Region of Cuba. *Open Journal of Marine Science*, 6, 49-65. doi: DOI: 10.4236/ojms.2016.61006
[24]. Tamayo Yero, Humberto Alejandro, Milanés Batista, Celene, & Milanés Clavijo, Vivian Aymée. (2019). Data Warehouse for the management of Hazard, Vulnerability and Risk studies in Cuba. *Cuban Journal of Computer Sciences,13(2),* 61-76. Recovered in July 12, 2020, from http://scielo.sld.cu/scielo.php?script=sci_arttext&pid=S2227-18992019000200061&lng=es&tlng=es
[25]. Pressman, R. 2002. Ingeniería del Software-Enfoque Práctico Mc Graw Hill 5ª. Edición Año.
[26]. Beck, K., & Gamma, E. 2000. Extreme programming explained: embrace change. Addison-Wesley Professional.
[27]. Milanés-Batista C. et al. (2020). Application of Business Intelligence in studies management of Hazard, Vulnerability and Risk in Cuba. 2020 I*OP Conf. Ser.: Mater. Sci. Eng.* 844 012033. IOP Publishing doi:10.1088/1757-899X/844/1/012033 Available in https://iopscience.iop.org/article/10.1088/1757-899X/844/1/012033
[28]. Perez Montero O & Milanes Batita C. (2020). Social perception of coastal risk in the face of hurricanes in the southeastern region of Cuba. *Ocean and Coastal Management.* Volume 184, 1 February, Article number 105010.https://doi.org/10.1016/j.ocecoaman.2019.105010
[29]. Nuñez Alvarez, J. R., et al. (2021). Design of a demilitarized network to secure the data stored in industrial networks. *International Journal of Electrical and Computer Engineering.* 11(1), pp. 611-619. DOI: 10.11591/ijece.v11i1.pp611-619
[30]. Milanés, C., Planas, J. A., Pelot, R., & Nuñez, J. R. (2020). New Methodology with Public Participation for Environmental Energy Planning within ICZM in Cuba. *Ocean and Coastal Management*, 186 (15) 105101. DOI: 10.1016/j.ocecoaman.2020.105101
[31]. Pereira, C.I., Carvajal, A.F., Milanés Batista, C., Botero, C.M. (2019). Regulating human interventions in Colombian coastal areas: Implications for the environmental licensing procedure in middle-income countries. *Environmental Impact Assessment Review* 79,106284. https://doi.org/10.1016/j.eiar.2019.106284
[32]. Milanés Batista, Celene, Pereira, Cristina y Botero Camilo Mateo. (2019). Improving a decree law about coastal zone management in a small island developing state: The case of Cuba. *Marine Policy* Volume 101, March 2019, Pages 93-107, doi.org/10.1016/j.marpol.2018.12.030.

IOP Conf. Series: Materials Science and Engineering 1154 (2021) 012006 doi:10.1088/1757-899X/1154/1/012006

Framework to Manage Software Quality on IIoT Apps

Yair Rivera Julio[1], Bertha Hernández Contreras[2], Samir Castaño Rivera[3], Clara Colón López[4], Angel Dario Pinto Mangonez[5], Harold Bula Herazo[6]

Department of Engineering System, [124]Corporación Universitaria Americana,[36]Universidad de Cordoba and [5]Universidad del Sinú, Colombia.

E-mail:yrivera@coruniamericana.edu.co

Abstract. The research presents a Framework based on CMMI and ITIL methodology. It complemented with an ASQ management type, which allows incorporating metrics that directly involve both the development process of embedded systems based on IoT architectures and oriented to industry 4.0 or IIoT. The methodology is based on the development of software for Embedded Systems using SysML modeling. The scheme to allows incorporates necessary mechanisms that enable the management of the System related to service continuity and easy adaptation to changes, considering service connectivity and compatible with an industrial environment. This scheme allows quantifying through indicators each one of the quality levels, providing at the same time the practicality and the systematization of the processes, which in the end translates into a consistent mapping for the standardization of the quality related to the lines of business that imply the treatment of IoT data at industrial level. The methodology permits are simplifying the need to deal with rapid software changes, one of the essential tasks of software systems development and maintenance teams. Due to the lack of standards and regulated work schemes for the development of IoT architectures; therefore, it is intended to establish implicitly the metrics and an essential work scheme to regulate the design, progress, and development of IIoT architectures, until defining an optimal state of implementation.

1.Introduction

Technological advances in software and embedded systems and trends in the industrial field have led to increases in process control and automation, controlled via the web. However, the lack of recognized standards for the design and development of IoT applications has caused the proliferation of private solutions that somehow limit the potential of such platforms. Besides, many of these wireless connectivity platforms are not compatible with a wireless environment with high variance, and that natively ensure a low delay for real-time connectivity. Currently, the solutions proposed with Bluetooth are not easily scalable, which makes connectivity and control of a long industrial process difficult. Connections developed through the WIFI protocol, do not allow to jump in frequency, which makes it challenging to apply in an environment with high interference[1]. Solutions established through the Wireless HART and 802.11ah protocol, thanks to the fact that they were developed openly, allow a native resolution, which reduces the interaction between layers of the open model and facilitates communication with the M2M (Machine-to-Machine) platforms and MQ Telemetry Transport (MQTT) in real-time[2]. One of the main advances in software engineering can be said to be the contribution of new quality assurance tools through control metrics for software validation at both the development and product stages, which partly guarantees the reliability, testing capabilities and maintenance of the product itself; all this is possible thanks to the development of activities that allow the quality assurance of the software throughout the development life cycle, events and activities that make up the monitoring and improvement of the process, which are linked to quality standards that allow the identification and treatment of errors as well as the management of their treatment, which implies the improvement of communication strategies for the treatment of errors at any stage of design and development; This document emphasizes a Framework for managing software quality in IoT architectures, used in this case for the development of a platform that interacts with thousands of devices for massive access and processing of information from embedded systems to a whole system for data storage and processing [3], [4]. However, to ensure a model that establishes best practices for the development of apps at the industrial level, not only is a technological platform necessary to ensure compatibility for optimal system performance at the wireless connectivity level. If necessary, the implementation of a model that guarantees from the design, efficiently facilitate its maintenance and operability, so that a common criterion is obtained to assess its level of maturity concerning its implementation processes through metrics that entail quality assurance. The scheme to address a quality environment on current methods and guide future ones, five levels of state, or maturity have been established through the CMMI model, where there is a variant that focuses on CMMI-DEV development. The Framework allows monitoring and administration on the development and maintenance process for a platform dedicated to the management and storage of data on IoT architectures. It is crucial for the organization of work, the incorporation of development processes grouped into four categories (Engineering - Project Management - Process Management and Support), where best practices are proposed at the level of team TSP (Team Software Process) and staff PSP (Personal Software Process). This design under structured stages what is sought as the only objects are successive and significant improvements in key development areas in the system; the document will analyze the necessary metrics for quality improvement under ASQ management, finally recreate the modeling of these metrics through a linear regression under linear parameters, with the aim of validating their behavior[5]–[8]. Finally, the model presented aims to establish an initiative based on mature models containing the best practices modified towards the development of IoT systems, which facilitates the final construction of this type of solution.

To make a detailed explanation of the Framework: Section 2, a general description of the methodology and its respective architecture under a 3-layer integral system through a component diagram is provided. In Section 3, the ASQ model is specified under the quality standards of ISO 9000-3, fields of action of the model, components, and factors are established. Section 4 defines a combination of technological resource management areas for the development of integrated systems. IoT refers to the CMMI-DEV standards in addition to a System Modeling Language (SysML) for virtual prototyping of components. To validate this architecture in some way. Sections 5 Simulation for development metrics, establish an experimental design through a linear model that characterizes the behavior of the development metrics under the ASQ standard. Sections 6 sets an analysis of the results and validation experimental. Finally, the model presented aims to establish an initiative based on mature models that contain the best practices modified towards the development of IoT systems, which facilitates the final construction of this type of solution.

2.Methodology and Architecture.

Quality management, as a guarantee of Quality, is one of the main concerns within the software industry. Quality must be quantifiable and even broader in the concepts of testing and measurement. It is vital to feedback the processes through indicators, which allows establishing an overview of each state. Due to the above, it is essential to build a QMS (Quality Management System) that structures the responsibilities and procedures necessary to achieve efficient management. This management allows establishing revisions of requirements, designs, review, and analysis of codes, verification of non-functional aspects of the products to develop metrics both in the logical part and in the development, to focus quantitative management and

IOP Conf. Series: Materials Science and Engineering 1154 (2021) 012006 doi:10.1088/1757-899X/1154/1/012006

thus define the parameters to work. Therefore, it is crucial to design a necessary Framework that allows the organization to reach a maturity level without leaving aside the synchronization with the commercial objectives and the priorities of the interest groups. Initially, measures are introduced to evaluate through metrics the entire software development life cycle (SDLC). The development must be aligned with the functional and performance requirements explicitly established. It is necessary to incur documented development standards and inherent characteristics expected from all the software developed; it is necessary to know the layers of the system and the type of architecture in which it is working. For this particular case, a 3-layer design based on IIoT applications and/or services is established, which is integrated by a middleware IIoT connectivity layer and complemented by an application layer, which provides efficient services for a large number of devices that send information through the data network, all of which gives rise to an integral system of components that involves (Processes, People, Data and Things with connectivity to the data network)[9]–[11] , see Figure 1.

Aplicartion	Framework Development IIoT App:HTTP-Request
Presentation	Format data Types and procedure
Session	
Transport	TCP/UDP , Negotiated Segment size
Network	Mesh-Redundant Path
Data Link	TDMA/CSMA, Master Slave Protocol
Physical	802.15.4-802.11 ah

Figure. 1. System architecture IIoT and Distributed component system under 802.11 ah[12].

It is essential to highlight the fact that ASQ given through the Framework is for the whole system, which goes from the data census through software embedded in the device drivers to the management of the information in the applications and the persistent system. it is necessary to highlight those other metrics go beyond those of development, such as those related to the maintenance and operation of the system, which becomes more extensive and complicated due to the resource established by the embedded systems.[13]–[15], the component diagram of this architecture is examined, see Figure 2.

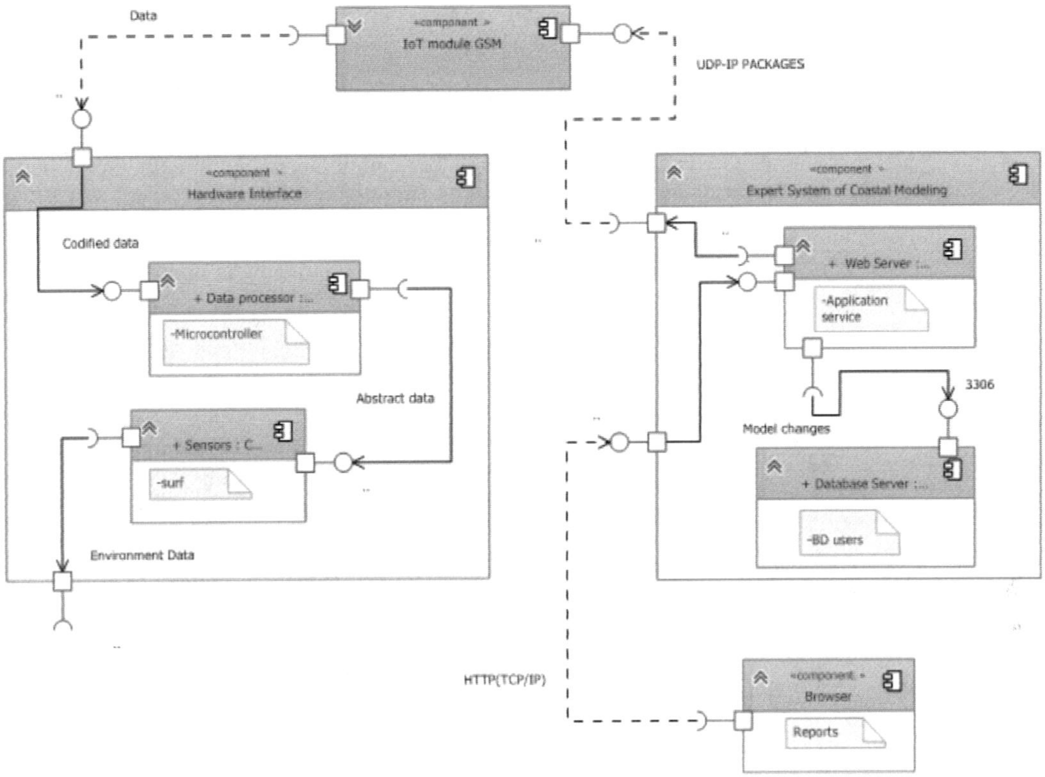

Figure. 2. Component diagram IoT architecture proposed.

3.Model ASQ

The ASQ model allows from the definition of requirements to guarantee reliability, test-ing and maintenance capabilities of the product, taking into account development factors such as the final product; If it is necessary to establish a Framework or planned systematic actions that guarantee an adequate development process, and that allows to reach the minimum established technical and operational requirements, it is necessary to establish a secure link with the quality processes of the ISO 9000 standards, more specifically with the ISO 9000-03, which is created through a maturity model, and that it is intrinsically linked to factors of aspects of the software that enter the life cycle of the system, these factors of respective form are a central part of the fundamental requirements and are grouped in specific categories that include attributes, thus establishing a hierarchy of components that attribute and focus on the characteristics of the system, see Table1.

Table 1. Table captions should be placed above the tables.

Software quality Components	Software quality Factors	Software quality criterion
Product Revision	Maintainability	Simplify
	Flexibility	Generality
	Testability	Instrumentation
Product Transition	Reusability	Machine independency
	Portability	Software system independency
	Interoperability	Data communality
Product Operation	Correctness	Consistency
	Reliability	Accuracy
	Usability	Communicativeness
	Integrity	Access audit
	Efficiency	Execution efficiency

The ASQ model to be worked on is the McCall's model. Also called the General Electric's Model, it was chosen for the structured hierarchy it has in terms of its evaluation levels. On the other hand, the type of metrics to be worked on according to the quality criteria can be obtained through complementary work with the ISO 9126-2 or ISO 9126-3 standards, whether the parameters to be worked on are external or internal, respectively [16].

4.SQA Architecture and Maturity Model

It is crucial to establish that the framework revolves around the classic SDLC model for software development with server-side architecture and complemented by SysML in the IoT devices, where at the beginning, building blocks are designed for the entire development process. These describe a linear sequence that goes from the definition of requirements to the operation and maintenance of services complemented by a distributed design, that is, a hybrid system made up of data connection devices to the network for the census and sending of data. Therefore, a component infrastructure is required according to this development methodology, which allows systems and devices to be easily integrated. It must also comply with SQA standards and be aligned with previously defined procedures, preventive actions, documentation, control of quality records, and tools used to avoid software errors. It is vitally important to consider an analysis based on quantifiable metrics within the same engineering standards since, in some way or another gives us a kind of accurate measure of our progress. [17], [18] ,see Table2.

IOP Conf. Series: Materials Science and Engineering 1154 (2021) 012006 doi:10.1088/1757-899X/1154/1/012006

Table 2. ASQ metrics in combination with technology resource management areas.

Categories	ITIL Standard	Metric	Severity Level
Quality Level Processes	PMC	CED (Code Error Density)	No
		DEV (Development Error Density)	No
		WCED (weighted code errors detected)	Yes
		WDED (weighted developed errors detected)	Yes
		WCEF (Weighted Code Errors per Function Point)	Yes
		WDEF (Weighted Development Errors per Function Point)	Yes
Metrics oriented to the quality of project time	MOCTP	TTO (Time table observance)	No
		ADMC (Average delay of milestone completion)	No
Development productivity-oriented metrics	MOPDEV	devP (Development productivity)	No
		FdevP (Function point Development productivity)	No
		Cre (Code Reuse)	No
		DocRe (Documentation Reuse)	No
Metrics oriented towards help desk management	MOGHID	HDD (HD Call Density)	No
		WHDD (Weighted HD calls per Function point)	Yes

The ASQ in the process and product area supports high-quality product delivery, provides the project manager with appropriate visibility into all levels of the project lifecycle, and ensures unified processes to meet specific form requirements. Therefore, it is necessary to carry out evaluations in a systematic and standardized way through a model of capacity and maturity such as the CMMI-DEV integrated into ISO 9001, here the quality model is considered in each of the processes. From the beginning, it allows us to establish plans, processes, standards, and procedures that, on the other hand, will enable us to add value to the project that satisfies the organization's policy. The model provides us with metrics that allow us to evaluate the work processes or products objectively, and whose results of the evaluations must be integrated into the work rhythms of each team. See the following table to examine some metrics [19]–[22] (see Table3).

Table 3. Quality assurance system (ISO 9001).

CCMI development process level 2 and 3			Metric
CCMI-3	Defined	Decision Analysis Resolution	% Level quantity of Analysis and resolution of problem associated with App
		Integrated Project Management	% Summarization Level. Application ($$, TT,Risk)
		Organizational Process Definition	% Associated document processes dev. Software, deployed and releases
		Organizational Training	% of persons certified or training tools
		Organizational Process Focus	% of processes documents associated with software development
		Product integration	% level of integration to app
		Requirements Development	% Development network
		Risk management	% Development solution (RQ-changes-disp and Integration)
		Technical solution	% Technical solutions during the second phase
		Validation	% defects of finished product
		Verification	% that meet the functional requirements
		Configuration Management	% Change in configuration elements (Baseline)
CCMI-2	Management	Configuration Management	% Change in configuration elements (Baseline)
		Measurement and Analysis	% identification and resolution of process-related problems
		Project Monitoring and Control	% Milestone review
		Project Planning	% from project life cycle phases
		Process and production Quality Assurance	% of processes and work products objectively evaluated according to the standard
		Requirements Management	% between project requirements, plans and outputs
		Supplier Agreement Management	% Suppliers' acquisition of products and services

Therefore, each of these results indicates the status of the intermediate work products against all the selected requirements in each development interaction; consequently, verification mechanisms must be defined incrementally. Besides, which must be aligned with the requirements, methods, and criteria available for product validation. the above shows whether the product as provided will meet its intended use, whereas verification examines whether the product establishes the specified requirements from its design [21], [23], [24]. All this scaffolding defined in CCMI-3 allows for the systematic elimination of process defects and the identification of any necessary changes. In the design of systems oriented to the development of IoT, verification includes the creation of prototypes through the metamodeling of a Systems Modeling Language (SysML), which is an extension of UML that takes into account the formalization of requirements and the virtual prototyping (VP) of complex systems, a construction framework focused on structured blocks to man-age the complexity of designs and reduce the development time for a modular implementation on a Hardware Description Language (HDL), This type of modeling allows an interaction of components that allows a simulation for the reading of data within tolerance zones according to specific standards, i.e., a parametric approach or mechanisms is required to cope with the complexity and unpredictability related to disturbances caused by various environmental conditions, in this case varied environmental requirements that lead to a range of data input for a nominal value, for example, a device could be designed that al-lows data reading from 20V to 29V for a nominal amount of 25V; Therefore, it is essential to establish an appropriate environment that allows for verification and verification methods, which for their connectivity environment is necessary to implement simulators and scenario generators that include the control of environmental variables in interaction with interfaces of electronic modules, which brings to consider another important aspect, as the intermittent connections caused by factors such as GPRS connectivity (General Packet Radio Service), battery consumption, radio interfaces, or simply that the link to the data network has been lost, therefore, when seen in some way all of the above simplified from the design side, finally allows us to focus our efforts on the server-side software ASQ metrics[25]–[29].

5.Development of the linear Model for metrics.

Having a clear concept about the metric, therefore, this data will be implemented to establish the level of progress through established stages of development with gradual improvement [30]–[32]. For this model, three levels (low, medium, and high) of importance ω were determined for each metric, since each model will have an associated set of parameters more critical than others [33]–[35], see table 4. For linear modeling, there is a dependent variable, and some independent variables x, the value of this variable is made up of checklists or processes completed. Each of these variables dependents is associated with their respective β weight or level of importance. An expert in the area defines this value of significance. Its process is assumed that a normal distribution provides the linear model of the data, see equation 1.

$$y \in \mathbb{R}^{N\times 1}, x \in \mathbb{R}^{N\times N}, \beta \in \mathbb{R}^{N\times 1}, \epsilon \in \mathbb{R}^{N\times N}$$
$$y = \beta_1 x_1 + \beta_2 x_2 + \cdots . + \beta_n x_n + \epsilon$$
$$y = \sum_{i=1}^{n} \beta_i . x_i + \epsilon, \epsilon \sim \mathrm{N}(0, \sigma^2) \quad (1)$$

Table 4. levels of Importance and weights.

Error severity Class	Relative weight
Low Importance	1
Medium Importance	3
High Importance	9
Extreme Importance	15

It is essential to clarify that within the model, there are variables that due to their compatibility, can be defined in a composite way. That is to say that a resulting value can be determined based on a set of metrics associated with a specific weight determined by the expert; a particular case could be to associate within the same vector the parameters of Efficiency, Portability, and Functionality and relate them to a value of importance. Establishing in this last way a linear combination of the parameters, see Table 5.

Table 5. table of metrics selected for the model[36]

Metris	Definition
Reliability	The possibility of ensuring the functionality of the system in claim time. The system functionality should not fail in that in claim period.
Integrity	Defend against attacks to its security.
Reusability	To use the accessible components of the system for the development of the software product.
Efficiency	The quality of a system to be used at different frameworks simultaneously.
Performance	The degree to which system or its component can be customized to correct errors and speedy recovery after a failure occurrence
Timeliness Cost	The degree to which the system is ready and released on time
Compatibility	The ability to which how a system component is with other components.

It is essential to define that each stage of project development leads to a maturity level set in an evolutionary way towards the final achievement of a mature software process. Each level of maturity provides a layer in the base so that it allows for continuous improvement of the process associated with the reduction of errors examined in table 2. There is a constant process of 4 states, where you want to bring all methods to an ideal state, in which they are managed systematically accompanied by continual improvement within the different states is possible to identify. The level of managed: processes documented in a minimum way is equivalent to fundamental and necessary training. The level Defined: everything documented seeks to focus as a standard process; it is crucial to identify the instructions of teachings and subject them to training processes. The Quantitatively Managed level: It is necessary to identify the relevant and agreed metrics. For this, it is essential to contemplate the standards related to the CDIO (Conceiving-Designing-Implementing-Operating). Finally, process management includes the Optimizing level: where each process is retained through continuous improvement through changes that imply the decrease of errors, see Figure 3.

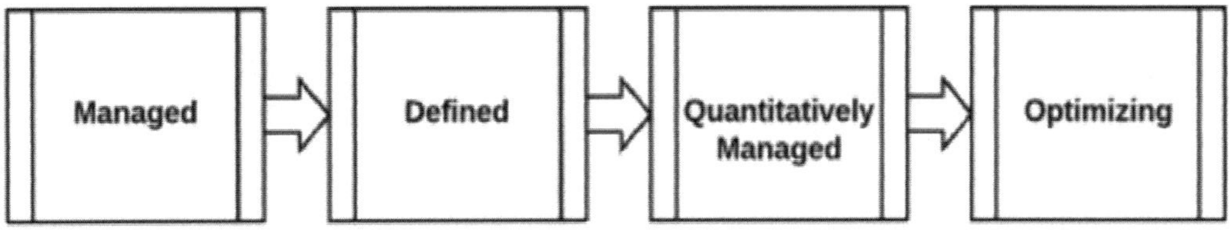

Figure. 3. Maturity Levels.

In particular, it is essential to reduce the cost of production without sacrificing the reliability and performance of the device, which allows in the long term to reduce maintenance costs by establishing a product with the highest quality standards. It is essential to determine that in the industrial environment, many variables randomly affect the operation and communication of IoT devices, it is desired from the design to skew those construction parameters to eliminate inherent problems that go against the functionality of mechanisms and sensors. For example, working with a wireless system for digital data transfer with the 802.11a/b/g/n standard is working with a communication system that tends to deteriorate quickly by a Gaussian noise produced by the industrial environment, which reset the communication. For such reason from the design, it is possible to implant another scheme of communication as 802.11 ah, which works with other frequencies more tolerant to the random interferences. Besides, it is compatible with a higher quantity of devices[37].

6.Simulation for development metrics.

An experimental design is established based on quantifiable metrics organized in a linear model to estimate a measure of performance through indicators, and whose simulation is based on a pseudo-random number generator with a normal distribution and a defined variance, which guarantees the independence of the data. Therefore, the parameters of the simulation are defined: GNA (Generator Numeric Aleatoric) and computational resources (S.O Windows 64 bits, Hardware Core IE5 processor, and MATLAB programming language). Consequently, the factors and levels of the experiment are established; in this case, each categorical metric is associated with three levels, where a value corresponding to the individual level is assigned, a priority to develop the value of each development metric with a specific associated weight. It's imperative establish that parameters to be simulated are those conformed in each level of the ASQ to conform to the linear model. Finally, it is necessary to carry out an average number of simulations sufficient to guarantee the normalization of each linear coefficient through the central limit theorem and to reduce the error E associated with the samples.[35], [38], [39]. For the next experiment, 100 simulations were taken from MATLAB to obtain synthetic data through the mvnrnd command. See the following table to examine the metrics to be simulated. (see Table5).

The metrics associated with severity levels are calculated, considering table n3 and Eq1. See the following table to examine a small simulation according to 3 severity levels. The model allows to model the projects with respect to a set of determined variables, which allows comparing the indicators with a predefined target with respect to a quality level through a Pareto diagram, with the aim of establishing continuous improvements over time, To carry out an implementation study or recent changes to prioritize the solution of problems concentrated on specific variables and to bring the value of the current indicators to a pre-established target in a progressive way, this analysis can be established through a Pareto diagram with respect to the last established measure, see Figure 4.

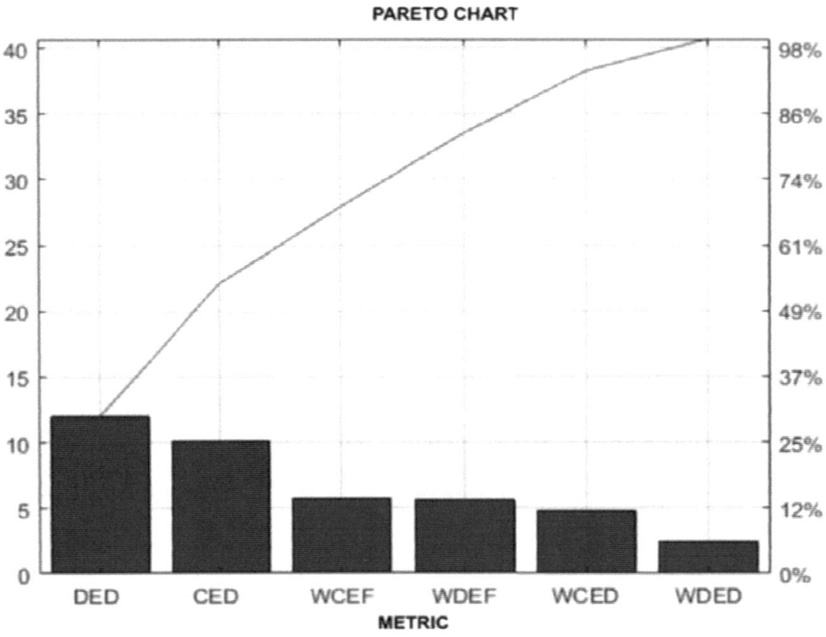

Figure. 4. Pareto diagram.

The diagram allows organizing the goals in the order of progress so that for each development interaction, the growth of the project can be evidenced through an index of progress. These elements allow the current status of the project to be quantified under a quality level, which allows achieving a considerable and optimal level of CMMI through an ASQ for IIoT devices, all this for a software factory service.

7.Analysis of Framework

Based on the estimation of the linear model, acute deviations from the objective values can be determined; such trends must be evaluated in each development cycle so that adjustments can be made as a result of error reduction. On the other hand, the model allows the establishment of activities that guarantee the quality of the product and the associated services through satisfactory operations that include full compliance with the client's requirements. Also, when the Framework is applied in an existing model, the organization's integrated software development processes allow an efficient approach to improve its development stage. The Framework as a systemic methodology seeks to integrate PSP / TSP / CMMI engineering theory and practice based on ASQ quality assurance.

8.Summary

What is sought from the framework, the characterization of a linear model and the extraction of the development metrics is to establish an estimation of indicators within a planned and systematic pattern of all the actions necessary to provide the quality assurance of the software within the development process within a maturity process, so that it meets the established minimum technical requirements, as well as the system management requirements; However, there are inherent shortcomings such as uncertainty regarding the validity of data rooted in information bias and others arising from human factors, especially the opposition of employees to evaluations of their activities. Finally, the framework aims to establish a robust solution on systems-oriented and compatible with the industrial environment, since many of the current solutions do not consider the physical scheme of data transmissions, which prevents a trade-off between components that intervene in the performance of the network and that at the same time are necessary for a more stable and secure application. It is desired to design an integrated engineering model for the development of distributed software, with an extension to the embedded software that represents the IoT terminal. All of the above according to the CMMI maturity model, with a systematic, disciplined and quantifiable approach to development, operation, and maintenance.

9.References

[1] J. Song et al., "WirelessHART: Applying wireless technology in real-time industrial process control", Proc. IEEE Real-Time Embed. Technol. Appl. Symp. RTAS, pp. 377–386, 2008.

[2] T. Yokotani y Y. Sasaki, "Comparison with HTTP and MQTT on required network resources for IoT", ICCEREC 2016 - Int. Conf. Control. Electron. Renew. Energy, Commun. 2016, Conf. Proc., pp. 1–6, 2017.

[3] S. Arshad y C. Tjortjis, "Clustering Software Metric Values Extracted from C# Code for Maintainability Assessment", Proc. 9th Hell. Conf. Artif. Intell. - SETN '16, pp. 1–4, 2016.

[4] M. Scotto, A. Sillitti, G. Succi, y T. Vernazza, "A relational approach to software metrics", Proc. 2004 ACM Symp. Appl. Comput. SAC 04, p. 1536, 2004.

[5] D. F. Barbosa, E. S. Furtado, y A. S. Gomes, "Uma proposta de institucionalização da usabilidade alinhada com práticas do modelo CMMI e foco nas necessidades da organização", Proc. VII Brazilian Symp. Hum. factors Comput. Syst. - IHC '06, p. 45, 2006.

[6] T. Guidinigoncalves, K. Oliveira, y C. Kolski, "La vérification et la validation de systèmes interactifs dans le modèle de maturité et d ' aptitude CMMI-DEV", pp. 48–58, 2016.

[7] M. Staples y M. Niazi, "Two case studies on small enterprise motivation and readiness for CMMI", Proc. 11th Int. Conf. Prod. Focus. Softw. - PROFES '10, pp. 63–66, 2010.

[8] S. Kusakabe y H. Lin, "Visualizing Centrality of Process Area Networks in", pp. 173–174, 2015.

[9] M. Raygoza, "Visual SDLC: Improving Requirements Engineering for Object-oriented Systems", Companion 18th Annu. ACM SIGPLAN Conf. Object-oriented Program. Syst. Lang. Appl., pp. 112–113, 2003.

[10] U. S. Shah, "An Excursion to Software Development Life Cycle Models", ACM SIGSOFT Softw. Eng. Notes, vol. 41, núm. 1, pp. 1–6, 2016.

[11] R. A. Majid, N. L. Noor, W. Adilah, W. Adnan, y S. Mansor, "A Survey on HCI Considerations in the Software Development Life Cycle : from Practitioner ' s Perspective", Proc. ICIS 2009, pp. 4–7, 2009.

[12] T. Adame, A. Bel, B. Bellalta, J. Barcelo, y M. Oliver, "IEEE 802.11AH: The WiFi approach for M2M communications", IEEE Wirel. Commun., vol. 21, núm. 6, pp. 144–152, 2014.

[13] N. S. Gill, "ACM SIGSOFT Software Engineering Notes March 2005 Volume 30 Number 2 Factors Affecting Effective Software Quality Management Revisited Nasib S . Gill Department of Computer Science & Applications , Haryana (India) ACM SIGSOFT Software Engineering Notes ", Softw. Eng. Notes, vol. 30, núm. 2, pp. 1–4, 2005.

[14] H. Yazbek, "A concept of quality assurance for metrics in CASE-tools", ACM SIGSOFT Softw. Eng. Notes, vol. 35, núm. 5, pp. 1–7, 2010.

[15] K. Iwanicki, P. Horban, P. Glazar, y K. Strzelecki, "Bringing Modern Unit Testing Techniques to Sensornets", ACM Trans. Sens. Networks, vol. 11, núm. 2, pp. 1–41, 2014.

[16] N. S. Nurul HASLINDA Ngah,Ziti Fariha,Syahrul Fahmy, "Refinement of the ISO 9126 Model for Evaluating Software Product Quality in e-Book", Aust. J. Basic Appl. Sci., vol. 8, pp. 29–34, 2014.

[17] R. Mo, Y. Cai, R. Kazman, L. Xiao, y Q. Feng, "Decoupling Level : A New Metric for Architectural Maintenance Complexity", ICSE '16 Proc. 38th Int. Conf. Softw. Eng., pp. 499–510, 2016.

[18] A. Sillitti, B. Russo, P. Zuliani, y G. Succi, "Deploying, updating, and managing tools for collecting software metrics", Proc. 2004 Work. Quant. Tech. Softw. Agil. Process - QUTE-SWAP '04, pp. 1–4, 2004.

[19] D. C. Boger y N. R. Lyons, "The organization of the software quality assurance process", SIGMIS Database, vol. 16, núm. 2, pp. 11–15, 1984.

[20] G. Zayaraz, D. P. Thambidurai, M. Srinivasan, y D. P. Rodrigues, "Software quality assurance through COSMIC FFP", ACM SIGSOFT Softw. Eng. Notes, vol. 30, núm. 5, p. 1, 2005.

[21] J. Abraham, J. Chapple, y C. Preve, "Improving quality of Ada software with range analysis", ACM SIGAda Ada Lett., vol. 31, p. 69, 2011.

[22] G. G. Gustafson y R. J. Kerr, "Some practical experience with a software quality assurance program", Commun. ACM, vol. 25, núm. 1, pp. 4–12, 1982.

[23] T. Hampp, "A cost-benefit model for software quality assurance activities", … Conf. Predict. Model. Softw. …, pp. 99–108, 2012.

[24] R. Nandakumar, A. K. Lal, y R. M. Parmar, "State of the art in software quality assurance", ACM SIGSOFT Softw. Eng. Notes, vol. 39, núm. 3, pp. 1–6, 2014.

[25] P. Patwari, S. R. Chaudhuri, A. Banerjee, S. Natarajan, y S. Pandey, "A complementary domain specific design environment aiding SysML", ISSE 2016 - 2016 Int. Symp. Syst. Eng. - Proc. Pap., 2016.

[26] D. Amyot, A. A. Anda, M. Baslyman, L. Lessard, y J. M. Bruel, "Towards Improved Requirements Engineering with SysML and the User Requirements Notation", Proc. - 2016 IEEE 24th Int. Requir. Eng. Conf. RE 2016, pp. 329–334, 2016.

[27] P. Bareiss, D. Schutz, R. Priego, M. Marcos, y B. Vogel-Heuser, "A model-based failure recovery approach for automated production systems combining SysML and industrial standards", IEEE Int. Conf. Emerg. Technol. Fact. Autom. ETFA, vol. 2016-Novem, 2016.

[28] H. Song, R. Srinivasan, T. Sookoor, y S. Jeschke, Smart cities : foundations, principles, and applications. .

[29] A. Abid et al., "A SysML based-methodology for modelling disturbances in manufacturing systems using ADACOR holonic control architecture", 2016 11th Fr. 9th Eur. Congr. Mechatronics, MECATRONICS 2016 / 17th Int. Conf. Res. Educ. Mechatronics, REM 2016, pp. 97–102, 2016.

[30] A. Díaz, "Principios del diseño estadistico", en Diseño estadistico de experimentos, 2009.

[31] H. Gutiérrez Pulido y R. De la Vara Salazar, Análisis y Diseño de Experimentos. 2013.

[32] A. . Fallis, Elementos de diseño de experimentos. 2013.

[33] D. Montgomery, Diseño y Análisis de Experimentos. 2004.

[34] H. Gutiérres Pulido y R. De la Vara Salazar, Análisis y diseño de experimentos. 2008.

[35] D. Monzón y D. Monzón Paiva, "Introducción al diseño de experimentos", Rev. la Fac. Agron. la Univ. Cent. Venez., 1992.

[36] M. Marwah, Q. Mateen, y M. Sirshar, "Software Quality Assurance in Internet of Things", Int. J. Comput. Appl., vol. 109, núm. 9, pp. 16–24, 2015.

[37] T. Pan y L. Zheng, "IPD-CMMI model of embedded software engineering under IEC-CDIO framework", J. Softw., vol. 7, núm. 11, pp. 2501–2510, 2012.

[38] D. C. Montgomery, Diseño y análisis de experimentos. 2004.

[39] Lauren H., "Diseño de Experimentos", en Diseño de Experimentos, 2000.

Bi-objective routing model with speed variation and consideration of emissions: Case study of solid waste collection in Coveñas, Sucre

J Ruiz-Meza[1], J R Montoya-Torres[1], D Mejía-Ayala[2], and S Navaja[3]

Faculty of Engineering, Universidad de La Sabana, Colombia.
Business Engineering School, Universidad César Vallejo, Perú.
Faculty of Engineering, Universidad Tecnológica de Bolívar, Colombia.

joserume@unisabana.edu.co

Abstract. Models for routing vehicles have been applied to solve a variety of transport route design problems. Also, it has taken an important trend on environmental care in consideration of the negative environmental impact generated by the emissions resulting from transportation. Another relevant aspect of environment care is the collection of solid wastes due to the high quantities generated by human activities. In this paper, we developed a bi-objective model for routing vehicles for solid waste collection that considers the minimization of costs associated with both vehicles and CO_2 emissions. The model considers heterogeneous fleet, capacity, hard time windows, and speed variation in an urban center. A case study in the Caribbean region of Colombia is studied, with a total of 50 nodes in the network. Because of the computational hardness, nodes are clustered in five sets of 10 nodes to obtain optimal solutions. The sum of costs showed a reduction of 9.447% compared to current costs (50 nodes without clustering). Subsequently, we reviewed and restructured some constraints that allowed to soften the complexity of the model and obtain improved solutions with reduced current costs by 16.55%.

1. Introduction

Nowadays, organizations need to make their processes more efficient to generate competitive advantages and improve their organizational logistics. For this purpose, the process of transportation is considered one of the most important in the supply chains in terms of reducing costs, distances, times, among other factors, when delivering goods [1].

The vehicle routing problem (VRP) is a well-known optimization problem dealing with decision-making at the operational level of the supply chain, usually employed for goods pick-up and delivery [2],[3]. This routing process may consider one or more stages of the transportation system, however, this type of operational planning generally includes only one level and must be carried out on a daily schedule [5], for which VRP has been highly efficient since its introduction by [6]. In this way, VRP translates into a model for organizations to minimize transportation costs, which can be between 20% and 55% of total logistics costs [7], [8].

The routing problem consists of assigning routes to satisfy customer demand in a given area starting from a depot and returning to it at the end of the route [9]. This model allows simulating real contexts [10], [11] presenting a large number of variants and restrictions, making it computationally complex. Indeed, the simplest version of the VRP is known to be NP-complete [12]. Among the variants addressed in the VRP are simultaneous pickups and deliveries [13], [14], time window [15], [16], multiple depots [17], capacity constraints [18], as well as variants considering the car of the environment [19]–[21].

In addition, the inclusion of vehicle speeds on the routing process is paramount since travel times depend on this magnitude. This is directly linked to fuel consumption and carbon emissions [22]. Also, speed is one of the main variables that specify the amount of energy a vehicle consumes while performing a job, in addition to load and distance traveled [23],[21],[24].

On another note, one of the relevant practical variants of the VRP is the solid waste collection problem [4]. Indeed, This type of problem has taken on greater relevance in response to the gradual increase in the generation of solid waste in terms of volume and composition, and which is linked to the growth in population. Hence, managing solid wastes is a problem that demands the attention of all countries and regions of the world, because it generates a direct impact that influences both public health and the environment [25]. To approach this problem, the academic community has formalized the so-called Waste Collection Vehicle Routing Problem (WCVRP). One of the first works is presented in [26], where the objective is the efficient collection of waste [27].

Several works associated with the solid waste collection have been developed by applying theoretical instances. In [28] develop a modified particle swarm optimization (PSO) algorithm to solve a capacitated vehicle routing problem. The routing is programmed for a week of collection. In [29] an improved non-dominated sorting genetic algorithm with direct local search (INSGA-dLS) is applied to solve a two-echelon multi-objective location routing problem (2E-MOLRP). Theoretical instances were developed due to the lack of publication of solid waste collection data that present the developed variants.

However, this type of problem presents more practice in real situations. In [30] a model of solid waste collection that considers container selection is developed as an inventory routing problem. The model is solved through a heuristic method for real instances of a company located in the Netherlands, generating cost savings of up to 40%. In [31] apply a two-stage method for solving a solid waste collection problem based on an assembly coverage problem and a VRP with split deliveries. The model is applied to the waste collection process in Kaohsiung City in Taiwan. In [32] apply an improvement approach to solve a problem that presents the combination of shift scheduling and generation of routes for waste collection. Moreover, in[33] a greedy randomized adaptive search procedure (GRASP) is applied to solve an eco-efficient routing model for waste collection on the island of La Palma, Spain.

In [34] the ORD software and dynamic programming are used for the construction of selective routes for the collection of recyclable elements in Rio de Janeiro with reductions of 27.72% in costs. Regarding scenarios under uncertainty, in[35] propose a model for the waste collection under uncertainty in a logistic network of three steps. The uncertain parameter corresponds to recyclable, biodegradable, and hazardous waste in southern Tehran, Iran, and is solved by the CPLEX solver. Finally, in [36] apply artificial intelligence in the development of an application that allows the identification of points and the construction of routes for the waste collection of electronic equipment in a city of Poland.

In this paper, we intend to show a bi-objective vehicle routing model with a heterogeneous fleet, capacity, hard time windows, and speed interval variation. The case of solid waste collection in the municipality of Coveñas, Colombia, is taken as a case study. In particular, the objectives presented by the model are: (i) minimization of the costs linked with the use of the vehicle in terms of distance traveled, and (ii) minimization of CO_2 emissions in terms of the fuel consumed during the collection process.

In addition to the above works, our model is also applied to a real case study. Likewise, CO_2 emissions are considered, which are less studied in this type of problem. On the other hand, our model applies the speed calculation for urban spaces considering that the processes of solid waste collection are carried out in this type of area worldwide. Finally, because this problem is NP-complete [37], we demonstrate the sensitivity in the constraint construction, which can make the model more complex or softer in terms of the number of nodes to be solved with exact methods.

The remainder of this document is organized as follows. Section 2 corresponds to the conceptual and mathematical development of the model. The computational results are summarized in Section 3. Finally, the conclusions are shown in Section 4.

2. Mathematical model

Formally speaking, the problem is modeled on a directed graph $G=(N,A,D,T,O,S)$, where N represents the set of nodes, being $N_n=\{n_1,n_2,n_3,\ldots,n_n\}$ customers and $N_o=\{0\}$ is the depot $N=N_o U N_n$. A is the set of arcs, the matrices of distance between nodes and time between nodes travelled at speed r are represented by $D=[d_{ij}]$ y $T=[T_{ijr}]$. Also, O corresponds to the collection vector in each i node$[d_i]$ and finally, S is the service time vector $[S_i]$. The fleet of vehicles k is heterogeneous $K=\{1,2,3,\ldots,k\}$, with limited capacity Q_k, which must perform service in a hard time window $[a_i,b_i]$. Each edge$(i,j)\in A$ where $i,j\in N$ and $i\neq j$, and is linked to a travel time T_{ijr} that is calculated through the Equation (1):

$$T_{ijr}=\frac{d_{ij}}{V_{ijr}} \tag{1}$$

Each vehicle can travel the arc (i,j) at a speed V_{ijr} which is calculated according to the speed equation for urban spaces adapted by [23], as shown in Equation (2):

$$V_{ijr}=l_r+\frac{dc_i+dc_j}{2\text{R}}(u_r-l_r) \tag{2}$$

A set of speeds r is considered , in addition to the speed limitations on each arc(i,j), where l_r is the minimum speed, and u_r is the maximum speed the vehicle can travel. We also have that dc_i is the distance from the node i to the center of the city and finally, R is considered the radius of the city.

Considering the environmental implications generated by land transport activities, we decided to work with an objective-based on minimizing fuel consumption by referencing works such as [20]. An objective function of fuel consumption based on mechanistic consumption models is proposed [38]. This model is based on the use of Equation (3):

$$W=F*C \tag{3}$$

Where C represents the distance, and F is the set of the four forces acting on the vehicle (the aerodynamic force fz_1, the rolling force fz_2, the slope force fz_3, and the inertia force fz_4) [39]; these forces are calculated as expressed in [40] and [41] obtaining a general equation for the calculation of the work, as shown in Equation (4):

$$W=\left[jCaRA_kV_{ijr}^{\,2}+m\left[\left(C_r cos\theta+sen\theta\right)g+a\left(u+\frac{n}{r_t^2}\right)\right]\right]C \tag{4}$$

Forces fz_2, fz_3, and fz_4, according to [20], are calculated from the distance traveled and the weight of the vehicle in the arc(i,j) (full and empty), so two equations are obtained. Equation (5) considers the weight of the empty vehicle, called "tare"[19], while Equation (6) calculates from the vehicle's load.

$$\left[\left(sen\theta_{ij}+Crcos\theta_{ij}\right)g+a_k\left(u+\frac{n}{rt_k^{\,2}}\right)\right]d_{ij}tare_k \tag{5}$$

$$\left[\left(sen\theta_{ij}+Crcos\theta_{ij}\right)g+a_k\left(u+\frac{n}{rpt_k^{\,2}}\right)\right]d_{ij}Load_{ik} \tag{6}$$

For the inertia force fz_1, which specifically considers vehicle parameters, it can be calculated by the travel speed V_{ijr}, the drag coefficient Ca, the frontal area of the vehicle A_k, and the air density R, as presented in Equation (7).

$$fz_1 = jCaRA_k V_{ijr}{}^2 d_{ij} \quad (7)$$

The cost of fuel cf is considered in liters, so the Equation (8) used to calculate the cost of emissions is applied according to [42].

$$e = \sum_{j=1}^{n} (\text{Spent fuel}_j * \text{Emission factor}_j * \text{Oxidized fraction}_j * \frac{44}{12}) \quad (8)$$

Therefore, the cost of emissions ce, considering an oxidized fraction of 0.99 [43], takes a value of 8.28 [41]. Then we have that the costs of CO_2 emissions are obtained by the sum of $(cf + ce)$.

Index and sets:

$i = j = h$: Index of clients or companies
k : Vehicle index
r : Speed index
N_n : Set of customers $\{1,2,3,\ldots,n\}$
N_0 : Deposit $\{0\}$
N : Set of customers and the deposit $N = N_0 \cup N_n$
K : Vehicle fleet $\{1,2,3\ldots k\}$
R : Speeds $\{1,2,3\ldots r\}$

Parameters:

Q_k : Vehicle capacity $k \in K$
Cf_k : Fixed cost per use of the vehicle $k \in K$
Cv_k : Variable cost per use of the vehicle $k \in K$
d_{ij}: Distance between nodes $i \in N, j \in N$
d_i: Offer to collect from customers $i \in N_n$
a_i : Earliest time of arrival to the customer $i \in N_n$
b_i : Late arrival time at the customer $i \in N_n$
S_i: Time of collection $i \in N_n$
M : Very large number to avoid negativity and customer-to-customer point selection
dc_i : Distance from point i to city center $i \in N$
l_r : Minimum speed $r \in R$
u_r : Maximum speed $r \in R$
V_{ijr} : Speed r in arc (i, j) $i \in N$, $j \in N_n$, $r \in R$
T_{ijr} : Travel time in the arc (i, j) at speed r $i \in N$, $j \in N_n$, $r \in R$
R : Radius of region or area

Variables:

$$X_{ijk} = \begin{cases} 1 \text{ vehicle } k \text{ goes from node } i \text{ to node } j \\ 0 \quad \text{otherwise} \end{cases}$$

$$Z_{ijr} = \begin{cases} 1 \text{ the arc } (i\ j) \text{ is traveled at speed } r \\ 0 \quad \text{otherwise} \end{cases}$$

$Load_{ik}$: Loading collected by the vehicle k at the customer i.

T_i : Time when the vehicle arrives at the customer i.

2.1. Formulation of the mixed integer linear programming model

The bi-objective model considers the minimization of costs by vehicle allocation and variable costs in terms of distance traveled, in addition to the costs of fuel consumption.

Bi-objective function:

$$\text{Min Z} = \alpha \left(\sum_{i \in N_0} \sum_{j \in N_n} \sum_{k \in K} Cf_k X_{0jk} + \sum_{i,j \in N: i \neq j} \sum_{k \in K} \left(Cv_k d_{ij} \right) X_{ijk} \right) + \beta \left(\begin{array}{l} \sum_{i,j \in N_n: i \neq j} \sum_{k \in K} (cf + ce) \partial_{ijk} d_{ij} tare_k X_{ijk} + \\ \sum_{i,j \in N_n: i \neq j} \sum_{k \in K} (cf + ce) \partial_{ijk} d_{ij} Load_{ik} + \\ \sum_{i,j \in N: i \neq j} \sum_{k \in K} (cf + ce) CaRA_k V_{ijr}^2 d_{ij} X_{ijk} \end{array} \right) \tag{9}$$

Constraints:

$$\sum_{j \in N: j \neq i} \sum_{k \in K} X_{ijk} = 1; \quad \forall i \in N_n \tag{10}$$

$$\sum_{j \in N_n} X_{0jk} \leq 1; \quad \forall k \in \text{K} \tag{11}$$

$$\sum_{i \in N: i \neq h} X_{ihk} = \sum_{j \in N: j \neq h} X_{hjk}; \ \forall k \in K, \forall h \in N \tag{12}$$

$$T_j \geq \sum_{r \in R} T_{0jr} * Z_{0jr} + S(j) - M * \left(1 - X_{0jk}\right); \ \forall i \in N, \forall j \in N_n, \forall k \in K \tag{13}$$

$$T_j \geq T_i + \sum_{r \in R} T_{ijr} * Z_{ijr} + S(i) - M * \left(1 - X_{ijk}\right); \ \forall i \in N, \forall j \in N_n, \forall k \in K \tag{14}$$

$$a_i \leq T_i \leq b_i; \quad \forall i \in N_u \tag{15}$$

$$Load_{ik} + d_j - Load_{jk} \leq M * \left(1 - X_{ijk}\right); \ \forall i \in N, \forall j \in N_n : i \neq j, \forall k \in K \tag{16}$$

$$Load_{ik} + d_j - Load_{jk} \geq -M * \left(1 - X_{ijk}\right); \ \forall i \in N, \forall j \in N_n : i \neq j, \forall k \in K \tag{17}$$

$$Load_{ik} \leq Q_k X_{ijk}; \forall j \in N_n, \forall k \in K \tag{18}$$

$$Load_{ik} \leq 0; \forall i \in N_0, \forall k \in K \tag{19}$$

$$\sum_{r \in R} Z_{ijr} = f_{ij}; \quad \forall i, j \in \text{N} \tag{20}$$

$$f_{ij} = \sum_{k \in K} X_{ijk}; \ \forall i, j \in N \tag{21}$$

$$X_{ijk}, Z_{ijr} \in \{0,1\} \tag{22}$$

$$Load_{ik} \in \mathbb{R}^{+} \tag{23}$$

$$T_i \in \mathrm{R}^{+} \tag{24}$$

The bi-objective function (9) minimizes vehicle usage and fuel consumption costs. Constraints (10) ensure that each customer is visited exactly once by a vehicle. Constraints (11) ensure that vehicles must leave the deposit. Constraints (12) ensure the flow of the network. Constraints (13) and (14) establish the time of arrival at each customer. Constraints (15) corresponds to the hard time windows. Constraints (16) and (17) ensure that pick-ups are met at each node. Vehicle capacity is restricted by Constraints (18). Constraints (19) ensure that, upon return to the depot, the vehicle must unload everything. These Constraints (16) and (17) are derived from the non-linear equation worked by [44], as shown in Equation (25), which ensures that a load is made on each node giving values to the variable, and also avoids the subtours in the routing:

$$\sum_{i \in N i \neq j} \sum_{k \in K} X_{ijk} (Load_{ik} + d_j - Load_{jk}) = 0; \ \forall j \in \mathrm{N} \tag{25}$$

Constraints (20) and (21) are equality artifices that assign the range of speeds and vehicle usage to the arcs made on the route. Constraints (22), (23), and (24) establish the nature of the decision variables.

3. Computational results

3.1. Data collection

The model was coded in GAMS (General Algebraic Model System) software and run on a computer with the following features: Windows 64-bit operating system, 4 GB RAM with 1 TB HDD, and processor Intel Core i7-5500 CPU@ 2.40 GHz. The model was evaluated with the data obtained in the characterization of the solid waste collection process in the municipality of Coveñas, Colombia. The data collection was done through the on-site monitoring of the current waste collection route by the company responsible for this process. Details of the coordinates of the points were recorded using a Garmin GPSMAP 64s; also, the service time and arrival and departure times at each point were registered. The average amount of waste was identified through data reported in the Plan for the Integral Management of Solid Wastes - (by its Spanish acronym PGIRS) of the municipality of the Coveñas - Department of Sucre [45].

A total of 49 waste collection points was identified along the current route. The vehicle starts the route from the landfill located in the municipality of Sincelejo, Sucre, and returns to it at the end of the tour. This process begins at 5:00 a.m., ends at approximately 12:00 a.m., and is performed by a vehicle with a capacity of 13 tons. The route presents 176,5 kilometers characterized by restrictions on vehicle speeds according to the tourist or urban area. These points are depicted in Figure 1.

Due to the computational complexity of the problem, it was not possible to work initially with the set of 50 nodes (49 collection points and the landfill). Therefore, a zoning is made based on the geographic conditions of the municipality. The total set of nodes was clustered into five zones, each one containing ten nodes; the initial node of a zone routing was the final node of the precedent zone, as shown in Figure 1. Additionally, we divide and allocate the fixed routing costs for each zone considering the cost of toll and of transporters.

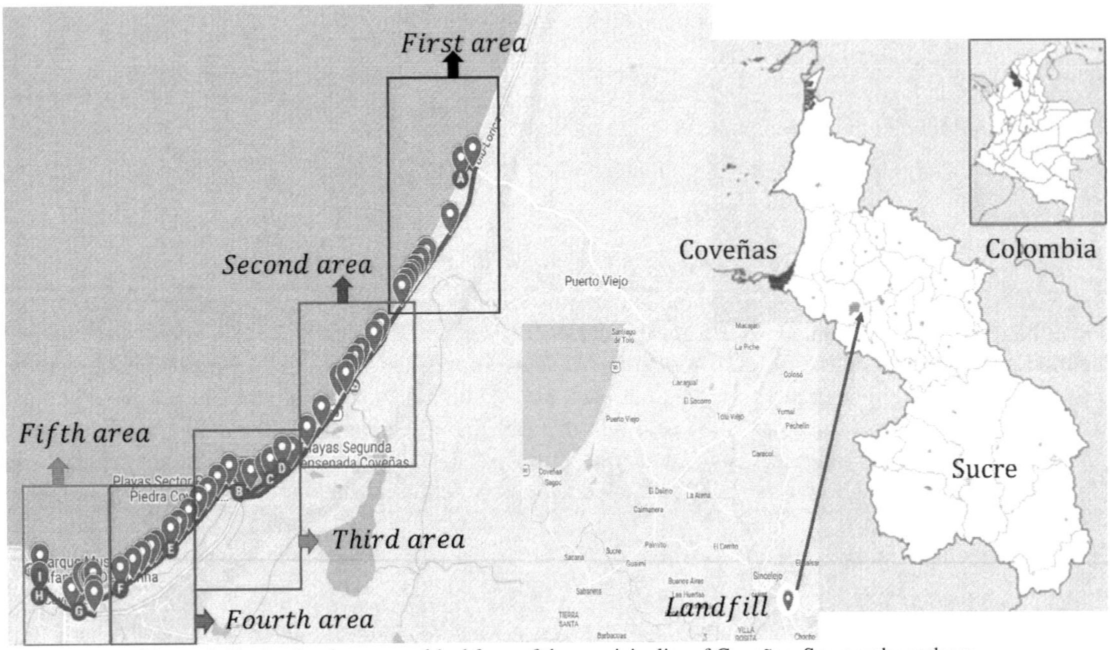

Figure 1. Zoning by the geographical form of the municipality of Coveñas. Source: the authors.

3.2. Results by zones

The zones include a total of 10 nodes each. In the first zone, the landfill is included, and three-speed ranges are established considering that the landfill is in a different city. For zones 2, 3, 4, and 5, two ranges of speeds were enabled according to those allowed in the urban center. Variable costs were calculated in terms of tire wear and oil consumption per kilometer traveled. In the third and fifth zones, the additional fixed cost is increased because these contain the location of tolls. The results of the zone are shown in Table 1.

Table 1. Results obtained by zones.

Zone	Fixed costs ($COP)	Variable costs ($COP/km)	Speeds (km/hr)	Total route costs ($COP)	Fuel consumption (Lt)
1	28.333,4		$r_1:[u_1=40,\ l_1=80]$ $r_2:[u_2=10,\ l_2=30]$ $r_3:[u_3=20,\ l_3=40]$	125.608,5	38,97
2	28.333,4	109,18		53.365,9	10,74
3	64.533,4		$r_1:[u_1=20,\ l_1=30]$	84.094,4	8,8
4	28.333,4		$r_2:[u_2=20,\ l_2=40]$	48.976,4	9,37
5	64.533,4			89.289	11,34

As a cumulative result, we added all the costs to obtain the total cost of routing in the municipality ($COP 401.333,3). We note that, despite the division of routes, the cost is still 9,45 % lower than the current total cost of $COP 443.206,5. The total fuel consumption calculated by the model is 79,22 Lt or

20,92 gallons. This, compared to the current consumption of 22 gallons, shows a difference of 1,08 gallons saved.

3.3. Constrains Analysis

Additionally, we performed an analysis of the constraints of the model due to the complexity of working with larger instances. We review other ways to build constraints to obtain the same functionality. We verified the constraints (16), (17), (18), and (19) that ensure the collection scheme and proposed the construction of the following constraints that fulfill the same function.

$$\sum_{i \in N: i \neq j} \sum_{k \in K} Load_{ijk} - \sum_{i \in N: i \neq j} \sum_{k \in K} Load_{jik} = d_i ; \forall j \in N_n \quad (26)$$

$$Load_{ijk} \leq Q_k X_{ijk} ; \forall i \in N_0, \forall j \in N_n, \forall k \in K \quad (27)$$

$$Load_{ijk} \leq MX_{ijk} ; \forall i \in N_0, \forall j \in N_n, \forall k \in K \quad (28)$$

$$Load_{ijk} = 0; \forall i \in N_n, \forall j \in N_0, \forall k \in K \quad (29)$$

For this proposition, the variable $Load_{ik}$ changes to $Load_{ijk}$ assuming the amount of waste collected in the arc ij with vehicle k. Constraint (26) ensures the collection of the amounts offered by each customer. Constraits (27) ensure that the capacity of the vehicle is not exceeded. Constraits (28) state that if a node is visited, a quantity must be loaded. Finally, Constraints (29) ensure that not everything is discharged into the depot.

3.4. Comparing results

For the new model, Constraints (26), (27), (28), and (29) replace Constraints (16), (17), (18), and (19), and the variable $Load_{ijk}$ replaces $Load_{ik}$. With this new building, it is possible to obtain model solutions for the 50 nodes corresponding to the solid waste collection in Coveñas, Sucre, without the need to make zoning. We obtained four solutions with a gap below 10%, of which α =0,6 y β =0,4 correspond to the lowest cost ($COP 369.860 with a gap of 9,98%). The total consumption was 18,7 gallons of fuel. This shows a 16,55% reduction in the current cost and a 15% reduction in the fuel consumed. Likewise, the percentage of reduction concerning the first model solution is 8,09% and 10,61% of the cost and fuel consumption, respectively.

4. Conclusions

In this paper, a bi-objective mixed-integer programming model was designed to provide decision-making for vehicle routing in the process of solid waste collection in urban centers. The model considers the variants of the heterogeneous fleet, capacity limits, hard time windows, and the implementation of speed ranges that allow calculating the optimal speed of travel within the city. The model aims at minimizing fixed and variable costs by vehicle assignment and distance traveled, as well as the costs for fuel consumption and equivalent CO_2 emissions, thus generating an environmentally friendly. Tests were carried out with real data on the solid waste collection in the municipality of Coveñas, Colombia, a coastal area dedicated to the sun and beach tourism.

Due to the complexity of the problem, it was initially not possible to generate solutions for the municipality's data set, so zoning was carried out based on the geography of the site. A total of five zones was constituted with 10 nodes each. The costs associated with the routing of each zone were obtained. The accumulated sum of the costs per zone determined the total cost of waste collection.

Total routing costs were $COP 401.333,3, representing savings of 9,447% compared to the current cost. This daily savings of $COP 41.873,2 can be translated into an annual savings of $COP 12.059.481,6. Additionally, fuel savings were estimated at 1,077 gallons. After modifying some constraints in the model to reduce the computational complexity, numerical results generated lower routing costs ($COP 369.860), which means a decrease of 16,55% over the current operational cost. That is, a reduction of 8,09% over the solution of the first model. Likewise, fuel consumption was reduced by 15% of the current value. This represents a daily saving of $COP 73.346,5.

In regards of future research opportunities, it is necessary to develop heuristic or metaheuristic algorithms that allow obtaining good solutions in low computational times. On the modeling approach, the waste collection problem can also be approached as an Open Vehicle Routing Problem (OVRP), where collecting vehicles do not leave from or finish the route at the same landfill.

References

[1] O. Belgin, I. Karaoglan, and F. Altiparmak, "Two-echelon vehicle routing problem with simultaneous pickup and delivery: Mathematical model and heuristic approach," *Comput. Ind. Eng.*, vol. 115, no. March 2016, pp. 1–16, 2018.

[2] J. R. Montoya-Torres, J. López Franco, S. Nieto Isaza, H. Felizzola Jiménez, and N. Herazo-Padilla, "A literature review on the vehicle routing problem with multiple depots," *Comput. Ind. Eng.*, vol. 79, pp. 115–129, 2015.

[3] A. Gharaei and S. H. R. Pasandideh, "Four-echelon integrated supply chain model with stochastic constraints under shortage condition," *Ind. Eng. Manag. Syst.*, vol. 3, no. 16, pp. 316–329, 2017.

[4] N. Chari, U. Venkatadri, and C. Diallo, "Design of a reverse logistics network for recyclable collection in Nova Scotia using compaction trailers," *INFOR Inf. Syst. Oper. Res.*, vol. 54, no. 1, pp. 1–18, 2016.

[5] E. Ruiz, V. Soto-Mendoza, A. E. Ruiz Barbosa, and R. Reyes, "Solving the open vehicle routing problem with capacity and distance constraints with a biased random key genetic algorithm," *Comput. Ind. Eng.*, vol. 133, no. August 2018, pp. 207–219, 2019.

[6] G. B. Dantzig and J. . Ramser, "The Truck Dispatching Problem," *Manage. Sci.*, vol. 6, no. 1, pp. 1–13, 1959.

[7] P. Toth and D. Vigo, "The Vehicle Routing Problem Monographs on Discrete Mathematics and Applications SIAM," *Philadelpia, Pennsylvania*, 2002.

[8] R. Pérez-Rodríguez and A. Hernández-Aguirre, "A hybrid estimation of distribution algorithm for the vehicle routing problem with time windows," *Comput. Ind. Eng.*, vol. 130, no. February, pp. 75–96, 2019.

[9] S. Aref and I. Kazemian, "A green perspective on capacitated time-dependent vehicle routing problem with time windows," pp. 1–15, 2015.

[10] P. Sombuntham and V. Kachitvichyanukul, "Multi-depot vehicle routing problem with pickup and delivery requests," *AIP Conf. Proc.*, vol. 1285, no. December, pp. 71–85, 2010.

[11] J. Ruiz-Meza, I. Montes, A. Pérez, and M. Ramos-Márquez, "VRP Model with Time Window, Multiproduct and Multidepot," *J. Appl. Sci. Eng.*, vol. 23, no. 2, pp. 239–247, 2020.

[12] C. B. Kalayci and C. Kaya, "An ant colony system empowered variable neighborhood search algorithm for the vehicle routing problem with simultaneous pickup and delivery," *Expert Syst. Appl.*, vol. 66, pp. 163–175, 2016.

[13] N. H. M. Wilson, J. M. Sussman, H.-K. Wong, and T. Higonnet, *Scheduling algorithms for a dial-a-ride system*. Massachusetts Institute of Technology. Urban Systems Laboratory, 1971.

[14] H. Min, "The multiple vehicle routing problem with simultaneous delivery and pick-up points," *Transp. Res. Part A Gen.*, vol. 23, no. 5, pp. 377–386, Sep. 1989.

[15] M. Solomon, "Algorithms for the vehicle routing and scheduling problem with timewindow," *Oper. Res.*, vol. 35, pp. 254–265, 1987.

[16] M. Solomon and J. Desrosiers, "Time Window Constrained Routing and Scheduling Problems," *Transp. Sci.*, vol. 22, no. 1, pp. 1–13, 1988.

[17] J. Renaud, G. Laporte, and F. F. Boctor, "A tabu search heuristic for the multi-depot vehicle routing problem," *Comput. Oper. Res.*, vol. 23, no. 3, pp. 229–235, Mar. 1996.

[18] G. Laporte, Y. Nobert, and D. Arpin, "An exact algorithm for solving a capacitated location-routing problem," *Ann. Oper. Res.*, vol. 6, no. 9, pp. 291–310, 1986.

[19] I. Kara, B. Kara, and M. Yetis, *Energy Minimizing Vehicle Routing Problem*. 2007.

[20] T. Bektaş and G. Laporte, "The Pollution-Routing Problem," *Transp. Res. Part B*, vol. 45, pp.

1232–1250, 2011.
[21] A. Palmer, "The Developement of an Integrated Routing and Carbon Dioxide Emissions Model for Goods Vehicle," *Sch. Manag.*, vol. Doctor of, p. 151, 2007.
[22] J. Zhang, Y. Zhao, W. Xue, and J. Li, "Vehicle routing problem with fuel consumption and carbon emission," *Int. J. Prod. Econ.*, vol. 170, pp. 234–242, 2015.
[23] W. Rao and C. Jin, "A Model of Vehicle Routing Problem Minimizing Energy Consumption in Urban Environment," *Asian Conf. Manag. Sci. Appl.*, pp. 21–29, 2012.
[24] W. Maden, R. Eglese, and D. Black, "Vehicle routing and scheduling with time-varying data: A case study," in *Journal of the Operational Research Society*, 2010, vol. 61, no. 3, pp. 515–522.
[25] S. Bashir and S. Goswami, "Tourism Induced Challenges in Municipal Solid Waste Management in Hill Towns: Case of Pahalgam," *Procedia Environ. Sci.*, vol. 35, pp. 77–89, 2016.
[26] J. W. Male and J. C. Liebman, "Districting and Routing for Solid Waste Collection," *J. Environ. Eng. Div.*, vol. 104, no. 1, pp. 1–14, 1978.
[27] M. Dotoli and N. Epicoco, "A Vehicle Routing Technique for Hazardous Waste Collection," *IFAC-PapersOnLine*, vol. 50, no. 1, pp. 9694–9699, 2017.
[28] M. A. Hannan, M. Akhtar, R. A. Begum, H. Basri, A. Hussain, and E. Scavino, "Capacitated vehicle-routing problem model for scheduled solid waste collection and route optimization using PSO algorithm," *Waste Manag.*, vol. 71, pp. 31–41, 2018.
[29] X. Yu, Y. Zhou, and X. F. Liu, "The two-echelon multi-objective location routing problem inspired by realistic waste collection applications: The composable model and a metaheuristic algorithm," *Appl. Soft Comput. J.*, vol. 94, p. 106477, 2020.
[30] M. Mes, M. Schutten, and A. P. Rivera, "Inventory routing for dynamic waste collection," *Waste Manag.*, vol. 34, no. 9, pp. 1564–1576, 2014.
[31] S. H. Huang and P. C. Lin, "Vehicle routing-scheduling for municipal waste collection system under the 'Keep Trash off the Ground' policy," *Omega (United Kingdom)*, vol. 55, pp. 24–37, 2015.
[32] P. De Bruecker, J. Beliën, L. De Boeck, S. De Jaeger, and E. Demeulemeester, "A model enhancement approach for optimizing the integrated shift scheduling and vehicle routing problem in waste collection," *Eur. J. Oper. Res.*, vol. 266, no. 1, pp. 278–290, 2018.
[33] A. Expósito-Márquez, C. Expósito-Izquierdo, J. Brito-Santana, and J. A. Moreno-Pérez, "Greedy randomized adaptive search procedure to design waste collection routes in La Palma," *Comput. Ind. Eng.*, vol. 137, no. September, p. 106047, 2019.
[34] L. S. Franca, G. M. Ribeiro, and G. de L. D. Chaves, "The planning of selective collection in a real-life vehicle routing problem: A case in Rio de Janeiro," *Sustain. Cities Soc.*, vol. 47, no. March 2018, p. 101488, 2019.
[35] H. Asefi, S. Shahparvari, and P. Chhetri, "Integrated Municipal Solid Waste Management under uncertainty: A tri-echelon city logistics and transportation context," *Sustain. Cities Soc.*, vol. 50, no. June, p. 101606, 2019.
[36] P. Nowakowski, K. Szwarc, and U. Boryczka, "Combining an artificial intelligence algorithm and a novel vehicle for sustainable e-waste collection," *Sci. Total Environ.*, vol. 730, p. 138726, 2020.
[37] D. V. Tung and A. Pinnoi, "Vehicle routing-scheduling for waste collection in Hanoi," *Eur. J. Oper. Res.*, vol. 125, no. 3, pp. 449–468, 2000.
[38] J. J. Posada Henao and C. A. González-Calderón, "Fuel Consumption in Vehicles for Highway Transportation - Predictive Models," *Rev. Ing. Univ. Medellín*, vol. 12, no. 23, pp. 35–46, 2013.
[39] R. Baxter, N. Hastings, A. Law, and E. J. . Glass, "Manual de conduccione eficiente para vehículos turismo," *Anim. Genet.*, vol. 39, no. 5, pp. 561–563, 2008.
[40] M. Yolanda, R. Morales, and A. H. Guzmán, "Manual De Conducción Técnica De Vehículos Automotores Diesel: Segunda Edición," 2012.
[41] J. Ruiz-Meza, "Diseño de un modelo multiobjetivo de VRP pick-up and delivery simultáneo (VRPSPD) para el aprovisionamiento de la leche en la cadena de suministros lácteos,"

Universidad Tecnológica de Bolívar, 2018.
[42] INECC and PNUD, "Actualización del Inventario Nacional de Emisiones de Gases de Efecto Invernadero 1990 - 2010 , para las Categorías de Energía y Procesos Industriales," 2012.
[43] IPCC, "IPCC Guidelines for National Greenhouse Gas Inventories." 2006.
[44] A. G. H. Kek, R. L. Cheu, and Q. Meng, "Distance-constrained capacitated vehicle routing problems with flexible assignment of start and end depots," *Math. Comput. Model.*, vol. 47, no. 1–2, pp. 140–152, 2008.
[45] Secretaria de Planeacion Municipal, "Plan de Gestión Integral de los Residuos Sólidos - PGIRS del municipio de Coveñas – departamento de Sucre," Coveñas, 2016.

Big Data and Artificial Intelligence in the Development of Industry 4.0; A Bibliometric Analysis

E González-Sarmiento[1,*], J Roa-Perez[1,] and L Ortiz-Ospino[2]**

[1] SI[2] Research and Innovation Seedbed. Universidad Simón Bolívar, Barranquilla, Colombia.
[2] Research Professor, School of Engineering. Universidad Simón Bolívar, Barranquilla, Colombia.

* egonzalez58@unisimon.edu.co
** jroa10@unisimon.edu.co

Abstract. Industry 4.0, also known as the Fourth Industrial Revolution, is characterized by process automation and digitization and the use of technologies such as the Internet of Things, cloud, big data, artificial intelligence (AI), and embedded systems, among others. In fact, AI and big data are two driving forces behind several technological innovations that have shaped the current digital environment and Industry 4.0. These two trends share a common goal, i.e., to obtain the maximal data out of the vast amount of data being generated today. Therefore, within the context of substantial scientific development in this subject, this work provides an overview of the development of Industry 4.0 based on big data and AI. The methodological design commenced from the development of a bibliometric exercise for the identification of scientific papers addressing this subject. The main results of this work include the identification of scientific development based on big data and AI in the advancement of Industry 4.0. Moreover, conclusions regarding the importance of these pillars in the new digitization era are obtained.

1. Introduction

Throughout history, the First Industrial Revolution has been regarded as an important step that led to the current technological development. Today, Industry 4.0, also known as the Fourth Industrial Revolution, is at its height, marked by numerous sophisticated technologies and the advent of several digital industries. In fact, Industry 4.0 focuses on the introduction and development of digital technologies, particularly the digitization of production processes to increase efficiency, quality, and safety within the industry. In other words, the integration of intelligent systems into the production environment based on concepts such as cloud, big data, artificial intelligence (AI), and 3D printing [1]. Some of these technologies have been in use for years but in an abandoned and reserved manner. However, their consolidation into current production models can boost the production industry with fully integrated, automated, and streamlined processes. Therefore, the technological evolution being fostered requires an effort from companies to adapt to the new networked society.

Consequently, in this new era of digital transformation, data have become one of the most important pillars of Industry 4.0. Actually, big data technology is the key for managing and transforming large amounts of data because such a magnitude of data cannot be processed using conventional processes and infrastructures. The role of big data in Industry 4.0 in terms of managing and assessing large volumes of data is critical for understanding the current technological context, market demands and requirements, and user information and their feedback regarding new products and services. Hence, data have not only become a necessity for organizations but also the foundation for conducting business in the 21st century. A large part of daily activities is performed or somewhat connected to the Internet; therefore, the number of data registered, accrued, and stored by organizations, both online and offline, is rapidly increasing [2]. Hence, every year, companies continue to face increasing numbers of challenges in terms of the management and assessment of these data. Still, what would big data be without a self-learning application? In fact, AI is advancing further because its concept includes facilitating machines or equipment with cognitive learning for conducting tasks in an "intelligent" manner and providing support to users when making operational decisions. AI applied to data management and transformation is another complement of this revolution, wherein new algorithms allow the development of predictive and behavior analyses, detect frauds, assist in medical diagnoses, and improve sales and supply processes. Furthermore, currently, there is a wide field of applications that employ AI coupled with big data, which will continue to be of great use in all industrial sectors and society [3].

Our new normal assumes that the Fourth Industrial Revolution is identified by its permissiveness and focuses on involving customers in the whole process, thus assimilating the constant change in demands inherent to the current market. In most parts of the world, this work is conducted as part of a government strategic policy to disseminate these technologies to become more competent in the market.

Therefore, because abundant scientific development has been achieved in this subject, this work provides an overview of the development of Industry 4.0 based on big data and AI. However, the approach used in this work differs from that used in other reviews or research papers addressing this topic. In other words, this study performed a bibliometric analysis based on the principle that the quantity and quality of published papers available in international databases are indicators of the contributions by different countries and institutions to this research area [4]. In fact, bibliometrics have been employed in the literature for many years. In this tool, contents from publications are analyzed using statistical methods to determine patterns [5], thus providing researchers with a clear overview of the research field. This tool can be used to study big data and AI for advancing Industry 4.0. Based on the generated maps and bibliometric information, we can unveil trends observed in this research area. Moreover, educators and researchers can use this information to identify problems or evaluate unanswered questions in the literature to indicate the orientation for future studies on industry digitization based on big data and AI within the universe of Industry 4.0.

2. Theoretical Framework

As a theoretical support for this work, the following concepts were used as a referential framework for our bibliometric assessment: Industry 4.0, AI, big data, bibliometrics. These concepts allowed us to select the appropriate approaches, authors, and concepts as the theoretical support for the research.

Over time, technological development has exerted an impact on the production field. For this reason, Navarro & Sabalza [6] reported that different Industrial Revolutions have emerged throughout history. The First Industrial Revolution started at the end of the 18th century, and it was prompted by the introduction of mechanical equipment characterized by steam engines. The Second Industrial Revolution was a consequence of industrialization and the new line production system. The Third Industrial Revolution integrated technologies such as electronics, automation, and Information and Communication Technologies (ICTs). Finally, the most recent, the Fourth Industrial Revolution, also known as Industry 4.0, brought about the Internet of Things (IoT), big data, AI, digital coordination, cyberphysical systems, and robotics.

Industry 4.0 has emerged to address the challenge of different industry issues. According to Martin [7], companies need to deploy new techniques, such as implementing technologies, adopting new management strategies, and developing new business models.

Industry 4.0 was first discussed was in Germany in 2011 during the Hannover Messe Industrial Technology Show. In fact, this term was promoted by three German engineers: Henning Kagermann, Wolfagang Wahlster, and Wolf-Dieter Lukas. Industry 4.0 is characterized by process automation and digitization, which was generated through the combination of digital and physical systems, in addition to the use of electronics and information technologies in manufacturing [8]. Moreover, service provision, production customization, and the creation and development of new value-added business models promote the exchange of information between humans and machines [9, 10].

Thus, there are several definitions that include the common points. Neugebauer et al. [11] states that Industry 4.0 is "The integration of human resources, physical elements, and systems in a dynamic, self-organized way in real time with autonomously optimized value-added systems." Alternatively, according to Sipsas et al. [12], "Industry 4.0 is a paradigm that promotes the connection of physical elements, such as sensors, devices, and business assets, with each other and with the Internet."

The Fourth Industrial Revolution is known for the integration and birth of new technologies, including robotics, AI, big data, quantum computing, nanotechnology, biotechnology, and IoT. To a significant extent, the digital evolution of industries and their survival in the market are currently dependent on the data they generate and their influence in decision-making strategies. They also depend on the correct implementation of AI. Moreover, it is the aforementioned technologies that allow organizations to rely on facts, data, and forecasts, instead of only assumptions when making decisions [13].

The concept of big data involves a collection of data obtained from both systems and objects, such as the reading of a sensor [14]. Alternatively, Lycett [15] stated that this technology has an advanced predictive capacity to identify the events that may affect production before it is conducted. In other words, it identifies patterns.

As reported by Stoicescu [16], using big data, it is possible to know which data are missing. Furthermore, big data can be considered as a perfect tool for achieving specific results and increased profits for a company within the production sector. Additionally, big data allows access to a number of valuable information related to customers, as well as any process that is directly related to performance along with unforeseen correlations and connections that can be used in the industry.

The concept of AI is defined [17] as a "branch of computational science that introduces the simulation of intelligent behavior in computers" and "the ability of a machine to imitate intelligent human behavior." Hence, these technologies have emerged to stay in our lives. They have been implemented in sectors that were never considered before. AI systems independently assimilate, learn, and adapt to organizations. Over time, an increasing number of applications will be developed, which will render the consumer the core of the entire marketing process. In turn, services would be highly personalized to suit the user's demands. Thus, the implementation of AI in industries leads to a high level of production through process automation, thus providing significant values and competitive advantages to organizations [18].

Today, AI and robotic technologies are adopted around the world, including in manufacturing sectors that use industrial robots and in other economic activities, such as transportation via autonomous vehicles, trade in the financial markets, customer relationship management based on Chatbots, legal provisions, and medical diagnostics and surgeries [19]. AI and big data are the two driving forces behind various technological innovations that have shaped the current digital environment and Industry 4.0. These two trends have a common goal, i.e., to obtain maximal data from the data generated in companies today.

In this research, a bibliometric analysis was conducted, which allowed the examination and quantitative analysis of the literature and academic research. This analysis was conducted using statistical and mathematical techniques, which have been integrated in multiple fields and disciplines over time [20]. Kostoff et al. [21] stated that these bibliometric techniques provide an evaluative analysis of publications, patents, citations, and other elements that provide potential informative for developing indicators, which assist in evaluating the performance achieved in science and technology. In other words, bibliometric analysis identifies the infrastructure (authors, journals, institutions, keywords, etc.) of a technical officer through whom experts are identified. This tool also allows the development of web visit strategies for evaluating organizations that are in the process of rapid growth worldwide. Additionally, it identifies the impacts (bibliographic citations) of people, research units, organizations, and countries, as well as all information regarding the subject of research.

Despite all positive qualities expressed by the authors, it is appropriate to mention that the application of a bibliometric analysis is not always convenient for the development of new research, as noted by Krauskopf [22], Johnson [23], and Johnson et al. [24] when discussing the impact of the bibliometric analysis on the transparency and benefit of the production of scientific articles. They mention how it is the function of the editors to choose the methods using which they can employ appropriate measurements for each field of research. Therefore, the benefits and weaknesses offered by the application of a bibliometric analysis to a quantitative and qualitative research can be easily identified. In this regard, Carvalho et al. [25], Bounes et al. [26], and Urquhart and Dunn [27], highlight how the quantitative refinement of bibliometrics strengthens the qualitative analysis of research results.

3. Methodology

In this section, we focus on obtaining a vision with respect to the digitization trend of industries based on big data and AI in the development of Industry 4.0. Therefore, the methodological design commenced from the development of a bibliometric exercise for the identification of scientific articles on the subject of big data, AI, and Industry 4.0.

To synthesize the existing analysis, determining patterns, themes and problems, as well as recognizing the conceptual content of the field contributing to the development of a theory, an adequate literature review was conducted. This process involved a methodical, organized, specific, and reproducible design for identifying, evaluating, and interpreting a body of existing documents. Hence, the process relied on the use of quantitative methods that analyze publications according to authors, countries, and institutes as well as the methods that map the science employing bibliometric software [28]. For this particular case, the analysis was limited to the purification and processing of scientific development based on the parameters established in the following methodological design subsection.

3.1. Methodological design

The methodological design of the present study included a documentary-type research design and correlational study by consulting documents (database, patents, scientific articles, reports, and journals) [29]. This approach allowed us to obtain an overview of the digitization trends of industries based on big data and AI in the world of Industry 4.0. The analysis performed in this study involved three phases, as detailed in Figure 1.

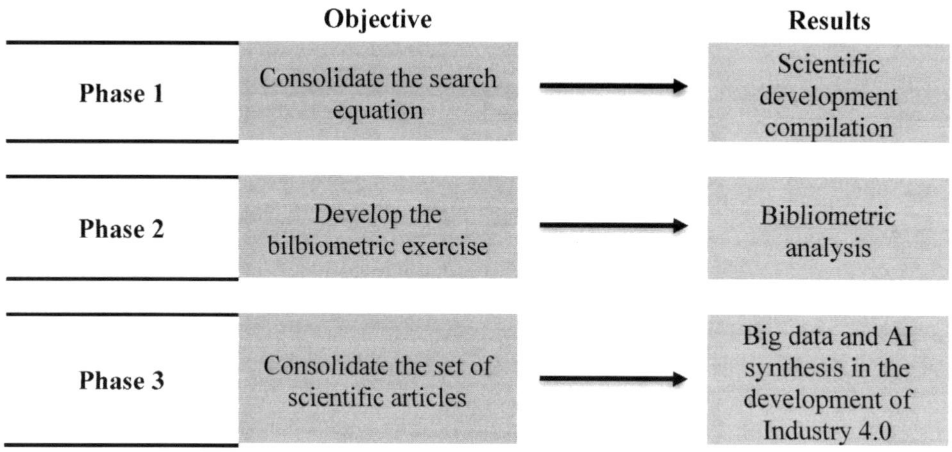

Figure 1. Study of the methodological design.

Figure 1 shows three methodological phases. In phase 1, the search equations were proposed and tested based on a combination of keywords characterizing the profile of the articles to be selected according to the study topic. Consequently, the search expression consigned in equation (1) was selected:

(TITLE-ABS-KEY ((Big Data) OR (Artificial intelligence)) AND ((Industry 4.0) OR (Smart manufacturing) OR (Fourth Industrial Revolution) OR (Cyberphysical system) OR (Smart factory)) (Equation (1)).

This equation was entered in the Scopus database (www.scopus.com), which yielded 2,155 results. This database was selected based on an information criterion of availability and accessibility. Moreover, it contains validated references in the development of bibliometric analyses.

In the second phase, the collected documents were refined using the information provided by the database, such as the publication year, keywords, research field, the type and source of document, language, institutions, and countries. This process was performed using an Excel spreadsheet.

Finally, the information gathered from the selected documents was processed using the cross-content analysis, which suggested a digitization trend of big data and AI for the development of Industry 4.0.

3.2. Content analysis

VOSviewer software (version 1.6.15) [30] was used to analyze the information. The analysis involved the determination of the co-occurrence of keywords and terms in the titles and abstracts of the publications using a co-occurrence method, revealing only the elements connected with others. The complete count method was established with several records for each term $\geq$15 and a minimum cluster size of 15 [31]. Based on different terms presented in the software, the maps of the network visualization were prepared. The algorithm was designed so that the terms exhibiting the highest correlation were positioned closer to each other with large bubbles that suggested a significant frequency. Those terms irrelevant to the map were not considered; therefore, they were removed.

4. Results

According to the Scopus database, 2,155 documents were identified between 2012 and August 5, 2020. The five-year period (2016–2020) was the period with the highest record of publications associated with big data and AI toward the development of Industry 4.0. The year 2019 showed the highest number of references. From 2012 to 2014, a poor performance was observed with an average of 1–9 publications per year. In 2015, the publication trend significantly improved compared with the previous years, reaching a total of 800 published documents in 2019 with an average of 354 publications per year. In fact, this topic has acquired further prominence within different areas of research, particularly in the field of computer science (27.7% of the publications classified in this area) and engineering (25.3%). To a small extent, this topic was prominent in business science, management and accounting (7.1%), mathematics (6.9%), decision sciences (6.2%), and 20 other areas of research.

Regarding the research fields and keywords, the vast majority of the collected documents originated in the field of computer science, where keywords such as Industry 4.0, big data, AI, IoT, and embedded systems showed the highest number of repetitions. In addition to the analysis of keywords, according to the documents identified by type and source, most of the obtained results were articles published in conferences (49.8%) and magazines (37.3%).

Following this analysis, the countries of origin of the documents were described. In general, the results reflect that there was a high concentration of scientific development in the United States, China, and Germany. At an institution level, South China University of Technology, Universidade do Minho, and University of Johannesburg were the most prominent. With reference to America, in addition to the USA, other countries such as Brazil, Colombia, and Mexico and institutions such as the Universidade de Sao Paulo, Tecnologico de Monterrey, and Universidad Distrital Francisco José de Caldas showed prominent research.

As far as the language of the documents, 96% of these are originally published in English and a small percentage in Chinese, Russian, Spanish, and German. In terms of the identification of the writings, using the search equation, it was easy to characterize scientific development associated with the digitization of industries based on big data and AI in the world of Industry 4.0 as a group of articles in magazines that mainly originated in institutions of higher education in the United States, and China, growing from various fields of knowledge according to computer science and engineering.

4.1. Content analysis

The keyword co-occurrence analysis provided an overview of trends in research studies by indicating the topics covered. The analysis was performed using VOSviewer software, which yielded a network visualization of big data and AI in Industry 4.0.

This bibliometric analysis is a map based on distance, where the distance between one element and another indicates the relationship strength among the two elements. A small distance reflects a stronger relationship. In this way, the frequency or level of occurrence and strength exiting between the words was measured. Hence, the words within the center of the map were as important as those on the periphery.

The results produced by the VOSviewer software reflect 205 terms, in which only those that had an occurrence greater than 15 were considered, thus delimiting the terms. Therefore, the terms were organized into six clusters with 9832 links.

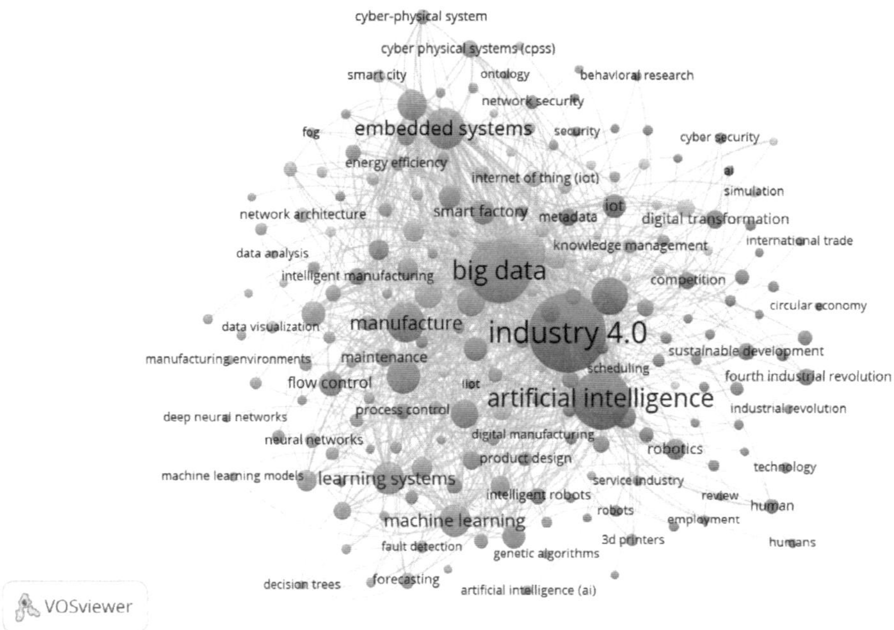

Figure 2. Visualization of the big data network and artificial intelligence in Industry 4.0.

A clustering method was implemented, wherein the unit of analysis was the keywords (author's words and indexed words) obtained from the dataset. The colors express the similarity and the level of relationship between the elements or nodes of the cluster. In this way, this map of the co-occurrence of keywords helped to determine the relationships between keywords of the research based on proximity within the map. In Fig. 2, clusters are observed to be associated with Industry 4.0, AI, big data, IoT, smart factories, and machine learning. All previous terms are interrelated topics; however, they cannot be consolidated as an individual cluster. Note that the clusters with the highest occurrence were INDUSTRY 4.0, ARTIFICIAL INTELLIGENGE, BIG DATA.

The aforementioned words are the nodes that have the most influence within the network. Therefore, the central-word sets reflect the semantic structure of the research field. Because it is built using keywords, this map reflects descriptors that have been selected by both the authors and journals.

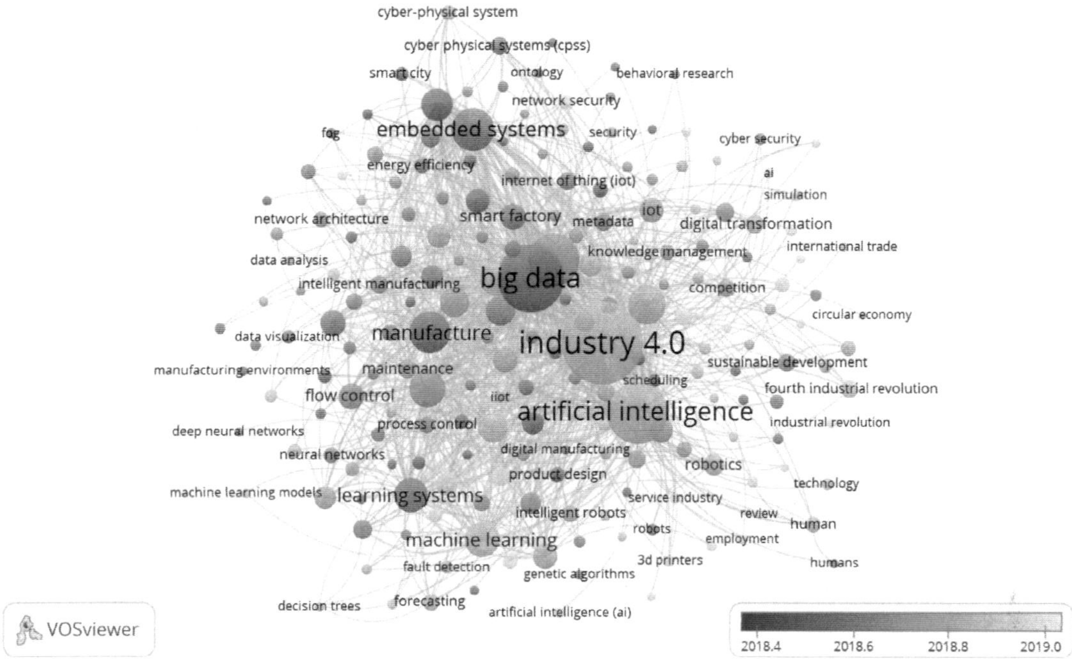

Figure 3. Visualization of the network overlap of big data and AI in Industry 4.0.

In Figure 3, the overlay visualization shows the transition and evolution of articles, reports, journals, and patents over the years. Moreover, the relationships and trends between them are shown. The themes Industry 4.0 and AI had more relevance at the end of 2018. The term big data showed higher relevance at the beginning of 2018, and the term Industry 4.0. was also included.

5. Discussion

The study conducted revealed that there is a significant co-occurrence of big data and AI in the development of Industry 4.0. From the previous analysis, it could be observed that the key paradigms in the development of digitization of industries are Industry 4.0, big data and AI. Moreover, the digital evolution of industries and their survival in the market depends, to a great extent, on the good use of data (big data) and AI. Because the use of these technologies allows organizations to make strategic decisions based on data, forecasts, information, and facts and not on hunches or strokes of wits, it must also be borne in mind that AI and big data projects are neither easy nor are the expected objectives achieved in a single phase because they require an iterative and careful approach that yields the expected results in continuous stages. The co-occurrence of the variables big data, IA and Industry 4.0 is evidenced by the bibliometric analysis and the number of documents found in the implemented databases. In addition, it is observed that it has a growing projection every year. Hence, it is said that we are currently living in the Fourth Industrial Revolution, an era where digital has taken on enormous importance. For this reason, big data and AI play a fundamental role in the businesses developed in Industry 4.0 because using these techniques, large volumes of data are managed and analyzed. This is crucial for understanding market demands, user profiles, and their response to new products or services.

The continuous technological development that marks this digital age offers an abundance of possibilities in the different areas of our lives. That is why the technological trends are currently being marked are diverse, but two particularly stand out: Big Data and AI. Admitted as the new Industrial Revolution, Industry 4.0 related research has garnered undue attention from academia, government, and industries. Furthermore, it is the development of Industry 4.0 characterized by the development of big data and AI that are necessary skills in this era and include advanced analysis, the IoT, and digital security.

As per Muktiarni et al. [32] every industrial sector in the world will be affected by the development of the Industrial Revolution 4.0. Because it is a new pattern, it is spreading so rapidly that all circles, especially industries, should be able to adapt immediately to industrial development 4.0. That is why big data and AI are born in this new digital era as tools that allow organizations to develop and evaluate strategies that benefit their businesses through data analysis. In addition, it is important to emphasize that the handling of the data, the interpretation, and what the company does with this information is more important than the quantity of data. That is why a good data analysis will help organizations in their productivity, the optimization of their processes, growth and economic development and, above all, good decision-making. This is reflected in the increase in its usefulness, which for any business is beneficial and important because, in this way, it maintains and/or increases its competitiveness in the market. In this new digital era, industries are betting on greater automation of their processes, connectivity, and globalization. We observe that the correlation between different areas, such as products, services, processes, and business models, has penetrated the industrial world, bringing with it the IoT, big data, and analytics.

The information analysis proposed for the Fourth Industrial Revolution allowed to have a broader perspective of the benefits, scope, challenges, and implications of Industry 4.0. These discoveries coincide with previous studies on this same topic, in which it has been reported that this new digital era encompasses a set of technologies that allow the integration of products into interconnected digital and physical processes at the same time. Likewise, these technologies such as big data and AI have contributed to the development of Industry 4.0 and are considered two fundamental pillars. Because data collection in industries has been continuously increasing, it is achieved through the incorporation of new data sources. Therefore, organizations can detect failures in real time, and in turn find points of improvement and forecast workloads. Consequently, big data is one of the essential technological tools for the digital transformation of organizations.

For this reason, companies are currently fighting to win the data battle because whoever does it will win the future. Through this strategy, they benefit since they understand the behavior of the market that is increasingly dynamic and competitive, as well as their clients. These new clients are increasingly demanding, and essential to progress on the path of maximizing profits. The industries, in addition to big data and AI, have complemented utilities such as IoT and machine learning, since a large amount of data is continuously generated by means of it. The abovementioned, together with those already collected, represent a new source of knowledge and that over time has taken on great importance for organizations that create competitive advantage through these tools and are in continuous transformation and innovation. To define the strategies for developing these technological tools, considerations regarding the storage and handling of data must be taken as well as compliance with the legislation and all technological support necessary to perform these strategies.

However, not everything is as easy as it seems. As industries present an increasing amount of hassles for the management and analysis of data, due to the large amount of data generated, AI arises as another important pillar of Industry 4.0. Although many years ago seemed unimaginable, AI has managed to replicate human intelligence in machines. Even so, robots will not become successors, or will replace people, but undoubtedly, they can contribute to streamline and optimize some activities, tasks, and processes, which are time and costs consuming resulting into savings for companies and organizations. Furthermore, it is precisely in the AI way of working where the key to its correlation with big data can be found since it is mainly this type of intelligence that feeds on the data generated by companies. Yes, AI feeds on data, and it uses it to develop algorithms and constitute the 'logic' of machines. In short, data can also be used to acquire information from the environment and interact with it. Big Data and AI work increasingly closely, and this is achieved to continue advancing in the progress of the industry digitization. Even so, it is true that prejudices continue to appear about the possible contradictory effects that robots can produce in humans. Moreover, there is a certain fear that working with intelligent machines will replace working with people and, precisely for this reason, it is essential to work on creating an ecosystem in which both actors, robots and human beings, can coexist and complement each other [33].

The results obtained in this study can help as a guide to learn to publicize the importance of big data and AI in the development of Industry 4.0. In addition, the industries may manage according to the characteristics, needs, and skills of digitization. Big Data and AI in the world of Industry 4.0. allow us to have a vision of the trend of digitization in industries, making industries increasingly profitable and durable, and meeting the needs of the Fourth Industrial Revolution.

6. Conclusion

This study shows the evolution of scientific research on AI and big data published in the Scopus database as part of the development of Industry 4.0. Within this context, 2155 documents were published between 2012 and August 5, 2020. The bibliometric analysis showed the results of the number of publications per year, the productivity of countries, research area, journals, and the analysis of the most frequent keywords and the relationship between them.

As presented in this work, based on the description of the texts on big data and AI in Industry 4.0 and its bibliometric analysis, we conclude that these research fields are generally developed within the scientific literature. Despite the fact that this literature focuses on aspects related to companies and industries, the importance of big data and AI was prominent as two pillars in the development of organizational digitization. In the forthcoming years of the Fourth Industrial Revolution, machines will be able to communicate with each other to absorb or transfer information and establish actions.

Moreover, considering that we are living in an era of digital transformation, which involves the fusion of the physical and digital systems, information transformation in a format that is understandable to all people has reinforced collaboration and the shared use of data relying on the automation of processes based on big data and AI. This new Industrial Revolution is characterized by the speed in facilitating visible results and the degree to which it affects people. Therefore, this new era, together with its pillars, is causing a stir today because of the advantages it offers.

Regarding the methodological aspects, the obtained results demonstrate how bibliometrics allow the collection, purification, and analysis of scientific development of studies associated with the subject, thus enriching this new research despite the limitations of the method.

References

[1] Del Val Román J L 2016 Industria 4.0: la transformación digital de la industria *In Proceedings of the Conferencia de Directores y Decanos de Ingeniería Informática, Informes CODDII, Valencia, Spain* 10
[2] Bonami B, Piazentini L and Dala-Possa A 2020 Education, big data and artificial intelligence: mixed methods in digital platforms *Comunicar: Revista Científica de Comunicación y Educación* **28** 65. DOI: 10.3916/c65-2020-04
[3] Ibrahim A, Asmawaty Abdul Kadir T and Kamaludin A 2020 Industry 4.0: eyeing the future via simulation *IOP Conf. Ser.: Mater Sci. Eng.* **769** 012001. DOI: 10.1088/1757-899x/769/1/012001
[4] Peykari N, Djalalinia S, Kasaeian A, Naderimagham S, Hasannia T, Larijani B and Farzadfar F 2015 Diabetes research in middle east countries; a scientometrics study from 1990 to 2012 *J. Res. Med. Sci.: Off. J. Isfahan Univ. Med. Sci.* **20** (3) 253
[5] Hood W W and Wilson C S 2001 The literature of bibliometrics, scientometrics, and informetrics. *Scientometr.* **52** (2) 291
[6] Navarro M and Sabalza X 2016 Reflexiones sobre la Industria 4.0 desde el caso vasco. *Ekonomiaz: Revista vasca de economía* **89** 142–173
[7] Martin L 2015 Industria 4.0: una industria como servicio (I) *AUSAPE* **25** 16–17
[8] Roblek V, Meško M and Krapež A 2016 A complex view of Industry 4.0 *SAGE Open* **6** (2) 215824401665398. DOI: 10.1177/2158244016653987
[9] Sommer L 2015 Industrial revolution - industry 4.0: are German manufacturing SMEs the first

victims of this revolution? *J. Ind. Eng. Manag.* **8** (5) 1512–1532. DOI: 10.3926/jiem.1470
[10] Cooper, J., & James, A. 2009. Challenges for Database Management in the Internet of Things. *IETE Technical Review*, **26** (5), 320 doi: 10.4103/0256-4602.55275
[11] Neugebauer R, Hippmann S, Leis M and Landherr M 2016 Industrie 4.0 - from the perspective of applied research *Procedia CIRP* **57** 2–7. DOI: 10.1016/j.procir.2016.11.002
[12] Sipsas K, Alexopoulos K, Xanthakis V and Chryssolouris G 2016 Collaborative maintenance in flow-line manufacturing environments: an Industry 4.0 approach *Procedia CIRP* **55** 236–241. DOI: 10.1016/j.procir.2016.09.013
[13] Schuh G, Anderl R, Gausemeier J, Ten Hompel M and Wahlster W 2017 Industrie 4.0 maturity index: managing the digital transformation of companies *Utz, Herbert*
[14] Porter M E and Heppelmann J E 2014 How smart, connected products are transforming competition *Harv. Bus. Rev.* **92** (11) 64–88
[15] Lycett M 2013 'Datafication': making sense of (big) data in a complex world. *Eur. J. Inf. Syst.* **22** (4) 381–386. DOI: 10.1057/ejis.2013.10
[16] Stoicescu C 2016 Big data, the perfect instrument to study today's consumer behavior. *Database Syst. J.* **6** 28–42
[17] Merriam-Webster. (n.d.) Artificial intelligence. *https://www.merriam-webster.com/dictionary/artificial%20intelligence*
[18] Gao S 2020 Innovative teaching of integration of artificial intelligence and university mathematics in big data environment *IOP Conf. Ser.: Mater. Sci. Eng.* **750** 012137. DOI: 10.1088/1757-899x/750/1/012137
[19] Webster C and Ivanov S 2020 Robots in travel, tourism and hospitality: key findings from a global study *Zangador*
[20] Thelwall M 2009 Bibliometrics and citation analysis: from the science citation index to cybermetrics *Libr. Inf. Sci. Res.* **31** (4) 268–269. DOI: 10.1016/j.lisr.2009.04.002
[21] Kostoff R, Tshiteya R, Pfeil K, Humenik J and Karypis G 2005 Power source roadmaps using bibliometrics and database tomography *Energy* **30** (5) 709–730. DOI: 10.1016/j.energy.2004.04.058
[22] Krauskopf E 2012 The uses and abuses of bibliometrics *Reprod. Biomed. Online* **25** (4) 434. DOI: 10.1016/j.rbmo.2012.07.005
[23] Johnson M 2012 Editors' response: the uses and abuses of bibliometrics *Reprod. Biomed. Online* **25** (4) 435. DOI: 10.1016/j.rbmo.2012.07.012
[24] Johnson M, Cohen J and Grudzinskas G 2012 The uses and abuses of bibliometrics *Reprod. Biomed. Online* **24** (5) 485–486. DOI: 10.1016/j.rbmo.2012.03.007
[25] Carvalho M, Fleury A and Lopes A 2013 An overview of the literature on technology roadmapping (TRM): contributions and trends *Technol. Forecast. Soc. Change* **80** (7) 1418–1437. DOI: 10.1016/j.techfore.2012.11.008
[26] Bounes V, Dehours E, Houze-Cerfon V, Vallé B, Lipton R and Ducassé J 2013 Quality of publications in emergency medicine *Am. J. Emerg. Med.* **31** (2) 297–301. DOI: 10.1016/j.ajem.2012.07.026
[27] Urquhart C and Dunn S 2013 A bibliometric approach demonstrates the impact of a social care data set on research and policy *Health Inf. Libr. J.* **30** (4) 294–302. DOI: 10.1111/hir.12040
[28] Tang M, Liao H, Wan Z, Herrera-Viedma E and Rosen M 2018 Ten years of sustainability (2009 to 2018): a bibliometric overview *Sustain.* **10** (5) 1655. DOI: 10.3390/su10051655
[29] Hernández Sampieri R, Fernández Collado C and Baptista Lucio P 2014 Metodología de la investigación *México: McGraw Hill Interamericana*
[30] Centre for science and technology studies 2020 VOSviewer Vol. 1.6.15. *Leiden Univ., Neth.*
[31] Van Eck N J and Waltman L 2010 Software survey: VOSviewer, a computer program for bibliometric mapping *Scientometr.* **84** (2) 523–538
[32] Muktiarni M, Widiaty I, Abdullah A, Ana A and Yulia C 2019 Digitalisation trend in education during industry 4.0 *J. Phys.: Conf. Ser.* **1402** 077070. DOI: 10.1088/1742-

6596/1402/7/077070
[33] Teng X 2019 Discussion about artificial intelligence's advantages and disadvantages compete with natural intelligence *J. Phys.: Conf. Ser.* **1187** (3) 032083. DOI: 10.1088/1742-6596/1187/3/032083

Energy savings in compressed air systems a case of study.

H Hernández Herrera[1], D Patiño Villalba[1], E Noriega Angarita[2], J I Silva Ortega[2] and C A Caraballo Echavarría[1].

[1]Facultad de Ingeniería, Universidad Simón Bolívar, Barranquilla, Colombia
[2]Departamento de Energía, Universidad de la Costa, Colombia.

Abstract. The compressed air applications are highly used in the industrial sector due to its easy transportation, safety, purity, cleanliness and storage capacity. In many regions, compressed air system accounts for about 10% of the industrial electricity energy bill. However, compressed air is expensive because around 30% of the consumed energy is used for these systems in an industrial facility. Energy used to be lost as leaks, pressure drops, heat, misuse, among others. Several energy efficiencies measures help to improve energy savings in companies such as: pressure reduction, reduce the inlet air temperature, use a well-calculated capacity tank for storage, control heat recovery and reduce leaks, giving potential energy savings in a range between the 20% to 60%, with a return of investment not higher than two years. As result, this document analyzes the energy efficiency measures that can be considered for compressed air systems applications, calculates the energy saving potentials, and assesses a case of study in a company in the city of Barranquilla.

Keywords: Compressed air systems; energy efficiency; energy savings.

1. Introduction.

Global energy consumption has doubled over the past four decades, and this trend will continue with an average annual increase of 1.5% in the period from 2018 to 2050 [1]. The industrial sector is consuming about 50% of the world's energy, while other sectors such as the residential and commercial are consuming between 29 and 21% respectively [2]. In Colombia, the manufacturing sector is consuming around 30% of the total energy, making it the second most consuming sector after transportation with 40% [3, 4].

The compressed air systems (CAS) in many industries worldwide represent around a 10% of the total electricity consumption as shown on figure 1. Commonly this system is used to operate plants and instrumentation equipment due to its easy transportation, safety, purity, cleanliness, and storage capacity [5-9]. Implementing these systems differs from the activity of every organization and its adequate usage guarantees great energy savings. The major cause of energy loss are leaks, inappropriate pipeline settings, small storage, pressure drops, and inadequate usage of the system. If energy efficiency measures (EEM) are applied in these systems, it could translate into a reduced energy consumption between 20%-60%.

These factors prove that CAS can be one of the most important objectives for the EEM application in the industrial sector [10-12]. A major energy saving, and the increase of efficiency could assure other benefits such as the increase of production with a capital investment reduction, improvement of the product, decrease maintenance costs; these advantages commonly are more representative than the energy-saving itself [13 - 15]. This research main purpose is to analyze the main EEM that could be applied for CASs and its evaluation in a case of study in the city of Barranquilla.

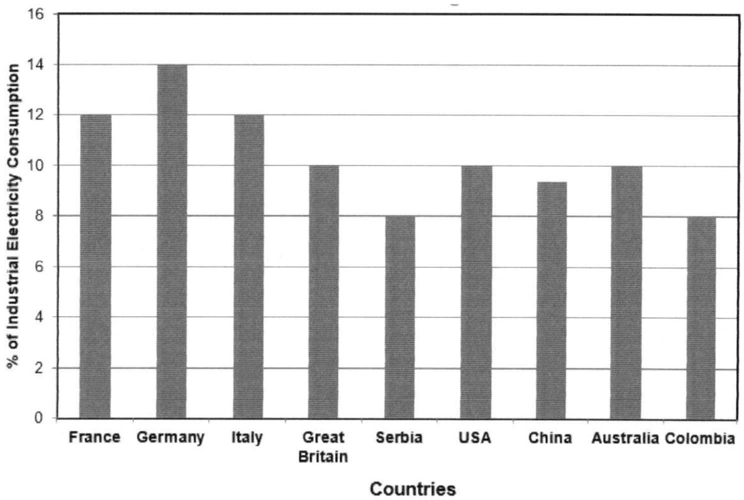

Figure 1. Percentage of electricity consumption of CAS in different countries. Data Sources based on: [5, 16-20].

2. Typical EEM in CASs.

2.1. Components of a CASs.

CASs has two 2 sides defined as supply and demand sides. The compressor (which oversees the increasing pressure), a storage humid tank, a dryer, a dry storage tank, and the filters located on the supply side. While, on the demand side, the flow and pressure controls and the transportation lines deliver the amount of air to every machine based on its consumption specifications. Figure 2 shows the key elements inside a CAS.

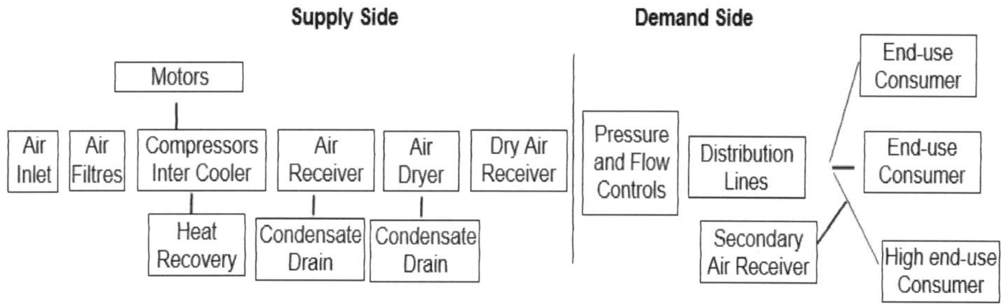

Figure 2. Description of the CAS sides and elements. Data source based on: [13, 14, 21].

2.2. Leaks: Identification and measurement.

The most significant cause of energy loss in the CAS is leakages. In an adequate system they are between 5% to 10% of the compressed air production [10, 22]. However, in the industry the regular range for leaks are in the range of 20% to 40% and without proper maintenance and use, it could reach up to 60% [23-25]. Commonly, leaks are found in the joints and connections of the compressed air pipelines, elbows, among others. The leakages value could be registered using two procedures: the first one is for compressors with a start/stop control and it comprises turning on the compressor when the system is without loads, the leaks can be able to pressure drops, by which the compressor will run into a load-unload cycle; according to [17, 26], the total leakage (T_L) measured in percentages are:

$$T_{L\%} = \frac{T_{load}}{T_{load} + T_{download}} \cdot 100 \quad (1)$$

Where:

T_{load} – Time the compressor is in load mode.

$T_{download}$ - Time the compressor is in download mode.

Other control strategies can be as follow: a pressure gauge (manometer) is placed downstream of the receiver tank. The compressor is turned on without loads until it reaches the working pressure P_1. Then, the compressor is then turned off measuring the time (t) until that the system drops of until a half of the working pressure P_2. It is necessary to know the system volume (V) including the storage tank capacity. The value of the loss flow (L_F) measure in (m^3/s) and can be calculated using the equation 2 as follow:

$$L_F\left(\frac{m^3}{s}\right) = V \cdot \frac{(P_1 - P_2)}{P_0 \cdot t} \cdot 1{,}25 \quad (2)$$

Where:

P_0 - atmospheric pressure

In equation 2, the multiplier factor 1,25 corrects the leak to the normal system pressure. The annual energy saving in a start/stop control can be expressed as:

$$\mathrm{ES_L} = \mathrm{AEC} \cdot \mathrm{T_{L\%}} \quad (3)$$

Using other control strategies

$$\mathrm{ES_L} = \mathrm{L_F} \cdot \mathrm{SEC} \cdot \mathrm{OPH} \quad (4)$$

Where,

OPH: is the operating hours in year (h/year).

SEC: is the specific energy consumption, (kWh/m^3).

AEC: Annual energy consumption in the CAS.

2.3. Storage Tanks Volume

The storage tanks in a CAS have several functions including: providing storage capacity to avoid short start/stop cycles, to condense air humidity, to cover periods of pressure peaks, to keep the system's pressure and to allow the systems control to work with more efficiency [24]. The compressed air provider companies recommend the use of two storage tanks, a humid and a dry tank. The first is placed between the dryer and the compressor while the other is located next to the dryer [25]. Sometimes, this makes more sense to have additional storage tanks near the applications with high consumption of air and intermittent usage [24]. Some authors recommend the tank volume should be between 12-120 m^3 for each cubic meter per second of air delivered by the compressor at full load as shown in figure 3 [21, 27, 28].

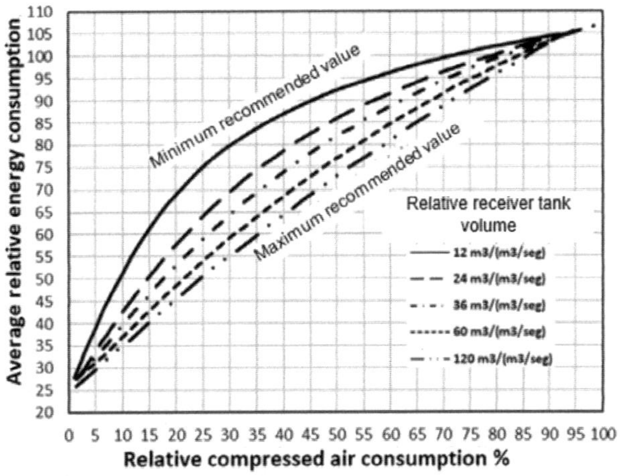

Figure 3. Recommended relative receiver tank volume in compressed air systems.

The annual energy saving for an adequate air receiver volume is:

$$ES_{TV} = AEC \cdot S_F \quad 5$$

2.4. Suction Temperature

Compressors are in a separate room inside the factory, or adjacent shelters, built specifically for this equipment. Most of them suck the air in that same room, at temperatures higher than the normal environment temperature, due to the heat generated by the compressor itself. The compression of the air becomes more inefficient as the suction temperature increases [29, 30], the potential savings from this concept can be calculated as:

$$T_S = \frac{T_1 - T_0}{T_1 + 273} \quad (6)$$

Where:

T_1 - average compressor air suction temperature.

T_0 - average environment air temperature.

The savings produced by the reduction of the air intake temperature are calculated as:

$$ES_{TR} = AEC \cdot T_S \quad (7)$$

3. Case Study

The plant analyzed is in the city of Barranquilla and it use is the glasses manufacture for the construction sector. To determine the potential savings in the compressed air systems, the efficiency measures previously discussed will be evaluated.

3.1 Leakage

The compressor has a start/stop control, so the leakage is calculated using equation (1). The corresponding measurements were carried out where it was determined that in the company the values of the leaks correspond to 33%. If we consider that a system with a proper functioning, the percentage of leakage allowed is 5%, we can state that there is a considerable saving potential in this case of 28%.

Using equation (3) potential savings could be determined, considering that the monthly electricity consumption is 10,852 kWh/month.

$$ES_L = AEC \cdot T_{L\%}$$

$$ES_L = 10852\text{kWh/month} \cdot 0{,}28\%$$

$$ES_L = 3038{,}56 \text{ kWh/month}$$

If the kWh cost is 451,28 COP/kWh, the company has an annual potential saving of 16.454.896 COP.

3.2 Storage Tanks.

The company has an Atlas Copco GA55P compressor, which has 56 kW and handles a flow (Q) of 0.128 m^3/sec, the system has a storage tank with a volume (V) of $1m^3$. The value of the relative receiver tank volume is.

$$Q^* = \frac{V}{Q} = \frac{1m^3}{0,128\frac{m^3}{seg}} = 7,81\,\frac{(m^3)}{\left(\frac{m^3}{seg}\right)}$$

In figure 4 load/unload compressor cycles are shown. The relative consumption conditions are at 53%.

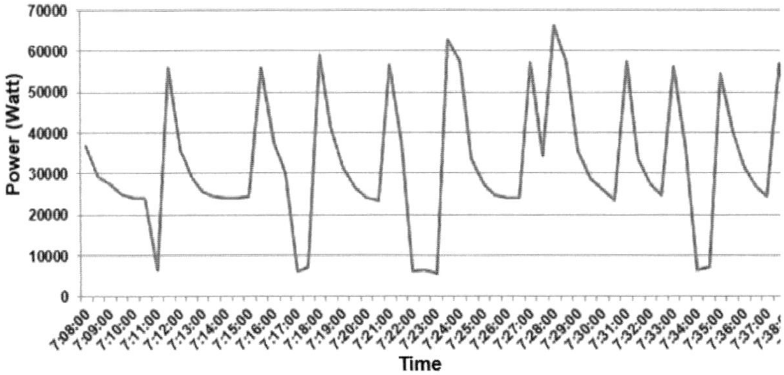

Figure 4. Power consumption of compressor.

Figure 5 shows the relative receiver tank volume for the plant. As seen that it is below the recommended value giving by manufacturers.

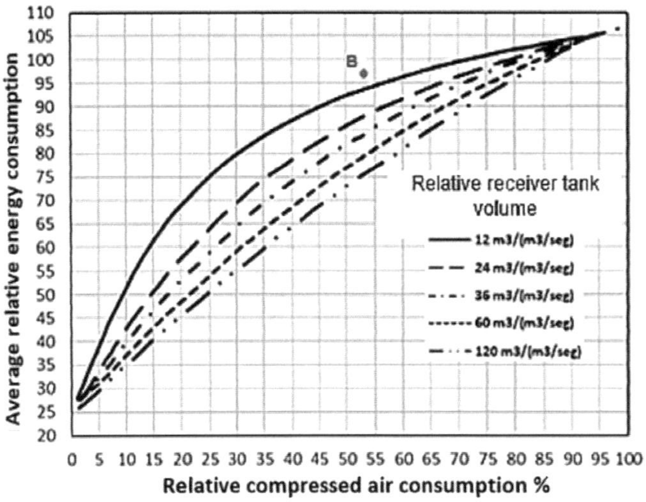

Figure 5. Relative receiver tank volume for the plant.

If the company buys a storage tank to achieve the relative volumes recommended by the manufacturers there would be savings in energy consumption (ES_{PC}) caused by increasing the relative storage values of 12 % figure 6.

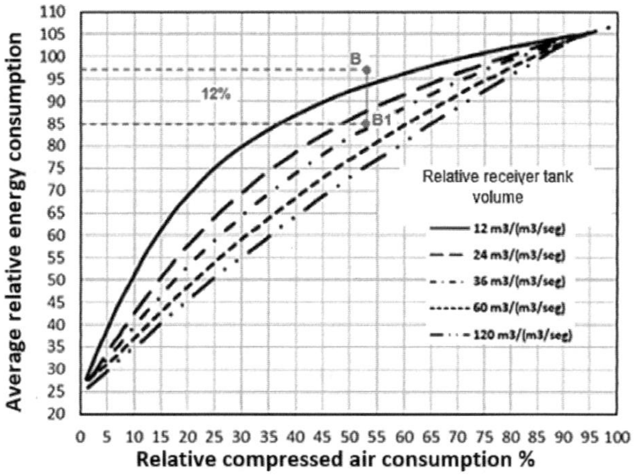

Figure 6. Potential savings with the required receiver tank volume.

Using equation (5) potential savings are determined as follow:

$$ES_{TV} = AEC \cdot S_F$$

$$ES_{TV} = 10852kWh/mes \cdot 0{,}12$$

$$ES_{TV} = 1302.24\ kWh/mes$$

Considering the kWh cost, the company has an annual savings potential of 7.052.098 COP.

3.3 Temperature.

For the case study analyzed, the compressor location is not appropriated because it is located on a second floor, above the oven and next to the electric motor of the oven's fan. Temperature measurements were taken in unison inside the compressor room and the ambient temperature for a week, resulting in an average suction temperature of 10 degrees above an ambient temperature at the company, reaching a difference of 15 degrees at the most critical times. The reduction in energy consumption as a result of the reduction in air temperature is calculated as follows:

$$T_S = \frac{42 - 32}{42 + 273}$$

$$T_S = 3{,}1\ \%$$

We calculate the savings from reducing the inlet temperature according to equation (7).

$$ES_{TR} = AEC \cdot T_S$$

$$ES_{TR} = 10852\text{kWh/mes} \cdot 0{,}031 = 336\ kWh/mes$$

Considering the cost of kWh, the company has an annual savings potential of 1.821.792 COP.

4. Conclusion.

The CASs in industry commonly consumes about 10% of the billed electricity, highlighting that they are inefficient systems where only between 10% to 30% of the produced air is used, implementing EEM such as reducing the improper use of compressed air, reducing the pressure of the compressors, lowering the air intake temperature, ensuring an adjusted storage capacity, obtaining a residual heat recovery from the air compressor and reducing leaks can provide savings in a range between 20% to 60% with a return of investment not higher than two years which makes it one of the main systems to be used by companies and countries when are considered strategies to reduce energy consumption. Also, this measure provides other benefits such as the increase of production with a lower capital investment, maintenance reduction, quality in final product improvements, and maintenance reduction; otherwise it is important to emphasis that these advantages are better than common electricity bill savings.

The company analyzed is dedicated to glass manufacture for the building sector where was evaluated all the compressed air systems located in the production area giving all possible recommendations that could be applied energy savings. According to this research could be estimated that exist a potential energy saving of 28% if the company applies measure to reduce leakages. If a storage tank is purchased to bring the relative volumes to the recommended values by the manufacturers it can be reached potential energy savings reaching 12% and if the temperature in the compressor can be reduced to the environment it can be reached potential energy savings equal to 3.1%. The potential energy savings of the company can reach a total energy saving of 43.1% which is equal to 25.328.786 COP per year.

5. References

[1] Wu, X. D., Guo, J. L., Ji, X., and Chen, G. Q. (2019). Energy use in world economy from household-consumption-based perspective. *Energy policy*, 127, 287-298.

[2] BP Energy. (2019), BP Energy Outlook. Available from: https://www.bp.com/content/dam/bp/en/corporate/pdf/energyeconomics/energy-outlook/bp-energy-outlook-2018.pdf. [Last accessed on 2020 Jun 28].

[3] Chikunov, S. O., Gutsunuk, O. N., Ivleva, M. I., Elyakova, I. D., Nikolaeva, I. V. and Maramygin, M. S. (2018), Improving the economic performance of Russia's energy system based on the development of alternative energy sources. *International Journal of Energy Economics and Policy*, *8*(6), 382-391.

[4] UPME (2018), Balance Energético Colombiano. BECO. Available from: http://www1.upme.gov.co/InformacionCifras/Paginas/BalanceEnergetico.aspx.[Last accessed on 2019 Feb 14].

[5] Benedetti, M., Bertini, I., Introna, V. and Ubertini, S. (2018). Explorative study on Compressed Air Systems' energy efficiency in production and use: First steps towards the creation of a benchmarking system for large and energy-intensive industrial firms. *Applied energy*, 227, 436-448.

[6] Radgen, P. and Annegret, C. (2003), Efficient Compressed Air a Successful Campaign for Energy Efficient Compressed Air Systems in Germany. Saint-Raphaël, France: ECEEE 2003 *Summer Study Proceedings.*

[7] Dos Santos Mascarenhas, J., Chowdhury, H., Thirugnanasambandam, M.,Chowdhury, T. and Saidur, R. (2019), Energy, exergy, sustainability, and emission analysis of industrial air compressors. *Journal of Cleaner Production*, 231, 183-195.

[8] Taheri, K. and Gadow, R. (2017), Industrial compressed air system analysis: Exergy and thermoeconomic analysis. CIRP, *Journal of Manufacturing Science and Technology*, 18, 10-17.

[9] Yin, Y., Zheng, B., Yang, C. and Zhang, X. (2015), A proposed compressed air drying method using pressurized liquid desiccant and experimental verification. *Applied Energy*, 141, 80-89.

[10] European Commission. (2009), Reference Document on Best Available Techniques for Energy Efficiency. Available from: http://www.eippcb.jrc.ec.europa.eu.

[11] Hernandez-Herrera, H., Silva-Ortega, J. I., Martínez Diaz, V. L., García Sanchez, Z., González García, G., Escorcia, S. M., and Zarate, H. E. (2020). Energy savings measures in compressed air systems. *International Journal of Energy Economics and Policy* 10 (3), 414-422

[12] Bonfàa, F., Salvatorib, S., Benedettic, M., Intronad, V. and Ubertinie, S. (2017), Monitoring compressed air systems energy performance in industrial production: Lesson learned from an explorative study in large and energy-intensive industrial firms. *Energy Procedia*, 143, 396-403.

[13] Nethler, T. (2018a), Linking energy efficiency measures in industrial compressed air systems with non-energy benefits-a review. *Renewable and Sustainable Energy Reviews*, 89, 72-87.

[14] Nethler, T., Parra, R. and Thollander, P. (2018), Implementation of energy efficiency measures in compressed air systems: Barriers, drivers and non-energy benefits. *Energy Efficiency*, 11(5), 1281-1302.
[15] Fleiter, T., Hirzel, S. and Worrell, E. (2012), The characteristics of energy efficiency measures a neglected dimension. *Energy Policy*, 51, 502-513.
[16] Šešlija, D., Ignjatović, I., Dudić, S. and Lagod, B. (2011), Potential energy savings in compressed air systems in Serbia. *African Journal of Business Management*, 5(14), 5637-5645.
[17] Saidur, R., Rahim, N. A., and Hasanuzzaman, M. (2010). A review on compressed-air energy use and energy savings. *Renewable and sustainable energy reviews*, 14(4), 1135-1153.
[18] UPME Corpoema. (2014), Determinación y Priorización de Alternativas de Eficiencia Energética Para los Subsectores Manufactureros Informe Final Códigos CIIU 19 a 31. Vol. I. Colombia. Available from: http://www.upme.gov.co/estudios/2014/informefinalvolumen_1.pdf. [Last accessed on 2020 Jul 2].
[19] UPME Corpoema. (2014a), Determinación y Priorización de Alternativas de Eficiencia Energética Para los Subsectores Manufactureros Informe Final Códigos CIIU 19 a 31. Vol. II. Colombia. Available from:http://www1.upme.gov.co/demandaenergetica/determinacioneficiencia/informe_final_volumen_2.pdf. [Last accessed on 2020 Jun 17].
[20] UPME Incombustion. (2013), Determinación del Potencial de Reducción del Consumo Energético en los Subsectores Manufactureros Códigos CIIU 10 a 18 en Colombia. Available from: http://www.upme.gov.co/demandaenergetica/INFORMEIII_caracterizacion_energeticaVerpub.pdf. [Last accessed on 2020 Jun 12].
[21] DoE, U. S. (1998). Improving Compressed Air System Performance, a Sourcebook for Industry. Prepared for the US Department of Energy, Motor Challenge Program by Lawrence Berkeley National Laboratory (LBNL) and Resource Dynamics Corporation (RDC).
[22] Castellanos, L. M., Hernández Herrera, H., Silva Ortega, J. I., Martínez Diaz, V. L., and García Sanchez, Z. (2019). Potential energy savings and CO_2 emissions reduction in Colombia compressed air systems. https://doi. org/10.32479/ijeep. 8084.
[23] Abdelaziz, E.A., Saidur, R. and Mekhilef, S. (2011), A review on energy saving strategies in industrial sector. Renewable and Sustainable Energy Reviews, 15(1), 150-168.
[24] Yang, M. (2009). Air compressor efficiency in a Vietnamese enterprise. *Energy Policy*, 37(6), 2327-2337.
[25] Dudić, S., Ignjatović, I., Šešlija, D., Blagojević, V., and Stojiljković, M. (2012). Leakage quantification of compressed air using ultrasound and infrared thermography. *Measurement*, 45(7), 1689-1694.
[26] Dindorf, R. (2012). Estimating Potential Energy Savings in Compressed Air Systems". *Procedia Engineering*. 39, 204-211.
[27] Compressed Air System Guide Designing Your Compressed Air System Provided as a Service by Kaeser Compressors, Inc. 2010. http://www.kaeser.com.co.
[28] Olszewski, P., and Borgnakke, C. (2016). Influence of volumetric capacity on energy consumption in oil-lubricated compressed air systems. *Journal of Thermal Science and Engineering Applications*, 8(4), 041007.
[29] UPME (2016), Plan de acción indicativo de eficiencia energética 2017-2022, una realidad y oportunidad para Colombia (PAI Proure 2017-2022), Available from:http://www1.upme.gov.co/DemandaEnergetica/MarcoNormatividad/PAI_PROURE_2017-2022.pdf. [Last accessed on 2018 Jan 29].
[30] Kluczek, A., and Olszewski, P. (2017). Energy audits in industrial processes. *Journal of cleaner production*, 142, 3437-3453.

The discharge and shell temperatures in hermetic compressors during its operation in tropical weather conditions

K.J. Arencibia Avila[1], J.M. Mendoza Fandiño[2], J.J. Cabello Eras[3], A. Sagastume Gutiérrez[3]* and F. Fernández Periche[1]

1. Universidad Antonio Nariño. Facultad de ingeniería Mecánica, Electrónica y Biomédica. Calle 58a bis # 37 – 94, Bogotá, Colombia.

2. Universidad de Córdoba. Facultad de Ingeniería. Carrera 6 No. 77- 305 Montería, Colombia

3. Universidad de la Costa. Departamento de Energía. Calle 50 No 55-66, PBX 336 22 00, Barranquilla, Colombia

* Corresponding author: asagastu1@cuc.edu.co

Abstract. Hermetic compressors are essential components of domestic refrigerators and freezers and are usually designed to operate in ambient temperatures under 32°C. This study measures the discharge and shell temperatures of hermetic compressors at different ambient temperatures. To this end, calorimetric tests are conducted in three different hermetic compressors to define the impact of tropical ambient temperatures on the discharge and shell temperatures. Environmental temperatures of 35, and 38°C and voltages of 93, 115, and 127 V were considered during the tests. Thi results are compared to the nominal parameters at 32°C and 115V. Furthermore, to identify if the wear-out of the different components affect the discharge and shell temperatures, the three compressors were tested in the calorimeter after an accelerated life test. In all cases, the evaporation temperature was controlled at -23 °C. Results show that the discharge and shell temperatures of hermetic compressors increase with the ambient temperature. Additionally, it is evidenced that the technology of hermetic compressors strongly influences the discharge and shell temperatures.

Keywords: hermetic compressors, discharge temperature, shell temperature, domestic refrigerators

1. Introduction

Hermetic compressors are essential components of the vapor-compression refrigeration systems used in domestic refrigerators and freezers. In 2019, some 201.11 million of new compressor units were produced, for a market value of 111,130 million USD, estimating the growth for the annual production of 2.5% (Statista, 2020). In total, the electricity demand of domestic refrigerators and freezers account for an estimated 6% of the global electricity production (Negrão and Hermes, 2011). Thus, defining the best approach to operate them in a tropical climate towards higher efficiency and reliability standards is essential to reduce their electricity use. Therefore, to reduce the demand for electricity in domestic refrigerators and freezers the adequate discharge and shell temperatures of hermetic compressors must be defined (Arencibia, 2004). The discharge temperature is the temperature of the exit flow of refrigerants in hermetic compressors, while the shell temperature is the temperature of the compressor's shell. The discharge temperature is used to control the flow of refrigerant in refrigerators to optimize its performance (Huang et al., 2019).

The performance of hermetic compressors is usually assessed using the coefficient of performance (COP). The COP of domestic refrigerators and freezers is strongly dependent on ambient temperature (Harrington et al., 2018), which in turn affects the discharge and shell temperatures in hermetic compressors. There is evidence that several domestic refrigerators worldwide are running at temperatures higher than the recommended values (James et al., 2017). This results in higher condensation temperatures, lower efficiency of compressors, and increased temperatures of the motor winding (Masjuki et al., 2001; Todescat et al., 1992). Another factor affecting the performance of hermetic compressors is the requirement that it must be properly ventilated, to facilitate the heat transfer by convection (Meza et al., 2017). Otherwise, the heat it rejects combined with the heat rejected by the condenser increases the ambient temperature, which in turn reduces the COP of the cycle (Bassiouny, 2009). Adequate ventilation needs over 200 mm of free space around the condenser, which seems unlikely during domestic exploitation (Bassiouny, 2009).

The recirculation of hot air, causing high temperatures, is accountable for over half of the burnout hermetic compressors (Alarcón, 1998). Furthermore, high temperatures reduce the thermal stability of refrigerants, while the oil reduces its lubrication capacity (Brignoli et al., 2017; Karnaz and Seeton, 2018). Extended periods of high temperatures affect the performance of the entire cycle (Mohd Yunus et al., 2016; Yin et al., 2014). Additionally, increasing the ambient temperature at 1°C increases in 5.6% of the electricity use of domestic refrigerators (Anjana et al., 2015). Ambient temperature is the main factor affecting the electricity demand of domestic refrigerators (Geppert and Stamminger, 2013). Additionally, some studies point that in tropical weather hermetic compressors operate at higher compression ratios, leading to higher discharge temperatures, and thus to lower COP. Domestic refrigerators operating between 15 and 35 °C in Sri Lanka showed differences in the electricity demand of up to 54% (Anjana et al., 2015).

Although the discharge and shell temperatures affect the lifespan of hermetic compressors and the electricity demand of domestic refrigerators and freezers, they are however overlooked in the assessment of these units (Hundy et al., 2016). Even catalogs from manufacturers fail to include this information for the operation of domestic refrigerators and freezers for tropical temperatures (Negrão et al., 2011). Until recently, most tests of refrigerating units were developed at a single ambient temperature (Harrington et al., 2018), frequently at 32°C. Therefore, reliable data on the electricity demand of domestic refrigerators and freezers as a function of the ambient temperature is limitedly available. Tropical regions are characterized by mean ambient temperatures between 20 and 30°C, with a maximum of 40°C (Meteoblue, 2020). In several tropical regions temperatures during the day are higher than 32°C, thus affecting the performance of hermetic compressors. Consequently, this study aims at measuring the discharge and shell temperature of hermetic compressors at different ambient temperatures, considering temperatures values typically from tropical regions. Following a literature review is developed to identify the state of the art on this topic. Next, the materials and methods used in this investigation are depicted, and next the results are presented and discussed.

2. Literature review

Measuring the performance of scroll compressors for two values of ambient temperature (i.e. 18 and 25 ° C), it was found that the pressure ratio is the main factor affecting the temperature change of refrigerants during the process (Winandy et al., 2002). Moreover, the heating-up experienced by refrigerants at the suction of compressors is the main factor affecting the mass flow rate. Moreover, during an experimental study developed in a small domestic refrigerator, measuring the refrigerating demand and the ambient temperature, an overheating was detected in different components of the compressor (Torchio and Anglesio, 2004). While the results of a survey of domestic refrigerators show that factors like ambient temperature and opening frequency strongly impact the performance of the refrigerator (Rahman et al., 2005). Likewise, experimental results showed that the power demand in hermetic compressors increases linearly with the discharge temperature (Cuevas and Lebrun, 2009). A semi-empirical model can be used to predict the mass flow, power, and discharge temperature for hermetic compressors considering different refrigerants and ambient temperatures (Li, 2012).

The energy demand of standard-size domestic refrigerator and freezers was forecasted to increase around 102,575 GWh, based on a survey of over 1,500 domestic refrigerators and freezers (Greenblatt et al., 2013). The main factor affecting the energy demand of domestic refrigerators is mostly influenced by the ambient temperature, which was highlighted during the experimental study of four domestic refrigerators (Geppert and Stamminger, 2013). In fact, for domestic refrigerators operating at ambient temperatures between 35 and 38°C for most of the year, the energy loss of hermetic compressors increases from 7 up to 15%, reducing its cooling capacity by some 10% (Arencibia and Tricio, 2014). Furthermore, some 90% of the refrigerators operated at voltages with differences of ± 10 % as compared to the nominal voltages of compressors.

Ambient temperatures over 32°C result in higher discharge temperatures. High discharge temperatures in hermetic compressors can cause a breakdown of refrigerant oils, reducing its lubrication capacity (Myszka et al., 2014). However, there are limited studies in this area, and a lack of standardization of these studies (Almeida et al., 2016). Domestic refrigerators are essential in the cold chain, and its inadequate operation is often a factor in food poisoning, yet, worldwide domestic refrigerators operate at temperatures higher than recommended (James et al., 2017). Regardless of the domestic refrigerator technology, they show a similar response to variations in the ambient temperature (Harrington et al., 2018).

Overall, most studies focus on optimizing the design and operation of domestic refrigerators. However, the discharge and shell temperatures of hermetic compressors resting from the operation of domestic refrigerators and freezers in tropical weather, are seldom considered.

3. Materials and methods

3.1. Compressors

Three different hermetic compressors (i.e. different make and models) were tested during the experiments. The main characteristics of the compressors are depicted in Table 1.

Table 1. Characteristics of hermetic compressors

Compressors	Frigorific Capacity (W)	Coefficient of Performance (W/W)
I	95.4	1.04
II	94.7	1.17
III	100.7	1.03

3.2. Calorimeter

A calorimeter permits to simulate different ambient temperatures during the operation of the hermetic compressor. Additionally, it permits to measure the discharge and shell temperatures of hermetic compressors.

The calorimeter was properly isolated to guarantee heat losses under 5% of the compressor power (Masjuki et al., 2001). The tests were developed according to the ASHRAE standard (UNE, 2000), which includes a subcooling process. The calorimeter used during the experiments is described in Figure 1.

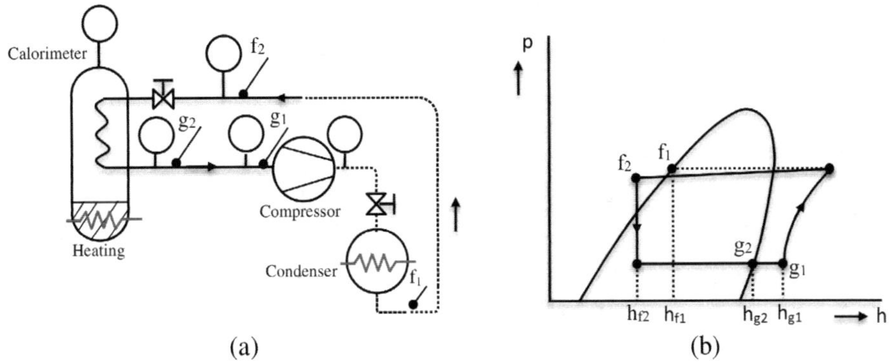

Figure 1. Calorimeter. a) Experimental layout b) Thermodynamic cycle

During the experimental tests, the calorimeter operates as the heat source in the refrigeration cycle (see Figure 1a) to measure the operating parameters of the hermetic compressors under defined operating conditions. The thermodynamic cycle is depicted with subcooling and without subcooling in Figure 1b. The precision and uncertainty of measurements are according to UNE-EN 13771-1: 2017.

3.3. Methods
Calorimetric tests are conducted in three different compressors to define the impact of ambient temperatures typical from tropical weather on the discharge and shell temperatures in hermetic compressors. Moreover, to identify if the wear-out of the different components affect the discharge and shell temperatures, the compressors were tested in the calorimeter after an accelerated life test. The accelerated life test, which is used to assess the reliability of hermetic compressors, was developed at a defined temperature and pressure during 678 h.

Based on the increased inefficiency of domestic refrigerators operating at ambient temperatures between 35 and 38°C for most of the year (Arencibia and Tricio, 2014), two environmental temperatures (i.e. 35 and 38 °C) were considered during the tests in the calorimeter. Additionally, considering the influence of voltage that was measured between 93 and 127 V in tropical regions (Arencibia and Tricio, 2014), three voltages levels were considered during the experiments (i.e. 93, 115, and 127 V) to assess the influence of low and high voltages on the compressors, and compare it to the nominal operation. The discharge and shell temperatures measured during the experiments are compared to the nominal operation of the hermetic compressors (i.e. at 32° and 115 V). Therefore, the operation of the hermetic compressors is measured at 32°C and 115 V (UNE, 2000). In all cases, the temperature of evaporation was controlled at -23 °C.

4. Results and discussions
Figure 2 depicts the shell temperatures measured during the experimental tests for the three compressors, for the different voltages and ambient temperatures considered. Additionally, it shows the shell temperature measured at nominal conditions (i.e. 32°C and 115 V).

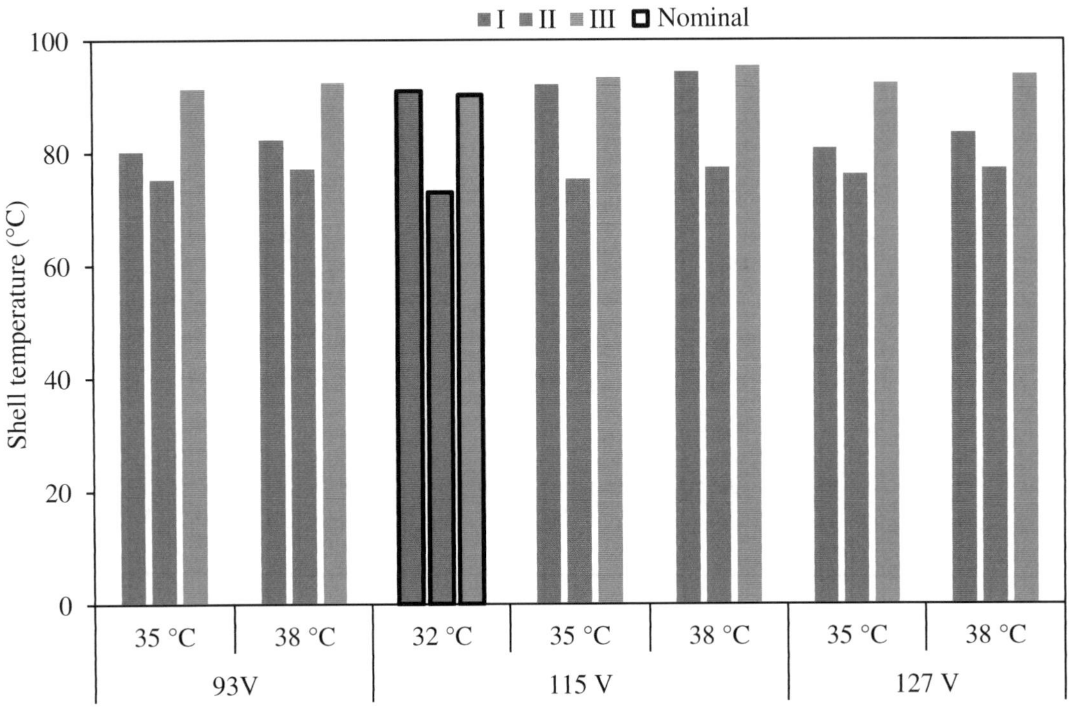

Figure 2. Shell temperature

The results show that, at the same ambient temperature, the voltage variation causes an increase in the shell temperature of compressors between 11.6 and 12.9% in compressor I, between 0.3 and 0.9% in compressor II, and between 1.3 and 2.9% in compressor III. Moreover, as compared to the nominal operation, the temperature reduces from 1 to 12% in compressor I, increases from 3 to 6% in compressor II, and from 1 to 6% in compressor III. Thus, the ambient temperature has more influence on the shell temperature than the voltage. As shown by the variation of the shell temperature in compressor II as compared to compressors II and III, the design and materials used during manufacturing also influence the shell temperature. While compressor I shows the highest shell temperature at 115 V, compressors II and III shows a similar behavior at the three voltages for the different ambient temperatures. Although the three compressors have similar characteristics, compressor II operates at lower shell temperatures, which might be a result of its higher efficiency. Furthermore, compressor I has lower shell temperatures at 93 and 127 V as compared to its shell temperatures at 115 V. While compressor III shows the highest temperatures, with little variations for the different conditions. Overall, except for compressor I, the voltage variations have limited influence on the shell temperature. Other factors, like designs, lubrication systems, and materials used for each compressor, which influences the performance, were excluded from the assessment. Overall, the results of the calorimetric tests permit to deduce that for ambient temperatures of 35 and 38 °C the heat rejection from compressors increases, which reduces its operation efficiency. Furthermore, the results show that the technology of the compressor influences its discharge temperature for given operating conditions.

Likewise, Figure 3 depicts the discharge temperatures measured during the experimental tests for the three compressors, for the different voltages and ambient temperatures considered.

IOP Conf. Series: Materials Science and Engineering 1154 (2021) 012010 doi:10.1088/1757-899X/1154/1/012010

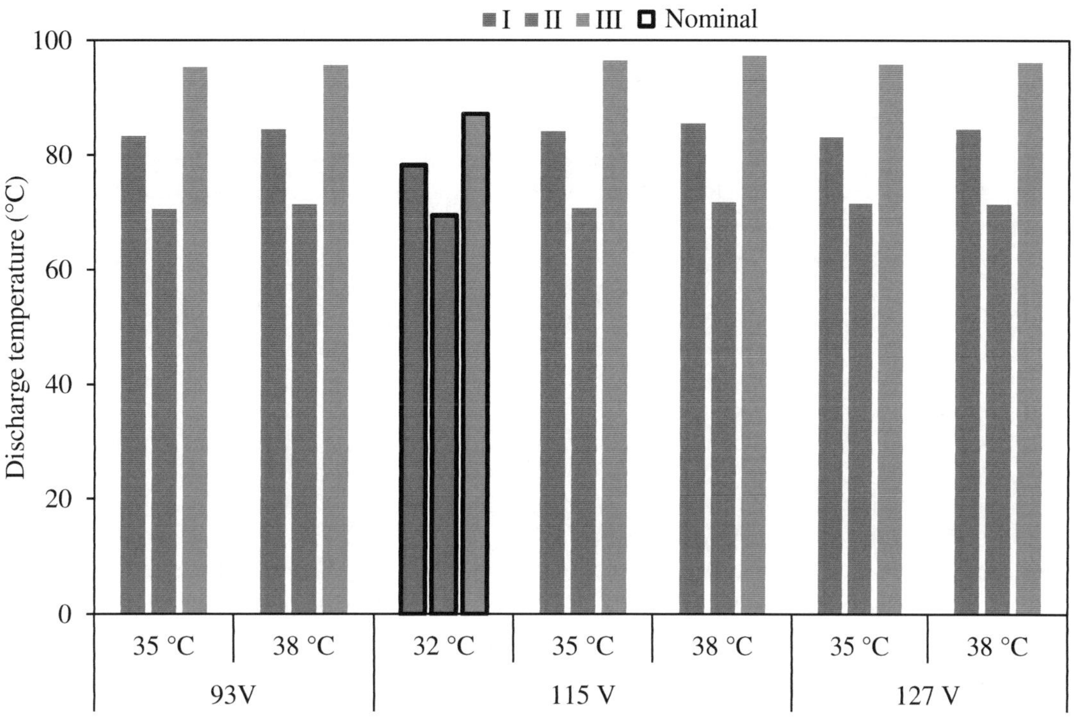

Figure 3. Discharge temperature

The results show that, at the same ambient temperature, the voltage causes an increase in the discharge temperature of compressors between 0.3 and 1.6%. Moreover, as compared to the nominal operation, the temperature increases from 6 to 9% in compressor I, from 1 to 3% in compressor II, and from 9 to 12% in compressor III. The results show that the discharge temperature of compressors depends on both the voltage and ambient temperature. However, the variations of the discharge temperature with the voltage and ambient temperatures are lower as compared to the variations of the shell temperature. Likewise, the compressor technology influences its discharge temperature for given operating conditions. Similarly, compressor II operates at lower discharge temperatures, which might be a result of its higher efficiency, while compressor III operated at the highest temperatures. Compressors III operates at temperatures over 90 °C, which is the maximum temperature recommended for hermetic compressor manufacturers. Overall, the voltage variations have a limited influence on shell temperature.

The temperatures in hermetic compressors are influenced by heat loss, which is mainly influenced by the fluctuation of the environmental temperature. Additionally, the heat losses because of the friction of the moving components and the lubrication system in the compressor have a significant influence on the shell and discharge temperatures, evidenced in the different performances observed for compressors from different manufacturers.

The calorimetric tests results show that in tropical weather conditions, the performance of compressors deteriorates, significantly increasing its shell and discharge temperature, because of the reduction of its dissipation capacity. The increase of the discharge temperature over the limits recommended by the manufacturer increases the condensation temperature, reducing the COP of the cycle. Additionally, these temperatures accelerate the deterioration of the lubricant oil and the coolant, reducing the lifespan of the compressors. Furthermore, the reduction of the COP in the cycle causes a reduction of its cooling capacity, because of the higher enthalpy at the input of the evaporator, moving the performance of compressors away from nominal conditions. These results highlight the need to adequate the current compressor tests to include tropical weather conditions and different electricity quality parameters, which is closer to the actual operating conditions of hermetic compressors in developing countries with tropical conditions.

5. Conclusions.

Although different studies have discussed the influence of environmental conditions on the energy demand and performance of hermetic compressors in domestic refrigerators and freezers, these are usually evaluated at nominal operation conditions (i.e. at 32°Cand 115 V). The results of the calorimetric tests show that for environmental temperatures between 35 and 38° C, the operation of hermetic compressors is affected by higher discharge and shell temperatures. This is influenced mostly by the ambient temperature, and to a lower extent by the voltage variations. Furthermore, the technologic characteristics of hermetic compressors strongly influence its performance, as shown in the calorimetric tests for the compressors assessed, where differences over 15 °C were observed for the discharged and shell temperature for the different compressors. These results highlight the need to address the operation and performance of hermetic compressors operating under tropical temperature conditions, to control the variations of the discharge and shell temperatures within adequate values.

References

[1] Alarcón, J. (1998). Tratado practico de refrigeracion automatica (12th ed.). Marcombo.
[2] Almeida, I. M. G., Barbosa, C. R. F., & Fontes, F. de A. O. (2016). Methods of Wear Tests for Hermetic Reciprocating Compressors: an Overview. Holos, 4, 86. https://doi.org/10.15628/holos.2016.2078
[3] Anjana, H. M.., Nimarshana, P. H. ., & Attalage, R. . (2015). Steady state performance variation of domestic refrigerators under different ambient conditions of Sri Lanka. MERCon 2015 - Moratuwa Engineering Research Conference, 177–181. https://doi.org/10.1109/MERCon.2015.7112341
[4] Arencibia, K. (2004). Sistema de pruebas de control para la comprobación de la fiabilidad en compresores herméticos y criterios para su selección. Universidad de Holguín.
[5] Arencibia, K., & Tricio, V. (2014). Evaluación experimental de compresores herméticos en condiciones de clima tropical. Eficiencia y capacidad frigorífica. Informacion Tecnologica, 25(2), 103–110. https://doi.org/10.4067/S0718-07642014000200012
[6] Bassiouny, R. (2009). Evaluating the effect of the space surrounding the condenser of a household refrigerator. International Journal of Refrigeration, 32(7), 1645–1656. https://doi.org/10.1016/j.ijrefrig.2009.03.011
[7] Brignoli, R., Brown, J. S., Skye, H. M., & Domanski, P. A. (2017). Évaluation De La Performance Du Frigorigène, Y Compris Les Effets Des Propriétés De Transport Et Des Échangeurs De Chaleur Optimisés. International Journal of Refrigeration, 80(September 2018), 52–65. https://doi.org/10.1016/j.ijrefrig.2017.05.014
[8] Cuevas, C., & Lebrun, J. (2009). Testing and modelling of a variable speed scroll compressor. Applied Thermal Engineering, 29, 469–478. https://doi.org/10.1016/j.applthermaleng.2008.03.016
[9] Geppert, J., & Stamminger, R. (2013). Analysis of effecting factors on domestic refrigerators' energy consumption in use. Energy Conversion and Management, 76, 794–800. https://doi.org/10.1016/j.enconman.2013.08.027
[10] Greenblatt, J., Hopkins, A., Letschert, V., & Blasnik, M. (2013). Energy use of US residential refrigerators and freezers: Function derivation based on household and climate characteristics. Energy Efficiency, 6(1), 135–162. https://doi.org/10.1007/s12053-012-9158-6
[11] Harrington, L., Aye, L., & Fuller, B. (2018). Impact of room temperature on energy consumption of household refrigerators: Lessons from analysis of field and laboratory data. Applied Energy, 211, 346–357. https://doi.org/10.1016/j.apenergy.2017.11.060
[12] Huang, L., Tao, L., Wang, C., & Yang, L. (2019). Theoretical and experimental research on using quasi saturation isentropic compression discharge temperature to control refrigerant mass flow rate. Heat and Mass Transfer/Waerme- Und Stoffuebertragung, 55(2), 489–500. https://doi.org/10.1007/s00231-018-2437-9
[13] Hundy, G., Trott, A., & Welch, T. (2016). Refrigeration, Air Conditioning and Heat Pumps (Elsevier (Ed.); 5th ed.). Butterworth-Heinemann.
[14] James, C., Onarinde, B. A., & James, S. J. (2017). The Use and Performance of Household

Refrigerators: A Review. Comprehensive Reviews in Food Science and Food Safety, 16(1), 160–179. https://doi.org/10.1111/1541-4337.12242

[15] Karnaz, J., & Seeton, C. (2018). Evaluation of Lubricant Properties and Refrigerant Interaction. International Compressor Engineering Conference.

[16] Li, W. (2012). Simplified steady-state modeling for hermetic compressors with focus on extrapolation. International Journal of Refrigeration, 35(6), 1722–1733. https://doi.org/10.1016/j.ijrefrig.2012.03.008

[17] Masjuki, H. H., Saidur, R., Choudhury, I. A., Mahlia, T. M. I., Ghani, A. K., & Maleque, M. A. (2001). The applicability of ISO household refrigerator-freezer energy test specifications in Malaysia. Energy, 26(7), 723–737. https://doi.org/10.1016/S0360-5442(01)00028-7

[18] Mohd Yunus, H., Nasution, H., Abdul Aziz, A., Sumeru, & Dahlan, A. A. (2016). The Effect of Ambient Temperature on the Performance of Automotive Air-Conditioning System. Applied Mechanics and Materials, 819(March), 221–225. https://doi.org/10.4028/www.scientific.net/amm.819.221

[19] Myszka, D., Guan, C., Murray, A., Hodapp, T., Myszka, D. H., Murray, A. P., & Hodapp, T. P. (2014). A Semi-Empirical Prediction Model for the Discharge Line Temperature of Hermetic Compressors. https://docs.lib.purdue.edu/icec

[20] Negrão, C. O. R., Erthal, R. H., Andrade, D. E. V., & Silva, L. W. Da. (2011). A semi-empirical model for the unsteady-state simulation of reciprocating compressors for household refrigeration applications. Applied Thermal Engineering, 31(6–7), 1114–1124. https://doi.org/10.1016/j.applthermaleng.2010.12.006

[21] Negrão, C. O. R., & Hermes, C. J. L. (2011). Energy and cost savings in household refrigerating appliances: A simulation-based design approach. Applied Energy, 88(9), 3051–3060. https://doi.org/10.1016/j.apenergy.2011.03.013

[22] Rahman, S., Sidik, N. M., Hassan, M. H. J., Rom, T. M., & Jauhari, I. (2005). Temperature performance and usage conditions of domestic refrigerator-freezers in malaysia. HKIE Transactions Hong Kong Institution of Engineers, 12(2), 30–35. https://doi.org/10.1080/1023697X.2005.10668000

[23] Statista. (2020). Refrigerators - worldwide | Statista Market Forecast. https://www.statista.com/outlook/16010100/100/refrigerators/worldwide

[24] Todescat, M. L., Fagotti, F., Prata, A. T., & Ferreira, R. T. D. S. (1992). Thermal Energy Analysis in Reciprocating Hermetic Compressors. Proceedings of the International Compressor Engineering Conference at Purdue, April 2014, 1419–1428.

[25] Torchio, M. F., & Anglesio, P. (2004). Refrigerant charge and ambient temperature effects on the refrigeration cycle of a small capacity food freezer. International Refrigeration and Air Conditioning Conference, 697. http://docs.lib.purdue.edu/iracc/697

[26] UNE. (2000). Ensayos de compresores para fluidos refrigerantes. Norma Internacional UNE 86-2020-84.

[27] Winandy, E., Saavedra O, C. S., & Lebrun, J. (2002). Experimental analysis and simplified modelling of a hermetic scroll refrigeration compressor. Applied Thermal Engineering, 22(2), 107–120. https://doi.org/10.1016/S1359-4311(01)00083-7

[28] Yin, P., Pate, M., Sweeney, J. ., & Kading, J. (2014). The effect of high ambient temperature on the performance of an R410a air conditioner. 1st International Conference on Energy and Indoor Environment for Hot Climates, 350–357.

Mathematical modeling for collaborative planning of two-echelon in the pineapple supply chain: A case study in San Marcos, Colombia

J Ruiz-Meza[1], Y Cárdenas-Díaz[2], M K Royeth-Angulo[2], and D Mejía-Ayala[3]

Faculty of Engineering, Universidad de La Sabana, Colombia.
Industrial Engineering Program, Corporación Universitaria del Caribe, Colombia.
Business Engineering School, Universidad César Vallejo, Perú.

joserume@unisabana.edu.co

Abstract. In the fruits sector supply chains, several decisions must be made about logistics operations that directly affect their productivity and competitiveness. Therefore, it is necessary to work collaboratively so that all actors and echelons of the chain benefit equally in strategic, tactical, and operational decisions. In this paper, we developed a mixed-integer linear programming model for the collaborative planning of the pineapple supply chain. The model allows the articulation and planning of the production and distribution echelons, considering the production planning and the transportation process, to maximize profits and increase the efficiency of the sector. We applied the model to the pineapple supply chain in the municipality of San Marcos, Sucre, Colombia. We carried out a characterization of the current state of pineapple production and distribution. Then, the mathematical model was developed and finally, the model was evaluated by developing analysis scenarios to make efficient decisions. Optimal computational results were obtained in a 10.53% increase in utility in the pineapple supply chain.

1. Introduction

Agricultural crops represent a high degree of importance for the sustenance of the people [1]. However, many of the supply chains for these types of crops located in rural areas require the incorporation of innovation to reduce costs and transfer the products from the production site to consumers [2]. Therefore, it is necessary to implement an efficient decision-making process to achieve a good position in the fruit market and avoid decreases in the supply chain income due to disruption, waste or loss of freshness in the products [3]. This approach involves a great deal of effort in coordinating actors, resources and activities to meet customer requirements and gain customer confidence in the value chain [2], [4]. This ensures that the chains act in a multidisciplinary manner, taking into account that they bring together different actors such as suppliers, producers, distributors, retailers and wholesalers, and most importantly, the consumer [5], [6].

Colombia has several areas with great wealth and agricultural potential [7]; despite these advantages, there is still a lack of use of technologies to enhance the sector and thus become a global agricultural powerhouse in contrast to the interests in improving the performance of agri-food supply chains [8]. Thus, it is necessary to organize logistics to contribute to the growth of this market [2].

Among fresh agricultural products, pineapple is one of the most demanded fruits nationally and internationally; in Colombia, the departments with the highest pineapple production are Santander, Valle del Cauca, and meta with 43%, 14%, and 12%, respectively [9]. One of the most closely monitored aspects throughout the agri-food supply chain concerns the ripening stage of the fruit so that there are no losses and it is consuming for the customer [10]. The pineapple producer has a close relationship with suppliers that provide inputs and services that facilitate the cultivation process, but at the same time generates higher costs and more need to organize each aspect of the chain [11]. Therefore, to improve the pineapple price and avoid economic and product losses, it is necessary that the new concept of supply chains must replace the traditional concept of segmented businesses. In this sense, the supply chain concept covers all aspects of production planning, inventories, transport, customer service, and integrated logistics [12].

Supply Chain Management (SCM) has received special attention from all organizations and researchers in the field of agriculture. A key aspect of the supply chain is collaboration in planning, coordinating and controlling the flows of information, material, customers, and suppliers [13]. The application of collaborative models leads to logistical responsiveness, cooperation, and guidance throughout the chain. It also allows for a more efficient organization of production capacity and better respond to fluctuations in demand. Benefits include reduced inventory, faster turnaround, unique capacity development, increased customer satisfaction, and increased profits [14].

The predominant concepts of collaborative planning include multiple stages with sites that can be connected by material flow, which allows a complete optimization throughout the network. Considering that there is a planning horizon, where great benefits are obtained in the processes of acquisition, production, storage, and sales [15]; this makes planning a difficult process due to the complexity of the chains [8] and opens up the application of decision-making models such as optimization models to improve strategic planning [6], [16].

Mathematical models for collaborative planning in the supply chain are widely applied through linear programming and mixed-integer programming models [6]. The main objectives are to minimize overall costs and maximize profits. The formulation of the model allows for periodic decisions on production, inventory and sales quantities, as planning challenges arise from the need for coordination in operations [17]. This type of model has been applied to agricultural production in terms of improving operations and decision making by considering aspects such as cultivation times, types of land, yield, the number of inputs, among others [18].

On the other hand, collaborative supply chains help fruit industries to get the most benefit from the product. In this type of chain, optimization models can also be proposed with the different echelons that make it up, so that each of its members achieves the proposed objectives [19]. These mathematical models can be applied to any type of fruit or Agri-food company, and even to SMEs, providing satisfactory support to incorporate them into the market [20]. Also, it is important to mention that the mixed-integer programming models manage to efficiently optimize all those operations that are carried out within the fruit crops. Likewise, it helps to determine the number of machines and workers needed for the entire production process and the allocation of vehicles for transport to achieve good results in the average percentage of cost reduction [21]. Certainly, one of the most important aspects within these models is the reduction of transport costs and their willingness to meet the demand for fruit, considering the processing capacities and the quantities of fruit in the storage centers [22].

In [23] propose a mathematical scheduling model to integrate the planning decisions of production and procurement, transport in manufacturing systems in a single optimization model. They simultaneously consider materials, production resource capacities and transportation planning decisions in the supply chain to avoid sub-optimal results, which are usually generated by the application of sequential and independent plans.

Additionally, in [24] propose a mathematical model designed to integrate and collaborate with planning, allowing to find optimal allocations and maximize the level of customer service, especially when production processes are intensive and capacity is limited. Therefore, they ensure compliance with new challenges that require competition in the chain; the flow of information is a crucial factor for efficiency and collaboration requires a great capacity for communication between different information and compatibility of their practices [25].

Via the development of this paper, mathematical modeling is applied through operational research to improve collaborative and competitive schemes in the pineapple supply chain. Considering that in our review we did not find a mathematical modeling approach applied to pineapple crops, our model presents a framework for decision making that allows us to better address the problems that may arise along the decisions involving the actors and echelons of the chain. The model is applied in a real case study about the pineapple supply chain in the municipality of San Marcos, Colombia. Additionally, a scenario analysis is developed to improve the decisions of the actors in specific situations.

The model considers two echelons of the chain, corresponding to agricultural production and distribution to wholesale and retail customers. In addition, transport aspects are included in the model. For the application of the model with real data, a current diagnosis of the pineapple production and distribution stables in the municipality of San Marcos was made. Subsequently, the mathematical model was built to maximize the profits of all the actors in the chain. The model is coded and solved using the GAMS CPLEX solver.

The remainder of this document is structured as follows. Section 2 corresponds to the development of mixed-integer mathematical modeling. Section 3 presents the computational results and scenario analysis for the study area. Finally, section 4 summarizes the conclusions.

2. Mathematical model

Mathematical model of collaboration for the pineapple supply chain considers two echelons (agricultural production and distribution to wholesale and retail customers). It considers a set of producers that require inputs for the cultivation of pineapple seeds and later, of the pineapple. Initially, the process of preparation, cultivation V_i of the seeds, and harvest X_i of the seeds (colinos) are carried out. Then, the farmers carry out the crop Y_i and harvest W_i of the pineapple considering the factors of loss of both seeds bx_i, and pineapple by_i.

It is considered initial storage of the pineapple harvest in each producer and the distribution from producers to wholesale customers H_{ict}, to retail customers HH_{imt}, finally, from wholesale customers to retailers Zx_{cmt}. Transport is carried out in t-type vehicles, which have a maximum capacity of Q_t or each trip Vx_{ict}, Vy_{imt}, and Vz_{cmt}. Additionally, the travel times $ttcx_{ic}$ and $ttcy_{im}$ as well as the weight loss factor fpp of the pineapple are considered. Travel costs are calculated as shown below.

$$Cvx_{ict} = ss_t 2CTNK_t Distx_{ic} \quad (1)$$

$$Cvy_{imt} = ss_t 2CTNK_t Disty_{im} \quad (2)$$

$$Cvz_{cmt} = ss_t 2CTNK_t Distz_{cm} \quad (3)$$

Equations (1), (2) and (3) determine the costs associated with the transportation process.

The times required for the individual processes of seed tts_i and pineapple ttp_i are calculated, in addition to the overall crops time $ttcp_i$. The representation of the supply chain is shown in Figure 1.

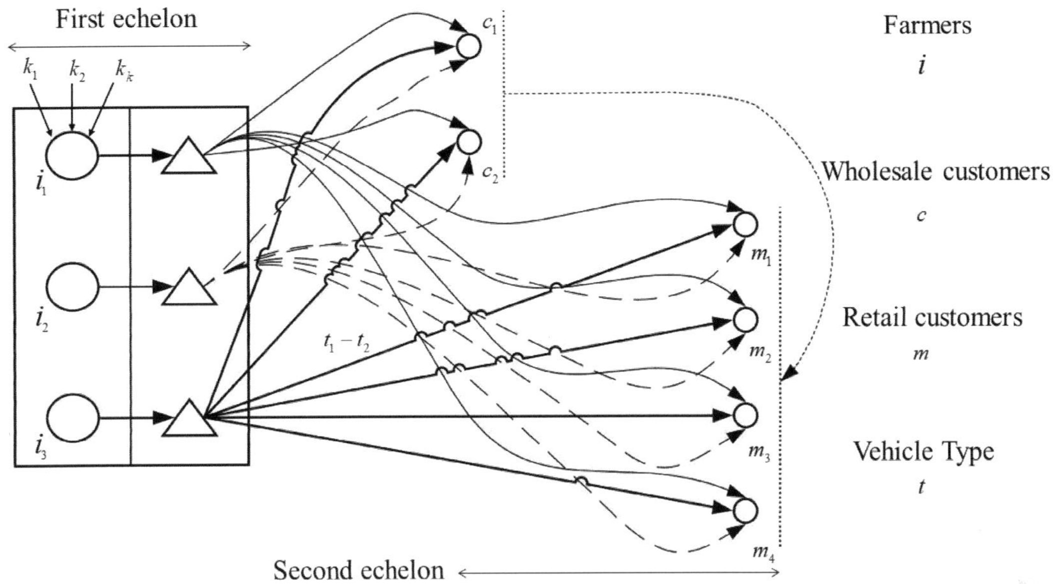

Figure 1. Conceptual model of the pineapple supply chain in the Municipality of San Marcos.

2.1. *Index and sets*

I Set of farmers
C Set of wholesale customers
M Set of retail customers
K Set of agricultural inputs
T Set of vehicles
n Seed supply set

2.2. *Parameters*

Ci_{ik} Cost of inputs per m^2 used
Cai_{ik} Quantity of inputs required per hectare
ts_i Drying time in months
trs_i Seed collection time in months
tss_i Seed sowing time in months
tcs_i Growing time of seed in months
bx_i Percentage of seed loss in months
hcs_i m^2 available for seed crops
Cs_i Cost of seeds
Cpt_i Cost of land preparation
nt_i Number of seeds allocated per worker

hcp_i m^2 available for pineapple crops
tcp_i Growing time in months for pineapple
tsp_i Sowing time in months for pineapple
$tspx_i$ Harvest time in months for pineapple

by_i Percentage of product loss in months
$cmop_i$ Labor cost
ntk_i Number of pineapples per worker

cal_i Storage capacity by volume
tal_i Pineapple storage time
$calpx_i$ Cost of storage

$demx_c$ Wholesale customer demand
$demy_m$ Retail customer demand
pcx_c Price paid per wholesale customer
pcy_m Price paid per retail customer
alm_c Wholesale customer storage capacity
$calpy_i$ Cost of storage at wholesale customer

Nmx_t Maximum number of trips per vehicle type
NK_t Number of vehicles per type
Q_t Maximum capacity of each type of truck per trip
ss_t Cost of preparation and enlistment
Cvx_{ict} Shipping cost from producer to wholesale customer
Cvy_{imt} Shipping cost from producer to retail customer
Cvz_{cmt} Shipping cost from wholesale customer to wholesale customer
tdx_{ic} Distribution time to wholesalers
tdy_{im} Distribution time to retailers
tdz_{cm} Distribution time from wholesalers to retailers
$Distx_{ic}$ Distribution cost from producer to wholesale customer
$Disty_{im}$ Cost of distribution from producer to retail customer
$Distz_{cm}$ Distribution cost from wholesale customer to retail customer

Jrn Cost of a daily wage
P Pineapple weight in pounds
as Seed area
fpp Pineapple weight loss factor
vp Volume of the pineapple
CT Transport cost per unit kilogram

2.3. *Variables*

H_{ict} Quantity of pineapple sold to the wholesale customer
HH_{imt} Quantity of pineapple sold to the retail customer
Zx_{cmt} Quantity of pineapple sold to the retail customer from the wholesale customer
nts_i Number of workers per seed
ntp_i Number of workers per pineapple
$tntp_i$ Total number of workers required
Vx_{ict} Number of trips to be made by the vehicle in the arc (ic)

Vy_{imt} Number of trips to be made by the vehicle in the arc (im)
Vz_{cmt} Number of trips to be made by the vehicle in the arc (cm)
V_i Quantity of seeds to be ordered
X_i Quantity of seeds to be cultivated
Y_i Quantity of pineapples to be cultivated
W_i Quantity of pineapples to be harvested
$Invx_i$ Quantity of pineapple sent to storage
$Invy_c$ Quantity of pineapple sent to storage
$Caif_i$ Quantity of inputs for seeds and pineapple
tts_i Total seed time
ttp_i Total time for growing pineapple
$ttcp_i$ Overall crop time
$ttcx_{ic}$ Transport time to reach the wholesale customer
$ttcy_{im}$ Transport time to reach the retail customer
Bn_i Binary variable to give openings for preparation costs if land is used

2.4. Objective function

Maximize =

$$\begin{aligned}
&\sum_i^I\sum_c^C\sum_t^T pcx_c PH_{ict} + \sum_i^I\sum_m^M\sum_t^T pcy_m PHH_{imt} + \sum_c^C\sum_m^M\sum_t^T pcy_m PZx_{cmt} \\
&-\sum_i^I\sum_c^C\sum_t^T Cvx_{ict} Vx_{ict} - \sum_i^I\sum_m^M\sum_t^T Cvy_{imt} Vy_{imt} - \sum_c^C\sum_m^M\sum_t^T Cvz_{cmt} Vz_{cmt} - \sum_i^I Cs_i V_i \\
&-\sum_i^I\sum_k^K Caif_{ik} Ci_{ik} - \sum_i^I Cpt_i Bn_i - \sum_i^I \left(Calpx_i Invx_i + 1000 Invx_i\right) - \sum_c^C \left(Calpy_c Invy_c + pcx_c Invy_c\right) \\
&-\sum_i^I 30 Jrn\left(tsp_i + tcp_i + tcpx_i\right) Npt_i - \sum_i^I 30 Jrn\left(tss_i + trs_i + tcs_i + ts_i\right) Npt_i
\end{aligned} \tag{4}$$

2.5. Constraints

2.5.1. Seed production

$$V_i as \le hcs_i Bn_i \,;\ \forall i \in I \tag{5}$$

$$tss_i + trs_i + tcs_i + ts_i = tts_i \,;\ \forall i \in I \tag{6}$$

$$V_i - \left(V_i Bx_i\right) = X_i \,;\ \forall i \in I \tag{7}$$

$$V_i / nt_i = nts_i \,;\ \forall i \in I \tag{8}$$

$$Cai_{ik} V_i as = Caif_{ik} \,;\ \forall i \in I, \forall k \in n \tag{9}$$

2.5.2. Pineapple product

$$Y_i as \le hcp_i Bn_i \,;\ \forall i \in I \tag{10}$$

$$Y_i \le X_i \,;\ \forall i \in I \tag{11}$$

$$tsp_i + tcp_i + tspx_i = ttp_i \,;\ \forall i \in I \tag{12}$$

$$Y_i - (Y_i By_i) = W_i;\ \forall i \in I \tag{13}$$

$$Cai_{ik} Y_i as = Caif_{ik};\ \forall i \in I, \forall k \in K - n \tag{14}$$

$$Y_i / ntk_i = ntp_i;\ \forall i \in I \tag{15}$$

$$\sum_c^C H_{ict} + \sum_m^M HH_{imt} + Invx_i = W_i;\ \forall i \in I, \forall t \in T \tag{16}$$

$$nts_i + ntp_i = tntp_i;\ \forall i \in I \tag{17}$$

$$ttcp_i = ttp_i + tts_i;\ \forall i \in I \tag{18}$$

2.5.3. Storage

$$tal_i + tdx_{ic} = ttcx_{ic};\ \forall i \in I, \forall c \in C \tag{19}$$

$$tal_i + tdy_{im} = ttcy_{im};\ \forall i \in I, \forall m \in M \tag{20}$$

$$Invx_i Vp \le cal_i;\ \forall i \in I \tag{21}$$

2.5.4. Major Customers

$$\sum_i^I (H_{ict} P) - (fppH_{ict} tdx_{ic}) - demx_c \le alm_c;\ \forall c \in C, \forall t \in T \tag{22}$$

$$\sum_m^M (Zx_{cmt} P) - (fppZx_{cmt} tdz_{cm}) \le Invy_c;\ \forall c \in C, \forall t \in T \tag{23}$$

$$Invy_c = \sum_i^I \sum_t^T (H_{ict} P) - (fppH_{ict} tdx_{ic}) - demx_c;\ \forall c \in C \tag{24}$$

2.5.5. Small Customers

$$\sum_i^I \sum_t^T (HH_{imt} P) - (fppHH_{imt} tdy_{im}) + \sum_c^C \sum_t^T (Zx_{cmt} P) - (fppZx_{cmt} tdz_{cm}) = demy_m;\ \forall m \in M \tag{25}$$

2.5.6. Transportation

$$\sum_i^I \sum_c^C Vx_{ict} + \sum_i^I \sum_m^M Vy_{imt} + \sum_c^C \sum_m^M Vz_{cmt} \le NK_t Nmx_t;\ \forall t \in T \tag{26}$$

$$H_{ict} P \le Q_t Vxx_{ict};\ \forall i \in I, \forall c \in C, \forall t \in T \tag{27}$$

$$HH_{imt} P \le Q_t Vxy_{imt};\ \forall i \in I, \forall m \in M, \forall t \in T \tag{28}$$

$$Zx_{cmt} P \le Q_t Vxz_{cmt};\ \forall c \in C, \forall m \in M, \forall t \in T \tag{29}$$

2.5.7. Nature of variables

$$H_{ict}, HH_{imt}, Zx_{cmt}, nts_i, ntp_i, tntp_i, Vx_{ict}, Vy_{imt}, Vz_{cmt}, V_i, X_i, Y_i, W_i, Invx_i, Invy_c, Caif_i, tts_i, ttp_i,$$
$$ttcp_i, ttcx_{ic}, ttcy_{im} \ge 0; \forall i \in I, \forall c \in C, \forall m \in M, \forall t \in T \tag{30}$$

$$Bn_i \in \{1,0\}; \forall i \in I \tag{31}$$

Equation (4) maximizes the chain's profit as a function of income from sales and costs of cultivation, harvesting, storage and transport. Constraints (5) and (10) ensure that the capacity of seeds and pineapples to be grown does not exceed the available hectares. Constraints (6) and (12) calculate the total time for obtaining colinos and pineapples. Constraints (7) and (13) calculate the fraction of seeds and pineapples that are lost due to externalities. Constraints (8) and (15) calculate the number of workers. Constraints (9) and (14) determine the amount of inputs required to grow and obtain the seeds and pineapple. Constraints (11) ensure that the amount of pineapple to be grown does not exceed the amount of colinos. Constraints (16) calculate the amount of pineapple to be produced and the farmer's inventory. Constraints (17) and (18) calculate the total number of workers and the total time for the pineapple harvest, respectively. Constraints (19) and (20) calculate transport times from the farmer to wholesale and retail customers, respectively. Constraints (21) determine the inventory at the farmer. Constraints (22) ensure shipment from the grower to the wholesale customers. Constraints (23) and (24) ensure inventory at the wholesale customers. Constraints (25) ensure compliance with the demand of retail customers. Constraints (26) calculate the number of trips. Constraints (27), (28), and (29) ensure that vehicle capacity is not exceeded on each trip. Finally, constraints (30) and (31) define the nature of the variables.

3. Computational experiments

3.1. Study area

San Marcos is one of the municipalities with the highest production of pineapple in the Department of Sucre, Colombia [26]. The presence of large wetlands in the municipality generates a high agricultural potential, being an important habitat for the development of economic activities of its inhabitants. The emerging crop of pineapple is characterized by the level of wealth and improved profitability in the area. This generates profits for the producers and the municipality [27]. With regard to production, in San Marcos, there are small and medium producers of pineapple with an approximate production of 330 tons. The marketing of this product is also driven by retail and wholesale customers located in the region; the production is sold to local customers and others located mainly in the municipalities of Sincelejo and Planeta Rica. However, these crops do not meet the logistical standards that form a well-structured supply chain. This prevents the efficient fulfillment of demand under the right delivery conditions of the fruit, which limits competitiveness [28].

The data collection was done through the construction of an instrument to evaluate the current logistic process of production, storage, and commercialization of pineapple. The information was obtained from three agricultural producers, who together have 20 hectares of total crops, harvest 309 tons, and distribute to two wholesale and for retail customers. Additionally, there are two types of vehicles with capacities of 10000 kg categorized as type 1 vehicle and 20000 kg for type 2 vehicle.

3.2. Computational results

Model was solved using the GAMS (General Algebraic Modeling System) CPLEX solver on a PC with 20 GB RAM, 512 MB DDSS and Intel Core i7 processor. With the actual data obtained in the process of characterizing the chain in San Marcos, a profit of $COP 80'684438 was obtained. That represents an increase of 10.53% over the current profits of the municipality. Likewise, within this solution, it was observed that from producer 2 to wholesale customer 2, 5014.3 kg of pineapple are sent in a type 1 vehicle and 105000 kg in a type 2 vehicle. A total of 110014.3 kg of pineapple is distributed. Shipments are made from producers to retail and wholesale customers, and wholesale customers to retail customers (see Figure 2a). Similarly, it is evident that only two producers grow and harvest, meeting the entire demand.

A total of 242828 seeds are required, of which 211260 are available for obtaining pineapple; applying the loss factor, there is a loss of 31567 colinos. Therefore, the total amount of pineapple produced is 209182, with a loss of the product of 2078. Another important feature is the amount of product stored at the wholesale customer equivalent to 890. As for the trips made in the transport process, 14 trips with type 1 vehicles and 22 trips with type 2 vehicles are made in the producer-wholesale customer arc. For the producer-retailer segment, only vehicle type 2 is used, which makes two trips. Finally, the number of trips to be made is 16 for vehicle type 2 and 7 for vehicle type 1 in the arch wholesaler-retailer customers. A total of 41 workers were required (initially 31 for the process of growing and obtaining the seeds and later, 10 workers were added for the cultivation of the pineapple). Overall time was 16.66 months.

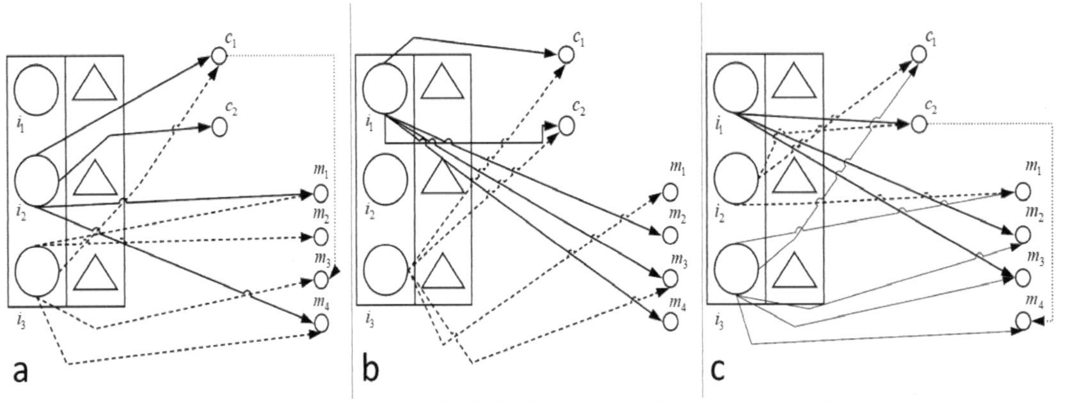

Figure 2. Results obtained in the initial model and the proposed scenarios.

3.3. *Scenario analysis*

It is evident from the results obtained in comparison with the current process, that the model suggests that with two agricultural producers, customer demands can be met. However, we considered carrying out an analysis through test scenarios that allow us to evaluate the behavior of the model by considering (i) the increase in demand to verify if it is possible to activate the production of all the available producers and (ii) to add another restriction that allows the activation of all the producers and to verify the behavior of the chain.

3.3.1. Increased demand. In this first scenario, a 30% increase in demand for retail customers is considered to real demand, to analyze the behavior of the model. The results show a profit on $COP 94'393000 higher by 16.99% concerning the initial profits shown by the model. The quantity of seeds requested is 271396 of which 236114 are obtained, i.e. there is a loss of seed of 35281. Subsequently, the total quantity of pineapple produced is 234692, extracting a loss of 1422.

Figure 2b shows the solution provided by the model for this scenario. Farmers 1 and 2 send the pineapple to wholesale and retail customers; unlike the original model that sends from farmers 2 and 3. This is because producer 1 has more production capacity than producer 2 because it has more land and therefore can supply the demand. In this scenario, 39 workers are required to obtain the seeds and 46 in the growth of the pineapple. Overall time to obtain the pineapple is the same as the initial model (16.66 months).

3.3.2. Production by all pineapple producers. In this scenario it is intended that all farmers produce pineapple. Therefore, constraints (32) is added.

$$Y_i \geq L;\ \forall i \in I \tag{32}$$

In this case, L can take value from any positive integer. The solution obtained generates profits of $COP 80'128792 which corresponds to a decrease of 0.69% of the initial solution obtained with the model. The amount produced of pineapple is 209182, which is sent from the producers to the retail and wholesale customers, and from the wholesale customers to the retail customers. There is an inventory generated in the second wholesale customer that is equivalent to 890 pineapples. A total of 34 trips is made from producers to wholesale customers (14 with type 1 vehicles and 22 with type 2 vehicles). From the products to the retail customers there are 5 trips with type 1 vehicles and 19 trips with type 2 vehicles. Finally, two trips are made using both types of vehicles from the second wholesale customer. Up to the fourth retail customer, as shown in Figure 2c above. For the latter scenario, 52 workers are required for the seed and 61 for the pineapple crop. The cultivation time is the same as in the previous scenario.

4. Concluding remark

In the concept of supply chain collaboration, strategic, tactical, and operational decisions play a major role. The articulation of these types of decisions can guide all actors and echelons of the supply chain to a generalized benefit. For this reason, the models and strategies applied for efficient decision-making are a relevant field in the academic and business level.

In this sense, operational research presents an opportunity for the development of mathematical models to improve the general scheme of the supply chain based on the achievement of common objectives; in this document we develop a two-echelon collaborative mathematical planning model for the pineapple supply chain. The mixed-integer linear programming model was validated with actual data from the municipality of San Marcos, Colombia. In this case, the supply chain presents problems of competitiveness due to the lack of knowledge of the general scheme of production and distribution, causing logistical failures.

With the application of this model, the general utility in the supply chain constituted by farmers, internal storage, wholesale, and retail customers was maximized. The computational results showed a 10.53% increase in profits without the need to allocate production to all farmers. Therefore, two test scenarios were developed to (i) observe the behavior of the supply chain in the face of increased customer demand and (ii) forces all producers to grow and harvest the fruit. The first scenario showed an increase in profits concerning the initial model of 16.99% when demand increased by 30%; this situation is easy to explain, because, as demand increases, so do the production and sale of the fruit. However, the variation is not proportional to 1:1. With the second scenario a decrease in profits of 0.69% was observed concerning the real data (i.e. by obliging all farmers to produce the costs associated with the production and distribution processes of the fruit).

For future research, we consider it important to add periods, consider storage that is not included in the producer and consider operational decisions on routing vehicles for the distribution process.

References

[1] R. Sharma, S. S. Kamble, A. Gunasekaran, V. Kumar, and A. Kumar, "A systematic literature review on machine learning applications for sustainable agriculture supply chain performance," *Comput. Oper. Res.*, vol. 119, p. 104926, 2020, doi: 10.1016/j.cor.2020.104926.

[2] L. Liu, J. Cavaye, and A. Ariyawardana, "Supply chain responsibility in agriculture and its integration with rural community development: A review of issues and perspectives," *J. Rural Stud.*, no. July, 2020, doi: 10.1016/j.jrurstud.2020.07.003.

[3] B. Yan, X. Chen, C. Cai, and S. Guan, "Supply chain coordination of fresh agricultural products based on consumer behavior," *Comput. Oper. Res.*, vol. 123, p. 105038, 2020, doi: 10.1016/j.cor.2020.105038.

[4] A. L. Macready, S. Hieke, M. Klimczuk-Kochańska, S. Szumiał, L. Vranken, and K. G. Grunert, "Consumer trust in the food value chain and its impact on consumer confidence: A model for assessing consumer trust and evidence from a 5-country study in Europe," *Food Policy*, vol. 92, no. February, p. 101880, 2020, doi: 10.1016/j.foodpol.2020.101880.

[5] Z. Vianchá, "Models and configurations of supply chains in perishable goods," *Ing. y Desarro.*, vol. 32, no. 1, pp. 138–154, 2014, doi: 10.14482/inde.32.1.4577.

[6] M. Nematollahi and A. Tajbakhsh, "Past, present, and prospective themes of sustainable agricultural supply chains: A content analysis," *J. Clean. Prod.*, p. 122201, 2020, doi: https://doi.org/10.1016/j.jclepro.2020.122201.

[7] R. Tordecilla-Madera, A. Polo, D. Muñoz, and L. González-Rodríguez, "A robust design for a Colombian dairy cooperative's milk storage and refrigeration logistics system using binary programming," *Int. J. Prod. Econ.*, vol. 183, pp. 710–720, 2017, doi: https://doi.org/10.1016/j.ijpe.2016.09.019.

[8] A. Castro-Nunez, A. Charry, F. Castro-Llanos, J. Sylvester, and V. Bax, "Reducing deforestation through value chain interventions in countries emerging from conflict: The case of the Colombian cocoa sector," *Appl. Geogr.*, vol. 123, no. July, p. 102280, 2020, doi: 10.1016/j.apgeog.2020.102280.
[9] Asohofrucol, "Panorama de la produccion de piña en Colombia," 2019. .
[10] A. M. Garrapa, "The Corporate Food Regime and immigrant farm workers in California strawberry harvests," *Norteamerica*, vol. 12, no. 1, pp. 233–264, 2017, doi: 10.20999/nam.2017.a008.
[11] E. B. Rúa, A. I. Barrera, and B. Pinzón, "Caracterización y Diagnóstico de la Cadena Productiva de la Piña en el Departamento del Casanare," *Rev. Estrateg. Organ.*, vol. 5, no. 0, pp. 29–47, 2016, doi: 10.22490/25392786.2099.
[12] H. Singh, R. K. Garg, and A. Sachdeva, "Supply chain collaboration: A state-of-the-art literature review," *Uncertain Supply Chain Manag.*, vol. 6, no. 2, pp. 149–180, 2018, doi: 10.5267/j.uscm.2017.8.002.
[13] G. Kumar, R. N. Banerjee, P. L. Meena, and K. K. Ganguly, "Joint planning and problem solving roles in supply chain collaboration," *IIMB Manag. Rev.*, vol. 29, no. 1, pp. 45–57, 2017, doi: 10.1016/j.iimb.2017.03.001.
[14] J. Oh and B. Jeong, "Tactical supply planning in smart manufacturing supply chain," *Robot. Comput. Integr. Manuf.*, vol. 55, no. April 2017, pp. 217–233, 2019, doi: 10.1016/j.rcim.2018.04.003.
[15] T. Nishi, T. Tsuboi, and M. Matsuda, "A simultaneous optimization framework for product family configuration and supply chain planning," *Procedia CIRP*, vol. 81, pp. 1266–1271, 2019, doi: 10.1016/j.procir.2019.03.305.
[16] H. Ge, R. Gray, and J. Nolan, "Agricultural supply chain optimization and complexity: A comparison of analytic vs simulated solutions and policies," *Int. J. Prod. Econ.*, vol. 159, pp. 208–220, 2015, doi: 10.1016/j.ijpe.2014.09.023.
[17] W. Albrecht and M. Steinrücke, "Assessing site integration into semi-continuous production, distribution and liquidity planning of supply chain networks," *EURO J. Transp. Logist.*, vol. 9, no. 1, p. 100002, 2020, doi: 10.1016/j.ejtl.2020.100002.
[18] J. Carvajal, W. Sarache, and Y. Costa, "Addressing a robust decision in the sugarcane supply chain: Introduction of a new agricultural investment project in Colombia," *Comput. Electron. Agric.*, vol. 157, no. April 2018, pp. 77–89, 2019, doi: 10.1016/j.compag.2018.12.030.
[19] S. S. Sana, G. Herrera-Vidal, and J. Acevedo-Chedid, "Collaborative Model on the Agro-Industrial Supply Chain of Cocoa," *Cybern. Syst.*, vol. 48, no. 4, pp. 325–347, 2017, doi: 10.1080/01969722.2017.1285160.
[20] J. Jonkman, A. P. Barbosa-Póvoa, and J. M. Bloemhof, "Integrating harvesting decisions in the design of agro-food supply chains," *Eur. J. Oper. Res.*, vol. 276, no. 1, pp. 247–258, 2019, doi: 10.1016/j.ejor.2018.12.024.
[21] M. M. Morales-chávez, J. A. Soto-mejía, and W. Sarache, "A mixed-integer linear programming model for harvesting , loading and transporting sugarcane . A case study in Peru Modelo de programación lineal entera mixta para el corte , carga y transporte de caña de azúcar . Un caso de estudio en Perú," vol. 83, no. 195, pp. 173–179, 2016.
[22] V. M. Albornoz, M. González-araya, M. C. Gripe, and S. V Rodríguez, "A Mixed Integer Linear Programming model for planning at operative level in a meat packing plant," no. Wfsc, pp. 1–9, 2014.
[23] M. Díaz-Madroñero, J. Mula, and D. Peidro, "A mathematical programming model for integrating production and procurement transport decisions," *Appl. Math. Model.*, vol. 52, pp. 527–543, 2017, doi: 10.1016/j.apm.2017.08.009.
[24] B. A. Mousavi, R. Azzouz, and C. Heavey, "Mathematical modelling of products allocation to customers for semiconductor supply chain," *Procedia Manuf.*, vol. 38, no. 2019, pp. 1042–1049, 2019, doi: 10.1016/j.promfg.2020.01.190.

[25] A. Badea, G. Prostean, A. Hutanu, and S. Popa, "Competency Training in Collaborative Supply Chain Using KSA Model," *Procedia - Soc. Behav. Sci.*, vol. 191, pp. 500–505, 2015, doi: 10.1016/j.sbspro.2015.04.352.
[26] Secretaria de Desarrollo Económico, "Encuesta Nacional Agropecuaria de los periodos 2007 a 2017," 2018.
[27] R. Ayazo-Toscano, *Caracterización ecológica de los humedales en la región de La Mojana*. Bogotá D.C., 2018.
[28] J. A. Orjuela-Castro, M. M. Herrera-Ramírez, and W. Adarme-Jaimes, "Warehousing and transportation logistics of mango in Colombia: A system dynamics model," *Rev. Fac. Ing.*, vol. 26, no. 44, p. 71, 2017, doi: 10.19053/01211129.v26.n44.2017.5773.

IOP Conf. Series: Materials Science and Engineering 1154 (2021) 012012 doi:10.1088/1757-899X/1154/1/012012

Proposal for an automated greenhouse to optimize the growth of hydroponic vegetables with high nutritional content in the context of smart cities.

S A Sánchez [1], A D Morales[1], J C Castillas [1], C A Martínez [1], A Z Meza [1]

[1]Corporación Universitaria Antonio José de Sucre, Sincelejo, Colombia, Faculty of Engineering Science.

sergio_sanchez@corposucre.edu.co

Abstract. Hydroponics is a cultivation method that has been gaining ground for a lo ng time worldwide, due to its high nutritional content, low water consumption, scalability and zero use of pesticides, in addition to the versatility that this technique has to be implemented in almost any space. This is why the Antonio José de Sucre University Corporation, taking advantage of the characteristics of this cultivation technique, proposed the construction of an automated hydroponic greenhouse that serves as a model in the sustainable generation of food from the cities, following the example of smart cities ; Within the construction of this proposal, it was necessary to study the requirements for planting the vegetables under hydroponic conditions in order to establish the technology that should have been used in their automation, for which the basic chemical and climatological needs in production were identified. of plant organisms, selecting those critical factors that intervene in productivity, in order to establish the appropriate types of sensors and an action protocol that maintain the values of these variables in optimum, tending to obtain healthy fruits and vegetables and superior in taste and nutrition. As a result, the design of a metallic hydroponic farm was obtained, for the location of the vegetables, recyclable materials such as PVC pipes were used and for the automation of the farm a raspberry pi electronic card was used complemented with PH, conductivity, temperature and humidity sensors. and lighting with the aim of obtaining quality products.

1. Introduction

Globally, there is widespread concern about a future food and water crisis due to the demographic explosion that is taking place in urban areas, and the effects of climate change; Thus, the cities of tomorrow will be increasingly populated and will require a greater quantity of food and efficient water management [1]. That is why it has been innovating in the improvement of crops, towards more productive ones and less prone to climatic and terrain variables, leading to the emergence of the hydroponic cultivation method, since the traditional planting presents various limitations such as the availability of fertile land and the use of water; limiting factors present in many nations when trying to ensure the food basket, so you have defined strategies for agricultural strengthening based on modernization and technological innovation, leading to food production through various techniques and methodologies, proving to be one of the hydroponic sowing is more effective, since from this a higher yield of crops is obtained, increasing their productivity above the traditional ones, since the availability of water, nutrients, radiation levels, environmental temperature, and the density of sowing or disposition of plants are not limiting, so they can be easily controlled. In addition, that the action of pathogens or pests is easily reduced, thanks to the control of the development conditions of the crop, allowing to optimize the crop performance, and achieve a clean and organic production [2].

Greenhouses and hydroponics are not a novelty in the agricultural technological environment, but they pose great opportunities to improve its advancement, an example of this are the plant factories, a method used in the industrial production of food of plant origin in greenhouses hydroponics; This production method was invented in the 80's, and it has been evolving in conjunction with the development of electronics and computing, so that today this industry has more efficient processes thanks to the automation and control of production parameters [3].

Thus, in order to avoid the nutritional vulnerability to which people with limited economic resources of the municipality of Sincelejo are exposed, it is necessary to explore alternatives that contribute in a sustainable way to the increase of the food stock of the region in the production of balanced sources of food so that the economy of the most vulnerable actors is not harmed.

All of the above is described within public policy at the international, national, regional and local levels, from the United Nations Sustainable Development Goals (SDG), which in turn is immersed in the new science, technology and innovation guidelines. for sustainable development set out in the Green Book 2030 and presented by the national government, in the guidelines for Responsible Production and Consumption, Zero Hunger, Well-being and Health, and Sustainable Cities and Communities, given the continuous expansion of the urban environment and with This increases the resources for the support of the population; likewise, the government of Sucre, who prioritizes food and nutritional security in the region, through the sustainable development of agricultural production, guaranteeing the consumption and sustainable production of the resources that the population requires, such as food; The Mayor of Sincelejo also assumes its commitment to these specific sustainable development objectives through some of the subprograms of its Comprehensive Health for All program [4][5]. Thus, ensuring a rich and balanced nutrition for the well-being and health of the communities, framed in a sustainable production scheme, is a priority for both the national and local government.

1.1 General aspects of hydroponics

The term hydroponics is based on the Greek hydro (water) and pónos (labor), and refers to the cultivation of plants in aqueous solutions or in a support medium, which can be gravel, sand, coco peat, pellets of expanded clay, a mixture of these elements, among others; It is for this reason that this method of agriculture is also known as "soilless agriculture", since plants do not require this medium for their development [6].

The development of this production method did not occur by chance, since it is due to the evolution and experimentation of agricultural practices due to the serious problems that it presents with the use of soil and water by some countries, such as they are the loss of fertility, the availability of water resources, extreme climatic conditions, in addition to the problems associated with soil diseases and pests that end up affecting crop yield. Sowing in hydroponics allows cultivating without the problems associated with land production, but despite the benefits that this technology has in productivity, pathogen control and resource management, it presents challenges, such as the initial cost of these projects, which is high. , greater care of the plants to avoid infections caused by the recirculation of water, and the need for higher standards in the handling and management of these, but despite all this, the instrumentation in this production technique has contributed to reducing the effect of these limiting factors, thus achieving excellent results compared to other sowing methods;

It is so much so that in the nations of northwestern Europe in recent decades the application of hydroponics has been increasing as a result of unfavorable soil conditions and monoculture in fertile areas, in addition to the legislation in countries such as the Netherlands and Germany it prohibited greenhouse production by direct administration of soil substrate due to the risk of groundwater contamination, being hydroponic sowing or sowing without soil the most viable option in the development of agricultural activity in greenhouse [7].

Thanks to the different technological advances, ways have been developed that facilitate the use of this cultivation method, among which the optimization of the same through automation is presented, as is the case of Fernandez, Costa, and Lemos [8], who built a small prototype of a transparent hydroponic greenhouse, where the internal environmental conditions of temperature, nutrients, and luminosity were monitored, controlling the first two in order to achieve optimal plant growth; The device also carried out the monitoring of plant growth through a camera and image processing software, which not only showed the progress of plant growth, but also the environmental conditions under which it was developed, it also allows the user can access information and images in real time of what happens in the greenhouse from anywhere on the internet of things, which facilitates the care of the conditions in the prototype and therefore of the crop.

Thus, from the boom that this technique has had together with the advances made in it, that today there are industrial facilities known as green factories, where in highly controlled environments due to the set of sensors and actuators arranged in variable such as lighting, temperature, nutrients, water quality and CO2 levels, such as those available in green factories located in Tokyo Bay in Japan, which produces up to 1,000 heads of lettuce organized per day, but all this does not possible without the heating, ventilation and air conditioning ducts and fans that keep the room in optimal climatic conditions and a concentration of 1,000 ppb (parts per million) of carbon dioxide, along with LED tubes that go on from 16 to 17 hours a day, and constant measurements of the quality of the water circulating through the greenhouse [9].

It should be noted that due to the versatility that this cultivation method has been implemented in areas with extreme climatic conditions that range from quasi-desert areas to very low temperatures; It is noteworthy that the international space station has its own hydroponic module in the development of experiments on the effect of zero gravity on the development of plants.

1.2 Requirements and care of plants in hydroponics

Production in a hydroponic medium is not exempt from the effect of variables that limit the growth and development of plants, and when it refers to variables, reference is not only made to the chemical part of the process, since the state of the plant both in the medium Hydroponic as in land is linked to the conditions of the environment to which the plant is exposed, that is why it is in this method where extra precautions must be taken so that the productivity of the crop is not affected; therefore it is necessary to talk about what these variables are and what precautions should be taken with them.

1.2.1 Support medium. The medium in which the plant is to be cultivated in the hydroponic state directly intervenes in the periodicity with which they must be irrigated, since the porosity and permeability of these media is directly linked to the transport and water retention capacity, and with it to the passage of oxygen and essential minerals to the roots. The water retention capacity of the medium is a factor that influences the type of crop that can be planted in it, since some plants have roots that are susceptible to high levels of humidity; In conclusion, when selecting the sowing medium, it is good to keep in mind the type of culture with which to work, water retention, oxygenation, structural integrity and sterility.

The last two conditions are related to the fact that the medium is maintained despite the growth of the plant and to avoid the incidence of pests and diseases in the roots as a result of a poorly sterile medium.

1.2.2 Water. Water is one of the essential factors in agriculture, since it is the source of all life and this does not exclude plants, which are composed of between 80% - 95% of this liquid and the rest of dry material. It is important to understand the importance of water when talking about photosynthesis, a process in which the energy from sunlight, nutrients and carbon dioxide from the air are transformed into oxygen and a building block for the plant in the form of sugars; In this, the water is responsible for transporting the nutrients from the soil to the root, and from the root into and out of each of the cells of the plant, to end its cycle in the evapotranspiration process, where through transpiration of water the vegetable cools. It should be noted that, the greater the perspiration, the greater the amount of water that the plant will need to adsorb from its roots in order not to lose its rigidity or turgor, otherwise when the body does not have enough water, it must slow down the transpiration process by closing their stomata in order to avoid the loss of water, which leads to its loss of turgor and begins to wilt, while essential minerals are lost. This type of stress causes growth to slow down and crop yields are reduced, because the stomata are responsible for capturing carbon dioxide (CO2), so if they are not completely open, photosynthetic activity is reduced. and it can even be stopped [7].

1.2.3 Illumination. Light is one of the requirements of photosynthetic activity in plants, but not all light is involved in it, since chlorophyll only uses the visible light spectrum in metabolic activity. It is known that chlorophyll is what gives the green color to plants and that in these there are organelles called chloroplasts, which contain chlorophyll-a, chlorophyll-b and carotenoids. Plants have evolved millions of years to better convert the energy in sunlight into carbohydrates and sugars, so they mainly absorb energy from the violet-blue and red light band of the visible spectrum, since these wavelengths coincide. With the range that goes from 400nm to 700nm, where photosynthesis is stimulated, in addition to this factor it is necessary that the quality and light intensity received are optimal, see figure 1.

The effective wavelengths in photosynthesis are within the range described above, since they are responsible for stimulating the metabolic processes of the plant, such as phototropic activity, stem and leaf growth, flowering, fruit production. , the generation of antioxidants, the formation of chlorophyll, and the photosynthetic activity; These processes are characteristic of specific wavelengths, Ultraviolet light (10nm-400nm) that at wavelengths of 350nm promotes the accumulation of phenolic compounds, improves the antioxidant activity of plant extracts, but has no significant effect on the increase; Blue light (430nm-450nm) is responsible for promoting vegetative growth, essential for the germination of seedlings and young plants during the vegetative stage of growth, limits the growth of mature plants, stimulates the production of secondary pigments that improve colors and encourages the production of aromatic compounds; Red light (640nm-680nm) has a very strong reaction to the photosynthetic action, which stimulates the generation of carbohydrates and sugars in the crop, allowing them to have a better flavor and a greater amount of nutrients within their composition and fruits, it also has a strong impact on the germination and flowering processes [8].

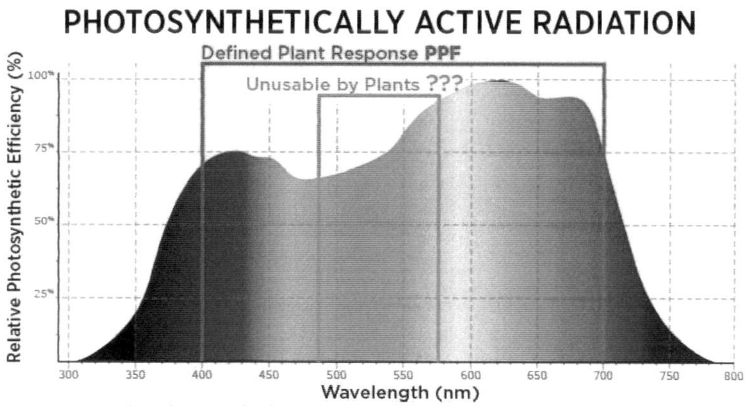

Figure 1. Active photosynthetic range (Source: Light spectrum and plant growth, p.2).

IOP Conf. Series: Materials Science and Engineering 1154 (2021) 012012 doi:10.1088/1757-899X/1154/1/012012

In addition to the wavelengths perceived by the leaves, another important factor with regard to lighting is the light intensity to which the crop is exposed, which is represented as the amount of energy the plant receives per day, and is given By the time the plant is exposed to the active photosynthetic range, figure 2 shows the effective saturation range for most plants, since those that are from very cold climates or from very hot climates are located on the outside of the same, since they are used to a lesser or greater solar irradiation. In conclusion, the effect of lighting is directly linked to photosynthetic activity, which is intrinsically related to plant growth, nutritional content, flowering and production cycles.

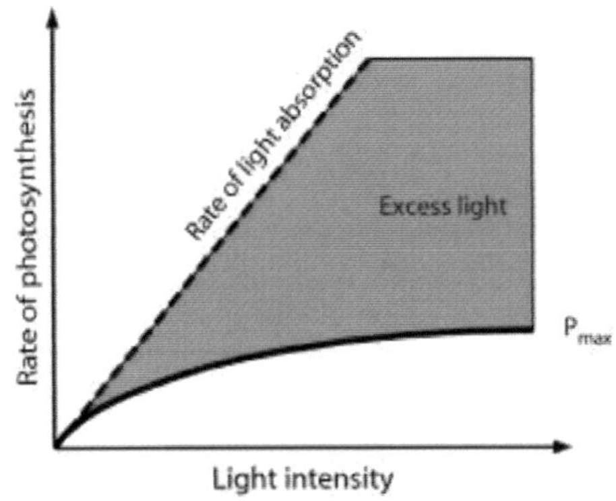

Figure 2. Photosynthetic activity vs light energy (intensity). (Source: Hydroponics for the home grower, p. 26).

Temperature and relative humidity of the air. The relationship between plant organisms and temperature is given based on their transpiration, since the higher the temperature, the greater the amount of water they will need to evaporate to keep their leaves fresh, however, as in other living beings if the relative humidity of the air is high, this interferes with the perspiration cycle reducing the cooling effect and therefore this slows or completely stops photosynthesis as seen in figure 3. It should be added that at high radiation loads UV plant leaves tend to burn, so in greenhouses UV filter coatings are used in order to remedy the consequence of this rays.

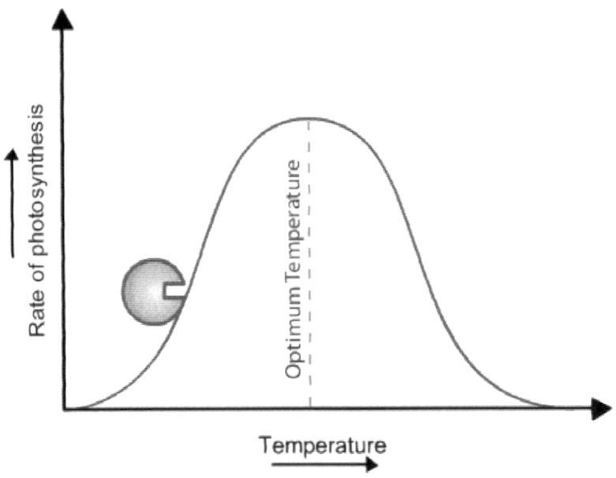

Figure 3. Effect of temperature on photosynthesis. (Source: Hydroponics for the home grower, p. 27).

Carbon dioxide (CO2) levels. Gaseous CO2 is the main source of carbon used by plants in the process of generating carbohydrates and sugars, essential blocks in obtaining energy for their growth and development, as can be seen in figure 4; That is why when growing in the greenhouse, the concentration of this gas within it must be known, since the environmental levels (400ppm) of the same are not sufficient to reach the optimal concentration (800ppm - 1200ppm), the which is 2- 3 times higher than the environmental one, as shown in figure 5 [9].

IOP Conf. Series: Materials Science and Engineering 1154 (2021) 012012 doi:10.1088/1757-899X/1154/1/012012

$$6CO_2 + 6H_2O \xrightarrow{\text{Sunlight}} C_6H_{12}O_6 + 6O_2$$

Carbon Dioxide Water Glucose (Energy) Oxygen

Figure 4. Photosynthesis equation. (Own source)

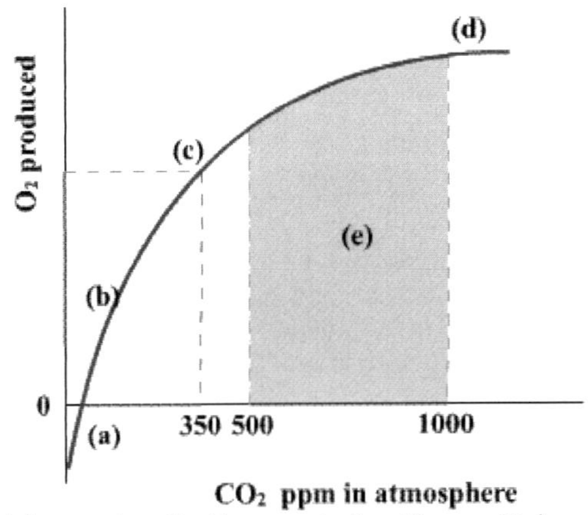

Figure 5. Photosynthetic activity vs carbon dioxide concentration. (Source: Hydroponics for the home grower, p. 28)

1.2.4 Oxygen concentration in the roots. The oxygen is in the roots of the plants is a critical factor in the active transport of the nutrients dissolved in the water towards the plant.

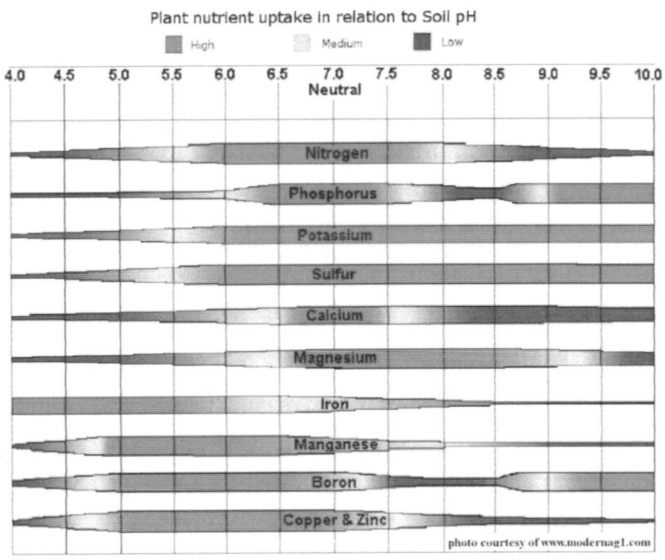

Figure 6. Effect of pH on the availability of nutrients for the plant. (Source: Hydroponics for the home grower, p. 36).

1.2.5 Essential nutrients. The minerals necessary for plants to grow and develop are known as essential, some of these are obtained through water and air, such as carbon (C), hydrogen (H) and oxygen (O), while others They are present in the soil and are transported to the organism by the water that the roots capture; These essential nutrients taken from the soil are divided into two categories, those that are required in a higher proportion and others in a smaller or much smaller proportion. Macronutrients include Nitrogen (N), Phosphorus (P), Potassium (K), Sulfur (S), Calcium (Ca) and Magnesium (Mg); and 8 micronutrients, Iron (Fe), Manganese (Mn), Zinc (Zn), Copper (Cu), Boron (B), Molybdenum (Mo), Chlorine (Cl) and Nickel (Ni). The deficiency or excess of these minerals can be seen directly in the disorders of the color and shape of the plant and others not so visible in the taste and nutritional content of the same.

1.2.6 pH. In nature, minerals are found in an insoluble and soluble form, those corresponding to the macronutrients and micronutrients necessary for plants are soluble and available; However, if the medium has a very acidic or basic pH, this will affect the solubility of the minerals and with it their availability, generating a deficit in the plant, said behavior is presented in Figure 6 [10].

2. Methodology

This document corresponds to a proposal developed in order to identify the requirements and instruments necessary in the automation process for a hydroponic greenhouse that was built at the Antonio José de Sucre University Corporation. In the process of identifying these requirements, a study was carried out of the factors that intervene in the growth and development of plants in the aforementioned environment, to later identify in which of these variables measurement and control action could be carried out, establishing the itself the type of possible sensors to use according to the weather conditions of the area where the project was developed, this being the city of Sincelejo, and the control actions to be developed in the management of said factors. The following research was developed in 3 stages, as evidenced below:

2.1 Stage 1: Research development.

In this first stage of the project, a study of an investigative nature was carried out with the purpose of stipulating the physical-chemical requirements that plants require in a hydroponic production method in order to guarantee their efficient production. Where the following activities were carried out.

- Bibliographic review of the methodologies and technologies applied in hydroponic planting in order to determine their advantages and disadvantages based on the criteria of production, space and necessary resources.
- Selection of the ideal hydroponic seeding technique for its application in the urban environment of the city of Sincelejo based on the spatial and climatological conditions of the area.
- Analysis of the electronic components for measurement and control of physicochemical variables that the hydroponic greenhouse requires for its automation.
- Inquiry about the types of plants to be used in hydroponic production for their selection according to their growth characteristics and nutritional contribution.
- Study of the effect of physical and chemical variables on the development and production of the selected plants.
- Establishment of the existing relationships between the physical-chemical parameters studied with the level of growth, production and food quality of the study plants in hydroponics.

2.2 Stage 2: Technological development.

This stage corresponds to the construction of the hydroponic greenhouse prototype for the Antonio José de Sucre University Corporation based on the automation operational criteria stipulated in the previous stage. Where the following activities were carried out.

- Design of the schematic circuit of the automated hydroponic greenhouse based on the measurement sensors and controllers defined in the previous stage.
- Coding of the hydroponic greenhouse monitoring and control system based on the correlational factors determined in the previous stage.
- Construction of the automated hydroponic greenhouse based on the technological and spatial criteria analyzed in its development.
- Analysis of the technical functionality of the prototype based on the testing of the monitoring and control systems of its physical-chemical variables.
- Evaluation of the productive capacity of the prototype.

2.3 Stage 3: Implementation of the system.

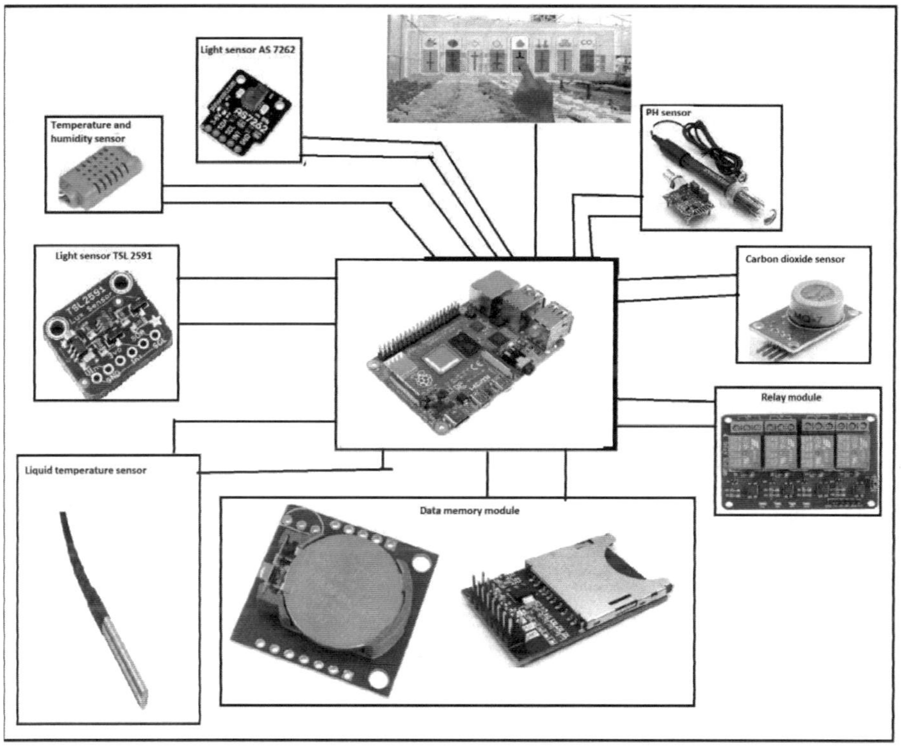

Figure 7. Block diagram of the proposed system.

3. Results

In the results phase, the main factors studied in the automation process in hydroponics were initially analyzed, which corresponds to the optimization of the cultivation space, the control of nutrients with precision, the elimination of pests and diseases transmitted by the soil, the quality of the light to which the plants should be exposed, since this factor has a direct inference with the development of the vegetables, mainly influencing the profitability of the crops, their nutrient content and flavor profile. Other factors that were taken into account were the pH of the irrigation water, a variable that can only be adjusted by adding acidic or alkaline substances as the case may be, the oxygenation of the water, the level of carbon dioxide, the conductivity of the water and the time of the macronutrients and micronutrients in the system.

To measure the level of carbon dioxide inside the greenhouse, a CO2 monitor was used in order to know its concentrations and, if necessary, look for auxiliary methods that help to increase the concentration of the gas inside the greenhouse, for For this reason, a CO2 sensor MG811 was used to control this variable within the greenhouse.

In the proposed greenhouse, a light intensity sensor and a visible light sensor were used to calculate the lighting, the ones chosen being the TSL2591 and the AS7262 respectively, since both devices have good sensitivity to the physical variables detected and operate. within the temperature range of the system.

The temperature and relative humidity of the air are factors involved in the transpiration process of the plant, directly intervening in photosynthesis, which is why a sensor is required that has the ability to measure these two variables, and more so in weather conditions. from the city of Sincelejo, whose temperatures range between 20 ° C and 36 ° C, these being the most extreme conditions. The sensor chosen for this task was the AMT1001 Humidity and Temperature Sensor, which presents a very accurate reading of these environmental variables.

The conductivity in the aqueous solution that contains the nutritional substrate and is linked to the level of mineral salts present in it, this indicator does not allow to individually detect the levels of each of the mineral compounds dissolved in the storage tank, however, gives an idea of when the levels of these are decreasing, serving as an indicator of when substrate must be supplied to the system again. As a monitoring device for this variable, an analog electrical conductivity sensor compatible with the Raspberry pi platform is used, used to measure the levels of electrolytes in water, and mainly applied in agriculture in order to establish the levels of fertilizers within of the aqueous solutions that are supplied to the greenhouse.

The pH of the aqueous solution is a factor that can limit the assimilation capacity of mineral salts by the plant, so this variable must be known at all times in order to avoid a deficiency of nutrients in it. The sensor used for this task is an analog pH sensor, whose temperature specifications and measurement precision determine the state of the water quality at all times.

During the transpiration process of the plants, fluid will be lost by evaporation, so in order to maintain operational water levels, a water level sensor in the tank will be added to the system, since the lack of said liquid can affect drastically the development of these.

The use of photovoltaic solar energy to power these systems is advisable, since it helps to reduce the energy resources used, relying on the free energy that the sun presents and even more so in the Colombian Caribbean, where it has a great photovoltaic potential.

Finally, in figure 8, the final design of the hydroponic greenhouse prototype is evidenced, conformed with dimensions of 4 meters wide by 7 meters long with a height of 2.6 meters high, fully metallic, lined with transparent paper caliber number 3, with 6 tubes of 4 inches, which contain 50 holes each, with their respective baskets where the seeds will be deposited for the growth of vegetables, on the other hand, the greenhouse has a 250-liter tank, which houses a PH, conductivity and water level sensor, a 10 W aeration pump, a 15w ejection pump and a 1 HP electric pump. As can be detailed in figure 8 and 9.

Figure 8. Final external structure of the hydroponic greenhouse prototype.

Figure 9. Final internal structure of the hydroponic greenhouse prototype.

In figure 10, we can see the distribution of the sensors.

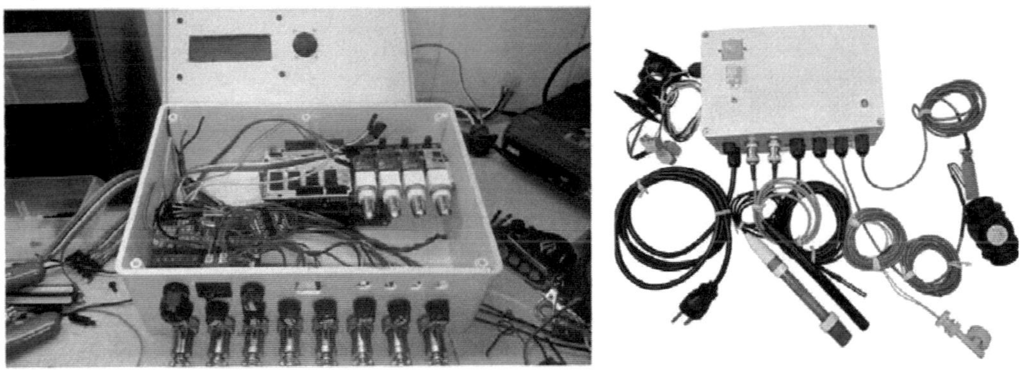

Figure 10. Organization and location of the hydroponic farm sensors.

4. Conclusions

The automation in the production of hydroponic crops entails multiple benefits in the generation of healthy organic food, rich in nutritional content and with a high yield, the initial investment in these processes may be expensive, but it is rewarded in a higher productive capacity Since you are working all year round and without the restrictions imposed by the weather. These climatic restrictions are overcome thanks to the set of control measures, which are more efficient by having a set of sensors and actuators that automatically contribute to keeping production in optimal conditions and at the same time keeping the farmer informed, supporting the taking of decisions of the same.

By automating this process,it also optimizes the use of resources and energy, since it helps to reduce the man-hours invested in the care of the plants, only the necessary amount of inputs that the system requires would be used, the total energy used in the development of agricultural activity is reduced thanks to the use of a less amount of fossil fuels and there would be a greater homogeneity in production by having the whole system under the same conditions.

5. References

[1] Yamaguchi, S., Takahashi, Y., & Hayashi, T. (2018). Small Indoor Hydroponic System with Renewable Energy. 18th International Conference on Control, Automation and Systems (ICCAS) (págs. 313-318). IEEE.

[2] Beltrano, J., & Gimenez, D. O. (2015). Cultivo en hidroponía. Buenos Aires: Editorial de la Universidad de La Plata.

[3] Goldstein, H. (2018). The Green Promise of Vertical Farms. IEEE SPECTRUM, 50- 55.

[4] Alcaldía de Sincelejo. (2016). Plan de Desarrollo 2016 – 2019. "Ciudad con Visión".

[5] COLCIENCIAS. (2018). Libro Verde 2030. Bogotá.

[6] Real Academia española – RAE. https://dle.rae.es/?id=KLUycbV.

[7] Poorter, H., & Nagel, O. The role of biomass allocation in the growth response of plants to different levels of light, CO2, nutrients and water:a quantitative review. Functional Plant Biology, 27(12): 1191-1191. (2000). doi: 10.1071/PP99173_CO.

[8] Li, Y., Chen, H., Ji, H., Wang, S., Zhu, Z., & Wang, X. Effect of led supplemental illuminat ion on the growth of strawberry plants. In: 2012 Symposium on Photonics and Optoelectronics: 1-4. IEEE. (2012). doi: 10.1109/SOPO.2012.6270919.

[9] H. M. Resh, Hydroponics for the home grower. Anguilla: CRC Press. (2015). doi: 10.1201/b18069.

[10] M. Saaid, A. Sanuddin, M. Ali y I. Yassin. Automated pH controller system for hydroponic cultivation. In: 2015 IEEE Symposium on Computer Applications & Industrial Electronics (ISCAIE): 186-190. IEEE. (2015). doi: 10.1109/ISCAIE.2015.7298353

IOP Conf. Series: Materials Science and Engineering 1154 (2021) 012013 doi:10.1088/1757-899X/1154/1/012013

Artificial intelligence model for the prediction of malignant tumors using a set of medical images from mammography studies.

S A Sánchez [1], A D Morales[1], M F Arroyo [1] y H D Asís [1]

[1]Corporación Universitaria Antonio José de Sucre, Sincelejo, Colombia, Faculty of Engineering Science.
sergio_sanchez@corposucre.edu.co

Abstract. Currently, the diagnosis of tumors and malignant cells through imaging studies is a great challenge for expert medical personnel, due to the complexity of achieving an early prediction of cancer cells, which would allow to accelerate early medical treatments. Today, technologies have become a fundamental ally for the health sector, specifically the area of artificial intelligence, which has permeated many disciplines, generating important advances. Advances in parallel computing, GPU technology, and deep learning have made real-time image processing easier. The main objective of this research was to generate a deep learning model for the prediction of malignant cells in medical images of diagnosed mammograms. Using the previously trained model based on Faster R-CNN, with the ResNet function extractor. This model works in the Python programming language, using the Tensorflow framework and the OpenCv library. The algorithms were previously trained through the DDSM and MIAS open medical image databases, published on the web. This model not only focuses on recognizing and classifying malignant cells in the image, but also on the location of objects within it, appropriately drawing a bounding box. One of the latent challenges of these models since their inception has been the consumption of computing, but today they have been optimized so much that they allow freezing the pre-trained models by loading them in the memory of the devices, managing to use them in computers without GPUs. As a result, it was found that the Faster R-CNN method with the Resnet101 extractor offers great advantages of precision and speed when it comes to detecting malignant tumors, studies that can serve as a great contribution to the bets of this algorithm in the health sector.

1. Introduction

Today there is a great public health problem worldwide, many countries face great challenges due to high mortality rates, the spread of infectious and non-communicable diseases and poor reproductive health. In recent decades, great strides have been made in increasing life expectancy and reducing some of the most common causes of death. As a result, globally, 170 countries have set 17 sustainable development goals in order to have a better planet for future generations, specifically for this research we will be working on objective number three, health and wellness, which aims to decrease the rate of mortality in the countries by 2030 [1].

According to the Pan American Health Organization, in the Americas, cancer is the second leading cause of death, the types of cancer most frequently diagnosed among men are: prostate (21.7%), lung (9.5%), colorectal (8%), bladder (4.6%) and stomach (2.9%). Among women, the types of cancer with the highest incidence are: breast (25.2%), lung (8.5%), colorectal (8.2%), thyroid (5.4%) and cervical (3.9%).

The types of cancer with the highest mortality in men are: lung (19.6%), prostate (12.1%), colorectal (9.3%), liver (65) and stomach (5.4%). The cancers that cause more deaths among women are: lung (17.4%), breast (15.1%), colorectal (9.5%) and cervical (5.2%) [2].

According to the World Health Organization [3], each year in the Americas, more than 462,000 women are diagnosed with breast cancer, and almost 100,000 die from this disease, if current trends continue, by 2030, the number of women diagnosed with breast cancer is projected to increase by 34%. Several countries in Latin America and the Caribbean have some of the highest rates of risk of death from breast cancer, highlighting health inequities in the Region. Prevention strategies in developing countries are carried out in very advanced phases, due to the high costs of medical equipment, the absence of trained medical personnel and the scarce resources to acquire these studies.

Despite notable advances in diagnostic technology in recent years, as well as an increase in campaigns aimed at promoting self-examination, many breast cancer patients continue to receive the diagnosis in advanced stages of the disease. Some risk factors for breast cancer are known, however, in most affected women it is not possible to identify specific risk factors [4][5]. Comprehensive cancer control encompasses prevention, early detection, diagnosis and treatment, rehabilitation and palliative care. In Figure 1, we can see the proportion of new cases and deaths from cancer in the Americas.

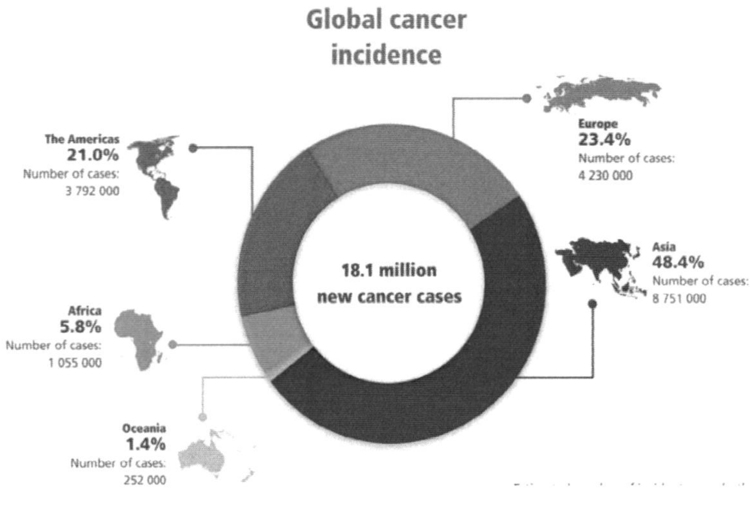

Figure 1. Incidence and mortality from breast cancer in the Americas

From the national level, according to the Ministry of Health with the support of the National Cancer Institute, the objectives of cancer control have been established, specifically in risk control, early detection, comprehensive treatment, palliative care, epidemiological surveillance, and talent development. human in oncology [6]. Presenting public policy and the 10-year cancer plan as an essential instrument. The number of sick and dead people from this cause has been increasing in recent years; the figures dictate that about 96 people die every day in Colombia from cancer. According to 2010 mortality figures, among women breast cancer was the leading cause of death (2,381), followed by cancer of the cervix (1,892), stomach (1,709), lung (1,606) and colon and rectum (1,456).

Among men, cancer mortality for the same year was led by malignant tumors of the stomach (2,796), followed by those of the prostate (2,431), lung (2,357), colon and rectum (1,261) and leukemias (890) [7].

From the local level, the population of the department of Sucre, like that of Latin America, has a significant incidence of chronic non-communicable diseases, in particular cardiovascular diseases, cancer, diabetes and chronic respiratory diseases. From the departmental administration they propose the initiative to make use of science, technology and innovation, which allow identifying and overcoming the social and environmental determinants related to the appearance of these diseases. Through alliances that include the public and private sectors, as well as NGOs, professional associations, the academic sector and international organizations [8].

In the department of Sucre, there are different health entities that have medical equipment to carry out radiological studies, but most of them are digitized, where the image is captured in analog form on a photostimulable phosphor plate (barium halide fluoride activated with europium impurities), compared to the most sophisticated radiology equipment, which comes with digital technology, which have introduced changes in the identification of suspicious images, digital storage to monitor patients, among others. This has allowed the generation of complementary decision-making tools when it comes to expert medical personnel making an accurate assessment.

Cancer is an aggressive disease with a low median survival rate. Treatment is long and very expensive due to its high recurrence and mortality rates. Early diagnosis and prediction of cancer prognosis are essential to improve the patient's survival rate. It exhibits a wide spectrum of behaviors, some are clinically indolent and others are aggressive, requiring prompt treatment. Today, doctors face different obstacles and challenges, such as the existence of less defined tumors, generating little visibility and distinction of cancerous tissues. This leads to false negative diagnoses, erroneous evaluations by medical personnel, when diagnosing a sick person as healthy, consequently, this can worsen the patient's health and delay treatment. Studies show that more experienced radiologists visually scan CT images more systematically [9]. The human field of vision may be short when making these diagnoses, since it only covers the equivalent of a circle of about 2.5 centimeters in diameter with high definition, and also for a defined time [10]. In figure 2, we can detail the width of the human visual field.

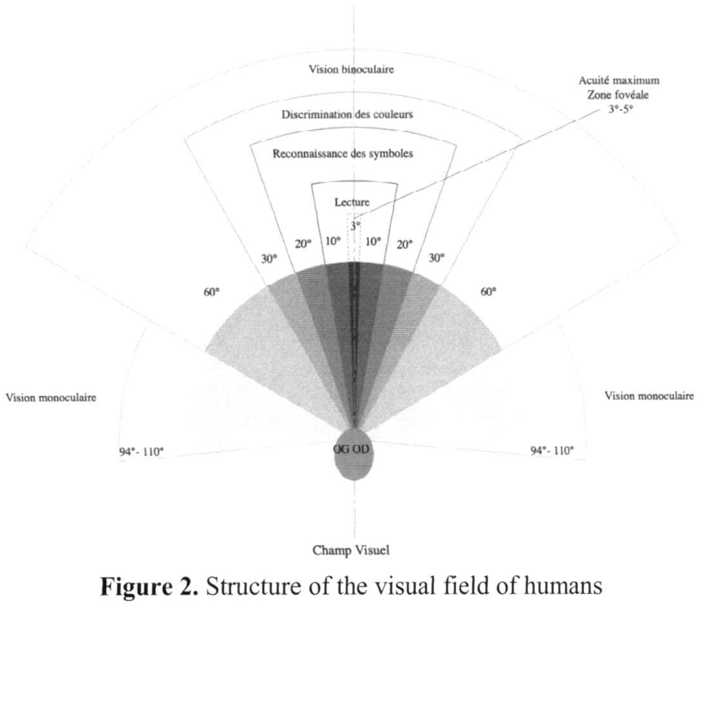

Figure 2. Structure of the visual field of humans

In figure 2, we detail the visual field of each individual where the objects located in the space can be seen. In which, there are differences in the more defined or less defined surface qualities depending on their distance and the quality of the colors is appreciated; if they are close, they are saturated; if they are far away, they are modulated and not modulated.

In relation to the above, technological advances will be a crucial alternative to develop methodologies that reduce mortality in cancer patients. This disease is difficult to diagnose in the early stages or can easily relapse after treatment. Furthermore, the precise predictions of the disease prognosis with high certainty are very complex. Some cancers are difficult to detect in their early stages due to their vague symptoms and distinctive indications from mammograms and scans [11]. Therefore, it is imperative to develop better predictive models using multi-variable data and high-resolution diagnostic tools in clinical cancer research. Artificial intelligence will be one of the best allies of medical professionals in the diagnosis and treatment of many diseases, such as cancer. Studies have shown that artificial intelligence algorithms have the ability to identify cell by cell and detect if it is a cancer cell or if it is a normal cell. And, through spatial correlation, it is able to make much more reliable predictions [12].

On the other hand, the creation of imaging techniques such as ultrasound, mammography, x-rays, computed tomography (CT), fMRI, magnetic resonance, or nuclear medicine, histology, and histopathology have marked a crucial point in the development of diagnoses of malignant cells in the medical sciences. The images generated by the medical teams of these modalities have two fundamental characteristics, compared to the analog radiography images, these being the spatial resolution (number of pixels per inch or cm) and the density or depth (gray levels that can be represent), who owns the image [13].

Within imaging tests for cancer diagnosis, artificial intelligence finds great utility in the execution of 3 main clinical tasks: (Detection, characterization, and monitoring of tumors).

Detection refers to the location of objects of interest on radiographs, collectively known as computer-assisted detection. The characterization broadly captures the segmentation, determining the shape or volume of the lesion, the histopathological diagnosis, and disease status or molecular profile. Segmentation defines the extent of an abnormality. Monitoring; It allows capturing a large number of discriminating characteristics, to verify the progression or regression of the disease and the monitoring of the appearance of a target or cancer mutations associated with resistance in near real time [14].

Although this technology is still in the testing phase and cannot be widely used at the moment, its implementation gives an idea of what the future may hold for the field of Artificial Intelligence applied to medicine. The researchers caution that these findings must be clinically validated in different types of patients.

The images today play a fundamental role in the process of medical care of patients, imaging studies have transformed modern medicine, to such an extent, from making diagnoses in acute pathologies.

Radiological and histopathological studies, which are part of imaging diagnosis, have become complementary tools that allow early detection of cancer cells in imaging tests, but today these evaluations are still short, due to the fact that medical opinions are highly dependent of specialist expertise.

The images provide valuable information for personalized medicine, becoming an important input for non-invasive medical procedures that do not involve instruments that break the skin or physically penetrate the body, which do not affect the integrity of people. These tools have strongly complemented the early detection of breast cancer patients.

The main causes of mortality in patients diagnosed with breast cancer lie in late diagnosis, high examination costs, extensive administrative processes, poor visibility and distinction of cancerous tissues.

Today, the integration of technology and computational methods have optimized early diagnosis, applications involving artificial intelligence technologies in medicine, imply a revolution in the way medical studies are perceived, this technology has left behind traditional prediction methods that are based on patient variables such as age, family history of cancer, or tissue density. However, these factors are weakly correlated with the possibility of developing cancer.

Collaboration between medical pathologists, oncologists, systems, electronic, and biomedical engineers has enabled the improvement of early cancer diagnoses, through the integration of technology and computational methods, in a new area known as digital pathology that optimizes traditional pathology.

The importance of carrying out this research lies in a non-invasive proposal related to a deep learning model for the detection of cancer cells in real time in breast tissues, with the aim of achieving early treatments in patients suffering from this pathology.

2. State of the art

A review of the state of the art was made from the international to the national level, of the investigations related to the use of deep learning algorithms for the prediction of malignant cells in mammographic images, finding the following studies:

Authors Scott Mayer McKinney, Marcin Sieniek, and Shravya Shetty of the Department of Surgery and Cancer, Imperial University of London in January 2020, conducted research titled "International Assessment of an Artificial Intelligence (AI) System for Cancer Detection breast", presenting a proposal as a complementary tool for expert medical personnel in the area, with the aim of making predictions of cancer cells using imaging studies. The authors evaluated the performance of the algorithm in a clinical setting with a representative dataset from the UK and the US They conducted an independent study with six radiologists, to assess the efficiency of the AI system, where it outperformed all expert medical personnel. The area under the receiver operating characteristic curve (AUC-ROC) for the AI system was greater than the AUC-ROC for the average radiologist by an absolute margin of 11.5%. Where they concluded that this technology can have a great impact on the health sector, specifically on the detection of different types of Cancer, which could help mitigate the high mortality rates from this cause [15].

On the other hand, the authors Rui Yana and Fei Ren from the area of Informatics and Technology of the University of Anhui, Hefei of China, in the year 2019, carried out an investigation entitled "classification of histopathological images of breast cancer using a deep neural network hybrid ", generating an artificial intelligence model to support medical personnel in histopathological diagnosis, due to the complexity of this type of study and the dramatic increase in workload, making this task time consuming, as a consequence, Results may be subject to the subjectivity of the pathologist. In this article, they propose a new hybrid recurrent convolutional deep neural network for the classification of histopathological images of breast cancer.

Based on multilevel feature representation of histopathological image patches, the method integrates the advantages of recurrent and convolutional neural networks, and short and long term spatial correlations between feature patches are preserved. The experimental results showed that the technique presented an efficient method in contrast to those of the state of the art with an average precision obtained of 91.3% for the classification task of 4 classes. Model training was performed with a dataset of 3,771 histopathological images of breast cancer, with variability in subclasses spanning different age groups, which provided sufficient diversity of data to alleviate the problem of relatively accurate classification. Low benign images [16].

The authors Andrea Duggento, Manuel Scimeca, Nicoletta Urbano and Elena Bonanno from the department of biomedicine and prevention at the University of Rome in Italy, developed an investigation entitled “a deep neural network of random initialization to discriminate malignant lesions from breast cancer”. In this context, the researchers affirm that the rates of false positives and false negatives when making a cancer diagnosis, commonly achieved by radiologists, are extremely high, and some authors have estimated figures of up to 20% of the total diagnosis or more. Hence, the great importance of computer-assisted diagnosis that has been driven by the introduction of deep learning techniques in general and convolutional neural networks (CNN) in particular. In this article, we propose a design and validation of an ad-hoc CNN architecture specialized in the classification of breast lesions and heuristic exploration on possible combinations of parameters and architecture styles to propose a model selection criterion that can emphasize reducing false negatives while maintaining acceptable accuracy. They achieved regular classification performance in the validation and test set, "Model 1" and "Model 2" achieved an AUC of 0.785 and 0.774, respectively. Demonstrating a moderate classification, demonstrating how an ad-hoc random initialization CNN architecture can provide practical help in the classification and staging of breast cancer [17].

Furthermore, the authors Zhen Zhang, Yaping Wang and Jiankang Zhang from the Zhengzhou University of China, developed an investigation entitled “comparison of multiple feature extractors in the fastest RCNN architecture for detecting breast tumors” The learning algorithm Deep shows great ability in pattern recognition tasks like object detection and speech recognition. Compared to typical machine learning methods that require manual feature extraction. The deep learning algorithm has more powerful feature learning ability. In this document, the deep learning associated object detection method is applied in the localization and classification of lesions for the detection of medical masses of mammary cells. At the same time, five network function extractors are used, which are ResNet101, V2 start, V3 start, Mobilenet and ResNet V2 start, which were investigated to explore the impact of the model. The dataset from the Digital Mammography Detection Database (DDSM) was used for research, with the objective of determining the performance of the models associated with the five feature extractors and comparing them separately in the detection of mammographic images. with benign and malignant tumors, according to the ROC compensation curves. The simulation results demonstrate that the classification model with the Inception ResNet V2 feature extractor had the best performance, compared to the other four feature extractors [18].

2.1 Comparison between deep learning algorithms for object detection

A state of the art review was performed to validate the performance of the Fast-RCNN, Faster-RCNN, R-FCN and SSD algorithms using different free databases on the web, where MS COCO, IMAGENET and PAS-CAL VOC stand out. with the purpose of reviewing speed and precision parameters of these methods using different image resolutions in different contexts.

It is important to note that technology is constantly evolving, any comparison can quickly become obsolete. An exhaustive review of scientific and academic documents on the performance of different object detection models with the Keras and TensorFlow framework was carried out, where the following studies were found;

In Table 1 the advantages and disadvantages of methods Fast R-CNN, Faster R-CNN, R-FCN, SSD and YOLO for detecting objects in images will be high- lighted by various experiments conducted by a variety of authors investigated in Deep Learning area.

Table 1. Advantages and disadvantages of some methods for detecting objects in images.

Method	Authors	Advantage	Disadvantages
Fast R-CNN	(Girshick, 2015) [19]	The calculation of the characteristics of CNN is done in a single iteration, achieving that the detection of objects is 25 times faster than the RCNN method (it requires 20 seconds on average to analyze an image).	The use of an external candidate region generator creates a bottleneck in the detection process.
Faster R-CNN	(Renet al., 2015)[20]	The RPN method allows object detection to be almost real-time, approximately 0.12 seconds per image.	Despite the efficiency of the algorithm, it is not fast enough to be used in applications that require real time, as would be the autonomous vehicles.
R-FCN	(Dai et al., 2016)[21]	The test time of R-FCN is much faster than that of R-CNN	R-FCN has a competitive mAP but lower than that of Faster R-CNN.
Mask R-CNN	(He et al., 2017)[22]	The location of the objects is more precise, when making a segmentation of the objects in the images.	Its execution time is greater than that used by the Faster-RCNN method, therefore, it can not be be implemented in applications that require real time.
YOLO	(Redmon et al., 2015)[23]	The location of objects is very efficient, allowing its use in real-time applications.	The method has difficulties to correctly detect small objects.
SSD	(Liu et al., 2016)[24]	The use of a single network, makes the location of the objects faster than the Fast-RCNN and Faster-RCNN methods.	The detection accuracy of the objects is lower compared to the Fast-RCNN and Faster-RCNN methods.

3. Methodology

In recent years, the architecture of the deep networks has been a significant progress for the moment Keras and TensorFlow dominates with different-pre- trained models already included in libraries, among these include VGG16, VGG1, ResNet50, Inception V3, Xception, MobileNet. The VGG and AlexNet 2012 net- works follow a typical pattern of classical convolutional networks. MobileNet is a simplified architecture Xception architecture, optimized for mobile applications. The following architectures; ResNet, Inception and Xception have become a reference point for subsequent studies of artificial vision and learning for its versatility Deep.

There are many factors that explain the revolution of deep learning, among these factors is highlighted; availability of sets of huge data and quality, parallel computing (GPU) features efficient activation for backpropagation, new architectures, new regularization techniques that allow train more extensive networks with less danger of overshooting, robust optimizers and software platforms with large communities like TensorFlow, Theano, Keras, CNTK, PyTorch, Chainer and Mxnet. All this has allowed solving problems easier. Today the Python programming language has great importance in Machine Learning compared to other languages because of its support for Deep learning framework.

Within this framework include TensorFlow which is a library of open source software for machine learning that allows you to deploy computing in CPU or GPU, developed by Google, using graphs flow data, PyTorch uses Python language and has the support of Facebook, Theano is a Python library that supports mathematical expressions involving tensioners, CNTK are a set of tools developed by Microsoft, open for Deep learning code, Keras is a library of neural networks high level created by Francis Chollet, member of Brain google equipment that lets you choose whether the models that are built will be executed in Theano, TensorFlow or CNTK. Keras and TensorFlow can construct models of three different ways; using a sequential model, a functional API and pre-trained models.

Earlier we talked about the different architectures (MobileNet, Inception, ResNet, among others), now we discuss models for object recognition and Keras TensorFlow; Faster-CCN R, R-FCN, SSD and YOLO. These models are classified based detectors in the region (Faster R-CNN, R-FCN, FPN) and single shot detectors (SSD and YOLO), start from different paths, but they look very similar now fighting for title faster and more accurate detector.

There are different metrics that can improve object detection algorithms based on more accurate positioning, faster speed and more accurate classification; metrics that stand out are: Intersection over Union (IoU), mean average precision (MAP) and rendered frames per second (FPS). Intersection over Union (IoU)It is an indicator that determines how close the predicted picture of the real picture. The average metric average accuracy (MAP) is the product accuracy and recovery detection bounding boxes. It's a good combined measure of how sensitive the network to objects of interest and how well it avoids false alarms. The higher the score the map, the more precise the network, but this has a cost of execution speed. Processed frames per second (FPS) is used to judge how fast is the system.

3.1 Datasets

For the training of the selected RCNN and SSD models, free databases of mammography medical images were used, already diagnosed by specialized medical personnel, specifically the digital database for detection mammography (DDSM) and the Society for the Analysis of Mammographic Imaging (MIAS).

3.1.1 DDSM - Digital Database for Screening Mammography. The Database (DDSM) is a resource for use by the mammographic imaging research community. Primary support for this project was a grant from the U.S. Army Medical and Materiel Research Command Breast Cancer Research Program.The DDSM project is a collaborative effort involving colleagues at Massachusetts General Hospital (D. Kopans, R. Moore), the University of South Florida (K. Bowyer), and Sandia National Laboratories (P. Kegelmeyer). The primary purpose of the database is to facilitate sound research in the development of computer algorithms to aid in detection. Secondary purposes of the database may include the development of algorithms to aid in diagnosis and the development of teaching or training aids. The database contains approximately 2,500 studies. Each study includes two images of each breast, along with some associated information about the patient and the diagnosed pathology [25].

3.1.2 Mammographic Image Analysis Society (MIAS). It is an organization of UK research groups interested in understanding mammograms and has generated a database of digital mammograms. Images taken from the UK's National Breast Detection Program have been digitized to a 50 micron pixel edge with a Joyce-Loebl scanning microdensitometer, a linear device in the 0-3.2 optical density range and representing each pixel with an 8-bit word. The database contains 322 digitized images. All images are diagnosed by expert medical personnel, with all locations of any abnormalities that may be present. All images have a dimension of 1024x1024 [26].

4. Results

For the development of this research, the deep learning method Faster-RCNN was used with the ResNet50 architecture, supported under the TensorFlow framework, the Python programming language and the OpenCV library. This model was trained on a Linux server with 32 cores and 93 GB RAM, where 10,000 medical images of mammograms diagnosed with malignant and binary tumors from the DDSM and MIAS databases were used, with different resolutions. The cross-validation technique was 85% of the images for training and 15% for validation. Deep convolutional neural networks were successfully applied to localize calcifications and masses on mammography images using the BI-RADS criteria. During the training there was a problem with the images with high resolutions, to solve this problem, smaller patches of the main images were formed, with the objective of improving the detection of the algorithms and increasing the data. As can be seen in figure 3 and 4.

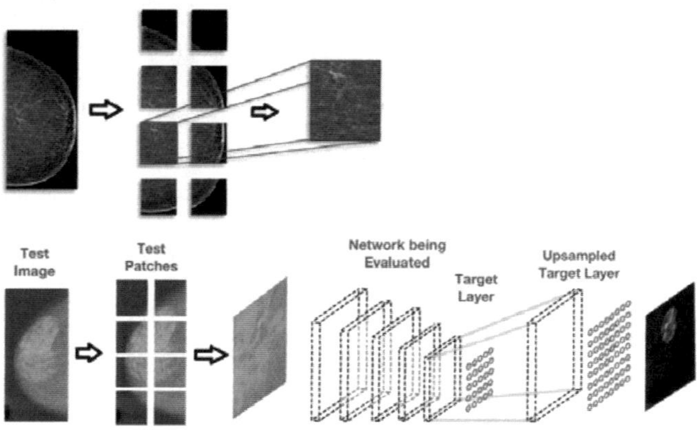

Figure 3. Evaluation of neural networks in image patches [27].

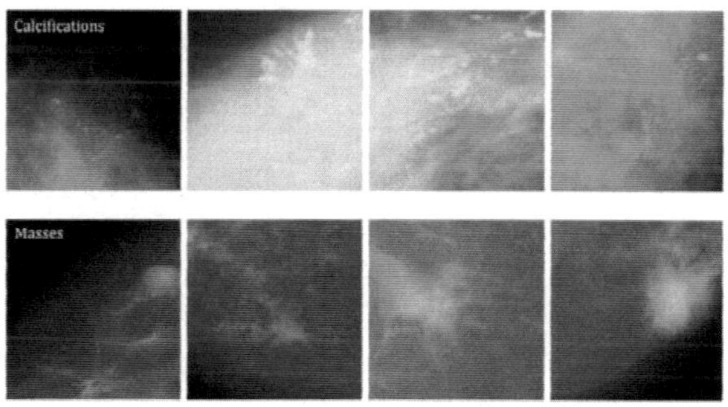

Figure 4. Examples of highest scoring patches for malignancy [27].

For the following investigation, the following block diagram was used to detect malignant tumors in mammography images.

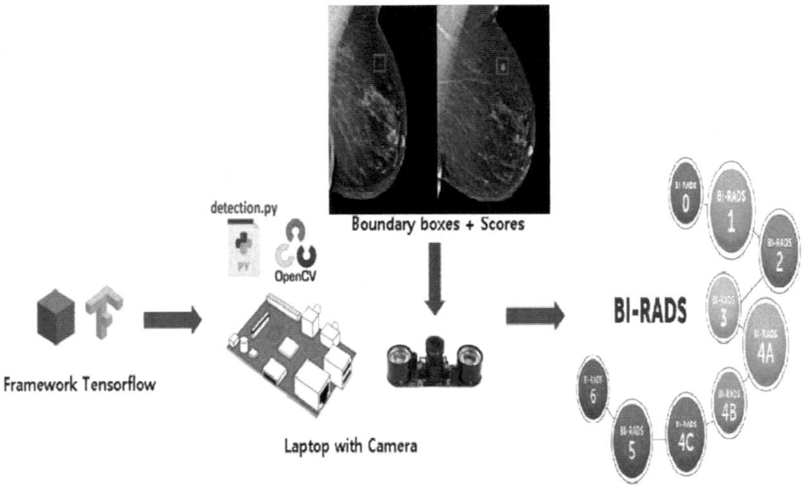

Figure 5. Block diagram of the system

The trained algorithm is capable of detecting masses and classifications in order to verify the probability of detecting a malignant or benign tumor, as we can see in Figures 6-8.

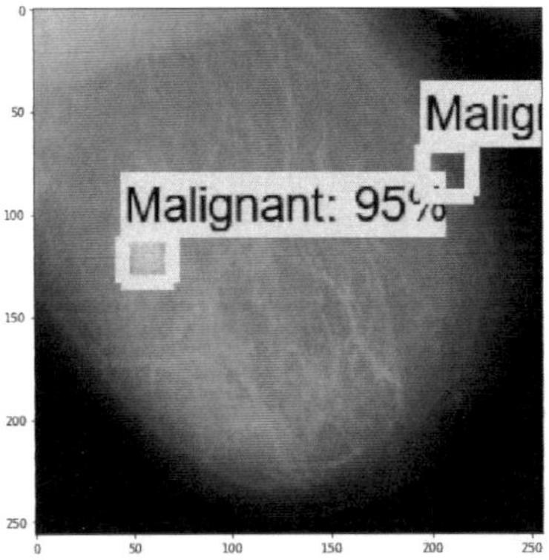

Figure 6. Detection of malignant tumors with the Faster-RCNN model and the Resnet50 extractor.

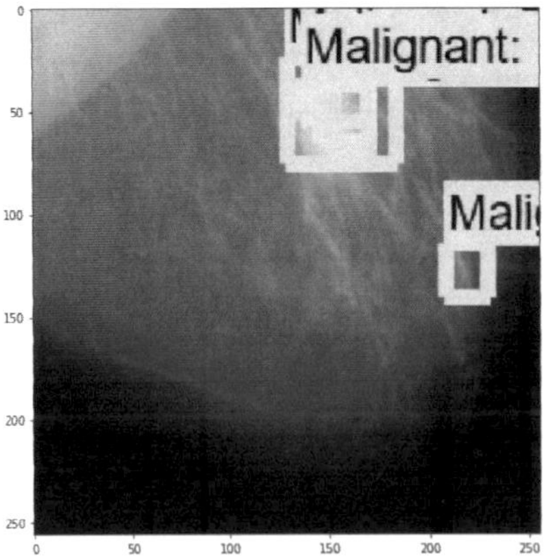

Figure 7. Detection of malignant tumors with the Faster-RCNN model and the Resnet50 extractor.

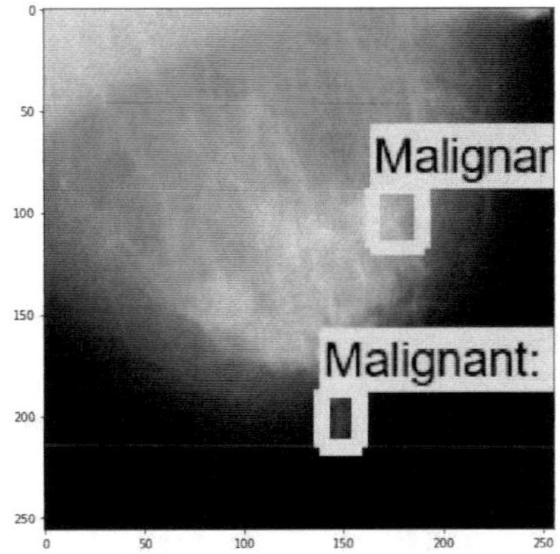

Figure 8. Detection of malignant tumors with the Faster-RCNN model and the Resnet50 extractor.

In figures 6,7 and 8 we can see the effectiveness of the Faster-RCNN model with the Resnet50 extractor, at the time of detecting malignant tumors, for this case images already previously diagnosed in the DDSM and MIAS databases were used. It is important to highlight that the more diagnosed images used to train the model, the better the algorithm's behavior.

5. Conclusion

We can conclude that the Faster-RCNN algorithm with the Resnet50 feature extractor is a good predictor of tumors in mammography images, but its functionality depends on the quantity and quality of the images with which the algorithm is trained, which shows us the great importance of this type of research, to complement the medical diagnoses of breast cancer, which has a high incidence and deaths worldwide.

It is important to highlight from this research that, depending on the type of application, a faster, slower or more precise model should be selected. In addition, combinations can be made between feature extractors and other parameters mentioned above, which will allow better performance depending on the application.

The difference between the detectors is narrowing. Single shot detectors use more complex designs to increase their accuracy, and region-based detectors speed up operations to be faster. There are many challenges to be overcome in object detection methods, deep learning will have a more prospective future in a wide range of applications in the field of medicine.

As future work, it is proposed to train other models and architectures using boxes and masks not only to locate the tumor within the mammogram, but also to create a mask with its shape. In this way, the tumor size can be estimated more precisely and taken into account in the decision-making task.

6. References.

[1] PNUD, 2019. Control del cáncer: conocimiento en acción: guía de la OMS para programas efectivos: detección temprana.
[2] PAHO, 2020. Organización Panamericana de la Salud: programa de cáncer.
[3] OMS, 2019. Control del cáncer: guía de la OMS para programas efectivos: detección temprana.
[4] IARC, 2018. Informe mundial sobre el cáncer, Agencia Internacional para la Investigación del Cáncer.
[5] Lacey JV Jr. y col, 2019. Epidemiología del cáncer de mama según factores de riesgo de cáncer de mama reconocidos en la cohorte de ensayos de detección de cáncer de próstata, pulmón, colorrectal y ovario (PLCO). BMC Cancer, 9, 84.
[6] Minsalud, 2019. Plan Decenal para el Control del Cáncer en Colombia, 2012-2021. Bogotá, Colombia.
[7] PND. Pacto por Colombia, pacto por la equidad: Plan nacional de desarrollo 2018-2022.
[8] Sucre. Plan departamental de desarrollo: 2016-2019.
[9] Alcalde, 2019. Inteligencia Artificial contra el cáncer de pulmón.
[10] Pubmed, 1981. Defectos del campo visual. PMID: 7258077. https://www.ncbi.nlm.nih.gov/pubmed/7258077.
[11] Huang, S., Yang, J., Fong, S., & Zhao, Q, 2019. Artificial intelligence in cancer diagnosis and prognosis: Opportunities and challenges. Cancer Letters. https://doi.org/10.1016/j.canlet.2019.12.007.
[12] Novartis, 2019. La inteligencia artificial decodifica las imágenes de patología del cáncer. [Recuperado el 14 de feb. de 2020]. https://www.novartis.es/stories/saber-mas/la-inteligencia-artificial-decodifica-las-imagenes-de-patologia-del-cancer.
[13] Castro, 2015. Algoritmo para la identificación de nódulos pulmonares solitarios en imágenes de tomografía de tórax. RCIM vol.7. versión On-line ISSN 1684-1859.
[14] Bi, W. L., Hosny, A., Schabath, M. B., Giger, M. L., Birkbak, N. J., Mehrtash, A., … Hoffmann, U, 2019. Artificial Intelligence in Cancer Imaging: Clinical Challenges and Applications, 69(2), 127–157. https://doi.org/10.3322/caac.21552.
[15] Mckinney, S. M., Sieniek, M., Godbole, V., Godwin, J., Antropova, N., Ashrafian, H., … Karthikesalingam. A, 2020. International evaluation of an AI system for breast cancer screening. Nature, 577(January). https://doi.org/10.1038/s41586-019-1799-6.
[16] Yan, R., Ren, F., Wang, Z., Wang, L., Zhang, T., Liu, Y., & Rao, X, 2019. Breast cancer histopathological image classification using a hybrid deep neural network. Methods, (February), 1–9. https://doi.org/10.1016/j.ymeth.2019.06.014.
[17] Duggento, A., Scimeca, M., Urbano, N., Bonanno, E., Aiello, M., Cavaliere, C., … Member, S, 2019. A random initialization deep neural network for discriminating malignant breast cancer lesions. 41st Annual International Conference of the IEEE Engineering in Medicine and Biology Society (EMBC), 912–915.
[18]Zhang, Z., Wang, Y., Zhang, J., & Mu, , 2019.. Comparison of multiple feature extractors on Faster RCNN for breast tumor detection. 8th International Symposium on Next Generation Electronics (ISNE), 1–4.
[19] R. B. Girshick, 2017. Fast R-CNN. CoRR, abs/1504.08083. 4, 5, 6

[20] Ren, S., He, K., Girshick, R., & Sun, J, 2017. Faster R-CNN: Towards Real-Time Object Detection with Region Proposal Networks. IEEE Transactions on Pattern Analysis and Machine Intelligence, 39(6), 1137–1149. https://doi.org/10.1109/TPAMI.2016.2577031.

[21] Dai, J., Li, Y., He, K., & Sun, J, 2017. R-FCN: Object Detection via Region-based Fully Convolutional Networks. Retrieved from http://arxiv.org/abs/1605.06409.

[22] Kaiming et tal, 2017. Mask R-CNN. https://arxiv.org/abs/1703.06870.

[23] Redmon, J., & Farhadi, A, YOLO9000: Better, Faster, Stronger. Pp-7.

[24] Liu, W., Anguelov, D., Erhan, D., Szegedy, C., Reed, S., Fu, C., & Berg, A. C, 2016. SSD : Single Shot MultiBox Detector, pp-5.

[25] USF, 2018. Computer Vision and Pattern Recognition Group, College of Engineering.Access: http://marathon.csee.usf.edu/Mammography/Database.html.

[26] J Suckling et al, 1994. The Mammographic Image Analysis Society Digital Mammogram Database Exerpta Medica. International Congress Series 1069 pp375-378, Access: http://peipa.essex.ac.uk/info/mias.html

[27] Wu et tal, 2018. Expert identification of visual primitives used by CNNs during mammogram classification. [Recuperado el 14 de feb. de 2020]. https://arxiv.org/abs/1803.04858

IOP Conf. Series: Materials Science and Engineering 1154 (2021) 012014 doi:10.1088/1757-899X/1154/1/012014

Application of 316L stainless steel coating by Directed Energy Deposition process

G A Barragan[1,2], F Mariani[2] and R T Coelho[2]

[1]Grupo de investigación en Ingeniería Aeroespacial, Universidad Pontificia Bolivariana, Circular 1 # 70 -01, Medellín - Colombia.
[2]Laboratory for Advanced Process and Sustainability - LAPRAS, Engineering School of Sao Carlos, University of Sao Paulo, 400 Trabalhador Sao Carlense ave, Sao Carlos, Brazil.

E-mail: german.barragan@upb.edu.co

Abstract. The corrosion problem faces a challenge for components in different industries. This research project aims to produce AISI 316L coatings in AISI 1045 plates of steel through additive manufacturing (AM) using a laser powder-fed system Direct Energy Deposition (DED) method. The coatings produced will be characterized using microstructural techniques (SEM/EDX and laser confocal), as well as microhardness Vickers. The average hardness value of the coating surface was 220 HV. The use of DED process for the coat of AISI 1045 material with AISI 316L has been proved as an efficient and viable operation.

1. Introduction

The materials employed during the build of a process plant or petrochemical equipment require different properties, among which are mechanical performance, resistance to corrosion, fabricability, and availability stand out due to the characteristics of its operation (highly corrosive and abrasive media) [1]. In many cases, the application itself dictates what materials are needed. Commonly employed materials include, between others, cast iron, alloyed steels, aluminium, titanium alloys, ceramics, and polymers.

The corrosion problem will be given more priority concerning materials for chemical process industries, than in other industries [2]. Carbon steels are widely used for either structural purposes or process equipment purposes [3]. Nevertheless, corrosion is its primary cause of failure [4], [5] with a substantial economic and environmental impact [6].

Directed Energy Deposition (DED), defined as "a category of additive manufacturing (AM) processes in which material is deposited onto a substrate employing the fusion and formation of a melt pool caused by a focused energy source" [7]. It is also used for coating metals with a variety of corrosion resistant or refractory materials [8]. The utilization of such technique in coating and repairing has many advantages, such as a reduced heat-affected zone and part distortion [9], [10].

AISI 316L manufactured by AM methods have been the focus of multiple studies covering aspects such the mechanical properties [11], [12], [13], [14], [15] and corrosion behaviour [16], [17] founding properties comparable with the cast and annealed wrought materials [18], [19], [20], [21].

During this research, a 316L stainless steel powder was deposited over a substrate of AISI 1045 steel employing the DED technique with laser as an energy source. The coating quality and the viability of its utilization for the corrosion improvement method for carbon steels in different applications such as the petrochemical industry was explored.

2. Materials, Equipment, and Methods

2.1. Materials

AISI 1045 drawn bar was selected as a substrate on the presented tests. As feedstock material, LPW 316, an atomized AISI 316L stainless steel powder, was employed with a normal distribution of the particle sizes, varying from 47.9 µm to 108µm. The materials were selected because the coating of carbon steels with stainless steels is a potential application of the DED process to improve its corrosion behaviour. Figure 1 shows an image of the powder used obtained by a laser confocal microscope Olympus LEXT OLS4100. It is possible to notice that spherical is the predominant particle geometry.

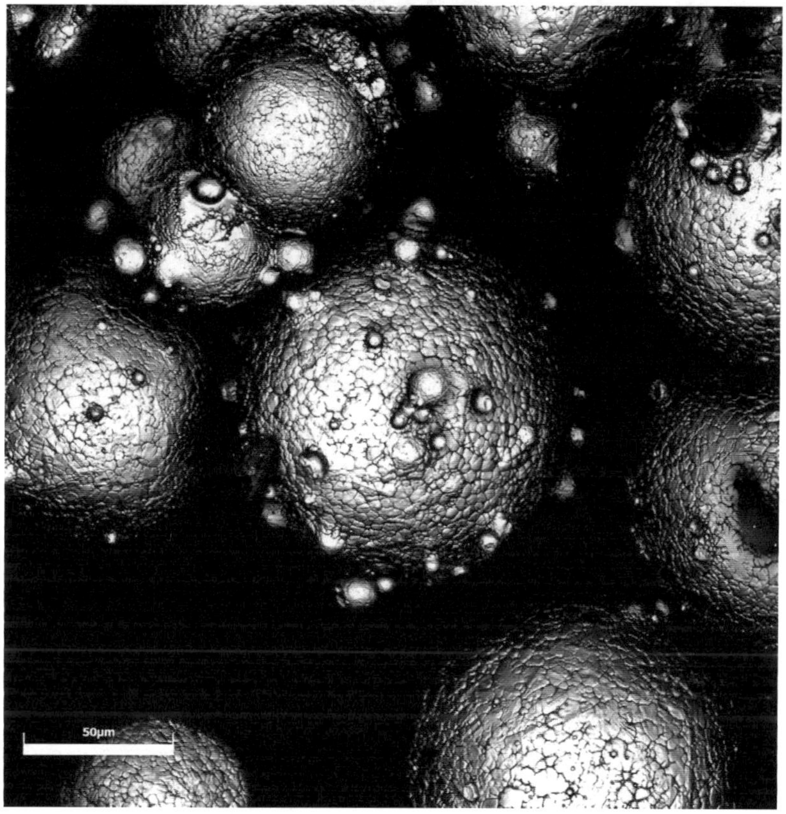

Figure 1. 316 L Stainless steel powder

Table 1 describes a typical chemical composition of the materials used in the here presented research.

Table 1. Material composition (wt.%) [22].

	Fe	C	Cr	MN	Ni	Cu	Mo	Si	S	P	N
AISI 1045	Bal.	0,43 – 0,5		0,06 – 0,09							
AISI 316L	Bal.	0,03	18 - 19	1,0 – 2,0	0,1	1,0	2,25 – 2,5	0,20 -0,75	0,02	0,04	0,1

2.2. Equipment
For the deposition of the coating, the ROMI DCM 620-5X Hybrid was used. It is a typical machining center, capable of metal cutting and equipped with an AMBIT metal deposition head, supplied by Hybrid Manufacturing Technologies, installed besides the regular machine spindle. The laser head is powered by a 500 W Nd: YAG continuous laser operating with 1.070 nm of wavelength. The material was delivered by a powder feeder Metco Twin 150. Powder feed rate was controlled by a rotation metering disk. Additional technical information available in [23], [24].

Figure 2. Deposition head - ROMI DCM 620-5X Hybrid [24]

2.3. Methods
The substrate material was face milled by the machine to obtain a flat surface and remove any existing corrosion film, avoiding possible interference on the coating results. Next, using the deposition head, the focused laser beam produced the melt pool of the carbon steel material at the surface of the workpiece. Then, stainless steel powder coating material is added simultaneously and melted by the laser to form a deposition track.

The deposition experiments consisted of a square surface coating of 225 mm2 employing multiple tracks to obtain a 316L thickness material of 1,5 mm over the substrate. At the present experiments, zigzag strategy was used to cover the entire area, with 90° rotations for each successive layer deposition. A similar strategy was employed by Gong et al, [25]. Using the experience of preceding experiments; the process parameters were selected to obtain an adequate track geometry for the coating procedure. Feed speed 350 mm/min, the feedstock material delivery rate of 5,5 g/min and laser power 300 W, they were not altered during layer stacking. The process was carried with a protective argon flux delivered by the deposition head in ambient atmospheric conditions.

Microhardness test, optical and scanning electron microscopy evaluation were carried out on the samples. The metallographic workpiece was mechanically polished and etched with Aqua Regia reagent.

3. Results and discussions
The sample analysed to assess the quality of the coating and the viability of its utilization for the corrosion improvement method on carbon steels in different applications such as the petrochemical industry. Figure 3 shows an SEM image of a sample cross-section. Three distinctive regions were observed: the substrate, HAZ zone and the coating. The coating tracks seems to be well attached to each other and with the substrate material in a metallurgical way. No significant defects or porosity could be found.

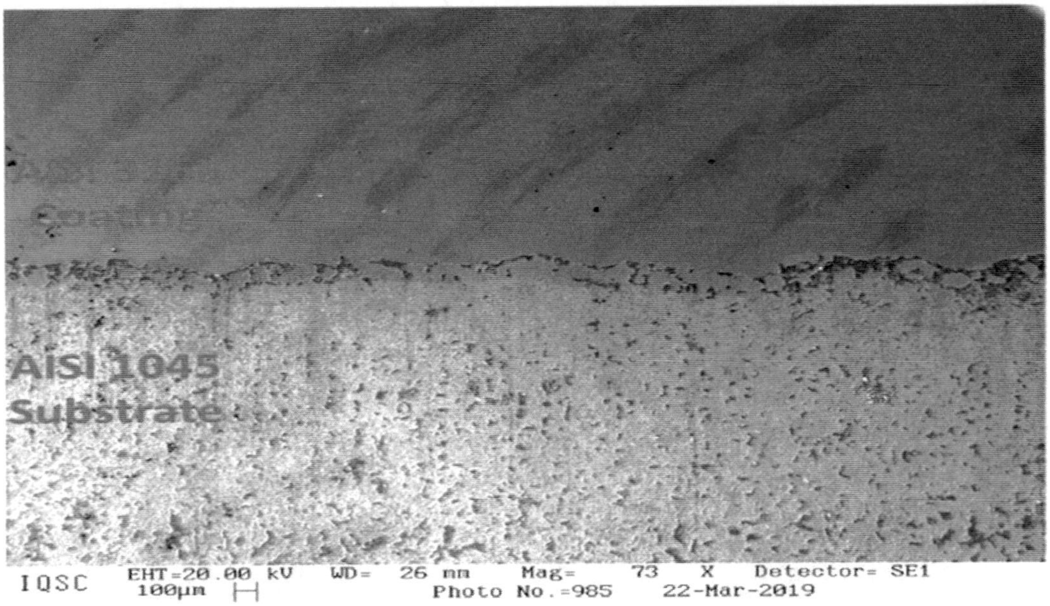

Figure 3. SEM image sample cross-section

An X-ray energy dispersive spectroscopy (EDX) equipment attached to the SEM enabled the chemical analysis quantification of the coating (Figure 4). The objective of such analysis was to check if there were any chemical element lost or diffusion during the deposition process that can reduce the corrosion resistance of the coating material. A stable condition of Cr and Ni was found as an important situation when consider that they are responsible to provide the good corrosion behavior of the 316L stainless steel. Lost, or dilution, of those elements must be avoided.

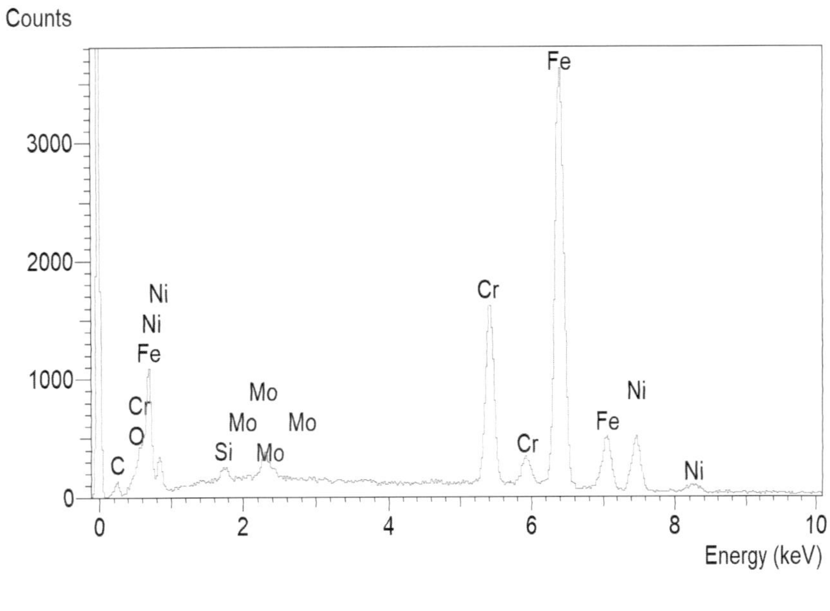

Figure 4. AISI 316L coating chemical analysis

The grain size and microstructure features show a direct relationship with the thermal history occurred during the deposition process under the DED technique. The laser-material interaction time is very short and high cooling/solidification rates are to be expected. Figure 5 shows a typical microstructure optical image of the deposited coating. The observed solidification is the type A. The microstructure is totally austenite, exhibiting cells and dendrites solidification structures.

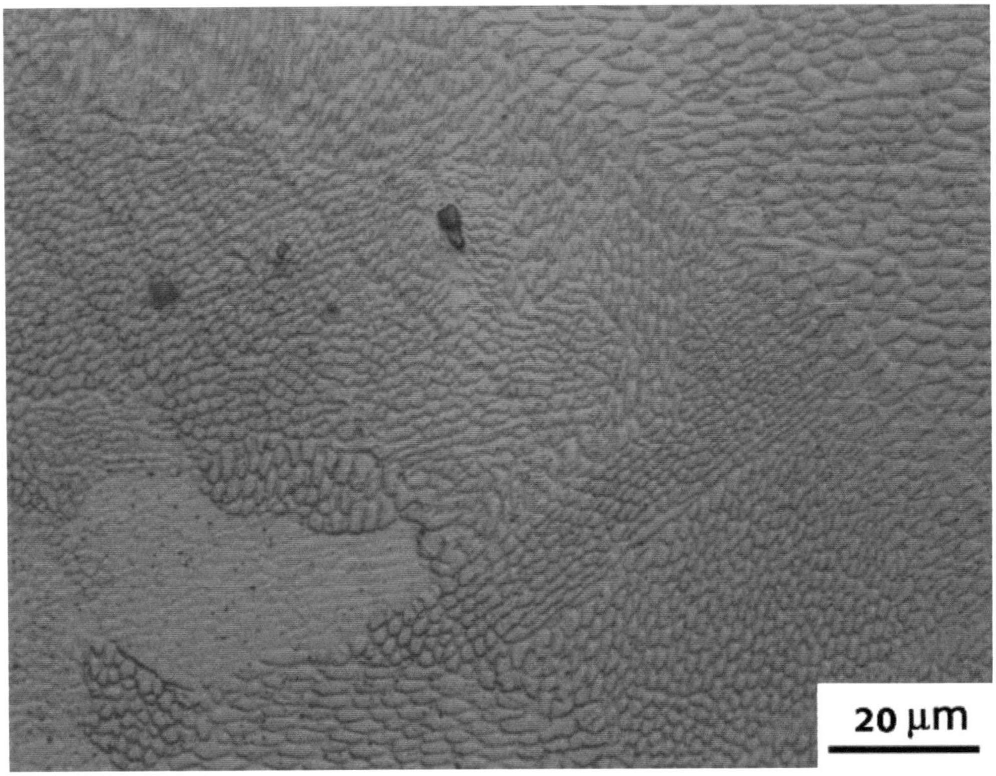

Figure 5. Typical microstructure image of the 316L coating

Microhardness Vickers is an essential mechanical performance indicator. The average workpiece hardness was 220 HV, acquired through averaging 10 measurements at different locations of the coating cross-section. Ahsan, et al. [26] reported a similar hardness value for AISI 316L laser deposited material. A substantial increase of hardness value was evidenced in the HAZ reaching 240 HV. Due to the high cooling rate, in this region, it is plausible to find a hard-martensitic microstructure.

4. Conclusions

The AISI 316L coating was mainly motivated to find a method that offers corrosion protection to the AISI 1045 metal substrate when come into contact with harmful agents, aiming at improvements in workpiece useful life.

The use of Directed Energy deposition process for coating of AISI 1045 material with AISI 316L has been proved to be a viable and feasible process. No considerable defects or changes in coating chemical composition was found, specifically on Cr and Ni contents.

The coating presented a typical microstructure containing austenite and the hardness values show a slight increase, compared with the traditional manufactured 316L.

Low porosity rates presented in the coating could be considered as a good indication of the protective capacity provided by the applied application method.

Acknowledgments
This study was partially financed by the São Paulo Research Foundation - (FAPESP) – grants 2016/11309-0; 2019/26362-2; Coordenação de Aperfeiçoamento de Pessoal de Nível Superior - Brazil (CAPES) - Finance Code 001; UPB training agreement - CDG 19.

References

[1] B. Raj, U. K. Mudali, T. Jayakumar, K. V. Kasiviswanathan and R. Natarajan, "Meeting the challenges related to material issues in chemical industries," *Journal of the Indian Academy of Sciences, SADHANA,* vol. **25**, no. 6, p. 519 – 559, 2000.

[2] L. Garverick, Corrosion in the petrochemical industry, ASM international, 1994.

[3] K. Elayaperumal and V. S. Raja, "Materials of Construction for Chemical Process Industries," in *Corrosion Failures: Theory, Case Studies, and Solutions*, John Wiley & Sons, Inc, 2015, pp. 75 - 110.

[4] K. Kidam and M. Hurme, "Analysis of equipment failures as contributors to chemical process accidents," *Process Safety and Environmental Protection,* vol. **91**, no. 1 - 2, pp. 61-78, 2013.

[5] A. S. Makhlouf and M. Aliofkhazraei, Handbook of Materials Failure Analysis with Case Studies from the Chemicals, Concrete and Power Industrie, Butterworth-Heinemann, 2015.

[6] G. Schmitt, "Global needs for knowledge dissemination, research, and development in materials deterioration and corrosion control," World Corrosion Organization, 2009.

[7] ASTM, "ASTM F2792-12a, Standard Terminology for Additive Manufacturing Technologie," 2015.

[8] W. Li and M. Soshi, "Modeling analysis of grain morphologies in Directed energy deposition (DED) coating with different laser scanning patterns," *Materials Letters,* vol. **251**, no. 15, pp. 8 - 12, 2019.

[9] X. Zhang, W. Cui, W. Li and F. Liou, "Effects of tool path in remanufacturing cylindrical components by laser metal deposition," *The International Journal of Advanced Manufacturing Technology,* pp. 1 - 11, 2018.

[10] X. Zhang, W. Li, K. M. Adkison and F. Liou, "Damage reconstruction from tri-dexel data for laser-aided repairing of metallic components," *The International Journal of Advanced Manufacturing Technology,* vol. **96**, p. 3377–3390, 2018.

[11] I. Tolosa, F. Garciandía, F. Zubiri, F. Zapirain and A. Esnaola , "Study of mechanical properties of AISI 316 stainless steel processed by "selective laser melting", following different manufacturing strategies," *The International Journal of Advanced Manufacturing Technology,* vol. **51**, p. 639–647, 2010.

[12] A. Zhukov, A. Deev and P. Kuznetsov, "Effect of Alloying on the 316L and 321 Steels Samples Obtained by Selective Laser Melting," *Physics Procedia,* vol. **89**, pp. 172 - 178, 2017.

[13] P. A. Kuznetsov, A. Zisman, S. Petrov and I. Goncharov , "Structure and mechanical properties of austenitic 316L steel produced by selective laser melting," *Russian Metallurgy (Metally),* p. 930 – 934, 2017.

[14] C. Wang, X. Tan, E. Liu and S. B. Tor, "Process parameter optimization and mechanical properties for additively manufactured stainless steel 316L parts by selective electron beam melting," *Materials & Design,* vol. **147**, pp. 157-166, 2018.

[15] M. Mukherjee, "Effect of build geometry and orientation on microstructure and properties of additively manufactured 316L stainless steel by laser metal deposition," *Materialia,* vol. **7**, 2019.

[16] M. Lodhi, K. M. Deen and W. Haider, "Corrosion behavior of additively manufactured 316L stainless steel in acidic media," *Materialia,* vol. **2**, pp. 111 - 121, 2018.

[17] M. Lodhi, M. Deen, M. C. Greenlee-Wacker and W. Haider, "Additively manufactured 316L stainless steel with improved corrosion resistance and biological response for biomedical applications," *Additive Manufacturing,* vol. **27**, pp. 8 -19, 2019.

[18] J. Yu, M. Rombouts and G. Maes, "Cracking behavior and mechanical properties of austenitic stainless steel parts produced by laser metal deposition," *Materials & Design,* vol. **45**, pp. 228 - 235, 2013.

[19] T. Amine, J. W. Newkirk and F. Liou, "An Investigation of the Effect of Direct Metal Deposition Parameters on the Characteristics of the Deposited Layers," *Case Studies in Thermal Engineering,* vol. 3, pp. 21-34, 2014.

[20] P. Guo, B. Zou, C. Huang and H. Gao, "Study on microstructure, mechanical properties and machinability of efficiently additive manufactured AISI 316L stainless steel by high-power direct laser deposition," *Journal of Materials Processing Technology,* vol. **240**, pp. 12 - 22, 2017.

[21] M. Akbari and R. Kovacevic, "An investigation on mechanical and microstructural properties of 316LSi parts fabricated by a robotized laser/wire direct metal deposition system," *Additive Manufacturing,* vol. **23**, pp. 487-497, 2018.

[22] ASM , ASM Handbook, Volume 1, Properties and Selection: Irons, Steels, and High Performance Alloys, ASM International Handbook Committee, 1990.

[23] OERLIKON METCO, "Product Data Sheet Metco Twin 150 Powder Feeder," 2017.

[24] ROMI, "ROMI DCM 620-5X HYBR D Catalogue," 2019.

[25] Y. Gong, Y. Yang, S. Qu, P. Li, C. Liang and H. Zhang , "Laser energy density dependence of performance in additive/subtractive hybrid manufacturing of 316L stainless steel," *The International Journal of Advanced Manufacturing Technology volume,* vol. **105**, p. 1585–1596, 2019.

IOP Conf. Series: Materials Science and Engineering 1154 (2021) 012015 doi:10.1088/1757-899X/1154/1/012015

Designing a system with machine learning to formative research projects management

C. Pineda[1], A. Morales[1], L. Lopez[1], C. Benitez[1] and S. Sanchez[1]

[1]Corporación Universitaria Antonio José de Sucre (Colombia).

carlos_pineda @corposucre.edu.co

Abstract. Artificial intelligence is marked among emerging technologies and is one of the most projected expectations for future dates. In the last decade this technology has been implemented in various sectors, without leaving the education and research sector, generating positive results. Among the most demanded algorithms are machine learning, because it gives the ability to "learn" to machines and information systems through experience and not rely only on data from system inputs in a given context. For this reason and also because of the difficulty that students and teachers express in the elaboration of training research documents, the objective of this research is to design a system that assists the user in the construction of the different sections of a formative research project. The research is descriptive and technological development. Among the results found is the behavior of the proposed system and identifies the different technologies to be used for future development. It is concluded that the system is structured by the MVC(Model View-Controller) architectural pattern which facilitates the scalability and maintenance of the system in the future, in addition to the implementation of the libraries of Python, NLTK(Natural Language Toolkit), Scikit-learn, Tensor Flow and Keras, allow the system to fully use machine learning for scientific and academic text prediction.

Keywords: *supervised machine learning, unsupervised machine learning, artificial intelligence, intelligent tutoring systems and formative research.*

1. Introduction

The Artificial Intelligence or AI, is considered to be the cognitive process simulation that influence the human decision making by computers or informatics systems [1]. For this these systems integrate mathematical modellings which through information banks detect patterns and this way these "understand" the behavior of context for which they were developed. In the last decades this technology has taken center stage in different sectors, generating expectations for the development and technology innovation of future years.

With regard to the field of education, AI(Artificial Intelligence) facilitates the automation of different processes supporting teaching and facilitating the learning process for students, ensuring that both the teacher and the student focus understand and generate new knowledges and they set aside operational and repetitive processes that do not contribute to vocational training. This aspect can benefit mainly to the field of formative and scientific research which are considered as one of the fundamental processes in the training of professional career and postgraduate students, because this wake up in the student curiosity and the searching and information classification skills for implement of innovation and development of new solutions for daily problems that allows the society progress. For this reason the Corporacion Universitaria Antonio José de Sucre, recognizes the importance of scientific and formative research, relying on bodies that manage this activity within the programs it offers. Among these bodies are the research and innovation center that relies directly on the research committees of the faculties, which in turn relies on the training research activities developed by teachers, which carry out pedagogical strategies such as the creation of research hotbed in an extracurricular way and development of classroom projects within some subjects, with the intention of fostering an investigative culture within the academic community.

Based on the above, the Corporacion Universitaria Antonio José de Sucre seeks to innovate in the creation of an information system that allows not only the management of research and academic documents generated by teachers and students, but also that the system is supported by Artificial Intelligence to automate processes of elaboration of the sections of research documents, for this reason the central objective of this document is to analyze and investigate articles research results of different authors of the last 5 years, all these related to artificial intelligence, and propose the architecture of a system and identify the technologies with which it should be developed. The document is initially organized by setting out the theoretical framework, then describes the research methodology and finally disseminates the results, discussions and conclusions.

2. Theoretical Framework

In this section sets out concepts and theories according to authors related to artificial intelligence, machine learning supervised and unsupervised systems.

2.1. Artificial Intelligence

The Artificial Intelligence (AI) is understood as the ability of computers or informatics systems to learn to make decisions. This technology allows the automation of various process that are carried out in different sectors in which the human being performs, which has spread rapidly in recent years, because more technology devices are incorporated every day that simplify the tasks that human beings develop. [2]

The AI is supported in mathematical models which make use of databanks with similar characteristics, which this technology detects by means of different techniques of identification of patterns using linear or nonlinear multivariate functions. These characteristics can be exploited through techniques that make use of a regression model of the process, and then be contrasted with the real process and get a residue; and techniques that directly model data output for both normal and abnormal operating conditions. [3]

2.2 Machine Learning

The Machine Learning, is a discipline in the field of Artificial Intelligence, that it has as purpose, generate the ability of learn for machines and informatics systems to resolve daily problems. Learning involves obtaining information to improve performance skills and abilities in a context, by identifying patterns in multi-data analysis, allowing machines to perform operations autonomously as they gain experience in the process. In the context of machines, learning is a set of algorithms that allow you to make decisions based on accumulated experiences to solve problems [4].

That is, machine learning is a discipline that arises to analyze data in an automated way, using the technique of recognizing patterns [5], generating the ability for realize generating the ability to make predictive judgments and make the best decisions by analyzing a large amount of existing data, for this reason it is argued that the key to machines acquiring artificial intelligence is through machine learning, the use of this technique has been developed in various sectors, from aspects such as Internet search, tasks that are commonly performed, to political procedures such as the United States presidential elections, to being applied in military decision assistants of developed countries; in other words, always that exist the necessity for data analysis, the machine learning can be used. [6]

From another perspective it can be said that machine learning does not depend of a programming for respond in a certain way according to the inputs received, but rather that utilize techniques to extract same characteristics of behavior from data input, and based on that information, carry out the evaluation of new entries. The intern algorithms that constitute the base of this machine learning have a strong statistic and algebraic component, with the consequent calculus ability [7].

2.3 Supervised and unsupervised machine learning algorithms
Among the machine learning algorithms the supervised and unsupervised algorithms were found. The unsupervised machine learning algorithms realize input-only processing. Unsupervised machine learning makes use of a dataset, which is not tagged or classified. The system is tasked with identifying patterns in pieces of data by recognizing similarities and grouping the data into categories. The unsupervised machine learning algorithms have the ability to update in relation to processing based only on inputs [7].

On the other hand, supervised machine learning algorithms have a manually classified dataset. On this particular set, the algorithm carries out two processes: find the best parameters for the algorithm, and evaluate the probability level with those parameters. Supervised learning is a technique for deducing a function from training or learning data [7].

2.4 Natural Language Processing
Natural Language Processing is one of many branches of knowledge from Artificial Intelligence, where it basically seeks to optimize and discover more thoroughly the way that the machines communicate with humans, using many different languages that exist in the world.

For the elaboration of the Natural Language Processing, it is necessary to eliminate what is known as stop words to then lemmatize the remaining words in the dataset. The stop words makes references to words that have no meaning, for example: prepositions (to, during, according to) or pronouns (he, she, and we) etc. While lemmatize makes references to the process for remove words that are in plural, female or conjugated, the purpose of this is to look for a word that represents all these words in the same form, which are the ones that will appear in the corpus. Based on this a dictionary is developed, with a number of keywords that will serve for the development of natural language. [7]

3. Methodology
This research is experimental design, this type of research has absolute control of the study variables considering that the research topic is projected to design the system architecture with machine learning. As the type of research is concerned, it is descriptive and technological development. Descriptive because the features and functionalities of the different Python libraries are described to suit the study context. According to Mario Tamayo and Tamayo in his work "the process of scientific research", he defines this type of research as the registration, analysis and interpretation of the current nature and the composition or processes of phenomena. The approach is made on dominant conclusions or on how a person, group or thing is conducted or works in the present [8]. Also, the research is technological development type, according to MinCiencias this type of study implements research results, for the design or development of new technological products, as well innovation in technological products. This activity will include the materialization of research results in a blueprint, scheme or design, as well the creation of non-marketable prototypes [9]. The purpose of this research is to generate the design of a computer system that integrates machine learning to support the development of the sections of formative research documents.

To realize the research, two phases were developed:

In the initial phase, a documentary analysis of the information was carried out by implementing the analytical matrix instrument of content where academic research results of topics related to artificial intelligence and machine learning were grouped identifying the functional and non-functional requirements of the proposed system.

Among the selected technologies is the Python programming language, a language that has taken center stage in the development of intelligent systems or algorithms, because it is intuitive and compatible with multiple platforms. Python integrates the NLTK (Natural Language Toolkit), Scikit-learn, Tensor Flow and Keras libraries, which have methods that allow to integrate machine learning activities.

In the second phase, the system architecture was designed by developing block diagrams, class diagrams, and flowcharts, which allow you to schematize the behavior that you want the proposed machine learning system to have, in addition to illustrating the relationship or integration of the different technologies that make up the system.

4. Results

Research has been conducted linking machine learning in artificial intelligence to education and analyzing the results of these researches, it has been proposed the design of a research project management system with artificial intelligence which includes machine learning. In Figure 1, is represented with a class diagram the structure in which the entire coding of the research project management system is composed, which it can also call Back End. In Figure 2, it was decided to represent using a block diagram, the architecture of the entire system, in this you will be able to observe the communication that exists between the different technologies used. Finally, it is shown in Figure 3, the behavior and processes that the artificial intelligence executes represented in a flowchart.

4.1 Defining how the Web Prototype Backend works

The following diagram illustrates the behavior of each of the Model-View-Controller sections that is implemented in the Web Prototype Backend (see Figure 1):

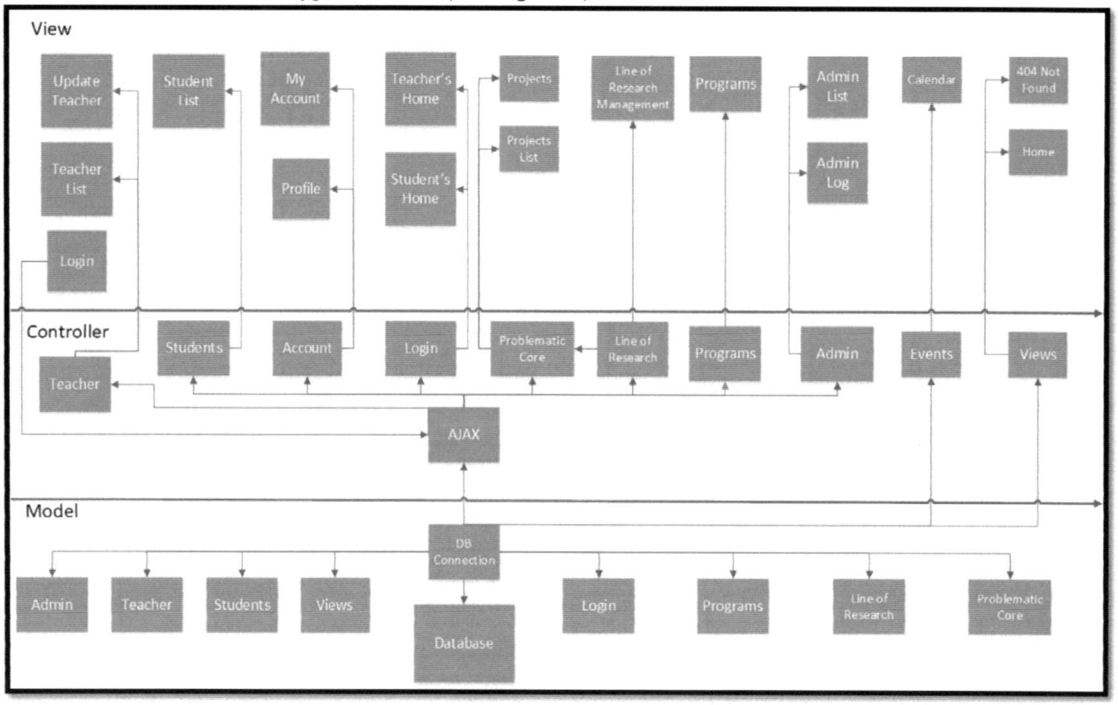

Figure 1. Back End class diagram.

The initial data flow when typing the URL (Uniform Resource Locator) in the browser, where the browser request runs to the web server, which immediately communicates with the view controller, which in turn is responsible for verifying that all data is properly hosted on the web server, redirecting to the page not found or the main interface of the website, from which the User Login can be accessed, at the time of login by user, the login controller communicates with AJAX (Asynchronous JavaScript

And XML), technology that is responsible for executing the actions of each of the main controllers, in this case it queries in the database and checks if the data is correct, it also checks the type of user that has been connected, and then redirects you to the home interface either the teacher or the student, if incorrect data is typed system would show the message "wrong user and/or password" and stay on the same interface.

AJAX (Asynchronous JavaScript And XML) is responsible for executing the tasks of each main controller, that is to say, AJAX has internal controllers for each of the controllers associated with this, and it also communicates directly with the database and with the client, AJAX is also responsible for running the interactive application, which will be the text writing section, this component allows interaction with the application and artificial intelligence asynchronously in the background, so the user would have no interruption when they are using the application.

Finally, there are multiple controllers for the different sections of the website, as they are, the controllers for teachers, students, programs, problematic cores, line of research, among other sections, so there is the possibility to find different views or interfaces depending on the section that the user wants to access.

4.2 Defining the Operation of the Web Prototype System

The overall operation of the system is as well as the task of each technology within the system below (see Figure 2):

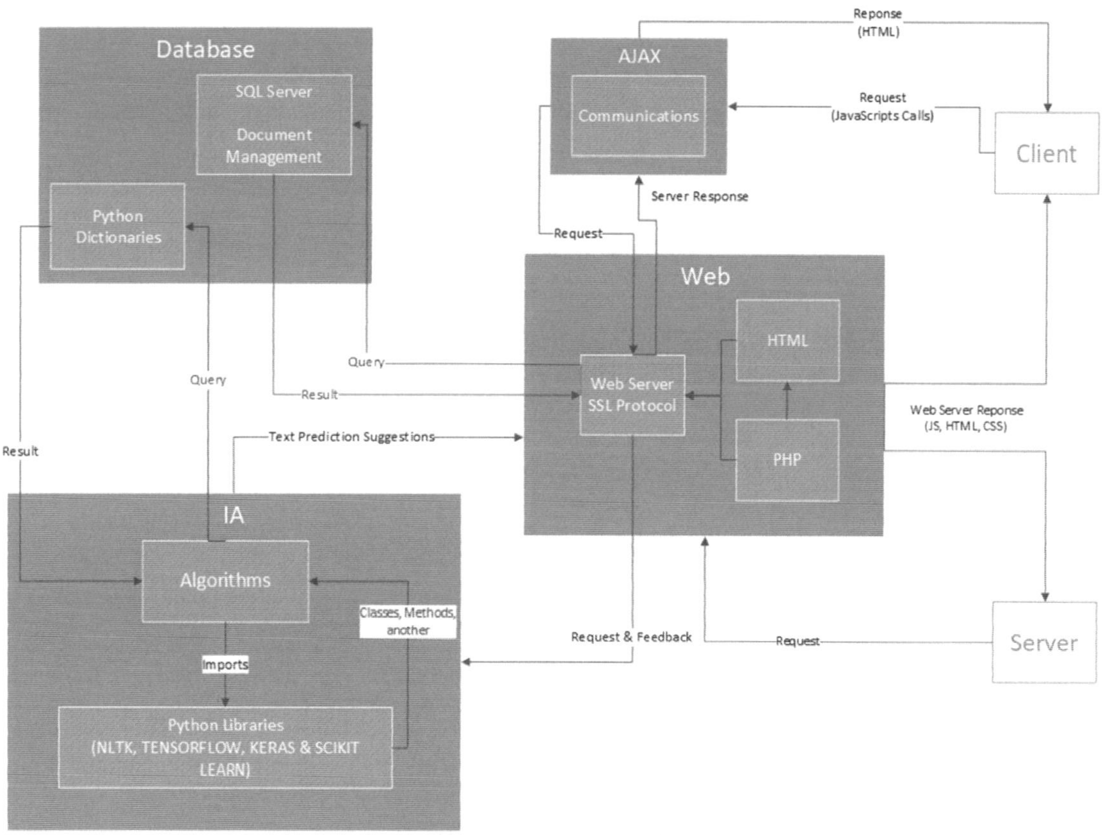

Figure 2. System block diagram.

Server: it is responsible for providing website services to the client and responding to requests that the client demands.

Client: it is the user who normally executes requests and performs activities on the website.

HTML: It is a markup language that is used for the development of Internet pages. This is the acronym for Hyper Text Markup Language.

PHP: (recursive acronym for PHP: Hypertext Preprocessor) it is a very popular open source language especially suitable for web development and that can be embedded in HTML.

Web Server: It is responsible for containing all the information of the website from the server side, it also takes care of the transfer of information by implementing the SSL protocol (Secure Sockets Layer) for the protection of information.

SSL Protocol: These are cryptographic protocols that provide privacy and integrity in communication between two points in a communication network. This ensures that the information transmitted by that network cannot be intercepted or modified by unauthorized elements, so ensuring that only the senders and legitimate recipients are the ones who have access to the communication in full.

AJAX (Asynchronous JavaScript and XML): It is a web development technique for creating interactive applications or RIA (Rich Internet Applications). These applications run on the client, that is to say, in the browser for users while these maintain asynchronous communication with the server in the background. In this way it is possible to make changes to the pages without reloading them, improving the interactivity, speed and usability in the applications.

AI (Artificial Intelligence): It is artificial intelligence that will be responsible for providing the text prediction tool in the section of project writing by users, the artificial intelligence algorithm has a connection to Python libraries and dictionaries, this way real-time queries are performed to provide the text prediction service to these users, it has also machine learning, which gives artificial intelligence the ability to increase the content of word dictionaries and thus not just leave learning manually for developers through data updates.

SQL Server: It is the database manager that will contain all the information of the website referring to research documents, user and administrator data, among others.

4.3 Defining the functioning of artificial intelligence

The following is how Artificial Intelligence works from the point where the user within the repository decides to start writing their research document (see Figure 3):

IOP Conf. Series: Materials Science and Engineering 1154 (2021) 012015 doi:10.1088/1757-899X/1154/1/012015

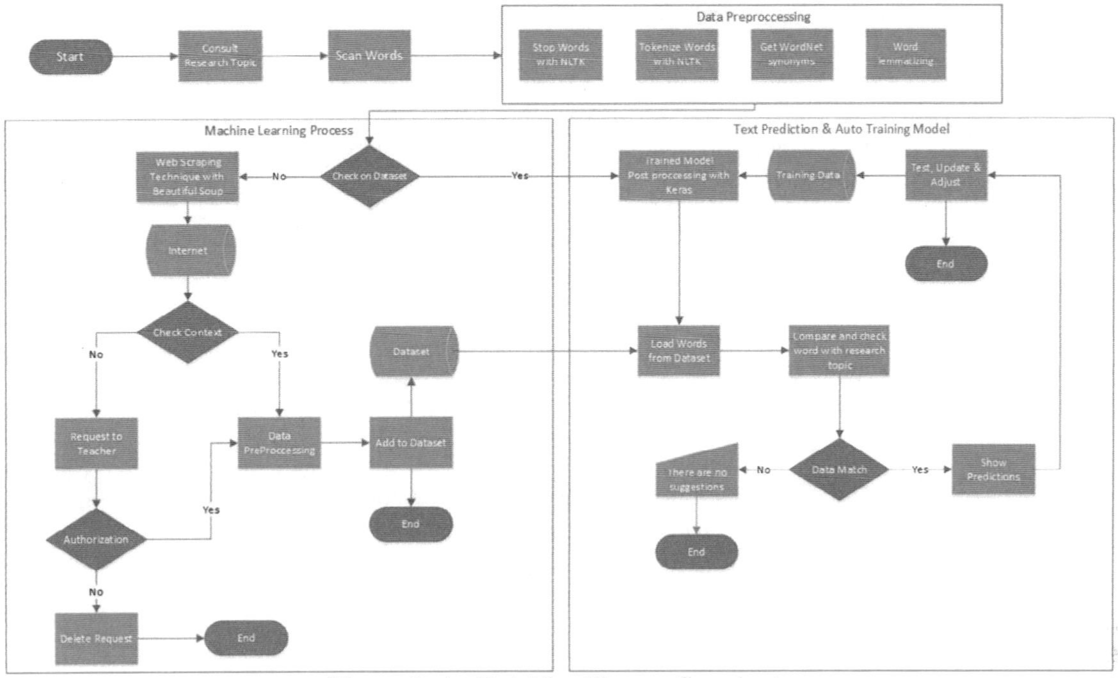

Figure 3. Artificial Intelligence flowchart.

- At the beginning, artificial intelligence conducts a query to determine the research topic or core of the problematic that will be working with, in order to have a better reception and response of the data, that is to say, from the beginning of the process artificial intelligence will already be able to preselect words and phrases that go in context with the problem.
- Having completed the above process, artificial intelligence will begin to capture all the words that are being written within the application area, this process could be carried out in different ways either by means of a key listener or some tool that helps store words in data structures like vectors.
- later, the Pre-processing stage of the data was initiated with the data that it has being received, this process is utilized to clean up data and that avoid all kinds of conflicts or error when running the processes of the text prediction model, for the pre-processing of the data is chosen to use the NLTK (Natural Language Toolkit) library for Natural Language Processing NLP and account is taken of 4 main tasks which are:
    - Deleting the stop words, this process is to remove words that could affect the results in this case the stop words would be connectors and helpers.
    - Tokenize words, Tokenize text is important because text cannot be processed without tokenizing it. The tokenization process means dividing larger parts into small parts. That is to say, convert paragraphs into sentences and make sentences into words according to your needs. NLTK contains a phrase tokenizer and a word tokenizer.
    - Get synonyms, WordNet is a database created for natural language processing. Includes synonym groups and a brief definition. It will be used to load synonyms of words for better search and response options.
    - Lemmatize Words means removing sharpening words and finding the word root. Search engines use this technique when indexing pages, since many people write different versions for the same word and all derive from the root word.
- After completing the Pre-Processing stage, the query is made in the dataset which is the set of words and sentences stored by artificial intelligence, so two decisions are made, in case

the words entered at that moment are related in the dataset, the process of loading the text prediction model immediately begins, otherwise it would initiate a machine learning process to quickly add the necessary words to provide a response to the user.

- If it starts the machine learning process, the first thing that is done is an internet search using the Web Scraping technique for which a tool called Beautiful Soup will be used that is responsible for executing this process, after searching on internet, it has to check if it belongs to the context and the theme that is being addressed in that moment, if it belongs to the theme the pre-processing stage is executed and then added to the dataset and that it is immediately available for use in the trained text prediction model, in case the query made on the internet does not match with the topic that is being worked on, the notification contains a request to the teacher which is that you must make the decision to add the words or phrases found on the internet or if you want to delete them that are not added.
- On the other hand, the process of text prediction and self-training of the model, consists of performing a post-processing of data using the Keras library, in this post processing stage the words are sorted in order of probability that they can be used, that is to say, these words are labeled with a probability obtained with a function called Softmax, the artificial intelligence immediately loads the dataset data into the state that is currently in and checks whether the words captured at the moment are related to some of the words in the dataset, in case of no coincidences this throws a message that says "no suggestions", otherwise if it finds matches, it loads the words in order of probability and they are displayed on the screen to the user, after this, a feedback stage is performed to the text prediction model with the results in order to generate a self-training of the model, That is to say, this text prediction model will be in constant training and updating your data to deliver the best possible results.

5. Conclusions

From the bibliographic review it is concluded that there are a number of techniques, technologies and algorithms that allow machine learning to be integrated into the software development process, allowing developers to select the most appropriate solution to solve the problem to be addressed.

Also, it is concluded that machine learning is one of the main areas of artificial intelligence because it allows the generation of knowledge autonomously in computer machines and informatics systems, by using pattern identification models that use mathematical functions that are supported in statistical and algebraic techniques for performing calculations and procedures.

It is concluded that the proposed system is a comprehensive system that integrates various technologies that achieve the proposed objective, this system is structured by the MVC architectural pattern which facilitates the scalability and maintenance of the system in the future, in addition to the implementation of Python's libraries, NLTK(Natural Language Toolkit), Scikit-Learn, Tensor Flow and Keras, allow the system to fully use machine learning for scientific and academic text prediction, finally the integration between the components and technologies of the system give reliability, creating positive expectations in the development of the system.

References

[1] «Inteligencia artificial en la docencia médica,» *Educ Med Super,* vol. 33, nº 3, 2019.

[2] «Tecnologías avanzadas e Inteligencia artificial: oportunidades para una Enfermería con competencias ampliadas,» *Index Enferm,* vol. 28, nº 3, 2020.

[3] L. Hurtado, E. Villarreal y L. Villarreal, «Detección y diagnóstico de fallas mediante técnicas de inteligencia artificial, un estado del arte,» *Dyna rev.fac.nac.minas,* vol. 83, nº 199, pp. 19-28, 2016.

[4] J. Ríos, G. Ulla y A. Gianni, «APRENDIZAJE AUTOMÁTICO EN UNA PLATAFORMA ONLINE CON FINES SOCIALES Universidad Tecnológica Nacional de Argentina,» *Concurso de Trabajos Estudiantiles,* pp. 30-39, 2019.

[5] A. Urbina y J. De la Calleja, «BREVE REVISIÓN DAPLICACIONES EDUCATIVAS UTILIZANDO MINERÍA DE DATOS Y APRENDIZAJE AUTOMÁTICO,» *REDIE,* vol. 19, nº 04, pp. 84-96, 2017.

[6] W. Sun y X. Gao, «The Construction of Undergraduate Machine Learning Course in the Artificial Intelligence Era,» *International Conference on Computer Science & Education (ICCSE),* pp. 1-5, 2018.

[7] T. Baviera, «TÉCNICAS PARA EL ANÁLISIS DEL SENTIMIENTO EN TWITTER: APRENDIZAJE AUTOMÁTICO SUPERVISADO Y SENTISTRENGTH,» *Revista De Comunicación Digital,* pp. 1-18, 2017.

[8] Tamayo y T. Mario, El proceso de la investigacion cientifica, Mexico: GRUPO NORIEGA EDITORES., 2003.

[9] MinCiencias, «Tipología de proyectos calificados como de carácter científico, tecnológico e innovación.,» MinCiencias, Bogota, 2016.

[11] I. Martín, RECONOCIMIENTO DE VOZ BASADO EN CARACTERÍSTICAS DNN BOTTLENECK, Madrid: Universidad Autonoma de Madrid, 2019.

[12] L. Ferreira, G. Spadon, . C. Carvalho and . F. Rodrigues, "UN ANÁLISIS COMPARATIVO DEL MODELADO AUTOMÁTICO DE ESTILOS DE APRENDIZAJE A TRAVÉS DE TÉCNICAS DE APRENDIZAJE AUTOMÁTICO," IEEE Frontiers in Education Conference (FIE), 2018.

[13] H. Pardede, E. Suryawati, R. Sustika and V. Zilvan, "APRENDIZAJE DE FUNCIONES BASADO EN EL CODIFICADOR AUTOMÁTICO CONVOLUCIONAL NO SUPERVISADO PARA LA DETECCIÓN AUTOMÁTICA DE ENFERMEDADES DE LAS PLANTAS.," International Conference on Computer, Control, Informatics and its Applications (IC3INA), pp. 158-162, 2018.

[14] R. Tommy, G. Sundeep and H. Jose, "DETECCIÓN AUTOMÁTICA Y CORRECCIÓN DE VULNERABILIDADES MEDIANTE APRENDIZAJE AUTOMÁTICO.," Robin Tommy, Gullapudi Sundeep, Hima Jose, pp. 1062-1065, 2017.

[15] K. Bengoetxea, A. Atutxa and M. Iruskieta, "UN DETECTOR DE LA UNIDAD CENTRAL DE UN TEXTO BASADO EN TÉCNICAS DE APRENDIZAJE AUTOMÁTICO EN TEXTOS CIENTÍFICOS PARA EL EUSKERA," Red de Revistas Científicas de América Latina, el Caribe, España y Portugal, pp. 37-44, 2017.

[16] O. Santa, L. Ramírez and F. Trujillo, "TÉCNICAS DE APRENDIZAJE AUTOMÁTICO APLICADAS A ELECTROENCEFALOGRAMAS.," Research in Computing Science 113, 2016.

Airfoil Shape Optimization: Comparative Study of Meta-heuristic Algorithms, Airfoil Parameterization Methods and Reynolds Number Impact

José D Hoyos, Jesús H Jímenez, Camilo Echavarría, Juan P Alvarado

Aeronautical Engineering Faculty, University Pontificia Bolivariana, Medellín, Colombia (050031)

E-mail: `jdaniel.hoyos@udea.edu.co`

Abstract. The aerodynamic efficiency in airfoil theory is defined as the ratio between the lift and drag force, which is the main objective function to be maximized in a wide kind of vehicle design due to its strong relationship between fuel consumption and range. This work employs the 4-digits NACA parameterization, a recently developed 6-parameters method, and the PARSEC technique with a correction of the matrices available in the literature, to compare the computational cost and the ability to achieved higher efficiency of these parameterizations. A genetic algorithm and particle swarm optimization routines are developed and implemented in Matlab, also a sine-cosine algorithm is tested, where Xfoil and the open-source computational fluid dynamic software OpenFOAM are coupled with the optimization algorithms. Finally, a Reynolds number impact study is performed related to the airfoil shape and the angle of attack which maximizes the aerodynamic efficiency. The results showed a faster convergence for the particle swarm optimization and the highest aerodynamic efficiency achieved by the 6-parameter method. Furthermore, with a higher Reynolds number, a higher angle of attack for the optimum lift-to-drag ratio as well a less camber is obtained.

1. Introduction

Nowadays one of the major concerns on airfoil implementation is to assure suitable aerodynamic characteristics to a defined operating condition, the lift-to-drag ratio is commonly one of the most dominant indicators which are highly correlated and affects directly aircraft performance. So far, there are several already designed airfoils that can be adopted, however those can present an efficiency peak on a particular and different application. Airfoil parameterization methods appear to present a novel formulation of both camber and thickness line in terms of several parameters where an original aerodynamic surface can be developed, In this sense, control and management of such variables must be applied in order to avoid useless airfoil generation,this can be translated establishing conscientiously limits in the parameters of the formulation. On the other hand, an optimization routine could be presented as an implementation attempting to search and reach an optimum value associated with a defined variable during any process with a set of constraints. The previously mentioned allows to establishment a convenience to link these two implementations where the employment of a computational tool to extract the aerodynamic coefficients will be required in order to find finally an optimum new airfoil shape for a defined operating condition.

This work aims to compare some conventional and novel methods of airfoil parameterization coupled with different meta-heuristic optimization algorithms, that is, algorithms that work like a black box, where a possible constraint optimization problem enters and an optimum is obtained, the nature of the problem does not affect the operation of the meta-heuristic method. The objective function to optimize is the lift-to-drag ratio, also called aerodynamic efficiency, at the certain wind and angle conditions. The search space is the airfoil shape.

The parameterization methods are 4-digit NACA [1] which has three parameters, another with six parameters [2] and the PARSEC method [3, 4] which employ eleven parameters. The ranges for the parameters are found and some corrections to the PARSEC method are proposed. The tested algorithms are the Particle Swarm Optimization (PSO) [5], the Genetic Algorithm (GA) [6] and the Sine-Cosine Algorithm (SCA) [7].

Moreover, a comparison between the optimum airfoil shapes achieved by Xfoil and OpenFOAM is carried out, where mesh independence and turbulence study is performed. The automatic routines are implemented in Matlab coupled with Xfoil and OpenFOAM. Finally, the Reynolds (Re) number impact in the airfoil shape is studied with the NACA parameterization coupled with Xfoil and PSO also varying the angle of attack.

2. Airfoil parameterization

2.1. Three parameters method

The airfoil shape generator method with three parameters selected is the 4-digit NACA family [1]. The shape is described using a series of four digits that are entered into equations to precisely generate the shape. The first digit specifies the maximum camber m in chord percentage, the second indicates the position of the maximum camber p in tenths of the chord, and the last two numbers provide the maximum thickness t of the airfoil in chord percentage. The thickness distribution for a closed trailing edge is shown in equation 1.

$$\pm y_t = \frac{t}{0.2}(0.2968\sqrt{x} - 0.126x - 0.3516x^2 + 0.2843x^3 - 0.1036x^4) \tag{1}$$

The mean camber line y_c coordinate is defined in two equations (2 and 3) as function of x coordinate.

$$y_c = \frac{m}{p^2}(2px - x^2) \quad \text{for} \quad 0 \leq x \leq p \tag{2}$$

$$y_c = \frac{m}{(1-p)^2}(1 - 2p + 2pc - x^2) \quad \text{for} \quad p < x \leq c \tag{3}$$

Finally, the coordinates of upper surface (x_u, y_u) and lower surface (x_l, y_l) are calculated from equation 4 to equation 7

$$x_u = x - y_t \sin(\theta) \tag{4}$$

$$x_l = x + y_t \sin(\theta) \tag{5}$$

$$y_u = y_c + y_t \cos(\theta) \tag{6}$$

$$y_l = y_c - y_t \cos(\theta) \tag{7}$$

where $\theta = arctan(\frac{dy_c}{dx})$.

Some combinations of the parameters generate intersecting geometries or very thin or thick shapes which Xfoil cannot solve, so the following ranges are established for the variables: $m \in [0, 0.12]$, $p \in [0.2, 0.7]$ and $t \in [0.07, 0.25]$.

2.2. *Six parameters method*

A recent parametric model for airfoil shape with six parameters is presented in [2]. The model has some interesting characteristics, the parameters have an intuitive meaning, and their change predictably impacts the airfoil. The generated curve is continuous and smooth, reducing the design space to realistic shapes. The x and y coordinates are described by equations from 8 to 10.

$$y(\theta) = \frac{T}{2}\frac{|\sin\theta|^B}{\sin\theta}(1 - x^P) + C\sin(x^E\pi) + R\sin(x2\pi) \tag{8}$$

$$x(\theta) = 0.5 + 0.5\frac{|\cos\theta|^B}{\cos\theta} \tag{9}$$

$$\theta \in [0, 2\pi] \tag{10}$$

Where B is the base shape coefficient. For B=2 the base shape is an ellipse. T is the thickness as a chord fraction. P s the taper parameter. C is the camber as a fraction of the chord. E is the camber parameter. R is a reflex parameter, a positive value generates a reflexed trailing edge. An example of the wide range of shapes that can be generated by this method is shown in figure 1.

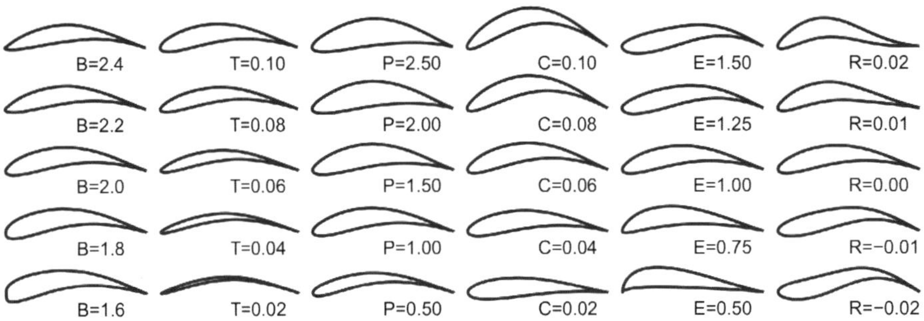

Figure 1: Airfoil for B=2, T=0.1, C=0.05, P=1, E =1, and R=0, where one of the parameters is changed, from [2].

The selected ranges for the optimization meet the same thickness and camber restrictions as the NACA parameterization: $B \in [1.95, 2.3]$, $T \in [0.085, 0.2]$, $P \in [0.85, 1.5]$, $C \in [0, 0.12]$, $E \in [0.65, 1.4]$ and $R \in [-0.015, 0.015]$. Some of the limits are imposed because Xfoil has problems working outside these values.

2.3. *Eleven parameters method*

The method selected is the PARSEC parameterization [3]. This parameterization has previously been coupled with genetic algorithms [4]. It uses eleven basic parameters to completely define the airfoil shape as shown in figure 2.

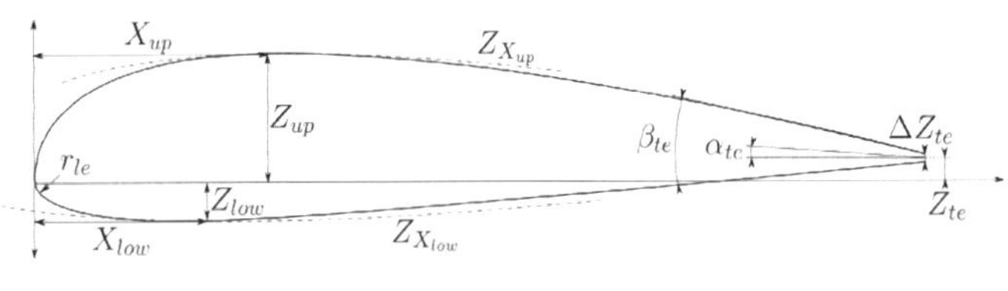

Figure 2: Airfoil shape defined by eleven PARSEC parameters.

$Z_{X_{up}}$ and $Z_{X_{low}}$ are the upper and lower values of curvature.

The parameter ΔZ_{te} is fixed, hence the parameterization is summarized to 10 variables. In this method, a linear combination of shape functions describes the aerofoil shape as 11 shows.

$$z_{up,lo} = \sum_{i=1}^{n=6} a^i_{up,low} \cdot x^{i-\frac{1}{2}} \tag{11}$$

$$a_{up} = C_{up}^{-1} b_{up} \tag{12}$$

$$a_{low} = -C_{low}^{-1} b_{low} \tag{13}$$

$$C_{up,low} = \begin{bmatrix} 1 & 1 & 1 & 1 & 1 & 1 \\ X_{up,low}^{\frac{1}{2}} & X_{up,low}^{\frac{3}{2}} & X_{up,low}^{\frac{5}{2}} & X_{up,low}^{\frac{7}{2}} & X_{up,low}^{\frac{9}{2}} & X_{up,low}^{\frac{11}{2}} \\ \frac{1}{2} & \frac{3}{2} & \frac{5}{2} & \frac{7}{2} & \frac{9}{2} & \frac{11}{2} \\ \frac{1}{2}X_{up,low}^{-\frac{1}{2}} & \frac{3}{2}X_{up,low}^{\frac{1}{2}} & \frac{5}{2}X_{up,low}^{\frac{3}{2}} & \frac{7}{2}X_{up,low}^{\frac{5}{2}} & \frac{9}{2}X_{up,low}^{\frac{7}{2}} & \frac{11}{2}X_{up,low}^{\frac{9}{2}} \\ -\frac{1}{4}X_{up,low}^{-\frac{3}{2}} & \frac{3}{4}X_{up,low}^{-\frac{1}{2}} & \frac{15}{4}X_{up,low}^{\frac{1}{2}} & \frac{15}{4}X_{up,low}^{\frac{1}{2}} & \frac{63}{4}X_{up,low}^{\frac{5}{2}} & \frac{99}{4}X_{up,low}^{\frac{7}{2}} \\ 1 & 0 & 0 & 0 & 0 & 0 \end{bmatrix}$$

$$b_{up} = \begin{bmatrix} \frac{\Delta Z_{te}}{2} + Z_{te} & Z_{up} & \tan\left(-\frac{\beta_{te}}{2} - \alpha_{te}\right) & 0 & Z_{X_{up}} & \sqrt{2r_{le}} \end{bmatrix}^T$$

$$b_{low} = \begin{bmatrix} \frac{\Delta Z_{te}}{2} - Z_{te} & Z_{low} & \tan\left(-\frac{\beta_{te}}{2} + \alpha_{te}\right) & 0 & Z_{X_{low}} & \sqrt{2r_{le}} \end{bmatrix}^T$$

A slight variation for this method is proposed. This parameterization lacks on camber control, besides, the parameter Z_{te} moves up or down the trailing edge, tilting the chord line. The airfoil shape is rotated after generated in such a way that the trailing edge is located exactly at the horizontal axis, giving a little more control of the airfoil curvature and making this parameterization some wider.

The following ranges are selected to generate the airfoil shapes: $r_{le} \in [0.008, 0.025]$, $X_{up} \in [0.317, 0.45]$, $Z_{up} \in [0.03, 0.12]$, $Z_{X up} \in [-2, 0]$, $X_{low} \in [0.3, 0.45]$, $Z_{low} \in [0.01, 0.06]$, $Z_{X low} \in [-2, 1]$, $Z_{te} \in [-0.25, 0]$, $\Delta Z_{te} = 0.006$, $\alpha_{te} \in [0, 0.436]$ and $\beta_{te} \in [0, 0.35]$.

3. Polar airfoil data

The aerodynamic coefficients are computed through the potential flow based software Xfoil and finite volume software OpenFOAM to test different models formulation where computational cost and extracted aerodynamic data could differ, resulting in a slight variation of the optimum airfoil shape.

3.1. Xfoil

Xfoil is a computational implementation of a panels method that incorporates a e^n laminar to turbulent transition technique coupled with an integral boundary layer formulation [8], which allows to predict C_l and C_d at small angles of attack. A Matlab code is developed to extract aerodynamic data for each mutated airfoil set by the optimization algorithm combined with the airfoil parameterization. A recursion within the code is implemented increasing the airfoil surface nodes as the maximum number of iteration, attempting to guarantee convergence of the solution avoiding random Xfoil issues during the calculation process. The airfoil is modeled with 250 points over the top and bottom surfaces applying a smoothing step fixing any high variation of the coordinates, the calculation is set with 300 available iteration and the viscous mode is activated to compute correctly the drag coefficient.

3.2. CFD: OpenFOAM

OpenFOAM is an open-source CFD software under the GNU General Public License which solves the main fluid governing equations through the finite volume method within the discretized computational domain [9], with the possibility to include additional fluid variables associated to already defined turbulence model. A Matlab code is developed to connects with OpenFOAM and compute automatically the aerodynamic coefficients for each airfoil shape established.

3.2.1. Mesh topology The Mesh is generated from the open-source software construct2d [10], which implements an elliptical or hyperbolic algorithm to develop a completely structured either C-Grid or O-grid topology, commonly used for airfoil analysis on CFD simulation. This tool allows us to interact through the operating system terminal command line, allowing us to link a Matlab script to generate automatically the mesh for any airfoil.

The O-grid topology is selected to guarantee a suitable refinement in any flow direction with a domain diameter of $25c$ with c being the chord length [11, 12]. The elliptical mesh algorithm is selected which perform additionally local smoothing steps near the airfoil wall, obtaining a higher cells refinement near the airfoil surface where the fluid variables gradients are more notables, accomplishing well transfer fluid variables information between the cells, this results on less number of cells closer the domain boundaries where fluid properties present small changes. In terms of mesh quality a $y^+ < 1$ is guaranteed [13] employing the flat-plate boundary layer theory [14, p. 467], where y^+ is the non-dimensional wall distance, finally the average cell skewness angle is less than 0 and the average cell aspect ratio is under 30.

3.2.2. Simulation Setup The Matlab code coupled to OpenFOAM allows us to establish any desired setup, varying the main simulation parameters like convergence criteria, numerical schemes, turbulence model, etc. For this particular case, the simulation is performed with a steady-state time scheme and Gauss linear divergence scheme [15] at standard sea level condition, the SIMPLE algorithm is implemented and the convergence criteria are set on 1×10^{-4} for the residuals of the fluid variables. The upstream domain side is defined as a velocity inlet boundary condition, the downstream domain side is set as a pressure outlet boundary condition with gauge pressure equal to zero and the airfoil is set as a wall boundary with no-slip condition. The angle of attack is guarantee setting the respective wind speed V_x and V_y components in order to use one single mesh for any angle. The normal and axial aerodynamic forces over the airfoil are computed, hence, must be transformed into the relative wind reference frame and then be normalized to obtain the lift and drag coefficient.

3.2.3. Grid independence and turbulence model To establish a number of cells that assure accurate data with the lowest computational time, a grid independence studied is performed, varying the number of grid cells between a range of 1.5×10^4 and 15×10^4 volumetric elements. The NACA 0015 airfoil is selected which has reported wind tunnel test data by Sheldahl and Klimas [16]. The Re number is selected as 1×10^6 with an angle of attack $\alpha = 0°$, results are shown in figure 3.

To select the most suitable turbulence model three options are considered. The Spalart-Allmaras which implements only one turbulence equation but is often used for low Re number applications [17], the $\kappa-\omega$ SST and a four-equation model Langtry-Menter $\kappa-\omega$ SST which adds two other equations: one for the intermittency γ and other for the laminar-turbulent transition with $Re_{\theta t}$ criteria, which links empirical transition data with intermittency equation [18].

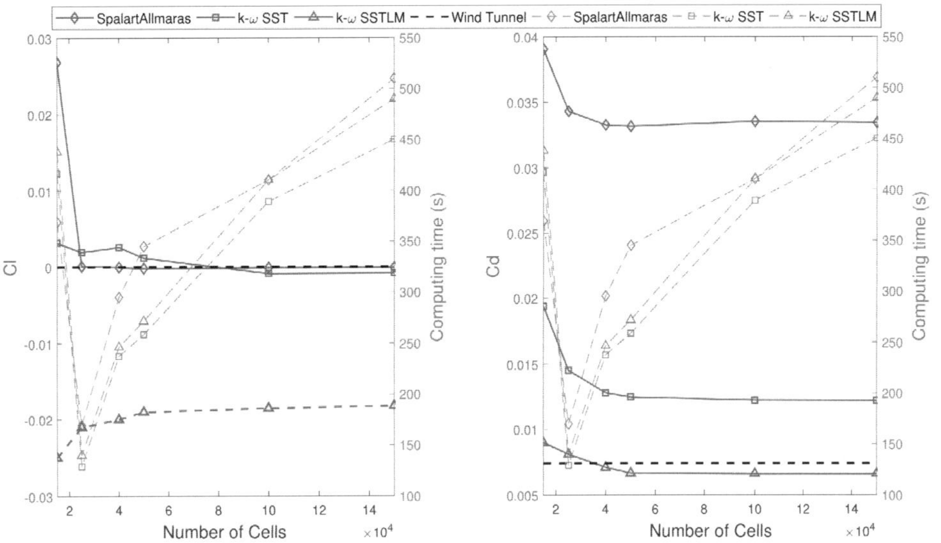

Figure 3: Grid independence and computational time analysis for O-Grid topology, lift coefficient (a) and drag coefficient (b).

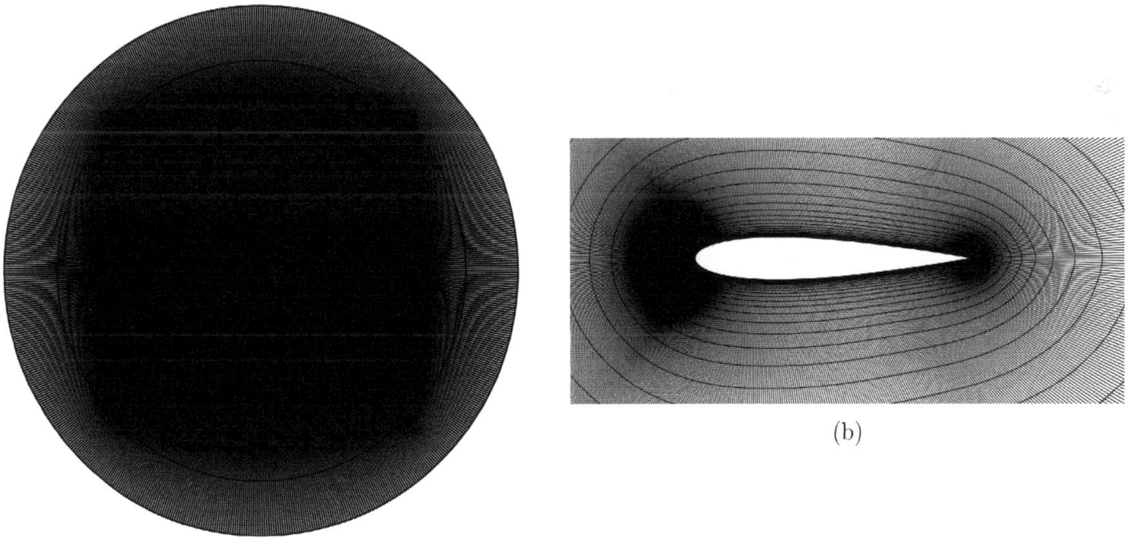

(a) (b)

Figure 4: O-grid topology generated by Construct2d with 5×10^4 cells.

It is shown from the results that C_l and C_d presents a good match in comparison with the experimental data. The C_l is close enough to zero in term of magnitude order from 5×10^4 number of cells for all turbulence models implemented, the C_d show a better fit employing the Langtry-Menter-$\kappa-\omega$ SST, also termed as the $\gamma-Re_\theta-$SST model [19], with a minimum relative error of 11% at a 5×10^4 number of cells, value where the mesh convergence is reached, so this number of cells is employed in the optimization.

3.3. Validation

Aerodynamic coefficients are computed for the NACA 0015 airfoil from $\alpha = 0°$ to $\alpha = 15°$ at a Re number of 1×10^6 through Xfoil and CFD OpenFOAM and are compared against experimental data, the results are shown in figure 5.

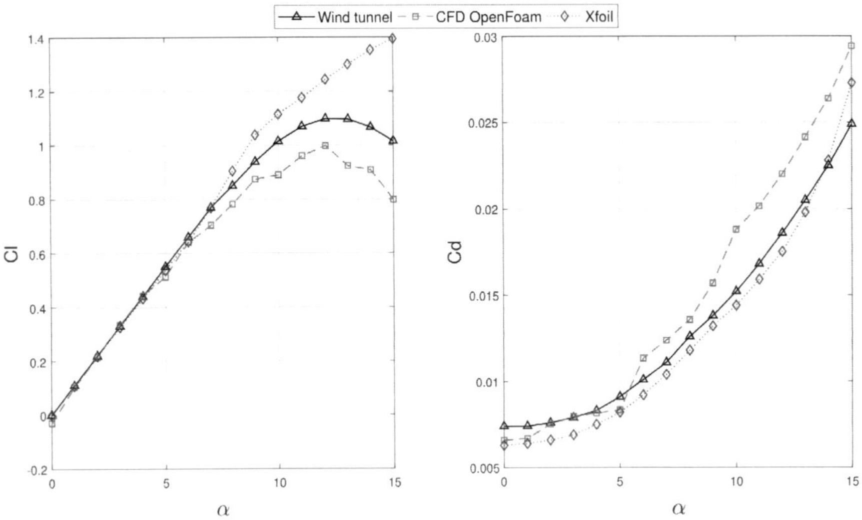

Figure 5: Wind tunnel data, CFD OpenFOAM and Xfoil Lift coefficients (a) and Drag coefficients (b) validation for a $Re = 1 \times 10^5$

From the results obtained, the lift coefficient extracted from CFD OpenFOAM and Xfoil present an average error of 8.21% and 8.23% respectively, highlighting the stall prediction by the finite volume software which is an important indicator on aerodynamics coefficient calculation, on the other hand, the average error associated to C_d is 8.09% and 6.78% for CFD OpenFOAM and Xfoil respectively, resulting on a quite good match in comparison with the wind tunnel data for both tools in this Re number range.

4. Meta-heuristic algorithms

4.1. Genetic algorithm

The GA method mimics the natural evolution, where some initial population is selected, then a crossover and mutation operations are performed usually based on the score of each individual by the objective function.

4.1.1. Initial population The initial population has a certain number of individuals and for each individual, the values for the variables are randomly distributed in the design space taking into account the maximum and minimum limits for each design variable.

4.1.2. Selection of parents The selection operator must choose with higher probability those individuals who present a better value of the fitness function, also including the possibility that individuals who are not as good have a certain probability of appearing. The latter tries to keep a certain diversity in regarding the information that the chromosomes contain. The objective is that the quantity of each one of the individuals in the new population is given by the proportion given in equation 14.

$$P(x_i) = \frac{f(x_i)}{\sum_{j=1}^{N_{ind}} f(x_j)} \tag{14}$$

where N_{ind} is the number of individuals, $P(x_i)$ is the probability of the selection for the x_i individual and $f(x_i)$ is the value of the function to be optimized. In this work, the Stochastic Universal Sampling (SUS) is employed to select the parents, which fulfill the proportion by equation 14. In order to avoid convergence in local optimums, a linear ranking function is employed, in this method, the individuals are ordered from highest to lowest in an array according to the fitness function value, and their position in the array is the new value of the fitness function. An example of the SUS methodology with and without ranking function is shown in figure 6.

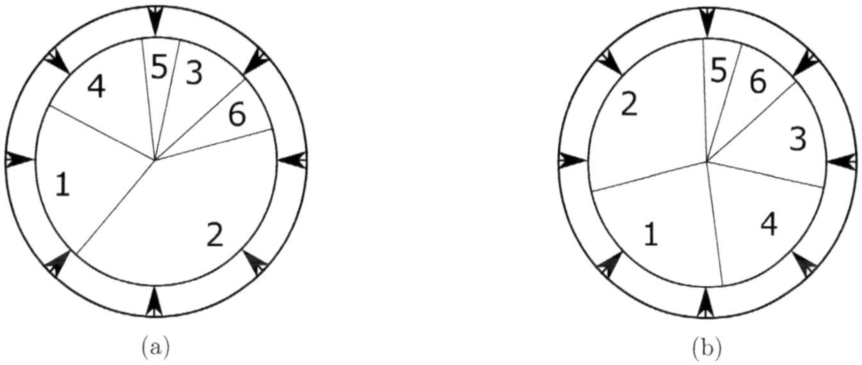

Figure 6: Stochastic Universal Sampling 8 parents selection from 6 individuals without (a) and with ranking function (b).

4.1.3. Cross over and mutation Cross over is performed to combine the desirable characters of two different parents selected by the SUS-linear ranking method. In this work, uniform cross over is employed. In this approach, a cross over probability P_c is defined and a probability test is performed for each chromosome. Since the coding of the problem is not binary, it is possible to perform the linear recombination to the chromosomes of the real variable to increase the diversity of the individuals shown in equations 15 and 16.

$$C_1 = \alpha_1 P_1 + (1 - \alpha_1) P_2 \tag{15}$$

$$C_2 = \alpha_2 P_2 + (1 - \alpha_2) P_1 \tag{16}$$

where C_n is the child n, P_n is the parent n and α_1, α_2 are random numbers between -0.5 and 0.5, which extends the linear combination beyond the limits set by the parents.

The mutation is performed to increase the diversity of genes and thus have a greater probability of success by finding global optima. Here a mutation probability P_m is defined and the probability test is carried out on each chromosome. If passed, the mutation is performed adding a low random number relative to chromosome value.

4.1.4. Parameters setting The main parameters to adjust are the mutation probability P_m and the cross over probability P_c. Due to the unknown nature of the function, parameters are chosen that have shown a good result for all types of function tested in [6], $P_m = 0.1$, $P_c = 0.6$. Moreover, the number of individuals in one generation N_{ind} is chosen as $10 \cdot n$ and the maximum number of generations as $25 \cdot n$ where n is the number of variables.

4.2. Particle swarm algorithm
The particle swarm optimization (PSO) method attempts to find an optimum design by moving a given number of possible solutions (the particles) along the design space of the optimization problem, which can be subjected to some constraints.

The movement of the particles is not completely random. The positions taken by particles are evaluated in order to define their further movement; the trajectory and velocity of an specific particle depend on the best position that has reached any particle at any time. A simple variant for the method establishes that the trajectory of a specific particle is affected also by its individual best position so local optimal values are avoided. Equations 17 and 18 are used to define the movement of an individual particle at a given time step (t) [5].

$$X_{(t+1)} = \mu X_{(t)} + V_{(t+1)} \tag{17}$$

$$V_{(t+1)} = \mu V_{(t)} + \phi_1 R_2 \otimes (G_{(t)} - X_{(t)}) + \phi_2 R_2 \otimes (L_{(t)} - X_{(t)}) \tag{18}$$

In this case, the operator $\otimes$ denotes the array element-by-element product. The design parameters are stored at the array X, the V array represents how the particle will be relocated at the next time step. G stores the best design reached by any particle while L represents the best design the individual particle reached. Look that speed depends on how far the particle is from the best one as well as from the local best, constants ϕ_1 and ϕ_2 are used to manipulate how strong the particle is attracted by the best one or by its individual best position. R_1 and R_2 are random vectors with many elements as many design variables the problem has. These random values vary between 0 and 1 giving randomness to the movement of the particles in order to improve the search. Eventually, after some time steps (or iterations) most of the particles will converge at an optimal design. The algorithm can be stopped by setting convergence criteria or well a maximum number of iterations. For this case the particles are randomly located on the design space; when a particle goes out of the boundary, a penalty function is applied so it will not attract the other particles to an undesired position.

4.2.1. Parameters setting A fixed inertia weight of $\mu = 0.5$ is used as a variant for the method to restrict a little the particle´s movement so they do not move unsteadily within the design space; ϕ_1 and ϕ_2 are both fixed to 2, meaning that local and global bests can have the same impact over a particle's trajectory. The number of particles is selected to be $10 \cdot n$ and the maximum number of iterations is set as $25 \cdot n$.

4.3. Sine cosine algorithm
An interesting optimization method called Sine Cosine algorithm (SCA) is chosen for the analysis due to the fact that it is an algorithm that easily allows limits to be used for the design variables, and it was previously employed for geometric airfoil optimization [7].

The SCA creates multiple initial random search agents (solutions) and requires them to oscillate outwards or towards the best candidate employing a sine and cosine functions. The algorithm employs several random and adaptive variables integrated to emphasize exploration and exploitation of the search space, however, this method is employed as is discussed in [7], no emphasis is placed on details. The number of search agents and the maximum number of iterations are also select as $10 \cdot n$ and $25 \cdot n$, respectively.

5. Results

The optimizations algorithms are compared for each parameterization method, the objective function is the aerodynamic efficiency $\frac{L}{D}$ for a Reynolds number of Re=1e6 and 5° angle of attack. The results show a higher optimum achieved by the GA and PSO algorithms compared to the SCA for all parameterizations as is shown in figures 7a, 7b and 8a. Moreover, the SCA shows a staggered search for the optimum, while PSO and GA quickly and smoothly reach the global optimum.

Although the PSO and GA achieve virtually the same optimum, the PSO achieves the optimum in a slightly smaller number of iterations as shown in figures 7a and 8a. Furthermore, the PSO does not evaluate the particles generated outside the limits of the variables, so it performs fewer evaluations of the objective function for the same number of iterations, considerably reducing the computation time as evidenced in the data in figures from 7a to 8a. If the number of evaluations of the objective function is compared, the rapid convergence of the PSO with respect to the GA is more marked.

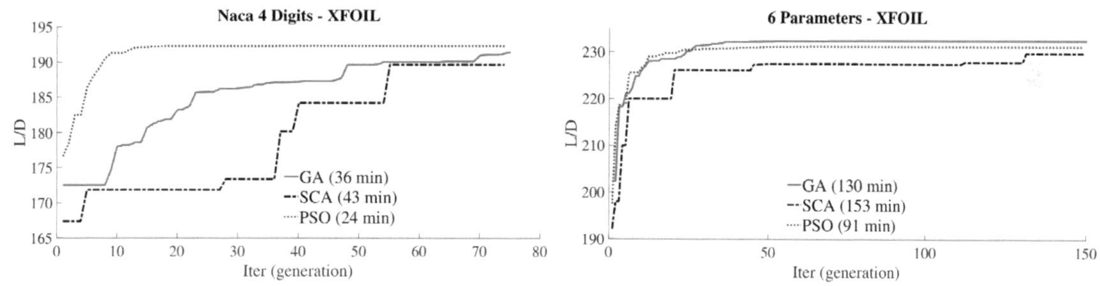

(a) (b)

Figure 7: Evolution of best aerodynamic efficiency employing Xfoil for 4-digit NACA parameterization (a) and 6 parameters method (b).

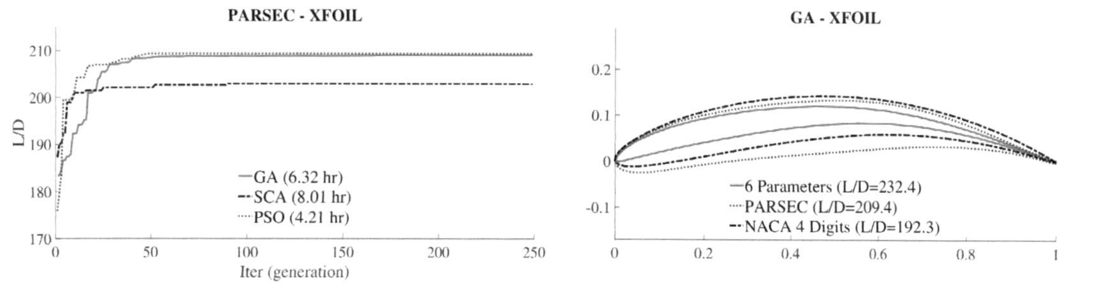

(a) (b)

Figure 8: Evolution of best aerodynamic efficiency employing Xfoil with PARSEC parameterization (a) and optimum airfoil shape with GA (b).

The best airfoil shape from NACA parameterization is described by the values $t = 0.1000$, $m = 0.0992$ and $p = 0.5475$. From the limits imposed on these variables, it is observed that no parameter reached the limit value of the variable.

Similarly, the six parameters method achieved the optimum with the values $B = 1.9501$, $T = 0.0850$, $P = 0.8500$, $C = 0.1047$, $E = 0.7821$ and $R = -0.015$, taking into account the limits set for the parameters, it is observed that B, T, P and R takes the lower limits. According to these results, an airfoil with a lower thickness would achieve more efficiency, however, the optimization obtained limit values, which means that Xfoil limits the design for this 6-number parameterization.

The PARSEC method achieved an optimum with the following values $r_{le} = 0.0169$, $X_{up} = 0.4108$, $Z_{up} = 0.0973$, $Z_{Xup} = -0.0101$, $X_{low} = 0.3937$, $Z_{low} = 0.0180$, $Z_{Xlow} = -1.3071$, $Z_{te} = -0.0783$, $\Delta Z_{te} = 0.006$, $\alpha_{te} = 0.3588$ and $\beta_{te} = 0.0025$. From this values, any parameter reached the limit values.

The optimal thickness according to NACA parameterization is higher compared to the parameterization of 6 values, however both methods reach an optimum for the same camber value of 10%.

The norm of the subtraction of the vectors that contain the optimal parameters is employed to calculate the distance between the solutions found by each optimization algorithm, the results are shown in table 1. From the results, it is found that for each airfoil parameterization method, the lower difference is found comparing the GA and PSO algorithms.

Method	PSO-SCA	GA-PSO	GA-SCA
NACA-4 Digits	0.0148	0.0008	0.0152
6 Parameters	0.0416	0.0186	0.0432
PARSEC	1.6311	0.3070	1.3428

Table 1: Solution vectors distance.

Finally, the PSO algorithm coupled with 4-digit NACA parameterization is employed to optimize the airfoil through OpenFOAM CFD code. The optimum parameters was $t = 0.0787$, $m = 0.1055$ and $p = 0.3556$, as it shown in figure 9a, less thickness than Xfoil method, with the maximum camber position closer to the leading edge, but about the same maximum camber of 10%. This airfoil is simulated at various angles of attack to compare Xfoil and OpenFOAM efficiency curves. It is shown in figure 9b that the airfoil was designed unintentionally by the PSO with the best aerodynamic efficiency at the arbitrarily selected angle for the optimization.

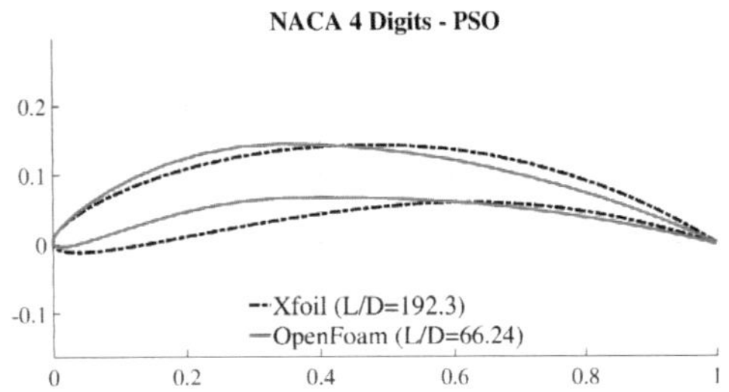

(a)

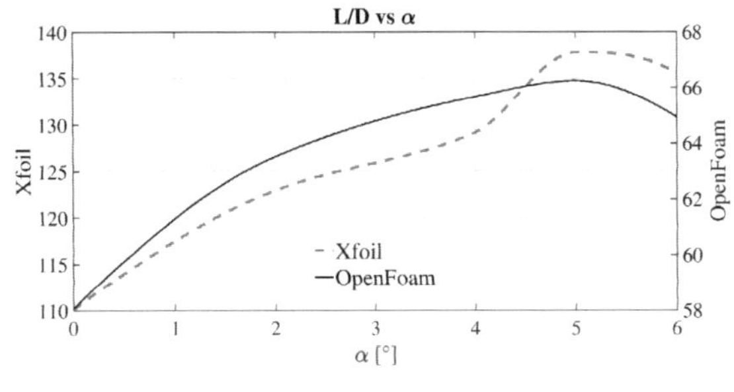

(b)

Figure 9: Optimum airfoil comparison between Xfoil and OpenFOAM (a) and efficiency polar for OpenFOAM optimum airfoil comparison between Xfoil and OpenFOAM (b).

A 4-digits NACA airfoil optimization process varying the angles of attack with the constraints $m \in [0, 0.12]$, $p \in [0.25, 0.65]$ and $t \in [0.08, 0.2]$ was carried out for different Re numbers. The optimum characteristics are shown in table 2 and the optimum shape for the lowest and highest

Re number are shown in figure 10. The camber has an increasing trend with the Re number, meanwhile the best efficiency angle of attack has the opposite behavior. Finally, the thickness achieved the lower limit from the constraint, which is the opposite behavior when optimizing by setting an angle of attack.

Re number	2e4	5e4	1e5	2e5	4e5	1e6
Thickness	0.0820	0.0800	0.0800	0.0807	0.0800	0.0802
Maximum camber	0.0385	0.0444	0.0654	0.0609	0.0729	0.0808
Maximum camber position	0.5720	0.4422	0.4570	0.5340	0.5669	0.5618
α best efficiency	7°	7°	7°	5°	4°	3°
Aerodynamic efficiency	22.4	43.3	68.5	203.1	148.9	222.2

Table 2: Optimum airfoil characteristics for different Re numbers.

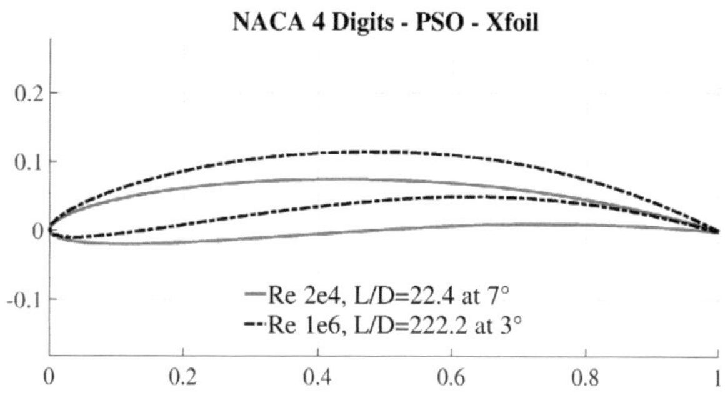

Figure 10: Re number impact in airfoil shape optimization.

6. Conclusions

The GA and PSO algorithms reached almost the same optimum in all tested parameterizations, while the SCA reached a lower optimum for the same cases. However, the PSO algorithm achieved a convergence in fewer iterations in NACA parameterization, even achieving a considerably lower computation time for all parameterizations.

The 4-digit NACA parameterization reaches an optimum 17% lower in comparison to the parameterization of 6 values, however, it requires approximately 3.7 times less due to the fact that algorithms were run with populations and iterations depending on the number of variables. According to this, whether it is desired to achieve maximum efficiency regardless of the computational cost, the 6-value parameterization is suitable, otherwise the 4-digit NACA method achieves enough efficiency in a short time.

The NACA optimization obtained an optimum inside the limits, while the optimization of 6 values obtained four limit values, which were imposed to avoid errors in the convergence and reading of the airfoil by Xfoil. It is concluded that a more optimal airfoil can be achieved by the parameterization of 6 values using another tool to compute the aerodynamic coefficients such as CFD tools, while the NACA parameterization has an adequate match with Xfoil capabilities.

A remarkable result is that both the NACA and the 6-value parameterization, including Xfoil and OpenFOAM methods, obtained an optimal camber of around 10%, while the thickness did not remain constant in the solutions.

The PARSEC method reached an intermediate aerodynamic efficiency for the longest computation time, so this method is not recommended for optimization from a random population.

The optimization was carried out by setting an angle of attack without a guarantee that the airfoil presents its peak efficiency at this angle, however, the results show that the airfoils generated have the maximum efficiency at the design value. Noticeably, this does not mean that a higher optimum cannot be achieved if the angle of attack is added as an optimization parameter, which actually happened in the Re number study.

The optimum airfoil shape shows an inverse behavior between the Re number and the angle of attack at which the best aerodynamic efficiency is achieved, concurrent, the camber increases with Re number.

References

[1] Eastman N Jacobs K E W and Pinkerton R M 1935 The characteristics of 78 related airfoil sections from tests in the variable-density wind tunnel Tech. rep. Washingto,D.C.
[2] Ziemkiewicz D 2017 *AIAA Journal* **55** 4390–4393 (*Preprint* `https://doi.org/10.2514/1.J055986`) URL `https://doi.org/10.2514/1.J055986`
[3] Sobieczky H 1997 *Geometry Generator for CFD and Applied Aerodynamics* (Vienna: Springer Vienna) pp 137–157 ISBN 978-3-7091-2658-5 URL `https://doi.org/10.1007/978-3-7091-2658-5_9`
[4] Della Vecchia P, Daniele E and DAmato E 2014 *Aerospace Science and Technology* **32** 103 – 110 ISSN 1270-9638 URL `http://www.sciencedirect.com/science/article/pii/S1270963813002046`
[5] Parsopoulos K E and Vrahatis M N 2010
[6] Ferragud F X B 1999 *Control predictivo basado en modelos mediante técnica de optimización heurística. Aplicación a procesos no lineales y multivariables* Ph.D. thesis URL `http://hdl.handle.net/10251/15995`
[7] Mirjalili S 2016 *Knowledge-Based Systems* **96** 120 – 133 ISSN 0950-7051 URL `http://www.sciencedirect.com/science/article/pii/S0950705115005043`
[8] Drela M 1989 *Low Reynolds number aerodynamics* (Springer) pp 1–12
[9] Open C 2011 *OpenFOAM Foundation* **2**
[10] 2012 Open source software construct2d URL `https://sourceforge.net/projects/construct2d`
[11] Morgado J, Vizinho R, Silvestre M and Páscoa J 2016 *Aerospace Science and Technology* **52** 207–214 ISSN 1270-9638
[12] Günel O, Koç E and Yavuz T 2016 *2016 IEEE International Conference on Renewable Energy Research and Applications (ICRERA)* pp 628–632
[13] Salim S and Cheah S 2009 *Proceedings of the International MultiConference on Engineerings and Computer Science, Hong Kong*
[14] White F 2003 *Fluid Mechanics* 5th ed McGraw-Hill series in mechanical engineering (New York: McGraw Hill)
[15] Suvanjumrat C 2017 *Engineering Journal* **21** 207–221
[16] Sheldahl R E and Klimas P C 1981 Aerodynamic characteristics of seven symmetrical airfoil sections through 180-degree angle of attack for use in aerodynamic analysis of vertical axis wind turbines Tech. rep. Sandia National Labs., Albuquerque, NM (USA)
[17] Islam M, Amin M R and Shariff Y M 2009 *ASME International Mechanical Engineering Congress and Exposition* vol 43826 pp 867–874
[18] Wang S, Ingham D B, Ma L, Pourkashanian M and Tao Z 2012 *Journal of Fluids and Structures* **33** 191 – 209 ISSN 0889-9746
[19] Menter F R, Langtry R and Völker S 2006 *Flow, Turbulence and Combustion* **77** 277–303

IOP Conf. Series: Materials Science and Engineering 1154 (2021) 012017 doi:10.1088/1757-899X/1154/1/012017

Use of augmented reality technology and GPS to encourage tourism in the Sucre department

D. Salgado[1], D. Benitez[1] A. Morales[1], M. Gonzalez[1] and N. Huertas[2,]

[1]Antonio José de Sucre University Corporation (Colombia). [2] Technological University of Bolívar (Colombia).

dsalgado012@gmail.com

Abstract. Globalization has boosted the tourism sector, making it one of the most important economic activities in developing countries in the 21st century, creating the need for urban growth through the construction of new physical infrastructures to guarantee accommodation, permanence , the security and mobility of large masses of people, putting the integrity of cultural heritage at risk. For this reason, the objective of this research is to design a system that integrates various emerging technologies, among which Augmented Reality and the Global Positioning System stand out, in order to develop digital content of cultural heritage and preserve its immaterial value. , in addition to allowing the tourist to live a different experience when visiting the host place, superimposing digital information in the places they visit. The type of research is descriptive and technological development. Among the results found, the communication between the different technologies of the proposed system is outlined, as well as the flow of data that allow the functionality of the system. It is concluded that the proposed system contributes to the conservation of cultural objects and expressions through the generation of multimedia content and that it allows tourists to live an innovative experience through the proposed technologies.

1. Introduction

Given the confinement situation, in the department of Sucre, there is a high impact in the tourism sector especially in the coastal areas of the department [1]; with a view to the search for options for tourism, the Antonio José de Sucre University Corporation and the Technological University of Bolívar came together to formulate a project, in relation to the Special Plan of Safeguarding, living paintings of Galeras-Sucre (among others). The project called the "technological tool with digital content for the dissemination of cultural objects or expressions of the departments of Sucre", which is in operation in 2020, is based on the line of action of the Research Programme, which becomes a device of memory and understanding of social phenomena, which contributes to the solution of practical problems [2] and according to Goal 11.4 Sustainable Development Goals. To highlight the need to redouble efforts to protect and safeguard the world's cultural and natural heritage [3].

An earlier project, between these two institutions, allowed to identify the possibilities of involving tourism, promoting creative and tourism vocations within the associative sector, in order to improve productivity and that could benefit communities and/or actors in the territory, establishing the need to promote technological tools [4].

The development of computer sciences offers a wide range of solutions to various problems that arise on a day-to-day basis in all sectors in which human beings operate, creating important changes in the various activities that are developed in these sectors. Among these activities is tourism, an activity that has taken center stage in recent decades for the drive of globalization, tourism is seen as one of the engines for the economic development of countries, because it generates significant jobs and incomes to the inhabitants of the host places; WTO states that tourism's contribution to the economy depends on the quality and innovative strategies tourism offers, WTO insists that developing countries must practice sustainable tourism in particular in order to obtain the different benefits it offers [5]. For this reason and in order to preserve the cultural and natural heritage of different communities the use of emerging technologies can contribute; among these technological trends are the so-called Augmented Reality or AR and Global Positioning System or GPS, which integrated have the ability to improve the experience of the user's perception of reality, by overlaying multimedia digital content supported by an electronic device with digital camera and GPS sensors.

Therefore, the objective of the following research is to design a system that integrates augmented reality with GPS, to allow to visualize objects and cultural expressions of the department of Sucre. The document is initially organized by exposing the tourist inventories of the department of Sucre (Colombia) and the contribution of AR with GPS to the tourism sector, then describes the methodology used in this study, then the architecture of the system is illustrated by block diagrams and flowcharts, and finally, the discussions and conclusions of the research are generated.

2. Tourism sector and emerging technology

2.1. The tourist inventories, valuation and importance of the Sucre department

Based on the methodology for the elaboration of tourist inventories of Colombia, the collection, respective qualification and diligence of the formats established for this purpose, were obtained from the governorate of the department of Sucre the results taking into account the established by the Ministry of Commerce, Industry and Tourism.

It is important to emphasize that attractive inventories along with other information inputs produced in tourism planning processes constitute important elements for decision-making for both the public and private sectors at the regional and national levels.

The updated categorisation sheets of the municipalities of Sucre correspond to the tourist corridor Golfo de Morrosquillo and Sabana: Coveñas, Corozal, Sampués, San Antonio de Palmito, San Marcos, San Onofre, Sincelejo and Tolú; prepared by a team and the Tourist Magister Edgar Villarraga Amaya, coordinator of the tourism area of the Governorate of Sucre, attached to the Secretary of Economic Development and Environment of the department, updated to May 2020. For the towns Sincé, Galeras, San Juan de Betulia, Los Palmitos, correspond to the project FNT-288 "Design of the tourist product Cultura Sabanera for the department of Sucre" in December 2015, by Jorge Gómez.

Therefore, from the inventories as an official source of consultation, the following results are established for the inclusion of a Sucre tourism system: the total resources are 173, of which 156 are the greater than 60 points and up to 70 (the highest score), the other 17 classified obtained a score of 58 and 59 and were considered as potentially susceptible to improvement of which 9 correspond to immovable cultural heritage; 2 movable material heritage; 4 intangible cultural heritage and 2 natural heritage.

The cultural heritage in terms of craft object techniques, classified from previous inventories [6], is diverse in hammocks woven from woven yarn in loom, bags and table runners. Popular traditional craftsmanship: artisanal technique in totumo; craftsmanship of wooden furniture, cabinet making; artisanal technique in tannery. Traditional popular handicraft in tannery: sandals-saddlery. Indigenous craftsmanship with artisanal seed painting technique and handcrafted techniques of making fabrics in arrow cane with headgear: vueltiao hat, the latter is a cultural symbol of Colombia and around it are identified ethnographic and anthropological investigations of the social and cultural manifestations of the Zenú ethnic group [7].

There is also the classification of cultural heritage, in the category of festivities and events, in addition to the sound and dance musical expressions. For this research, the living paintings of the municipality of Galeras stand out since they constitute a manifestation of popular art and festive and recreational events. In addition, they meet the requirements of belonging, representativeness, relevance, nature and collective identity, validity, equity and responsibility [2]. It is also a "community artistic expression" that encompasses events, allegory, religious, moral, historical, satirical and everyday events. For this research project within its classification will take into account the first and second place of the traditional and contemporary paintings, from 2014 to 2020.

2.2. *Emerging technologies*

In the 21st century, emerging technologies are considered to be any technology little known but that creates high expectations in the fields in which they are applied, in terms of the tourism sector can be highlighted among these technologies augmented reality or AR and the Global Positioning System or GPS, which have taken center stage thanks to the arrival of mobile cellular devices that integrate multiple sensors, allowing to detect actions and / or external stimuli to the device that integrates them. The AR and GPS features provide answers to the needs of the tourist in being informed about both historical and cultural data and to be located in unknown places without fear of being lost.

AR is defined as the integration of digital information within physical environments, by overlaying text, 2D and/or 3D animations, videos, graphics among other computer-generated multimedia content displayed through the screen of a technological device [8], such a device must have minimal digital cameras, software that has the properties of reflecting the information on the screen when activating the device camera and a trigger or information activator, which can be 2D images, 2D codes (bar code and QR code), real objects (chairs, doors, among others) and/or geolocation points [9] (see figure 1).

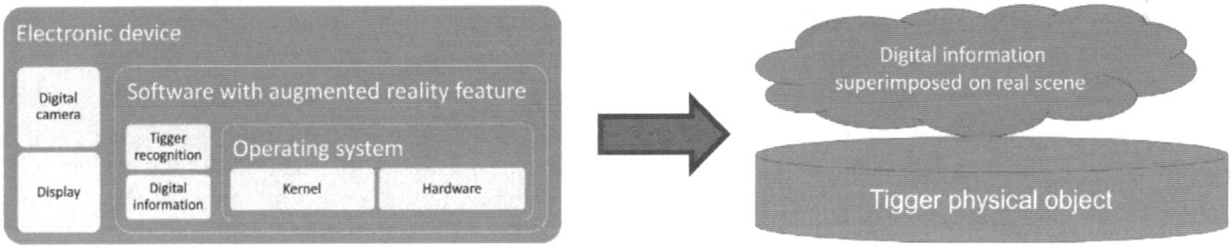

Figure 1. Structure of an AR system. Source authors of the article

With regard to GPS technology, it provides the possibility to determine the geographical location of an electronic device by identifying the network devices to which the mobile electronic device (Smartphone and Tablet) is connected or directly by the internal GPS receiver, communicating directly with the satellites orbiting the Earth, which are able to locate the devices located somewhere on the planet with a minimum margin of error [10].

It should be noted that the positioning is given by the integration of software technologies associated with the GPS device, which can also be supported by IP location or by geometry of wireless networks. Electronic devices, in addition to geographical coordinate location, can orient cardinal points towards the situation and synchronize other georeferenced information using orientation sensors, magnetometers, inclinometers and inertial sensors, which allows to increase the accuracy of the location of the device [11].

Now, in what concerns AR and GPS studies in the tourism sector, the latest findings reflect that this technology allows the user or tourist to take full advantage of the exploration of the scenarios and attractions of the host place, promoting historical, cultural, social data, entertainment and commercial and economic information in real time and in different multimedia formats, making it an attractive technology especially for new generations that are more likely to use electronic devices [12]. Likewise, AR supports integration with other technologies both at the software level such as the use of databases, applications for photography, artificial intelligence for process automation, software for generating GPS points, among others, and at the hardware level with the use of sensors and actuators such as the accelerometer to measure the movement of the device, gyroscope for reading the orientation of the device, capacitive sensors for the detection of events on the touch screen, GPS sensors for the geographical location of the device, among others, increasing the autonomy of the tourist, allowing him to explore the destination in an enriched and entertaining way; this integration with other technologies enhances the use of AR and complements functions that especially benefit tourism companies. Another advantage of AR is given in the educational field, because it allows to visualize information from different perspectives, intuitively, interactive, personalized and attractive in real time [13], capturing the attention of the tourists, allowing them to appropriate knowledge of the places, structures and / or tourist spaces, creating awareness of the cultural and heritage wealth of place, managing to convince the tourist that frequent the places visited and protect those riches; It should be noted that through the elaboration of multimedia contents such as photographic, videos and 3D modeling it is contributed to the conservation of cultural heritage, in addition to adding the function of giving life to static elements, allowing the visitor to live a different experience, bringing this group of people closer to cultural spaces and themes, which are regularly ignored by them.

3. Methodology

Research is technological development type, according to MinCiencias, this type of study is given in the application of the results of research or any other type of scientific knowledge, for the design or development of new processes, systems or provision services, as well as substantial technological improvement of materials, products, processes or pre-existing systems, this activity will include the materialization of research results in a plan, scheme or design, as well as the creation of non-marketable prototypes [14]. The purpose of this research is to generate the design of a computer system that integrates augmented reality and GPS technology for the preservation and dissemination of expressions and cultural objects of the department of Sucre located in Colombia.

3.1 Instruments

Based on the methodology for the elaboration of tourist inventories in Colombia, the collection, respective qualification and completion of the forms established for this purpose, were obtained from the government of the department of Sucre; which are based on what is established by the Ministry of Commerce, Industry and Tourism [15]. Los inventarios de atractivos junto con otros insumos de información son de trascendencia para la toma de decisiones para el sector público y privado en los niveles regional y nacional [16].

On the other hand, the bibliographic matrix and content analytical matrix instruments were used to organize the information of the consulted theoretical references and analyze the requirements of a software with augmented reality technology using GPS and physical Target, in addition to the application of this technology. in the tourism sector. Finally, unstructured interviews were conducted with experts in cultural heritages of the Sucre - Colombia region, which serve as a communicative process for the collection of information, which takes place through meetings between subjects, previously agreed and planned [17].

To conduct the research, two phases were developed using the waterfall software development methodology (see Figure 2):

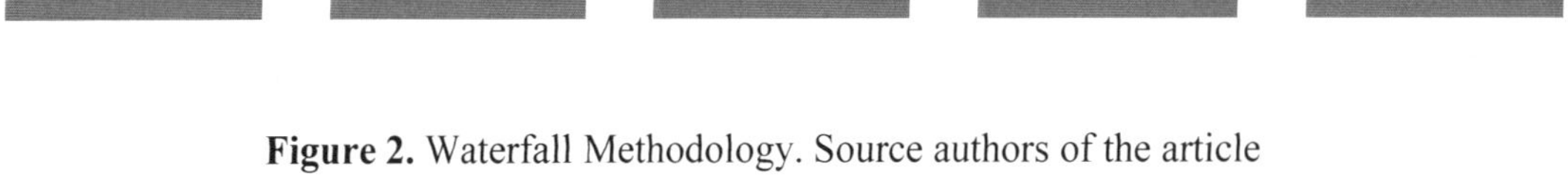

Figure 2. Waterfall Methodology. Source authors of the article

In the requirements phase, the documentary analysis of the information provided by the Sucre Governorate and academic documents research results on tourism-related topics in Sucre and the use of AR and GPS technologies in the tourism sector was carried out, in order to identify the functional and non-functional requirements of the system proposed in this research.

For the design phase, the requirements identified in the previous phase were taken into account and from there, through the use of block diagrams and flowcharts proceeded to the design of the system architecture relating the communication between the different technologies that integrated the operation of the system.

4. Results

4.1.1. Descripción de Resultados

Initially, a bibliographic review of documents, research results, referring to issues of AR and GPS integrated with the tourism sector was made, which were organized with the help of the bibliographic matrix and later they were analyzed with the help of the analytical matrix of content.

Subsequently, semi-structured interviews were made and developed with teachers M. González and M. Huertas, experts in the area of tourism, which provided information for functional and non-functional requirements. (see table 1).

Table 1. System requirements

Functional requirements	Non-functional requirements
The System must provide a means that allows adding Points of Interest (POI).	The Graphic Interface must be simple and intuitive so that the user does not get lost in the use of the application.
The System must superimpose digital content with the use of the web camera, cultural objects and expressions in multimedia format.	
The System must be able to communicate with the web server.	
The system must recognize points of interest (POI)	

For the development of an application of this type, there are 2 methods to do it, which are Online and Offline, below, the operation of the application is exposed using both methods (see figure 3).

4.1.2. Offline operation via TAG. The following diagram describes the operation of the application and how it reacts to the different actions of the user, it starts by requesting the POI parameters, this process is supported in a web API, if these are not accessible the application will activate the camera and start a recognition cycle of TAG offline, when a TAG is found, the content of this is rendered and then viewed by the user through the device screen, if the user decides to exit it will start to dismount the previously requested resources, otherwise the POI parameters will be requested again.

4.1.3. Online operation via GPS or TAG. When starting, the application makes an AJAX request to the web server, to obtain the JSON control list where it specifies the components that make up each of the POI entities that are uploaded to the web server (the reload function to make the request again, is available as a button on the interface). The application pre-loads the model meshes and the geographical references obtained from the server, and for each geographical reference a single POI entity is created. An algorithm proceeds to generate a geographic matrix with the previously defined POI entities. Then, a cycle begins where the system performs two essential functions in the application and in a process that repeats itself indefinitely:

It captures the GPS data of the device and compares it with the geographical matrix of the POIs by means of an algorithm of approximation and correlation of points, which allows to show a POI as "close as possible", In case of finding a geographical match between the GPS position and one of the POIs in the matrix, the pertinent information of the POI, the textures of the 3D model and associated multimedia content are downloaded from the server to be displayed in the Augmented Reality Viewer.

The feed data of the camera is validated in search of a valid QR TAG within the system. In case of finding a match, the pertinent information of the POI, the textures of the 3D model and associated multimedia content are downloaded from the server to display it in the Augmented Reality Viewer.

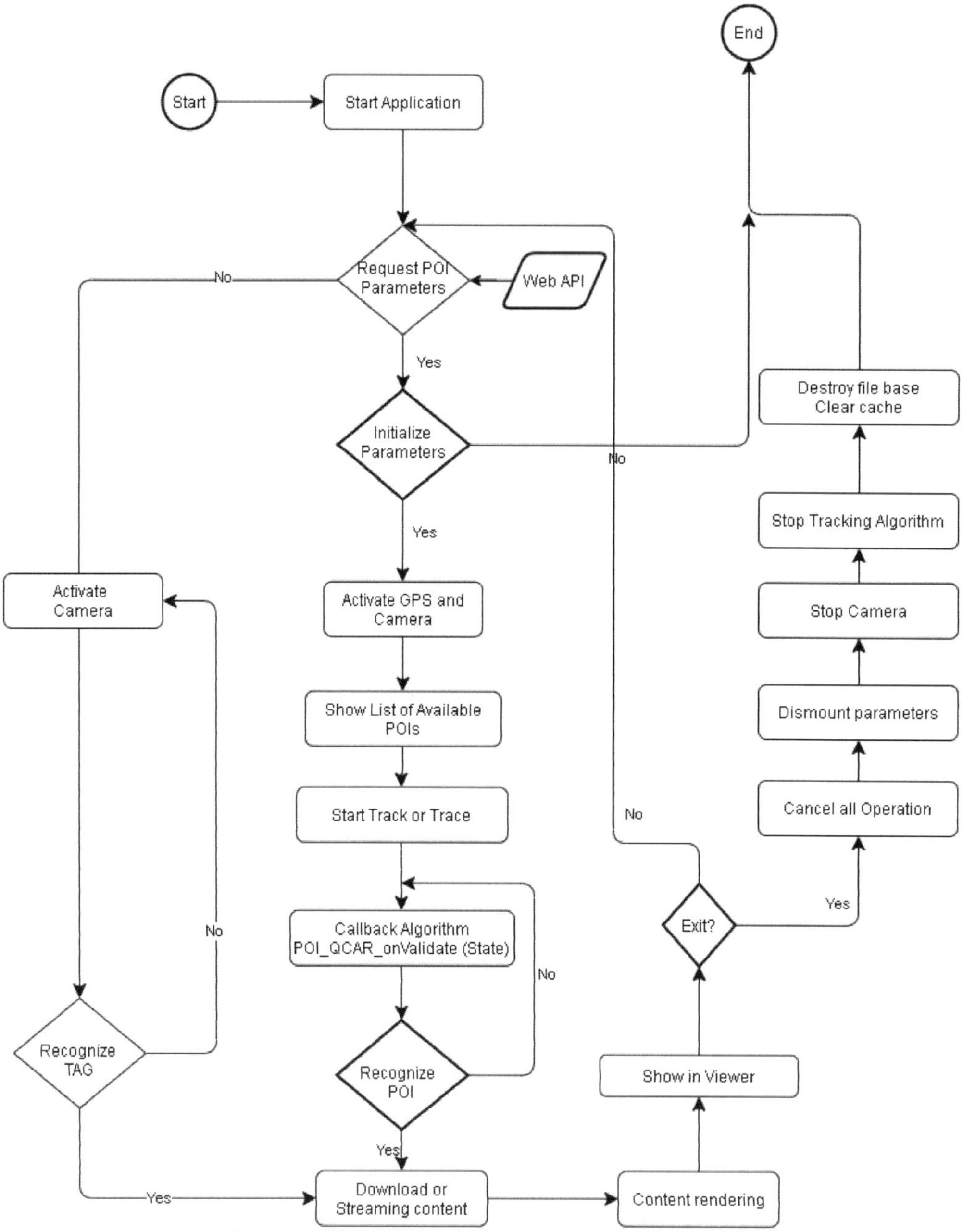

Figure 3. Sequence diagram - customer. Source authors of the article.

Next, the operation of the application is exposed starting from the point where the Mobile App makes a request to the API. (see figure 4):

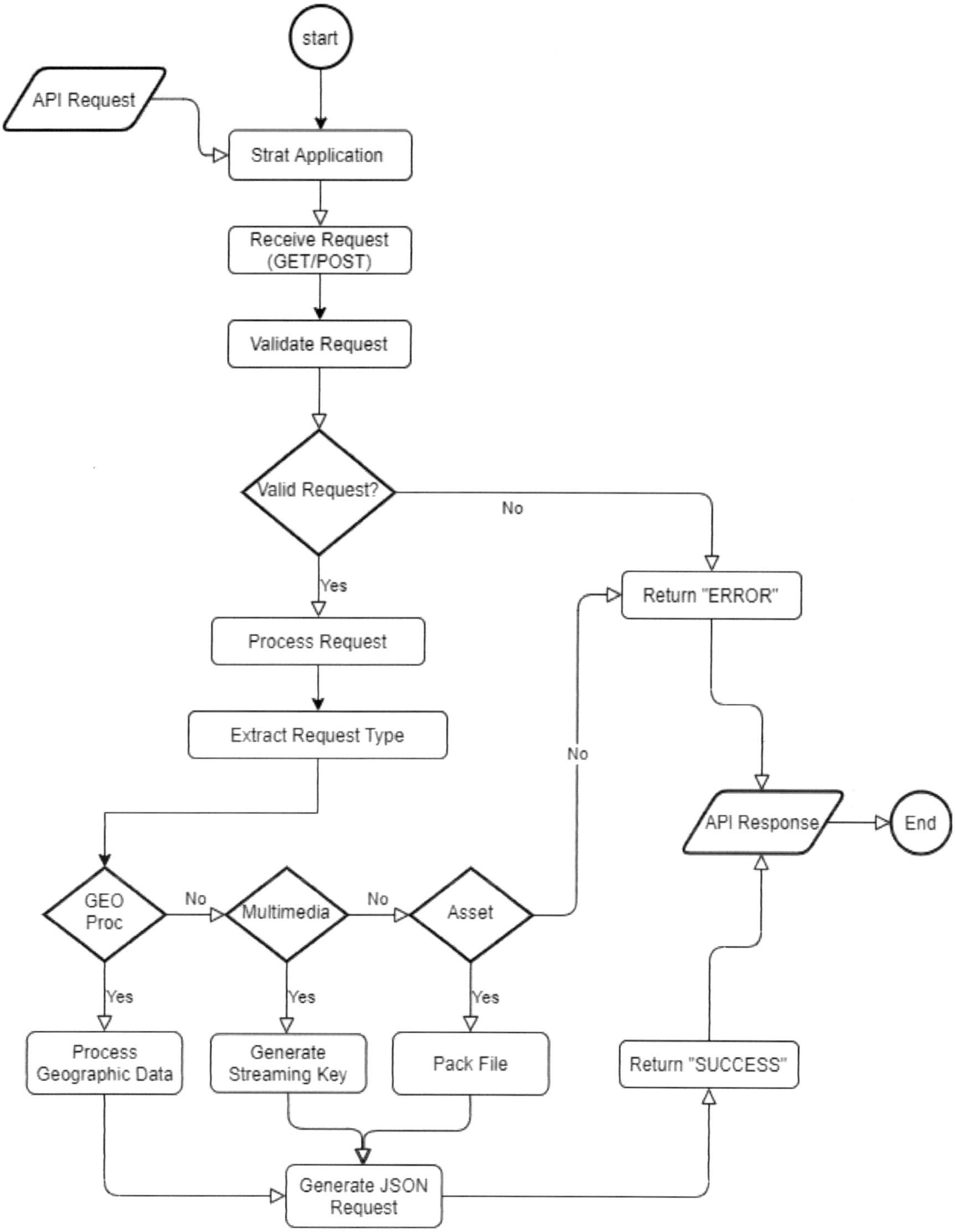

Figure 4. Sequence diagram - Server. Source authors of the article.

Pre-requisite:

(Constant connection to the Internet and the App must have made a Request to the API). The process described below is a general view of a request to the API, because given the nature and type of request, a more specific process must be carried out to deliver a response according to the request.

The process carried out by the application is as follows:

When starting, the application receives an AJAX request from the mobile application, this request must be validated, seeing that it comes from a valid source, that it does not contain wormStrings and that it is correctly composed within the valid patterns of the API. After validating the request in case of not passing the test, an "ERROR" response is returned that will be processed by the corresponding Mobile App and the request to the API is completed. Otherwise, if the request is correctly validated, we proceed to process the request and verify the type of request, and then process the specific data of the request, generate a JSON Response as a result of the algorithm and return "SUCCESS" so that The Mobile App processes the Chord response.

Next, the general operation of the system is exposed noting the way in which the various technologies that compose it are related, such as the vuforia SDK, the AR application and the web API. (see figure 5):

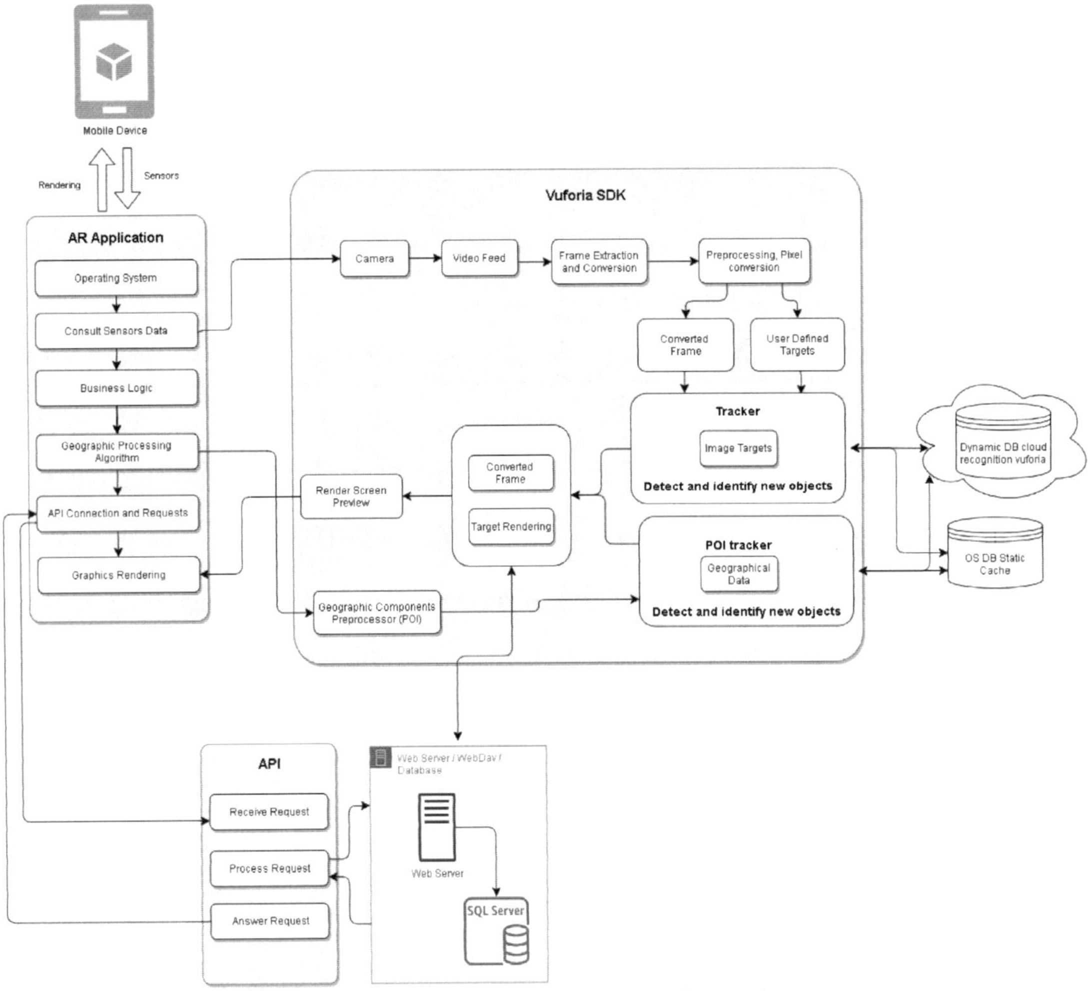

Figure 5. Block diagram - Application interaction.

Vuforia: It is a Framework for the development of Augmented Reality applications, which runs on Unity, so that it integrates with the development of the application with Unity and Vuforia provides tools for the detection of markers, or POIs, based on geographic data, Likewise, it performs image processing to display a composite video feed in the application.

Augmented Reality (AR) Application: This application developed using the Unity + Vuforia Framework, and the application is responsible for processing the geographic data obtained from the GPS of the mobile device, and making inquiries to the API to display relevant information to the end user during the AR process.

API (Web Server / WebDav / Database): The API is an interface that allows the mobile application to connect with the web server, so that the application can request the processing of certain data from the server side, as well as request multimedia content or specific assets.

The following describes the process of how the aforementioned applications interact

The mobile device has a constant interaction with the AR application, which performs different operations, one of which is the query of sensor data, this process makes a request to the Vufuria SDK, which through the camera performs different processes for the correct operation of the application, during this process the recognition algorithms are found, which work in 2 types, one with TAG and the other with POI points, both use the cache memory of the operating system to carry out this task.

In addition, it has Cloud Recognition of vufororia which is used as an administrative platform for the management of "tigger" triggers. When some of the 2 aforementioned methods recognize a point of interest, the content is obtained and rendered, this is done through the web server, which contains all the multimedia content hosted in a database. Then this content already rendered is shown on the device screen, this process is carried out by the AR application, the requests to the server are also made by said application which maintains constant communication with the web server.

5. Conclusions

Tourism, like other economic sectors, has been affected by the decrees issued by the leaders in recent months, where for reasons of the Sarv-Cov. 2 governments have had to implement contingency measures through resolutions to prevent its spread, and based on this, the economy of many countries has been drastically affected, seeing how these measures gradually affect tourism, it has been necessary to implement mechanisms that help to face this situation, with technological advances being one of the options that offer us the most tools when it comes to promoting mechanisms to impart culture in a didactic and creative way, in order to represent the heritage and cultural expressions of certain places.

On the other hand, it is concluded that thanks to the contribution and ideas that are given in the research sector, the problems imposed in the document can be mitigated, where through technological development projects, the AR technology with GPS contributes in what today it is known as sustainable tourism, and providing advance tools for the sector where they are established, in this case and specifically the tourism sector.

The applications proposed for the development of the project provide innumerable tools to be able to carry out completely integral systems, in addition to having a free license, which facilitates the use in certain processes.

By using augmented reality and integrating it with GPS and having a web server where all multimedia content is stored, it is possible to achieve an optimal system, since the content load being hosted on a server, the AR application will consume less resources on the device where it is running.

Finally, it is concluded that the AR integrated with GPS represents great benefits for the tourism sector, representing the cultural part and keeping it intact with the passage of time, giving an enrichment of the entire history to the tourist, involving him in a striking and creative way, diversifying his model representative for all audiences.

References

[1] E. Villarraga, Interviewee, *Coordinador de turismo, adscrito a la Secretaría de Desarrollo y Medioambiente de la Gobernación de Sucre.* [Interview]. 23 junio 2020.

[2] Ministerio de Cultura, "Plan Especial de Salvaguardia. Cuadros Vivos de Galeras, Sucre," diciembre 2013. [Online]. Available: https://www.mincultura.gov.co/prensa/noticias/Documents/Patrimonio/14-Cuadros%20vivos%20de%20Galeras,%20Sucre%20-%20PES.pdf.

[3] CGLU, "LA CULTURA EN LOS OBJETIVOS DE DESARROLLO SOSTENIBLE: GUÍA PRÁCTICA PARA LA ACCIÓN LOCAL," Barcelona, 2018.

[4] M. Gonzalez, N. Huertas and E. Lugo, "Diagnóstico de las organizaciones solidarias del sector artístico-cultural en Córdoba y Sucre," *Revista Venezolana de Gerencia,* vol. 25, no. 89, pp. 174-188, 2020.

[5] OMT, "Desarrollo Sostenible," [Online]. Available: https://www.unwto.org/es/desarrollo-sostenible. [Accessed 11 Agosto 2020].

[6] M. González Vergara and A. Baquero Tobías, "PRODUCTO TURÍSTICO CULTURAL ARTESANAL EN MORROA Y SAMPUÉS SUCRE-COLOMBIA," *International Journal of Scientific Management and Tourism (2018) 4-1: 7-28,* pp. 7-28.Volumen 4-1, 2018.

[7] P. Trocha, "Sombrero Vueltiao: Transformaciones de un objeto artesanal," *Cuadernos del Centro de Estudios en Diseño y Comunicación [Ensayos],* no. 87, pp. 165-220, 2020.

[8] J. Malca, B. Carrasco, V. Guamán, B. Guevara, F. Ruiz, J. Delgado and P. Vinueza, "REALIDAD AUMENTADA APLICADA AL TURISMO - CASO DE ESTUDIO EN RIOBAMBA, ECUADOR.," *International Conference on Information Systems and Computer Science (INCISCOS),* pp. 116-123, 2019.

[9] D. R. Dela Cruz, J. S. Sevilla, J. W. San Gabriel, A. J. Dela Cruz and E. J. Caselis, "Design and Development of Augmented Reality (AR) Mobile Application for Malolos Kameztizuhan (Malolos Heritage Town, Philippines).," *IEEE Games, Entertainment, Media Conference (GEM),* pp. 15-19, 2018.

[10] C. Rodríguez, "ETOURISM APLICANDO TECNOLOGÍA DE GEOLOCALIZACIÓN, RECORRIDOS VIRTUALES Y MÓVILES DE REALIDAD AUMENTADA," *IEEE Trigésima Quinta Convención Centroamericana y de Panamá (CONCAPAN XXXV),* pp. 1-6, 2015.

[11] J. Fombona, M. Del Valle and E. Vázquez, "Análisis de la geolocalización y realidad aumentada en dispositivos móviles, propuestas socio-educativas relacionadas con el entorno y las salidas de campo," *Revista de curriculum y formación del profesorado,* vol. 22, no. 4, pp. 197-222, 2018.

[12] N. López, E. Loredo and J. Sevilla, "REALIDAD AUMENTADA EN DESTINOS TURÍSTICOS RURALES: OPORTUNIDADES Y BARRERAS," *International Journal of Information Systems and Tourism (IJIST),* pp. 25-33, 2019.

[13] J. Azevedo and B. Alturas, "The Augmented Reality in Lisbon Tourism," *Ieeexplore,* pp. 1-4, 2019.

[14] MinCiencias, "Tipología de proyectos calificados como de carácter científico, tecnológico e innovación.," MinCiencias, Bogota, 2016.

[15] MinCIT, *Metodología para la elaboración del inventario de atractivos turísticos,* Bogota, 2020, p. 26.

[16] MinCIT, *Metodología para la elaboración del inventario de atractivos turísticos,* Bogota, 2010, p. 3.

[17] Septiembre 2007. [Online]. Available: http://sedici.unlp.edu.ar/handle/10915/64407.

Design of a telemedicine system for the provision of primary health care (PHC) services

D. Jaramillo[1], M. Mondragon[1], A. Morales[1], S. Sanchez[1] and H. Romero[1],

[1]Antonio José de Sucre University Corporation (Colombia).

alex_morales@corposucre.edu.co

Abstract. Currently, technological progress has allowed health-related issues to be worked on collaboratively, expanding their learning, managing to provide health services in remote places, guaranteeing one of the fundamental rights of every human being, among these technological advances are Telemedicine stands out, which has been playing a leading role especially in these times of pandemic. For this reason, the objective of this research is to design an information system that integrates multiple technologies to support the provision of primary health care services. The research is descriptive and technological development in scope. As a final result, block diagrams and flow diagrams were obtained to represent the set of technologies to be used and the information data flow of the proposed system. It is concluded that the proposed system would facilitate the provision of health services in areas with or without internet connectivity, as well as support decision-making by executives of health organizations, in addition, telemedicine is a new technological trend for what projects like this allow Colombia to be at the forefront in health and technology issues.

1. Introduction

The interaction of technological advances, government regulation and market dynamics in the telecommunications industry has been a booming topic in the context of research [1]. In the last decades the generations of telecommunications called 1G, 2G, 3G, 4G and now 5G have experienced extremely rapid growth and promising deployment of use in all sectors that benefit from telecommunication [2]. Among these sectors is the health sector, which must ensure the provision of high-quality services to its users, because health is a fundamental right of every human being, and thanks to the increase in the use of mobile phones and access to high-speed mobile networks, an expansion has been created in the offer of activities that link to health. Among these activities we can highlight telemedicine, which is the focus of this document.[3]. Telemedicine allows the health ecosystem to be able to visualize, communicate and transfer patient data through a communication network, contributing to collaborative learning in the health area and to the timely treatment of patients who are located in neighboring areas or who, for Disability reasons make it difficult for them to travel to health centers.

Nowadays in different countries they make use of this new technological trend through simplex or duplex communication, the first is characterized by the fact that there is a single transmitter and receiver in the communication scheme, allowing data and information to be transmitted in a single direction, while the second consists of a two-way communication scheme during telemedicine sessions, in which tele-consultation, tele-monitoring, live surgical skills training, tele-mentoring, among other techniques that require two-way interactive sessions can be applied. [4].

Colombia presents a challenge in matters of health service coverage, due to the fact that a large percentage of Colombians live in areas near the urban area, which often do not have a complete road infrastructure, making it difficult for these populations to move to the benefits they provide. They are located within the urban area, it should also be noted that in Colombia few rural areas have health centers, so that the urban area becomes the epicenter of the provision of health services.

For this reason, the research proposes the design of an information system that allows telemedicine to be carried out in rural areas of Colombia, taking into account the little technological infrastructure and internet coverage in these places. The document is divided into the following parts, initially the theoretical framework of telemedicine is proposed, then the methodology of the study is exposed and finally the results and conclusions are described by the researchers.

2. Theoretical framework

2.1 Telemedicine

According to Daniela Chueke [5], Wilson Giovanni, Jhoanna Jiménez and Sareth Acuña [6], affirm that telemedicine is the exchange of medical information between two sites through information and communication technologies, wired and wireless, such as video conferences, smartphones, emails, among others, to your Once the use of these devices improves the clinical health of a patient, providing education and care.

Likewise, Reyna M. Durón, Néstor Salavarría, Heike Hesse, Andrea Summer, Kenton Holden [7] Imayasil Castellón Rodríguez [8], Olga Stolik Lipszyc, Claudia Jáuregui Izquierdo, Laura Galeano Zaldivar [9], establish that telemedicine is the provision of remote media services, through electronic telecommunications means, which includes the diagnosis, treatment, education and medical research of patients, guaranteeing the quality of health services provided in any geographic environment that requires medical attention for its inhabitants.

For her part, Jenny Paola Buitrago Aguilar [10], defines telemedicine as the use of information and communication technologies to bring health services to sectors that are far from the urban case, generating the optimization of the provision of these services.

Unquestionably, telemedicine is distance medicine that relies on the constant evolution of information and communication technologies to provide optimal health services in hard-to-reach areas, guaranteeing the well-being of patients and the quality care provided to these through the valid exchange of information on diagnosis, treatment and prevention of diseases, given through doctor-patient interaction, through the use of technological tools such as: the internet, videoconferences, smart phones, portable ultrasounds, videophones , radio, among others.

2.2 Telemedicine services and classification

Today, telemedicine offers tele-consultation, tele-monitoring, tele-diagnosis, tele-assistance, teleradiology, health emergencies, tele-surgeries, among others, which are carried out through various classifications of the telemedicine.

For Wilson Giovanni, Johanna Jiménez, Sareth Acuña Gómez[6] and Jenny Paola Buitrago Aguilar [10], telemedicine is classified into three groups:

- *Storage and shipping*, are the processes that are carried out through an asynchronous communication route, for example, email, by which exam documentation or other documentation can be sent, to be studied by medical personnel, to later send the respective results to the birthplace.
- *Remote monitoring*, allows health service providers to monitor patients in settings outside health centers, that is, it is done at home.
- *Interactive telepresence*, this type of telemedicine consists of the provision of health services in real time, where the doctor-patient interacts synchronously, either by videoconferencing

or another ICT tool, which allows simultaneous communication between the patient and the health professional.

2.3 Conditions and elements for the implementation of telemedicine

Every technological advance at the time of being implemented or executed requires conditions and elements for its proper functioning and evolution, for telemedicine is no exception, in the case of conditions Reyna M. Durón, Néstor Salavarría, Heike Hesse, Andrea Summer, Kenton Holden [7], establish that for the implementation of telemedicine it is necessary to ensure that:

- *Network security,* establish passwords and protection software that ensure tampering with the network.
- *The privacy and confidentiality of patient information*, You must work and store patient information in a prudent manner, without compromising their integrity.
- *Informed consent* It is important to use this instrument and have the monitoring of the ethics committees of the participating institutions.
- *The benefits to the patient*, the patient will always benefit from the case discussions. However, the legal responsibility for second opinions should be noted.
- *Ethical training of suppliers*, all participants in the tele-consultations must be trained on the privacy and confidentiality of the service.

For their part, Olga Stolik Lipszyc, Claudia Jáuregui Izquierdo, Laura Galeano Zaldivar [9], determine that the conditions to keep in mind for the management of telemedicine are:

- *ICT infrastructures*, to have multimedia network services for healthcare.
- *E-Health infrastructures*, the general aspects of ICT necessary for the development of the telemedicine service must be determined and ensured.
- *Support from research and education networks*, this aspect refers to the link with specialized providers of high-speed internet services, dedicated to supporting the exchange of data.
- *Ensuring the usability and technological scalability of the service*, it is necessary to ensure the usability and technological scalability to guarantee the correct implementation and development of the telemedicine service.

From another perspective, Jenny Paola Buitrago Aguilar[10], recommends that for the implementation of Telemedicine it is necessary to take into account elements such as: the determination of the demographic and epidemiological factors of morbidity and mortality, the economic capacity of the region and the resources available for the implementation, visit and prepare a detailed questionnaire of all the hospitals object of the implementation as infrastructure and to establish the services that are provided in the institution.

2.4 Advantages and Disadvantages of Telemedicine

The implementation of telemedicine brings with it advantages and disadvantages that are reflected at the time of its execution, starting from the perspective of different authors. In figure 1, several conceptions expressed about the advantages of the implementation of telemedicine are reflected. Similarly, in Figure 2, the disadvantages of this new health technology are analyzed.

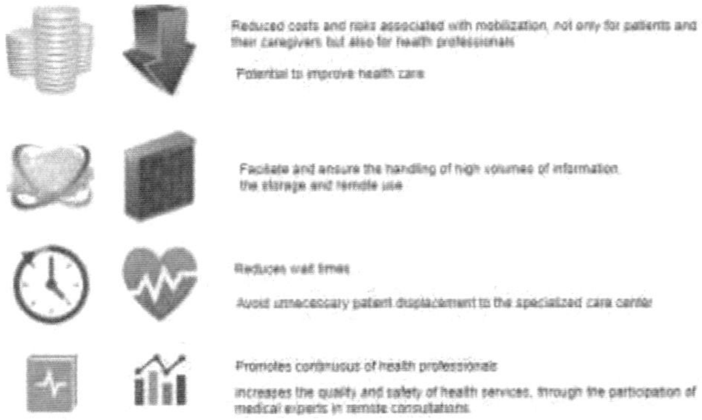

Figure 1. Advantages of Telemedicine. (Source: Own) referenced information[5], [6], [7], [8], [9], [10], [11]

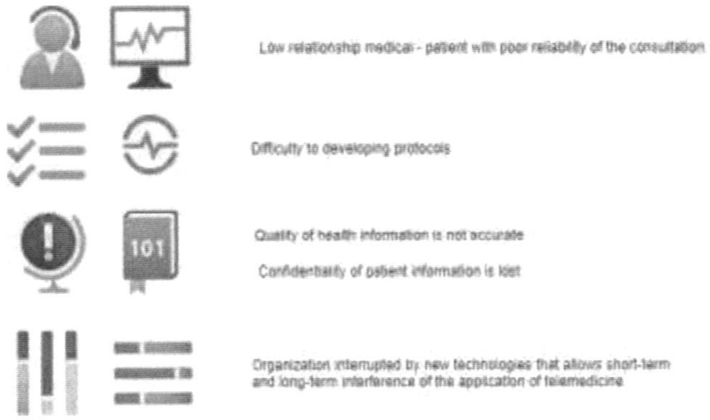

Figure 2. Disadvantages of Telemedicine. (Source: Own) referenced information from[10]

3. Methodology

The research is initially descriptive according to Hernández Sampieri, this type of research seeks to specify the properties, characteristics and profiles of people, groups, communities or any other phenomenon that is subject to analysis. [12]. Within the research process, an analysis of the state of the art was carried out to know the latest telemedicine findings and determine the minimum technical requirements and technologies that the system to be designed must have.

The final phase of the research is of a technological development type, according to MinCiencias this type of study occurs in the application of the research results, or of any other type of scientific knowledge, for the design or development of new processes, systems or provision of services, as well as the substantial technological improvement of pre-existing materials, products, processes or systems. This activity will include the materialization of the research results in a plan, scheme or design, as well as the creation of non-marketable prototypes[13]. The final purpose of the research is to generate the design of a computer system that implements telemedicine activities using technologies to meet the minimum needs for rural populations in Colombia.

The research is embodied in two phases, taking into account the software development methodology called Cascade.

In the requirements phase, the documentary analysis of the information organized with the bibliographic matrix and analyzed with the content analytical matrix was carried out, in addition to the application of semi-structured interviews with health personnel, in order to identify the functional requirements and non-functional of the system proposed in the present investigation.

For the design phase, block diagrams and flow diagrams were used to outline the architecture of the system and illustrate the communication between the different technologies that make up the system.

4. Results

Initially, a survey of requirements was made (see table 1) through documentary analysis, using the bibliographic matrix and content analytical matrix instruments, in addition, semi-structured interviews were implemented with health personnel from the Cruz Rojas Sucre Section in Sincelejo. Based on these requirements, the proposal of the system architecture is made, through block diagrams and flow diagrams.

Table 1. Requirements.

Functional	Not functional
Registration, consultation, elimination and updating of data related to patients, doctors, auxiliaries and administrators of the System.	Intuitive interface
First level data record in patient health or physical examination.	Terms and conditions for the treatment of information
Data storage on the web server (online) or locally on the device (offline)	Security in the transmission of data over the network.
Allow synchronization of device data with web server data (online)	Use of technologies such as HTML, PHP, Ajax, SQL, Chart.js library and SSL certificate
Elaboration of graphs of circular, bar and linear type using information stored in the database	MVC architectural design
Registration, consultation, elimination and updating of diseases and causes	Use of data entry techniques.

By drawing up the block diagram (see figure 3), the behavior of the functional requirements of the telemedicine prototype is illustrated, where the systems based on both the cloud and the local are shown in each block, the scheme of which presents the following technologies, a) structure of the view or interface (HTML), b) PHP, c) Database (SQL), d) chart.js library, e) Ajax and f) 4G technology, Communication protocols and protocols are also displayed in each layer. security, such as: a) HTTPS protocol, b) IP protocol, c) TCP protocol and b) SSL protocol.

Subsequently, the general operation of the telemedicine prototype is defined, and the activities carried out by each of the technologies mentioned above.

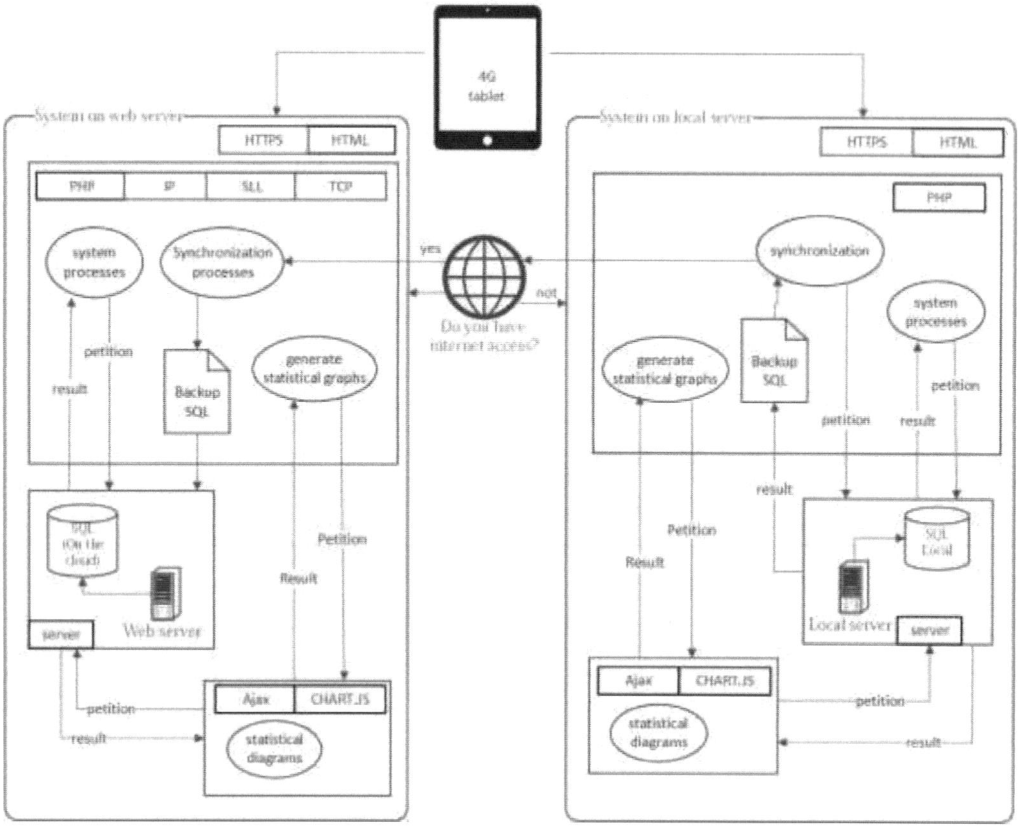

Figure 3- Block diagram - telemedicine prototype. (Own source).

When starting the telemedicine system 4G technology is used, the system verifies if there is internet coverage in the area, since there is no internet, the system will proceed to work offline, recording the data registered in the local database.

Once the data records are finished and the internet has access, the synchronization process is carried out, which makes a request to the database (local SQL), which allows the download of the SQL backup to carry out the upload and update in the database (SQL) that is in the cloud.

After performing the synchronization (whether or not the web server database (Cloud SQL) is updated) the system will have the ability to make Ajax request to the database (Cloud SQL or Local SQL), where it generates a query, later returns a response that is processed with the chart.js library, generating statistical graphics, which benefits when making decisions. The technologies that make up the system are:

- *HTML*: It is the hypertext markup language, which is responsible for the interface or visual for the client side, this is used for the creation of web pages.
- *PHP*: This is an open source programming language, based for web development, which processes on the server side, performs all kinds of processing such as the connection with the database, and performance of internal system processes, this on the other hand it can be embedded with HTML.
- *SQL*: This is the structured query language, where it facilitates the manipulation and integrity of the information established in the database stores.
- *Chart.js*: It is a library based on JavaScript to make graphics of all kinds, where it represents data based on a given information, either through a file, data structures such as lists, stack and array or

in the form of a query to the stores of database, displaying the data graphically (data bar, pie and donut pie charts, line charts, radar chart, among others).

- *Ajax (Asynchronous JavaScript and XML):* It is a technique that combines several programming languages, it consists of requesting data from the server, it returns the data, while the page has a response time, to obtain the results on the client side.
- *4G technology*: It is the fourth generation of mobile technologies (wireless modems, Smartphone, among other mobile devices), in data transmission, it is based on the IP protocol, for processing access to the network, it also has a capacity of 100Mbit / s in motion and 1Git / s idle with respect to internet access speed.

The flowcharts, each of the following components, system data log, physical exam log, system stat graphs, and timing are described in detail below.

4.1 Data logging component in the system.

Next, the data recording process in the telemedicine prototype is defined, using a flow diagram (see figure 4).

Initially the system has a login to allow user access, for this it makes a verification of the database, depending on whether it has internet access or not, then it returns a result of existence, if the result is no, it returns to login with a login error message, if the result is yes, it is entered into the telemedicine system.

Once the entry has been validated, internally the system must validate the access level of the user, whether he is an administrator, doctor or assistant.

If you give him administrator user permission, he has the functions of a) access to the registry of information about patients, doctors, assistants and system administrators and b) access to the registry of diseases and their causes.

If you give permission from a doctor or assistant, you are allowed to a) register patients and b) medical records of patients.

The administrator records the information of the patient, doctor, assistant or system administrator, if the system is connected to the internet the data is stored in the database that is in the cloud, if not, the data is stored in the local database. It also records diseases, for them a search query is made in the database, if the disease does not exist, it proceeds to register it, if the disease exists, the user can give the causes of the diseases, at the final if the system has internet access the data is stored in the cloud database, if not, it is stored in the local database.

The doctor and assistant, verify the patient's stock in the system, if the patient does not exist, it is added to the system, if the patient exists, make a new patient consultation.

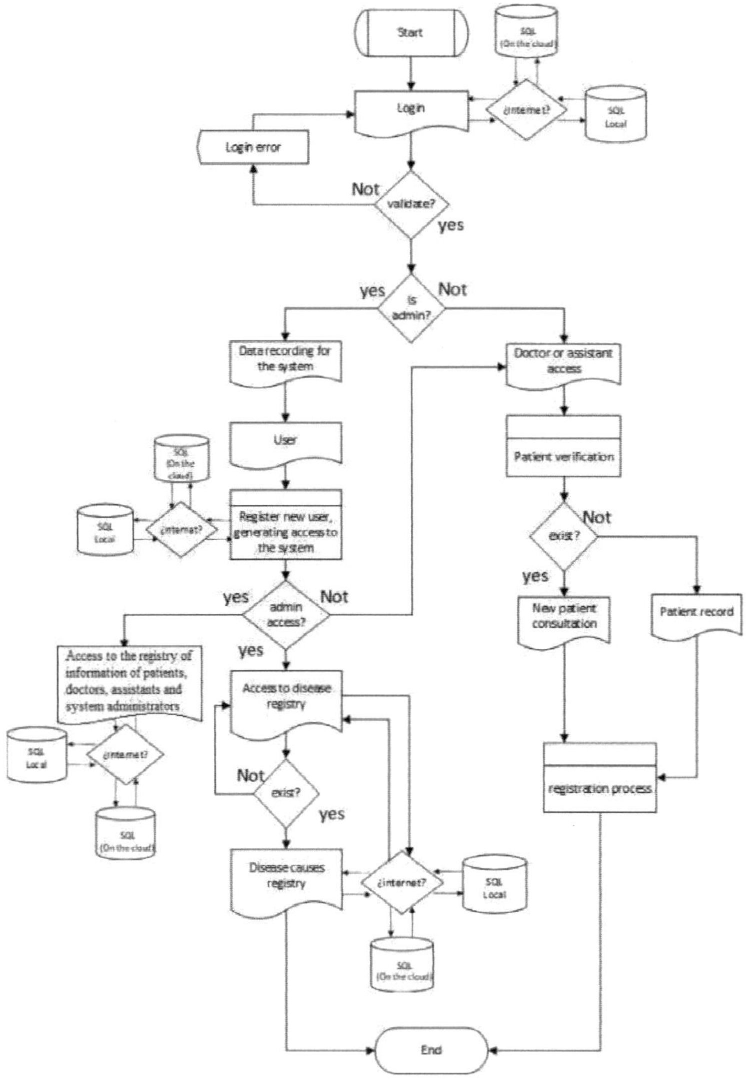

Figure 4 Flow chart, data recording in the system. (Own source)

4.2 Physical examination record component.

In this section, the existing and non-existent patient physical examination record diagram is defined in the telemedicine system (see figure 5).

Initially, the patient process performs an existence search, for them, if it has internet access, it will search the cloud database, if it does not search the local database.

If the patient does not exist, a patient not found message will be displayed, and it will be redirected to enter data for the new patient, this will ask for the personal data of the new patient, data from the physical examination, doctor or assistant in charge, current location of the consultation, if the patient has any disease or not, and reason for the consultation.

If the patient exists, a new consultation is added, where you must enter the following information, data from the physical examination, doctor or assistant in charge, current location of the consultation, if the patient has any disease or not, and reason for the consultation.

Once all this data has been collected, depending on the internet access, if you have the data it is stored in the cloud database, if it is not stored in the local database.

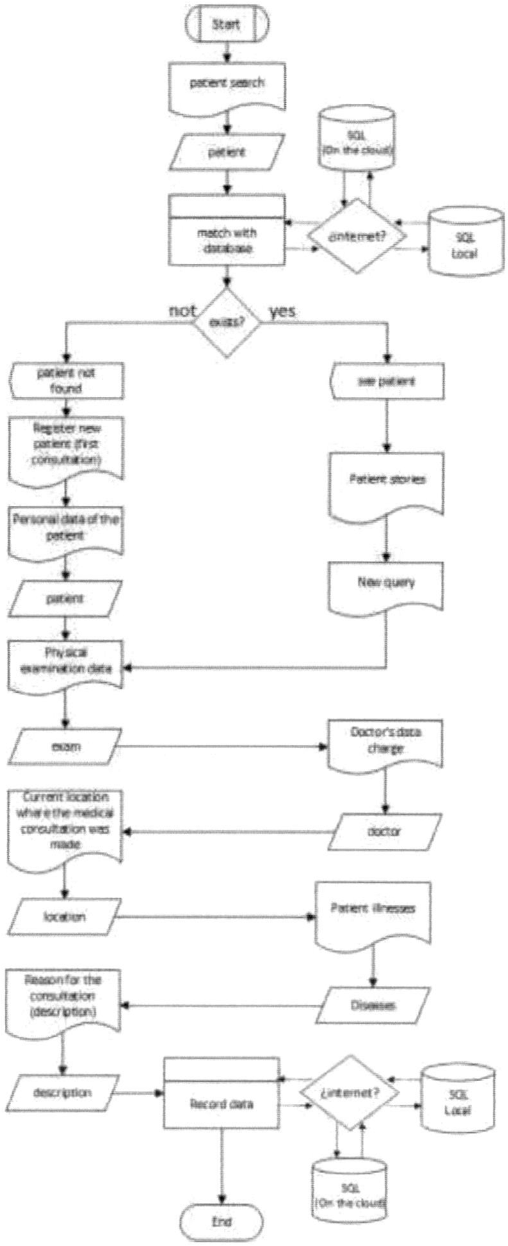

Figure 5 Flow chart, physical examination record. (Own source)

4.3 Statistical graphic components of the system

Subsequently, the flow diagram was elaborated (see figure 6), where the statistical generator process of the telemedicine system was determined.

It starts with a simple question, what statistical graph do you want to generate? Gradually, you enter the query to be generated.

Then an internal process is carried out which makes a request through Ajax, if the system has internet then the query is made to the cloud database, if it is not, the query is made to the local database.

Once a result of the request has been obtained, it generates a statistical graph with the query generated previously.

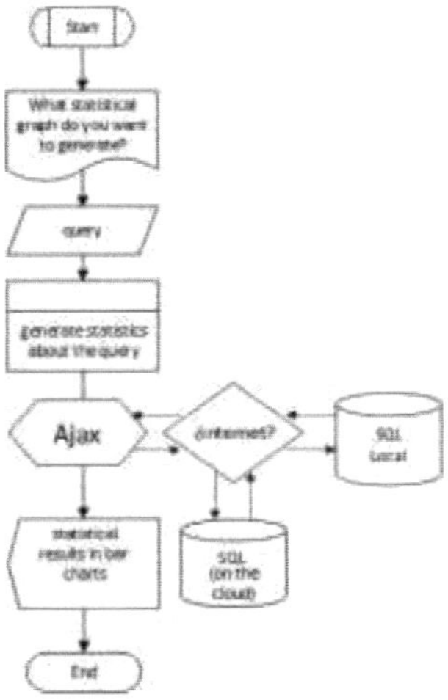

Figure 6 Flow chart of the statistical graphics component. (Own source)

4.4 Synchronization components.

Finally, the synchronization flow diagram of the local database with that of the cloud is described (see figure 7).

When starting the synchronization process, it must be ensured that the mobile device has internet access.

If you do not have access to the internet, it can carry out the registration processes on the local system, in the certain processes they are saved in the local database.

To achieve synchronization, this download a backup of the local database, once this action is performed, it is asked again if it has access to the internet, if so, it proceeds to the cloud system, where the process of upload the backup to the cloud database.

If it has always been on the Internet, what it does is save all the records and processes carried out in the cloud database.

Figure 7 flow diagram of synchronization of the local system with the cloud system. (Own source)

4.5 Discussion

Telemedicine is one of the technological trends in the field of health that has been studied by different researchers in the area of health and engineering in the last two decades, expanding its role in the provision of health services; which leads to continue developing research and innovations in this field of technology and engineering. Reviewing what is involved in this research, it is detailed that there are two types of telemedicine according to the authors consulted that stand out for the provision of health services in places of difficult access, among them there is remote monitoring or home consultations through technologies and Interactive telepresence or consultations through videoconferences or other technological tools that allow synchronous communication between patient and doctor.

On the other hand, the proposed system seeks to offer representations of important data for health organizations, such as the most frequent diseases, characterization of rural populations in terms of health, segmentation between genders and their health states, segmentation between age and their health states. health, among other possible categories in order to be able to make more accurate decisions when planning future health brigades in these communities.

This system seeks to support the provision of first-level health services in places far from the urban area or areas of difficult access, mitigating the different disadvantages raised by the authors consulted, among these advantages is the low interaction between patient and doctor, difficulty in the development of health protocols, confidentiality of patient information, quality of the information provided by the patient in the consultation and interruptions due to technical difficulties, which are solved with the home consultants, where the health personnel guarantee compliance with all these tasks; If it focuses on the last disadvantage, interruptions due to technical difficulties, this is where the synchronization system can support the lack of technological infrastructure in the place of residence of these communities.

5. Conclusions

At present, telemedicine has become a strategy to break down barriers to physical access to health services, supporting its development in new information and communication technologies, which allow tele-consultation to tele-surgery. In the same way, telemedicine has allowed asynchronous and synchronous communication between the health professional and the patient.

In the same vein, it can be concluded that the implementation of telemedicine brings benefits to society and mainly to the field of health, such as the reduction of cost and time to patients, improves the quality of services health, greater opportunity to access the right to health, among others.

On the other hand, it is concluded that the integration of statistical graphics in the proposed telemedicine system provides organization of the data, through these various variables can be compared, yielding various results, which facilitate and favor decision-making for positions administrative, to establish strategies to improve the provision of health services.

It should also be noted in conclusion that the operation of synchronization brings as an advantage access to the telemedicine system, without the need for an internet connection, that is, it favors health brigades that are carried out in areas without internet coverage, ensuring the provision of the health service and the registration of data that will later be updated with the web server through synchronization.

To conclude, it should be noted that telemedicine is a new technological trend that is being implemented in different countries of the world, and projects like these allow Colombia to be at the forefront in health and technology issues.

References

[1] X. Lin, T.-J. Lv and X. Chen, "The coevolutionary relationship of technology, market and government regulation in telecommunications," China Communications, vol. 15, No. 8, pp. 152173, 2018.

[2] Y. Xiaodai, W. Xu, and M. Yuen, "Cellular and WiFi Co-design for 5G User Equipment," IEEE 5G World Forum (5GWF), pp. 256-261, 2018.

[3] CR-RJA-PPR-CA Mariano-Melo, "Telemedicine Acceptance in Brazil: Explaining behavioral intention to move towards internet-based medical consultations," 15th Iberian Conference on Information Systems and Technologies (CISTI), pp. 1-4, 2020.

[4] R. Dhaman, A. Kumar, A. Kumar, P. Tiwari, R. Rajnish and S. Kanta, «Advanced Communication Technologies for Collaborative Learning in Telemedicine and Tele-care,» 9th International Conference on Cloud Computing, Data Science & Engineering (Confluence), pp. 601-605, 2019.

[5] D. Chueke, "Latin America Telemedicine Panorama," 2015. [Online]. Available: https://www.teleiberoamerica.com/publicaciones/TelemedicinaAmericaLatinaEyeforPharma04-16-2015.pdf.

[6] W. Giovanni, J. Jiménez and S. Acuña, «Advances in telehealth and telemedicine: strategy to bring health services closer to users,» 2019. [Online]. Available: http://www.bdigital.unal.edu.co/60180/.

[7] NSHHASKH Reyna M. Durón, "Prospects for telemedicine as an alternative for health care in Honduras," 2017. [Online]. Available: https://www.camjol.info/index.php/INNOVARE/article/view/3184.

[8] I. Castellón Rodríguez, "Proposal for network architectures for telemedicine," 2016. [Online]. Available: http://www.auditoriamedicahoy.com.ar/biblioteca/TELEMEDICINAfinalDra.%20Jenny%20Buitrago.pdf.

[9] CJILGZ Olga Stolik Lipszyc, "TELEMEDICINE: HEALTH AND ICT SERVICES," 2019. [Online]. Available: https://www.researchgate.net/profile/Laura_Galeano_Zaldivar/publication/337740803_REVISTA_CUBANA_DE_ECONOMIA_INTERNACIONAL_TELEMEDICINA_SERVICIOS_DE_SALUD_Y_TIC_Telemedicine_health_services_ICdANAlinks3VISTA_SERVICES6_ICD28 / cclinks3-VISTA_SERVICES6_ICdTUBA7 / ccTELEMEDICINE_Y_TIC_Telemedicine_health_services6_ICDANA-

[10] JPB Aguilar, «Telemedicia,» 2019. [Online]. Available: http://www.auditoriamedicahoy.com.ar/biblioteca/TELEMEDICINAfinalDra.%20Jenny%20Buitrago.pdf.

[11] MC Moreno, «REMOTE DIAGNOSTIC SYSTEM FOR RURAL HEALTH CENTERS FROM,» 2017. [Online]. Available: http://repositorio.puce.edu.ec/bitstream/handle/22000/13707/TESIS%20MARIA%20CECILIA%20MORENO%20BRAVO.pdf?sequence=1&isAllowed=y.

[12] R. Hernandez, Research Methodology, Mexico: McGRAW-HILL, 2014.

[13] MinCiencias, "Typology of projects classified as scientific, technological and innovation in nature.," MinCiencias, Bogota, 2016.

Evaluation of a virtual learning object with augmented reality technology for teaching of the computer parts.

I J Romero [1], H J Romero, K A Sierra[2], A D Morales[1], S A Sánchez [1],

[1]Corporación Universitaria Antonio José de Sucre, Sincelejo, Colombia,
Faculty of Engineering Science, GINTEING Research Group
[2]Institución Educativa COLMERCES, Sincelejo, Colombia
ingrid_romero@corposucre.edu.co

Abstract. This article refers to the evaluation of the Virtual Learning Object (OVA) Computer AR, OVA with augmented reality technology to facilitate the teaching-learning of the concepts of computer maintenance, related to the internal parts of the computer, with the sixth grade students of the COLMERCEDES educational institution. The methodological design with a quantitative and descriptive approach involved the application of the OVA quality assessment instrument called LORI (Learning Object Review Instrument), under OVA quality assessment dimensions and criteria, which determine that it is a quality digital teaching resource. The quantitative descriptive analysis focuses on the assessments of students and teachers when exploring the tool and its contents, aspects such as motivation, usability, accessibility, among others, are measured, so that the relevance of designing and using interactive virtual objects with augmented reality technology for teaching and learning the basic concepts related to the internal parts of the computer is concluded. Among the results obtained were positive, it was evident that the OVA designed met the quality criteria to obtain a high assessment by the population under study, concluding that it was appropriate to design an OVA with augmented reality to generate contributions to the teaching and learning processes on concepts related to the internal parts of the computer. The data in the dimensions of evaluation with a low tendency indicate that it should be complemented with a contextualized setting, so that it can be adapted to the characteristics and needs of the students.

1. Introduction

Information and communication technologies (ICT) make it possible to access information in such a way that it is available in any space and time, and they are a support tool in different professional fields such as social, health and education. In education, for example, they serve as a pedagogical support [1], ICTs s are recognized as innovative resources that allow teachers to implement various teaching strategies [2], and they provide environments to monitor the learning of each student, improving the performance of students and teachers [3]. The demand for technological resources in educational environments has generated in teachers who teach subjects such as technology the need to seek strategies that enhance the work of teachers so that students have more than one alternative for learning, so in education with face-to-face mode is important to include new technologies and the application of quality educational material standards adapted to emerging trends every day, Among the strongest is mobile learning and augmented reality, in short, the trend towards mobility, ubiquity and permanent access to information in a virtual, interactive and closer to reality, so that the classroom is not the only environment available for learning.

The teaching of technology has become one of the needs in educational environments, where the most relevant issues are those related to recognizing the difficulty of teaching the parts of the computer in educational institutions that lack the physical resources to guide such classes. The Computer AR application provides a battery of opportunities to teach students the different parts of the computer, visualizing in real time and virtually the components such as: the processor, RAM, power supply, hard disk, memory card reader, etc. The students will be able to see and hear through the "app" each of these parts as well as their concept. It is worth mentioning that in the academy the teaching and learning of concepts related to computer parts is not easy given the scarce access to such components and that these internal parts of the computers are very delicate and prone to damage with constant or inadequate manipulation, especially in students between the ages of 9 and 14 years who belong to the sixth grade of educational institutions such as COLMERCEDES, whose computer room does not have this resource as opposed to external parts that allow greater manipulation without incurring damage.

In response to this perspective and problem, the objective of this research is to evaluate the Virtual Learning Object (VLO) with students in the sixth grade at COLMERCEDES using the LORI (Learning Object Review Instrument), a validated tool that allows for the evaluation of learning objects [4], which measures the quality of an OVA in nine dimensions. The OVA Computer AR was designed to teach concepts related to the internal parts of the computer, under the OVA design model called MOVA (Model for the design of virtual learning objects for the subject of computer maintenance). This model was the result of research carried out by a group of thesis students at the Antonio José de Sucre educational institution in 2018. In this context, Computer AR is an OVA with augmented reality technology with the potential to provide a virtual environment for teaching and learning that is appropriate for the needs of the context, offering students the opportunity to learn about the internal parts of the computer. In this sense, it is pertinent to validate it in order to identify some characteristics that contribute to the redesign or adaptation of the presentation of the virtual object. At the end, the results of the present investigation and together with the diagnostic evaluation, it will be tried in a future proposal to implement and to adapt the OVA, so that this one adapts to the own characteristics, the needs and the interests of the population object of study.

1.1 Learning through mobile devices

Movil Learning: The use of mobile tools such as laptops and tablets, MP3 players, smart phones and mobile phones. As more and more students are bringing their personal mobile devices to higher education institutions, this trend offers an unprecedented opportunity to use these devices for teaching and learning [5], [6]. Mobile, personalized, portable, cooperative, interactive and contextualized learning has unique characteristics not found in traditional e-learning [7]. 7] The former emphasizes access to information at the right time [8], since instruction through it can take place anywhere and at any time. As an aid to formal and informal learning, it therefore has enormous potential to transform educational provision and training. Mobile learning is becoming one of the solutions to the problems facing the education sector. This is why UNESCO's programme of activities is based on a growing number of joint initiatives to explore how mobile technologies can support the achievement of Education for All. Its partners include Nokia and the United States Department of State [9].

1.2 Augmented Reality

Augmented reality in education: Augmented reality -AR is a technology that provides seeing the digital world in the real world, through a technological device that supports such technology, the AR what it does is to superimpose images that are generated through a computer, a smart phone or augmented reality glasses in a real physical space so that the user sees a better perception of things. Therefore, AR adds virtual information in an existing physical environment, i.e. digital information is superimposed on a surface with scarce information and combined with other types of technology it is possible for the object to be displayed to become interactive. The applicability of AR is in any field because it allows to add layers of visual information about the real world around us [10].

1.3 Learning Object and Virtual Learning Object

Virtual Learning Object in education: Recognizing that existing definitions are intended to meet the particular needs of the bodies issuing them and can be very ambiguous in relation to the educational value, the Ministry of National Education, with the support of experts from various higher education institutions, has agreed on the following definition, within which the Ministry's initiatives on the subject are framed. A learning object is a set of structural elements. The structure of an object of learning has taken several important turns over time that demonstrate the analysis that the academic community has made of the subject. At the beginning, without using the term of object of learning, we talked about resources that could be reused in different contexts, such as documents or images, whose structure was self-contained in the document's summary, key words or simply the name. Later on, it evolved towards interoperability, where the structure of an object of learning had to contain all the necessary aspects, both technical and pedagogical, in order to "connect" two or more objects, where the initiatives of object description and metadata standards emerged, thus integrating the virtual learning objects - OVAs. The pedagogical value is present in the availability of objectives, contents, activities, and contextualization elements of the OVAs [11].

According to the user manual of the instrument for the evaluation of OVAs, LORI, an OVA: "is an information resource or interactive software used in online learning. A single image, a page of text, an interactive simulation, or a complete course are examples of learning objects" [12].

1.4 OVA Quality

OVA Quality in education: Authors [13] affirm that it is possible to determine the quality of an OVA in a specific context by taking into account the user's perspective. They also describe that the quality of OVAs focuses on identifying the various ways in which the student perceives information and is measurable through the evaluation of the OVA. The LORI OVA evaluation instrument includes 9 criteria or dimensions necessary for quality OVAs, each with specific quality characteristics [12]. Content-quality OVAs have a positive impact on the teaching-learning process and help to strengthen use in educational settings [13].

1.5 Computer AR

Characteristics of the Computer AR Virtual Learning Object: Computer AR is a mobile application that incorporates augmented reality technology, uses markers designed based on the sketch of each component of the computer, as seen in the image, so that when the user uses the application and focus a marker is displayed 3d model of the associated component, with information in audio and text, allowing zoom in and out for better exploration of the component, images 1 and 2, illustrating the internal components of the computer

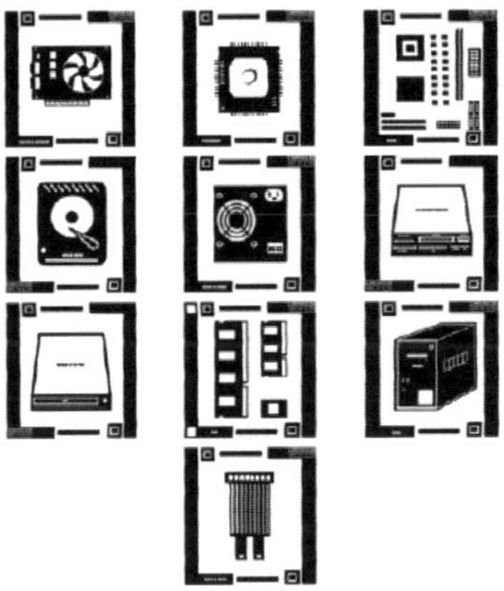

Figure 1. Marcadores

IOP Conf. Series: Materials Science and Engineering 1154 (2021) 012019 doi:10.1088/1757-899X/1154/1/012019

Figure 2. Component RAM memory

2. Materials & Methods

2.1 Methodological Design

This research has a quantitative and descriptive approach that seeks to understand the students' perspective and to deepen their experiences, perspectives, opinions and meanings [16].

For the evaluation of the OVA Computer AR, the LORI (Learning Object Review Instrument) was applied, which is a validated tool that allows for the evaluation of learning objects [4], with characteristics according to evaluation criteria to determine the quality of the OVA, in terms of the quality of the technological and didactic design, and which applies dimensions according to the general criteria for evaluating the quality of OVAs [17]. In this context, it should be considered a valid educational tool for future implementation in the population under study

In applying the LORI instrument, the parameters of the noncomparative continuous rating scale were taken into account, as they allow the measurement of the opinion of the interviewees, presenting various response alternatives in which numerical ratings can also be used [18].

The procedures used to interpret and analyze the data collected from the instrument were the organization of the data using descriptive statistics, specifically central tendency measurement techniques [16]. The use of the mean allowed the categorization of the variables under study (OVA quality level, student and teacher perspective) [17], taking into account the dimensions according to the criteria for evaluating the quality of the OVAs.

A complete inductive logic research method was used, in which the final reflections and conclusions were generated from the study of the results obtained from the population under study, i.e. from each and every member of the [4], [19].

2.2 Participants

The evaluation of the OVA Computer AR was carried out with 80 students of the subject Technology and computer science of the sixth grade of the educational institution COLMERCEDES, with ages between 9 and 11 years and 43 students who studied the subject computer maintenance, in both subjects the learning of the parts of the computer is oriented. In total there were 123 students and 3 teachers.

2.3 Application of the instrument

2.3.1 Virtual Learning Object assessment tool -LORI

For the evaluation of the OVA Computer AR, students and teachers responded to a checklist on functional, technical and pedagogical aspects of the Virtual Learning Object, according to the quality standards of the learning objects, in which the pedagogical, technological and human-computer interaction dimensions should be considered, as proposed by the scientific studies of [20] and [4],

Applying the LORI (Learning Object Review Instrument), LORI is a tool for evaluating learning objects [21], which applies the dimensions for evaluating the quality of virtual learning objects, as demonstrated in a study by the University of Antioquia, which compared it with a model from the University of Salamanca proposed by Morales and others, LORI shows and includes characteristics according to the criteria for evaluating the quality of OVAs [4], [17], [20].

Training was provided on the administrative management and operation of the OVA Computer AR. Students were also trained in the use and operation of the OVA. The OVA evaluation process was carried out using the LORI (Learning Object Review Instrument), which evaluates the relevant aspects of a Virtual Learning Object as contemplated by the Ministry of Education [22]. The LORI tool has been researched and applied with validated results since 2002 [21].

This tool LORI proposes to measure the following aspects of the OVA:

1. Quality of content: truthfulness, accuracy, balanced presentation of ideas and level of detail appropriate.
2. Alignment of learning objectives: coherence between objectives, activities, assessments and student characteristics.
3. Feedback and adaptability: adaptive content directed at each student's response student and his learning style.
4. Motivation: Ability to motivate and create interest in learners.
5. Design and presentation: the design of audiovisual information favours the adequate processing of the information.
6. Usability and interaction: easy navigation, predictive user interface and quality of interface help resources.
7. Accessibility: the design and presentation of the information is adapted for people with some type of disability and for mobile devices.
8. Reusability: ability to be used in different learning scenarios and with students of various contexts.
9. Compliance with standards: Consistency with international standards and specifications

Each aspect was evaluated on a scale of 1 to 5, with 1 being the Low level and 5 the High level, as follows: 1 - Low, 2 - Medium Low, 3 - Medium, 4 - Medium High, 5 - High.

Although the tool offers flexibility on the choice of items, it is important to highlight that none of the items proposed by the LORI were excluded since they are all considered to be aimed at the correct evaluation of the educational material. However, it should be taken into account that the Accessibility item should be considered only from the point of view of accessibility through desktop or mobile devices and the adaptability to the use by disabled people should be excluded, since this aspect is not within the scope of the present project.

3. Results

3.1 Teachers' evaluation

3 teachers participated in the evaluation, who were guiding the subject of Technology. The average OVA score given by teachers was 4.3 and by students was 4.5, which corresponds to a High score on the LORI scale [21]. This rating serves as an indicator that the designed OVA largely meets the objectives set, as well as the quality standards required for this type of educational material.

This shows that the OVA design largely satisfies the expectations of the teachers in the Teacher Survey, and they agree that it meets the pedagogical and technical elements required to achieve the goals set in terms of encouraging student motivation and providing a platform that contributes to improving information accessibility, under quality standards.

Table 1. Teachers' evaluation

Item	Evaluation
Quality of contents	4.6
Appropriateness of learning objectives	4.7
Feedback and adaptability	4.5
Motivation	4.4
Design and presentation	3.8
Usability	4.8
Accessibility	3.8
Reusability	4.2
Standards compliance	4.6
Total	4.3

3.2 Students' evaluation
The aspects best evaluated by the students were the quality of the contents (4.7 - High), the adaptation to the learning objectives (4.6 - High), Design and presentation (4. 6 - High), the other dimensions were at a Medium-High level. This showed that the design of the OVA Computer AR greatly satisfies the expectations of the students and meets the quality criteria of OVAs, which is why it is considered a useful and relevant tool in the learning processes of the computer parts of the Technology and Information Technology subject in the sixth grade of COLMERCEDES and that the students manage to identify that the OVA reflects the themes of the subject and offers feedback and adaptability to their learning process. It is also observed that the motivational component is considered high and that the OVA is in line with current quality standards. 123 students participated in the evaluation, 80 students in the subject Technology and Computer Science in the sixth grade and 43 students who had previously taken the subject

Table 2. Estudents' evaluation

Item	Evaluation
Quality of contents	4.7
Appropriateness of learning objectives	4.6
Feedback and adaptability	4.4
Motivation	4.4
Design and presentation	4.6
Usability	4.4
Accessibility	4.5
Reusability	4.5
Standards compliance	4.0
Total	4.5

4. Conclusions & Discussion

4.1 Conclusions

The research showed that the OVA Computer AR was accepted and highly rated, 69% in High and 21% in Medium High, the quality of content was rated 75% in High and 22% in Medium High, the adequacy of learning objectives, The design of the presentation and reusability showed a high and medium-high level rating of over 90% by students and 70% by teachers, showing that the topics, activities and evaluations proposed were in line with those proposed by teachers and students at the COLMERCEDES educational institution, followed by the item motivation by 61% at the high level.

The evaluation showed that the OVA was adapted to international standards and specifications, and that it was highly valued as a technological and educational tool that can be used in different learning scenarios and with students from different contexts. The usability, feedback and accessibility, as well as compliance with standards tended to be in the range of 60% to 70%, showing that future implementations should take into account improving accessibility to the disabled and adapting the OVA to the ages of the sixth grade. Some of the observations were to include the OVA Computer AR in a digital primer so that the presentation would be more age-friendly, to include help text, and to improve audio quality, which is not within the scope of this research but is important to consider for future implementation.

4.2 Discussion

The role of the new information technologies in the processes of social and cultural change is particularly relevant in the field of education. Certain conceptions of education system reform attribute to the incorporation of new information technologies a decisive effect in improving the quality of teaching and learning processes. But these cannot be determined without an analysis of the political and social conditions that structure teaching practices. The economic, social and political transformations that took place in the seventeenth century stimulated scientific development, setting in motion the technological production processes that characterized modernity, in which knowledge began to respond predominantly to an instrumental or technical rationality". 23], with this it can be understood that the development of computing and its influence on economic and social sectors contributed greatly to the evolution of societies and to the form of communication and interaction with the environment. Taking into account the evolution of information technologies in education, it can be stated that the construction of strategies mediated by ICT has been the result of the work of researchers interested in increasing knowledge regarding the application of strategies that improve and facilitate current teaching methodologies, the principle included audio-visual media, then programmed instruction, as is the case with Skinner when he implemented his teaching machines based on laboratory research in working condition, triggering a powerful movement with the purpose of revolutionizing teaching by trying to base it on scientific principles [24].

Worldwide, the learning of the internal parts of the computer has been one of the aspects that has received the most attention from teachers who teach this area of knowledge because of the complexity of teaching with few resources. At the end of 2010, the National Government of Colombia, with the presentation of the Education Policy for Prosperity, assumed the commitment to close educational gaps and affirmed that in order to advance in the consolidation and implementation of innovative projects, individual and collective actions and practices are required that respond to a planned, intentional, deliberate and systematized process for both their gestation and their development. The National Innovation System aims to make innovation a condition and aspect of educational practice, strengthen conditions and capacities for the educational use of ICTs in the Colombian education sector and meet the needs of educational communities [25].

Information and communication technologies are a fundamental component of citizen training in this millennium, which is why this field has become an element of interest for scientists and computer engineers and a complementary component for various areas of knowledge. However, research has shown that there are difficulties in the teaching-learning process of technology in secondary education [26], and that it is necessary to generate computer tools to try to solve the many problems that arise in this process,

Of the needs to integrate ICT tools in education, especially in the teaching-learning process of the subject of Technology, it is important to include Virtual Learning Objects, since they allow the student to have moments of significant learning, so the Virtual Object behaves as an additional aid to the teacher, knowledge and learning that the student must acquire, in which the teacher promotes and induces the development of innovative skills aimed at significant learning [26], [27], and through them it is possible to reuse educational processes [28]. OVAS is used worldwide as a tool to support the learning process: "The global Learning Object initiative allows any teacher or student to access the material published in the different Learning Object Repositories (LOR)" [22].

Learning Objects could be designed and developed for any kind of subjects. The Object is a resource or tool different from the traditional or conventional ones, with educational games, that is, elements with playful and didactic characteristics that contribute and allow improving the processes involved in the teaching and learning of knowledge [29]. It can be pointed out that the game can be used as an effective teaching-learning strategy to be applied in our educational environments [30]. It has been demonstrated through research that incorporates the game as a pedagogical strategy, based on the educational point of view, appreciating its virtues and designing, implementing, applying and validating a pedagogical proposal in a formal educational context [31].

Technologies make it possible to take advantage of various possibilities for interaction between students and knowledge using resources such as augmented reality, OVAs, serious video games, or social games or gaming [33]. These technologies are the basis of m-learning and e-learning, they are communication methods that enable a learning experience and will transform all forms of education and learning in the 21st century, since communication is at the centre of all forms of educational interaction, which proves that it has an enormous potential worth exploiting by including it in teaching and learning strategies [34]. Virtual educational resources such as OVAs with RA influence the development of cognitive skills in students and the promotion of their capacity for analysis, synthesis and understanding, and also favor motivation, accessibility and appropriation of knowledge. It is important to evaluate OVAs with instruments such as LORI that measure quality characteristics to ensure a good teaching-learning environment [4], [35].

5. References

[1] Río, M. L. (2014). Diseño de un software para la enseñanza de la asignatura Programación y Gestores de Bases de Datos en la carrera de Tecnología de la Salud. Rev. Ciencias Médicas, 3.

[2] Capuano, V. (2011). El uso de las TIC en la enseñanza de las Ciencias Naturales. Virtualidad, Educación y Ciencia 2, 79.

[3] UNESCO (2013). En foque estratégico sobre las TICS. América Latina y el Caribe. Oficina Regional de Educación para América Latina y el Caribe.

[4] Hernandez, Y. S. (2012). Instrumento de evaluación para determinar la calidad de los objetos de aprendizaje combinados abiertos de tipo práctica. Comunidad Latinoamericana de Objetos de Aprendizaje, 11, 3.

[5] Scornavacca, E., Huff, S., & Marshall, S. (2009). Mobile phones in the classroom: If you can't beat them, join them. Communication of the ACM.

[6] Santos, I. (2013). Use of students' personal mobile devices in the classroom: Overview of key challenges. E-Learning in Corporate, Government, Healthcare, and Higher Education 2013. Las Vegas, NV, United States.

[7] Tatar, D., Roschelle, J., Vahey, P., & Penuel. (2003). Handhelds go to school: Lessons learned. IEEE Computer & Education.

[8] Trifonova, & Rochetti. (2003). eprints.biblio.unitn.it. Retrieved from http://eprints.biblio.unitn.it/493/1/A _General_Architecture_for_M-Learning__m-ICTE2003_.pdf

[9] La Organización de las Naciones Unidas para la Educación, l. C. (2015). unesco.org. Retrieved from http://www.unesco.org/new/es/unesco/themes/icts/m4ed/

[10] Bello, C. R. (2017). La realidad aumentada: lo que debemos conocer. Tecnología Investigación y Academia, 5(2), 257-261.

[11] Colombiaaprende. (2015). colombiaaprende.edu.co. Retrieved from http://www.colombiaaprende.edu.co/html/directivos/1598/article-172369.html
[12] Otamendi, A. Belfer, K. Nesbit, J. Leacock, T. & Álvarez, J. (2003). Instrumento para la evaluación de objetos de aprendizaje (LORI _ ESP). Manual de usuario, 1–12.
[13] Velázquez, C. Muñoz, J. Álvarez, F. & Garza, L. (2006). La determinación de la calidad de objetos de aprendizaje. Avances en la ciencia de la computación, 346–351.
[14] Cáceres, A., Hernández, J. y Merlano, J. (2018) . Modelo para el diseño de objetos virtuales de aprendizaje de la asignatura mantenimiento de computadores (Tesis de grado).Antonio Jose de Sucre, Colombia.
[15] Kurt, S. "ADDIE Model: Instructional Design," in Educational Technology, August 29, 2017. Retrieved from https://educationaltechnology.net/the-addie-model-instructional-design/
[16] Hernández, S., Fernández, C. R, Baptista, L. P. (2010) Metodología de la investigación, 5ª Edición, México, Mc Graw Hill.
[17] Callejas, M., & Pinzon, J. (2011). Objetos de aprendizaje un estado del arte. REDALYC, 5.
[18] Baena, P. G. (2014). Metodología de la investigación Serie integral por competencias Derechos, Grupo editorial PATRIA, ISBN ebook: 978-607-744-003-1
[19] Carvajal, B. A. (2012). Elementos de investigación social aplicada, 1a reimpresión de la 3ª. Edición, Cali, Escuela de Trabajo Social y Desarrollo Humano-Facultad de Humanidades, Universidad del Valle.
[20] Villodre, S., & Llarena, M. (2009). sedici.unlp.edu.ar. Obtenido de Universidad de San Juan, Argentina. Retrieved from: https://www.google.com.co/url?sa=t&rct=j&q=&esrc=s&source=web&cd=1&ved=0ahUKEwjKlNHK8IvLAhXCqx4KHXoCBA8QFggaMAA&url=http%3A%2F%2Fsedici.unlp.edu.ar%2Fbitstream%2Fhandle%2F10915%2F18856%2FDocumento_completo.pdf%3Fsequence%3D1&usg=AFQjCNHG5SVVD6yWBlAGXoxzGe
[21] Nesbit, J., Belfer, K., & Leacock, T. (2002). transplantedgoose.net. Retrieved from transplantedgoose.net: https://www.google.com.co/url?sa=t&rct=j&q=&esrc=s&source=web&cd=1&cad=rja&uact=8&ved=0ahUKEwj08eXs9ovLAhVFLB4KHVBrBU8QFggdMAA&url=http%3A%2F%2Fwww.transplantedgoose.net%2Fgradstudies%2Feduc892%2FLORI1.5.pdf&usg=AFQjCNFRqkqphJ0XR5keLRfZFnxSuAXgBg&bvm=bv.1
[22] MEN, M. d. (2016). colombiaaprende.edu.co. Retrieved from http://www.colombiaaprende.edu.co/html/directivos/1598/article-172366.html
[23] Litwin, E., & et. (16 de Mayo de 2011). aprendiendomediantetecnologias.blogspot.com. Retrieved from http://aprendiendomediantetecnologias.blogspot.com.co/2011/05/conceptos-basicos-sobre-tecnologia.html
[24] Ariza, A. O. (2007). http://aprendiendomediantetecnologias.blogspot.com.co. Retrieved from http://aprendiendomediantetecnologias.blogspot.com.co/2011/05/conceptos-basicos-sobre-tecnologia.html
[25] Calderón, G. G., Buitrago, B., Acevedo, M. A., & Tobón, M. I. (2013). Colombia Aprende. Retrieved from Colombia Aprende: http://www.colombiaaprende.edu.co/html/micrositios/1752/articles-318264_recurso_tic.pdf
[26] Pantoja, A., & Huertas, A. (2010). Integración de las TIC en la asignatura de tecnología de educación secundaria. Revista de Medios y Educación. 37, 225-337.
[27] Valverde, B., Garrido, M., & Fernández, R. (2010). Enseñar y aprender con tecnologías: un modelo teórico para las buenas prácticas con tic. Teoría de la Educación. Educación y Cultura en la Sociedad de la Información. 11(1), 203-209. Retrieved from: https://www.redalyc.org/pdf/2010/201014897009.pdf
[28] Hodgins, W. (2000). cudi.edu.mx. Retrieved from http://www.cudi.edu.mx/diplomadoOA/materiales/modulo_01/evolucion-conceptual-OA.doc+&cd=1&hl=es-419&ct=clnk&client=firefox-b-ab
[29] Geist, E. (2011). The game changer: Using iPads in college teacher education classes. En E.

Geist, The game changer: Using iPads in college teacher education classes. College Student Journal.
[30] Torres, C. M. (2002). El juego: Una estrategia importante. Redalyc, 3(4), 10-47.
[31] Rocha, M. C., Espinoza, I. C., & González, P. G. (2006). Repositorio Universidad de Chile. Retrieved from Repositorio Universidad de Chile: http://repositorio.uchile.cl/tesis/uchile/2006/campos_m/sources/campos_m.pdf
[32] Zamora, H. D. (2013). acofipapers.org. Retrieved from acofipapers.org: http://www.acofipapers.org/index.php/acofipapers/2013/paper/viewFile/52/11
[33] Sánchez, A. B., & Salvador, C. C. (2010). Los entornos virtuales como espacio de enseñanza y aprendizaje. Revista Mexicana de Investigación Educativa, 4, 19.
[34] Anderson, D. R. (2010). e-learning en el siglo XXI. Barcelona: Ediciones Octaedro , S.L.
[35] Vera, R. G. (2005). Objetos de aprendizaje en la educación virtual: una aproximación en bibliotecología. Scielo, 157, 175.

Bibliometric review of emerging technologies in Gas Insulated Substations SF6

K Y Gómez-Sandoval[1,a], I A Marriaga-Márquez[1], J I Silva-Ortega[1], J W Grimaldo-Guerrero[1], H Hernandez-Herrera[2]

[1] Department of Energy, Universidad de la Costa, Barranquilla-Colombia.
[2] Universidad Simón Bolivar, Facultad de Ingeniería, Barranquilla-Colombia.

[a] kgomez17@cuc.edu.co

Abstract. The gas insulated substations (GIS) are made up of a set of equipment submerged in SF6 gas and a metal enclosure in which all the equipment associated with the substation is located. The GIS are being widely used in the high voltage electrical sector, due to this, developments and a high degree of research corresponding to new trends or emerging technologies that allow advances in this equipment are being presented. for this reason, this report contains a documentary investigation on the subject of isolated gas substations, through which a bibliometric analysis was carried out that allowed observing the high interest that GIS present in the scientific community in different parts of the planet. The objective of this work is to provide information corresponding to GIS such as some technical characteristics, a regulatory framework and new emerging technologies.

1. Introduction

A gas insulated substation is the set of electrical devices and appliances immersed in the dielectric gas Sulphur hexafluoride (SF6), contained within an aluminum alloy enclosure [1]. These substations can be used in industrial areas to meet high demands for electrical power through a space-saving design with minimal cost in maintenance and operation. These technologies are typically modular in design and can be used both indoors and outdoors [2], [3]. The GIS have advantages such as small volume, less maintenance and good technical performance, being a type of advanced high voltage electrical distribution equipment that is developing rapidly [4].

In GIS, the two main insulating media used are SF6 gas and solid insulating supports called spacers. The most used materials for spacer insulation functions are alumina and epoxy resin. The preferred shapes of spacers are disc, pole and cone type [5].

The most important element for gas insulated substations is the SF6, which is used as a means of isolating GIS busbars and equipment, and in the switches it is used as a means of extinguishing the electric arc produced by the opening or closing of the interrupting device contacts [6], [7].

Considering that gas is an important factor in ensuring safety in both construction and operation, it is important to have constant monitoring of this element [8], through the use of measuring instruments capable of detecting lack of SF6 gas [9], because the gas in the presence of moisture has by-products that have a high degree of toxicity [10], being moisture one of the biggest problems in the presence of failures in the insulation systems of electrical equipment [11], which compromises the reliability of the substation, affecting the quality indicators of the electric service supplied [12], The substation operating conditions must be adapted to future changes in the regulations of the transmission networks, since they can generate a technical-economic impact in the sector [13]. In addition, this gas is also classified as a gas with high greenhouse potential by the KYOTO protocol [14]. Therefore, in this work a documentary search was carried out with respect to emerging technologies that consider regulatory frameworks and allow significant improvements in these equipments.

This document is organized as follows, Section II presents the concept of gas insulated substation, in Section III gas quality and leak prevention, in section IV the technical standards corresponding to encapsulated substations, section V presents a bibliometric review with respect to emerging GIS technologies, and section VI presents the conclusions of this paper.

2. Gas insulated substations

Gas insulated substations (GIS) consist of a compact set of several components, enclosed in a grounded metal enclosure in which the primary insulating medium is a compressed gas. Within this envelope are the busbar and equipment associated with the substation such as circuit breaker, disconnectors, busbars and instrumentation transformers [3], [15], [16], [17], [18].

A gas insulated substation uses a superior dielectric gas, SF6 Sulphur hexafluoride, at a moderate pressure for phase-to-phase and phase-to-ground insulation [19].

Gas insulated substations are used when little installation space is required or there is a high impact on the environment. Sulphur hexafluoride is the gas used as an isolation medium for this type of substations, because dielectric stiffness this atmospheric pressure gas is three times that of air, and the space required for installation is only 10% to 25% of those required in an air-insulated substation (AIS) [6], [20], [21]. Figure 1 illustrates the scheme of an isolated substation in ABB brand gas.

The gas insulated substation has the following main characteristics [2]:

- Units assembled and tested from manufacturing.
- Operational life over 50 years.
- Major inspection not before the age of 25.
- Self-lubricating motor-operated mechanisms.
- Minimum cleaning requirement.
- Corrosion resistant.
- Low probability of failure/high availability.
- Protected against aggressive environmental conditions.
- Seismic-resistant.

The GIS envelope must be electrically grounded. In the middle of the envelope is the conductor (bar) attached through cones called separators. The area between the bar and the driver is filled with SF6 gas at a pressure between three to 7 bar [6]. The basic design of the envelope is illustrated in figure 2.

Encapsulation is the main protection of electrical components against external influences such as environmental effects [21]. In addition, the metal envelope reliably prevents electrical parts under voltage from being touched, providing protection to service personnel from the direct consequences of the arches due to internal dielectric faults [22].

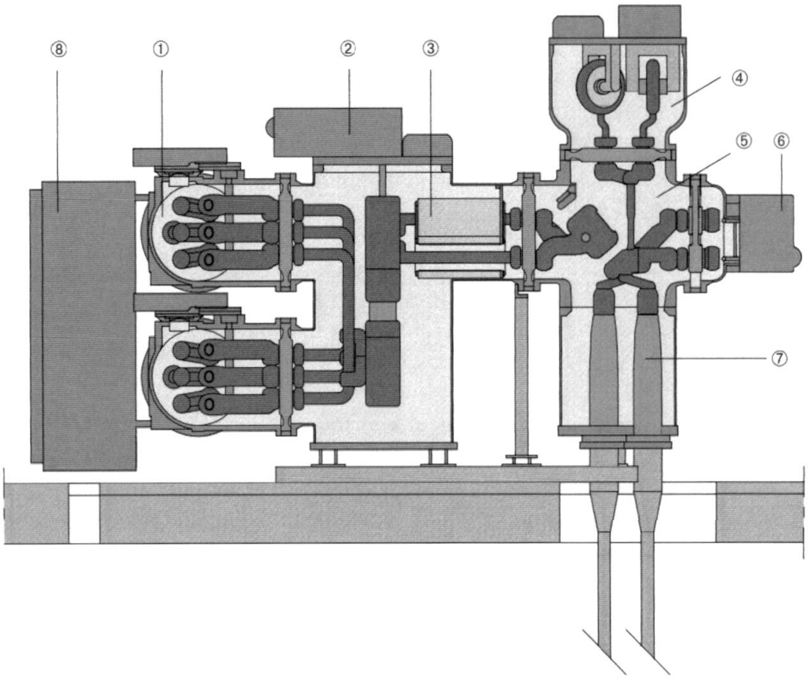

Figure 1. ABB's GIS. 1) busbar with high voltage switches combined and grounding switches, 2) circuit breaker, 3) current transformer, 4) voltage transformer, 5) combined high voltage switches and grounding switches, 6) high-speed grounding switches, 7) cable terminal, 8) control cabinet. By: [23].

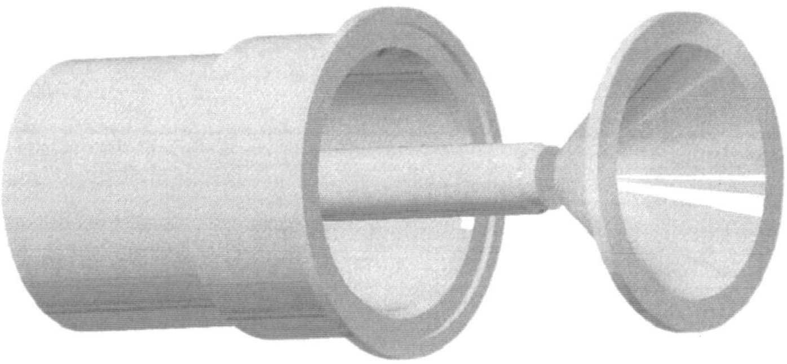

Figure 2. Basic design of a GIS envelope.

3. SF_6 Gas Quality and leak prevention

The most important element for GIS technologies is the sulphur hexafluoride gas which is used as an insulator of busbars and equipment and as a means of arc extinction. Use requires a high degree of purity because otherwise the dielectric coefficient of the gas is reduced [6].

A gas with 100% purity has some physical and chemical properties such as thermal conductivity and sound propagation rate, properties that can be verified with portable instruments. Contamination causes changes in gas properties [9]. To determine contaminants, the comparison between contaminated gas and a pure SF6 gas [15], [24].

When a leak occurs, it is important to find the exact location of each vanishing point, which in most cases is difficult to detect. To locate leaks an instrument that is used to perform SF5 leak detection issued, after inspection the maintenance team can take the necessary steps to eliminate the problem [9]. GIS during construction and operation are closely monitored and regulated, but there is always the possibility of accidents, because a dangerous situation could arise if SF6 gas escapes and effective measures are not given to eliminate contaminants on time [8]. While SF6 is a chemically stable gas (non-toxic, inert, non-flammable gas), it can cause suffocation when its concentration in the air exceeds 12 $^{mg}/_{m^3}$ [10], [25]. Leakage is due, in their normality to insufficient resistance of the GIS structure, caused mainly by defects in design and operation [26], [27].

Since Sulphur hexafluoride gas has high dielectric performance and interrupt capacity, it is widely used in the electrical energy industry [28]. When subjected to an electric arc and when the gas temperature exceeds 600 °C, it decomposes and recomposes extremely quickly when the temperature drops again, allowing the gas to regain its electrical resistance [29], [30]. This recovery property makes it suitable for high voltage switches. However, at 140oC, after operating for a period longer than 25 years, approximately 5% [25]. Can be broken down. It is known that some by-products are of high toxicity, particularly those that occur during discharges in the presence of moisture [10].

4. Regulatory framework

In this section the main rules applicable to gas-insulated substations are presented (table 1 and table 2).

Table 1. Regulations applicable up to GIS.

Source	Standard	Description
[31]	IEEE Std. 1125-1993, IEEE Guide for Moisture Measurement and Control in SF6 Gas-Insulated Equipment	It describes guidelines for measuring moisture level, interpreting moisture data, and controlling moisture in gas-insulated transmission class equipment.
[32]	IEEE Std. C37.123-1996, Guide to Specifications for Gas-Insulated, Electric Power Substation Equipment	Technical requirements for the design, manufacture, testing and installation of a gas-insulated substation (GIS) are described
[15]	IEEE Std. C37.122.1-1993, IEEE Std. Guide for Gas-Insulated Substations.	This guide provides information of special relevance for the planning, design, testing, installation, operation, and maintenance of substations and gas-insulated equipment (GIS).
[33]	IEEE Std. 693-1997, IEEE Recommended Practice for Seismic Design of Substations	Recommendations for seismic substation design, including the qualification of each type of equipment, are discussed.
[34]	IEEE Std. 1127-1998, IEEE Guide for the Design, Construction, and Operation of Electric Power Substations for Community Acceptance and Environmental Compatibility	Significant elements of community acceptance and environmental compatibility are identified and will be considered during the planning and design phases, the construction period and operation of power substations, and document ways to address these concerns in order to obtain community acceptance and environmental compatibility.
[35]	IEC 62271-1 2007, IEC High-voltage switchgear and control gear - Part 1: Common specifications	IEC 62271 applies to AC distribution devices designed for indoor and outdoor installation and for operation at service frequencies up to 60 Hz even in systems with voltages greater than 1 000 V.
[36]	IEC 62271-203:2003, IEC high-voltage switchgear and control gear - Part 203: Gas-insulated metal-enclosed switchgear for rated voltages above 52 kV	Specifies the requirements for gas-insulated substations, enclosed in metal where insulation is obtained, at least in part, by an insulating gas other than atmospheric pressure air, for alternating current of rated voltages greater than 52 kV, for indoor and outdoor installation, and for service frequencies up to 60 Hz.

Table 2. Regulations applicable up to GIS.

Source	Standard	Description
[37]	IEC 60137: 2017, Insulated bushings for alternating voltages above 1 000 V	Specifies the characteristics and tests for insulated bushings. This standard applies to bushings, as defined in Clause 3, intended for use in electrical appliances, machinery, transformers, substations and installations for three-phase AC systems, which have the highest voltage for equipment above 1000 V and power frequencies of 15 Hz. up to 60 Hz.
[38]	IEC 60376:2018, Specification of technical grade sulphur hexafluoride (SF6) and complementary gases to be used in its mixtures for use in electrical equipment.	Defines the quality of technical grade Sulphur hexafluoride (SF6) and complementary gases such as nitrogen (N2) and carbon tetrafluoride (CF4) for use in electrical equipment.
[24]	IEEE Std 1416-1998, IEEE Recommended Practice for the Interface of New Gas-Insulated Equipment in Existing Gas-Insulated Substations	Recommendations for connecting a gas-insulated substation to a gas-insulated substation of a different brand are described.
[37]	IEC 60137: 2017, Insulated bushings for alternating voltages above 1 000 V	Specifies the characteristics and tests for insulated bushings. This standard applies to bushings, as defined in Clause 3, intended for use in electrical appliances, machinery, transformers, substations and installations for three-phase AC systems, which have the highest voltage for equipment above 1000 V and power frequencies of 15 Hz. up to 60 Hz.

5. Bibliometric review

5.1. Bibliometric analysis

The methodology applied consisted of the analysis of 576 documents in the subject of study, obtained through the search in the Web of Science database, using as a keyword "gas insulated substations", with this input were proposed bibliometric indicators, allowing to analyze the temporal evolution of the subject between 2011 and 2020.

The countries with the greatest scientific contribution to gas-insulated substations are illustrated in Figure 3. Where it can be seen that China is the country that stands out the most because it has the highest number of publications with a total of 276 publications between 2011 and 2020, followed by Japan with 117 publications in the same time period. Figure 3 illustrates the number of country publications in the 2011-2020 period.

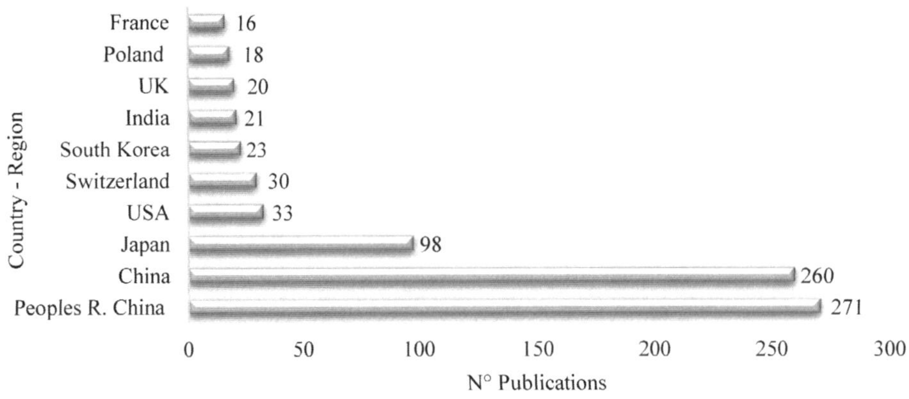

Figure 3. Number of publications by country in the 2011-2020 period.

In figure 4 shows the publications recorded in the 2011-2020 period, allowing to identify that the years with the highest number of publications were 2017 with 75 and 2018 with 71 publications, respectively.

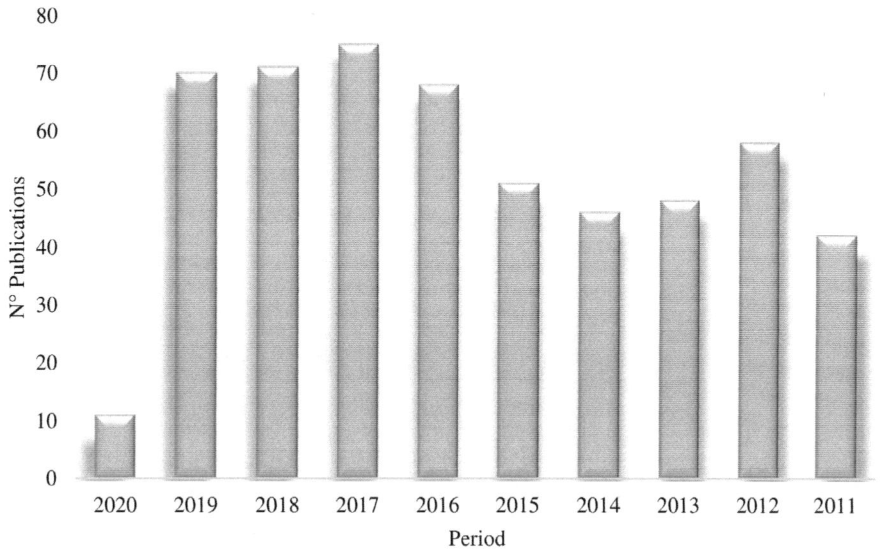

Figure 4. Number of publications per year in the 2011-2020 period.

Figure 5 shows the publication register of the 10 universities with the highest number of publications regarding GIS in the world, among which Xi an Jiaotong University stands out the most with the highest number of publications, with a total of 14 publications regarding GIS.

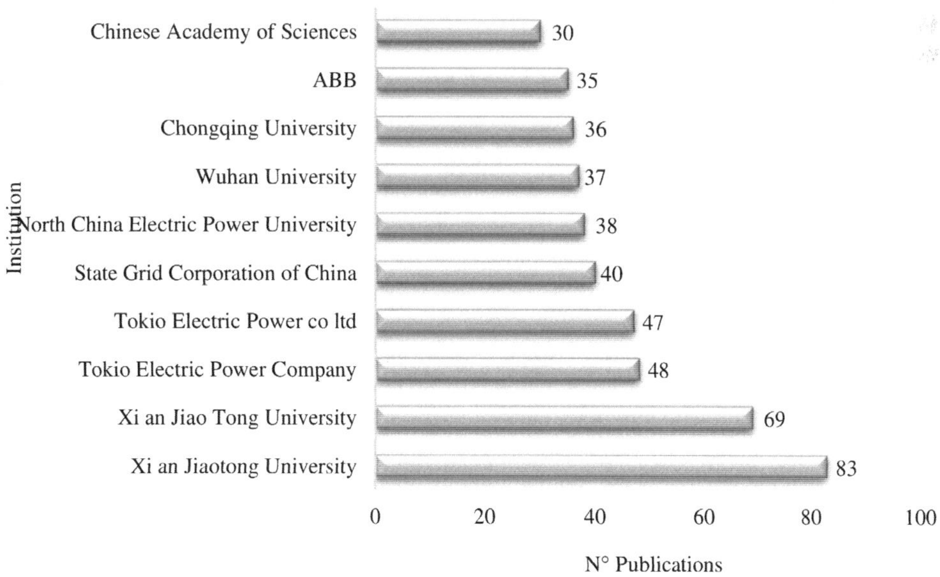

Figure 5. Number of publications by institutions in the 2011-2020 period.

Table 3 features the 10 authors with the highest number of publications between 2011-2020, including the author Okabe S. with a total of 14 publications.

Table 3. Number of publications per author

Authors	Nº publications
Okabe S.	49
Ueta G.	34
Zhang XX.	29
Xiaoxing Zhang	28
Chen WJ.	26
Thang j.	26
Ju Tang	24
Li CR.	24
Wada J.	22
Liu WD.	20

Table 4 shows the number of publications in the different IEEE journals indexed on The Web of Science. In which the high contribution of IEEE in terms of the research trend regarding gas-isolated substations can be identified, the journal IEEE transactions on dielectrics and electrical insulation is highlighted for having the highest number of publications in the 2011 to 2020-time period.

Table 4. Number of IEEE indexed journal publications on web of science

Journal	Nº publications
IEEE transactions on dielectrics and electrical insulation	182
IEEE transactions on power delivery	61
IEEE transactions on plasma science	17
IEEE electrical insulation magazine	11
IEEE access	10
IEEE sensors journal	5
IEEE transactions on components packaging and manufacturing technology	5
IEEE transactions on industry applications	4
IEEE transactions on electromagnetic compatibility	3
IEEE transactions on magnetics	3
IEEE industry applications magazine	2
IEEE transactions on applied superconductivity	2
IEEE transactions on instrumentation and measurement	2
IEEE antennas and wireless propagation letters	1
IEEE electron device letters	1
IEEE transactions on electron devices	1
IEEE transactions on power electronics	1

5.2. Emerging technology for gas insulated substations
Gas insulated substations present a strong research trend in different parts of the planet, opening a door to the search for emerging technologies for the improvement of these equipment. In table 5, some emerging technologies for the improvement of GIS can be seen.

Table 5. Regulations applicable up to GIS.

Source	Title	Year	Scientific contribution
[39]	Research and experiments on an external miniaturized VFTO measurement system	2020	A. P. Purnomoadi, A. Rodrigo Mor, and J. J. Smit propose a miniature sensor system that uses a capacitive voltage system and a differentiation-integration circuit to increase the bandwidth of the very fast front transient surge measurement system (VFTO), so that researchers can reduce the damage of equipment caused by VFTO.
[40]	Emerging technologies in high voltage gas insulated switchgear-clean air gis and NCIT	2019	The authors propose a new unconventional instrumentation transformer, which offers an advantage over the conventional instrumentation transformer, because it has less copper wiring and therefore allows to open the door to digital substations, presenting better functionality in both measurement and protection.
[41]	Highly Sensitive Acoustic Emission Sensor Based on Polarization-maintaining and Absorption-reducing Polarization-maintaining Fiber	2018	X. Li, Y. Zhang, H. Cao, X. Zhang, and R. Shen propose a new type of acoustic emission sensor (AE) based on a polarization maintenance fiber (PANDA) that maintains polarization (PANDA) to detect ultrasound information induced by partial discharge accurately in GIS.
[42]	Electric field computation of epoxy- nano and micro composite conical spacer in a gas insulated bus duct	2018	The authors propose an alternative for existing insulating materials, based on the mixture of nano and micro composite polymers that exhibits excellent electrical, thermal, and mechanical properties.
[43]	Application of a multi-parameter sensor system for monitoring dielectric insulation of gas mixtures	2017	M. Porus, T. A. Paul, and A. Kramer propose a robust, low-cost method for deriving concentration into a binary gas mixture using simultaneous pressure, temperature, and density measurement to detect perfluorinated ketones
[44]	Research on the microscopic state parameters of SF6/N2 and SF6/CO2 circuit breaker in arc formation process	2017	The authors L. Jing, Z. Hao, C. Yundong, L. Shuxin, and Y. Longbin propose to complete a process of formation of SF6/N2 and SF6/CO2 mixtures as an electrical insulation gas
[45]	Comparison of CF3CHCl2 Gas with SF6 Gas as an alternative substitute for gas insulated switchgear equipment	2017	Authors T. Juliandhy, T. Haryono, Suharyanto, and I. Perdana say cf3CHCl2 has a high potential as an insulating gas because it has a higher insulation capacity than SF6 gas and is recommended by the KYOTO protocol.
[46]	Development of 72/84 kV dead-tank VCB with rated normal current of 3000 A	2013	Author H. Saito et al says that insulated dead tank has been developed in dry air of 72/84 kV type VCB with a nominal normal current of 3000A, which is shown as a feed because the rated normal current of the conventional type is limited to 2000 A.
[47]	A survey on the potential of CF3I gas as an alternative for SF6 in high voltage applications	2010	Authors M. S. Kamarudin, M. Albano, P. Coventry, N. Harid and A. Haddad say trifluoroyodomethane (CF3I) has recently been considered as a candidate to replace SF6. Because it has been used as a fire suppressor and many researches is being conducted to assess its capacity in high voltage applications.

5.3. Patents in the optimization of gas-insulated substation technologies

Taking into account the importance of isolated gas substations and the impact that SF6 generates due to its potential as greenhouse gas, the scientific community and companies dedicated to the invention of this technologies have developed new patents capable of providing alternatives for the use of sf6. table 6 and table 7 shows some patents for the optimization of GIS technologies.

Table 6. Patents in the optimization of gas-insulated substation technologies.

Source	Title	Year	Scientific contribution
[48]	Switchgear having earth switch.	2019	A sideboard is provided that includes: a fixed electrode that is arranged inside a cabinet; a moving electrode that is provided corresponding to the fixed electrode and rotates between a first position that is in contact with the fixed electrode and a second position that is separated from the fixed and grounded electrode.
[49]	Use of a linear octafluorobutene as a dielectric compound in an environmentally safe dielectric-insulation or arc-extinction fluid.	2019	A combination, comprising: a linear octafluorobutene used as a dielectric compound in an arc-extinguishing fluid or environmentally safe dielectric isolation for a device for the generation, transmission, distribution and/or use of electrical energy, where the ratio of the linear octafluorobutene dielectric compound in dielectric or arc-extinguishing isolation fluid is at least 3%.
[50]	Transformer for measuring currents in a gas-insulated substation.	2019	An electronic current transformer for measuring currents, comprising: a Resistive Magnet Tunnel sensor (RMT); a driver; an amplification circuit; an armour structure; a circuit board; where the RMT sensor and amplification circuit are arranged on the circuit board; where the circuit board is arranged between the conductor and the shielding structure; and where the RMT sensor is configured to receive driver data and transmit the data to the amplification circuit, which is configured to amplify the data and release the data as a transformer output.
[51]	Closed loop interferometric sensor using loop gain to determine interference contrast.	2018	Sensor to measure interference contrast in an interference-based closed-loop phase modulation optical sensor device, the gain of the feedback circuit is evaluated on a feedback controller. This gain is found to be a measure for contrast. Contrast evaluated in this way can be used, for example, for periods to change when determining the measurement and sensor device. The sensor device can be, for example, a high voltage sensor or a current sensor. This device can be used in HVDC in GIS.
[52]	Apparatus containing a dielectric insulation gas comprising an organofluorine compound.	2017	The present invention refers to a device for the generation, transmission, distribution and / or use of electrical energy. The appliance comprises a housing enclosing an insulating space and an electrical component arranged in the insulating space. The insulating space containing a dielectric insulation gas comprising an organ fluorinated compound. The appliance also comprises a desiccant ready to meet the insulating gas. The desiccant contains or consists essentially of lithium bromide, ideal for gas encapsulated substations.

Table 7. Patents in the optimization of gas-insulated substation technologies.

Source	Title	Year	Scientific contribution
[53]	Temperature measurement in switchgear stations.	2017	The model refers to a device for measuring temperature at a high-voltage part of a power station, which is characterized by comprising: at least one temperature sensor located at one point in the high-voltage serving of which the temperature must be monitored, at least one antenna connected to at least one temperature sensor, a control module located in a low voltage part of the switching station and at least one antenna connected to the control module. This invention refers to temperature measurement in high voltage switching stations, type AIS (Isolated Air Substation) or GIS (Isolated Gas Substation)
[54]	Method of extracting a jacket of a gas insulated substation (GIS) under a metal casing.	2017	This method refers to a removable stabilizer for a spacer housing of an electrical installation of the GIS type, the spacer housing comprising a first opening as well as a conductor fitted with a hole, the removable stabilizer comprises a plug configured to be inserted through the first opening in the separator housing, being designed the stopper to penetrate the driver's drill to block that conductor against translation movement after such housing spacer has been put under atmospheric pressure.
[55]	Control of spring(s) type for a high- or medium-voltage breaker furnished with a pawled free wheel coupling device.	2017	The model of a spring controller for high or medium voltage electrical appliance, with the controller having a freewheel coupling device between a crank or motor and the drive shaft to operate a switching contact of the appliance and respectively providing coupling while loading the springs and decoupling while releasing the springs. The mechanical freewheel device is incorporated into a controller sprocket and includes at least one ratchet that engages or does not engage with an internal set of sprocket teeth. The invention applies more particularly to high or medium voltage switches and particularly to high or medium voltage automatic switches, either for those used in gas-insulated substations (GIS) or those used with air insulation.

6. Conclusions

Electrical substations play an important role within electrical power systems, thus having to meet safety and reliability requirements within the system, because of this since 1960 the isolated substations with SF6 gas were resulted, to meet the demand for electrical energy reliably and safely, also providing advantages such as little space for installation and lower operating and maintenance costs. These substations have a higher insulation medium than air-insulated substations because it employs a gas with excellent dielectric properties within a metal envelope, ensuring easy operation and greater safety for personnel engaged in installation activities.

However, it is necessary to have constant monitoring of the state of the GIS metal envelope to check that there are no dents, also, constant monitoring should be carried out through the use of sensors or gas measuring instruments SF6 to detect the existence of gas leaks, which may cause accidents in the facility by suffocation if the gas exceeds a concentration and their of 12 $^{mg}/_{m^3}$ [10], [25]. for this reason, we must be persistent in the search for new technologies that can identify these problems quickly and effectively, through scientific research.

On the other hand, SF6 gas is a gas with high greenhouse efficiency according to the KYOTO protocol [14], so there is a need to look for alternatives that can replace the use of SF6 in high voltage applications, with environmentally friendly elements.

7. References

[1] Aragón Quesada G, Leyva Paz E and Vigil Sánchez D 2015 Análisis comparativo entre subestaciones eléctricas encapsuladas en SF6 y subestaciones convencionales (Instituto Politécnico Nacional)
[2] Bolin P and Koch H 2006 Gas insulated substation GIS 2006 IEEE Power Engineering Society General Meeting, PES
[3] Koch H, IEEE and Ltd J W & S 2014 Gas Insulated Substations (IEEE, John Wiley & Sons Ltd)
[4] luo X 1999 SF6 Gas Insolated Switchgear (GIS) China Electr. Power Press 83–135
[5] Adari J, Gadi V S K R and Gundavarapu V N K 2018 Mitigation of field stress with metal inserts for cone type spacer in a gas insulated busduct under delamination Eng. Sci. Technol. an Int. J. 21 850–61
[6] S.A. M V H M . 1989 Subestaciones de Alta y Extra Alta Tensión (Medellin, Colombia)
[7] Díaz S, Nuñez J, Berdugo K and Gomez K 2020 Study of technologies implemented in the operation of SF6 switches IOP Conf. Ser. Mater. Sci. Eng. 872
[8] Han O, Zhang Y, Li A, Li J, Li Y and Liu H 2020 Experimental and numerical study on heavy gas contaminant dispersion and ventilation design for industrial buildings Sustain. Cities Soc. 55
[9] Ogiboski L, Cunha A C, Kajikawa C O, Magrini L C and Jardini J A 2012 Development and installation of a data acquisition and monitoring system to gas insulated substation Proceedings of the International Conference on Optimisation of Electrical and Electronic Equipment, OPTIM pp 1134–9
[10] Beroual A and Haddad A 2017 Recent advances in the quest for a new insulation gas with a low impact on the environment to replace Sulfur Hexafluoride (SF6) gas in high-voltage power network applications Energies 10
[11] Marriaga-Márquez I A, Gómez-Sandoval K Y, Grimaldo-Guerrero J W and Nuñez-Álvarez J R 2020 Identification of critical variables in conventional transformers in distribution networks IOP Conf. Ser. Mater. Sci. Eng. 844
[12] Mantilla Florez B H, Silva Ortega J I and Candelo Becerra J E 2018 Fault Effect Analysis Based on Elements Loadability to Evaluate Reliability in Power Systems IEEE Lat. Am. Trans. 16 2649–56
[13] Andrade-Becerra A, Marenco-Rua S, Grimaldo-Guerrero J W, Noriega-Angarita E M and Silva-Ortega J I 2019 Economic assessment of changes in the regulation of the transmission activity in Colombia J. Eng. Sci. Technol. Rev. 12 11–6
[14] Montillet G 2004 Implementation Of Treatment & Recovery Of the SF6 Gas Containing A High Amount Of Decomposition Products Due To High Voltage Electrical Interruptions
[15] IEEE 1993 IEEE Guide for Gas-Insulated Substations. IEEE Std. C37.122.1-1993
[16] Bautista E 2013 Subestación Encapsulada en SF6 ElectroIndustria 4–6
[17] Kondalu M 2011 Estimation of Transient over Voltages in Gas Insulated busduct from 220kv Gas Insulated Substation vol 20
[18] Rao M M and Kumar M 2019 Ultra-High Frequency (UHF) Based Partial Discharge Measurement in Gas Insulated Switchgear (GIS) Proceedings of 2019 International Conference on High Voltage Engineering and Technology, ICHVET 2019 (Institute of Electrical and Electronics Engineers Inc.) pp 1–5
[19] McDonald J D 2012 Electric Power Substations Engineering (New York)
[20] Harper G E 2002 Elementos de diseño de subestaciones eléctricas (Editorial Limusa)
[21] Nagarsheth R and Singh S Study of gas insulated substation and its comparison with air insulated substation
[22] Bolin P and Koch H 2008 Basic information on gas insulated substation (GIS) IEEE Power and Energy Society 2008 General Meeting: Conversion and Delivery of Electrical Energy in the 21st Century, PES

[23] ABB Subestación aislada en gas tipo ELK-04 Sistema modular hasta 170 kV, 4000 A, 63 kA
[24] IEEE 1999 IEEE Recommended Practice for the Interface of New Gas-Insulated Equipment in Existing Gas-Insulated Substations IEEE Std 1416-1998
[25] SÉGUR P 1990 Gaz isolants Tech. l'Ingénieur, Trait. Génie Électrique; Tech. l'Ingénieur D2530–D2531
[26] Zhang H, Jian Y, Guo Q, Deng H, Jian M, Zhang G and Du B 2019 Vibration Analysis of GIS Pipe System in Transformer Substation IOP Conference Series: Earth and Environmental Science vol 300 (Institute of Physics Publishing)
[27] Ohshita Y, Hashimoto A and Kurosawa Y 1989 A diagnostic technique to detect abnormal conditions of contacts measuring vibrations in metal enclosures of gas insulated switchgear IEEE Trans. Power Deliv. 4 2090–4
[28] Kawamura T, Matsumoto S, Hanai M and Murayama Y 1998 SF6/N2 Mixtures for HV Equipment and Practical Problems Gaseous Dielectrics VIII (Springer US) pp 333–43
[29] Acevedo Vasquez D, Chacón Pedraza G and Santamaría Piedrahita F 2016 Metodología para la medición de parámetros de sobretensiones transitorias en redes de distribución de 11.4 kV INGE CUC 12 65–72
[30] di Mauro G, Ferreyra R, Suárez J and Jurado A 2016 Overvoltage by Ferroresonance on a Rural Distribution Feeder : Case Report and Simulation Inge Cuc 11 34–47
[31] IEEE 1993 IEEE Guide for Moisture Measurement and Control in SF6 Gas-Insulated Equipment IEEE Std. 1125-1993
[32] IEEE 1996 IEEE Guide to Specifications for Gas-Insulated, Electric Power Substation Equipment IEEE Std. C37.123-1996
[33] IEEE 1997 IEEE Recommended Practice for Seismic Design of Substations. IEEE Std. 693-1997
[34] IEEE 1997 IEEE Guide for the Design, Construction, and Operation of Electric Power Substations for Community Acceptance and Environmental Compatibility IEEE Std. 1127-1998
[35] IEC 2007 IEC High-voltage switchgear and controlgear - Part 1: Common specifications IEC 62271-12007
[36] IEC 2003 IEC High-voltage switchgear and controlgear - Part 203: Gas-insulated metal-enclosed switchgear for rated voltages above 52 kV IEC 62271-2032003
[37] IEC 2017 Insulated bushings for alternating voltages above 1 000 V IEC 601372017
[38] IEC 2018 Specification of technical grade sulphur hexafluoride (SF6) and complementarygases to be used in its mixtures for use in electrical equipment IEC 603762018
[39] Wang L, Zhang W, Tan X, Chen W, Liang S and Suo C 2020 Research and experiments on an external miniaturized VFTO measurement system Sensors (Switzerland) 20
[40] Thummapal D, Kothari S and Thirumalai M 2019 Emerging technologies in high voltage gas insulated switchgear-clean air gis and ncit Proceedings of 2019 International Conference on High Voltage Engineering and Technology, ICHVET 2019 (Institute of Electrical and Electronics Engineers Inc.)
[41] Li X, Zhang Y, Cao H, Zhang X and Shen R 2018 Highly sensitive acoustic emission sensor based on polarization-maintaining and absorption-reducing polarization-maintaining fiber Sensors Mater. 30 1145–53
[42] Swamy Naidu Y and Nagesh Kumar G V. 2018 Electric field computation of epoxy- nano and micro composite conical spacer in a gas insulated busduct Advances in Intelligent Systems and Computing vol 668 (Springer Verlag) pp 575–83
[43] Porus M, Paul T A and Kramer A 2017 Application of a multi-parameter sensor system for monitoring dielectric insulation of gas mixtures IEEE Trans. Dielectr. Electr. Insul. 24 847–51
[44] Jing L, Hao Z, Yundong C, Shuxin L and Longbin Y 2017 Research on the microscopic state

IOP Conf. Series: Materials Science and Engineering 1154 (2021) 012020 doi:10.1088/1757-899X/1154/1/012020

parameters of SF6/N2 and SF6/CO2 circuit breaker in arc formation process ICEPE-ST 2017 - 4th Int. Conf. Electr. Power Equipment- Switch. Technol. 2017-Decem 247–50
[45] Juliandhy T, Haryono T, Suharyanto and Perdana I 2017 Comparison of CF3CHCl2 Gas with SF6 Gas as an alternative substitute for gas insulated switchgear equipment International Conference on High Voltage Engineering and Power Systems, ICHVEPS 2017 - Proceeding vol 2017-January (Institute of Electrical and Electronics Engineers Inc.) pp 198–203
[46] Saito H, Nagatake K, Yamauchi K, Takeshita Y, Kobayashi T, Cho T, Katsumata K and Sakaki M 2013 Development of 72/84 kV dead-tank VCB with rated normal current of 3000 A 2013 2nd Int. Conf. Electr. Power Equip. - Switch. Technol. ICEPE-ST 2013 1–3
[47] Kamarudin M S, Albano M, Coventry P, Harid N and Haddad A 2010 A survey on the potential of CF3I gas as an alternative for SF6 in high voltage applications Proceedings of the Universities Power Engineering Conference
[48] Seog, Won; Jeong, Young Woo; Jin Y 2019 Switchgear having earth switch
[49] Doiron, Charles; Paul, Thomas Alfred; Radisavljevic B 2019 Use of a linear octafluorobutene as a dielectric compound in an environmentally safe dielectric-insulation or arc-extinction fluid
[50] Huang, Qi; Khawaja, Arsalan Habib; Chen Y 2018 Transformer for measuring currents in a gas-insulated substation
[51] Gu, Xun; Marchese, Sergio Vincenzo; Bohnert K 2018 Closed loop interferometric sensor using loop gain to determine interference contrast
[52] Di-Gianni, Anna; Porus M 2018 Apparatus containing a dielectric insulation gas comprising an organofluorine compound
[53] Stella, Eros; Rayón, Jean-Luc; Girlando V 2017 Temperature measurement in switchgear stations
[54] Gerber, Heinz; Lüscher R 2017 Method of extracting a jacket of a gas insulated substation (GIS) under a metal casing
[55] Von Allmen, Peter; Boiteux R 2017 Control of spring(s) type for a high- or medium-voltage breaker furnished with a pawled free wheel coupling device

Recommendation system with graph-oriented databases for repository of open educational resources

A. Morales[1], J. Gonzalez[1], K. Alquerque[1], J. Reyes[1] and S. Sanchez[1]

[1]Antonio José de Sucre University Corporation (Colombia).

jaider_reyes @ corposucre.edu.co

Abstract. The advancement of telecommunications has stimulated individualized study and collaborative learning, allowing the creation of pedagogical innovations by the teaching and administrative staff of educational institutions. Among these innovations are the Open Educational Resources (OER), which since its inception have contributed to teaching practices and contributed to the free learning of students. For this reason, the objective of the research is to design a repository of open educational resources integrating a recommendation system with graph-oriented databases. The study is descriptive and technological development. The results are the schematization of the system architecture by means of a block diagram and flow diagram.

1. Introduction

Globalization contributes to the arrival of an interactive society, but it can also cause secondary effects, especially with regard to access to information and knowledge in cyberspace. Therefore, international institutions and the competent national authorities of each country must develop strategies to mitigate this problem. UNESCO proposes that public access to information and its free circulation should be promoted without geographic, economic or social discrimination, through appropriate public policies that guarantee its reliability. [1]

The advancement of telecommunications has stimulated individualized study and collaborative learning, allowing the creation of pedagogical innovations by the teaching and administrative staff of educational institutions. Among these innovations are the Open Educational Resources (OER), which since its inception have contributed to teaching practices and contributed to the free learning of students [2].

The development of a repository for educational resources is undoubtedly an excellent tool for the teaching-learning process of educational institutions, but it should be noted that the growth of information is exponential and, therefore, better approximate and match the information that it is consulted of interest, it becomes more complex and tedious. Today there is the need to select between several alternatives without having a fixed knowledge of each of them, in these situations, the final decision may depend on the recommendations made by other people, for this reason currently companies such as Spotify, YouTube, Amazon guide users through recommendations.

These systems have been called "Recommendation Systems". These systems are in charge of providing users with personalized and differentiated information on certain topics, products and / or services that may be of interest. That is, they are responsible for guiding the user through recommendations in the search for those topics, services or products that may be more attractive, modifying the browsing or search process. The use of these systems is becoming increasingly popular due to their usefulness in evaluating and filtering the large amount of information available on the Web. This is undoubtedly a great advantage for users, who will find what they need in a faster, more comfortable and easier way within electronic repositories on the Internet and also allow them to discover new topics,

Therefore, the main objective of this research is to design a repository that integrates a recommendation system with the use of graph-oriented databases, which facilitate the recognition of patterns in the relationships that join the nodes. The database will relate user searches by topics of interest in order to make personalized recommendations.

2. Recommendation Systems

With the arrival of Big Data, the massive amount of information available on the web creates the need to make decisions when making inquiries and selecting products, content or topics of interest, however frequently due to not having the necessary information or sufficient experience staff on the subject, use is made of recommendations from other people (family or friends) or by the media such as Radio or TV [3].

At the beginning of the 90's, developers and computer science researchers, faced with this situation, developed recommendation algorithms, which in the last decade have taken center stage in many of the popular computer applications such as, for example, online stores, movie repositories, video repository, music repositories, academic databases, among others. The purpose of these recommendation systems is to satisfy the search needs of users in a personalized way [4].

Recommendation systems have been added to e-commerce websites (such as Amazon.com) and leisure applications such as Netflix, Spotify, YouTube, Facebook, Instagram, Telegram, Snapchat, among others that have made these systems a prominent part of the provision of services [5].

Recommendation systems are tools whose main objective is to help users in making decisions to select the item that best suits their preferences, tastes and interests. [6], the recommendation systems can also be defined as common solutions in the personalization of content, they identify elements that are of interest to the user using a recommendation component that compares the characteristics defined and stored in the user profiles, to predict the score of the Internet user towards items you have not yet considered [7].

These systems have characteristics in which the system presents elements that different users with similar tastes like, the system forms a user profile according to the interaction that the user has had with the elements and recommends content based on the knowledge acquired. [3], so that the system already has the ability to select the items that the user might be interested in and the user is recommended to recommend topics, products or services that resemble other products that they have liked. [8].

2.1 Recommendation Technique

It can be said that these systems select the useful elements for the user among a large set of existing search elements, that is why these systems must anticipate the possible decisions that the user is going to select. For this, the system uses different methods and algorithms to calculate the recommendations, among the different algorithms there are the content-based, the collaborative filtering and the hybrid hybrids. [9].

2.1.1. Content-based algorithm. In this type of algorithm, the context is any information that can be used to characterize the situation of the entities (person, place or object) that are considered relevant for the interaction between a user and an application, including the user and the application itself. themselves. Context is typically the location, identity, and state of people, groups, and physical or digital objects. [10]

2.1.2 Collaborative Filtering Algorithm. This type of algorithm is a method to make predictions about the interests of a user based on information about the preferences of other users. Collaborative filtering is the process of filtering information or models, which uses techniques that involve collaboration between multiple agents, data sources, among others.[11]

2.1.3 Hybrid algorithm: The hybrid filtering technique combines various recommender techniques to obtain a better optimization of the system to avoid some limitations and difficulties of the pure recommender systems. The idea behind hybrid techniques is that a combination of algorithms will provide more accurate and effective recommendations than a single algorithm, since the disadvantages of one algorithm can be improved by another algorithm. The use of multiple recommendation techniques can suppress the weaknesses of an individual technique in a combined model [12].

3. Graph-Oriented Database

Graph-oriented data stores information through the structure of graphs, where the information is grouped into nodes and these in turn are related by memory pointers, which allows these databases to apply graph theory for navigation ; These are very useful for storing information in models with many relationships such as social networks and connections, where each node is assigned a grade that indicates the number of connections it has[13].

Since a high demand for information is required when storing it and when searching for information in the application when making inquiries, it leads to investigating new technological tools that facilitate data management. Taking into account that the data provided by the user must be studied and interpreted in order to be useful, therefore, if we could represent the data and the relationships between objects as a single data set, knowledge patterns could be generated that describe the set of data contained in these elements together. For this purpose, the databases are presented in graphs that are flexible and powerful enough to represent these elements.[3].

3.1 Neo4j

Neo4j is a native graphics database, built from the ground up to take advantage of not only data but also data relationships. Neo4j connects data as it is stored, enabling queries never before imagined, at speeds never thought possible [14].

It is a robust (fully ACID) highly scalable graph-oriented database. Neo4j is used in mission critical applications by thousands of leading companies and governments around the world.

Below in the figure. 1 presents a diagram of what would become "NEO4J", as an implementation in a graph-oriented database [3].

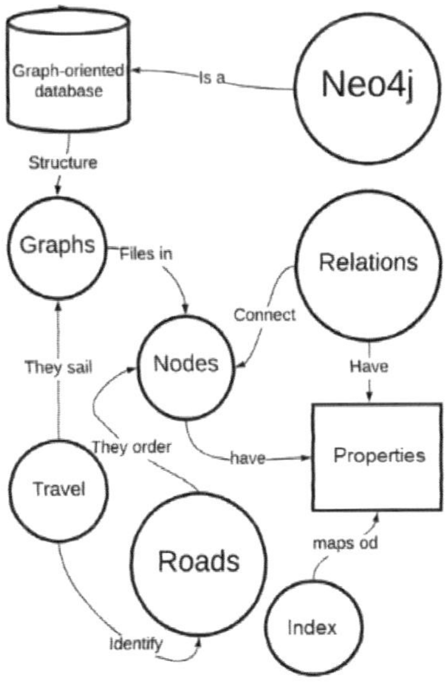

Figure 1. Function representation of neo4j.

Neo4j Main Features

- Intuitive: use a graphical data model for data representation.
- Reliable - Complete ACID transactions.
- Durable and fast: native storage engines.
- Highly scalable: thousands of nodes, relationships and properties.
- High availability: distribution across multiple machines.
- Expressive: powerful query language.
- Embedded
- Simple: accessible with a REST interface or an object-oriented API (JAVA).

4. Methodology

Research is of a technological development type, according to MinCiencias this type of study occurs in the application of the results of the research, or of any other type of scientific knowledge, for the design or development of new processes, systems or provision of services, as well such as the substantial technological improvement of pre-existing materials, products, processes or systems. This activity will include the materialization of the research results in a plan, scheme or design, as well as the creation of non-marketable prototypes [fifteen]. The purpose of this research is to design an OER repository for the Antonio José de Sucre University Corporation to offer educational content to the student community and teachers of the department of Sucre located in Colombia.

To carry out the research, two phases were developed using the Cascade software development methodology (see figure 2):

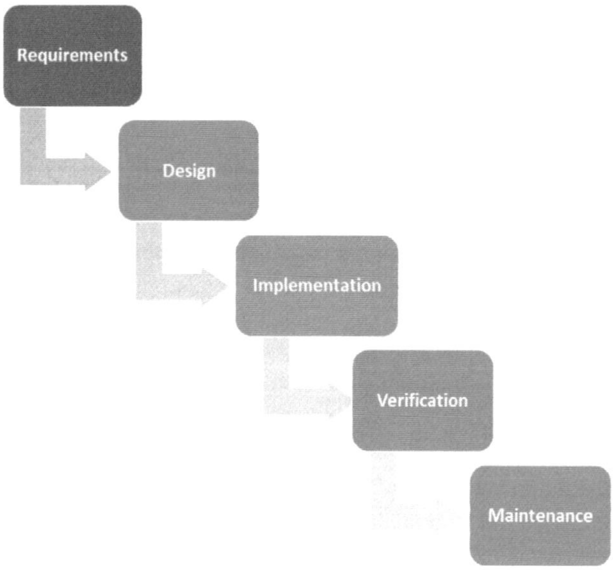

Figure 2. Cascade Methodology.

In the requirements phase, a bibliographic review was carried out and with it a documentary analysis of the information provided by the expert entities in charge of the development and directors of virtual learning environments, and of academic documents results of the investigation of related topics recommendation systems in any field, be it tourism, commerce, entertainment or education, in order to identify the functional and non-functional requirements of the system proposed in the research.

For the design phase, the requirements identified in the previous phase were taken into account and from then on, through the use of flow diagrams and block diagrams, the system architecture was designed, relating the communication between the different technologies that integrated the system. system operation.

5. Results

Due to the continuous development of digital content, it is necessary for the repositories databases of these contents to have scalability characteristics as they grow, in addition to having the capacity to facilitate the storage and retrieval of information. However, the massive amount of information leads to users not always finding the desired information, and if they do, the task becomes complex and tedious. That is why recommendation systems play an important role in solving and managing this information within these repositories.

Within the research, the design of the recommendation system for a repository of educational resources for the Antonio José de Sucre University Corporation, was made initially a bibliographic review was made organized in the bibliographic matrix and analyzed by means of an analytical content matrix, later The requirements survey was carried out through semi-structured interviews applied to experts in the area, these requirements are listed in table 1.

Table 1. Functional and non-functional requirements

Functional	Not functional
The system should allow searching for topics of interest to the student.	The system interfaces should be easy or intuitive.
The system must allow modification of the search.	
The system should provide assistance when conducting a search.	
The system must list the users according to the topics of interest.	
The system must identify which topics are of interest to the user or not.	
The system must make recommendations according to the user's profile.	
The user must view the results of the recommendation according to interests.	
The user must view independent results in case of no history.	

As a proposal after the analysis of the proposed requirements, the following software architecture was designed, which expresses each of the processes that are necessary for the proper functioning of the recommendation system. The tests will be carried out with a NEO4J graph-oriented database engine.

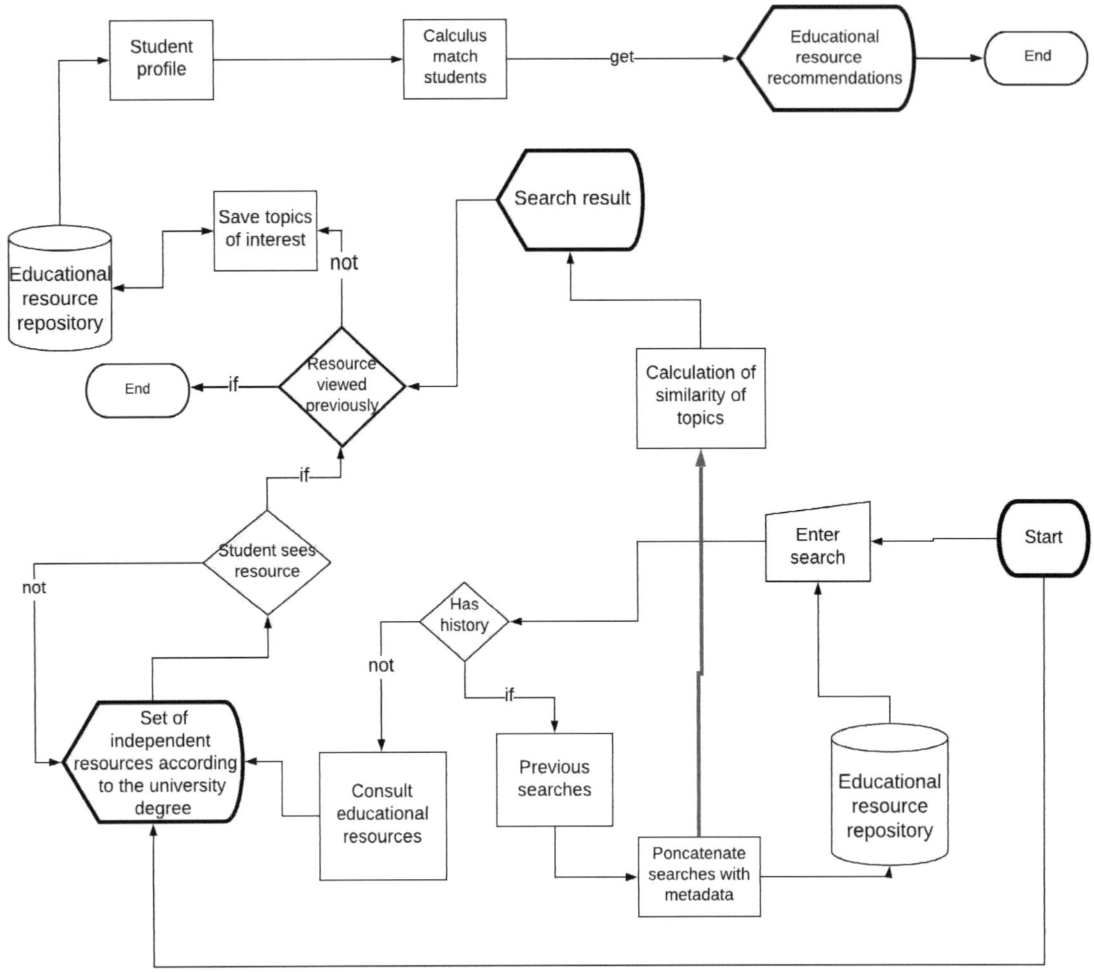

Figure 3. Software architecture scheme

This system will give recommendations of topics to each user individually according to the interest of each one of them, thus optimizing searches, permanence on the site and most importantly, it is the optimization in the performance of the same system, since This affects the user's perception of their learning process.

As can be seen in the flow diagram figure (see figure 3), the system starts with a search, which returns a list of results that concatenates with search history, in addition the system provides recommendations of the most viewed or consulted topics by their peers. The system identifies a new user who provides independent results according to the search topic, results that will be saved in their profile for a future recommendation.

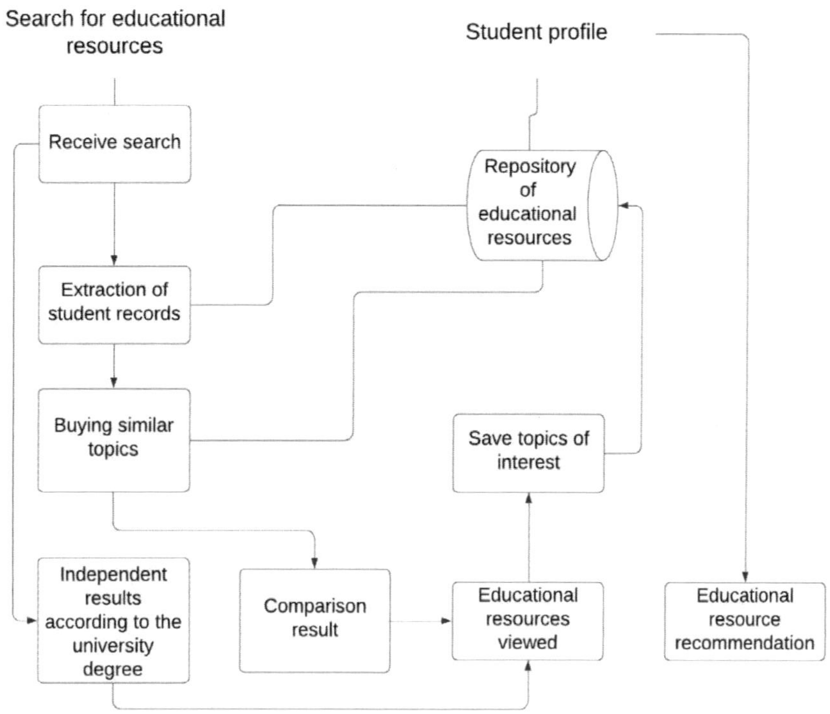

Figure 4. Scheme of more generalized recommendation system above.

As can be seen in figure 4 "block diagram", it is a more generalized representation of the recommendation system previously analyzed. The process begins in the search for educational resources, then goes to the first block, which receives the search data, to later continue to the next block, where the data of queries made in the user's history are extracted, followed by this a comparison of the data from the new search with the history, checking similarities between the two. If you do not have any history as a user, it will present you with independent results according to the topic consulted, the system makes a comparison of the resources seen and not seen to record it as history and topics of interest for future comparisons and then generates new recommendations.

6. Conclusions

It is concluded that the implementation of recommendation systems has had a boom in recent years, because they allow the handling of information to be more precise for end users, that is, these systems prevent the user from being frustrated and lost on the web due to the massive amount of information that exists in cyberspace.

It is also concluded that recommendation systems have been a great pivot for electronic commerce and marketing, to such an extent that they have proven to be a tool of great value for organizations that have implemented them within the Internet and their consumers, with the aim of that users continue to visit the site and increase the probability that they carry out some commercial activity. Due to these positive results in this sector, it is necessary to carry out research on the implementation of these systems in other areas to evaluate their impact.

Regarding the results of the research, it was possible to analyze and design an architecture capable of generating recommendations for educational resources from a historical set of data, information that will be in constant growth due to the relationships that will be formed according to the interests of each user, thus creating patterns that help the system to better predict in each recommendation process; It is worth mentioning that, to make a prediction with a minimum degree of error, it is inevitable to analyze a large amount of data.

References

[1] UNESCO, "Access to Information," [Online]. Available: https://es.unesco.org/themes/acceso-informacion. [Last access: August 2020].

[2] E. Nila, C. Juárez, J. Sánchez and A. Soberanes, "Open educational resource design with an expert system for database teaching-learning," Research in Computing Science 111, pp. 23-32, 2016.

[3] WAV Vargas, «Context-sensitive recommendation algorithm for reusable educational elements with Graph-Oriented storage,» Revista Tecnológica ESPOL - RTE., Vol. 30, No. 1, p. 101, 2017.

[4] VHMDAZG Rodrigo Vences Nava, «Performance optimization of a text document recommendation system based on server configuration,» COMPUTACIÓN E INFORMÁTICA, nº 2, 2016.

[5] SGMB Ajay Kaushik, "A Movie Recommendation System using Neural Networks," International Journal of Advance Research, Ideas and Innovations in Technology, p. 425, 2018.

[6] CAR Morales1, «SVD algorithm applied to recommendation systems in commerce,» Tecnología, Investigación y Academia, vol. 6, no. 1, p. 19, 2017.

[7] DVDYML José Carlos Hernández Nieto, "NAPI: NEWS PERSONALIZATION SYSTEM," VI INTERNATIONAL SYMPOSIUM ON COMMUNITY, ENTERTAINMENT AND SOCIAL MEDIA ", 2018.

[8] LDM-CLCR-LV and. S.-AMA Ibarra-Báez, «Book Recommendation System,» Applied Computational Technology, vol. 1, No. 1, 2019.

[9] P.-y. LRW and. Z.-f. Z. Pei-pei Wang, "A Recommendation Algorithm Based on Density Peak Clustering and Key Users," 13th International Conference on Natural Computation, Fuzzy Systems and Knowledge Discovery (ICNC-FSKD 2017), 2017.

[10] TGJCABCAMTGVB Paola Gómez, «Recommendation Systems: an approach to filtering techniques,» Revista Ibérica de Sistemas e Tecnologias de Informação, p. 288, 2019.

[11] XLWDRZ FULIAN YIN, "TAG-BASED COLLABORATIVE FILTERING RECOMMENDATION ALGORITHM FOR TV PROGRAM," 13th International Computer Conference on Wavelet Active Media Technology and Information Processing (ICCWAMTIP), 2016.

[12] W. Wenzhen, "Personalized Music Recommendation Algorithm Based On Hybrid Collaborative Filtering Technology," International Conference on Smart Grid and Electrical Automation (ICSGEA), p. 281, 2019.

[13] C. Pinilla, M. Bello and C. Peña, «Graph-oriented databases,» Tecnología, Investigación y Academia, vol. 5, no. 2, pp. 153-160, 2017.

[14] Neo4j, «Neo4j,» August 13, 2020. [Online]. Available: https://neo4j.com/. [Last access: 13 August 2020].

[15] MinScience, Typology of Projects Classified as Scientific, Technological and Innovation Characteristics, Bogota: Government of Colombia, 2019.

IOP Conf. Series: Materials Science and Engineering 1154 (2021) 012022 doi:10.1088/1757-899X/1154/1/012022

Operation of a Dish Stirling Heliothermic System with Directly Illuminated Tube Receiver

G E Carrillo[1,3], Y Cárdenas[2]

[1]Universidad Tecnológica de Bolívar, Research Group EOLITO, Cartagena – Colombia
[2]Universidad de la Costa, Research Group GIOPEN, Barranquilla – Colombia
[3]Universidade Federal de Itajubá, Research Group NEST, Itajubá-Brasil

gcarricaballero3@hotmail.com

Abstract. The generation of electricity from solar energy is one of the options within the renewable technologies that have gained much importance in recent years. Within this area, Dish Stirling systems stand out among solar concentration technologies, as they have the highest efficiency in converting thermal energy into electricity. Unfortunately, there are few publications and information on operating parameters of this type of technology. For this reason, this paper presents a methodology to characterize the behavior of this type of system, considering project parameters of a system installed in the city of Itajuba-MG Brazil. In the initial stage, a geometric and optical analysis is presented, which allows us to determine parameters such as the parabolic disk opening area and optical parameters that define the performance of this component. In the next stage, a thermal analysis is elaborated as a way to quantify the losses in the receiver system and to be able to calculate the energy supplied to the Stirling engine fluid. Finally, relationships are established to characterize the behavior of the Stirling engine working with a DIR type receiver. The results show that for speeds below 8 m/s, the heat loss due to radiation is higher than the loss due to convection, representing up to 90% of the total heat loss in the system, for solar irradiation of 1000 W/m2, the maximum electrical power obtained by the system was0.99 kW.

1. Introduction

The continuous development of society and the industrial sector has required the implementation of various technologies and processes based on the use of conventional energy sources. Throughout history, the use of these technologies has grown considerably due to their high potential to meet energy needs [1]. Unfortunately, this behavior has contributed to an increase in the level of pollutants, with harmful effects on the environment, human health, the deterioration of the ozone layer, and the greenhouse effect [2]. In this global scenario, renewable energies and technologies, besides gaining more importance as options to supply thermal energy or generate electricity, constitute new options for the diversification of the global energy matrix, with a positive impact on people's health and the environment [3]. We can highlight wind energy, biomass energy, geothermal energy, tidal and solar energy [4].

Emphasizing that, for regions with high intensity of solar irradiation, heliothermal or solar concentration technologies (Linear Fresnel, Parabolic Cylinder, Central Tower, Dish Stirling) represent one of the first options to generate electricity in a sustainable and friendly way [5]. In the case of Dish

Stirling systems, it is currently considered to be one of the most promising technologies, due to its high capacity to concentrate solar radiation, conversion efficiency, and compact configuration [6].

Some research on this type of system is presented below: Vahidi y Abbassi [7] introduced a dynamic analysis of a Dish Stirling system, using the TRNSYS simulation platform. For the conditions evaluated, the maximum efficiency obtained by the system was 24 % and electrical power of 2.1 kW, for an engine operating with helium as working fluid at a pressure of 50 bar. Sandoval et al. [8] developed a thermal analysis of a Dish Stirling system for climatic conditions in Natal, Brazil. As the main results of this system, system conversion efficiencies of 19% to 26% were obtained for the period April 14-15, 2015. Lai et al. [9] developed a mathematical model of a Stirling solar engine, which considers the climatic conditions and project parameters of this type of engine. For the conditions evaluated, the maximum power obtained was 25 k, at a conversion efficiency of 40 %. Shabanpour et al. [10] presented the analysis of a cogeneration system using a Dish Stirling system for electricity generation. The results obtained show that the implementation of this type of system allows considerable reductions in CO2 emissions to be achieved, and a period of return on investment suitable for the proposed method.

Because of the above and considering the global panorama, the chief scientific contribution of this work consists in the implementation of a methodology through a mathematical model to develop the analysis of the operation of a Dish Stirling system with a directly illuminated tube receiver, considering the project and operational parameters of a real system installed in the city of Itajubá-MG/Brazil, and the climate conditions of this city.

2. Analysis of the Dish Stirling system.

This section will present the stages implemented in this research work required to characterize the behavior of the Dish Stirling system. Figure 1 shows the calculation sequence defined in the methodology described in this paper.

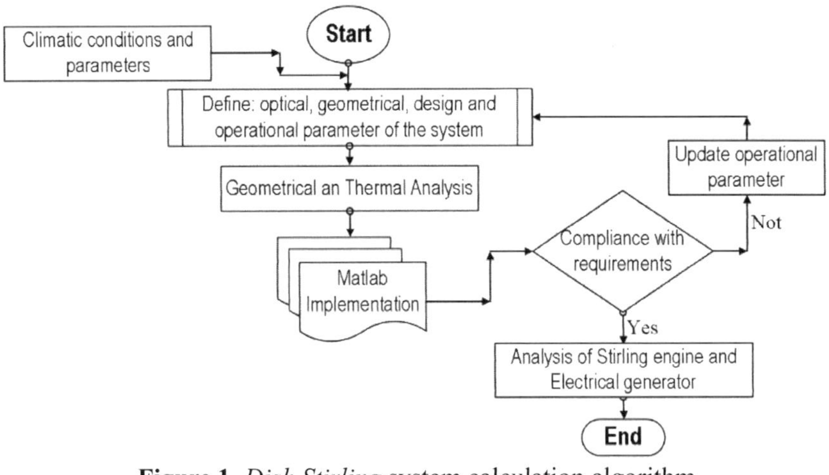

Figure 1. *Dish Stirling* system calculation algorithm

It is essential to mention that the calculation sequence defined in figure 1, allowed to set the equations and considerations necessary to characterize this type of technology. And from these, a programming code was developed in Matlab, to determine and analyze iteratively the different calculations required to create the analysis proposed in this work.

The project parameters and the operating conditions of the system modeled in this work were defined from the TRINUM system installed in the NEST/UNIFEI Heliothermal Energy Laboratory, shown in figure 2.

Figure 2. Dish Stirling system, [Taken from Heliothermal Energy Laboratory NEST/UNIFEI].

As can be seen in figure 2, the main components of this type of technology include the concentrator disk and its support structure, the two-directional (elevation and azimuthal) solar tracking system, the receiver assembly, Stirling engine, an electric generator, the system's control panel and sensors, and the system's support pedestal.

The operation of this type of technology starts when the irradiation reaches the surface of the collector disk, which is concentrated and directed on the surface of the system's receiver. This is a tube receiver, through which circulates a working fluid (can be air, helium, hydrogen), which is heated by the concentrated solar irradiation. When the fluid is heated, it expands and causes the pistons of the Stirling engine to move. This same working fluid, after this expansion stage, is subjected to a cooling process, so that the fluid is compressed, and in this way can cause an alternative movement of the pistons of the engine. In the final stage of this process, this reciprocating movement is used to drive the shaft of an electric generator[11].

During the operation of this technology, the control system, sensors, and solar tracking, are responsible for ensuring the monitoring of the regularly throughout the period of operation, or even are responsible for bringing the system to the safe position in the absence of solar radiation, presence of rain or high wind speed, as a way to protect the integrity of the system and its components [11].

Table 1 shows the main parameters of the two systems installed at the NEST/UNIFEI Heliothermal Energy Laboratory in the city of Itajubá, Brazil.

Table 1. Parameters of the Dish Stirling systems of the project TRINUM [12].

System Operation	Electrical power[kW]	1
	Overall efficiency [%]	17.37
Concentrator	The diameter of the	3.75
	Net reflective surface [m^2]	9.54
	Gross reflective surface [m^2]	11.23
	Focal distance [m]	2.3
	Focal ratio f/d	0.61
	Concentrator efficiency [%]	76.9
Receiver	Receiver Diameter [mm]	230
	Receiver temperature [°C]	525
Stirling Engine	Electric power [kW]	1
	Gas temperature (high)[°C]	497
	Stirling engine efficiency [%]	22

The following will define some of the main equations used to characterize the operation of the system shown in Figure 2. The calculation sequences implemented in this work, the calculation algorithm presented in Figure 1.

2.1. Thermal analysis of the system

Once the solar radiation is captured by the collector disk, it is concentrated and directed on the system's receiver, located at the focal point. In the system's receiver, we have the main heat losses of this type of technology; for this reason we focus the energy balance of the order on this component, as shown in figure 3.

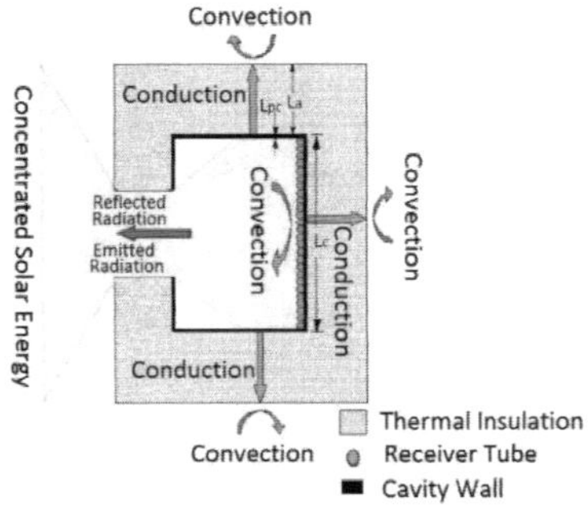

Figure 3. Heat loss mechanisms in the Dish Stirling system receiver [13].

From figure 3, it is possible to establish the heat balance relationships to determine the useful energy that is transferred to the working fluid of the Stirling engine, equation (1) [14]:

$$Q_{useful} = Q_{receiver} - \left(Q_{cov\_w} + Q_{cov\_aper} + Q_{r,em} + Q_{r,ref}\right) \quad (1)$$

Considering that all the concentrated energy reaches the surface of the system's receiver. Therefore the energy on the receiver can be determined with equation (2) [15]:

$$Q_{receiver} = I A_{DC}^{aper} \rho f_{int} \quad (2)$$

to ρ = 0,94 according to [16] and is between 0,9-0,99, according to [15].

From the solar radiation on the receiver, a fraction is transferred to the environment, through the external wall of this component, this heat loss can be calculated with the equation (3) [17]:

$$Q_{cc} = \frac{T_{receiver} - T_{amb}}{\frac{L_{PC}}{K_{ac}A_{ic}} + \frac{L_a}{K_a A_a} + \frac{1}{h_{ec}A_{ec}}} \quad (3)$$

Additionally, in this component it is possible to identify a loss of heat by convection through the opening where the concentrated solar radiation enters the receiver. This type of heat loss can be determined with equation (4) [17]:

$$Q_{cov\_aper} = h_{if} A_{ic} (T_{receiver} - T_{amb}) \quad (4)$$

In the case of heat losses by radiation, two main terms are identified, one for emitted and reflected radiation, which can be calculated with the equation (5) y (6) [18]:

$$Q_{r,em} = \varepsilon \sigma A_{cav}^{aper} \left(T_{receiver}^4 - T_{amb}^4\right) \quad (5)$$

$$Q_{r,ref} = \left(1 - \alpha_{ef}\right) Q_{receiver} \quad (6)$$

Once the useful energy of this type of system has been determined, it is possible to decide on the electrical power generated by this type of operation using the equation (7) [18]:

$$W_{DS} = Q_{useful} * n_{DS} \quad (7)$$

In equation (7) the term n_{DS}, represents the overall efficiency of the system, calculated with the equation (8):

$$n_{DS} = n_{con} * n_{rec} * n_{motor} * n_{gen} \quad (8)$$

3. Main results of the system's behavior

In order to determine the accuracy of the methodology presented, Figure 4 shows the comparative results of the behavior of the electrical power generated: in red by the system installed in the city of Itajuba/Brazil and in blue, the results obtained by the methodology proposed in this paper.

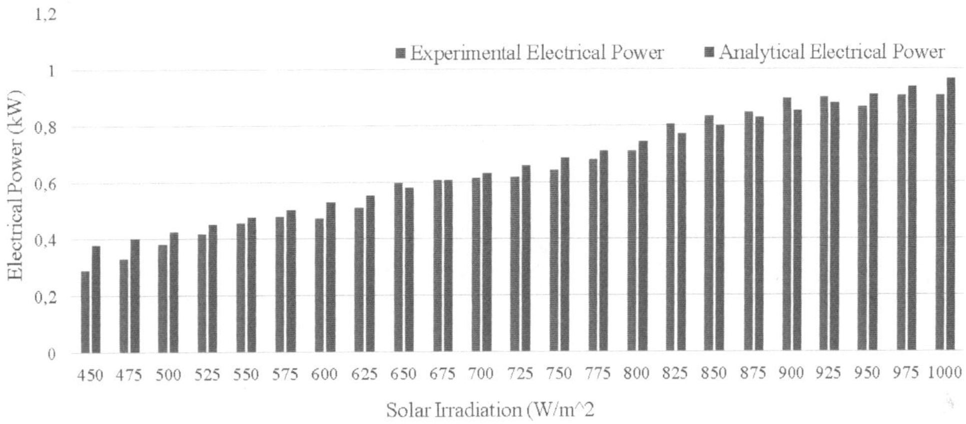

Figure 4. Comparative analysis of the electrical power generated by the Dish Stirling system.

In table 1, some of the leading project parameters of the Dish Stirling system, used to develop the analysis presented in figure 4, are described. It is essential to mention that for these analyses, it was necessary to assume the value of some parameters of the system, once these were not provided by the company that sold this technology, and likewise they could not be measured directly on the network. Some of these parameters and their values assumed in this work are collector disc reluctance (90%), receiver absorbance (96%), and the receiver tilt angle (80°).

In order to make these simulations, data of the behavior of the climatic conditions of the city of Itajubá MG, of day 06 of June of 2017, were used. These data were obtained from the meteorological station installed at UNIFEI University. For this day analyzed, it was possible to determine through the methodology implemented in this work, minimum electrical power of 0.3 kW, obtained for a solar irradiation value of 450 W/m^2, and a maximum value of 0.99 kW, for solar irradiation of 1000 W/m2.

When analyzing the experimental values and those obtained with the methodology proposed in this work, it is possible to identify the differences between the compared value at an irradiation value of 450-525 W/m^2. For values of solar irradiation higher than 525 W/m^2, minimum differences of the compared values are observed, with values close to 5 %. This behavior can be explained by the fact that at low solar radiation values, the Dish Stirling system has to use energy to overcome the inertia of the system and its components. This behavior cannot be replicated in the mathematical model proposed in this work. In addition to the system parameters that are not known or could not be determined.

Figure 5 shows the behavior of heat losses, present in the system's receiver, for which some parameters were considered as, a tilt angle of the receiver of 80 °, a value of solar irradiation of 906 W/m2, interception factor of 92%.

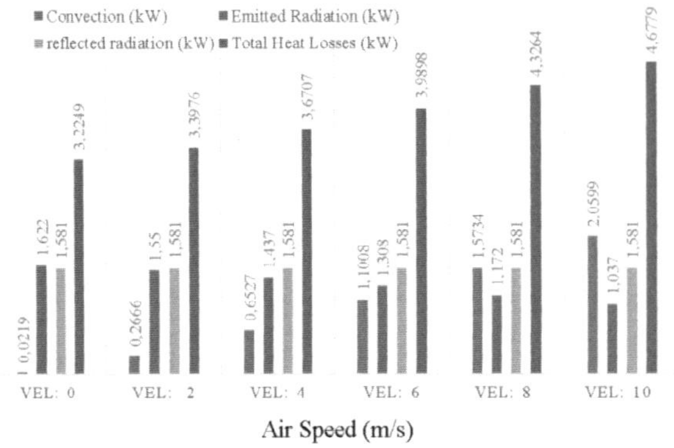

Figure 5. Heat loss behavior as a function of wind speed variation

From the results shown in figure 5, it is possible to observe that for the conditions evaluated, the heat loss by reflected radiation is observed to remain constant. In the case of the heat loss emitted, it is possible to keep a small variation for the different evaluated. Presenting a minimum difference of 4 % and a maximum difference of 36 %. For the loss of heat by convection, it is possible to observe a considerable increase with the increase of the wind speed. For the low-speed values, this heat loss has the best impact on the performance of the system. It is changing behavior, from a wind speed higher than 8 m/s, where the value of convection heat loss exceeds the values of radiation heat loss.

Figure 6 shows the behavior of heat loss by convection, by emitted and reflected radiation as a function of the system's operating temperature behavior, considering a wind speed of 5 m/s, an ambient temperature of 298.15 K and a receiver inclination angle of 80 °.

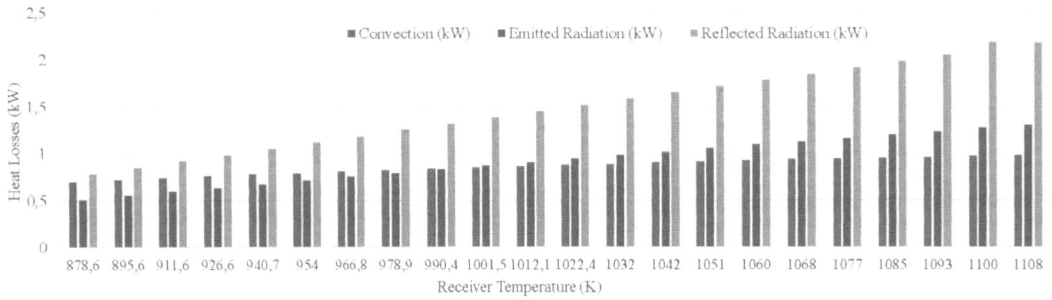

Figure 6. The behavior of system receiver thermal losses for different operating temperatures.

For the evaluated conditions and operating temperatures between 879.6 and 1108 K, it is possible to observe that the heat loss by reflected radiation has a more significant impact on the operating behavior of the Dish Stirling technology. In some of the conditions evaluated, up to a maximum difference of 55 % in relation to the other forms of heat loss. In the case of low operating temperatures, heat loss by convection shows higher values than the radiation emitted. For a temperature of 876.8 K, the heat loss by convection exceeds the maximum difference in relation to the radiation emitted, in a proportion of 27 %. In this analysis, it is observed that as the temperature increases, the emitted radiation increases. From the values of 1001 K, the heat loss exceeds the values of heat loss by convection.

Conclusions

The results obtained in this work allow us to conclude that the performance of Dish Stirling systems for cities or places with conditions similar to those of Itajubá/Brazil are affected to a greater extent by heat losses due to radiation. Once the data obtained by the weather station (June 06, 2017), the low values of wind speed present in this region of Brazil.

The analyses carried out in this research also allowed to determine that for some operating conditions, heat losses by radiation, considering the UNIFEI project parameters, can represent up to 90-95% of the heat losses present in the system's receiver.

From the comparison of the electrical power of the experimental system and that obtained in this work, it is possible to observe that there is a functional characterization of the behavior of this type of technology. For this reason, it can be concluded that the methodology implemented in this work can be used as a useful tool, which helps to have a better understanding of this type of technology and its operating parameters. And from which it is possible to make decisions on the system parameters that contribute to the improvement of its performance. In the particular case of Itajubá, the analysis shows that efforts should be made to investigate and invest in strategies to reduce the effects of heat loss from radiation.

Acknowledgments

The authors wish to thank the Federal University of Itajuba UNIFEI, CAPES, CNPq, FAPEMIG, ANEEL, for the support in the development of this project and the support of the students. To the working group of the meteorological station of UNIFEI, for providing the data of the weather conditions of the city of Itajuba/Brazil. And to the company INNOVA as this one is the representative of the Dish Stirling system installed at the NEST/UNIFEI Heliothermal Energy Laboratory.

References

[1] Heras J and Martín M 2020 Social issues in the energy transition : Effect on the design of the new power system (*Applied Energy* vol. 278) April, p. 115654.

[2] Strezov V and Cho H H 2020 Environmental impact assessment from direct emissions of Australian thermal power generation technologies (Journal of Clean Production) vol. 270, p. 122515.

[3] Cuesta M A Castillo T and Borges C. E 2020 A critical analysis on hybrid renewable energy modeling tools: An emerging opportunity to include social indicators to optimize systems in small communities (*Renewable Sustainable Energy Reviews* vol. 122) June 2019 p. 109691.

[4] Mahmud M Huda A N Farjana S H and Lang C 2020 Life-cycle impact assessment of renewable electricity generation systems in the United States (Renewable. Energy, vol. 151) pp. 1028–1045

[5] Pearre N and Swan L 2020 Combining wind, solar, and in-stream tidal electricity generation with energy storage using a load-perturbation control strategy (*Energy*, vol. 203) p. 117898,

[6] Guarino S Buscemi A Ciulla G Bonomolo M and Lo V 2020 A dish-Stirling solar concentrator coupled to a seasonal thermal energy storage system in the southern Mediterranean basin : A cogenerative layout hypothesis (Energy Conversion and Management vol. 222) March p. 113228.

[7] Vahidi M and Abbassi Y. 2020 Exploring dynamic operation of a solar dish-Stirling engine : Validation and implementation of a novel TRNSYS type (Sustainable energy technology and assessments vol. 40) March, p. 100765.

[8] Sandoval O R Castro B Ungaretti M García J J and Molina R 2019 Modelling, simulation and thermal analysis of a solar dish / Stirling system : A case study in Natal, Brazil (Energy Conversion and Management vol. 181) September 2018, pp. 189–201.

[9] Lai X Yu M Long R Liu Z and Liu W 2019 Dynamic performance analysis and optimization of dish solar Stirling engine based on a modi fi ed theoretical model (Energy vol. 183) pp. 573–583.

[10] Shabanpour R Sayyaadi H and Hosseinzade H 2013 Sizing a solar dish Stirling micro-CHP system for residential application in diverse climatic conditions based on 3E analysis (Energy Conversion and Management vol. 75) pp. 348–365,

[11] . Carrillo G. E 2013 Estudo e Modelagem dos Componentes de um Sistema Dish Stirling Visando à Otimização da Potência e a Eficiência do Sistema. Dissertação de Mestrado,"

Universidade Federal De Itajubá.
[12] INNOVA, "Solar Dish System Cogenerative and Thermal Module. Trinium Technical Manual for Installation and Use.," Pescara. Italia, 2014.
[13] Nepveu F Ferriere A and F. Bataille 2009 Thermal model of a dish/Stirling systems (*Solar Energy*, vol. 83, no. 1, pp. 81–89.
[14] Wu S Y Xiao L Cao Y and Li Y. R 2010 A parabolic dish/AMTEC solar thermal power system and its performance evaluation (Applied Energy vol. 87 no 2) pp. 452–462.
[15] Hafez A. Z Soliman A El-Metwally K A and Ismail I M 2017 Design analysis factors and specifications of solar dish technologies for different systems and applications (Renewable and Sustainable Energy Reviews vol 67) pp. 1019–1036,
[16] Ni M 2016 Improved Simple Analytical Model and experimental study of a 100 W b-type Stirling engine (Applied. Energy vol. 169) pp. 768–787.
[17] J. Ruelas, N. Velazquez, and J. Cerezo, "A mathematical model to develop a Scheffler-type solar concentrator coupled with a Stirling engine," *Appl. Energy*, vol. 101, pp. 253–260, 2013.
[18] Fraser P. R 2008 Stirling Dish system performance prediction model Master thesis. The University of Wisconsin-Madison.

Exergetic Analysis of a Parabolic Trough Collector System Operating with Different Heat Transfer Fluids

G E Carrillo[1,3], Y Cárdenas[2]

[1]Universidad Tecnológica de Bolívar, Research Group EOLITO, Cartagena – Colombia
[2]Universidad de la Costa, Research Group GIOPEN, Barranquilla – Colombia
[3]Universidade Federal de Itajubá, Research Group NEST, Itajubá-Brasil

gcarricaballero3@hotmail.com

Abstract. Solar concentrating systems are one of the most promising technologies to supply process heat or generate electricity. Responding to the global problem of increasing pollutants in the atmosphere and contributing to the diversification of the world's energy matrix. In this sense, the Parabolic Trough Collector system represents the technology with the greatest technological maturity and hours of commercial operation. Thus, this paper presents an exergetic analysis of this type of system, based on the configuration of a system installed in the city of Itajuba-MG, Brazil. As a way to identify the improvement potentials of this type of system and thus obtain performance improvements. In the initial stage, the exergetic balance on the system's receiver is developed. The effect of the loss of exergy and the destruction of exergy on the exergy of the fluid inside the receiver tube is determined. In the next stage, the required parameters were defined to calculate the terms defined in the first stage. Among these parameters is the effect of exergy of solar irradiation, heat losses, the optical efficiency of the collector, and heat absorbed by the heat transfer fluid. The results obtained show that the optical losses of the collectors of this type of plant represent one of the main causes of exergy losses. Additionally, it was possible to determine that for low-temperature applications, water allows the best exergetic use.

1. Introduction

The high impact caused by the increase in the level of pollutants worldwide, as a consequence of the operation of systems and processes based on conventional fuels [1], has promoted the development of projects and research on technologies based on renewable energy, such as solar energy [2]. Where it is possible to highlight technologies such as Dish Stirling, Linear Fresnel, Central Tower, and Parabolic Trough Collector systems [3], these are characterized by generating a positive impact on the environment, people's health, and the diversification of the countries' energy matrix in the life cycle analysis [4].

There are various projects implemented and in the development phase worldwide, but without a doubt, this whole panorama shows that the solar thermal plants with the greatest maturity and commercial development are those of Parabolic Trough Collectors (PTC) [5]. Many are the hours of operation that this type of system, however, it is necessary to find strategies to improve performance and resource utilization that can have this type of system, for the various operating conditions that may occur throughout the world [6].

Within these strategies, we find the exergetic analysis and its diverse tools as a way to identify the potentials of improvements that can be had in a system or process [7]. Some of the research that has been carried out for this type of technology is presented below: Akbarzadeh et al. [8] presented an energy and exergetic analysis of a PTC system, with the purpose of determining the performance of this type of system using pipes with special arrangements inside. This investigation showed that the use of helical piping allows improvements in the performance of these systems. Babaelahi et al. [9] by simulating an electricity generation plant, in the conventional PTC and Rankine Cycle configuration, establish a system control methodology based on energy and exergetic analysis. At the end of the system analysis, improvements of 36.06 % were obtained in the overall efficiency of the system. Ghazouani et al. [10] perform thermo-economic, exergetic, and impact optimization analysis of operational and design parameters, a small PTC system. This analysis allowed us to determine that a reduction in the , an increase in the receiver tube, and the diameter of the collector are considerable improvements for this system. Zare y Moalemian [11] propose the energy, exergetic and economic analysis of an integrated PTC system with a Kalina cycle. The main results are a power cycle efficiency of 64% and an overall system efficiency of 14%.

Based on the review of the literature and the search for better use of the resources by these emerging technologies, the main scientific contribution of this work consists of the implementation of an exergetic analysis of a system of Parabolic Trough Collectors, based on the configuration of the system installed in the NEST/UNIFEI Heliothermal Energy Laboratory, in the city of Itajubá/Brazil. For this analysis, the operation of this system with different heat transfer fluids is considered, an aspect that is not taken into account by the publications found in the literature review.

2. Exergetic Analysis of a Parabolic Trough Collector System

To present a structured methodology for calculating this type of system, figure 1 shows the calculation algorithm designed and implemented in this work, using the Matlab. Programming tool.

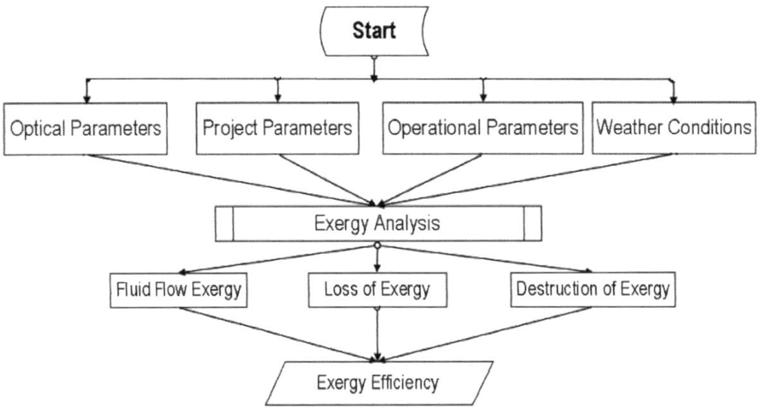

Figure 1. Calculation algorithm of a Parabolic Trough Collector System.

Based on the calculation algorithm defined in figure 1, and the configuration of the PTC system installed in the NEST/UNIFEI Heliothermal Energy Laboratory, shown in figure 2, the different conditions and mathematical relations were raised, to elaborate the analysis proposed in this study.

IOP Conf. Series: Materials Science and Engineering 1154 (2021) 012023 doi:10.1088/1757-899X/1154/1/012023

Figure 2. PTC system, [Taken from NEST/UNIFEI Heliothermal Energy Laboratory].

The Parabolic Trough Collector system is shown in figure 2, the system installed at UNIFEI, supplied by the Canadian company RACKAM, and its main characteristics are described in table 1. Among the main components of this type of system, we find The reflective mirrors or collectors, the system's receiver (glass cover and receiver tube), the solar tracking system, the control panel, and sensors.

The operation of this system begins when the solar radiation reaches the surface of the collectors, and this is concentrated and directed on the system's receiver, located at the focal point of the system (linear type focus). The concentrated radiation generates the progressive heating of the heat transfer fluid located inside the absorber tube. Fluid, which is transported to a power block or an energy demand system, such as an industrial process, a commercial, residential or hospital sector application, among other applications. In these technologies, the tracking system is responsible for moving the collectors constantly following the sun to ensure that the sun's rays reach the surface of the collectors almost perpendicularly at all times, reducing losses due to radiation overflow.

Table 1. Main parameters of the PTC system, RACKAM project [12].

	Parameter	**RACKAM System**
Collector	Length [m]	24.4
	Width [m]	1.2 m
	Height [m]	1.5
	Surface installed [m^2]	112
	Orientation	East-West
	No. of rows	4
	Thermal efficiency [%]	65
Receiver	Type of fluid	Water
	Maximum pressure [bar]	10
	Maximum temperature [°C]	160
	Absorber tube material	Stainless steel, ASTM A312-TP
	Absorber diameter [m]	0.0254

Based on the configuration of the system installed at UNIFEI and some of the parameters of this project, the considerations and mathematical relationships required to develop the exergetic analysis of this type of technology were defined.

2.1. Exergetic Analysis of the System

Once the PTC technology is in operation, the concentrated solar radiation reaches the surface of the system's receiver (absorber glass tube cover). Since the most important heat transfer process occurs in this component, it is necessary to study the parameters and conditions that can affect energy interaction, losses, and destruction. A simplified diagram of the receiver of a PTC system is shown in Figure 3.

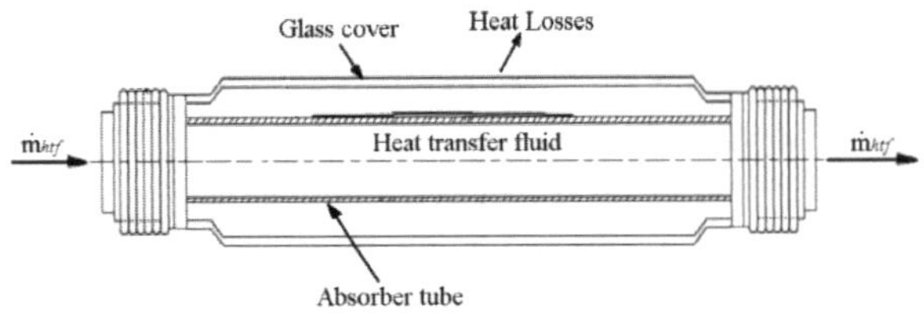

Figure 3. The receiver of a PTC system and its main components [13].

By applying an exergetic balance to the system's receiver (Figure 3), it is possible to obtain equation (1). Which we can call it as the basis of calculation of the methodology proposed in this study [14]:

$$E_{sol} = E_{htf} + E_{loss} + E_{dest} \tag{1}$$

Where the first term of equation (1), represents the exergy of the heat transfer fluid inside the absorber tube, calculated with equation (2) [14]:

$$E_{htf} = q_{ac} - \dot{m}_{htf} T_{amb} \, Ln \left[\frac{T_{Col\_out}}{T_{htf\_in}} \right] \tag{2}$$

In the case of loss of exergy, the second term of equation (1), this can be determined with equation (3) [14]:

$$E_{loss} = E_{thermal\_loss} + E_{optical\_loss} \tag{3}$$

As shown, equation (3) takes into consideration the effect of heat and optical losses, which can be determined with equations (4) and (5), respectively [14]:

$$E_{thermal\_loss} = q_{heat\_loss} \left(1 - \frac{T_o}{T_{htf}} \right) \tag{4}$$

$$E_{optical\_loss} = \left(1 - \eta_{optical} \right) E_{sol} \tag{5}$$

In the case of the loss of exergy due to the optical effect of the collector material, the optical efficiency of the collector and the exergy contribution of the solar radiation, calculated with equation (6) and (7), respectively, are taken into account [14]:

$$\eta_{\acute{o}tica} = \gamma_{fi} \rho_{er} K_{\theta} \tau_{gc} \alpha_{a} \tag{6}$$

$$E_{sol} = \left(I_{beam} (a - D_{absorver}) \right) \psi \tag{7}$$

Where γ_{fi} is the interception factor, ρ_{er} the reflectivity of the collector, K_{θ} is the modified angle of incidence, τ_{gc} is the transmittance, α_a is the absorbency of the absorber, a is the opening width of the Collector, I is the solar radiation, D is the diameter and ψ useful working potential of solar radiation. Now, to determine the exergy destruction term, the third thermal of equation (1), equation (8) is used [14]:

$$E_{dest} = E_{dest\_sol} + E_{dest\_ab\_f} \tag{8}$$

Where the first term of equation (8), represents the destruction of exergy between the sun and the absorber tube and the second thermal term represents the destruction of exergy between the absorber tube and the heat transfer fluid, terms that can be determined with equations (9) and (10) [14]:

$$E_{dest\_sol} = \eta_{optical} E_{sol} - q_{ac} \left(1 - \frac{T_{amb}}{T_{htf}} \right) \tag{9}$$

$$E_{dest\_ab\_f} = q_{ac} \left(1 - \frac{T_{amb}}{T_{htf}} \right) - E_{htf} \tag{10}$$

Once the terms of the exergetic balance, proposed in equation (1), have been calculated, it is possible to determine the exergetic efficiency of the system, with equation (11) [14]:

$$\eta_{exergy} = \left(\frac{E_{htf}}{E_{sun}}\right) \tag{11}$$

3. Main results of the system's behavior

The behavior of the decrease in exergy for three heat transfer fluids, such as water, Downtherm A, Downtherm Q and Syltherm 800, as shown in Figure 4.

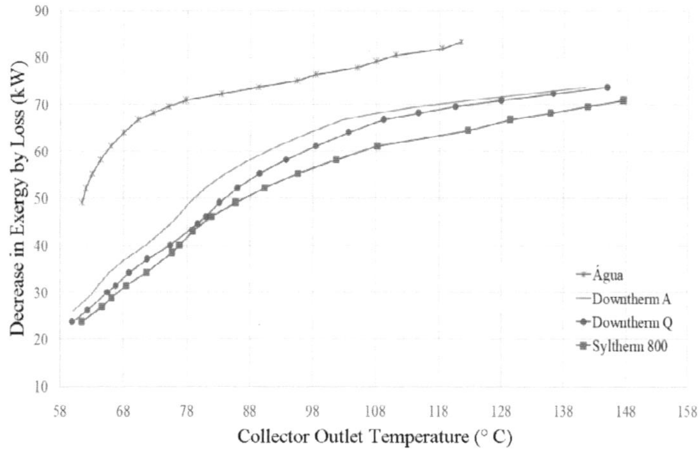

Figure 4. The behavior of the decrease of exergy by losses.

As it is possible to observe from figure 4, of the fluids evaluated in the solar field of Parabolic Trough Collectors, the configuration with water, presents the biggest decreases, followed by the configuration with Downtherm A, the Downtherm Q and the Syltherm 800. It is important to mention that the biggest contributions to this effect of the decrease of exergy, are caused mainly by the optical effects (99%) of the solar field and in smaller proportion by the thermal losses (1%). This behavior is due to the optical losses that depend on the optical efficiency of the collectors and the exergy of the solar radiation.

Another aspect evaluated for this type of system consists of analyzing the exergy destruction of the set of collectors, depending on the output temperature of the solar field, as shown in Figure 5.

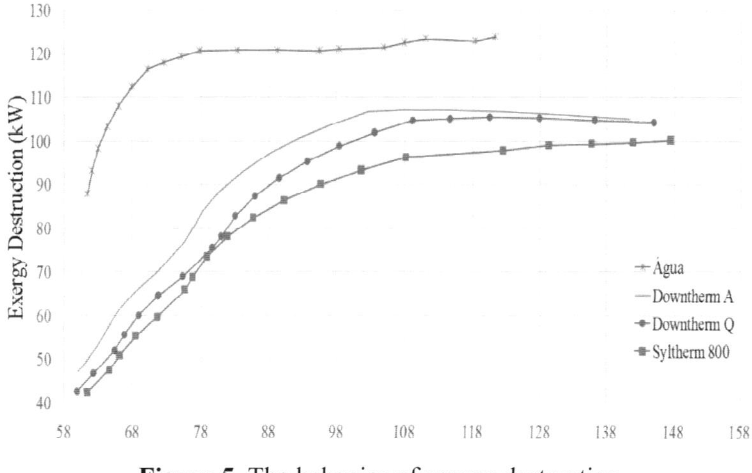

Figure 5. The behavior of exergy destruction.

From figure 5, it is possible to determine that the configuration with water as heat transfer fluid, presents the maximum values of exergy destroyed. Followed by the configuration with Downtherm A, Downtherm Q, and Syltherm 800. There is no considerable difference in the exergy destroyed between the configurations of the system operating with oils

Finally, this analysis presents the exergy behavior of the collectors as a function of the operating temperature (Figure 6).

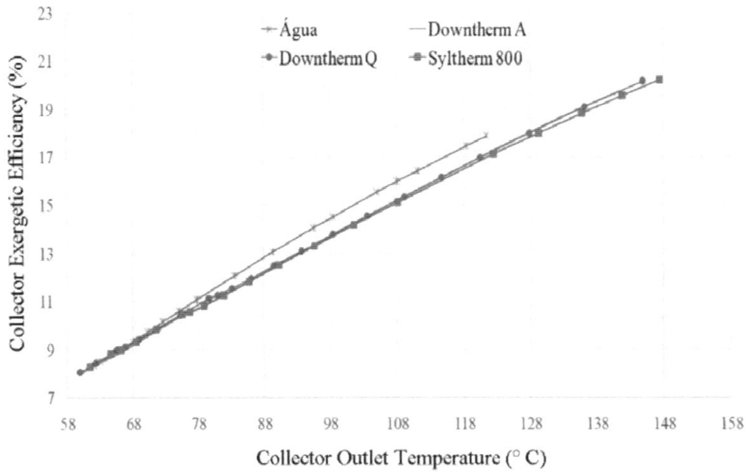

Figure 6. The behavior of the exergetic efficiency of the collector.

For the fluids chosen in this analysis, the highest efficiency values were obtained for the system working with water. For a temperature range of 70.2 ° C to 121.4 ° C, the efficiency values were between 9.7% - 17.9%, higher than those obtained with the other heat transfer fluids used. This analysis also shows that for the defined configuration, higher temperature values are obtained at the output of the solar field with the Syltherms 800, Downtherm A, and Downtherm Q. For these fluids, there is no significant difference between the exergetic efficiency values of the solar field. In the temperature range of 60° C to 147.5° C, the efficiency values determined are in the range of 8.2% - 20.2%.

Conclusion

From the results obtained in this work, it was possible to determine that in the case of the technology of PTC, it becomes important to concentrate the efforts in looking for ways to reduce the effects of optical losses, as well as to obtain better use of the solar irradiation on the receiver of the system. For this purpose, the use of new materials of the collectors and receiver, or even the possibility of using auxiliary components for a better distribution of the solar radiation concentrated on the system's receiver, can be considered.

It can also be concluded that, even though the configuration of the system working with water presents the highest values of loss and destruction of exergy, this configuration allows us to obtain the highest exergetic efficiency. The low difference in efficiency with the other configurations guarantees the possibility of using water for small system applications, contributing to reducing costs associated with the purchase of other types of fluids and final costs of the projects, which would have to be assumed in these other configurations.

Acknowledgments

The authors wish to thank the Federal University of Itajuba UNIFEI, CAPES, CNPq, FAPEMIG, ANEEL, for the support in the development of this project and the support of the students. And to the company RACKAM for being the representatives of the PTC system installed at UNIFEI.

References

[1] Ozturk M and Dincer I 2019 ScienceDirect Comparative environmental impact assessment of various fuels and solar heat for a combined cycle (International Journal of Hydrogen Energy vol 44 no. 10) pp. 5043–5053.
[2] Dixit S 2020 Proceedings Solar technologies and their implementations : A review (Materials Today: Proceedings vol 28) pp. 2137–2148, 2020.
[3] Kumar L Hasanuzzaman M and Rahim N A 2019 Global advancement of solar thermal energy technologies for industrial process heat and future prospects : A review (Energy Conversion and Management vol 195) May pp 885–908.
[4] Fan Y 2020 A unified ecological assessment of a solar concentrating plant based on an integrated approach joining cosmic exergy analysis with ecological indicators (Renewable and Sustainable Energy Reviews vol 129) May p 109934.
[5] Fuqiang W Ziming C Jianyu T Yuan Y Yong S and Linhua L 2017 Progress in concentrated solar power technology with parabolic trough collector system : A comprehensive review (Renewable and Sustainable Energy Reviews vol 79) February pp 1314–1328.
[6] Vouros A Mathioulakis E Papanicolaou E and Belessiotis V 2020 Modelling the overall efficiency of parabolic trough collectors (Sustainable Energy Technologies and Assessments vol 40) May p 100756.
[7] Hassan H Yousef M. S Fathy M and Ahmed M S 2020 Assessment of parabolic trough solar collector assisted solar still at various saline water mediums via energy, exergy, exergoeconomic, and enviroeconomic approaches (Renewable Energy vol 155) pp 604–616.
[8] Akbarzadeh S and Valipour M S 2020 Energy and exergy analysis of a parabolic trough collector using helically corrugated absorber tube (Renewable Energy vol 155) pp 735–747.
[9] Babaelahi M Mo E and Rafat E 2020 Combined Energy-Exergy-Control (CEEC) analysis and multi-objective optimization of parabolic trough solar collector powered steam power plant (Energy vol 201).
[10] Ghazouani M Bouya M and Benaissa M 2020 Thermo-economic and exergy analysis and optimization of small PTC collectors for solar heat integration in industrial processes (Renewable Energy vol 152) pp 984–998.
[11] Zare V and Moalemian A 2017 Parabolic trough solar collectors integrated with a Kalina cycle for high-temperature applications : Energy, exergy, and economic analyses (Energy Conversion and Management vol 151) July pp 681–692.
[12] RACKAM 2014 Solar system design and specification sheet of Parabolic Trough Collectors," Valcourt, Canada.
[13] Carrillo G E 2018 Modelagem do Comportamento Integrado de um Sistema de Coletor Cilíndrico Parabólico Operando com Ciclo Rankine Orgânico e Armazenamento Térmico de dois Tanques. Tese de Doutorado Universidade Federal De Itajubá,
[14] Bellos E. and Tzivanidis C 2017 A detailed exergetic analysis of parabolic trough collectors (Energy Conversion and Management vol 14) pp 275–292.

IOP Conf. Series: Materials Science and Engineering 1154 (2021) 012024 doi:10.1088/1757-899X/1154/1/012024

Analysis of the behavior of a parabolic trough collectors' plant for weather conditions of Cartagena - Bolívar

V Lopez[1], Y Cárdenas[2], G E Carrillo[1,3]

[1]Universidad Tecnológica de Bolívar, Research Group EOLITO, Cartagena – Colombia
[3]Universidad de la Costa, Research Group GIOPEN, Barranquilla – Colombia
[2]Universidade Federal de Itajubá, Research Group NEST, Itajubá-Brasil

ycardena6@cuc.edu.co

Abstract. Solar energy and concentration technologies, have gained importance in recent years due to their versatility and positive impact on the environment. Varios son las tecnologías que se pueden utilizar para generar electricidad, a partir de la irradiación solar, sin embargo, la tecnología de Colectores Cilíndrico Parabólico, recibe el reconocimiento por tener mayor desarrollo comercial. Before this panorama, this work analyzes the solar radiation potential of the city of Cartagena is located in Colombia with Elevation of 32 m, Latitude (10.57°), Longitude (-75.46), DNI 40.84KWh/m^2/day, DHI 2.26 KWh/m^2/day, Average wind speed 4.5 m/s for the use of parabolic trough collector to generate electricity. To achieve the purpose of this work, a methodology is implemented for the characterization of this type of plants. In the initial stage, the requirements of the solar field are considered and configured. Followed by the stage of characterization of the thermal storage system, and configuration of the system. To later characterize the power cycle, an operating parameter to respond to the energy demand defined in this research. In the final part of this work, an analysis of operating parameters is proposed, in order to obtain improvements in the configuration of the proposed system. The results obtained allowed us to determine that for the evaluated conditions and the configuration of the determined plant, it can operate with the Therminol PV-1 fluid, a solar multiple of 1.7 and solar salt, as the storage system fluid.

1. Introduction

Due to the high cost of electricity, energy independence, and increased demand, solar energy became an emerging form of end-use energy choice. After wind power, it is the second-largest source of energy for electricity generation [1]. Solar energy is available in abundance across the globe and can be used for solar thermal power generation applications and electricity generation [2].

Currently, the use of solar technologies has diversified, for example, energy generated through solar concentration collectors. [3]. In particular, the Parabolic Trough Collector (PTC) is a concentration technology that has proven to be useful in medium to high-temperature applications [4]. PTC technology is more straightforward and more feasible compared to similar types of solar collectors [5]. Like the ones in the Central Tower [6]and collector Fresnel [7]. A PTSC is a line-focus concentrator which converts concentrated solar energy into high-temperature heat [8]. Depending on the application, temperatures up to 550 °C is achievable in these systems [9]. As shown in Figure 1, the PTSC assembly necessarily has several subsystems to be functionally operated.

Figure 1. Parabolic Trough Collector (PTC).

The PTSC has a mirror or reflector curved in the shape of a parabola, which thus allows concentrating the sun's rays onto the focal line. The mirror is produced from different raw materials such as aluminum or low iron glass to lessen the absorption losses. The solar-weighted reflectivity of the mirror and its cost, durability, and abradable properties are essential factors during the production of the collector mirrors, after bending the mirror, a set of manufacturing processes such as silvering, protective coating, and gluing. In published scientific works, the following stand out Bilal et al. 2018 [10] investigated the PTC system under transient weather conditions. A detailed study and its validation with the experimental data has been carried out. The thermal efficiency of the system is reported to be 76%. In a research study, Agagna et al. 2018 studied [11]. The parabolic trough system has been investigated by Bellos et al. 2020 [12]. Yuehong Bi [13] In this paper, the composition and operation strategy is elaborated in detail. Given the measured data of the solar energy and the cooling load demand of buildings, the intelligent matching design of the solar collector, absorption refrigeration chiller, and the three-phase accumulator is carried out. On this basis, the dynamic performance of the solar air conditioning system with the independent-developed PTC is analyzed. In general, there has been a lot of research related to PTCs that has allowed demonstrating the efficiency of these systems.

In view of the above and considering the global panorama, the main scientific contribution of this work consists of in the implementation of a calculation methodology for a Parabolic Cylindrical Collector plant integrated to a thermal storage system and a conventional Rankine cycle. Taking as a case study the climatic conditions of Cartagena - Bolivar. Research that has not been carried out by other authors, so this work represents an important basis for evaluating the potential of the solar resource for this region of the Country and starting point for other research.

2. Place Description

Cartagena is located in Colombia with Elevation of 32 m, Latitude (10.57°), Longitude (-75.46), DNI 40.84KWh/m2/day, DHI 2.26 KWh/m2/day, Ambient temperature 27.3°C, Average wind speed 4.5 m/s. Because of its geographical and astronomical position, it has a high potential for solar radiation. Table 1 shows the solar radiation available in Colombian regions:

Table 1 Solar radiation in the Colombian areas [14]

Region	kW /m2/year
Guajira	1980-2340
Atlantic coast	1260-2340
Orinoquia	1440-2160
Amazonia	1440-1800
Andina	1080-1620

3. Methodology

This section will present the stages implemented in this research work required to characterize the behavior of the PTC system. Figure 2 shows the calculation sequence defined in the methodology described in this study.

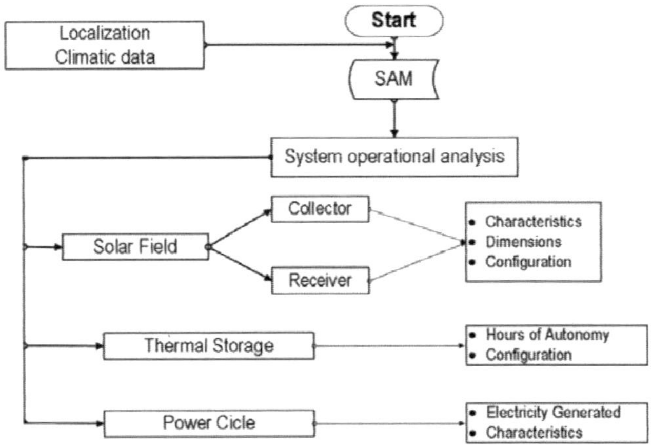

Figure 2. PTC system calculation algorithm.

It is important to mention that the calculation sequence defined in Figure 2 allowed the definition of the equations and considerations necessary to characterize this type of system. From these, a System Advisor Model (SAM) analysis was developed to determine and analyze in an iterative way, the different calculations required to create the report proposed in this study.

3.1 Characteristics of the System

For this analysis, parabolic cylinder diagrams with Physical Model specifications were used in the SAM software. This allows the study of the geometry and properties of the system components to be represented using parameters from the manufacturers of these devices. Table 2 shows the Description of the analysis properties of the solar plant for the SAM simulation, and later in each subsection of this stage will be detailing the established configurations.

Table 2. Description of the solar plant's properties.

Nominal potential (MW)	1
N° looping	3
N° total collector	12
Extension collectors (m^2)	6 acres
Solar multiple	1,7
Annual gross electricity production (MWh)	2 818 537
Collector type	Eurotrough ET-150
Thermal oil	Therminol VP1
Storage type	Indirect from two tanks
HTF storage	Sola Sal r
Rankine Cycle Performance	0,358
Overall plant performance	0,89

3.2 Mathematical components

3.2.1. Solar field

For this study, the configuration (Solar Multiple(SM)) was used, for which SM is defined as the quotient between the thermal power produced by the solar field at the design point and the thermal energy required by the power block (Rankine Cycle), a multiple of 1.70 is selected [15].

$$SM = \frac{Qthermal, solar\ field}{Q\ thermal, power\ block}$$

3.2.2. Configuration verification for PTCs

When verifying the sizing configuration in this section, it should be noted that the maximum performance of the Rankine (η_{pot}) of a solar thermodynamic power plant is from 38% [15], so.

$$P_R = \frac{P_{t,real}}{\%performance} = \frac{0{,}648MWt}{0{,}38} = 1{,}70734311MWt \quad (1)$$

$P_{t,real}$= thermal power consumed ;P_R= efficient thermal capacity.

3.2.3. Increase in oil temperature per collector

To calculate the temperature increase in the oil under the conditions of analysis, it is determined from the thermal losses that will be obtained with the overall coefficient ($U_{L)abs}$ W/m2 °C specified in the manufacturer's data sheet and used in another essential study [16].

$$U_{L)abs} = a + b(T_{abs} - T_{amb}) + C.(T_{abs} - T_{amb})^2 \quad (2)$$

The values of the coefficients a, b, and c of the equation are chosen for a working temperature in the range of 340°C as a reference for the next value of 342°C. The values of the coefficients are 2.8954; -0.0164 and 0.000065, respectively, according to the coefficient tables calculated by the Latin American solar platform for the Eurotrough [16].

$$U_{L)abs} = 2{,}8954 - 0{,}0164(342 - 27{,}3) + 0{,}000065\ .\ (342 - 27{,}3)^2$$

$$U_{L)abs} = 4.1716\ \frac{W}{m^2\ abs°C}$$

Taking into account the dimensions of the collector's absorber tube, the thermal losses are calculated (Q_{loss}) in the collector whose average temperature is the same in the absorber tube (T_{abs}) is of the 342°C [16].

$$Q_{loss} = U_{L)abs}.A_{abs}.(T_{abs} - T_{amb}) \quad (3)$$

$$A_{abs} = L.D.\pi \quad (4)$$

$$Q_{loss} = U_{L)abs}.32{,}65.(342-27{,}3) \quad (5)$$

$$Q_{loss} = 4286{,}67W$$

Incident solar energy is calculated (Q_{sol}) at the opening of the Eurotrough collector, taking into account the area of the collectors (Ac), the direct irradiation (Identical design), the angle of incidence considering the installation and handling space is increased to K=17,80°).
The following equations are taken from [17].

$$Q_{sol} =.\ A_c.\ I_{desing}\ cos(\theta) \quad (6)$$

$$Q_{sol} = 817{,}5\ (950)cos(10)$$

$$Q_{sol} = 386725{,}235991$$

4. Results

In table 3, the configured and obtained values are collected to proceed with the analysis of the expected results.

Table 3. Values obtained in the configuration and simulation of SAM.

T_{amb}(°c)	27,30
SM	1,70 Solar Múltiple
I_{design} (W/m²)	950
T_{in} (°c)	293
T_{out} (°c)	391
$T_{abs}=T_{media}$(°c)	342 Theoretical average value
N. Reynolds	1,37. 10^6
V_{max} (m/s)	4,96554 (Fluid velocity in the solar loop)
V_{min} (m/s)	0,35 (Fluid velocity in the solar loop)
$\dot{m}_{max}$ (kg/s)	12kg/s (Mass flow in the solar field)
$\dot{m}_{min}$ (kg/s)	1kg/s (Mass flow in the solar field)
▲P (P_a)	15767,18 (Pressure drop)
S_t (m^2)	33,18 .10^{-4} (Section of the pipe)
$\dot{m}_{max}$ (kg/s)	12,85kg/s (notional value)
$\dot{m}_{min}$ (kg/s)	0,95kg/s (notional value)
N_{lazos}	3
A_c (m^2)	817,50 (opening reflective area)
P_R *(MWt)*	1,701 (efficient thermal capacity)
$P_{t,real}$ *(MWt)*	0,64 (Actual thermal output)
$U_{L)abs}$ $\frac{w}{m^2\,abs°C}$	4,17 (overall rate)
A_{abs} (m^2)	32,65 8 (Absorber tube area)
Q_{Loss} (W)	4286,67 (thermal losses)
Q_{sol} (W)	386725,23 (Incident solar energy)
Q_{Useful} (W)	327 124,89 (Collector power)
T_{oil} (°C)	10.76 (theoretical value of the thermal increase of the oil)
$\eta_{c,solar}$ (%)	76 Theoretical value of the solar field yield)

4.1. Analysis of plant energy production

The first data provided by SAM when carrying out the simulation is the analysis of annual net electricity production for the 25-year life of the plant, which we will compare from the following table 4:

Table 4. Comparison of obtained vs. theoretical results.

Power rating SAM	0,671MW	**Net power with a safety factor**	108 KW
Power rating Req	0,648MW		
Annual Energy SAM	2 818 537 KWh	**Extra energy produced by SAM**	119 569 KWh
Energía anual Req	2 698 968 KWh		

4.2. Theoretical calculations to verify power output

The peak consumption value for the case study defined in this work is 224914 KWh/ in the month of August. Considering the operating parameters, the nominal power required to produce the plant is obtained:

$$224914\,\frac{\frac{KW}{h}}{month}\left(\frac{1month}{26\,days}\right)\left(\frac{1\,days}{16h}\right) = 540KW$$

1. To have an estimate of the energy that should be produced with this peak value:

$$224914\,\frac{\frac{KW}{h}}{month}\left(\frac{12month}{1\,year}\right) = 2698968KWh\,anual$$

2. As the value obtained can fluctuate to a higher or lower value, it has been considered a safety factor of 1.2. This will allow over-dimensioning the energy production of the plant and make use of the storage system for when the system requires it.

$$540KW\,(1{,}2) = 648\,KW = 0{,}648MW$$

This means that 108 KW is protected from the real peak value, which is why the solar plant is simulated with 1MW and not with 0.648MW.

The results of the annual production with SAM should be considered that the dimensioning of the installed power depends mainly on the limitations of the available area (9 acres) and critical parameters that have been assumed to affect the optimal performance that can be achieved by the plant Figure 3. Even so, the result obtained satisfies the demand for the peak value generated by the needs of the Universidad Technological de Bolívar.

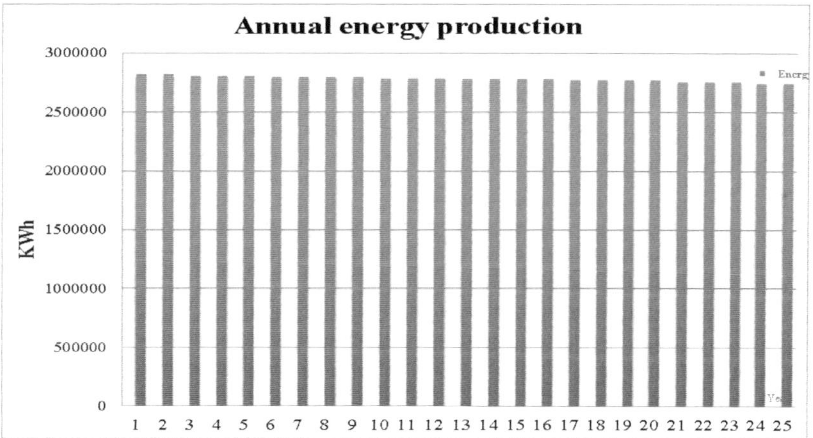

Figure 3. Annual energy production result SAM.

In figure 3 you can see that a value of energy higher than the nominal value of the power is obtained, and although it was estimated a drastic degradation in the performance of the plant of 20% per year, it is obtained for the final year of estimated useful life 2750000KWh.

In Figure 4, and can be seen that each month is obtained a value ranging from 224914Kwh, which corresponds to the maximum value consumed by the university.

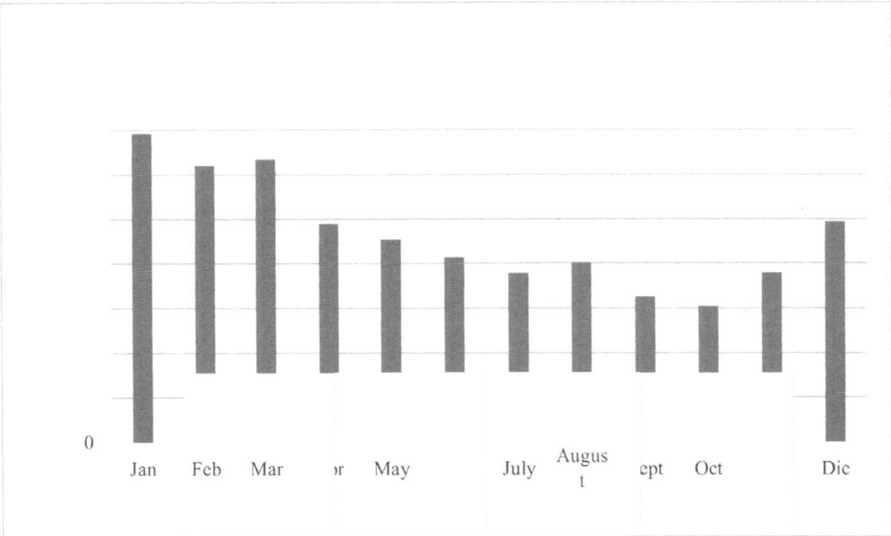

Figure 4. SAM monthly energy production result.

Figure 4 shows that in the first months approximately 121224KWh is produced in addition to the required peak value of 224914Kwh. This monthly surplus can be stored to supply the months detected by the program with less production, and this is possible because the surplus energy that is stored in the tanks can be used later months as if it were seasonal thermal energy storage, where commonly stores energy in the summer to supply during the winter.

Conclusions

At the end of this work, the results obtained were analyzed. It was concluded that the generation of electric energy using the installation of the parabolic cylinders, which corresponds to a thermoelectric plant of heliothermic technologies, has been viable due to the fact that 2,818,537 kWh are produced per year, which corresponds to 1MW of nominal power against the demand of required energy that is 0.648MW. This is mainly because there is a suitable location where radiation values are sufficiently high. There are no seasonal changes, allowing for much more stable solar irradiation than in other countries where these solar plants have been implemented.

There are several options that can be implemented in this type of system, in order to obtain improvements in the performance of this type of plant. As a result of a sensitivity analysis of project parameters and operation of the system proposed in this work. It was possible to determine that with the increase in the area available for the installation, going from 6 acres to 9 acres and using a solar multiple of 1.7, it allows to adequately guarantee the energy available in the thermal storage system, to guarantee autonomy system for 6 hours.

From the analysis carried out in this work, it was also possible to determine for this parameter analysis, which for the different heat transfer fluids of the solar field. Therminol PV-1 allows to obtain the best performances of the proposed system, in relation to other fluids used, such as: Dowtherm, Caloria HT 43 and Therminol 59.

References

[1] N. Kannan and D. Vakeesan 2019 Solar energy for the future world: - A review (*Renew. Sustain. Energy Rev.*, vol. 62) pp. 1092–1105

[2] S. Ben Romdhane, A. Amamou, R. Ben Khalifa, N. M. Saïd, Z. Younsi, and A. Jemni 2020 A

review on thermal energy storage using phase change materials in passive building applications (*J. Build. Eng.)*.

[3] S. Dixit 2020 Solar technologies and their implementations: A review (*Mater. Today Pro vol 28)*, pp. 2137–2148

[4] A. Kumar, S. Vyas, and D. N. Nkwetta 2020 Experimental study of single slope solar still coupled with parabolic trough collector (*Mater. Sci. Energy Technol).*,

[5] T. Akba, D. Baker, and A. G. Yazıcıoğlu 2020 Modeling, transient simulations and parametric studies of parabolic trough collectors with thermal energy storage (*Sol. Energy*, vol. 199) pp. 497–509, 2020

[6] M. Hakimi, E. Baniasadi, and E. Afshari 2020 Thermo-economic analysis of photovoltaic, central tower receiver and parabolic trough power plants for Herat city in Afghanistan (*Renew. Energy*, vol. 150) pp. 840–853

[7] A. Vouros, E. Mathioulakis, E. Papanicolaou, and V. Belessiotis 2020 Performance evaluation of a linear Fresnel collector with catoptric subsets (*Renew. Energy*, vol. 156), pp. 68–83

[8] H. Jamali 2019 Investigation and review of mirrors reflectance in parabolic trough solar collectors (PTSCs) (*Energy Reports*, vol. 5) pp. 145–158

[9] M. Imtiaz Hussain, C. Ménézo, and J. T. Kim 2018 Advances in solar thermal harvesting technology based on surface solar absorption collectors: A review (*Sol. Energy Mater. Sol. Cells*, vol. 187) pp. 123–139

[10] B. Lamrani, A. Khouya, B. Zeghmati, and A. Draoui 2018 Mathematical modeling and numerical simulation of a parabolic trough collector: A case study in thermal engineering (*Therm. Sci. Eng. Prog.*, vol. 8,) pp. 47–54,

[11] B. Agagna, A. Smaili, Q. Falcoz, and O. Behar 2018 Experimental and numerical study of parabolic trough solar collector of MicroSol-R tests platform (*Exp. Therm. Fluid Sci.*, vol. 98), pp. 251–266

[12] E. Bellos and C. Tzivanidis 2020 Solar concentrating systems and applications in Greece – A critical review (*J. Clean. Prod.*, vol. 272), p. 122855

[13] Y. Bi, L. Qin, J. Guo, H. Li, and G. Zang 2020 Performance analysis of solar air conditioning system based on the independent-developed solar parabolic trough collector (*Energy*, vol. 196) p. 117075

[14] IDEAM., *Atlas de Radiación Solar en Colombia*. 2018.

[15] I. Purohit, P. Purohit, and S. Shekhar 2015 Evaluating the potential of concentrating solar power generation in Northwestern India (*Energy Policy*, vol. 62) pp. 157–175, 2013

[16] R. Senthil and M. Cheralathan 2019 Enhancement of the thermal energy storage capacity of a parabolic dish concentrated solar receiver using phase change materials (*J. Energy Storage*, vol. 25) p. 100841

[17] H. Hijazi, O. Mokhiamar, and O. Elsamni 2020 Mechanical design of a low cost parabolic solar dish concentrator (*Alexandria Eng. J.*, vol. 55), pp. 1–11

Electronic distiller to improve learning processes in natural science laboratories

C P Chamat-Torres[1], G E Romero Cabarcas[1] and G Bracho Tovar[1]
[1] Grupo de investigación GINTICS, Universidad Popular del Cesar, Valledupar, Colombia

E-mail: cchamat@unicesar.edu.co

Abstract. With the continuous advance of technology, new challenges arise in the face of innovative learning models and their theoretical-practical representation, in which we seek to strengthen effective teaching methods and maximize academic and experiential performance. Therefore, this article describes the development of an electronic distiller for basic natural science laboratories as a result of the research process, which allows teachers and students to perform experimental practices by means of simple distillation. The team developed, involves the concepts of Electronic and Systems Engineering, to improve the efficiency of the simple distillation process, solving needs in terms of assembly time and resource utilization, and optimizing the learning experience. This allows for automatic control of the boiling point of the substances and the quantity of grams to be distilled as the final product. This, thanks to the instrumentation and control modules that interpret the sensor signals, in order to control the actuators such as, the heating, pumping and cooling water system. The processes are visualized thanks to an Android graphic interface application, which allows monitoring and configuring the Setpoint of the measured variables.

1. Introduction

Basic sciences such as Biology, Physics, Geology, Mathematics and Chemistry, seek to understand the phenomena present in nature and in some cases replicate them, thus deciphering laws that satisfy this end. Linus Pauling defines chemistry as the science that studies substances, their structure, properties and reactions that transform these into other substances in reference to time [1]. However, in the specific case of a student, the interest in understanding these concepts, replicating them through methods used for more than 15 years, is increasingly complex. The sequence of teaching and learning lies in the selection and use of means that among others allow the student to interact, visualize and manipulate their learning context, to build bridges between theory and practical experience [2]. Therefore, providing the student with a way to become familiar with this bridge is one of the main foundations of any learning method.

Understanding the concepts that encompass chemistry from reality has become a challenge in both the basic and higher education cycles; the lack of evolution in the use of theoretical-practical bridges, in comparison with new tools embraced by other branches of science, generated a need in view of the fascinating technological developments of this century.

Hence, that incorporating "Investigación, desarrollo e innovación" (I+D+i) technologies focused on pedagogical processes, reveals a new alternative to face challenges in educational models. Although technologies such as augmented reality, virtual reality and Information and Communication Technology (ICT) have been involved in these processes, very few educational establishments have become involved.

On the other hand, engineering is a multidisciplinary practice based on the above-mentioned sciences, applying concepts to develop technologies capable of satisfying needs and solving problems, providing an efficient and productive management of the resources used. Therefore, all these new technologies that have evolved the way of teaching and learning, are a product of engineering, so much so, that ICT are based on software engineering, computing, among others.

The American Institute of Chemical Engineers (AIChE) defines in [3] Chemical Engineering as a profession in which knowledge, such as the case of basic sciences such as Mathematics, Chemistry, among others, is acquired through study, experience and practice. In order to apply it in the development of economic ways to use materials and energy for the benefit of humanity.

Education is undergoing major changes influenced by the advance of ICTs, which have generated educational resources in all areas of knowledge. This has led to the emergence of what are now known as virtual laboratories, which are a safe and easily accessible form of education. They allow real experiments to be carried out through the Internet, being tools that support classroom teaching [4], and an ideal complement to distance education [5]. Although one should not make the mistake of thinking that this form of teaching can completely replace face-to-face learning.

The "Laboratorio Tradicional" (LT) or hands-on allows the direct manipulation of the equipment and devices, through observation and experimentation, in which the student develops and optimizes his cognitive abilities and practical skills [6] necessary in the learning of Chemistry and Physics. Therefore, implementing I+D+i technologies developed through engineering helps to awaken a new interest in acquiring learning experiences.

In this sense, the learning and teaching of chemistry with new technologies is having a strong growth trend, which leads us to think about what advantages can be offered by innovating these technologies to make a significant change in the student. On the other hand, it has prompted profound changes in the curriculum of some curricula. Some cases are exposed by [7]: 1) in Hungary, there is no science without communication technology; 2) in the Czech Republic, where a combination of Chemistry and Computer Science has been implemented for undergraduate students; 3) in India, where a general evaluation of Chemistry Education has been made; 4) in Ireland, mainly in Organic Chemistry through the program Organic Chemistry in Action.

According to the above, it can be reflected as one of the objectives that have been pursued with the incorporation of new technologies in education, is to optimize the student's attention and achieve learning that adapts to existing needs. Therefore, this research proposes the engineering development of an electronic distiller to improve learning processes in simple distillation in natural science laboratories.

2. Simple distillation

Distillation has been a widely used separation technique in the industry and in organic laboratories around the world, so it is necessary to analyze and study the performance of the most common methods: Simple Distillation and Fractionated Distillation. Distillation bases its separation efficiency on the differences in vapor pressures, or boiling points, of two or more substances in solution, since the technique consists of boiling the more volatile substance and recondensing it in a different vessel than the initial one [8]. Therefore, substances with close boiling points are more difficult to separate by this method [9]. However, a disadvantage present in the distillation process is the low efficiency when trying to separate substances with a very high boiling point and also those present with low concentrations in large volume carriers [10]. This has made it challenging to achieve a quality that can be considered high, even using non-linear controllers to do so [11].

This is because distillation is not an equilibrium process, since the mass is continuously removed from the system, it is based on states of liquid-vapor equilibrium, the boiling temperature, since the vapor phase is only removed when this equilibrium has been reached. Such balances happen to each other because the composition of the original liquid changes continuously [12].

In simple distillation the vapor is removed from the liquid sine and immediately passes into the refrigerant where it condenses and then the distilled liquid is collected. Using this procedure, mixtures of two components with a difference of 30-60° can be separated by repeated single distillations [12].

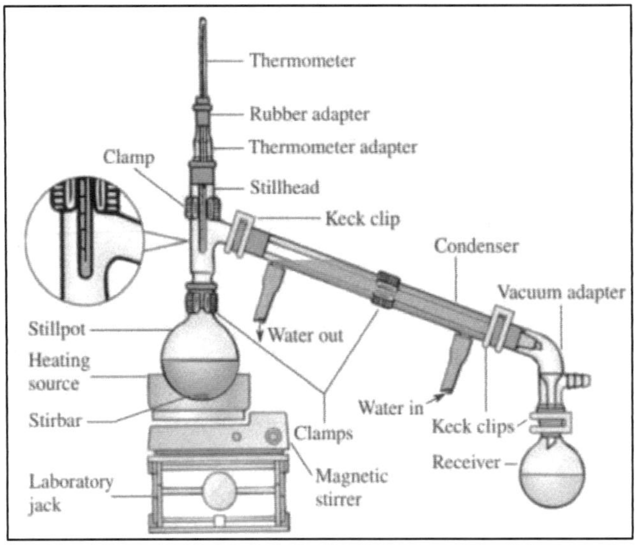

Figure 1: Simple distillation assembly [13]

3. Methodology

This research is based on the methodology set out in [14], which in turn is focused on the ICONIX software development method that is halfway between a Rational Unified Process (RUP) and an eXtreme Programming (XP). Its objective is based on the fact that each requirement is identified with some use case and in this way, to be able to verify in each instant of the system if the requirement has been fulfilled satisfactorily.

The ICONIX methodology is structured in four phases as described below:

- Phase I – Requirements Analysis: Collect information related to the research in books, documents and articles, as a basis or foundation for its development, decision making and generalities of the distillation process.
- Phase II – Analysis and preliminary design: Design a logical and organized structure of the different modules and the functions that each one must perform to obtain the final product.
- Phase III – Design: Design the modules in charge of operating each function in the system, such as the instrumentation module, which is responsible for operating, controlling and supervising the modules that make up the distillation process. As well as the heating, cooling and pumping modules, which allow the simple distillation process to be carried out.
- Phase IV – Implementation: Carry out the necessary tests to validate the correct functioning of the equipment in each of its stages, through the supervision of chemical engineers for the analysis of the results obtained, corroborating the fulfilment of the objectives previously set.

4. Results

The following is a description of the results obtained in the course of the research to meet the objectives set and to implement the selected methodology.

4.1. Requirements Analysis:

Examining the needs of the traditional method used to carry out simple distillation, the following four main problems become evident: a) Time and difficulty in carrying out the assembly, b) Fragility of the materials when handled by the students, given that most of them are made of glass, c) Deficiency in the use of resources such as the water used to cool the substance obtained, d) Difficulty in controlling the temperature of the substance. For this reason, it is necessary to implement innovative methods to solve these difficulties in natural science laboratories.

4.2. Analysis and preliminary design:

Once the existing difficulties in simple distillation processes in natural science laboratories have been raised, the preliminary design of the electronic distiller is started, which is divided into four main parts, such as: the heating, cooling and pumping modules, the instrumentation and control module in which the digital processing is carried out, the software and firmware, and finally the graphic interface application. This can be seen in Figure 2.

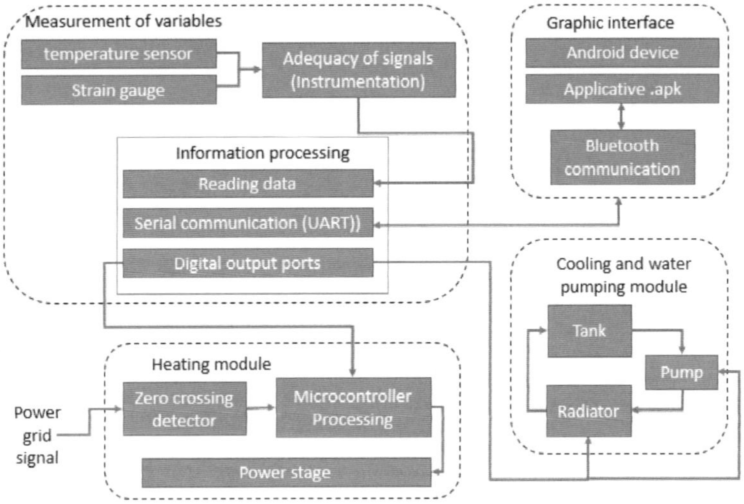

Figure 2: Electronic Distiller Design Diagram

4.3. Design:

For the design of the electronic distiller, in its first stage, a K type thermocouple was implemented with a 10cm probe for temperature measurement, since it is made of a material resistant to industrial environments, resistant to corrosion and has an ideal probe length; which makes it suitable for the application in which it is going to be used. It also offers a temperature range up to 800°C and a resolution of 0.25°C. It is connected to the main microcontroller through the MAX6675 converter (see Figure 3a) which performs the compensation and linearization of the sensor response with an ADC, and the output data is obtained through an SPI communication.

The measurement of the distillate quantity is carried out by means of a 10Kg (extensiometric) load cell (see Figure 3b), which varies its resistance according to the deformation it undergoes when a force is applied. It is connected thanks to the HX711 converter with a two-wire serial-data communication.

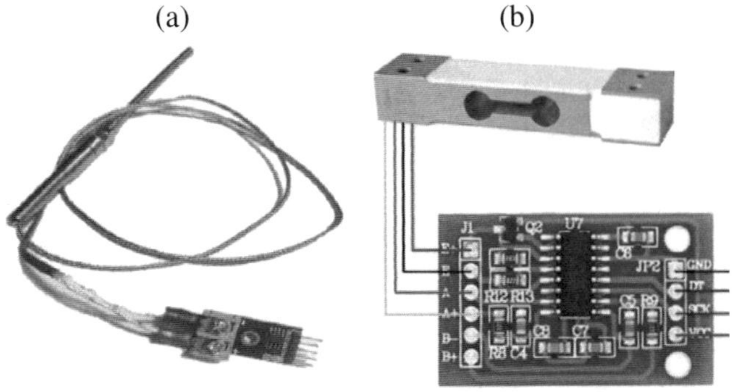

Figure 3: a) type k thermocouple and converter and b) load cell and converter

In the main data processing stage, a microcontroller of the ATMEL family (ATMEGA328) was chosen, because it has a great variety of fast applications in the market, due to countless libraries developed freely by programmers around the world. This microcontroller is present in the instrumentation and control module, which receives the input signals from the sensors and user interface, processes them and makes the decisions pre-conditioned in the programming code (firmware).

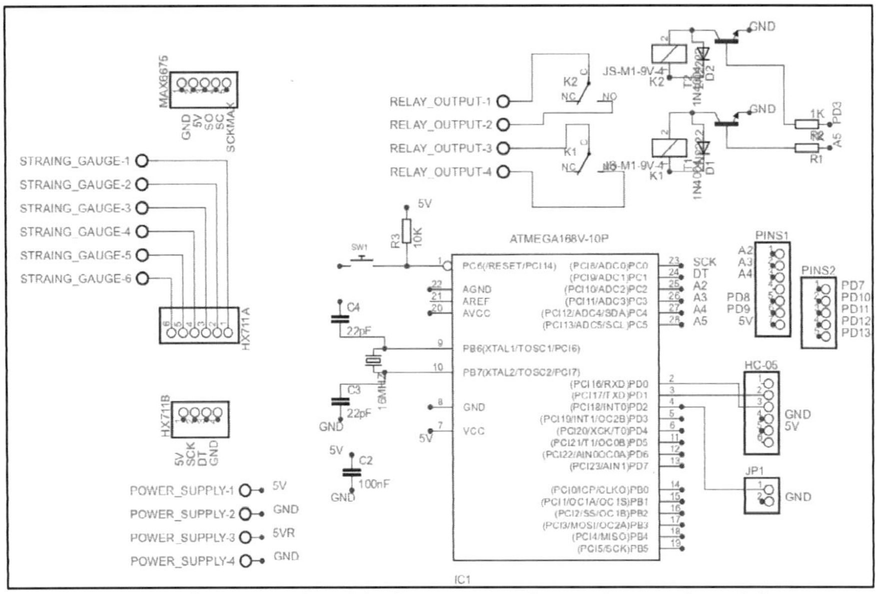

Figure 4: Schematic of the instrumentation and control module

For the heating module, a simple microchip microcontroller (PIC16F628P) was used, due to its easy availability in the market and great performance in terms of speed in its interruption ports. It captures the signal from the electrical grid and depending on the control signal from the main microprocessor, it injects different levels of power into the thermal resistor, exerting an ON/OFF control by levels; and thus, maintaining a stable substance temperature Setpoint.

The heater consists of a metal cylinder and on it a spherical base in stainless steel mesh that holds the distillation ball; a spiral-shaped thermal resistance of 120V AC/1000W located inside the metal cylinder was used, which is in turn connected to an AC power control board, which is activated by the control and instrumentation module. In the case of the cooling module which is used to recirculate the water, once it has fulfilled the task of condensing the vapor present in the distillation process, a 1300ml storage tank is used, connected to a small DC water pump, with a flow rate of 3litres/min. Once the condensation has been carried out, the liquid that gained heat passes to a radiator, which is linked to a Peltier cell and a dissipator that extract the previously gained heat. Finally, it goes back to the tank for a new cycle.

For the last stage corresponding to the display and configuration of variables, we designed a graphic interface application developed in the MIT App Inventor 2 environment, dedicated to the creation of Android applications with .apk extension, which is connected via wireless connection (Bluetooth) to the instrumentation and control module, allowing the reception and transmission of data through the serial port of the main microcontroller.

In the drop-down menu of the application there are six options, in which you can see Water distillation, Alcohol distillation and Advanced distillation respectively. Figure 5 shows the windows that open when selecting one of these three options. In each of them, it is possible to monitor, track and display the reception and emission of data and variables involved in each process, such as the amount in grams, the temperature in degrees Celsius and the time spent.

(a) (b)

Figure 5: Graphic interface (a) options menu and (b) Alcohol distillation

The development of the software constituted by the processing and graphic interface stage was named "AutodestLAB" and has software registration with the "Dirección Nacional de Derechos de Autor" (DNDA) with registration number No. 13-80-15.

4.4. Implementation:

Finally, in the implementation phase, the different tests were carried out in production in order to validate its functionality and performance. For the water pumping and cooling modules, their flow was verified through the entire system, in addition to monitoring the temperature in which the coolant was distributed by means of hoses, as shown in Figure 6. In addition, tests were carried out on the heating module, which consisted of verifying that the heat it supplies manages to reach the boiling point of the substance to be distilled, establishing voltage levels to reach the required temperature.

Figure 6: Water pumping and cooling system

With the assistance of professionals from the Natural Sciences Department of the "Universidad Popular del Cesar" (UPC), several tests were carried out to obtain alcohol at an ambient temperature of 33°C, by means of mixtures of 200ml of 2-propanol isopropyl alcohol and 200ml of water, as well as water distillation tests, with mixtures of dye and water. In Figure 7, the behavior of the temperature control of the distiller can be seen, which allows an optimization in the energy consumption during the process, since the heating system is capable of regulating the power supplied by means of control signals, depending on the temperature sensed and required.

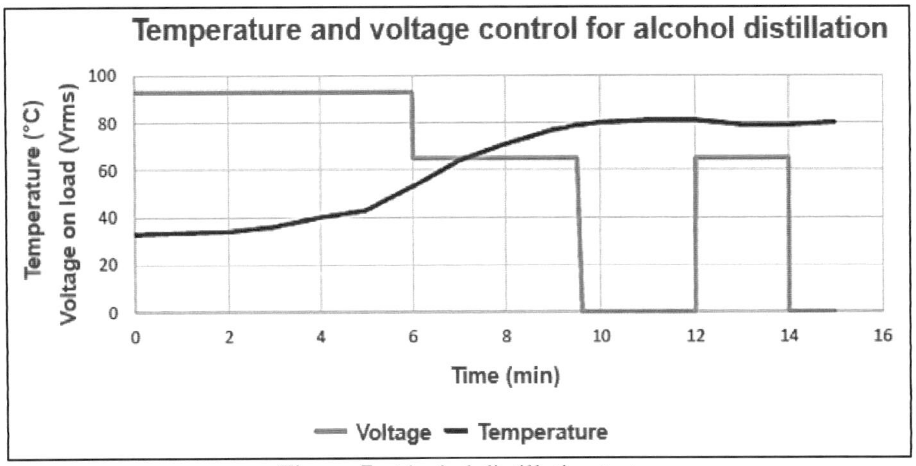

Figure 7: Alcohol distillation test

The tests previously exposed, allowed to validate the correct operation in production, of the electronic distiller for academic laboratories, evaluating performance, water and energy consumption, temperature control, amount of substance obtained and optimization in the execution time of the distillation process. In Figure 8, you can see the final result of the design and development of the equipment in its front part.

IOP Conf. Series: Materials Science and Engineering 1154 (2021) 012025 doi:10.1088/1757-899X/1154/1/012025

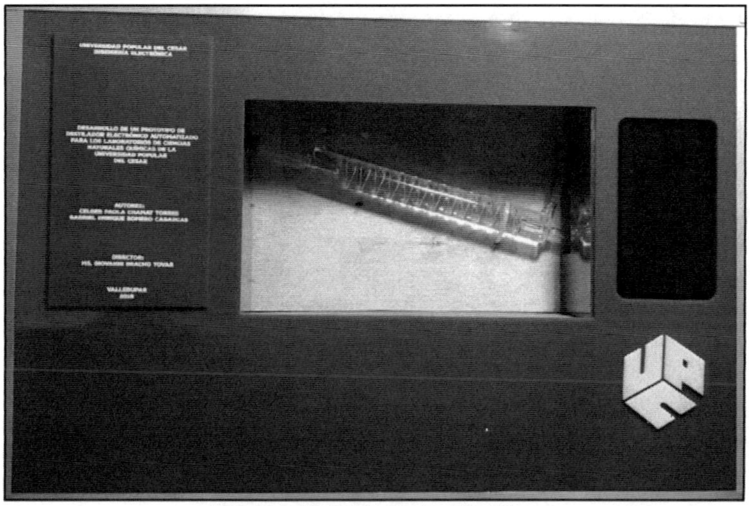

Figure 8: Electronic Distiller

5. Conclusions

As a result of research based on the ICONIX methodology, the objectives set out for the engineering development of an electronic distiller to improve learning processes in simple distillation in natural science laboratories were achieved. To this end, difficulties in the practical process and their repercussions on the students' learning experience were initially analyzed, resulting in the need to develop equipment that would allow for the automation of the stages involved in simple distillation, such as; the heating and temperature monitoring stage, the cooling stage, and the stage of obtaining the final product. In addition, to identify that these traditional practices lead to a deficiency in the use of resources such as water and gas, to a misuse of time, given the delay in the assembly of the components, and to the difficulty in the sequence of learning and teaching in the student, due to the lack of interaction, visualization and manipulation of their learning environment.

Once the requirements analysis was done, the heating, pumping and cooling modules were designed and built, with which the required temperature regulation and the reduction of the equipment's energy consumption were achieved, by means of an ON/OFF control. In addition to the optimal use of water in the process, due to the fact that it was possible to significantly avoid its waste, when carrying out the condensation of the vapors, thanks to the implementation of a closed-loop pumping system. Going from a waste of approximately 90 liters of water for 30 minutes of process, to a need of only 1 liter of it in the storage tank, and that could be reused without any problem.

In addition to the above, an Android graphic interface application was developed, for the visualization, monitoring and configuration of process variables, through a screen on the computer. Finally, the different modules developed were integrated, with a robust, eye-catching, user-friendly architecture that requires minimum manipulation, in order to facilitate the development of the laboratory experiences and contribute to the effective training of the students.

References

[1] Stapleton C B, Hughes C E, Moshell J M 2002 Mixed reality and the interactive imagination *Proceedings of the First Swedish-American Workshop on Modeling and Simulation* **1** 30-31

[2] Merino C, Pino S, Meyer E, Garrido J M, Gallardo F 2015 Realidad aumentada para el diseño de secuencias de enseñanza-aprendizaje en química *Educación química* **26(2)** 94-99

[3] Álvarez-Borroto R, Stahl U, Cabrera-Maldonado E V, Rosero-Espín M V 2017 Los paradigmas de la ingeniería química: las nuevas fronteras *Educación química* **28(4)** 196-201
[4] Navarro Arévalo E, Tizón Pulido J M 2016 Docencia presencial y laboratorio remoto: una unión idónea para las prácticas de motores alternativos *Modelling in Science Education and Learning* **9(1)** 129-138
[5] Angulo I, García-Zubia J, Rodríguez-Gil L, Orduña P 2016 A new approach to conduct remote experimentation over embedded technologies *13th International Conference on Remote Engineering and Virtual Instrumentation* (Madrid: IEEE)
[6] Infante Jiménez C 2014 Propuesta pedagógica para el uso de laboratorios virtuales como actividad complementaria en las asignaturas teórico-prácticas *Revista Mexicana de Investigación Educativa* **19(62)** 917-937
[7] Frías M V, Arce C, Flores-Morales P 2016 Uso de la plataforma socrative.com para alumnos de Química General *Educación química* **27(1)** 59-66
[8] Cesar G V J, Sebastián O B 2019 *Destilación simple y fraccionada: método de purificación de líquidos* (Universidad Industrial de Santander Facultad de ciencias Escuela de química: Colombia)
[9] Cerezo F, Sastrón F 2015 Laboratorios virtuales y docencia de la automática en la formación tecnológica de base de alumnos preuniversitarios *Revista Iberoamericana de Automática e Informática industrial* **12(4)** 419-431
[10] Espinoza J 2009 *Automatización de una columna de destilación* (Centro Nacional de Investigación y Desarrollo Tecnológico, Departamento de Mecatrónica: México)
[11] Jana A K 2010 A nonlinear exponential observer for a batch distillation *11th International Conference on Control Automation Robotics & Vision* (Singapur: IEEE) 1393-1396
[12] Lamarque A, Zygadlo J, Labuckas D 2009 *Fundamentos teórico-prácticos de química orgánica /Theoretical and practical organic chemistry* (Córdoba: Editorial Brujas)
[13] Gilbert J C, Martin S F 2011 *Experimental Organic Chemistry A Miniscale and Microscale Approach* (Boston: COLE Cengage Learning) **6**
[14] Figueroa R G, Solís C J, Cabrera A A 2008 *Metodologías tradicionales vs. metodologías ágiles* (Universidad Técnica Particular de Loja Escuela de Ciencias de la Computación: Ecuador) **9**

Acknowledgment
Special thanks to the Department of Natural Sciences of the "Universidad Popular del Cesar", Valledupar, Colombia, for supporting and validating the development of this research.

IOP Conf. Series: Materials Science and Engineering 1154 (2021) 012026 doi:10.1088/1757-899X/1154/1/012026

Simulation-aided walking pattern generation for a humanoid robotic platform

Sergio Zapata Velásquez[1], Iván Darío Mora Orozco[1], Gustavo Suárez Guerrero[2], Wuitman Garrafa Jara[2]

[1]Semillero de Investigación A+D, Universidad Pontificia Bolivariana, Medellín, Colombia.
[2]Grupo de Investigación Sobre Nuevos Materiales, Universidad Pontificia Bolivariana, Medellín, Colombia.

Email:gustavo.suarez@upb.edu.co,sergio.zapatave@upb.edu.co,ivan.mora@upb.edu.co,wuitman.garrafa@upb.edu.co

Abstract. Bipedal locomotion is one of the main research topics in humanoid robotics due to the complexity of generating dynamically stable gait patterns in robots that are generally very unstable. Good walking performance requires diverse and complex control strategies based on a mathematical model of the robot. Here, a methodology to easily generate stable bipedal locomotion patterns in a humanoid robotic platform is presented. Using a simulation environment and a simple programming method, the robot gait, stability and general behaviour are first simulated and then transferred to the actual machine. Three gait patterns are developed according to the biomechanics of human walking and validated using a humanoid robot platform. The robot exhibits good performance in terms of stability, displacement and velocity. The proposed methodology allows the study of locomotion in biped robots and can serve as a basis for developing controls in closed-loop schemes.

1. Introduction

Humanoid robotics has been under research for more than 40 years, since the pioneering study of Kato and Vukobratovic [1, 2]. Owing to the high complexity of such robots and the dynamics involved in performing tasks such as locomotion over uneven terrain, a preliminary simulation of the robot behavior is mandatory [3, 4]; tasks such as path planning, robot control, and teleoperation, are commonly developed and tested via simulation prior to their execution in actual machines [5, 7].

Similarly, machine learning algorithms and methods, such as reinforcement learning, genetic algorithms and neuroevolution, require several tests of the agent to achieve the expected behavior, and this process can be facilitated with an accurate robot simulation [8, 9]. Besides, the online training is susceptible to be extrapolated to the real robot [10, 12]. A promising field for this methodology is represented by legged robots and their applications to biped robots is still open to new developments due to the complexity of generating a control for locomotion on highly unstable machines generally operating in non-ideal environments.

The abovementioned applications have been allowed by the improvement in physical engines implemented in the simulation software. Nowadays, Gazebo, virtual robot experimentation platform (V-REP), and Webots are the most common simulators used in robotics [13, 14], facilitating the assessment of the kinematics and dynamics of a robot even without its mathematical model.

This paper presents a methodology to develop a stable bipedal locomotion with an open-loop control scheme in a humanoid robotic platform (HRP), by using a physical simulation and a simple programming method, which can allow an easy and fast extrapolation of gait patterns to be successively implemented in the hardware without the need for a robot mathematical model.

The primary purpose of the simulation is to assess the machine behavior while performing different tasks, especially bipedal locomotion, over a flat and known surface, implemented without any disturbance or noise; the robot is configured according to the real counterpart characteristics and the gait patterns are developed based on the biomechanics of human walking.

Two gait patterns are developed by simulation and their execution in HRP, which also serves as the basis for a third pattern that significantly improves the robot displacement, is described. The proposed methodology allows studying different locomotion patterns that can form the basis for developing closed-loop controls or used in control schemes with artificial intelligence

The rest of this paper is structured as follows, it is described the PRH project; then, the methodology proposed to configure and control the robot in a simulation environment and the hardware characteristics, is presented; next, the locomotion tests and both the simulated and experimental results, are reported; the last two sections present the discussion and conclusions of the study.

1.1 HRP Project

The HRP project is currently under development at the Semillero de Investigación en Automática y Diseño A+D, of the Universidad Pontificia Bolivariana (Colombia), and focuses on developing a low-cost humanoid robot for research and education. The platform has 56 degrees of freedom (DOFs), making it modular, easy to assemble and 3D printed. Its anthropometric design is one of its main differentiating fea-tures, which includes the same proportions, shape and joint ranges as those of an average human.

This robot is being developed primarily to study the human bipedal locomotion and implement stable gait patterns on a regular flat floor with an open-loop control scheme. For this purpose, it has 12 DOF legs providing the same mobility as that of a person. The robot is 90 cm tall and weighs 4.7 kg, resulting in a lightweight and robust platform [15]. It is sought to develop closed-loop controls and explore the machine learning application in locomotion. The robot is also designed for research on human/robot interfaces, object manipulation, robot interactions with the environment, basic limb control, etc.; its kinematic structure and design are shown in Figure1.

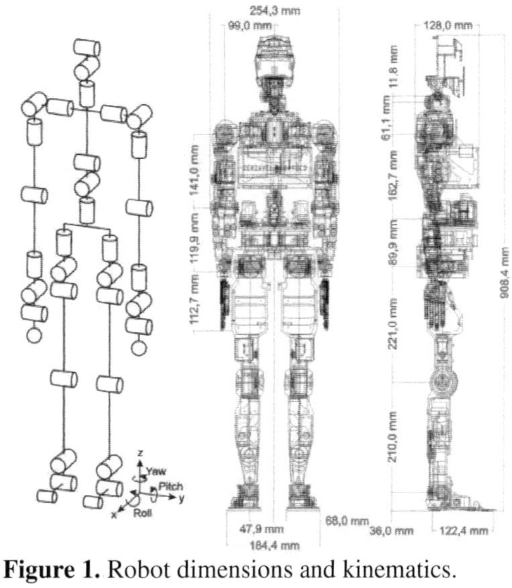

Figure 1. Robot dimensions and kinematics.

2. Methodology

2.1. Simulation environment

The 3D simulation is performed with the V- REP program, which presents a friendly environment for the development of rigid-body robots and allows the use of four physical tools (Bullet physics library, Open Dynamics Engine, Vortex Dynamics Engine and Newton Dynamics Engine) and the creation of controls in different programming languages, such as C/C++, Java, and Matlab [16, 17]. Moreover, in its distributed architecture, each object of the simulation can be individually controlled by a plug-in, an embedded script, or a remote application programming interface (API) client [18]

2.2. Robot configuration on V-REP

Each segment of the robot, designed on a CAD program, is exported into the simulation environment as a mesh piece and transformed into a convex hull configuration, which preserves its external shape and contact points without the detailed mechatronic design, simplifying the interactions between the parts and speeding-up the simulation. The pieces are configured with parameters defining the dynamics in the actual robot, such as mass, center of mass (CoM) position, and moments of inertia; collision detection is also included.

The kinematic chain and general structure of the robot are configured on a hierarchical tree. Revolute joints are used, whose parameters, including torque, speed, and proportional–integral–derivative control properties, are set according to the actuator characteristics. The robot in the simulation environment is shown in Figure 2.

2.3. Control Method

The robot is controlled via a Matlab script, which works as a remote API, communicates with the V-REP environment through a child script related to the HRP hierarchical tree, and controls the execution time of each simulation step.

The main task of the Matlab program is to deliver (write) the angles for each actuator; similarly, the actual joint angles must be read in the simulation to determine the possible deviations due to the robot dynamics and interaction with the environment.

The commands sent to the simulation are manually generated and indicate the set of angles to be executed by the robot for each time period. Then, these data are interpolated in the program with the shape-preserving piecewise cubic Hermite interpolating polynomial method to execute the commands every 50 ms.

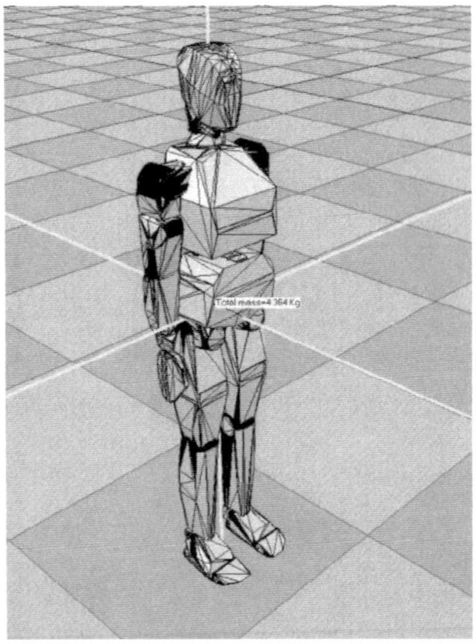

Figure 2. Humanoid robot platform in the simulation environment.

The four physics tools provided by V-REP have been successively tested with the robot (Figure 3), and the Bullet Physics Library, which presents a stable and reliable contact between the robot feet and floor, is selected. The other three show similar behaviors, but the contact with the ground exhibits slight vibrations and movements, destabilizing the machine.

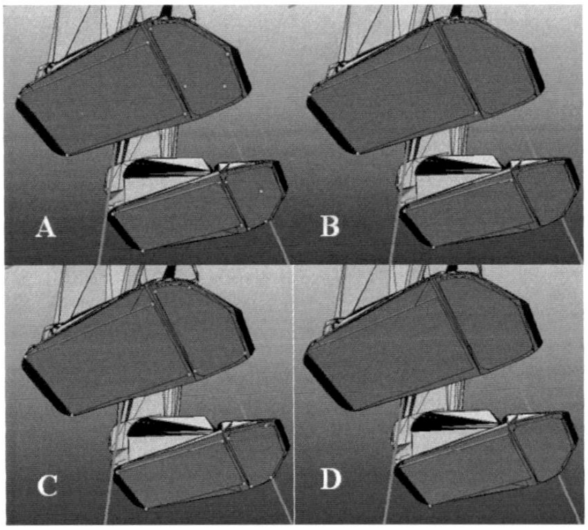

Figure 3. Contact points of the robot feet with a surface (dots), simulated with four tools: (A) Bullet Physics Library, (B) Open Dynamics Engine, (C) Vortex Dynamics Engine, and (D) Newton Dynamics Engine.

2.4. Joints torques

Servomotors are chosen as the robot actuators and connected to the joints directly or through a reduction system. To determine the static torque required in the joints for a stable locomotion, a kinematic analysis is carried based on the robot position; for this analysis, HRP is supported on one foot, whose ankle is considered the origin of the coordinate system and anchored to the ground.

As regards the frontal plane, the following five joints are analyzed: the roll movement of the supporting foot ankle, two joints of the legs with the hip, and two joints in the shoulders. For the sagittal plane, there are seven joints of the pitch movement corresponding to the one supporting the foot ankle, two knees, two legs, and two arms.

The total torque is calculated as the sum of individual torques, derived from the weight and distance between the origin and the robot nodes,

$$\tau_T = \sum_{i=0}^{n} \tau_i \quad (1)$$

$$\tau_i = F_i l_i(\theta_0, l_0, \dots, \theta_k, l_k)\cos\big(\theta_i(\theta_0, l_0, \dots, \theta_k, l_k)\big) \quad (2)$$

where τ_T, τ_i, F_i, l_i, θ_i are, respectively, the total torque on the joint, the individual torques, the force due to the node weight, and the node distance and angle, which are both functions of the θ_0 to θ_k and distances l_0 to l_k. The analysis is performed considering the main segments of the limbs not exceeding a weight of 375 g, for a total robot weight of 4.7 kg.

The projection of the robot COM on the ground is generally found, considering it as a particle system [19], where its extremities are represented by links whose mass is concentrated in a predefined node. The torque generated in the frontal and sagittal planes of the supporting foot ankle joint is determined based on (2). Then, the system CoM [20] is calculated as,

$$r_{cm}\hat{\imath} = \frac{1}{M}\sum_{i=0}^{n} \tau_i \quad (3)$$

$$r_{cm}\hat{\jmath} = \frac{1}{M}\sum_{i=0}^{n} m_i |r_i| \sin(\theta_i) \quad (4)$$

where τ_i, m_i, θ_i, r_i are the torque, mass, angle and distance, respectively, of each particle and M the total mass of the system; thus, the COM position in i^ and j^ is obtained. Based on the previous equations, a program is developed in Matlab to find the maximum torque in each joint and calculate the robot COM; in particular, two functions are elaborated for the sagittal and frontal planes, where robot characteristics such as its limbs weight and length, node position and general geometry are configured. The inputs are the desired joint angles.

Matlab Simulink is used to improve the user interface; specifically, a robot representation is generated by a SimMechanics module and sliders allow its position configuration based on the joint angles. The maximum torque generated in each joint is obtained from these simulations according to a set of fundamental positions in the locomotion process, mainly in the single-support stages.

3. Results

The results are divided into two main sections that present, respectively, the performance of locomotion patterns generated in the simulation environment and their validation in the real HRP.

3.1. Locomotion simulation tests

In a first method, the maximum and minimum tilts in the frontal and sagittal planes executed by the robot without losing balance, achieved through the gradual increase in the ankle and hip joint angles while keeping the trunk vertical, are determined. A second method with a graphical estimation is also elaborated based on the COM position and its projection in the floor.

The results obtained with the two methods are presented in Table 1; although the values are similar in most cases, the simulated ones tend to be smaller due to the dynamics involved when the robot is moving to the commanded angles. These results are useful to determine the basic limits of stability in some locomotion phases, but the dynamics involved in the walking cycle may further limit these values. The similarly determined angle, giving the COM projection at the center of the supporting foot polygon, is 7.74° for the frontal plane and 3.19° for the sagittal one.

Table I
Minimum and maximum angles obtained via the 2d method and the virtual robot experimentation platform (v-rep)

Actuator	2D Method (°)	V-REP (°)
Sagittal plane minimum inclination	4.19	4.50
Sagittal plane maximum inclination	10.96	7.20
Frontal plane minimum inclination	-2.47	-1.50
Frontal plane maximum inclination	8.89	8.88

For the locomotion tests, two patterns are generated based on the biomechanics of human gait, including the typical joint angles. Although the robot is anthropometrically designed, these angles cannot be extrapolated directly, but must be first adapted in a 2D environment and then edited in V-REP until achieving a dynamically stable gait. In these tests, only 10 of the 56 DOFs, corresponding to the pitch/roll rotation of the ankles, knees, and hip, are used.

The first gait pattern searches a very human-like locomotion, implying mainly knee-stretched, heel-contact and toe-off motion, i.e., the supporting leg is maintained with low knee flexion and the toes are used as additional support in the pre-swing stage. Figure 4 shows from the support phase, until terminal stance and pre-swing, also showing the COM projection remaining within the feet support.

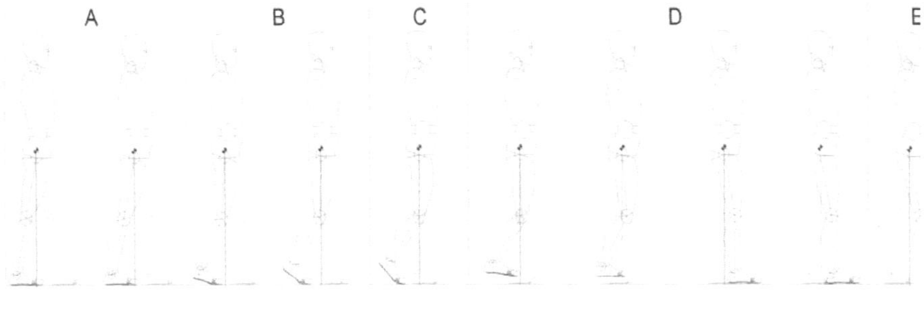

Figure 4. Stance phase (60% of the locomotion) in the first pattern showing the heel strike (A), loading response (B), mid-stance (C), terminal stance (D) and pre-swing (E).

Multiple tests are performed for adjusting the different angles according to the results obtained in the 3D simulation until achieving satisfactory and stable locomotion at all times. In Figure 5, the locomotion on V-REP, where 4 steps corresponding to 2 gait cycles are executed for 200 s, is presented. The motion of the toes is illustrated in Figure 6, showing their additional support to the robot at some gait phases, as the pre-swing stage.

The thus-generated pattern achieves a stable locomotion and a very human-like walking. However, slight changes in the angle commands can easily destabilize the machine due to the nature of this gait type; therefore, the sequence is executed slowly to counteract the effects of the overall dynamics. On the other hand, the execution of the initial oscillation (pre-swing), when the supporting foot is switched and a new phase of single support initiates, implies a very complex series of movements in the joints.

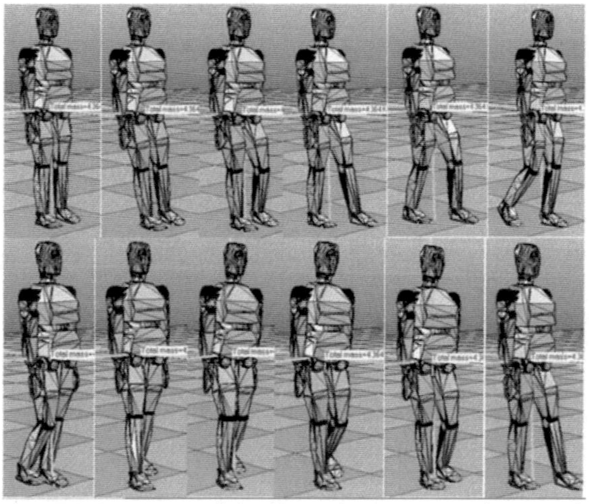

Figure 5. Sequence of the first gait pattern implemented.

The second gait pattern aims to reduce the machine instability via a bipedal locomotion performed with the knees flexed and without the additional support of the toes. Such gait maintains a low COM and allows better control of parameters such as the zero-moment point. The commands are also generated from the 2D environment and then edited in the 3D simulation.

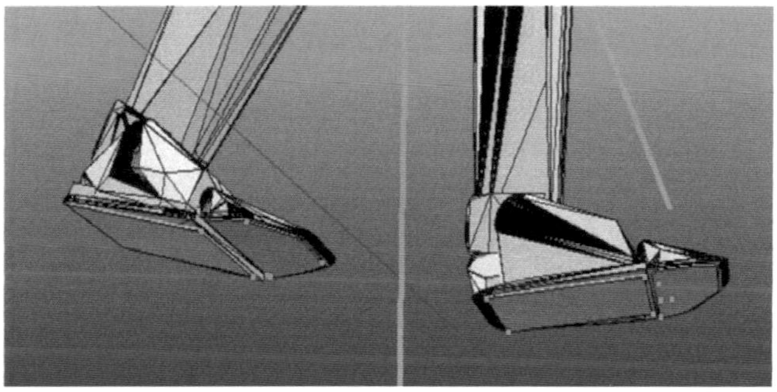

Figure 6. Additional support of the toes in the pre-swing stage.

The two gait cycles are executed successfully on V-REP during 195 s, as shown in Figure 7; the stability increases with a knee flexion of 25°, although the COM height is decreased by only 2.03 %. Moreover, with the second gait pattern, the process of COM transfer from one foot to another in the pre-swing phase is performed easily as compared to the first pattern, because the knee flexion allows a smoother and more stable transition.

The two as-generated gait patterns can be modified to enhance the locomotion, and hence, improve the stability and execution speed. These tests also confirm that the HRP is mechanically functional, since the torque in the joints allows the correct development of the walking movements.

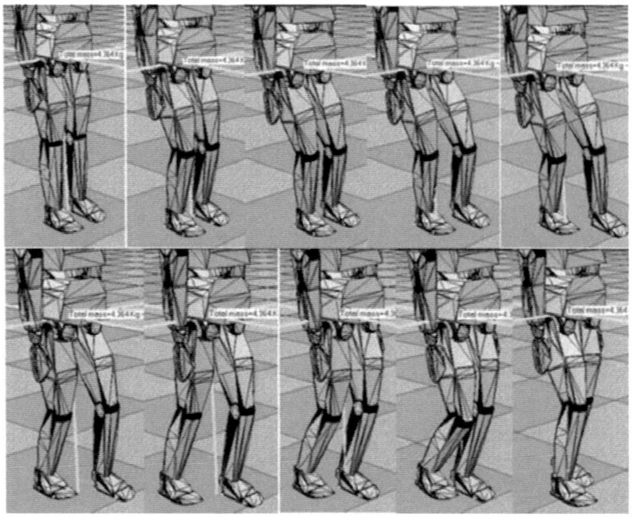

Figure 7. Sequence of the second gait pattern implemented on the virtual robot experimentation platform; the white line indicates the center of mass projection in the ground.

3.2. Locomotion tests with the real hardware

To validate the robot locomotion patterns, the actuation system, electronics, and control protocol are preliminarily tested. The two walking patterns, evaluated in the 3D simulation, are executed to assess the performance of complex movements in the platform. Figure 8 shows the acceleration exhibited by the femur in the hip flexion/extension, which does not exceed 4 m/s2, and results in a smooth and continuous movement of the joint.

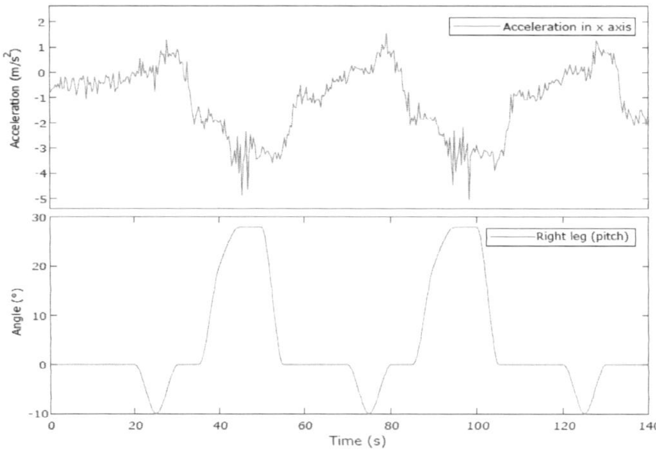

Figure 8. Right femur acceleration (up) corresponding to the angle commands sent to the hip joint (down).

In the bipedal locomotion process, the inclination movement in the frontal plane, where the robot COM is transferred to the supporting foot, is fundamental to maintain. The simulation gives a maximum inclination of 8.88°, with an optimum value of 7.74°; therefore, in a process similar to that in V-REP, the inclination is gradually incremented while keeping the robot supported by both feet; an optimum value ranging between 7° and 8° is found (Figure 9).

Figure 10 shows the response of FSRs, revealing how the inclination movement transfers the robot COM to one foot, since the total force on it increases and that on the other one decreases. Moreover, during this process, the COP is kept in an appropriate position in the foot receiving the COM, maintaining its balance (Figure 11).

Figure 9. Inclination test in the frontal plane of the humanoid robot platform.

Based on these preliminary results, the bipedal locomotion in HRP is explored without the full support by the walking frame, although this support has to be manually moved to prevent the hindering of the robot free movements by the cables, used as security elements.

First, the locomotion is tested with the first simulated gait pattern by editing only some angles referred to the inclination in the frontal plane. Figure 12 shows the frontal view of a four locomotion-step sequence, which allows an approximately 65 cm displacement in 200 s. The gait is correctly executed with sufficient spacing between each step, fulfilling the planned phases including the additional support provided by the toes. However, between the initial and final oscillation stages of the leg, the feet rise few millimeters from the ground, often causing a foot dragging, and hence, destabilizing HRP.

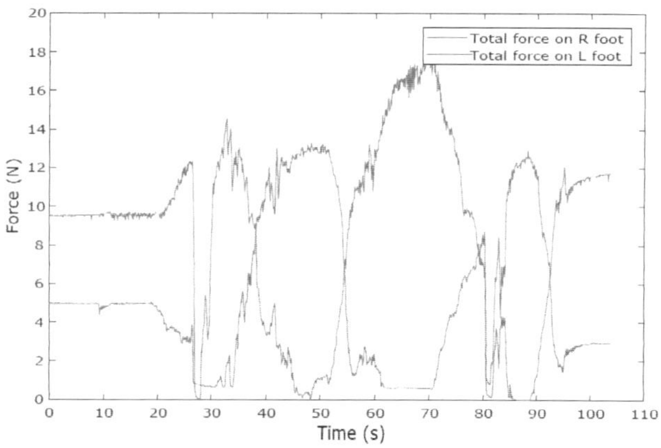

Figure 10. Total force received on each foot during the inclination test between -7° and 7° revealed by the FSRs.

First, the locomotion is tested with the first simulated gait pattern by editing only some angles referred to the inclination in the frontal plane. Figure 15 shows the frontal view of a four locomotion-step sequence, which allows an approximately 65 cm displacement in 200 s. The gait is correctly executed with sufficient spacing between each step, fulfilling the planned phases including the additional support provided by the toes. However, between the initial and final oscillation stages of the leg, the feet rise few millimeters from the ground, often causing a foot dragging, and hence, destabilizing HRP.

The data from the inertial sensors show no significant difference in the extremity acceleration between the supported and unsupported execution, as shown in Figure 13, for the case of the femur segment, suggesting that the actuators work similarly in both conditions, i.e., the locomotion on the ground does not require a greater effort from the joint; nevertheless, a slight increase in the lateral vibration is perceived due to the corresponding gait movement.

Then, the second simulated gait pattern is validated in the robot, and unlike the first one, it achieves a displacement of only 25 cm in 200 s for a total of two gait cycles and deficient stabilization is also observed.

Owing to the limitations exhibited by the two gait patterns, which perform well in the simulation, a new pattern is developed. The process to generate the gait pattern is carried out similarly to the previous ones, based on the simulation and angle commands.

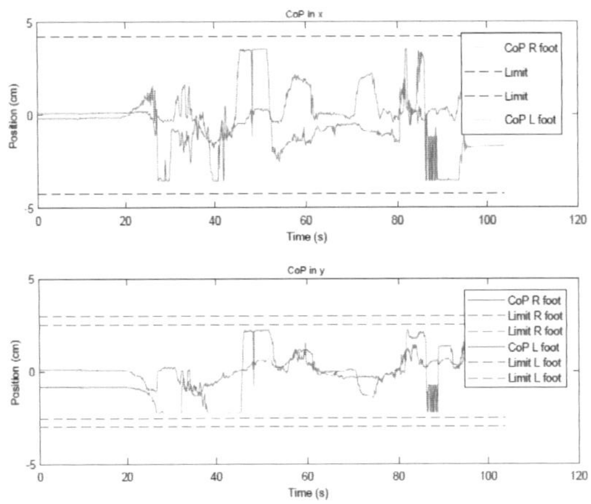

Figure 11. Center of pressure position in the inclination test between -7° and 7° over the foot boundaries.

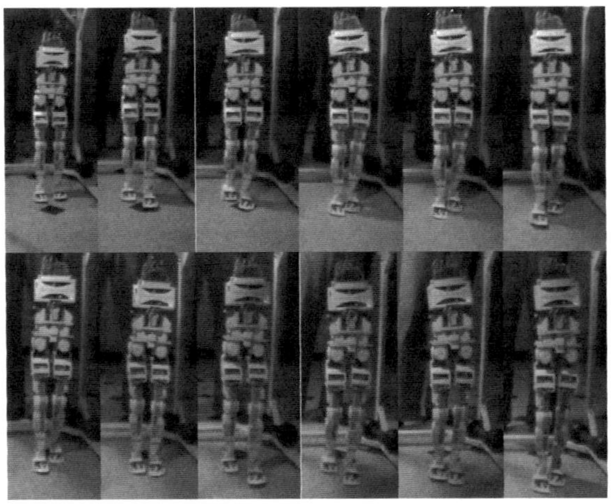

Figure 12. First locomotion pattern implemented on the HRP.

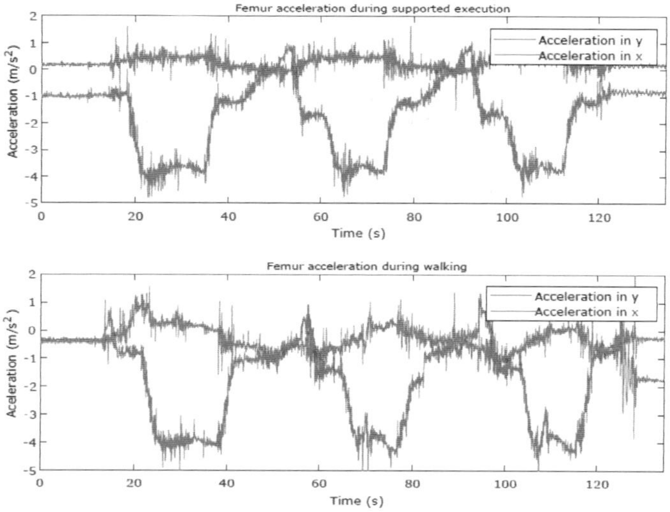

Figure 13. Acceleration in the femur segment during the supported and unsupported execution of the first locomotion pattern.

The new locomotion pattern uses a completely upright posture without pre-flexing the knees, which should allow a greater displacement despite the less stability, as well as additional support of the toes. The generated sequence comprises two gait cycles, i.e., a total of four steps in 140 s; it is successfully executed in HRP without the total support of the walking frame, as shown in Figure 14. A more stable gait is observed, with a step length of 20 cm. The transition in the pre-swing and initial swing phases is also significantly improved; a permanent contact with the toes is maintained (Figure 15), allowing the other foot to rise more than 1.5 cm above the ground. The robot covers a distance of 85 cm in the 140 s programmed, which is an improvement of approximately 30% in both displacement and speed compared to the previous patterns.

The new gait pattern is tested with different locomotion speeds, showing good performance for a speed up to 2.15 times the original, completing two gait cycles in 65 s. Despite a greater hip movement during these tests at increased speed, the acceleration at the extremities does not significantly change, as shown in Figure 16, for the femur and lower leg segments.

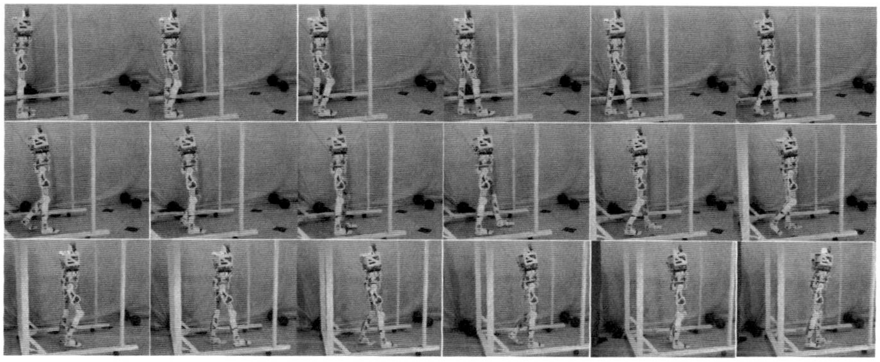

Figure 14. Locomotion implemented with the new (third) gait pattern.

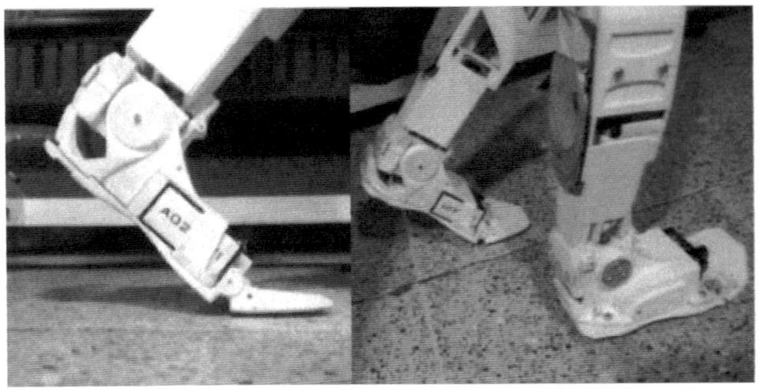

Figure 15. Toe support in the pre-swing phase.

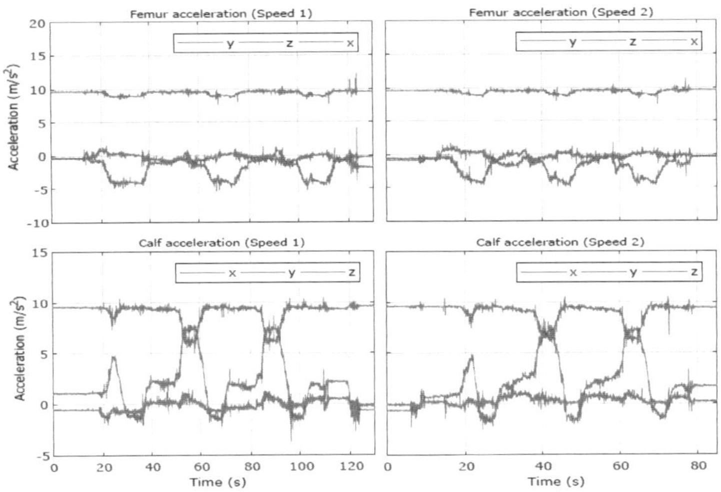

Figure 16. Acceleration in the femur and lower leg segments at the two tested speeds (Speed 1: 2 gait cycles in 140 s; Speed 2: 2 gait cycles in 65 s).

4. Discussion

Although the locomotion patterns are based on the human gait biomechanics, which gives the robot a human-like walking, they must be adjusted online in the simulation to improve the stability and increase the displacement. Nevertheless, the simulation environment and programming method provide an easy and quick way to edit and test the walking without the need for an explicit mathematical model of the robot. In this way the simulation is easily set up and the locomotion execution takes a few iterations to be accomplished.

The robot anthropometry results in a high that acts like an inverted pendulum, where small differences in the set point can totally destabilize the machine; hence, to reduce the overall robot dynamics, the gait is designed to be slow.

Simulation is a tool for generally testing the machine behavior and the simulated patterns confirm the possibility of a stable walking on HRP; however, it is not guaranteed that the simulated patterns always work correctly on the real robot and effectively achieve a stable locomotion due to the differences between simulation and real hardware and the unstable nature of the robot.

The execution of the locomotion patterns on the real HRP demonstrates an acceptable transfer from the simulation to the real hardware, validating the methodology proposed as easy and quick to implement in the robot. Although, the first two patterns do not achieve perfect walking, they provide, in conjunction with the simulation, a basis to develop the final locomotion pattern on HRP, which exhibits good displacement, stability, and speed performance.

As mentioned before, the walking patterns are developed in an open-loop control scheme and the good stability achieved can provide a fundamental basis for developing closed-loop control algorithms able to correct errors under perturbations. In addition, these generated patterns can further serve as a base to optimize the locomotion through machine learning, and thus, to produce a dynamically stable movement.

The interactions of the upper body movements should be analyzed in future works since it can significantly improve the stability and error correction as well as decrease unwanted accelerations in the hip sway and surge. In the long term, locomotion is expected to be generated by machine learning through neuroevolution or reinforcement learning.

5. Conclusions

A methodology to easily edit and develop bipedal locomotion patterns for a humanoid robotic platform using a physical simulation environment, where they are tested before being transferred to the real hardware, was presented.

Two gait patterns were generated through a simple programming method that facilitated their adjusting and validation in the simulation. Then, these patterns were tested on the real hardware and achieved good performance in the robot gait. Based on the results, a third pattern was developed, providing better performance, in terms of speed and displacement, of around 30 % compared to the previous patterns generated, a better stability is also observed.

Although the walking patterns were developed in an open-loop control scheme with preprogrammed sequences executed directly in the simulation, which required a previous configuration of the robot characteristics, no mathematical model of the machine was required.

References

[1] M. Vukobratovic and B. Borovac, "ZERO-MOMENT POINT — THIRTY FIVE YEARS OF ITS LIFE," *International Journal of Humanoid Robotics*, vol. 01, no.01, pp. 157–173, mar 2004. [Online]. Available: http://www.worldscientific.com/doi/abs/10.1142/S0219843604000083

[2] M. Sherman and C. Dembia, "DARPA Robotics Challenge", *DARPA*, pp. 1–42, 2012. [Online]. Available: https://www.darpa.mil/program/darpa-robotics-challenge

[3] Y. Kakiuchi, K. Kojima, E. Kuroiwa, S. Noda, M. Murooka, I. Kumagai, R. Ueda, F. Sugai, S. Nozawa, K. Okada, M. Inaba, "Development of humanoid robot system for disaster response through team nedo-jsk's approach to darpa robotics challenge finals", *2015 IEEE-RAS 15th International Conference on Humanoid Robots (Humanoids)*. IEEE, 2015. p. 805-810. [Online]. Available: https://ieeexplore.ieee.org/abstract/document/7363446/

[4] P. Oh, K. Sohn, G. Jang, Y. Jun, and B. K. Cho, "Technical Overview of Team DRC-Hubo@ UNLV's Approach to the 2015 DARPA Robotics Challenge Finals", *Journal of Field Robotics*, vol. 34, no. 5, pp. 874–896, aug 2017. [Online]. Available: http://doi.wiley.com/10.1002/rob.21686

[5] J. Englsberger, A. Werner, C. Ott, B. Henze, M. Roa, G. Garofalo, R. Burger, A. Beyer, O. Eiberger, K. Schmid, and A. Albu-Schaffer, "Overview of the torque-controlled humanoid robot TORO", *2014 IEEE-RAS International Conference on Humanoid Robots*. IEEE, 2014. p. 916-923. [Online]. Available: https://ieeexplore.ieee.org/abstract/document/7041473/

[6] Q. Huang, K. Yokoi, S. Kajita, K. Kaneko, H. Aral, N. Koyachi and K. Tanie, "Planning walking patterns for a biped robot", *IEEE Transactions on Robotics and Automation*, vol. 17, no. 3, pp. 280–289, 2001. [Online]. Available: https://ieeexplore.ieee.org/abstract/document/938385/

[7] M. Wahde and J. Pettersson, "A Brief Review of Bipedal Robotics Research," *Proceedings of the 8th UK Mechatronics Forum International Conference (Mechatronics 2002)*, pp. 480–

488, 2002.
[8] B. F. Allen and P. Faloutsos, "Evolved controllers for simulated locomotion", in Lecture Notes in Computer Science (including subseries Lecture Notes in Artificial Intelligence and Lecture Notes in Bioinformatics), vol. 5884 LNCS, 2009, pp. 219–230. [Online]. Available: http://link.springer.com/10.1007/978-3-642-10347-620
[9] J. Clune, B. E. Beckmann, C. Ofria, and R. T. Pennock, "Evolving coordinated quadruped gaits with the hyperNEAT generative encoding," in *2009 IEEE Congress on Evolutionary Computation, CEC 2009*, 2009, pp. 2764–2771.
[10] P. Manoonpong, T. Geng, T. Kulvicius, B. Porr, and F. Wörgöter, "Adaptive, fast walking in a biped robot under neuronal control and learning", *PLoS Computational Biology*, vol. 3, no. 7, pp. 1305–1320, 2007. [Online]. Available: http://dx.plos.org/10.1371/journal.pcbi.0030134
[11] R. Tedrake, T. Zhang, and H. Seung, "Learning to walk in 20 minutes", *Proceedings of the Fourteenth Yale Workshop on Adaptive and Learning Systems*. 2005. p. 1939-1412. [Online]. Available: http://groups.csail.mit.edu/robotics-center/public papers/Tedrake05.pdf
[12] J. Yosinski, J. Clune, D. Hidalgo, S. Nguyen, J. Zagal, and H. Lipson, "Evolving Robot Gaits in Hardware: the HyperNEAT Generative Encoding Vs. Parameter Optimization," *Proceedings of the European Conference on Artificial Life (ECAL)*, pp. 1–8, 2011.
[13] M. Freese, S. Singh, F. Ozaki, and N. Matsuhira, "Virtual robot experimentation platform V-REP: A versatile 3D robot simulator," in *Lecture Notes in Computer Science (including subseries Lecture Notes in Artificial Intelligence and Lecture Notes in Bioinformatics)*, vol. 6472 LNAI, 2010, pp. 51–62. [Online]. Available: http://link.springer.com/10.1007/978-3-642-17319-68
[14] N. Koenig and A. Howard, "Design and use paradigms for gazebo, an open-source multi-robot simulator," in *2004 IEEE/RSJ International Conference on Intelligent Robots and Systems (IROS)*, vol. 3, pp. 2149–2154. [Online]. Available: http://ieeexplore.ieee.org/document/1389727/
[15] S. Zapata, I. D. Mora, and G. Suarez, "Mechatronic design of PRH: A NEW 3D printed and anthropometric humanoid robotic platform," in *2017 IEEE 3rd Colombian Conference on Automatic Control, CCAC 2017 - Conference Proceedings*, vol. 2018-Janua. IEEE, oct 2018, pp. 1–6. [Online]. Available: http://ieeexplore.ieee.org/document/8276477/
[16] E. Rohmer and S.P. Singh, "V-REP: a Versatile and Scalable Robot Simulation Framework", in *Proc. of The International Conference on Intelligent Robots and Systems (IROS)*, 2013. [Online]. Available: https://ieeexplore.ieee.org/abstract/document/6696520
[17] Coppelia Robotics Switzerland, "Coppelia Robotics v-rep: Create. Compose. Simulate. Any Robot," 2018. [Online]. Available: http://www.coppeliarobotics.com/
[18] M. Mendez and S. Kannan, "V-REP & ROS Testbed for Design, Test, and Tuning of a Quadrotor Vision Based Fuzzy Control System for Autonomous Landing," *Proceedings of The International Micro Air Vehicle Conference and Competition 2014*, 2014. [Online]. Available: http://publications.uni.lu/handle/10993/19225
[19] J. B. Marion, Dinámica clásica de las partículas y sistemas, 1996.
[20] J. Millard F. Beatty, Principles of Engineering Mechanics Volume 2 Dynamics - The Analysis of Motion, 2006.

Computational analysis of the mechanical behavior of carbon nanotubes in a rail-wheel system

Carlos García[1], Mateo Echeverri[1], Jesus Carlos[2] and Gustavo Suárez[1]

[1]Grupo de Investigación sobre Nuevos Materiales, Universidad Pontificia Bolivariana, Medellín, Colombia
[2]Grupo de Investigación de Tribología de Superficies, Universidad Nacional de Colombia, Medellín

E-mail: carlos.garciac@upb.edu.co, mateo.echeverrip@upb.edu.co, iqcornelio@gmail.com, gustavo.suarez@upb.edu.co

Abstract. Currently, railway systems have presented many problems associated with friction and wear due to movement between the interface of the wheel-rail system, a situation that has been demanding large investments and railway costs in transport companies due to the frequent change of pieces of the system.As a possible solution, the addition of carbon nanotubes (CNTs) has been underway in some modified lubricants to increase the antifriction properties. In this investigation, computational simulations were developed to analyze the behavior of the CNTs as a reducing component of the friction of the material. A constitutive model was considered by Hooke's law for the wheel-rail system and the CNTs. Numerical lattices were designed to develop the computational simulations. The constitutive equations were solved by the Finite Differences Method (F.D.M.). The computational models were subjected to transverse loads emulating the movement of the system. Simulations were performed for one wheel-rail model with insert of 0.01 percent of CNTs. The results identified in the report of the insertions of the CNTs in the wheel-rail system show a dissipative-absorbent behavior of the deformation energy is detected, combined with rotation effects of the CNTs, a condition that may allow a reduction in friction and wear CNTs can contribute to improve the tribological properties of lubricants in affected railway systems and a longer service life of mechanical elements. The computational developments carried out in this research can facilitate the analysis of behaviors of nanostructured materials under conditions of extreme functionality where the experimental processes are complex.

1. Introduction

Rail systems for years have presented a great difficulty regarding the constant wear experienced by the wheel-rail interface, which leads to multiple maintenance costs, increased energy consumption and decreased useful time for the components involved due to damage caused to rails and wheels. This causes surface irregularities and can lead to potential accidents or increased energy consumption. (see Figure.1 and 2) [2]. The most important function of lubricants is the reduction of friction and wear, in some cases, the relative movement of two surfaces is only possible if a lubricant is present [3]. Modern lubricants are formulated from a wide range of base fluids and chemical additives. The base fluid has several functions, but it is mainly the lubricant that the fluid layer provides to separate moving surfaces. Also, it removes heat while minimizing friction. Many of the lubricant's properties are improved or created by adding special chemical agents to the base fluid. For example, the oxidation and degradation stability in a motor oil is improved by the addition of antioxidants, while the anti-wear properties in extreme pressure (EP) lubricants are given by the addition of special additives. The base fluid acts as the carrier for these additives and, therefore, this fluid must be able to remain in solution in all normal working conditions. [5]

Figure 1. Rail damaged by high friction in the Medellín metro.

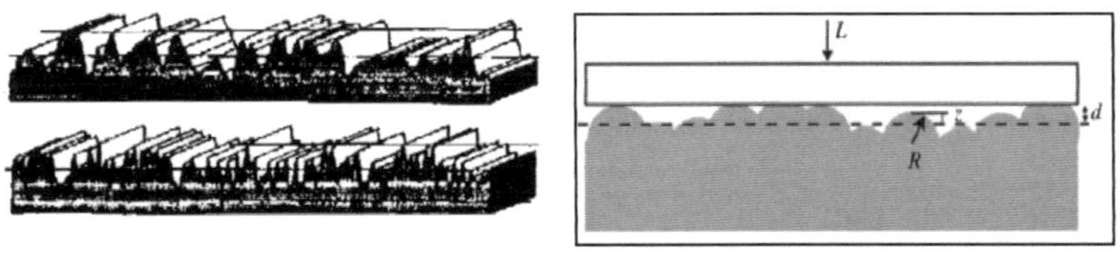

a b

Figure 2. Friction Damage approximation on a)Rail and b)Wheel . [2]

In search of counteracting this problem, multiple tests have been carried out in previous works with modified lubricants with carbon nanotubes to reduce the wear of the wheel-rail systems, specifically for the Medellín Metro (see Figure.3). [7]

CNTs are an allotropic form of carbon, like diamond or graphite. Its structure can be considered as coming from a graphite sheet rolled on itself [5]. Depending on the degree of winding and the way in which the original sheet is formed, the result can lead to nanotubes of different diameters and internal geometries. Nanotubes shaped as if the corners of a sheet were joined at their ends to form a tube, are called single-walled nanotubes, or SWNTs. There are also nanotubes whose structure resembles that of a series of concentric tubes, included one inside the other and logically of increasing thickness from the center to the periphery [3]. The latter are multi-walled nanotubes or MWNTs. Nanotubes usually have a high ratio length /

radius, since the radius is usually less than a couple of nanometers, however, the length can even be centimeters. [1]

Figure 3. Metro of Medellín city. [12]

The stability and robustness of the bonds between carbon atoms, of the sp 2 - sp 3 type, which give it the ability to be the strongest fiber that can be manufactured today. On the other hand, in the face of very intense deformation stresses, CNTs are able to deform noticeably and remain in an elastic regime. Young's modulus could range from 1.3 to 1.8TPa, although to date it has only been possible to obtain experimentally up to 0.8TPa. Furthermore, these mechanical properties could be improved. In this way, even if a nanotube breaks, since they behave as independent units, the fracture would not propagate to the other neighboring units. [2]

In short, they can function as extremely firm springs under small stresses and, under higher loads, can deform drastically and subsequently return to their original shape. Various studies have tried to measure the mechanical properties and the maximum tension supported by a nanotube, with heterogeneous results, although it could be assumed as a guideline that the maximum tension could be around 150GPa [8]. This data implies that a 1cm thick cable made up of CNTs could support a weight of around 1,500 tons. The potential that they have in the tribology area is related to their mechanical properties. [6]

According to what was mentioned in the previous paragraphs, the use of CNTs has been proposed as lubricant additives for sliding wheel rail rolling systems and it has been verified through a series of computational modeling based on the physical results obtained by J.A Carlos Cornelio et al, verifying that the carbon nanotubes in the lubricant contribute to the reduction of rail wear by studying their behavior at high pressures. [1]

The physical tests to evaluate this lubricant in real systems imply a risk for the structures, because the behavior of the nanotubes can alter the balance of the system and also the cost that these tests would precede would be very high without first being completely certain of the efficiency of this lubricant in the real system, therefore, simulations have been performed to describe the behavior of nanotubes in a real system, to verify the behavior and mechanical properties in a wheel-rail system.For this reason, this work is carried out to verify and describe the behavior of CNTs subjected to high pressures in a rail-wheel system as lubricating agents. [1]

2. Materials and Methods

The simulation was performed in Matlab which is a multi-paradigm numerical computing environment and a proprietary programming language developed by MathWorks, allowing matrix manipulations, the mapping of functions and data which is of total utility for the situation

IOP Conf. Series: Materials Science and Engineering 1154 (2021) 012027 doi:10.1088/1757-899X/1154/1/012027

performed. The values and data from the literature were taken into account and the variables that adjust to the real conditions that CNTs may have in lubricants were taken into account. [6]

2.1. Mathematical modeling

The constitutive model to describe the mechanical behavior of the CNTs was through the formulation of the hook law [6]. The equation 1 corresponds to that equation expressed in a general way and is [11] :

$$\sigma_{ij} = (C_{ijl})(\varepsilon_{kl}) \tag{1}$$

Where the left side of the equation represents the Cauchy stress tensor. The first term on the right side represents the constitutive tensor that depends on the mechanical properties of the material and the second one the deformation tensor. Matrix formulation can be described as:

$$\begin{pmatrix} \sigma xx \\ \sigma yy \\ \sigma zz \\ \sigma xy \\ \sigma xz \\ \sigma yz \end{pmatrix} = \frac{E}{1+v} \begin{pmatrix} \frac{1-v}{1-2v} & \frac{v}{1-2v} & \frac{v}{1-2v} & & & \\ \frac{v}{1-2v} & \frac{1-v}{1-2v} & \frac{v}{1-2v} & & & \\ \frac{v}{1-2v} & \frac{v}{1-2v} & \frac{1-v}{1-2v} & & & \\ & & & 1 & 0 & 0 \\ & & & 0 & 1 & 0 \\ & & & 0 & 0 & 1 \end{pmatrix} \begin{pmatrix} \varepsilon xx \\ \varepsilon yy \\ \varepsilon zz \\ \varepsilon xy \\ \varepsilon xz \\ \varepsilon yz \end{pmatrix} \tag{2}$$

The simulation was carried out in the Matlab program. For this simulation, a constitutive model defined by Hooke's law was established for the analysis of the materials of the Wheel-Lubricant-Rail system subjected to shear forces to model the conditions of the Medellín metro system. The values corresponding to the rheological properties of the lubricant are taken from the literature where the measurement of these properties had been sought through experimental tests [11].

2.2. Design and loads considerations

The geometry of the rail wheel system was designed based on Figure. 2 where the carbon nanotube structures were incorporated and a three-dimensional geometry is obtained as illustrated in Figure. 4. The concentration of carbon nanotubes in the system is 0.01 because this is the concentration limit where nanotubes exhibit the highest efficiency as possible lubricating agents. Beyond this concentration, the possible lubrication is not increased and therefore CNTs would be wasting unnecessarily. All this is given by the high surface area that these structures have. Loading conditions generated by the shear forces were established at the borders of the CNTs, due to the close interaction of the wheel-nanostructure-rail. Some roughness of the structures were considered. [2]

The mechanical properties of carbon nanotubes were determined and are presented below at Table 1.

Table 1. Mechanical and physical properties of MWCNTs. [1]

Property	Value	Unit
Density	2.6	g/cm^3
Young's modulus	1200	GPa
Tensile strength	150	GPa
Inside diameter	5 - 10	nm
External diameter	30 - 40	nm

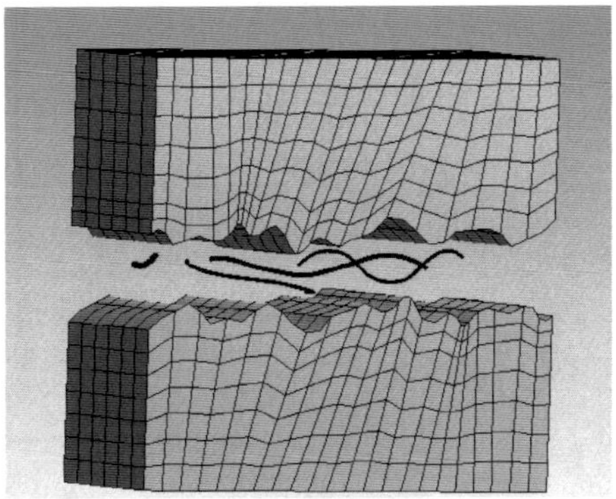

Figure 4. Three-dimensional approach to the study system with oriented CNTs.

2.3. Numerical model

Mathematical equations were solved using the finite difference numerical technique to analyze the mechanical behavior of CNTs. A computational algorithm was built to develop the system of differential equations using the taylor series (see Equation 3). [3]

$$U_{i+1,j} + U_{i-1,j} + U_{i,j+1} + U_{i,j-1} = 4U_{i,j} + h_x^2 U^{"}_{i,j} \mid_x + h_y^2 U^{"}_{i,j} \mid_y \quad (3)$$

After truncation lead to an F.D.M model, wich was applied to the hooke law as follows(see Equation 4):

$$-2\left(\frac{1}{h_x^2} + \frac{1}{h_x^2}\right) U_{i,j} + \frac{U_{i+1,j}}{h_x^2} + \frac{U_{i-1,j}}{h_x^2} + \frac{U_{i,j+1}}{h_y^2} + \frac{U_{i,j-1}}{h_y^2} = -\frac{p}{k} \quad (4)$$

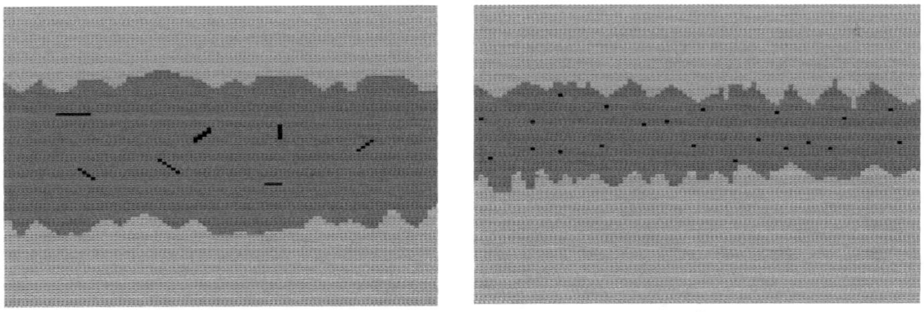

Figure 5. CNTs lattices a)oriented and b)unoriented.

The terms hx and hy are the dimensional steps of the mesh, k is the stiffness of the material and is made up of the cross sectional area and young's modulus. The equations were implemented using an algorithm made in matlab. Numerical lattices were constructed using rectangular elements and node densities equal to 10,000 to be used in the Matlab program. The numerical lattices obtained can see at Figure.5.

IOP Conf. Series: Materials Science and Engineering 1154 (2021) 012027 doi:10.1088/1757-899X/1154/1/012027

3. Results and discussion

The simulations of the behavior of the material were obtained for two different cases, for the conditions of Figure 6, the response of the lubricant was evaluated with the CNTs oriented, thus being able to see the energy absorption that these present and possible rotation; For the conditions in Figure 7, lubricant values were obtained with the unoriented CNTs with a concentration of 0.01, this also presents energy absorption. [1]

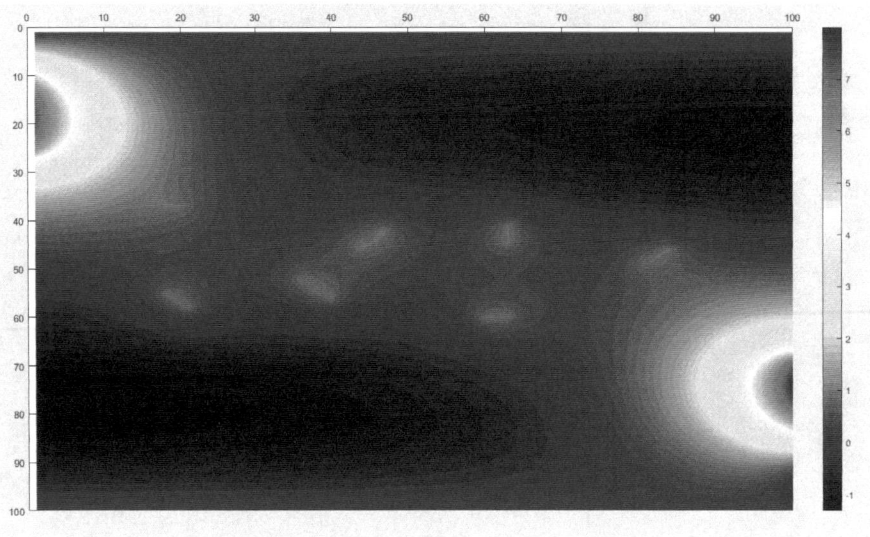

Figure 6. Results of the mechanical behavior of the oriented carbon nanotubes impregnated in the wheel-rail system. Scale 1.

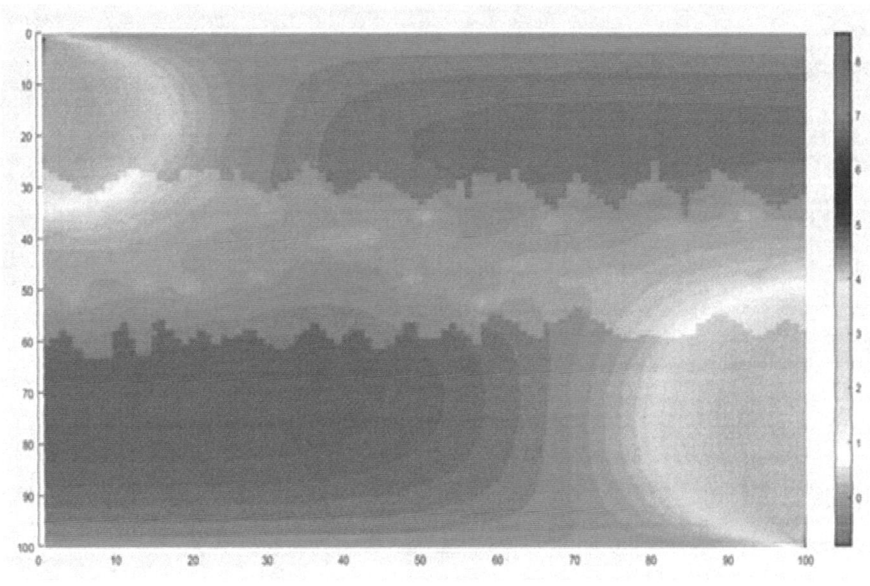

Figure 7. Results of the mechanical behavior of the unoriented carbon nanotubes impregnated in the wheel-rail system. Scale 1.

Reduced coefficient of friction and high wear resistance may be related to the formation of a carbon film amorphous transferred by carbon nanotubes [4]. The simulation results support this hypothesis, but further studies on the chemistry of worn surfaces are required to reveal the mechanisms real and improve the tribological properties of the tribo-system. Therefore, it is foreseen that the next step is to evaluate the properties in a system with real conditions. [5]

Based on the results of these simulations, it can be verified that the CNTs are fulfilling the function of absorbing as much energy as possible to make the lubrication properties optimal and the shear effects have the least possible wear on the wheel system -rail [7]. The increase in the lubricating properties of the oil occurs because the CNTs absorb a large amount of energy as evidenced in Figures 6 and 7 [9], causing in some cases the breakdown of the CNTs, which leads to the formation of a film of graphene that avoids contact between surfaces and this film continues to absorb energy just like CNTs. [8]

4. Conclusions

In this study the tribological properties of CNTs used as lubricant additives were investigated with the aid of a computational simulations.The best tribological response of the pair evaluated was obtained when CNTs were orientated.

Simulations show that nanotubes improve energy absorption, which contributes to improving the tribological properties of lubricants. The implemented algorithms provide the possibility of analyzing different material properties and working conditions with geometries in 3 dimensions, approaching the real system with higher resolution. The project is in the structural simulation phase in 3 dimensions.

The posible reduced coefficient of friction and high wear resistance may be related to the formation of a carbon film due to the CNTs damage under high pressures. The aspect of the surfaces supports should be studied to disclose the actual mechanisms of enhancing the tribological properties of the system wheel-rail becouse of the CNTs by using phisical experiments.

The results obtained by means of the computer simulation are of great help for future works and allowed us to predict how the behavior of this lubricant would be in real conditions. This allows for greater clarity when making the same assembly in a real environment, in order to be more certain of the results that can be obtained and to reduce the cost of said experiment and its risk factors.

References

[1] Carlos, J.A., Cuervo, P.A., Ardila, M.I., Orozco, W., Jaramillo, L.F., Hoyos, L.M., Lara, J.,Toro, A.: Tribological properties of carbon nanotubes as lubricant additive in oil and water for a wheel–rail system. (2015).10.15446/dyna.v83n199.54900

[2] Velásquez, P.A.: Análisis experimental del efecto del proceso de reperfilado sobre el desgaste y fatiga por contacto de rodadura de riel en el sistema rueda riel en el Metro de Medellín. (2014).

[3] Kim, H., Kim, D.: MD simulation of the frictional behavior of CNTs with respect to orientation : Tribology International (2011).10.1016/j.triboint.2012.01.007

[4] Heo, S., Sinnotta, S.: Effect of molecular interactions on carbon nanotube friction : Journal of Applied Physics (2007).10.1063/1.2784007

[5] Padilla, M., Espinosa, J., Velasco, J.: Mecánica molecular estructural para el cálculo del módulo de Young y los modos de vibración de nanotubos de carbono : Ingeniería y Competitividad (2012).10.25100/iyc.v14i1.2641

[6] Xu, Z., Kleinstreuer, C.: Mathematical Modeling and Computer Simulations of Nanofluid Flow with Applications to Cooling and Lubrication : Fluids (2016).10.3390/fluids1020016

[7] Olofsson, U., Lyu, Y, Zhu, Y.: Mapping the friction between railway wheels and rails focusing on environmental conditions : Wear (2015).10.1016/j.wear.2014.12.028

[8] Ajay, B.S., Amarnath, M., Ramkumar, J., Rai, P.K.: Experimental investigations to enhance the tribological performance of engine oil by using nano-boric acid and functionalized multiwalled carbon nanotubes:a comparative study to assess wear in bronze alloy : Journal of Materials Engineering and Performance (2018).10.1007/s11665-018-3384-9

[9] Chen, K., Wang, L., Chen, Y., Wang, Q.: Molecular dynamics simulation of microstructure evolution and heat dissipation of nanoscale friction : International Journal of Heat and Mass Transfer (2017).10.1016/j.ijheatmasstransfer.2017.01.105

[10] Reinert, L., Green, I., Gimmler, S., Lechthaler, B., Mücklich, F., Suárez, S.: Tribological behavior of self-lubricating carbon nanoparticle reinforced metal matrix composites : Wear (2018).10.1016/j.wear.2018.05.003

[11] Faura, F., Estrems, M., Pedrero, J.: Desarrollo de un metodo numerico para la determinacion del mapa tensional en la zona de contacto entre dos cuerpos : revista internacional de métodos numéricos para cálculo y diseño en ingeniería (2000).

[12] El Colombiano, http://www.elcolombiano.com. Last accessed 11 Jun 2020

IOP Conf. Series: Materials Science and Engineering 1154 (2021) 012028 doi:10.1088/1757-899X/1154/1/012028

Prototype virtual assistant as a tool to support research activities at CORPOSUCRE

C M Pineda-Pertuz[1], H J Romero[1], J A Vergara[1], A D Morales[1], S A Sánchez[1]

[1]Corporación Universitaria Antonio José de Sucre, Sincelejo, Colombia

Email: docente_ingenieria1@corposucre.edu.co

Abstract. Chatbots have become widely used tools in different fields of society, such as in industry, banking, health among other sectors. Their success is mainly that they can provide assistance to people quickly and in a timely manner no matter the time or place where the user is located. This project consisted of a virtual assistant (chatbot) based on artificial intelligence more specifically in the subfield of natural language processing (NLP). Basically what was intended with this prototype is to provide remote assistance to teachers and especially to students researchers of the Antonio José de Sucre Corporation answering their questions on research-related issues, particularly to support those who start in this field and wish to clarify some aspects related to the exercise of research within the institution. The system has been developed with open source tools in its latest versions, it is also modular in nature to facilitate its maintenance and easy integration with other programs, and in addition to being a web-oriented system can be deployed on a wide variety of devices which has greatly facilitated its validation.

1. Introduction

Research is considered one of the fundamental processes in university centers because it basically allows to further enhance knowledge in the educated and is also a source of creation and development of new creative ideas that properly aimed set the course for the achievement of the progress of society. Antonio José de Sucre Corporación Universitaria, recognizing the importance of this activity and with the contribution of the research groups of each of his faculties, is constantly promoting the development of training research within the academic community. Different research strategies have been implemented, such as classroom projects plus illustration provided by research seedlings, which has led the way for students to have their first approaches in this area and consequently through discipline, reflection and teamwork to forge a culture towards research.

In this sense from the faculty of engineering science, different technological projects have also been proposed to facilitate and support the research process through the use of artificial intelligence as a bastion of these developed tools. In this context, the branch of AI that excels in built applications is the processing of the natural language NLP that according to [1] deals with the communication capabilities of computers with humans using their own language.

Natural language processing seeks to help computers understand, interpret, and manipulate human language by borrowing from many disciplines, including computer science and computational linguistics, in order to bridge the gap between human communication and computer understanding [2].

NLP and artificial intelligence in general for some years has been disrupting a large number of social sectors through the application of many of its techniques and tools to address different problems of society, and education has not been the exception as the demands and challenges presented in educational institutions that justify the need to incorporate technology and innovation into their processes to provide more efficient and timely services.

In the university corporation Antonio José de Sucre has seen the need to have a web tool to provide virtual assistance, through which to answer all those questions that the research community may have. This application is designed to interact with research users in a similar way to human, being able in addition to responding to their concerns, performing some specific tasks. This purpose is fundamentally aimed at the number of concerns usually generated within the exercise of research, especially in those who start in this process within the corporation. Such questions arise mainly in processes such as drafting of undergraduate project documents, research reports and diligence of institutional formats. In this regard questions such as: What is the research group of the engineering faculty?, What are the lines of research that are handled in the engineering faculty?, What are the types of research?, Where do I radiate my undergraduate project? are some of the many doubts that the minds of the beginner students are assailed and that through the outcome of this project are intended to respond in an effective and accessible way. Although many of the aforementioned concerns can be resolved by a teacher or a collaborator of the investigative area is not always granted the necessary immediacy required by the user due mostly due to time limitations of the same working days, lack of documentation of the processes among others; therefore the need for this proposal to ensure much more timely responses to these requests for information.

Faced with this problem, the need arises the development of this web-based chatbot system through which we will seek to provide the necessary assistance for users from the reception and analysis of users' questions and giving them successful answers based on the appropriate context.
Chatbots are highly responsive and interactive, they are able to establish conversations like those that humans do using artificial intelligence techniques and tools being able to solve any question at any time by means of a chat. These applications can learn from expressions in the conversation and from previous dialogs to further customize and fine-tune their responses [3].

The virtual assistant proposed in this project will bring many benefits such as: improvements in response times, permanent attention 24 hours a day, better use of time by users by reducing administrative wear and more user satisfaction, who will see in this software the ideal tool to answer their doubts quickly and accurately.

As a fundamental library for this development Rasa NLU (Natural Lenguage Understanding) and Rasa Core were used which are very robust modules ideal for the creation of conversational systems in a simple and very efficient way. The first helps to determine the intentions and analyze the entities grammatically, while with the second we configure the domain and the stories are the way we want our robot to communicate with users [4].

One of the advantages offered by Rasa NLU in addition to being a project with a large community of collaborators actively participating in it, is that it allows you to have full control over the chatbot allowing you to train, refine and optimize the data in the way the developer wants, as well as supports testing with different machine learning algorithms to choose the one that best fits the available data set without having to rely on a fixed algorithm [5]. The machine learning algorithm is a prediction program that learns from data in order to find patterns or trends to understand what this data tells us, and thus build a model to predict or classify elements [6].

The development of this prototype was carried out in three phases that are generally stated below but will be further described during the article:

Planning: At this stage a search was carried out of more used open source environments and libraries for the development of virtual assistants both in academia and industry in general. Going so far as to choose RASA along with Python for facilitating the programming of applications with artificial intelligence due to the large amount of resources available to the programmer.

Design and Construction: In this phase a development architecture was defined based on the Model View Controller (MVC) pattern and the user interfaces and entity model relationship of the database were developed. The modular nature used allowed the integration of this system with other websites and applications of machine learning-based research projects that have been developed articulately with this project within the engineering faculty.

Evaluation: Performance of the predictive model generated with artificial intelligence was estimated using various controlled tests, assessing some assessment metrics such as *accuracy*, as well as the performance of the chatbot was deployed on a locally accessed web page to verify its operation.

2. Methodology

For the development of the prototype, the agile SCRUM methodology was used through which the system was built iteratively and detecting change needs so that new functionalities could be added.

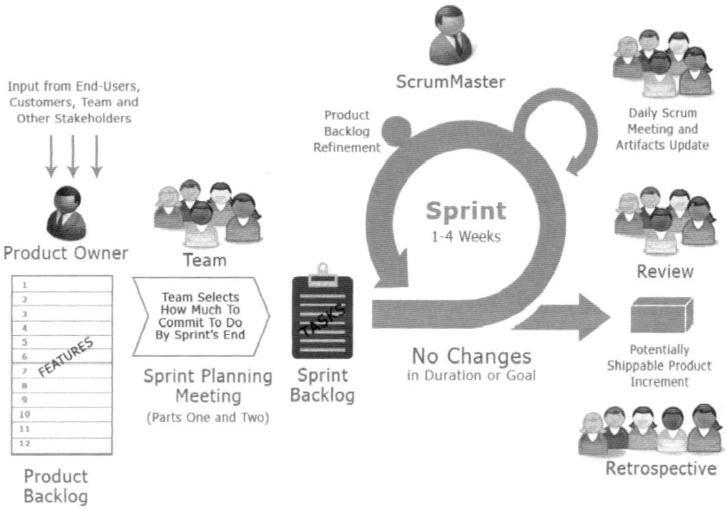

Figure 1. Elements of the Scrum methodology.

A key aspect of these agile methodologies like Scrum is the high visibility of all events in the evolution of projects. For this reason, it is important to track and discipline the activities that generate the necessary visibility. Son: Daily Scrum, Sprint Review, Spring Retrospectives, Product Backlog and Burndown chart [7]. First, the first three help to gain an understanding of the current state of development and advancement in the Springs at all times, as well as to verify improvement actions focused on achieving better group performance. On the other hand, the Product Backlog as one of the most important assets was the reason for permanent reviews to more accurately determine the highest priority stories for each Spring, leaving those in a pending state, reflected in the Burndown Chart.

The overall process using the Scrum methodology can be described in the following 3 stages:

2.1. Planning

For this stage, special attention was paid to the lifting of user requirements through the implementation of user stories, as well as defining the roles and those responsible for making each user story, and also defining the dates of the deliverables of the project.

2.2. Springs Development

At this stage, each responsible user story was encoded using the object-oriented programming paradigm. Each story as the programmer developed marked it as finished on an online dashboard created with Jira software. The Jira Scrum board is the tool that unraies teams around unique goals and promotes iterative and incremental delivery [8].

2.3. Evaluation of Springs

Finally, at this stage the validation of each of the Springs was carried out where the clients (commissioned teachers) reviewed that the proposed functionalities met the established requirements.

3. Proposed solution

For the proposed system stands out the Rasa framework which is a library based on artificial intelligence that has made it possible to create the virtual assistant. Which has two components that are: Rasa NLU (Natural Understanding of Language) which is used for the recognition of messages entered by the user and is responsible for finding a correspondence with an *intent*. The latter represent what the user can type or want and are used to describe how to categorize those messages. For example: If the user enters "Hello" or "Good morning" the intention in this case will be *greeting*. That component also extracts from the user's message keywords called *entities* with which the chatbot recognizes the structure of the sentence. Rasa Core, meanwhile, is a learning engine that uses machine learning, which defines what the next action to take based on the user's message. The input data for this engine for wizard training is defined in the stories [9].

Rasa Core specifies all the things that the chatbot can say or do, these are called *actions*. The actions defined in the proposed system were basically: greeting, farewell, FAQs and research_concepts, the latter makes connection to the database to obtain information related to research projects such as: Research lines, participants, directors, methodologies, among others. Training data trained a probabilistic model to predict what action will be taken based on the historical conversation of the user or users [5].

Where the main component of the model is a recurring neural network (RNN) type LSTM which maps the history of raw dialogs directly to the distribution of system actions [10]. LSTM (Long short term memory) is a special type of memory-ended neural networks. Allowing them to learn from important experiences they have spent before. LSMs allow RNNs to remember their inputs for a long time, using a memory that can be considered similar to a computer's memory in the sense that a neuron in an LSTM can read, write, and erase information from its memory [6].

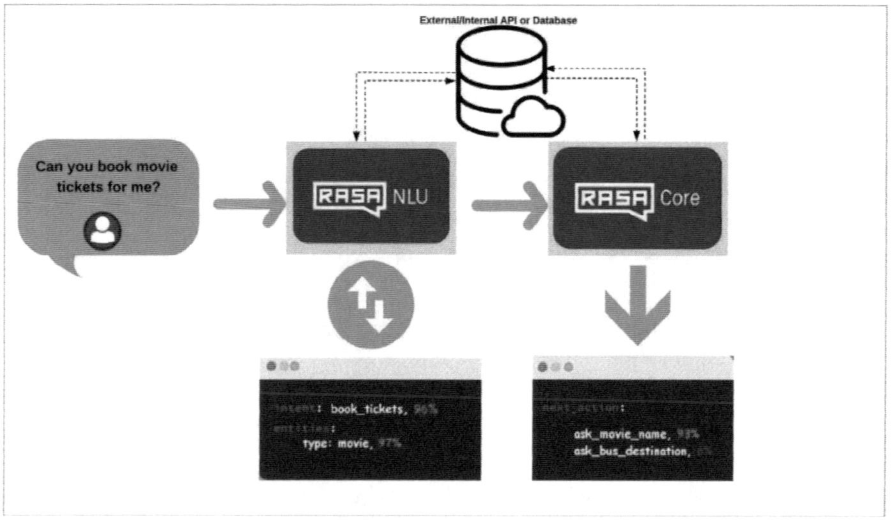

Figure 2. RASA Architecture.

To carry out the training you must clearly define intentions, entities, among others. To do this you must configure the following files: 1. config.xml (configuration for Rasa NLU), this basically specifies the language, the training pipeline that are a sequence of processing components that run over incoming messages and policies specifying what action to take at each point of a dialog, leaving by default MemoizationPolicyPolicy , MappingPolicy and TEDPolicy the latter with a maximum of 5 stories and a total of 100 eras or iterations. 2. nlu.md, here are defined the possible messages of the user that will be grouped according to an intention. 3. domain.xml, this is the file where you mainly define default intentions, actions and responses that the system should know and that define its scope.

To perform training with NLU based on the above files, the *rasa train* command is used, which at the end of its execution generates a compressed file in a folder called models. This helps the model understand what the user writes, to make it respond to the user's queries, and we must run the Rasa Core, this component helps decide the next action to execute in a conversation with the user based on previous intents, entities, and conversations. To do this work is based on stories where examples of possible conversations between a user and the chatbot were defined, those stories were written in a file called stories.md. Custom actions that were created are executed on the action server by using the rasa run actions command by making them accessible through the url "http://localhost:5055/webhook" which is specified in a file called endpoints.xml. These actions retrieve information from a database in Mysql using code written in Python by establishing the proper connection to the mysql-connector-python driver.

Finally, the generated model is deployed to the server with the *rasa run* command, making several functions available through several paths that can be reached using the GET and POST methods. For the point-in-time case of this prototype, "http://localhost:5005/webhooks/rest/webhook" was accessed using a request in Ajax and processing the response emanating from the server in JSON format.

```
function sendMessage(textoServ)
{
    var xhttp = new XMLHttpRequest();
    xhttp.onreadystatechange = function() {
        if (this.readyState == 4 && this.status == 200) {
            var dataJson = eval(this.responseText);
            insertText("res", dataJson[0].text);
        }
    };
    xhttp.open("POST", "http://localhost:5005/webhooks/rest/webhook");
    xhttp.setRequestHeader("Content-Type", "application/json");
    xhttp.send(JSON.stringify({message:textoServ}));
}
```

Figure 3. Access to the model deployed on the server. Source: Own elaboration.

To perform the evaluation of the model we obtain a test data group of the original dataset using the command: *rasa data split*. This action generates a file named conversation_tests.md in a folder called tests, with which you can run the test using the command: *rasa test -u tests-conversation_tests.md --model models-20200815-063926.tar.gz*, after which an *accuracy* of more than 70% is achieved. Accuracy is an evaluation metric that is defined as the sum of all correct predictions divided by the total number of predictions or mathematically: A = (TP/TN) / (TP+TN+FP+FN) [11].

The communication between the different elements of the prototype is shown below:

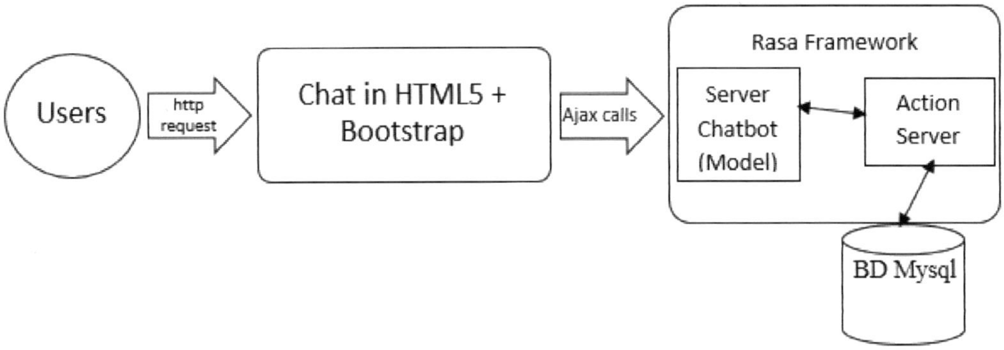

Figure 4. General chatbot architecture. Source: Own elaboration.

4. Results

The purpose of this project was to develop a smart assistant that would help the student researchers answer the questions that most afflict them when participating in some research activity, for which it was used to build the frontend the Html5, css and Javascript languages, thus making a page with an interactive chat so that the user can dialogue with the chatbot. For its part for the backend was used in addition to the aforementioned Rasa library, the Python programming language to program custom actions involving queries on the database in Mysql, whose model entity relationship shown below.

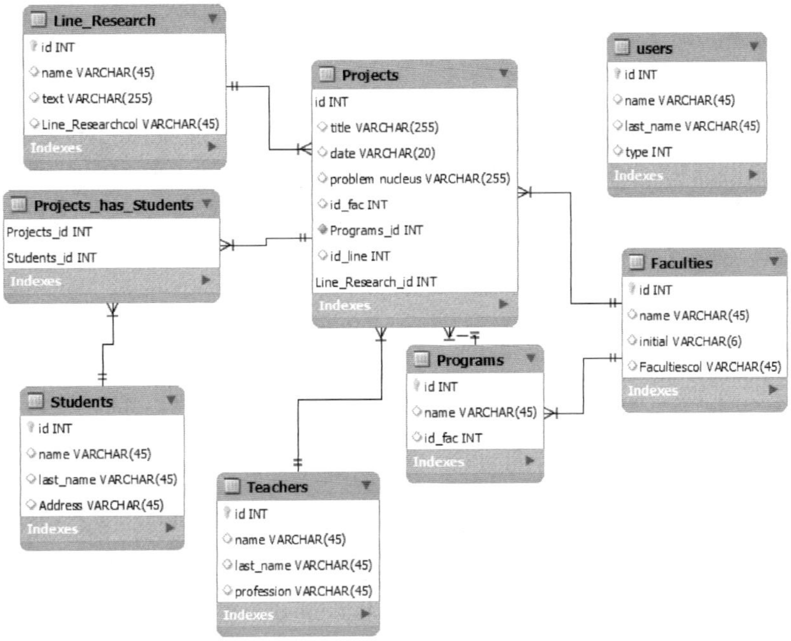

Figure 5. Relationship entity model. Source: Own elaboration.

The working wizard deployed in the web browser is also presented in the following figure.

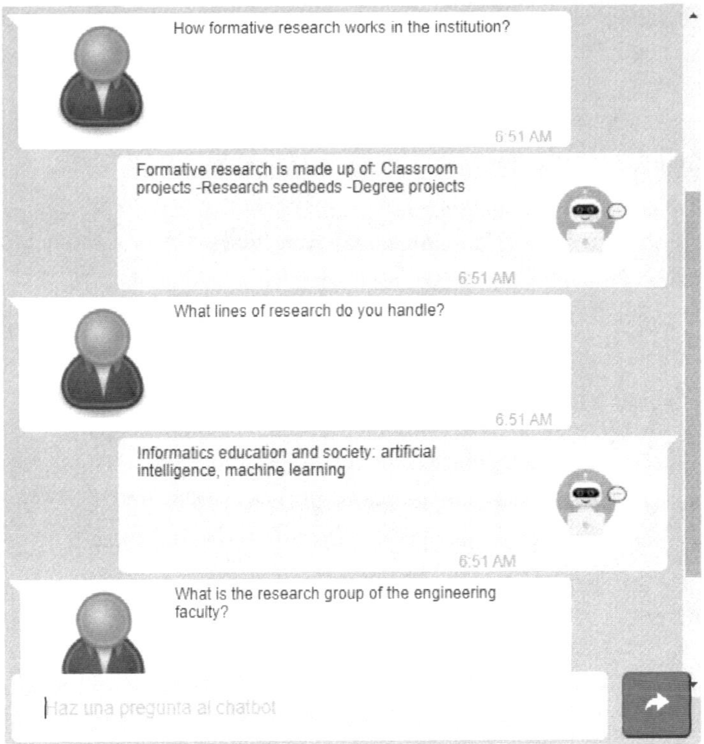

Figure 6. Chatbot in operation: Own elaboration.

5. Conclusions

Based on the objectives of the previous research work, the following can be concluded:
The needs presented in the students of the Faculty of Engineering were identified at the time of carrying out some activity related to research and from the collection, analysis and selection of useful information, the current situation of the faculty and the institution could be further understood by proposing an effective and efficient solution that responds to the concerns of these actors in an optimal way improving their satisfaction.

The use of an agile Scrum programming methodology has allowed to maintain a permanent communication between all those involved in the project including of course the clients, thus guaranteeing a better control and monitoring of it since the master tools of the team have been used to visualize and have a balance of the outstanding, developing and resolute tasks of Spring each and also the acceptance tests have been carried out at the end each one of these periods of development.

As a main recommendation, it is highlighted, first of all increasing the number of examples of dialogues in the stories, so that the machine learning model can have a greater repertoire and have better performance in conversations with users. Similarly, more tests should be performed based on different validation metrics to improve the accuracy of the results obtained.

References

[1] A. G. Serrano, Inteligencia artificial, Madrid: Alfaomega, 2018.

[2] SAS, "SAS," 2019. [Online]. Available: https://www.sas.com/es_co/insights/analytics/what-is-natural-language-processing-nlp.html. [Accessed 13 08 2020].

[3] T. S. E. C. Department of Information Technology, "Customer Support Chatbot Leveraging Machine Learning," *International Research Journal of Engineering and Technology (IRJET)*, 2020.

[4] B. S. Desikan, Natural Language Processing and Computational Linguistics: A practical guide to text analysis with Python, Gensim, spaCy, and Keras, Birmingham: Packt Publishing Limited , 2018.

[5] S. Raj, Building Chatbots with Python, Bangalore: Library of Congress, 2019.

[6] J. Torres, Python Deep Learning, Bogotá: Alfaomega, 2020.

[7] Atlassian, "Atlassian," [Online]. Available: https://www.atlassian.com/es/software/jira/scrum-boards. [Accessed 14 08 2020].

[8] G. P. y. R. Rinaudo, Ingeniería de software, Buenos aires: Alfaomega, 2015.

[9] J. C. Palma, "Desarrollo de un sitio web para público infantil para clasificar sonidos del cielo," ETSI Informáticos, Madrid, 2020.

[10] P. Pandey, "Planetachatbot," [Online]. Available: https://planetachatbot.com/tutorial-construyendo-chatbot-conversacional-slack-usando-rasa-y-python-part-1-7ff7eb92a9c6-7ff7eb92a9c6. [Accessed 14 08 2020].

[11] W. M. Lee, Python machine learning, Indianapolis: John Wiley, 2019.

Topology optimization and additive manufacturing in the building and construction industry

G A Barragan[1], J Perafan[1], and G Urrea[1]

[1]Grupo de investigación en Ingeniería Aeroespacial, Universidad Pontificia Bolivariana, Circular 1 # 70 -01, Medellín - Colombia.

E-mail: german.barragan@upb.edu.co

Abstract. A systematic review covering Topology optimization and Additive manufacturing has been conducted. Covering areas such as its application on the building and construction industry and the potential of its mixed-use. The results found shows the increased interest of the research in this area during the last decade and the vast potential of its full application for safety improvements and resource efficiency.

1. Introduction

During the construction of structural components, the best possible configuration is always desired to achieve superior performance characteristics . The architectural design process employs different techniques for this propose, ranging from purely empirical approaches to mathematical methods [1]. The latter being the most relevant for the case of structural optimization.

The optimization processes work to obtain the optimal component dimension, geometry, or material distribution, considering different criteria such as maximizing its rigidity, minimizing stresses, deformations, vibrations, and others. Topological optimization (TO) as a mathematical design tool has been widely used for several years with rapid growth as a research field.

The use of OT methods has spread to different industries with promising results. Nevertheless, in most cases, its potential has been limited because the geometries obtained were impossible or presented significant challenges to be manufactured using traditional construction techniques.

Additive manufacturing (AM) techniques were based on the construction employed successively adding material layers. They have done away with the typical constraints of conventional construction processes allowing the materialization of parts with high geometrical complexity or functional designs without significant cost effects.

The construction industry has been embraced AM as a tool in recent times [2], opening a wide range of application possibilities, including improvements in design flexibility [3] with a reduction in material and labor requirements [4].

This paper presents findings of a systematic review covering TO and AM. It is covering areas such as its application on the building and construction (B&C) industry and the potential of its mixed-use. The results presented are part of a study to identify technologies with potential application during a spatial colonization process.

2. Methodology

The research methodology adopted in this research was a systematic review. Considered an essential scientific research method could be used to analyze, summarize, and convey the conclusions and consequences of large quantities of research publications [5].

For this specific case, the process has been conducted to identify the evolution of Topology Optimization and Additive manufacturing techniques in the construction industry. Two databases were employed, Web of Science and SCOPUS. The reputation of these databases was the decisive factor for this selection.

The critical terms for the search were "Topology optimization", "Additive Manufacturing", and "3D printing". Only English published studies with construction as a focus were considered. A two-stage selection process was executed. During the first stage, a search employed any individual term plus construction was conducted, this phase was used to identify some main trends, close the search period, and to refine the search process. In the second stage, the three terms were used together. The manuscripts found were classified, examined, and reviewed. The relevant information contained was summarized and presented.

3. Results

Overall results of the first stage of the work were presented in table 1.

Table 1. Documents found during the first stage

Search term	Scopus	Web of Science
Additive Manufacturing	888	588
3D printing	1299	881
Topology Optimization	361	239

Through the last decade, the number of documents having within their scope, the selected research terms have increased significantly. Between the different types of documents available, the most representative were scientific articles be more of the 55%. Its information and results can bring a profound contribution to the body of knowledge for new construction technologies. In figure 1, it is possible to see the distribution by year and type of the found documents.

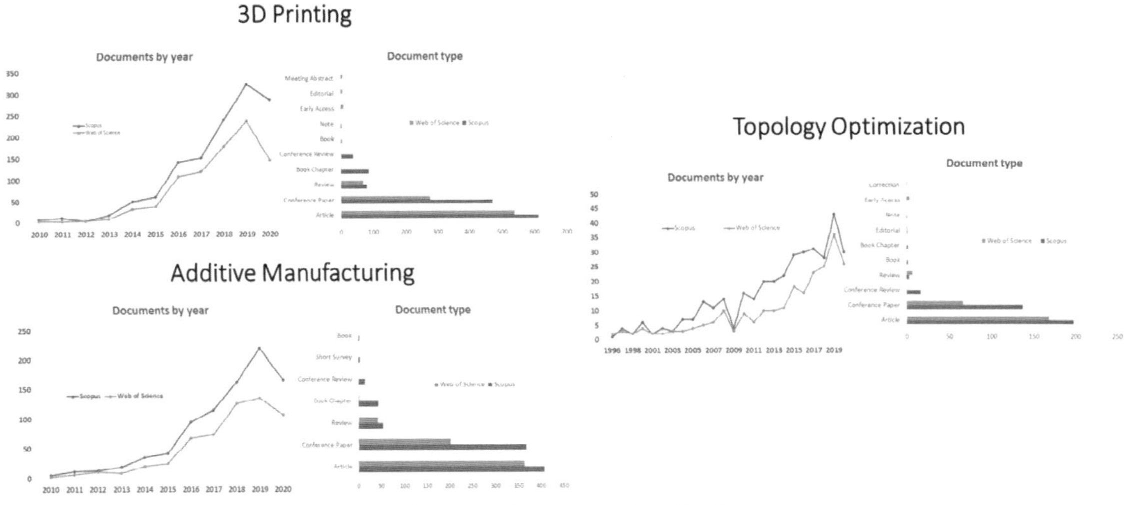

Figure 1. Firs stage documents distribution

For the second stage, the number of documents obtained in Scopus and Web of Science was 32 and 26, respectively. Duplicated papers were eliminated. The final analysis was developed with 44 selected documents.

3.1. Topology optimization

The development and application of numerous computer optimization tools were found. Its purpose is the cost or material reduction while attaining a specific design criterion. TO commonly appear as a gradient-based design instrument that establishes where put material is based on the load distribution, stress, and boundary conditions for a goal [6].

3.1.1. Methods. Different TO methods and approaches [7] have been introduced for the work with a continuum structure. These include the density approach [8], [9];the level set strategy [10], [11]; evolutionary approaches [12]; and several others [13].

Density-based methods form the basis for most recent topology optimization literature because of the degree of sophistication achieved. It represents the optimized structure by a set of density vary values at discrete nodes in the project domain [14]. Between these methodologies, the SIMP ("Solid Isotropic Material with Penalization") has demonstrated to be very successful and extremely skillful. A full description of this approach can be found in [15].

Adaptations of this along with other methods were used to integrate enhancements or modifications to be applied during the design phases of components to be manufactured using AM techniques such as [16].

3.1.2. Applications The development of proper applications of TO methods in the construction industry has gained an expanding consideration [17]. Some contributions are Staged construction [18]; Optimized strut-and-tie models for RC members [19]; Reinforced structures [20], [21], [22]; Prestressed girder (Figure 2) [14]; Concrete beams and structures [23], [24]; Dapped beams [25], Building vibration behaviour improvement [26] and Buildings outrigger placement [9].

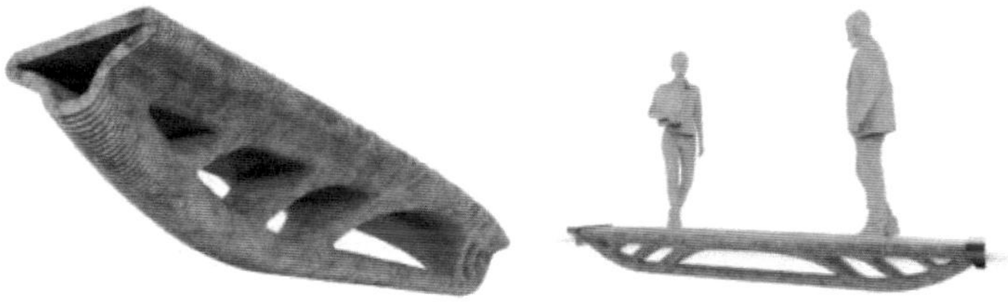

Figure 2. Topology optimized design for 3D-print construction. Source [14]

3.2. Additive manufacturing

AM also now as 3D printing is considered an innovative process that could create intricate shape geometries directly from a digital model [27]. The B&C industry have been explored AM [28] in academic [29] and practical applications (Figure 3), primarily focused on demonstrating viability [4].

Figure 3. AM constructed pillar. Source [30]

AM could allow the development of well custom-built components and inspire intricate and refined design [31]. Its ability to works in an autonomous or semi-autonomous way can reduce the requirement of workforce [28] situation that can be able to generate substantial diminishes on fatalities and injuries that are a severe problem of the B&C industry [32], [33]. Nevertheless, as with any new technology, to harness its full potential and achieve its extensive use, many faces remain.

3.2.1. Processes All the AM categories recognized by the ASTM [34] have a potential application in different aspects of the B&C industry, from architectural representation to full building construction. However, considering only the last case (Large-scale printing) is possible to find process names such as Contour Crafting [35]; Concrete Printing [36]; Freeform construction [23]; Cementitious paste extrusion [37]; Layered extrusion processes [38]; Material deposition method (MDM) [27]; Stick dispenser [39]; Digital construction platform [40]; Flow-based fabrication [41]; Mesh-mould [27]. Figure 4 shows an image of the concrete printing process.

Figure 4. Concrete printing process. Source [42]

3.2.2. Materials Limited materials are now available for AM in the B&C industry [43]. One of the major faces is the conversion of existing high-quality concretes to printable materials. Concrete or cement-based are considered as intricate material with a non-homogeneous composition. Intensive research has recently been done on the production and use of admixture to achieve the desired material performance [44], [45], [46].

4. Conclusions

Additive manufacturing and Topology optimization have a vast potential of application on the building and construction industry.

The use of these combined techniques could bring the industry's most significant benefits in terms of resource economy and safety.

A proper integration route for AM and TO is essential to take full advantage of its potential.

Limited materials are now available for AM in the B&C industry, but intensive research is happening in this knowledge area.

References

[1] K. Besserud, N. Katz and A. Beghini, "Structural emergence: Architectural and structural design collaboration at SOM," *Architectural Design,* vol. 83, no. 2, pp. 48-55, 2013.

[2] P. Wu, J. Wang and X. Wang, "A critical review of the use of 3-D printing in the construction industry," *Automation in Construction,* vol. 68, pp. 21 - 31, 2016.

[3] B. Khoshnevis, "Automated construction by contour crafting—related robotics and information technologies," *Automation in Construction,* vol. 3, no. 1, pp. 5 - 19, 2004.

[4] R. Buswell, R. Soar, A. Gibb and A. Thorpe, "Freeform construction: mega-scale rapid manufacturing for construction," *Automation in construction,* vol. 16, no. 2, pp. 224 - 231, 2007.

[5] O. Babalola, E. Ibem and I. Ezema, "Implementation of lean practices in the construction industry: A systematic review," *Building and Environment,* vol. 148, pp. 34-43, 2019.

[6] L. Beghinia, A. Beghini, N. Katz, W. Faker and G. Paulino, "Connecting architecture and engineering through structural topology optimization," *Engineering Structures,* vol. 59, pp. 716-726, 2014.

[7] O. Sigmund and K. Maute, "Topology optimization approaches," *Structural and Multidisciplinary Optimization,* vol. 48, no. 6, pp. 1031-1055, 2013.

[8] M. Bendsoe and O. Sigmud, Topology optimization: theory,method and applications, Springer, 2003.

[9] S. Lee and A. Tovar, "Outrigger placement in tall buildings using topology optimization," *Engineering Structures,* vol. 74, pp. 122-129, 2014.

[10] G. Allaire, F. Gournay, F. Jouve and A. Toader, "Structural optimization using topological and shape sensitivity via a level set method," *Control and Cybernetics,* vol. 34, no. 1, 2005.

[11] M. Wang, X. Wang and D. Guo, "A level set method for structural topology optimization," *Computer Methods in Applied Mechanics and Engineering,* vol. 192, no. 1-2, pp. 227-246, 2003.

[12] X. Huang and Y. Xie, Evolutionary Topology Optimization of Continuum Structures: Methods and Applications, John Wiley & Sons, Ltd, 2010.

[13] A. Takezawa, S. Nishiwaki and M. Kitamura, "Shape and topology optimization based on the phase field method and sensitivity analysis," *Journal of Computational Physics,* vol. 229, no. 7, pp. 2697-2718, 2010.

[14] G. Vantyghem, W. De Corte, E. Shakour and O. Amir, "3D printing of a post-tensioned concrete girder designed by topology optimization," *Automation in Construction,* vol. 112, 2020.

[15] E. Andreassen, A. Clausen, M. Schevenels, B. S. Lazarov and O. Sigmund, "Efficient topology optimization in MATLAB using 88 lines of code," *Structural and Multidisciplinary Optimization,* vol. 43, no. 1, pp. 1- 16, 2011.

[16] M. Zhu , Y. Yang , A. Gaynor and J. Guest, "Considering constructability in structural topology optimization," in *Structures Congress*, Boston, United States, 2014.

[17] T. Pastore, V. Mercuri, C. Menna, D. Asprone, P. Festa and F. Auricchio, "Topology optimization of stress-constrained structural elements using risk-factor approach," *Computers & Structures,* vol. 224, 2019.

[18] O. Amir and Y. Mass , "Topology optimization for staged construction," *Structural and Multidisciplinary Optimization,* vol. 56, p. 1679–1694, 2018.

[19] M. Bruggi, "Generating strut-and-tie patterns for reinforced concrete structures using topology optimization," *Computers & Structures,* vol. 87, no. 23-24, pp. 1483-1495, 2009.

[20] M. Bogomolny and O. Amir, "Conceptual design of reinforced concrete structures using topology optimization with elastoplastic material modeling," *International Journal forNumerical Methods in Engineering,* vol. 90, no. 13, pp. 1578-1597, 2012.

[21] M. Bruggi, "A numerical method to generate optimal load paths in plain and reinforced concrete structures," *Computers & Structures,* vol. 170, pp. 26-36, 2016.

[22] M. Buggi, "Analysis and design of reinforced concrete structures as a topology optimization problem," in *European Congress on Computational Methods in Applied Sciences and Engineering*, Crete, Greece, 2016.

[23] J. Jewett and J. Carstensen, "Topology-optimized design, construction and experimental evaluation of concrete beams," *Automation in Construction,* vol. 102, pp. 59-67, 2019.

[24] P. Martens, M. Mathot, F. Bos and J. Coenders, "Optimising 3D printed concrete structures using topology optimisation," in *Fib Symposium - High Tech Concrete: Where Technology and Engineering Meet*, Maastricht, 2017.

[25] R. Oviedo, S. Gutierrez and H. Santa Maria, "Experimental evaluation of optimized strut-and-tie models for a dapped beam," *Structural Concrete,* vol. 17, no. 3, pp. 469-480, 2016.

[26] V. Hoorickx, M. Schevenels and G. Lombaert, "Topology optimization of wave barriers for railway induced vibrations in buildings," in *European Congress on Computational Methods in Applied Sciences and Engineering*, Crete, Greece, 2016.

[27] Y. Tay, B. Panda, S. Paul, N. Mohamed, M. Tan and K. Leong, "3D printing trends in building and construction industry: a review," *Virtual and Physical Prototyping,* vol. 12, no. 3, pp. 261-276, 2017.

[28] R. Buswell, W. Leal de Silva, S. Jones and J. Dirrenberger, "3D printing using concrete extrusion: A roadmap for research," *Cement and Concrete Research,* vol. 112, pp. 37-49, 2018.

[29] Y. E. Kalay and M. J. Skibniewski, Eds., Automation in Construction, Special Issue: Rapid Prototyping, ELSEVIER, 2002.

[30] XtreeE, "Krypton, a column in Aix-en-Provence," [Online]. Available: https://xtreee.com/en/project/krypton/. [Accessed 30 7 2020].

[31] C. Gosselin, R. Duballet, P. Roux, N. Gaudillière, J. Dirrenberger and P. Morel, "Large-scale 3D printing of ultra-high performance concrete – a new processing route for architects and builders," *Materials & Design,* vol. 100, pp. 102-109, 2016.

[32] J. Meliá, K. Mearns, S. Silva and M. Lima, "Safety climate responses and the perceived risk of accidents in the construction industry," *Safety Science,* vol. 46, no. 6, pp. 949-958, 2008.

[33] I. Rudchenko, A. Bychkov and D. Levchenko, "Labor Safety in Construction," in *International Scientific Conference on Construction and Architecture: Theory and Practice for the Innovation Development*, Kislovodsk, Russian Federation, 2019.

[34] ASTM, "ASTM F2792-12a, Standard Terminology for Additive Manufacturing Technologie," 2015.

[35] B. Khoshnevis, "Automated construction by contour crafting—related robotics and information technologies," *Automation in Construction,* vol. 13, no. 1, pp. 5-19, 2004.

[36] S. Lim, A. Buswell, T. Le, S. Austin, A. Gibb and T. Thorpe, "Developments in construction-scale additive manufacturing processes," *Automation in Construction,* vol. 21, pp. 262-268, 2012.

[37] R. Duballet, O. Baverel and J. Dirrenberger, "Classification of building systems for concrete 3D printing," *Automation in Construction,* vol. 83, pp. 247-258, 2017.

[38] T. Wangler, E. Lloret, L. Reiter, N. Hack, F. Gramazio, M. Kohler, M. Bernhard, B. Dillenburger, J. Buchli, N. Roussel and R. Flatt, "Digital Concrete: Opportunities and Challenges," *RIMEL Technical Letters,* vol. 1, pp. 67 - 75, 2016.

[39] H. Yoshida, T. Igarashi , Y. Obuchi, Y. Takami, J. Sato, M. Araki, M. Miki, K. Nagata, K. Sakai and S. Igarashi, "Architecture-Scale Human-Assisted Additive Manufacturing," *ACM Transactions on Graphics,* vol. 34, no. 4, pp. 88:1-88:8, 2015.

[40] S. Keating, N. Spielberg, J. Klein and N. Oxman, "A Compound Arm Approach to Digital Construction," in *Robotic Fabrication in Architecture, Art and Design*, W. McGee and M. Ponce de Leon, Eds., 2014, pp. 99 - 110.

[41] J. Duro-Royo, L. Mogas-Soldevila and N. Oxman, "Flow-based fabrication: An integrated computational workflow for design and digital additive manufacturing of multifunctional heterogeneously structured objects," *Computer-Aided Design,* vol. 69, pp. 143-154, 2015.

[42] R. Wolfs, F. Bos and T. Salet, "Early age mechanical behaviour of 3D printed concrete: Numerical modelling and experimental testing," *Cement and Concrete Research,* vol. 106, pp. 103-116, 2018.

[43] G. De Schutter, K. Lesage, V. Mechtcherine, V. Nerella, G. Habert and I. Agusti-Juan, "Vision of 3D printing with concrete — Technical, economic and environmental potentials," *Cement and Concrete Research,* vol. 112, pp. 25-36, 2018.

[44] D. Marchon, S. Kawashima, H. Bessaies-Bey, S. Mantellato and S. Ng, "Hydration and rheology control of concrete for digital fabrication," *Cement and Concrete Research,* vol. 112, pp. 96-110, 2018.

[45] J. Cunha and L. Chaves, "The use of topology optimization in disposing carbon fiber reinforcement for concrete structures," *Structural and Multidisciplinary Optimization,* vol. 49, p. 1009–1023, 2014.

[46] E. Grigoryan, A. Babanina and A. Eropkin, "Concrete mixture grout for printing extruder," in *Innovative Technologies in Environmental Science and Education, ITESE*, "Raduga" Divnomorskoe Village, 2019.

Discrete event simulation: Case study in an extractive sector company

R Quintana-Londoño[1,2], **A García-Pérez**[1,3]

[1]Universidad Tecnológica de Bolívar, Faculty of Engineering, Department of Industrial Engineering. Cartagena de Indias.
[2]raul.quintana1997@hotmail.com, [3]agarcia@utb.edu.co

Abstract. At present, companies are obliged to continually improve their productivity to remain in a very competitive industry, that said, the development of processes must be done efficiently and due to this, meet the various requirements of their customers. In order to offer and execute all these services, it is necessary to carry out an analysis to each area of the company, that is, prior analysis and error processes must be implemented, in such a way, the present investigation will aim to determine the faults present in the system produced in the company and provide a viable alternative for the improvement of its system. The ProModel software will be used as the main tool for the development of the current system model and its analysis to interpret the difficulties of the system and look for an improvement alternative, later the development of the model of this alternative to compare with the statistical analyzes thrown by the software between models. As a result, a clear improvement was achieved with the proposed alternative, obtaining a significant increase in production and also a reduction in costs in the medium and long term thanks to the reduction in the amount of resources used.

Keywords: discrete-event simulation, system analysis, improvement proposal.

1. Introduction

A quarry is a mining industry of all kinds of raw materials for different industries. This industry works mainly with all types of heavy machinery such as backhoes, loaders, screens, and vehicles for transporting this raw material. All these interactions within a quarry can be simulated to analyze and seek improvements within the system to increase productivity. On the other hand, the simulation of discrete events is a "set of logical, mathematical, and probabilistic relationships that integrate the behavior of a system under study when a specific event occurs. The objective of the simulation model consists precisely in understanding, analyzing, and improving the relevant operating conditions of the system " [1].

Discrete event simulation modeling using software has proven to be an effective method for troubleshooting manufacturing process problems and service system problems [2][3][4] and, in turn, to be a reliable support for decision making. Decisions have been made based on the use of simulation in the search for the best possible scenario for companies. Emphasizing on improving the production process of this, employing a reasonable allocation of spaces from plant design to be more productive [5].

This article aims to develop the simulation model of the internal process of the company CIMACO SAS, using the ProModel software, where the construction of the current system in the software is proposed for the analysis of its productivity and in turn the identification of their problems. Subsequently, an improvement alternative is developed within the software to finally describe the comparison of these. Next, the company undergoes a process of analysis and decision-making in which the information provided with the model is validated and a decision is made regarding the execution of the proposal provided.

The procedure necessary for the development of the simulation model of the extraction and processing process of the CIMACO SAS quarry is as follows.

The simulation comprises an indispensable set of technological tools and methods for the successful implementation of digital manufacturing since it allows experimenta-tion and validation of the design and configuration of products, processes, and sys-tems. Especially in today's turbulent manufacturing environment, which is affected by megatrends such as globalization and the ever-increasing requirements for a greater degree of product customization and customization, the value of simulation is evident.

2. System

The quarry has multiple stations, transport vehicles, and heavy machinery of different types for various operations. One type is the excavators, they are in charge of extracting the raw material from the worked land, the company has four units of excavators of different sizes and brands such as CATERPILLAS and DOOSAN. Another type of machinery is the screens that are in charge of classifying the raw material between grading, crushing and foundation stone, the company has 2 TESAB brand screens. Another type of machinery is the loaders, these are in charge of handling the raw material between different stations, the company has three units of the brand VOLVO, LIUGONG and CATERPILLAR. The vehicles to transport the raw material inside and outside the company are dump trucks with bucketsof 15 mt2 capacity of multiple brands. And the central station of the company is the crusher that is in charge of processing and classifying the raw material to obtain the final products of the company such as dust, gravel, foundation stone, gravel, crushed ¾ and ½ inch.

2.1. Formulation of the model

Once the extraction and processing system of the quarry has been finalized, the flow diagram that fully represents the quarry system is proposed. In Figure 1, which corresponds to the flow diagram of the system, you can see the steps of the raw material extraction and processing system. The process that is carried out is as follows: The excavators are in charge of extracting the raw material from the ground in the extraction area and loading it onto the screens, then the filters are in charge of classifying the material, then a load collects the resulting material and It loads the dump trucks of the company or individuals, then the dump trucks of the company transport the material to the main area and unload the material, then another loader picks up the material and loads it to the crusher, and finally the crusher processes the material to obtain the resulting products.

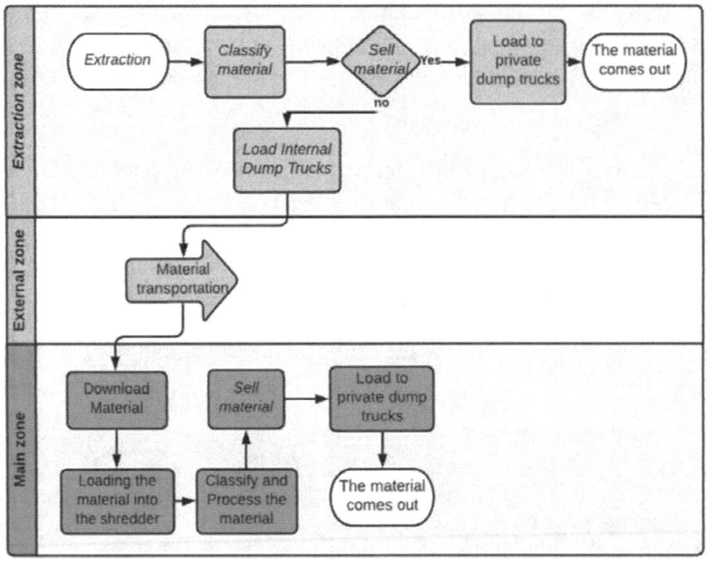

Figure 1. Flow diagram of the extraction and production system.

2.2. Data collection
The data collection of the duration of each step of the process was carried out directly in the work area. The time measurement was performed on a sample of 30 repetitions of each process and the mean and standard deviation of each process were identified. These times were implemented respectively in the model developed.

2.3. Implementation
The simulation of the mining-type raw material extraction and processing process begins with the analysis of the system, examining the interaction between different elements, seeking to create a model for decision making, with which the parameters can be modified and different scenarios. Said model can be developed by taking advantage of the computational capacity of specialized systems such as "ProModel, which is designed for modeling manufacturing systems ranging from small workshops to large-scale production machining cells" [6]. Within our model, we worked with different elements to represent the extraction and processing of raw material.

2.4. Entities
The first element defined as the entity called "Material" which represents a "Trip" of 15 m^3 since it is the unit of measure within the company and as can be seen in Figure 2., three graphs were assigned to it an entity to represent the changes that the material undergoes within the processing.

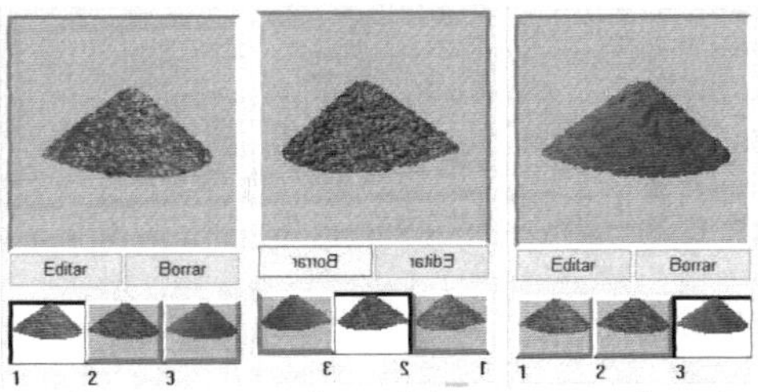

Figure 2. Graphic representation of the company's unit of measure.

2.5. Layout

Subsequently, the graphic representation where the entire model is worked was located, it can be seen in Figure 3. This layout was developed with images taken from the Google Maps digital platform and represented an area of 7.5 km^2.

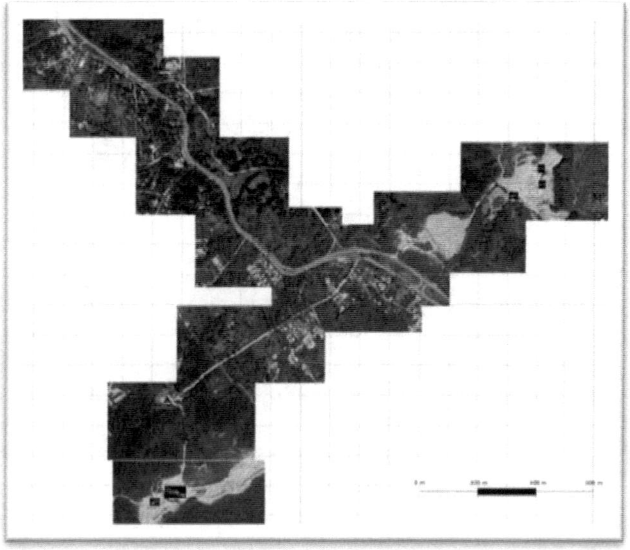

Figure 3. Layout of the study area.

The graphic representation of the model is divided into three main areas: The first area located in the lower-left part consists of the central area of the company where the office, the workshop, the dump truck parking lot and the shredder are located. The second zone located in the upper middle right part of the model that consists of the extraction zone. This area is where the raw material extraction area, the different screens and machinery parking are located. The last region of the model that makes up most of the model represents the entire path required for transportation between the two previous zones. This area is not part of the property of the company but instead constitutes a public road. Transportation is done through this area by dump trucks. The three zones can be seen in greater detail in Figure 3. Zone1 corresponds to the central region, zone 2 belongs to the extraction zone and zone 3 corresponds to the outer area of the company in greater detail in Figure 4.

IOP Conf. Series: Materials Science and Engineering 1154 (2021) 012030 doi:10.1088/1757-899X/1154/1/012030

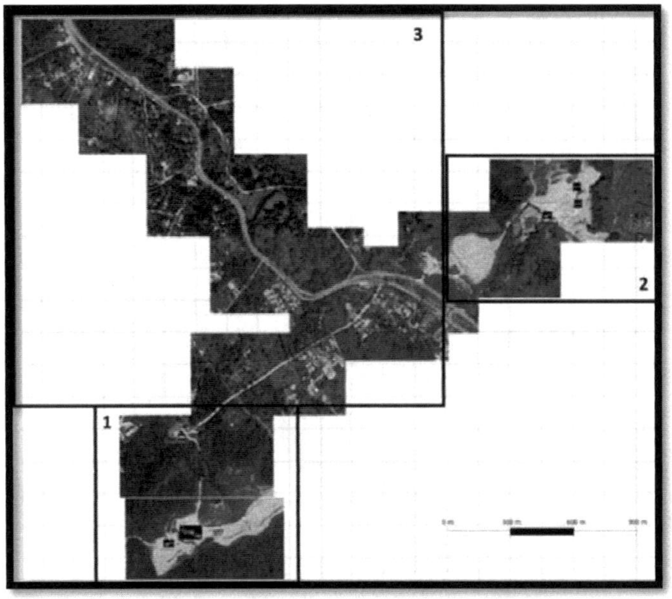

Figure 4. Layout with the different zones.

2.6. Locations

Afterward, he was given the step to define the different locations with their respective capacities and units. The areas were limited as follows, as shown in Table1.

Table 1. Locations.

Name	Capacity	Units
ZonaExtraccion	INF	1
ParqueaderoMaquinaria	4	1
ZonaCarga	INF	1
ParqueaderoVolquetas	10	1
ZonaDescarga	100	1
Trituradora	5	1
ZonaFinal	100	1
Zaranda1	3	1
Zaranda2	3	1

The placement of these locations in the different zones was as follows: The area of "ParqueaderoVolquetas" was located in the upper part of the central zone at point number 1, the locations of "ZonaDescarga", "Crituradora" and "ZonaFinal "Were located in the lower part of the main zone at point number 2, the locations of" Extraction Zone "," Zaranda1 "," Zaranda2 "," Loading Zone " and " ParkingMachinery "were located in the extraction zone at point number 3. You can see the points in greater detail in Figure 5.

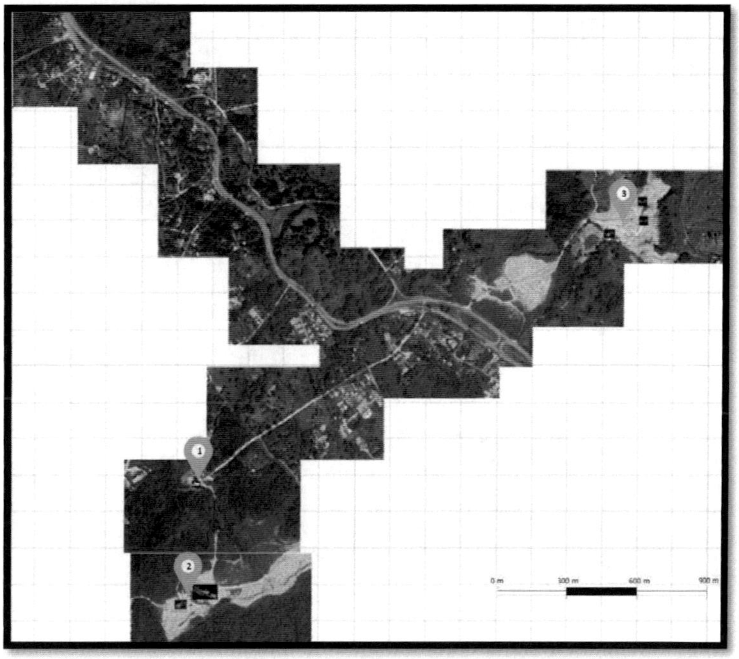

Figure 5. Layout with the different locations points.

Afterward, the input element of the entity "Material" was developed, as can be seen in table 2.

Table 2. Arrivals.

Entity	Location	Qty Per Arrival	Occurrences	Frequency
Material	*ZonaExtraccion*	1	INF	5

2.7. Route networks

Subsequently, the different routes where the resources worked with their respective nodes, routes and interfaces were developed. As can be seen in Table 3.

Table 3. Route networks.

Name	Routes	Interfaces	Nodes
RExcavadoras	4	4	4
RCargador	3	4	4
RCargador2	1	2	2
Rvolquetas	5	3	4

The route corresponding to "RVolquetas" is distributed throughout the model, as shown in Figure 6. And it was necessary to extend this route to the upper part since the dump trucks need to cross to connect the two areas of the company on the public highway and as it has two lanes in one direction, it is necessary to travel extensively to the nearest return that is in the upper part of figure.

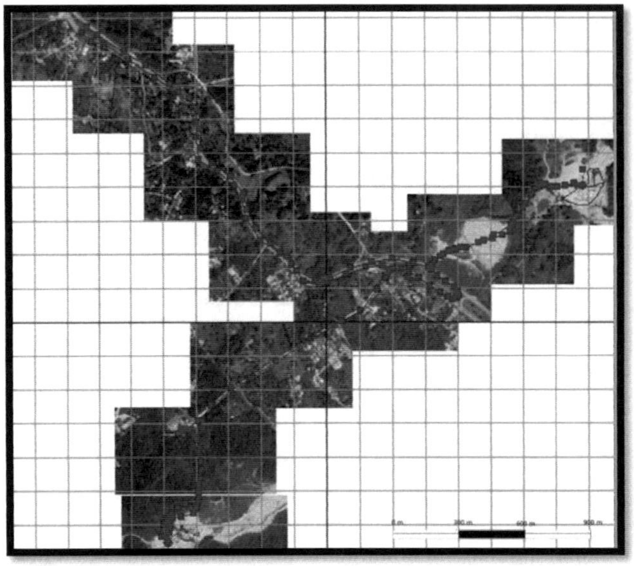

Figure 6. Route corresponding to "RVolquetas".

2.8. Resources

Subsequently, it was given step to establish the resources that will work in the already established route networks. These will be in charge of transporting the material between the different locations of the model. These resources were created with different specifications reflecting the current state of the company. In table 4 you can find the different resources.

Table 4. Resource

Name	Units	Speed without entity	Speed with entity
Excavadora	3	15	10
Cargador1	1	25	20
Cargador2	1	25	20
Volquetas	10	50	30

This difference in speeds is since as the entity "Material" represents 15 mt3 of quarry materials, this unit has a considerable weight for all these machines, which means that its transport activity is forced to be carried out with less speed when they count with the entity of "Material".

2.9. Processing

To finalize the main elements, it was given step to establish all the processing of the model that represents the behavior of the different stations with the entity and how the resources interact with the entity to perform the different transports within the model, all the processing can be found in more detail in table 5.

Table 5. Processing

Process			Routing				
Entity	**Location**	**Operation**	**Blk**	**Departure**	**Destination**	**Rule**	**Montion logic**
Material	*ZonaExtraccion*	- Wait N(20,3)	1	*Material*	*Zaranda1*	LU 1	- Move With Excavadoras Then Free
				Material	*Zaranda2*	LU	- Move With Excavadoras Then Free
Material	*Zaranda1*	- Wait N(5,1) - Graphic 2	1	*Material*	*ZonaCarga*	FIRST 1	- Move With Cargador1 Then Free
Material	*Zaranda2*	- Wait N(5,1) - Graphic 2	1	*Material*	*ZonaCarga*	FIRST 1	- Move With Cargador1 Then Free
Material	*ZonaCarga*	- Get Cargador1 And Volquetas - Wait N(10,3) - Free All	1	*Material*	*ZonaDescarga*	FIRST 1	- Move With Volquetas Then Free
Material	*ZonaDescarga*	- Wait N(5,2)	1	*Material*	*Trituradora*	FIRST 1	- Move With Cargador2 Then Free
Material	*Trituradora*	- Wait N(30,3) - Graphic 3	1	*Material*	*ZonaFinal*	FIRST 1	
Material	*ZonaFinal*	- Wait N(40,3)	1	*Material*	EXIT	FIRST 1	

With this, the entire model of the company's internal system is observed and by varying the parameters of the different elements, it will be possible to see how the results of the model will affect.

3. Analysis of the current situation

Subsequently, the statistical results on the entity worked are interpreted, shown by the model, shown in Table 6.

Table 6. Results.

Name	Total departures	Current quantity in system	Time in system	Time in motion logic	Waiting time	Time in operation	Lock time
Material	22	62	246,02	95,02	30,00	109,45	11,56

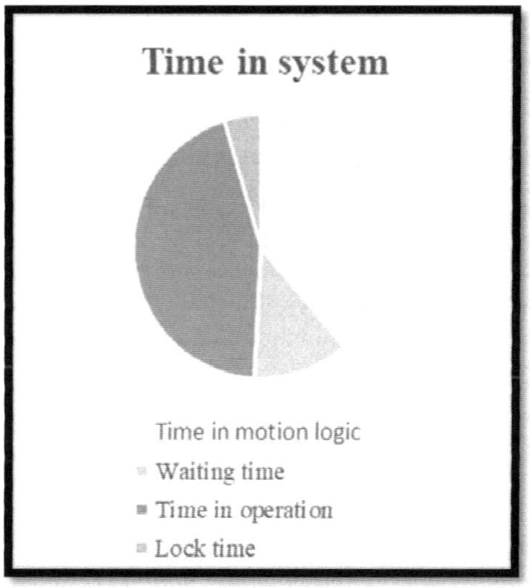

Figure 7. Baseline's pie chart.

According to the results obtained, it is possible to interpret that a large amount of time within the system is covered by the movement logic representing 38.6% as seen in Figure 7. This is identified as a bottleneck within the system since it causes the quantity within the system to be greater than the total output as shown in Table 5.

4. Improvement proposal

Taking into account the properties of the land owned by the company, it is identified that the extraction area is going to be the work area for the next 50 years, it is proposed to add a crusher in that area, in order to reduce the time in motion logic. But the current crusher cannot be neglected or moved, so a system is proposed where the two crushers work simultaneously in the different areas of the company. You can see a grass flow with the improvement implemented in Figure 8.

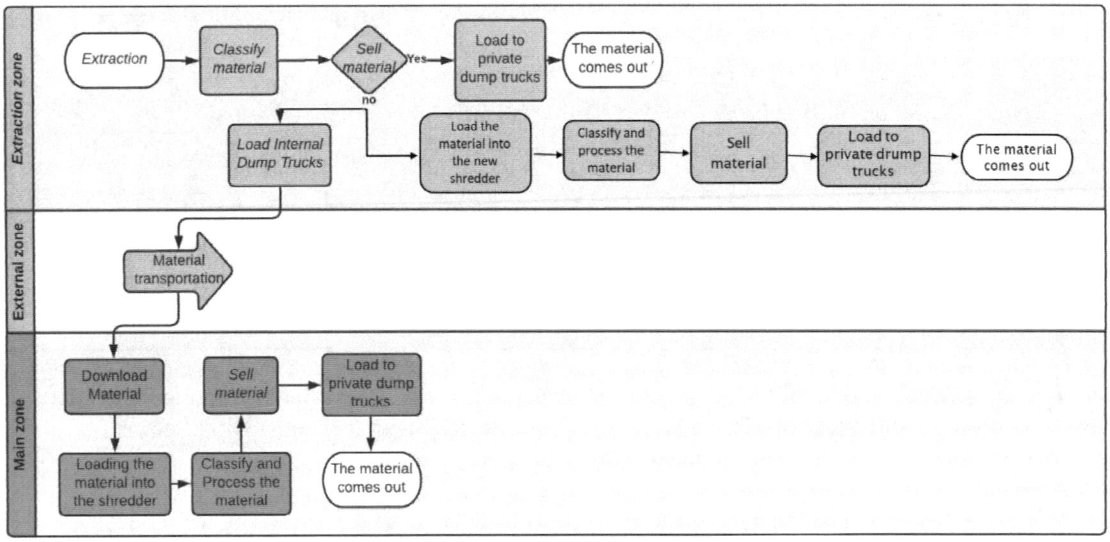

Figure 8. Proposed flow chart.

5. Implementation

First, the new locations are added in the extraction area with their respective capacities and units as shown in Table 7.

Table 7. Added locations

Name	Capacity	Units
Trituradora_2	5	1
ZonaFinal2	100	1

Subsequently, the route network "RCloader" was modified by adding the route for the new locations with their respective interfaces, later in the processing the registry of the "Loading Zone" was changed to implement the decision policy with priority in the new added crusher and also The records were added for the locations of "Trituradora_2" and "ZonaFinal2". Finally, the optimal number of units of the "Volquetas" resource is determined, for which a parameter is established in the units of said resource called "Quantity of Volquetas", which has a range from 3 to 10 units. Afterward, the SIMRUNNER statistical analysis tool is used, which is conFigure d with said parameter established for experiments lasting 7 hours. From this, it is possible to obtain the optimal solution of 6 units for the number of dump trucks, and in turn, developing the scenario with the said result.

Subsequently, the statistical results on the entity worked are deduced from the model of the improvement proposal, shown in Table 8.

Table 8. Results of the improvement proposal

Name	Total departures	Current quantity in system	Time in system	Time in motion logic	Waiting time	Time in operation	Lock time
Material	40	44	167,99	63,77	0,00	96,08	8,14

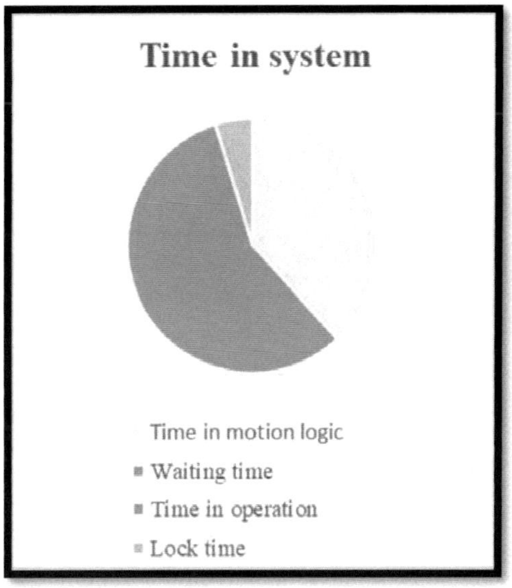

Figure 9. Entity pie chart of the improvement proposal

It can be seen that the results were obtained from the proposed alternative (see table 10.) in comparison with the results obtained by the model of the current situation (see Table 6.) a significant decrease of 78.03 minutes in the time in the average system of the entity and also a considerable increase of 18 units produced in a time of 7 hours.

6. Conclusion

In the results shown, a definite improvement in the production capacity of the materials in the system can be seen with the proposed alternative, as evidenced above, and this information provided is beneficial in the following processes in the execution of company decisions. In addition, the great utility in decision-making is demonstrated thanks to the simulation of all kinds of situations in the ProModel tool with which statistical analysis is obtained quickly.

7. References

[1] Dunna, E. G., Reyes, H. G., & Barrón, L. E. C. (2006). *Simulación y análisis de sistemas con ProModel*. Pearson Educación.
[2] W Septiani et al 2020 IOP Conf. Ser.: Mater. Sci. Eng. 847 012054.
[3] Muthanna Jamil and Noraini Mohd Razali 2016 IOP Conf. Ser.: Mater. Sci. Eng. 114 012049.
[4] J Fernandez 2015 J. Phys.: Conf. Ser. 622 012002-

IOP Conf. Series: Materials Science and Engineering 1154 (2021) 012030 doi:10.1088/1757-899X/1154/1/012030

[5] Orozco, E., Cervera, J.: Design and Installations Distributions of Industrial Facilities Supported by the Use of Process Simulation Diseño y Distribución de Instalaciones In-dustriales apoyado en el uso de la Simulación de Procesos," Investig. Innov. e Ing., vol. 1, pp. 6–12, (2014).

[6] Harrell, C. y Price R. (2003). Simulation modeling using ProModel technology. Proceedings of the 2003 Winter Simulation Conference.

[7] D Suhardini and S D Rahmawati 2019 IOP Conf. Ser.: Mater. Sci. Eng. 528 012062

[8] Sushma Rani et al 2019 J. Phys.: Conf. Ser. 1240 012002

Generation and Capture of Electric Energy Using Piezoelectric Materials: A Review

A D Rincón-Quintero[1,2], C L Sandoval-Rodriguez[1,3], N D Zanabria-Ortigoza[1], C G Cárdenas-Arias[1,4], J G Ascanio-Villabona[1,2] and M A Durán-Sarmiento[1,2].

[1] Design and materials research group (DIMAT), Automation and Control Energy Systems Research Group (GISEAC), Electromechanical engineering, Faculty of Natural Sciences and Engineering, Unidades Tecnológicas de Santander (UTS), Calle de los estudiantes No 9-82, 680005, Bucaramanga, Colombia.

[2] Doctoral Program in Energy Efficiency and Sustainability in Engineering and Architecture, Department of Machines and Thermal Engines, University of the Basque Country, UPV/EHU, 48013, Bilbao, Spain.

[3] Doctoral Program in Electronic and Communications, Department of Communications Engineering, University of the Basque Country, UPV/EHU, 48013 Bilbao, Spain.

[4] Doctoral Program in Engineering of Materials and Sustainable Processes, University of the Basque Country, UPV/EHU, 48013, Bilbao, Spain.

arincon@correo.uts.edu.co

Abstract. Piezoelectricity comes as a principle of transformation of mechanical energy into electrical energy, it is limited in terms of investment, time and research, due to this fact, the need arises to be able to innovate with data collection on the most used models to collect and generate electrical energy. In this research, the literature regarding the generation and collection of electrical energy using piezoelectric materials was analyzed, from this analysis, fifty innovative articles were determined in the last three years, which were reflected in a data matrix, in which It presents the generation and collection element, the applications, and shows the performance in terms of power or voltage that the prototypes supply. With the results of the table, a condensed panorama of current data is obtained, about the most used and outstanding of this form of little-used energy, but which is a competent and efficient alternative for the generation of electricity.

1. Introduction

Electric energy is currently used in order to develop any service in daily life, which ends up being more environmentally friendly, profitable when using energy in any work field and a priority for the development of many technologies in history [1][2][3].

The point is that, when wanting to generate electrical energy, there are not many energy devices that make the most of their configuration, or generation processes that are not polluting with the environment. Piezoelectricity, as the principle of producing electrical energy under a force that is applied in a certain area, is an innovative option for the generation of renewable energy, since it has a wide range of materials, such as piezoelectric transducers, for various availabilities, offering a good potential differential when used [4][5]

[6]. For this reason, it is necessary to carry out this research, to expose through research in the literature regarding the generation and collection of electrical energy, various sustainable models that, using piezoelectric materials, can provide efficient solutions for this form of energy, little used, clean and renewable. Also, it is very important to contribute this study to subsequent projects in the same framework, so that they have current antecedents of what was done, and thus further improve the efficiency obtained in the model that is being sought [7][8][9][10].

The objective of this article is to base research on the literature regarding the generation and collection of electrical energy using piezoelectric materials, and then make a selection of the fifty most outstanding technical-scientific articles in the last three years. The authors of the different sections focus mainly on selecting a difficulty in some field, when wanting to supply electrical energy to a process or equipment that needs it, but always prioritizing the use of piezoelectrics for this service. Some highlight the use of piezoelectric materials, such as Lead Zirconate Titanate (PZT) and Polyvinylidene Fluoride (PVDF), since due to their properties they are truly efficient in producing electrical power; Other authors emphasize that the use of these elements are polluting with the environment, since they contain Lead, deducing that it is necessary to carry out tests on other materials that allow obtaining an efficiency equal to or greater than those commonly used [11][12][13][14][15].

This review presents a methodological process that consists of the creation of a data matrix focused on the technological models already investigated, in order to summarize and visualize the analysis made. The table will show the most prominent devices or prototypes by the authors of the inquiries, highlighting the piezoelectric material used in each of them. Also, the application in which the prototype takes place and in which service it will be developed will be defined, in order to later determine how much each model fulfilled in the generation of this clean energy, making known the active voltage or power output that was reached to provide. Finally, an analysis of the results will be carried out with the creation of graphs to extract the information from the previous table, and thus be able to visualize the investigated models more easily [16][17][18].

2. Work development

2.1 Data matrix

To detail the investigation of the fifty technical-scientific articles, the elements, applications and performance of the investigated models within the framework of generation and collection of electrical energy are illustrated through a data matrix, using piezoelectric materials as a solution for transform mechanical energy into electrical energy through vibrations, in order to power a selected device or process. Next, it is presented in Table 1, of the aforementioned.

Table 1. Data matrix on the technical-scientific articles investigated.

Articles	Piezoelectric element generator and energy collector	Application	Performance
[16][17][4][5] [19][20][21][22][23][1][2][24][25][26][27] [28][29][30][31][32][33][34] [35][36][37][38][39][40][41] [42][43][44]	Energy harvesting devices based on Lead Zirconate (PZT) piezoelectric ceramic transducers, such as cantilever beam, springs or patches, are presented. Also, the PZT-5X, which is made up of hardened steel and ceramic, is used in two prototypes. A device is manufactured with the use of PZT-5A material, excellent in fields where the temperature is high and variable.	Installations are found in offshore buoys, asphalt pavements, tank pipes, buildings, broadband environments, in tanks that obtain chemical reactions and the human body. This at the rate of powering implantable medical electronic elements, LED lights and portable devices such as low-power wireless sensors.	Innovative piezoelectric energy collector designs are presented, reaching between 4.91 μW and 2.09 kW / day, as output power. Also, in terms of potential differential, between 400 mV and 20 V/s can be discovered.
[7][8][10][45] [46][47][48][49][50][51][52]	The prototypes present in these eleven articles consist of piezoelectric material in Polyvinylidene Fluoride (PVDF), either in folded sheets, as a cantilever, in the form of patches or more commonly used as a nanogenerator.	There are installations of the prototypes on local roads, bridges for vehicles or trains, tires for automobiles or bicycles, pipes with constant fluid flow and also, in replacement of batteries. This to supply the power supply of wireless sensors, LED diodes and microelectronic systems.	Optimum energy harvesting performance is obtained in the different designs, finding electrical powers between 1.2 μW and 16.5 W. Also, in terms of voltage, there are values between 29 mV and 7 V.
[11][13][14]	The authors of these three articles expose prototypes composed of piezoelectric macro fiber material (MFC), which is made up of rectangular PZT 5A fibers and are embedded with copper electrodes, in order to generate the necessary electrical energy. They are visualized in the form of a cantilever or as patches.	Installations of energy harvesting devices in wind generators, air flow pipes and in the human body are exposed. To achieve constant power to wireless sensors and portable mobile electronic devices.	Optimal electrical power generation is achieved using piezoelectric transducers with an output power between 700 μW and 13.6 mW.
[53]	A high performance piezoelectric elastic nanogenerator is manufactured, composed of piezoelectric nanowires (NW) with Zinc Oxide (ZnO). It also consists of an AC / DC converter and a DC/DC converter, for rectification of the transformed signal from environmental vibrations.	It is installed in the human body so that when exerting a movement, it generates electrical energy to provide self-powered touch sensors.	It features excellent performance with a maximum output power of 35 μW.
[54]	Device in the form of a tile, with sheets of substrates. Mainly, it is composed of a flexible, thin film, chemically stable and biocompatible piezoelectric material of Group III Nitride (III-N).	Implementation in tiles to power light-emitting diodes, and to charge commercial capacitors and batteries that operate optical pulse sensors.	It complies with the transformation of mechanical energy into electrical energy, collecting an active output power of up to 167 μW.
[55]	The prototype is based on the use of piezoelectric materials of Perovskite Oxide in sheet form; Also, it consists of a rechargeable battery and a rectifier. Everything named is located superficially on the heart to transform the heart rhythm into electrical energy.	It is established in portable electronic devices or implantable biomedical devices for a continuous feeding of the same.	It show excellent electrical properties, exhibiting good flexibility in generating electrical power up to 0.3V.
[56]	Power generator prototype located on the model; Composed of flexible piezoelectric material with thin films of Poly (Vinylidene Fluoride/Trifluoroethylene) P(VDF/TrFE)] ferroelectric polymer.	Device implanted in the heart, such as pacemakers, to obtain self-powered sensors implantable in the human body.	Power generation is obtained but with minimum voltages of 3V; recommending optimizing the capacity of the piezoelectric.

Source: Authors.

In the matrix produced, the most relevant research and applications of the fifty technical-scientific articles are condensed. The piezoelectric material, PZT, is used in 64% of the investigated sections, emphasizing its excellent piezoelectric properties and its efficiency in applications, reaching up to 2.09 kW/day in incrustations within roads. Also, 22% of the authors of technological articles present prototypes that mainly contain PVDF material, emphasizing its flexibility and effectiveness, reaching up to 16.5 W, in local road installations.

At the end, other types of piezoelectric materials that have fewer applications in recent years are presented, but this does not make them less important, since the authors of the sections work to eradicate the use of Lead, being a highly polluting chemical element environmentally friendly, but still widely used in piezoelectricity applications over time.

It is important to expose through research in the literature regarding the generation and collection of electrical energy, various sustainable models using piezoelectric materials, in order to provide efficient solutions for this form of energy, little used, clean and renewable, which is and will be a competent alternative for electricity production.

2.2 Analysis of results

For the generation and collection of electrical energy, it is very important to identify the piezoelectric element used, since it helps the transformation of mechanical energy by exerting a force into electrical energy through vibrations.

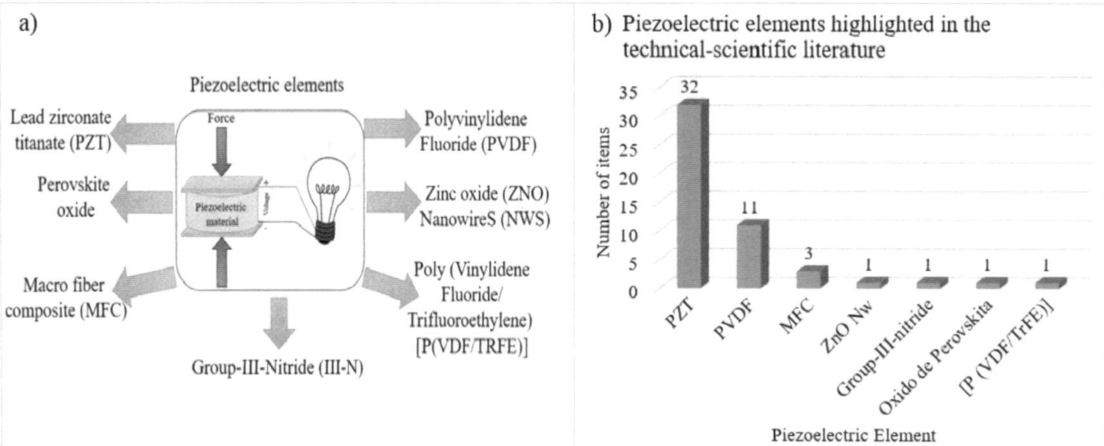

Figure *1.* Piezoelectric elements highlighted in the investigated technical-scientific literature. a) Identification of piezoelectricity and elements used in recent years; b) Change in the number of investigated articles depending on piecoelectric materials.

Source: Authors.

In Figure 1, the different piezoelectric materials used in the investigated sections can be observed, it is identified that Lead Zirconate Titanate is the most used at a rate of thirty-two articles, highlighting its flexibility and efficiency in applications. Polyvinylidene Fluoride (PVDF) was used in eleven articles, which although it did not produce power as in the PZT road application, it manages to generate 16.9 W in this same adaptation, complying with the collection of electrical energy. Finally, the MFC elements can be observed with three articles that use it and the piezoelectric nano cables (NW) with Zinc Oxide (ZnO), Group III Nitride (III-N), Perovskite Oxide and [P (VDF/TrFE)], which only have one application, but it does not make them less important, since they provide elements without Lead, this being very polluting with the environment.

The most prominent places for the installation of piezoelectric generators are illustrated below, which were found in the fifty articles investigated.

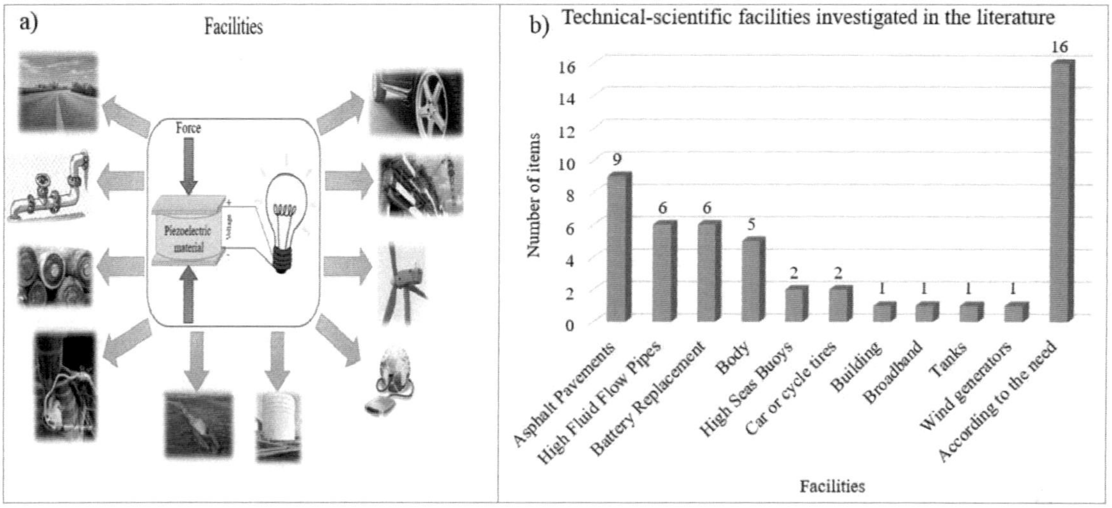

Figure *2*. Installation of piezoelectric generators featured in the technical-scientific literature investigated. a) Identification of piezoelectricity and facilities used in recent years; b) Change in the number of articles investigated depending on the instances.

Source: Authors.

In Figure 2, the different installations of piezoelectric generators in the last three years are observed. The technological development has been quite extensive, identifying 18% of the articles investigated from installations on local roads, 12% in high-flow pipes and the same, to achieve the replacement of conventional batteries. It is illustrated, the incursion of piezoelectricity into the human body in 10%, also, 4% in level buoys that use the waves of the sea as their main force and the same percentage for their placement on automobile or bicycle tires. Likewise, they are obtained from a single item in installations in buildings, broadband, tanks and electric generators. Finally, with 32%, there is the arrangement of piezoelectric generators in fields where it is required to produce energy to power a low-power device, thus taking advantage of any mechanical energy that is had.

It is illustrated below, the outstanding devices of the investigation in the articles, which will be fed by the electrical energy produced from the use of piezoelectric materials which first take some external force to transform it into the required energy.

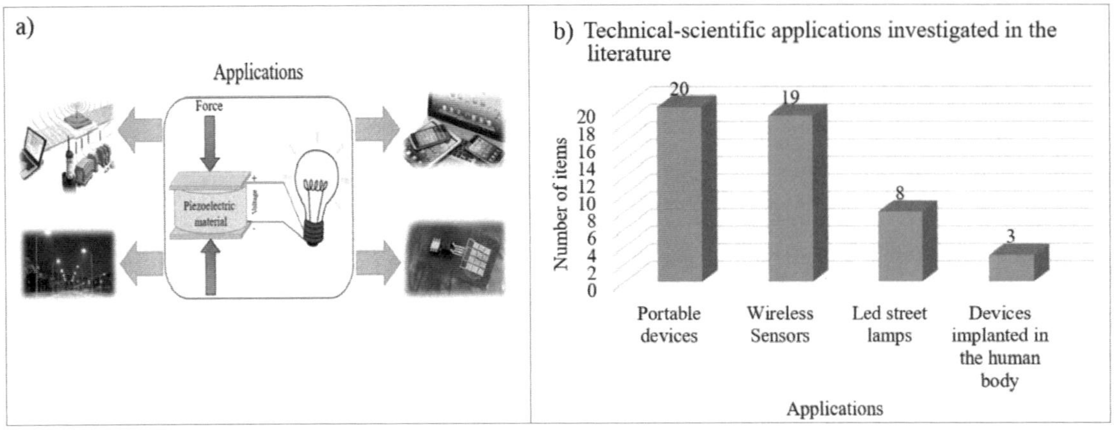

Figure 3. Applications of piezoelectric generators featured in the investigated technical-scientific literature. a) Identification of piezoelectricity and applications used in recent years; b) Change in the number of articles investigated depending on the applications.

Source: The authors.

In Figure 3, four types of applications found in all the investigated sections are identified, obtaining that twenty of them used the electrical energy generated to power low-power portable electrical devices, nineteen of them were applied to promote the sustainable and durable self-powering of wireless sensors. With eight articles, applications in Led street lamps are discovered, and with three the innovative application of devices implanted in the human body is detected, for the replacement of batteries, opening a wide panorama, in subsequent investigations within the same field.

3. Conclusions

It was presented the feasibility of carrying out a technical-scientific evaluation of applications in the framework of generating and collecting electrical energy using piezoelectric materials, obtaining as a result a matrix with a wide panorama of data on the most used and outstanding of this little-used form of energy, but that it is a competent and efficient alternative for generating electricity.

From a table of data on the research in the fifty models destined to generate and collect electrical energy, it is obtained that the main device to carry out the investigations were prototypes made by the authors of the same, since the few suppliers of piezoelectric devices Existing elements make these elements very expensive and a good economic viability of the service of this kind of technology is not obtained.

In the applications of the service to be supplied, their use in low-power electronic devices is appreciated in all the technical-scientific models, seeking their self-supply in implantations of the human body, wireless sensors or portable devices in a specific field.

Finally, it can be clarified that all the models fully meet their objectives, although some generate more than what is established in the sections and others lack very little to fulfill their purpose. From this, it can be inferred that it is very important to contribute this research to subsequent projects in the same framework, so that they have current antecedents of what was done, and thus, further improve the efficiency obtained in the model that is being sought.

References

[1] Y. Hu, B. Yang, X. Chen, X. Wang, and J. Liu, "Modeling and experimental study of a

piezoelectric energy harvester from vortex shedding-induced vibration," *Energy Convers. Manag.*, vol. 162, pp. 145–158, 2018.

[2] L. Nyamayoka, L. Zhang, and X. Xia, "Feasibility study of embedded piezoelectric generator system on a highway for street lights electrification," *Energy Procedia*, vol. 152, pp. 1015–1020, 2018.

[3] A. D. Rincón-Quintero *et al.*, "Manufacture of hybrid pieces using recycled R-PET, polypropylene PP and cocoa pod husks ash CPHA, by pneumatic injection controlled with LabVIEW Software and Arduino Hardware," *IOP Conf. Ser. Mater. Sci. Eng.*, vol. 844, no. 1, 2020.

[4] A. Jasim, G. Yesner, H. Wang, A. Safari, A. Maher, and B. Basily, "Laboratory testing and numerical simulation of piezoelectric energy harvester for roadway applications," *Appl. Energy*, vol. 224, pp. 438–447, 2018.

[5] M. Rajarathinam and S. Ali, "Energy generation in a hybrid harvester under harmonic excitation," *Energy Convers. Manag.*, vol. 155, pp. 10–19, 2018.

[6] B. E. Romero-Tarazona, C. L. Rodriguez-Sandoval, J. G. Villabonai-Ascanio, and A. D. Rincón-Quintero, "Development of an artificial vision system that allows non-destructive testing on flat concrete slabs for surface crack detection by processing of digital images in MATLAB," *IOP Conf. Ser. Mater. Sci. Eng.*, vol. 844, no. 1, 2020.

[7] S. Youn *et al.*, "Piezoelectric polymer-based roadway energy harvesting via displacement amplification module," *Appl. Energy*, vol. 216, pp. 741–750, 2018.

[8] P. Cahill, B. Hazra, R. Karoumi, A. Mathewson, and V. Pakrashi, "Vibration energy harvesting based monitoring of an operational bridge undergoing forced vibration and train passage," *Mech. Syst. Signal Process.*, vol. 106, pp. 265–283, 2018.

[9] J. G. Ascanio-Villabona, C. L. Sandoval-Rodriguez, A. D. Rincón-Quintero, B. E. Tarazona-Romero, and R. E. Paez-Castro, "Building a prototype for functional analysis of the energy potential of the water flow in pipe 1/2 "using microturbines applied to Unidades Tecnológicas de Santander," *IOP Conf. Ser. Mater. Sci. Eng.*, vol. 844, no. 1, pp. 0–26, 2020.

[10] D. Maurya *et al.*, "Energy harvesting and strain sensing in smart tire for next generation autonomous vehicles," *Appl. Energy*, vol. 232, pp. 312–322, 2018.

[11] R. Hua, H. Liu, H. Yang, Y. Wang, and J. Ferrante, "A nonlinear interface integrated lever mechanism for piezoelectric footstep energy harvesting," *Appl. Phys. Lett.*, vol. 113, no. 5, 2018.

[12] C. L. Sandoval-Rodriguez, J. G. Ascanio Villabona, C. G. Cárdenas-Arias, A. D. Rincon-Quintero, and B. E. Tarazona-Romero, "Characterization of the mechanical vibration signals associated with unbalance and misalignment in rotating machines, using the cepstrum transformation and the principal component analysis," *IOP Conf. Ser. Mater. Sci. Eng.*, vol. 844, no. 1, 2020.

[13] J. Ramirez, "Desarrollo de dispositivos recolectores de energia de fuentes vibratorias," Universidad Nacional del Sur, 2019.

[14] J. Wang, S. Zhou, Z. Zhang, and D. Yurchenko, "High-performance piezoelectric wind energy harvester with Y-shaped attachments," *ELSEVIER*, vol. 181, pp. 645–652, 2019.

[15] C. G. Cárdenas-Arias, A. D. Rincón-Quintero, A. Santos-Jaimes, C. L. Sandoval-Rodriguez, D. F. Rojas-Gomez, and S. J. Ardila-Galvis, "Elasticity modulus variation of the AISI SAE 1045 steel

subjected to corrosion process by chloride using tension test destructive," *IOP Conf. Ser. Mater. Sci. Eng.*, vol. 844, no. 1, 2020.

[16] S. Nabavi, A. Farshidianfar, and A. Afsharfard, "Novel piezoelectric-based ocean wave energy harvesting from offshore buoys," *Aplied Ocean Res.*, vol. 76, pp. 174–183, 2018.

[17] C. Wang, S. Wang, Q. Li, X. Wang, Z. Gao, and L. Zhang, "Fabrication and performance of a power generation device based on stacked piezoelectric energy-harvesting units for pavements," *Energy Convers. Manag.*, vol. 163, pp. 196–207, 2018.

[18] J. G. Ascanio-Villabona, A. D. Rincón-Quintero, C. G. Cardenas-Arias, and C. L. Sandoval-Rodriguez, "Incidence of corrosion in low voltage electrical conductor," *IOP Conf. Ser. Mater. Sci. Eng.*, vol. 844, no. 1, 2020.

[19] D. Wang, J. Mo, X. Wang, H. Ouyang, and Z. Zhou, "Experimental and numerical investigations of the piezoelectric energy harvesting via friction-induced vibration," *Energy Convers. Manag.*, vol. 171, pp. 1134–1149, 2018.

[20] Y. Wu, J. Qiu, S. Zhou, H. Ji, Y. Chen, and S. Li, "A piezoelectric spring pendulum oscillator used for multi-directional and ultra-low frequency vibration energy harvesting," *Appl. Energy*, vol. 231, pp. 600–614, 2018.

[21] Z. Hong, Z. Wen, L. Wen, W. Ke, H. Kai, and G. Meng, "Magnetically coupled flextensional transducer for wideband vibration energy harvesting: Design, modeling and experiments," *J. Sound Vib.*, vol. 416, pp. 55–79, 2018.

[22] G. Wang, L. Wei, B. Yang, X. Wang, W. Xu, and X. Li, "Dynamic and energetic characteristics of a bistable piezoelectric vibration energy harvester with an elastic magnifier," *Mech. Syst. Signal Process.*, vol. 105, pp. 427–446, 2018.

[23] S. Dhote, Z. Yang, and J. Zu, "Modeling and experimental parametric study of a tri-leg compliant orthoplanar spring based multi-mode piezoelectric energy harvester," *Mech. Syst. Signal Process.*, vol. 98, pp. 268–280, 2018.

[24] S. Ju and J. Chang, "Impact-based piezoelectric vibration energy harvester," *ELSEVIER*, vol. 214, pp. 139–151, 2018.

[25] K. Kyu, C. Si, K. Hyun, and S. Kyu, "Wave power generation by piezoelectric sensor attached to a coastal structure," *J. Sensores*, pp. 1–7, 2018.

[26] M. Safaei, M. Meneghini, and S. Anton, "Energy Harvesting and Sensing With Embedded Piezoelectric Ceramics in Knee Implants," *IEEE/ASME Trans. Mechatronics*, vol. 23, no. 2, pp. 864–874, 2018.

[27] Z. Xie, K. Kwuimy, Z. Wang, and W. Huang, "A piezoelectric energy harvester for broadband rotational excitation using buckled beam," *API Adv.*, vol. 8, no. 1, 2018.

[28] J. Chen, Q. Qiu, Y. Han, and D. Lau, "Piezoelectric materials for sustainable building structures: Fundamentals and applications," *Renew. Sustain. Energy Rev.*, vol. 101, pp. 14–25, 2019.

[29] F. Ali, W. Raza, X. Li, H. Gul, and K.-H. Kim, "Piezoelectric energy harvesters for biomedical applications," *Nano Energy*, vol. 57, pp. 879–902, 2019.

[30] M. Hamlehdar, A. Kasaeian, and M. Safaei, "Energy harvesting from fluid flow using

piezoelectrics: A critical review," *Renew. Energy*, vol. 143, pp. 1826–1838, 2019.

[31] G. Yesner, A. Jasim, H. Wang, B. Basily, A. Maher, and A. Safari, "Energy harvesting and evaluation of a novel piezoelectric bridge transducer," *Sensors Actuators A Phys.*, vol. 285, pp. 348–354, 2019.

[32] L. Guo and Q. Lu, "Numerical analysis of a new piezoelectric-based energy harvesting pavement system: Lessons from laboratory-based and field-based simulations," *Appl. Energy*, vol. 235, pp. 963–977, 2019.

[33] C. Wang, Z. Song, Z. Gao, G. Yu, and S. Wang, "Preparation and performance research of stacked piezoelectric energy-harvesting units for pavements," *Energy Build.*, vol. 183, pp. 581–591, 2019.

[34] J. Zhang and L. Qin, "A tunable frequency up-conversion wideband piezoelectric vibration energy harvester for low-frequency variable environment using a novel impact- and rope-driven hybrid mechanism," *Appl. Energy*, vol. 240, pp. 26–34, 2019.

[35] Z. Wu and Q. Xu, "Design and testing of a novel bidirectional energy harvester with single piezoelectric stack," *Mech. Syst. Signal Process.*, vol. 122, pp. 139–151, 2019.

[36] L. Min, K. Chang, P. Woon, C. Jeong, P. Jong, and Y. Jeong, "Energy harvesting performance of unimorph piezoelectric cantilever generator using interdigitated electrode lead zirconate titanate laminate," *Energy*, vol. 179, pp. 373–382, 2019.

[37] J. Chen, N. Nabulsi, W. Wang, J. Kim, M.-K. Kwon, and J.-H. Ryou, "Output characteristics of thin-film flexible piezoelectric generators: A numerical and experimental investigation," *Appl. Energy*, vol. 255, 2019.

[38] J. Scornec, B. Guiffard, R. Seveno, and V. Cam, "Frequency tunable, flexible and low cost piezoelectric micro-generator for energy harvesting," *Sensors Actuators A Phys.*, vol. 312, 2020.

[39] L. Dong *et al.*, "Cardiac energy harvesting and sensing based on piezoelectric and triboelectric designs," *Nano Energy*, vol. 76, 2020.

[40] Z. Zhang, S. Wang, J. Kan, W. Hu, Z. Chen, and H. Xu, "A pneumatic piezoelectric vibration energy harvester based on the compressed air-transducer-structure interaction," *Energy Convers. Manag.*, vol. 213, 2020.

[41] S. Velasco and F. Lopez, "Generadores de electricidad bioinorgánico. Conversión de energía química renovable a través de materiales piezoeléctricos funcionalizados con enzimas," *Tecnol. Cienc. y Educ.*, vol. 15, pp. 125–139, 2020.

[42] M. Beyaz, F. Tat, K. Ozkaya, and R. Ozbek, "Hybrid Magnetic-Piezoelectric Energy Harvester for Power Generation around Waistline During Gait," *J. Electr. Eng. Technol.*, vol. 15, pp. 227–233, 2020.

[43] C. Lesmes, J. Perez, and D. Juez, "Generadores Piezoeléctricos como Alternativa Energética: Evaluación de Factibilidad en la Implementación en Vías en Ciudades Intermedias," *L'esprit Ingénieux*, vol. 9 (1), pp. 69–82, 2020.

[44] L. Wang, L. Zhao, G. Luo, Z. Jiang, and R. Maeda, "System level design of wireless sensor node powered by piezoelectric vibration energy harvesting," *Sensors Actuators A Phys.*, vol. 310, 2020.

[45] J. Fu, Y. Hou, X. Gao, M. Zheng, and M. Zhu, "Highly durable piezoelectric energy harvester

based on a PVDF flexible nanocomposite filled with oriented BaTi2O5 nanorods with high power density," *Nano Energy*, vol. 52, pp. 391–401, 2018.

[46] S. Antiopi and A. Kalfas, "Experimental investigation of energy harvesting from swirling flows using a piezoelectric film transducer," *Energy Convers. Manag.*, vol. 171, pp. 1405–1415, 2018.

[47] D. Hu, M. Yao, Y. Fan, C. Ma, M. Fan, and M. Liu, "Strategies to achieve high performance piezoelectric nanogenerators," *Nano Energy*, vol. 55, pp. 288–304, 2019.

[48] J. Yan *et al.*, "Performance enhancements in poly (vinylidene fluoride)-based piezoelectric nanogenerators for efficient energy harvesting," *Nano Energy*, vol. 56, pp. 662–692, 2019.

[49] S. Kumar, H. Singh, and N. Khare, "Flexible hybrid piezoelectric-thermoelectric generator for harnessing electrical energy from mechanical and thermal energy," *Energy Convers. Manag.*, vol. 198, 2019.

[50] Y. Hwang, S. Choi, and H. Seong, "Highly flexible all-nonwoven piezoelectric generators based on electrospun poly (vinylidene fluoride)," *Sensors Actuators A Phys.*, vol. 300, 2019.

[51] S. Chauhan, U. Bhatt, P. Gautam, S. Thote, M. Joglekar, and S. Manhas, "Fabrication and modeling of β-phase PVDF-TrFE based flexible piezoelectric energy harvester," *Sensors Actuators A Phys.*, vol. 304, 2020.

[52] S. Yayla, S. Ayca, and M. Oruc, "A case study on piezoelectric energy harvesting with using vortex generator plate modeling for fluids," *Renew. Energy*, vol. 157, pp. 1243–1253, 2020.

[53] G. Poulin-Vittrant *et al.*, "Challenges of low-temperature synthesized ZnO nanostructures and their integration into nano-systems," *Mater. Sci. Semicond. Process.*, vol. 91, pp. 404–408, 2019.

[54] J. Chen, S. Kyu, N. Nabulsi, H. Johnson, W. Wang, and J.-H. Ryou, "Biocompatible and sustainable power supply for self-powered wearable and implantable electronics using III-nitride thin-film-based flexible piezoelectric generator," *Nano Energy*, vol. 57, pp. 670–679, 2019.

[55] W. Gao, Y. Zhu, Y. Wang, G. Yuan, and J.-M. Liu, "A review of flexible perovskite oxide ferroelectric films and their application," *J. Mater.*, vol. 6, no. 1, pp. 1–16, 2020.

[56] Y. Kondo, S. Horike, Y. Koshiba, T. Fukushima, and K. Ishida, "Directly monitoring and power generation from pulsating 3D heart model with organic flexible piezoelectric device," *Jpn. J. Appl. Phys.*, vol. 59, 2020.

IOP Conf. Series: Materials Science and Engineering 1154 (2021) 012032 doi:10.1088/1757-899X/1154/1/012032

A bibliometric analysis of the uses of the cocoa pod shell

A D Rincón-Quintero[1,2], L A Del Portillo-Valdés[2], C G Cárdenas-Arias[1,4], B E Tarazona-Romero[1,2], W L Rondón-Romero[1] and M A Durán-Sarmiento[1,2]

[1] Design and materials research group (DIMAT), Automation and Control Energy Systems Research Group (GISEAC), Electromechanical engineering, Faculty of Natural Sciences and Engineering, Technological Units of Santander (UTS), Calle de los estudiantes No 9-82, 680005, Bucaramanga, Colombia.
[2] Energy in Building research group (ENEDI), Doctoral Program in Energy Efficiency and Sustainability in Engineering and Architecture, Department of Machines and Thermal Engines, University of the Basque Country, UPV/EHU, Engineer Torres Quevedo Plaza, 1, 48013, Bilbao, Spain.
[3]Doctoral Program in Electronic and Communications, Department of Communications Engineering, University of the Basque Country, UPV/EHU, 48013, Bilbao, Spain.
[4]Doctoral Program in Engineering of Materials and Sustainable Processes, University of the Basque Country, UPV/EHU, 48013, Bilbao, Spain.

arincon@correo.uts.edu.co

Abstract. The research explores the use of cocoa pod shell (CPS) in rural areas and accessible to the cocoa farmer, applying VOSviewer software. The results show that the production of ethanol or biochar with CPS is not feasible, the first due to its low sugar content and the second, due to the requirement of processes with high temperatures for the extraction of different chemical compounds (i.e activated carbon). On the other hand, obtaining pellets represents a viable energy potential, since the calorific value is competitive with other biomass; Also, through anaerobic digestion under the ideal conditions, it can generate 55% in biogas of the solid matter used. Additionally, the ashes produced in direct combustion or pellets rep-resent an excellent catalyst, which can be used in transesterification, for the generation of biodiesel, used with other residual oils, giving a plus in the efficient use of resources.

1. Introduction

Cocoa (*Theobroma cacao*) is an economically important crop because it has been used mainly in the manufacture of chocolate. During the cocoa bean production process, the ripe fruits are harvested from the trees and then cut open to extract the almonds from the interior. After the fermentation of the pulp, the juice is drained, the beans are dried and bagged even more, which constitutes the cocoa bean [1].

The cocoa pod shell (CPS) represents more than 60% of the weight of the fruit, originating plant material that is normally discarded in situ for natural degradation, causing a negative impact on the environment, generating a bad smell and a source of disease inoculum, due to its lignocelluloses components and biological decomposition [2]. Therefore, cocoa farmers are encouraged to use SCP as feed for farm animals and as a source of alkali for soap making, in addition to its use as a potash fertilizer (rich in potassium) in the cocoa plantation u other cultivation fields, since around 10 tons of wet CPS are generated from one ton of dry cocoa beans produced, which represents a large volume without harvesting [3], but without neglecting a series of other uses that are can give this material, with techniques and methodologies accessible to the cocoa farmer, such as: the manufacture of pellets for different thermal applications due to its considerable calorific value [4], highlighting the use of residual ash [5] as a catalyst in the biodiesel production [6]. Furthermore, the production of biogas in up to 55% solid matter under ideal conditions that can be controlled [7][8]; among other uses at an industrialized level.

In this research, the Web Of Science (WOS) database is addressed, with the aim of identifying the energy potential of the residual biomass that is generated in the production of cocoa, highlighting the different methodologies used for the production of biofuels. For this, a bibliographic analysis methodology is implemented, which uses a network visualization tool (software) VOSviewer. This tool efficiently improves information processing and facilitates understanding of areas of greatest interest, along with trends in academic and technological development.

2. Formatting the title, authors and affiliations

In the planning of this research, two thematic centers were proposed, for the approach of the databases specifically speaking in the development of this with the Web Of Science (WOS). The first thematic center is cocoa (*Theobroma cacao*) which represents a self-sustainable development opportunity for the Santander region and Colombia in general, and the second is the use of residual biomass from the production of cocoa almonds, which is mainly represented by its bark or rind, which makes up approximately 70% of the fruit.

For this reason, to address the search for research focused on the use of this residual biomass, the following keywords are implemented: Cocoa, Biomass, Biogas and Biofuel, within which they are identified as the main forms of energy use of the residual biomass of cocoa. The first equation used to approach the database is: cocoa AND (biomass OR biogas OR biofuel *) with this equation a set of scientific articles consisting of 221 publications is obtained. This record is analyzed using the VOSviewer graphic visualization tool, which allows to efficiently recognize different fields of research in less time, among other things. Taking into account that within the first stage of research it is important to determine the data set to study is a practical way to filter those articles that are not within our field of interest. For this reason, the registry of the 221 articles is downloaded, obtaining information such as, title, abstract, authors, citations, bibliography, keywords, among other relevant data, to carry out a bibliographic analysis through a text file.

It should be noted that the main functionality of VOSviewer is the visualization of similarities, which allows to identify within a two-dimensional map, the strength and relationship that different nodes have, which can be represented by authors, keywords, entities, magazines, etc. To start, a co-occurrence map of keywords is made on the downloaded articles. Within these, the program identifies 1,567 words which must be revised to make a duplicity filter and within this it is possible to reduce 10 words which are corrected so as not to generate a distortion within the map by observing two separate components when they should be displayed. like one. Within these duplicities it is highlighted that the program does not recognize the pluralization of words such as cocoa pod husk and cocoa pod husks (see Figure 1).

Now with this first map it is possible to make several deductions, firstly the intense strength of the word "biomass" within this research field is observed, fully maintaining centrality, which implies that it is the main subject of study within this data set. Now looking at the entire map, two large fields are identified. The first one that is on the right side of the map, is mainly oriented to agroforestry, biodiversity and agriculture, which makes relevance to the fact that cocoa is a crop that stands out for its degree of environmental sustainability, because it has the advantage if implemented through the combination of trees (leftover) and cocoa, since it has a biodiversity within the crop, it allows mitigating the environmental impact of deforestation.

IOP Conf. Series: Materials Science and Engineering 1154 (2021) 012032 doi:10.1088/1757-899X/1154/1/012032

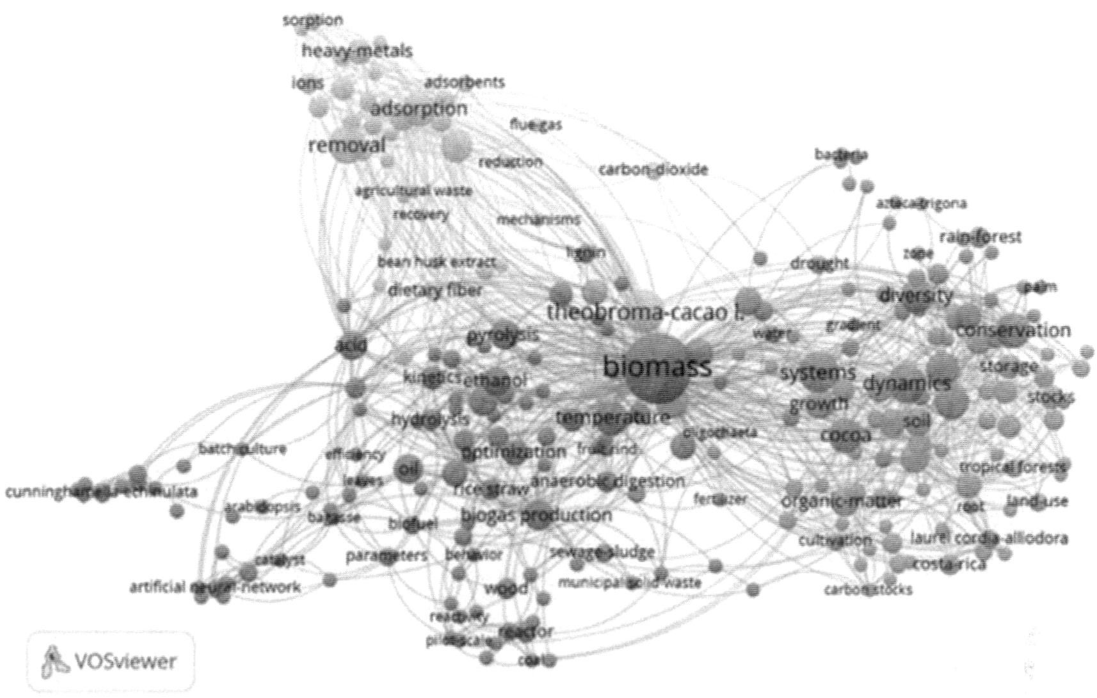

Figure 1. Keyword display - first record (252 nodes, 4 groups and 1752 link).

Now with this first map it is possible to make several deductions, firstly the intense strength of the word "biomass" within this research field is observed, fully maintaining centrality, which implies that it is the main subject of study within this data set. Now looking at the entire map, two large fields are identified. The first one that is on the right side of the map, is mainly oriented to agroforestry, biodiversity and agriculture, which makes relevance to the fact that cocoa is a crop that stands out for its degree of environmental sustainability, because it has the advantage if implemented through the combination of trees (leftover) and cocoa, since it has a biodiversity within the crop, it allows mitigating the environmental impact of deforestation.

The second field on the left side of the map made up of two groups, the first of which is yellow in the upper left, is defined by themes of organic waste, where the quantities of heavy metals found, the absorption processes of nutrients among others. The second group, represented in red, is made up of studies on the production of biogas, biofuels, biodiesel, combustion, improvement of thermal efficiency, temperatures and energy in general. This same represents the center of study that you want to address, so it is therefore necessary to generate a more precise search equation.

Taking into account the aforementioned, the equation is adjusted as follows: cocoa AND (biomass OR biogas OR biofuel * OR biodiesel OR bioethanol) AND (energy OR combustion OR temperature OR pyrolysis). With this equation, a record of 90 documents is obtained that allows us to more precisely address our field of interest.

The activities for the debugging of the key words within this registry are repeated. The program identifies 662 words of which only those with a minimum co-occurrence of 2 documents will be displayed, granting the following map (see Figure 2):

IOP Conf. Series: Materials Science and Engineering 1154 (2021) 012032 doi:10.1088/1757-899X/1154/1/012032

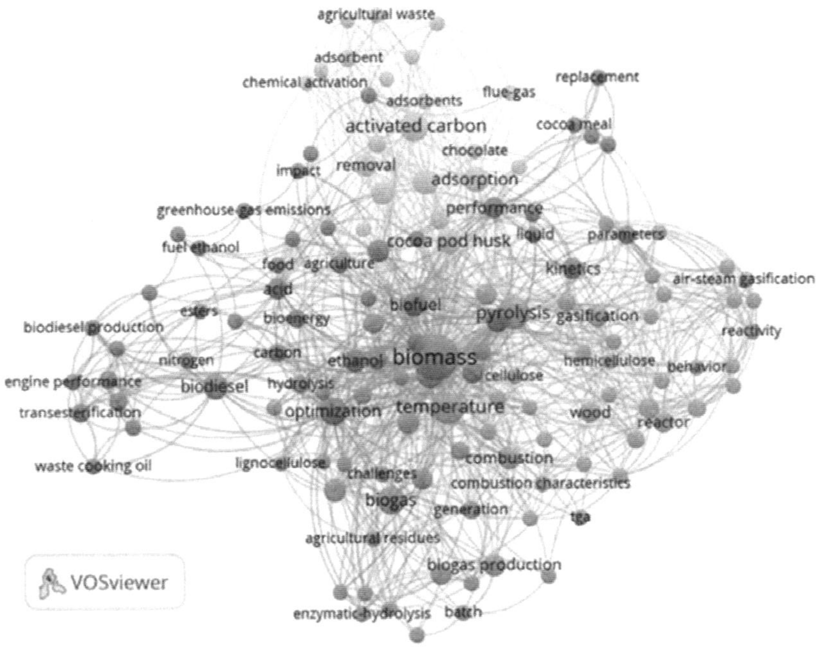

Figure 2. Keyword map - second record (135 nodes, 4 groups and 1031 links).

With this map of keywords, a network according to the research theme is identified, in which 4 groups stand out, red is centralized by biodiesel, which is why it is generalized by topics such as production, alcohols, yield and even Observe the incorporation of cooking oil waste, taking into account that it is one of the main components for its production. In the blue group, the centrality of biogas stands out, accompanied by topics such as generation, combustion, hydrolysis, lignocellulose and the use of different agricultural residues for its production [9].

The green group is mainly dominated by pyrolysis, where temperature is the main factor analyzed in this process and the use of pellets is taken into account. This group focuses on the direct combustion of plant material. Yellow represents a theme oriented to the study of activated carbon, where chemical processes are analyzed, the composition of the waste generated by the plant, ranging from the shell of the cocoa bean, to the shell of the almond among other organic wastes that Chocolate production is generated (see Figure 3).

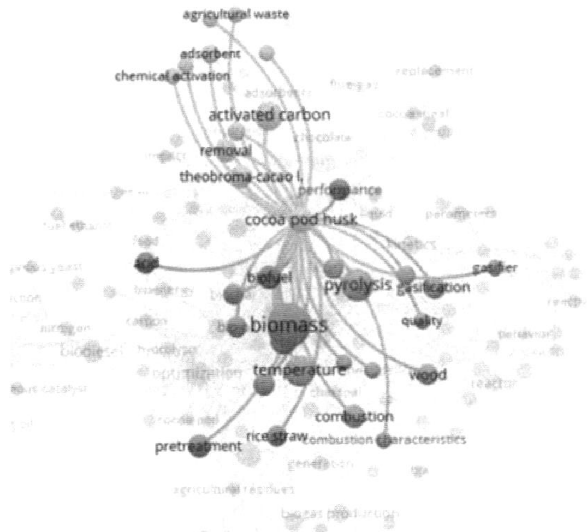

Figure 3. Links related to the cocoa pod shell (CPS).

When this research begins, it starts with a hypothesis about the cocoa pod shell (CVC), which consists of using it as a source of biomass, to be used in thermal processes. With this map it is possible to demonstrate the importance of this resource for the generation of biogas, pellets and biochar in general and its strength within this network of keywords, is a clear example of the relevance that it represents for the academic community on issues of energy use of renewable resources, taking into account that the biomass of the cocoa pod shell is normally discarded to be used as fertilizer (see Figure 4).

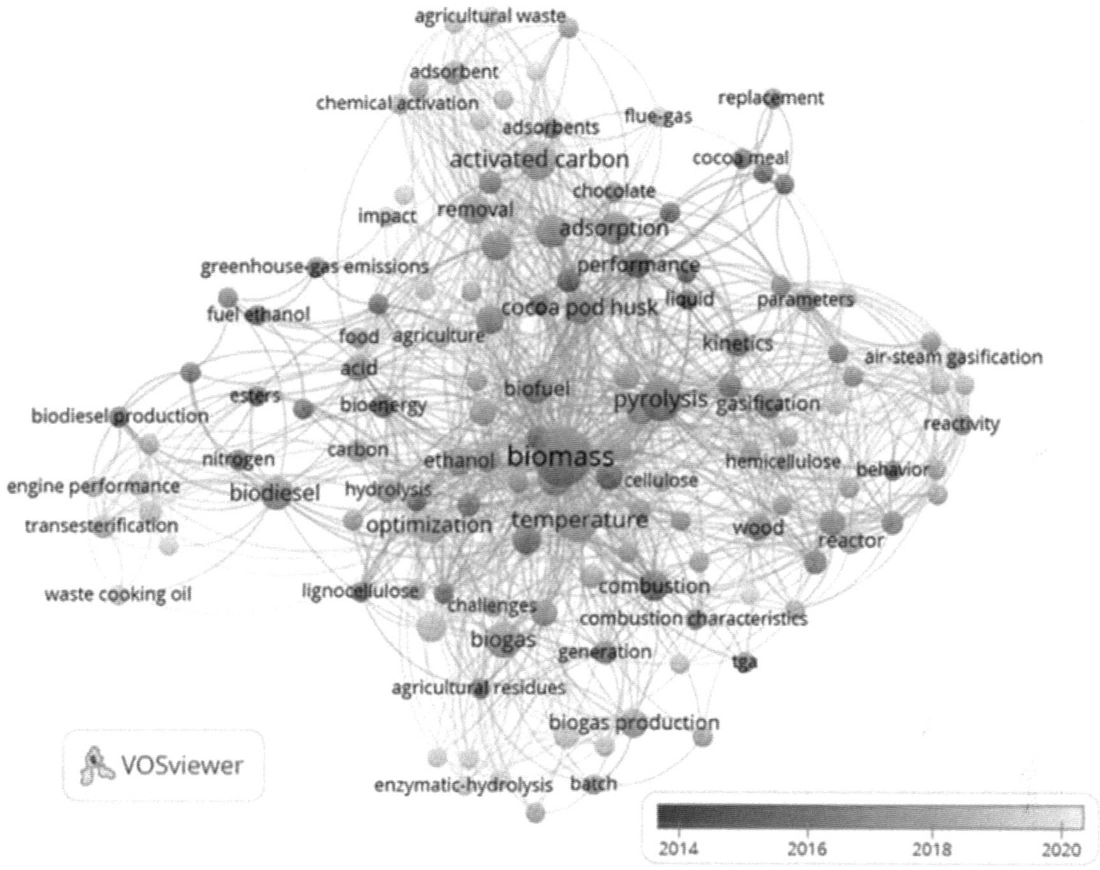

Figure 4. Temporary display.

With the VOSviewer tool, it is possible to graphically observe the use of keywords over time and this works as a trend analysis tool, in order to identify new research topics for decision-making in the continuity of this renewable energy field. For example, in this network it is observed that optimization begins to be used from 2018 approximately with recurrence, taking into account the links that this word has, it can be pointed out that this is achieved through anaerobic digestion processes, pre-treatments, biomass in pellets, heterogeneity catalysis, among other topics that will be addressed in more detail in the results section (see Figure 5).

Figure 5. Bibliographic coupling.

With the bibliographic link it is possible to see the different documents and their strength within the map thanks to the citations they have between them. The strength of the older articles is identified, which as they have been published for a longer time can obtain a higher number of citations, with this tool it is possible to access the document just by clicking on the node of interest, so a general scan of these documents to analyze their content taking into account those nodes with the greatest strength (co-citations) and with the highest number of citations, selecting the first 5 of each list in the table 1.

Finally, the documents processed by the VOSviewer tool are read and analyzed, which serve to have a more precise idea of the results obtained by the tools and the research approach.

Table 1. Selected documents.

Author	Citations	Links of strength	Documents
Moses Hensley Duku (2011)	106	1	[10]
L. Fryda (2010)	86	1	[11]
Cecile Bessou (2013)	82	5	[12]
Disney Ribeiro Dias (2007)	48	2	[13]
T G Chuah (2006)	47	6	[14]
J. Ratte (2011)	44	3	[15]
Wen-Tien Tsai (2020)	1	50	[2]
M. P. González-Vázquez (2018a)	15	40	[16]
Chi Hung Tsai (2018)	7	39	[17]
M. V. Gil (2019)	7	30	[18]
J. Daniel Martinez-Angel (2015)	10	29	[19]

To start reading and analyzing the documents processed by the VOSviewer tool, which serve to have a more precise idea of the results obtained. Firstly, the content of the work published by Chuah (2006) is read, this document is titled "Biomass as a renewable energy source in Malaysia: an overview", this document provides an overview of the types of biomass used, among which stands out the energy potential of the waste generated in the production of cocoa, the research work on the conversion of biomass into energy and the projects that at the time focused on the development of biomass energy in Malaysia.

The second work that is read is that of Dias (2007) which is entitled "Elaboration of a fruit wine from cocoa pulp (Theobroma cacao L.)", in this the alcohol levels generated in the process of fermentation, which have served as a documentary basis for research into the generation of different biofuels.

Consecutively, the document published by [11], which is entitled "Study on the deposition of ash in the combustion of oxyfuel from coal/biomass mixtures", exposes the behavior of the combustion of coal with cocoa flour additives, with the which seek to minimize CO_2 emissions, taking into account that biomass contributes to energy generation. This work identifies that the temperature and the levels of the fuel mixture are the main factors that affect emissions and the ash generated during combustion.

A document that represents a very interesting vision for this research and is incorporated due to its relevance, is the one published by [15] which is entitled "Mathematical modeling of a continuous biomass roasting reactor: TORSPYD ™ column", in her work it was describes the roasting of biomass, by depolymerization consisting of a "gentle" heat treatment (<250 ° C) that allows the definitive removal of biomass water, as well as the transformation of part of the organic matter from biomass. The biomass is heated evenly. The roasting process dries the biomass, makes it irreversibly non-hygroscopic and concentrates its energy potential in a solid fuel to facilitate its transportation and grinding. As part of their research they use the cocoa shell in their tests.

Taking into account the list of documents that is ordered by number of citations [10] is the leader, his work is entitled "A comprehensive review of biomass resources and biofuel potential in Ghana" and it is possible to demonstrate the potential energy from renewable resources from biomass, among which cocoa is named, as a fundamental part of them.

To finish in the list of the leading documents in citations there is the work of [12] entitled "LCA applied to perennial cultivation systems: a review focused on the farm stage" this focuses on the cultivation systems for the proportion of bioenergy raw materials, this is important to keep in mind, to maintain an efficiency margin from the point of production.

Now it is not only important to take into account those documents that have been most frequently cited, but also those that can generate a new field of research and represent an opportunity for development in a specific topic, which is why the first 5 documents ordered by their strength link level. To start, we talk about the work carried out by Daniel Martinez-Angel (2015), his work is entitled "Characterization and evaluation of the cocoa husk (theobroma cacao l.) As a renewable energy source" this work is carried out in Colombia, and It directly analyzes the energy potential of the cocoa shell, and concludes that the gasification method is the most appropriate to adequately take advantage of this renewable resource.

The next work chronologically speaking is the one published by [16] this study investigates the air-vapor gasification of ten commercial and alternative lignocellulosic biomass fuels (pine sawdust, chestnut sawdust, roasted pine sawdust, roasted chestnut sawdust, almonds, cocoa hulls, grape marc, olive stones, pine nut shells and pineapple shells) to assess the product gas composition and process performance in a bubbling fluidized bed gasifier with focus on the different properties of the biomass. The work published by [17] is entitled "Thermochemical characterization of biochar from cocoa pod shell prepared at low pyrolysis temperature." In this, the energy behavior and the waste produced by the cocoa shell when processed as biochar are studied very precisely, with this it is found that it has a calorific value close to 17.8 MJ / kg, using these data it is possible to start from comparative values in the results of the work.

Taking the continuity of the documents based on the strength of their links, there is the one published by [18] whose title is "Assessing the influence of biomass properties on the gasification process using multivariate data analysis" in this work use multivariate analysis to study the influence of biomass characteristics on the gasification process, within which they use different types of biomass such as the cocoa shell between them.

The last article with the strongest links and which has been recently published is the work carried out by [2]. This work focuses on the study of the use of cocoa beans for the preparation of activated carbon through the use of potassium hydroxide. Low contamination as activating agent.

As it could be intuited within the documents previously described, the first group that stands out for having been cited more frequently, are documents that describe an energy potential of the cocoa biomass, these documents were located approximately 10 years ago and were the first engines in the research of this renewable resource. Subsequently, the documents that stand out for the strength of their links, describe in a more specific and detailed way, the thermochemical capacities of the different wastes generated in the production of cocoa, such as the cocoa pod shell, the shell of the almond and the pulp.

In order to solve the research question, the different documents continue to be analyzed using the VOSviewer visualization tool and with this, clarify the continuity of the research aimed at the cocoa biomass.

3. Results and discussions

The objective of this research is to achieve a complete overview of the energy potential of the biomass that is discarded in the cocoa growing process (*Theobroma cacao*), especially to know those processes that are required to transform the biomass conditions into a fuel, whether solid, liquid or gaseous. All this in order to implement this knowledge in support of agricultural, regional and national activities, taking into account that they can benefit from implementing new technologies that generate new economic or energy resources for their ordinary work activities [20][21].

Taking into account the different ways in which the residual biomass of cocoa crops can be used, it is evident that fermentation for the production of biofuels such as ethanol, does not represent a sustainable opportunity in terms of energy use, this counting on the different pretreatments that have been explored in the different investigations, due to the levels of sugars present in the CVC, also when compared to other plant residues that have greater potential such as banana or corn waste [22][23].

On the other hand, the pyrolysis process, used to generate biochar, is analyzed, which is a methodology with a high energy requirement, when working with temperatures between 300 to 900°C, this represents an energy storage strategy, not very viable for rural-type implementations, bearing in mind that this technological process or implementation has other types of purposes, such as the extraction of different chemical compounds or as some researchers establish activated carbon [24][25].

Now within the aforementioned processes, it is necessary to clean, dry and crush the CVC, in order to extract the fuel through fermentation or pyrolysis, and this also applies to the production of pellets, bearing in mind that after being crushed they are sieved. or tablets, with the aim of increasing their density, facilitating their transport and combustion. In the investigations it is identified that the cocoa pellet is not explored as the main raw material for commercial reasons, taking into account that the pellet industry already has an important participation of other biomass such as pine. The calorific value of cocoa is approximately 17.5 MJ/Kg, so it is located on the average of different solid vegetable materials and represents an opportunity for energy use in different agricultural processes, such as cooking, food drying, generation of energy in non-interconnected areas, among others [19]–[21]. It is important to highlight that in terms of pellets it was mainly implemented as an additive for other solid materials, such as pine, sawdust, coal, among others, with the aim of giving greater consistency to the pellet or reducing costs as it is a waste from a food industry.

As mentioned above, cocoa does not have the potential to generate liquid fuels, but this does not completely displace it from belonging to a biodiesel production line [29][30], the ashes generated after the combustion of the CVC plant material is used in the transesterification process as a catalyst, these residues within a thermal process at a temperature close to 70°C. The results obtained by the researchers in this area highlight the catalytic capacity of this ash, which is really important considering that for these production processes the main raw material is cooking oil waste, animal fats and even research with other types of oil with palm oil, in which Colombia has a stake. In terms of generating new products, the incorporation of the use of all cocoa wastes in different industrial processes such as the biodiesel industry is quite innovative.

When talking about biogas generation, CVC has been quite remarkable as the main raw material, by achieving a transformation of close to 50% of its solid matter into gaseous, researchers through pretreatments manage to increase this percentage and thereby increase the energy use of this Biomass, through the anaerobic digestion process and implementing different inocula in controlled environmental conditions, it is possible to generate biogas efficiently. With this, a new viable energy product is opened to be used in the cocoa food industry, which generally continues to grow and represents one of the crops with the least environmental impact due to its diversity of trees, which maintain the consistency of the ground [18][31].

4. Conclusions and recommendations

The bibliographic analysis methodology implemented for this research allows us to advance efficiently in the search for specific information on the use of cocoa residues, the graphical network visualization tool, allows us to observe the links between the different topics, such as for example, pretreatments to improve the production of biogas, biodiesel, bioethanol, among others. It should be noted that these topics were identified before making a deep reading of the documents and investigations registered with the search equation.

With this research it is possible to point out that the main methods for the energy use of biomass in rural areas are the drying, crushing or milling of the biomass, to be used in two ways, the first as solid plant material in direct combustion, with a low humidity percentage, it becomes a fuel with great potential for different thermal processes. If it is compressed or sieved into pellets, it can be marketed for different applications. For the second option, this material can be used within a biodigester, through anaerobic digestion with pretreatments and, under the ideal conditions, it can generate 55% in biogas of the solid matter used, the researchers highlight it as a biomass with a potential to be used in industrial biogas generation plants, incorporating different raw materials in rural areas.

Additionally, the ashes produced in the combustion of the solid plant material of the cocoa pod shell are linked, as an excellent catalyst, which can be used in transesterification, for the generation of biodiesel, this biofuel uses residual oils such as that of the cooking, animal fats, among others that are commonly used in rural areas, this represents an opportunity to take advantage of the different wastes generated during cocoa production. Giving this industry a plus in the efficient use of its resources.

It is recommended in the first instance the recurrent use of the VOSviewer application or network visualization tools, to carry out an efficient analysis of the databases in those projects that require an overview of the level of research or technological progress, for a given subject, as is the case of the energy potential of cocoa. This improves efficiency at the time of understanding or directing certain investigations from which it is desired to generate a significant advance or that causes a greater impact at the academic and social level.

In terms of research, the implementation or development of a pilot plant in cocoa growing areas is recommended, in order to establish a cocoa biomass production line. With this, a production line must be founded that involves cleaning the hulls of the cocoa pod (CVC) to remove impurities such as dirt, drying looking for a humidity close to 10%, grinding to decrease the size of its particles. in the order of mm and pressing or sieving to increase the density of the material, facilitating its commercialization and transport. This production line does not represent a high investment and can lead to an improvement in the economic sustainability of cocoa farmers.

For the generation of biogas, a control of the environmental conditions of anaerobic digestion is commonly required, in order to improve its production, including temperature and monitoring of different thermochemical variables. In addition, it is important to implement suitable equipment for the extraction process of the same for its corresponding storage and transport. For this reason, it is recommended for the production of biogas with the CVC, to implement a central pilot plant that processes all the waste from a region.

5. References

[1] Z. S. Vásquez *et al.*, "Biotechnological approaches for cocoa waste management: A review," *Waste Manag.*, vol. 90, pp. 72–83, 2019.
[2] W. T. Tsai, Y. C. Bai, Y. Q. Lin, Y. C. Lai, and C. H. Tsai, "Porous and adsorption properties of activated carbon prepared from cocoa pod husk by chemical activation," *Biomass Convers. Biorefinery*, vol. 10, no. 1, pp. 35–43, 2020.
[3] S. O. Dahunsi, C. O. Osueke, T. M. A. Olayanju, and A. I. Lawal, "Bioresource Technology Co-digestion of Theobroma cacao (Cocoa) pod husk and poultry manure for energy generation : Effects of pretreatment methods," *Bioresour. Technol.*, vol. 283, no. March, pp. 229–241, 2019.
[4] G. Kilama, P. O. Lating, J. Byaruhanga, and S. Biira, "Quantification and characterization of cocoa pod husks for electricity generation in Uganda," *Energy. Sustain. Soc.*, vol. 9, no. 1, 2019.
[5] A. D. Rincón-Quintero *et al.*, "Manufacture of hybrid pieces using recycled R-PET, polypropylene PP and cocoa pod husks ash CPHA, by pneumatic injection controlled with LabVIEW Software and Arduino Hardware," *IOP Conf. Ser. Mater. Sci. Eng.*, vol. 844, no. 1, 2020.
[6] S. Y. Chua *et al.*, "Biodiesel synthesis using natural solid catalyst derived from biomass waste — A review," *J. Ind. Eng. Chem.*, vol. 81, pp. 41–60, 2020.
[7] N. Acosta, J. De Vrieze, V. Sandoval, D. Sinche, and I. Wierinck, "Bioresource Technology Cocoa residues as viable biomass for renewable energy production through anaerobic digestion," *Bioresour. Technol.*, vol. 265, no. May, pp. 568–572, 2018.
[8] S. Raj, S. Prasad, O. Kyung, K. Young, Y. Mo, and J. Woo, "Bioresource Technology Pretreatment of agricultural biomass for anaerobic digestion : Current state and challenges," *Bioresour. Technol.*, vol. 245, no. July, pp. 1194–1205, 2017.
[9] B. E. Romero-Tarazona, C. L. Rodriguez-Sandoval, J. G. Villabonai-Ascanio, and A. D. Rincón-Quintero, "Development of an artificial vision system that allows non-destructive testing on flat concrete slabs for surface crack detection by processing of digital images in MATLAB," *IOP Conf. Ser. Mater. Sci. Eng.*, vol. 844, no. 1, 2020.
[10] M. H. Duku, S. Gu, and E. Ben Hagan, "A comprehensive review of biomass resources and biofuels potential in Ghana," *Renew. Sustain. Energy Rev.*, vol. 15, no. 1, pp. 404–415, 2011.
[11] L. Fryda, C. Sobrino, M. Cieplik, and W. L. Van De Kamp, "Study on ash deposition under oxyfuel combustion of coal/biomass blends," *Fuel*, vol. 89, no. 8, pp. 1889–1902, 2010.
[12] C. Bessou, C. Basset-Mens, T. Tran, and A. Benoist, "LCA applied to perennial cropping systems: A review focused on the farm stage," *Int. J. Life Cycle Assess.*, vol. 18, no. 2, pp. 340–361, 2013.
[13] D. R. Dias, R. F. Schwan, E. S. Freire, and R. dos S. Serôdio, "Elaboration of a fruit wine from cocoa (Theobroma cacao L.) pulp," *Int. J. Food Sci. Technol.*, vol. 42, no. 3, pp. 319–329, Mar. 2007.
[14] A. Luqman Chuah, W. Ghani, R. Yunus, and R. Omar, "Biomass as the Renewable Energy Sources in Malaysia: An Overview," *Int. J. Green Energy - INT J GREEN ENERGY*, vol. 3, pp. 323–346, Sep. 2006.
[15] J. Ratte, E. Fardet, D. Mateos, and J. S. Héry, "Mathematical modelling of a continuous biomass torrefaction reactor: TORSPYD^{TM} column," *Biomass and Bioenergy*, vol. 35, no. 8, pp. 3481–3495, 2011.
[16] M. P. González-Vázquez, R. García, M. V. Gil, C. Pevida, and F. Rubiera, "Comparison of the gasification performance of multiple biomass types in a bubbling fluidized bed," *Energy Convers. Manag.*, vol. 176, no. September, pp. 309–323, 2018.
[17] C. H. Tsai, W. T. Tsai, S. C. Liu, and Y. Q. Lin, "Thermochemical characterization of biochar

from cocoa pod husk prepared at low pyrolysis temperature," *Biomass Convers. Biorefinery*, vol. 8, no. 2, pp. 237–243, 2018.
[18] M. V. Gil, M. P. González-Vázquez, R. García, F. Rubiera, and C. Pevida, "Assessing the influence of biomass properties on the gasification process using multivariate data analysis," *Energy Convers. Manag.*, vol. 184, no. December 2018, pp. 649–660, 2019.
[19] J. D. Martínez-Ángel, R. A. Villamizar-Gallardo, and O. O. Ortíz-Rodríguez, "Characterization and evaluation of cocoa (Theobroma cacao L.) pod husk as a renewable energy source," *Agrociencia*, vol. 49, no. 3, pp. 329–345, 2015.
[20] J. G. Ascanio-Villabona, C. L. Sandoval-Rodriguez, A. D. Rincón-Quintero, B. E. Tarazona-Romero, and R. E. Paez-Castro, "Building a prototype for functional analysis of the energy potential of the water flow in pipe 1/2 "using microturbines applied to Unidades Tecnológicas de Santander," *IOP Conf. Ser. Mater. Sci. Eng.*, vol. 844, no. 1, pp. 0–26, 2020.
[21] C. L. Sandoval-Rodriguez, J. G. Ascanio Villabona, C. G. Cárdenas-Arias, A. D. Rincon-Quintero, and B. E. Tarazona-Romero, "Characterization of the mechanical vibration signals associated with unbalance and misalignment in rotating machines, using the cepstrum transformation and the principal component analysis," *IOP Conf. Ser. Mater. Sci. Eng.*, vol. 844, no. 1, 2020.
[22] B. K. Koua, P. M. E. Koffi, and P. Gbaha, "Evolution of shrinkage, real density, porosity, heat and mass transfer coefficients during indirect solar drying of cocoa beans," *J. Saudi Soc. Agric. Sci.*, vol. 18, no. 1, pp. 72–82, 2019.
[23] K. Sandesh, R. K. Shishir, and C. V. Rao, "Optimization and comparison of induction heating and LPG assisted acid pretreatment of cocoa pod for ABE fermentation," *Fuel*, vol. 262, no. October 2019, p. 116499, 2020.
[24] F. Fernandes-Klajn, J. M. Romero-García, M. J. Díaz, and E. Castro, "Comparison of fermentation strategies for ethanol production from olive tree pruning biomass," *Ind. Crops Prod.*, vol. 122, no. March, pp. 98–106, 2018.
[25] C. G. Cárdenas-Arias, A. D. Rincón-Quintero, A. Santos-Jaimes, C. L. Sandoval-Rodriguez, D. F. Rojas-Gomez, and S. J. Ardila-Galvis, "Elasticity modulus variation of the AISI SAE 1045 steel subjected to corrosion process by chloride using tension test destructive," *IOP Conf. Ser. Mater. Sci. Eng.*, vol. 844, no. 1, 2020.
[26] R. García, M. V. Gil, F. Rubiera, and C. Pevida, "Pelletization of wood and alternative residual biomass blends for producing industrial quality pellets," *Fuel*, vol. 251, no. April, pp. 739–753, 2019.
[27] T. A. Yassen and H. H. Al-Kayiem, "Experimental investigation and evaluation of hybrid solar/thermal dryer combined with supplementary recovery dryer," *Sol. Energy*, vol. 134, pp. 284–293, 2016.
[28] E. Antwi, N. Engler, M. Nelles, and A. Schüch, "Anaerobic digestion and the effect of hydrothermal pretreatment on the biogas yield of cocoa pods residues," *Waste Manag.*, vol. 88, pp. 131–140, 2019.
[29] M. S. Sinaga, S. Pandia, T. F. Tarigan, and W. G. Tiffani, "Utilization of cacao peel waste to K2O heterogeneous catalyst in biodiesel synthesis by waste cooking oil : effect of catalyst calcination temperature," 2018.
[30] J. G. Ascanio-Villabona, A. D. Rincón-Quintero, C. G. Cardenas-Arias, and C. L. Sandoval-Rodriguez, "Incidence of corrosion in low voltage electrical conductor," *IOP Conf. Ser. Mater. Sci. Eng.*, vol. 844, no. 1, 2020.
[31] J. Jochum and F. E. Sierra, "Effect of particle size and addition of cocoa pod husk on the properties of sawdust and coal pellets," vol. 35, no. 1, pp. 17–23, 2015.

IOP Conf. Series: Materials Science and Engineering 1154 (2021) 012033 doi:10.1088/1757-899X/1154/1/012033

AISI SAE 1045 steel tenacity variation evaluation subjected to corrosion process by chloride using the Charpy pendulum machine

C G Cárdenas-Arias[1-2], A D Rincón-Quintero[1-3], C L Sandoval-Rodriguez[1-4], B E Tarazona-Romero[1-4] J S Muñoz-Gutierrez[1], J L Jímenez-Rueda[1], O A Acosta-Cárdenas[1]

[1] Design and materials research group (DIMAT), Energy Systems, Automation and Control Research Group (GISEAC), Electromechanical Engineering, Faculty of Natural Sciences and Engineering, Unidades Tecnológicas de Santander (UTS), Bucaramanga, Colombia.

[2] Doctoral Program in Engineering of Materials and Sustainable Processes, University of the Basque Country, UPV/EHU, Bilbao, Spain

[3] Doctoral Program in Energy Efficiency and Sustainability in Engineering and Architecture, Department of Machines and Thermal Engines, University of the Basque Country (UPV / EHU), Engineer Torres Quevedo Square, 1, Bilbao, BI, Spain

[4] Doctoral Program in Electronic and Communications, Department of Communications Engineering, University of the Basque Country, UPV/EHU, Bilbao, Spain.

ccardenas@correo.uts.edu.co

Abstract. The steel corrosion affecting study is important for industries and companies as well as for manufacturers of these material types. AISI SAE 1045 steel was used in this study, due to its frequent use. A specimen's series were prepared which were induced in an aggressive environment with the aim of degenerative process accelerating by increasing the material corrosion impact by altering the material mechanical properties. 15 specimens were exposed to the fog bank chamber, likewise 15 specimens in saline immersion chamber all with different exposure times 15. 30, and 45 days, subsequently the specimens are the impact test subjected to by the method of Charpy obtaining the tenacity value. To determine these values, the ASTM E23, ASTM A673 and ASTM E739 standards were used.

1 Introduction

Steel is considered one of the materials with the highest production worldwide due to its high demand in the different processes and applications within the industry.[1] It is also exposed to corrosion that a degenerative transformation generates directly affecting the material mechanical properties. The corrosion severity and speed will depend on the chemical or electrochemical reactions incidence on the exposed environment, as well as there are variations depending on the type of steel alloy.

AISI SAE 1045 steel it was decided to use to analyze how the corrosion degenerative process affects the material toughness, because it is widely used for the manufacture of automotive parts, the agricultural sector, metal structures, among others; this due to its outstanding characteristics such as hardness and toughness. The study method is experimental[2]; 35 specimens are subjected to saline solutions at different forms and exposure times, formed into groups of 5 test tubes according to the ASTM G-16 standard; the first 5 are the basis of the analysis since these will not be induced to no type of oxidizing agent, for this reason will give us the reference values of study material. 15 specimens will be subjected to an immersion process with sodium hypochlorite as an oxidant with demineralized water, a saline solution to which steel does not offer greater resistance, as they state [3] in their research, for a period of 15, 30 and 45 days respectively. At the same time, another degenerative process of 15 test tubes will be carried out, which will be introduced into a saline mist chamber, using sodium chloride with demineralized water as an oxidizing agent, for a period of 15, 30 and 45 days, they will be extracted. in groups of 5 test tubes respectively. Afterwards, the impact tests are carried out on the Charpy pendulum machine, to obtain the quantitative values and verify the AISI SAE 1045 steel toughness variation. This test consists of impacting a specimen located on a support pre-designed for this purpose in the machine, by means of a mass that is attached to one end of the pendulum that has a characteristic distance; it is dropped from a height (h), by means of which the energy of application of the load is controlled for the subsequent impact. After the test, the information obtained was collected to perform the calculations and subsequent analysis of the results.

2 Work development

2.1 Design and execution stage.

The ASTM E32 standard was applied to obtain the design parameters required to manufacture the specimens for the impact tests under the Charpy pendulum.

Figure 1 shows the specimens finished design for the Charpy impact test execution according to ASTM E-23, taking into account the testing machine type and all the characteristics established for the test correct development and a result obtaining true.

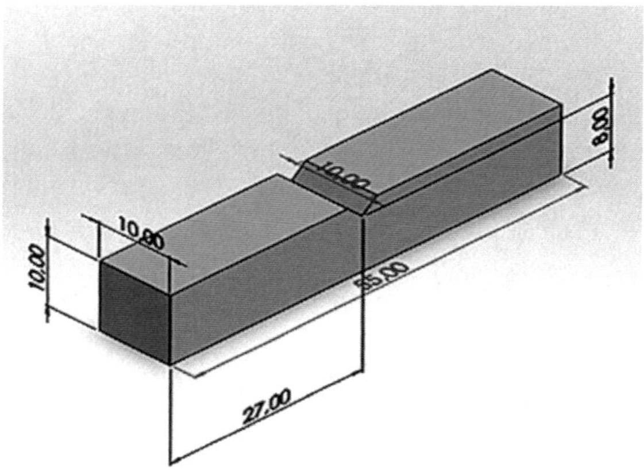

Figure 1 Specimen modeling.

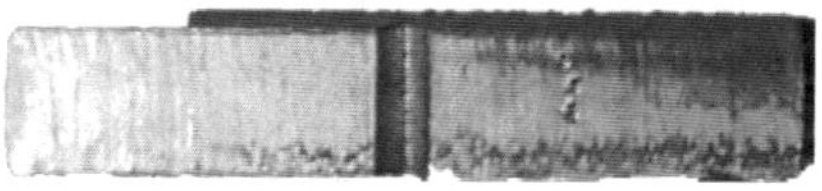

Figure 2 Machined specimen.

35 specimens are manufactured with the dimensions already specified by the water jet cutting method in order to make clean and uniform cuts that maintain the characteristics necessary for the test's correct development. Finally, they undergo a rectifying process any possible detail or imperfection that they may have after the previous method, thus obtaining 35 specimens made of AISI SAE 1045 sheet-type steel with a thickness of ½".

2.2 Corrosion stage

The exposure time quantification selected for corrosion by the methods of saline immersion and fog bank, is presented below in the Table 1

TEST	DESIGNATIÓN	ESPECIMEN´S QUANTITY	EXPOSITION TIME (DAYS)
Saline immersion (sodium hypochlorite + demineralized water)	INM-15	5	15
	INM-30	5	30
	INM-45	5	45
Salt spray chamber (sodium chloride + demineralized water)	NIE-15	5	15
	NIE-30	5	30
	NIE-45	5	45

Table 1 Corrosion test type and exposure times

2.2.1 Salt immersion corrosion. Sodium hypochlorite and demineralized water were used for the corrosion process by the saline immersion method, which were poured into a clean container previously prepared for this purpose[4]. Once the container and the sodium hypochlorite with demineralized water mixture were ready, the test tubes previously marked by groups were introduced, which were left at room temperature and protected to avoid the inclusion of any external agent that could alter the corrosion process.

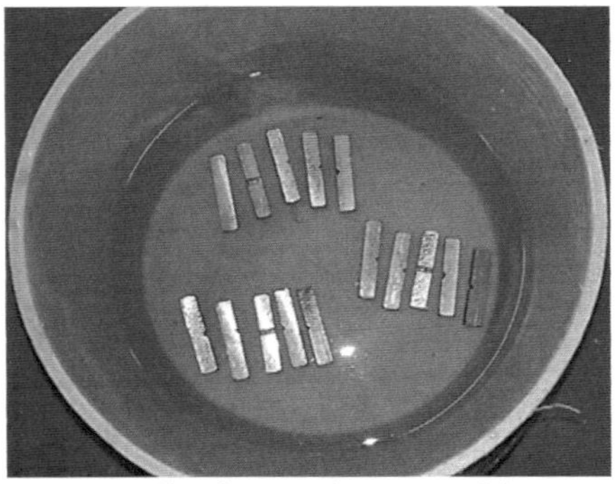

Figure 3 Saline immersion exposition

According to the schedule to the oxidizing agent exposure, they were extracted in groups of 5 specimens during a period of time comprised in 15 days intervals.

Figure 4 Saline immersion test specimens 15 days

Figure 5 Saline immersion test specimens 30 days

Figure 6 Saline immersion test specimens 45 days

2.2.2 *Corrosion in saline fog chamber.* For the corrosion test through the cloud chamber, sodium chloride and demineralized water were used, dissolving the substances and pouring them into the storage tank to generate the saline mist, prior to the tank a cleaning process was carried out with in order to eliminate external agents that could influence the development of the research as indicated [3] in their research.

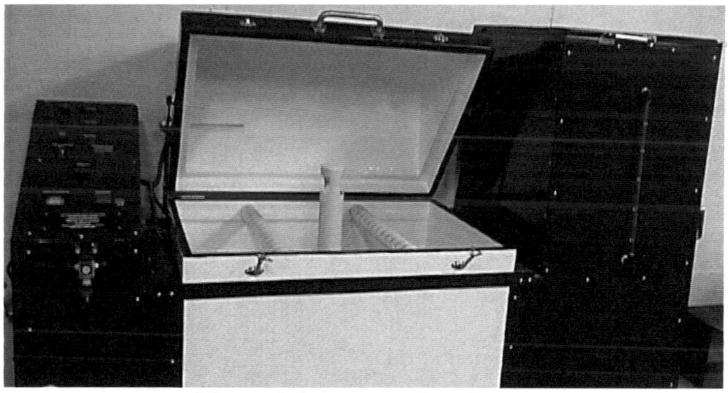

Figure 7 Salt spray chamber

The cloud chamber is prepared and the 15 properly marked specimens are introduced, with the objective the corrosion process starting; in order for the three groups of specimens to retain the characteristics necessary for the tests, they are placed at the same height, all this with the purpose that the only variable between them is the exposure time.

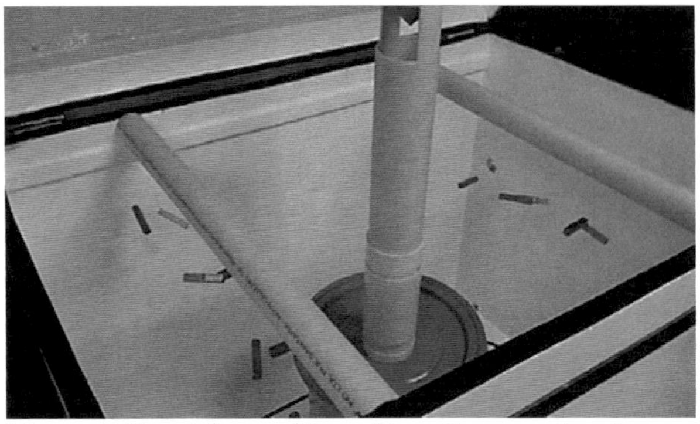

Figure 8 Test tube positioned in salt spray chamber

The extraction is carried out in groups of 5 specimens during the different time periods defined above of 15, 30 and 45 days respectively.

Figure 9 Corroded specimens in fog bank 15 days

Figure 10 Corroded specimens in fog bank 30 days

Figure 11 Corroded specimens in fog bank 45 days

2.2.3 Charpy impact test.

Finished the corrosion process, the Charpy pendulum machine is used based on bibliographic information [5] to obtain the quantitative values of the breakage resistance energy by impact mean (toughness).

The correct operation of the Charpy equipment is checked and the necessary data is taken to perform the toughness calculations of the steel as indicated by the ASTM-E23 standard, the eigenvalues of the Charpy pendulum machine used are presented below:

Test machine	**Charpy pendulum**
Arm length	0,95 m
Hammer weight	30 kg
Hammer starting angle	120°
Idle hammer end angle	118,5°

Table 2 Impact machine features

IOP Conf. Series: Materials Science and Engineering 1154 (2021) 012033 doi:10.1088/1757-899X/1154/1/012033

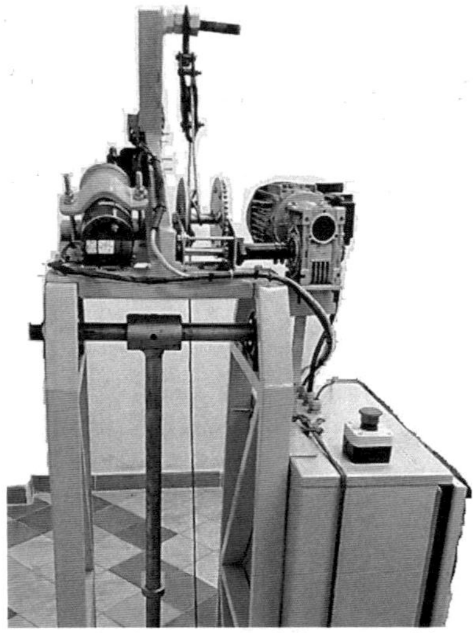

Figure 12 Charpy pendulum machine

The specimen´s first group test that were without corrosion were carried out, taking the reference to respective data obtained for the subsequent results analysis.

Figure 13 Test specimen position

Figure 14 Specimens without corrosion after the test

Figure 15 Corroded specimens for 15 days in fog chamber after impact test execution

Figure 16 Corroded specimens for 30 days in fog chamber after impact test execution

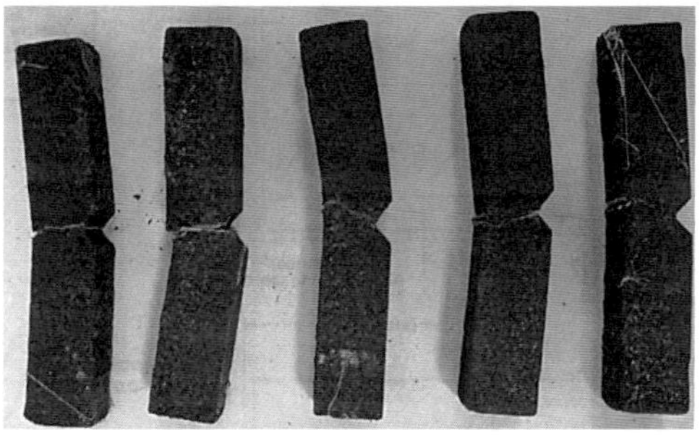

Figure 17 Corroded specimens for 45 days in fog chamber after impact test execution

Figure 18 Corroded specimens for 15 days in saline immersion after impact test execution

Figure 19 Corroded specimens for 30 days in saline immersion after impact test execution

Figure 20 Corroded specimens for 45 days in saline immersion after impact test execution

Once the impact test execution phase with the Charpy pendulum machine was completed, using the ASTM-E23 standard, the test data of each in the different groups of specimens were collected to perform the necessary calculations to obtain the results and investigation subsequent conclusions.

3 Results

At this research project stage, the necessary calculations were carried out to obtain the results based on the collected data in impact test execution phase on the test pieces different groups, following the ASTM G-16 standard criteria and the analysis was performed by means the different graphs obtained.

Based test without specimen of the Charpy pendulum equipment, the loss due to friction and other factors that the execution of the tests may entail was quantified. The final average angle obtained was 118.5 °.

The energy loss is obtained based on the height differential. For this reason, both the initial height with an angle of 120 ° and the final height with the angle of 118.5 ° are calculated

$$h=0,95+[0,95*\sin(\theta-90°)]$$
$$E=m*g*h$$
$$h=0,95+[0,95*\sin(120°-90°)]$$
$$h=1,425m$$
$$Et=30*9,81*1,425$$
$$Et=419,37\,J$$
$$h=0,95+[0,95*\sin(118,5°-90°)]$$
$$h=1,4033\,m$$
$$Ev=30*9,81*1,4033$$
$$Ev=412,99\,J$$
$$E_{loss}=E_t-E_v$$
$$E_{loss}=419,37-412,99$$
$$E_{loss}=6,38\,J$$

The lost energy of 6.38 J must be taken into account in tests carried each out to correctly determine the AISI 1045 steel toughness in the different corrosion degrees by the salt immersion and fog chamber methods.

IOP Conf. Series: Materials Science and Engineering 1154 (2021) 012033 doi:10.1088/1757-899X/1154/1/012033

Data obtained in the impact test shows below.

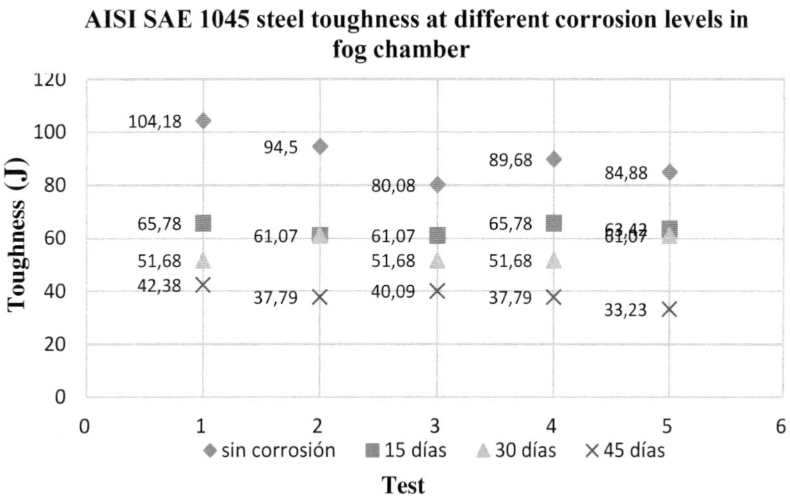

Figure 21 AISI SAE 1045 steel toughness at different corrosion levels in fog chamber

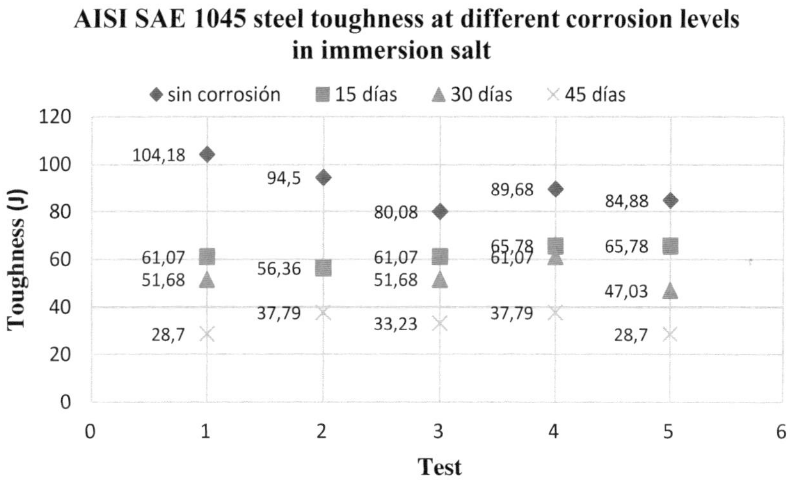

Figure 22 AISI SAE 1045 steel toughness at different corrosion levels in immersion salt

Material exposure times	Corrosion methods	
	Fog chamber	**Salt immersion**
Without corrosion	90,664 J	90,664 J
15 days	63,424 J	62,012 J
30 days	55,436 J	53,564 J
45 days	38,256 J	33,242 J

Table 3 AISI SAE 1045 steel average toughness according to applied corrosion methods and exposure times

With the specimen´s toughness average value that was not any corrosion type subjected, as a reference, it can be seen how this property is affected with time to exposure corrosion as indicated[6].

A toughness reduction of 30%, 38.9% and 57.8% was obtained due to in fog chamber corrosion in the periods of 15, 30 and 45 days, in the same way for saline immersion corrosion at exposure times 15, 30 and 45 days of 31.6%, 40.9% and 63.3% respectively, this being the most critical exposure.

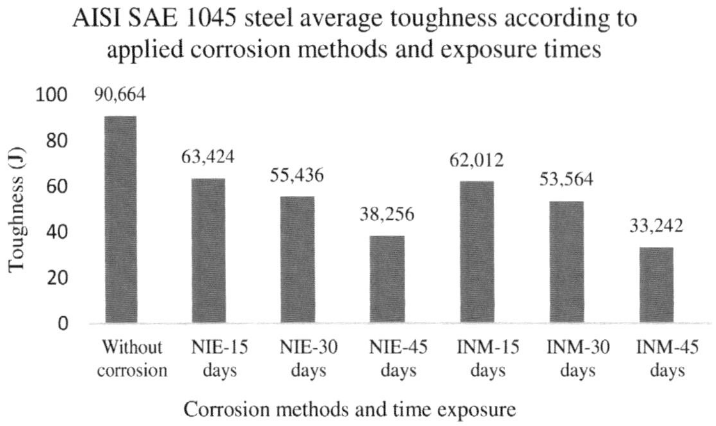

Figure 223 AISI SAE 1045 steel average toughness according to applied corrosion methods and exposure times

4 Conclusions

Corrosion affects the AISI 1045 steel toughness an ascending way according to the material´s exposure time to the presents oxidizing agent. for the immersion test case it´s evidenced that at the end of 45 days the material has lost approximately one 63% the ability to absorb energy by impact, which indicator clear of the deterioration level material suffered.

When observing the specimen subjected fracture with 45 days of corrosion, based on the fracture development collected information[7], it can be evidenced that the steel structure has deteriorated presenting a rough surface with signs of that corrosion has penetrated the specimen interior weakening its structure and making it less resistant to impact fractures.

Comparing the corrosion procedures, the immersion is more aggressive, degrading the material at a higher speed, which is observed in the toughness results.

A two-corrosion test representative characteristic used for this investigation is that after 30 days the material exposure, the material's structure deterioration seems to drastically increase, which has in toughness direct impact.

The longer exposure time, it´s see that, the tenacity difference increases, possibly due to the passivity phenomenon that can occur in a less aggressive environment.

References

[1] C. G. de A. S.A., "SAE 1020 Y SAE 1045.," 2013.

[2] E. A. Pérez Ruiz, A. C. Galeano Perilla, and L. I. Negrín Hernández, "Evaluación de la dureza y microestructura del acero 1045 templado en sustancias refrigerantes no tradicionales," *Sci. Tech.*, vol. 21, no. 3, p. 213, 2016.

[3] L. Sáenz and E. Patiño, "Impact toughness of the duplex stainless steel SAF 2304, with heating between 1100°C y 1300°C / Tenacidad de impacto de un acero inoxidable dúplex SAF-2304, con calentamientos entre 1100°C y 1300°C," 2018.

[4] C. G. Cárdenas Arias, A. D. Rincón Quintero, A. Santos Jaimes, and C. L. Sandoval Rodriguez, "Elasticity modulus variation of the AISI SAE 1045 steel subjected to corrosion process by chloride using tension test destructive," 2020.

IOP Conf. Series: Materials Science and Engineering 1154 (2021) 012033 doi:10.1088/1757-899X/1154/1/012033

[5] E. Medina Bejarano, "Ensayo De Impacto Charpy," no. October, 2017.
[6] P. Cortes, M. T.; Ortiz, "Corrosion," *Apunt. científicos uniandinos*, vol. 4, 2004.
[7] J. Niño, "Propiedades de escalamiento en la fractura de un acero hipoeutectoide," Universidad Autónoma de Nuevo León, 2003.

IOP Conf. Series: Materials Science and Engineering 1154 (2021) 012034 doi:10.1088/1757-899X/1154/1/012034

Implementation of a prototype water electrolysis system as an alternative to produce hydrogen

B E Tarazona-Romero[1 2], C L Sandoval-Rodriguez[1 3], J G Ascanio-Villabona[1 2], A Quintero-Ruiz[1 3], R E Páez-Castro [1] and O A Acosta-Cárdenas[1]

[1] Automation and Control Energy Systems Research Group (GISEAC), Faculty of Natural Sciences and Engineering, Electromechanical engineering, Unidades Tecnologicas de Santander (UTS), Student Street No 9-82, 680005, Bucaramanga, Colombia.
[2] Energy in Building research group (ENEDI), Doctoral Program in Energy Efficiency and Sustainability in Engineering and Architecture, Department of Machines and Thermal Engines, University of the Basque Country (UPV / EHU), Engineer Torres Quevedo Square, 1, 48013, Bilbao, BI, Spain.
[3] Doctoral Program in Electronic and Communications, Department of Communications Engineering, University of the Basque Country, UPV/EHU, 48013 Bilbao, Spain.

Abstract. The objective of the development of the prototype or scale model of production of hydrogen and oxygen through the electrolysis process of water is to generate an alternative with a sustainable approach, which proposes the supply or source of energy initially with a direct current power source. replaceable in the future by a photovoltaic solar system. For the development of the prototype, it began with the academic compilation of information, then the model was made in CAD, followed by the need to integrate an electronic control system and a mechanical system to obtain the final model, materials were selected and used premises for the development of the two systems, the electronic and physical schematization was carried out, chaired by the assembly and field experimentation. It is important to highlight that the constructed scale model generates hydrogen and oxygen with a low current consumption and a constant voltage, the maximum point of gas generation is the maximum point of energy consumption and its relationship is almost linear.

Keywords: Electrolysis of water, prototype, Hydrogen production, photovoltaic system and CAD modeling.

1 Introduction

Electrolysis[1] is an electrochemical process[2] that uses electrical conductivity [3] of some substances or materials to generate a non-spontaneous oxidation-reduction reaction [4] . Electrolysis derives from electrolytes that are on ionic conductors of positive or negative loads that have the ability to carry electrical energy, generate electrical conductivity in materials and substances [5] . Water electrolysis requires water not to be in its pure state, as pure water is not conductive so it must have some concentrations of salts and some minerals [6] [7].

Electrolysis consists of the chemical decomposition of a substance by means of electricity (electro-electricity and lysis - destruction) [7] . Due to the electric current circulating from the cathode causing reduction, towards the oxidation anode, provided that between them a conductive substance (electrolyte)is present[8]. For the production of electrolysis in water you need a dissolved electrolyte[9] (to increase its conductivity) to introduce two internal electrodes [10] (anode and cathode) being connected to a DC source, the electrolyte transformation of water into hydrogen and oxygen will occur requiring high energy consumption, so it is only used for practical purposes exceptionally [11] [13] [14].

It is important to define the term renewable energy[15] [16]: they are clean, endless energy sources that differ from fossil fuels[17] [18] especially in their diversity, abundance and willingness potential anywhere on the planet, but mainly in that they do not cause greenhouse gases[19] [20], nor polluting emissions[21] [22], therefore, it is proposed in this Paper how to produce hydrogen[23] from a small-scale electrolyze that is powered by the energy generated by a DC source that simulates power through photovoltaic solar panels[24]. The direct current source, will allow the fractionation of water molecules in oxygen and hydrogen, the latter, will be used as an energy vector or serve as an energy storage system.[25] [15].

Thus, the main objective of this document is to present a small-scale prototype as an alternative [26] to the production of hydrogen by water electrolysis[27]; simulating the power supply of a photovoltaic system through a conventional direct current source[28] [29]. The present technological development of an investigative nature, performs in an exploratory and descriptive way, the methodological development of the model[30], through the implementation of an electrolyze that takes the current of a switched power supply, which in the future can be replaced by a load regulator to receive power through photovoltaic solar panels[31].

Considering the This Paper begins with the methodological development[32] implemented for the construction of the prototype, as well as the description of the implemented materials. Followed by the experimentation presented in section 3 where tests of operability and operation of the prototype are carried out[33]. Subsequently, in section 4 the results are described and the functional behavior of the system is analyzed, and. Finally, the most significant conclusions of the development of the prototype are presented. Finally, this document will present a scale prototype of a hydrogen production system from the electrolysis of water[34], fed with a direct current source as an alternative of operation through photovoltaic solar panels as a sustainable and scalable prototype.

2 Methods and Materials

2.1 Methodology

For the development of a prototype at scale, **Figure *1*** presents the methodological workflow diagram, beginning with the system sizing process, which initially requires an academic information search where the initial need to develop CAD modeling, with two integrated systems for the proper functioning of the prototype: electronic control system and mechanical system. The first mentioned system must satisfy the needs of the project and control the scale model, selected as the central axis to carry out this task, the Atmega 328p microcontroller is identified, compatible with Arduino systems; simply remove the original microcontroller from the card and mount the Atmega 328p, it also highlights the flexibility to program the new microcontroller, since the free Arduino software allows its configuration. Also, a PCB board that connects to the microcontroller is required to have all the connections in one module.

On the other hand, it is necessary to build the mechanical section of the prototype that is fundamentally responsible for carrying out the physical electrolysis process and consists in the development of hydrogen cells (HHO); They are basically made up of two positive, three negative and sixteen neutral plates separated from each other by a neoprene gasket that separates the anode and the cathode, so that the gas generated in the anode (oxygen) does not mix with that generated in the cathode (hydrogen). This system will be fed by a water tank, connected through hoses. The prototype has temperature and level sensors, which send signals to the microcontroller, which in turn processes them and transmits them to a visual indicator or LCD screen. After the electrolysis process is run and the hydrogen and oxygen are obtained, all this will be deposited in a bubbling tank made of PVC. The hydrogen and oxygen generated by the electrolysis process will be measured by rotameters, these will be responsible for defining the flow generated

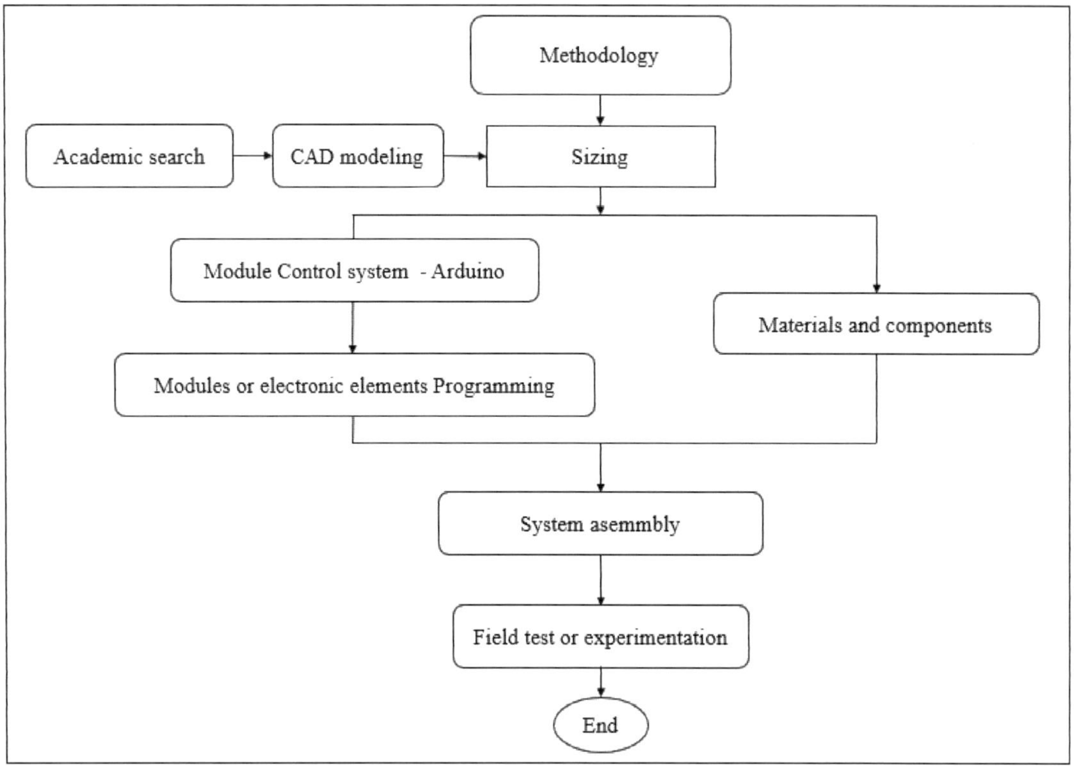

Figure 1. Working diagram

After developing the control system and mechanical system, the two are integrated into a compact module built with 1-inch structural tubes. Additionally, a satin stainless-steel plate was installed for mounting the aforementioned measuring instruments, indicator and two rotameters to measure flow. After assembling the prototype, the sensors or measuring instruments are calibrated. Finally, field tests or final experimentation are carried out, obtaining promising measurements of hydrogen and oxygen.

2.2 CAD Modelling

The 3D schematization of the module was developed after the collection of information using the CAD tool or SolidWorks Software, dimensioning it on a real scale in order to facilitate its construction. Modeling integrates mechanical and electronic control systems; Basically, the elements that compose it are: H2 cell, structure, water tank, instrument panel, power supply, control panel, hydraulic and gas system, among others (See **Figure *2***).

Figure 2. CAD modelling of the model

2.3 *Control System*

The materials required for the development of the electronic control system are described below:

2.3.1 PCB Board. A printed circuit board (PCB) as shown in Figure 3 is a sheet of rigid insulating material, covered by copper tracks on one or both sides to serve as connections between the various components to be mounted on it. Depending on the type of plate, copper can in turn be protected by a layer of photosensitive resin.*[35] [36]*. For our model it was necessary to develop it for integration with the microcontroller.

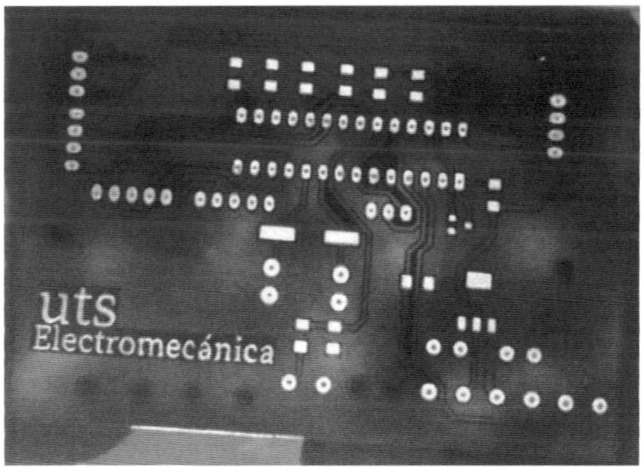

Figure 3. Printed circuit

2.3.2 *Atmega 328-p microcontroller.* Sensors and other electronic elements generate digital signals; these signals have to be received and mapped to the required values. To solve this is used the Atmega 328-p microcontroller, shown in **¡Error! No se encuentra el origen de la referencia.Figure 4**, this on a PCB board previously designed for operation and easy compatibility with its electronic components, simplifying connections and giving a better appearance to the module. It is important to note that this microcontroller is compatible with the Arduino platform, being thus easy to program.

Figure 4. Atmega 328P microcontroller.*[37]*

2.3.3 Programing IDE. The source code for the IDE is published under the GNU General Public License. The IDE supports C and C++ languages using special code structuring rules. The IDE supplies a software library of the Wiring project, which provides common I/O procedures, the user-written code requires only basic and simple functions. The IDE uses the avrdude program to convert the executable code into a hexadecimal-encoded text file that is loaded into board Atmega using a loading program into the board firmware.*[38]*

2.3.4 Current Sensor to Arduino. The hydrogen gas flow of an electrolyte cell is directly proportional to the intensity of the consumed current. In order to measure the current consuming the cell, it is necessary to implement the module an ammeter to know that variable. The maximum current to be measured in the cell is 20 amps, little less than the maximum current of the power supply. Therefore, the ACS712 current sensor of allegro (See **Figure 5**), which meets most requirements, will be used. This Hall effect current sensor provides the most economical solution with a fairly acceptable accuracy to measure the current intensity of the cell and with the appropriate size to adapt it to a PCB board being also compatible with it.*[39]*

Figure 5. ACS712 current sensor for Arduino. *[40]*

2.3.5 MAX6675 module for thermocouple (type K). The MAX6675 thermocouple sensor is a specialized analog-to-digital signal converter for type K thermocouples. With this module it is possible to easily connect a thermocouple to any microcontroller via a one-way SPI interface. Within this small circuit is the electronics needed to amplify, compensate and convert to digital the voltage generated by the thermocouple, which makes it very easy to connect a thermocouple to a microcontroller. The only problem is that this circuit is only achieved in SOIC encapsulation, so it is not so easy to use it in the protoboard. However, in this module we find the MAX6675 with all the necessary electronics and the appropriate terminals to facilitate its use. *[41]*

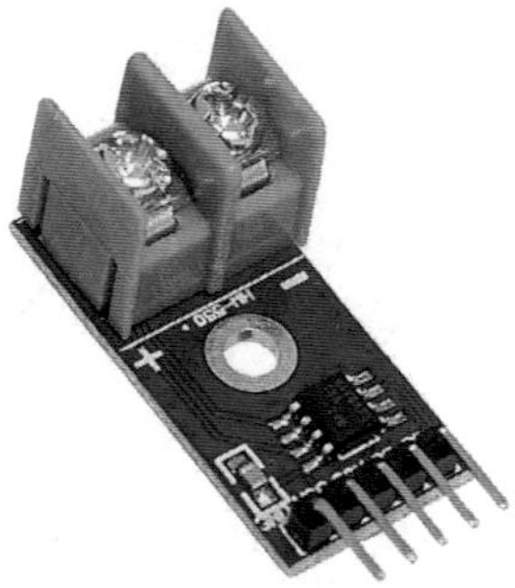

Figure 6. Max6675 module for thermocouple (type K) *[41]*

2.3.6 *Type K thermocouple.* This sensor supplies an electrical voltage depending on the temperature. A thermocouple doesn't directly measure temperatures, but the temperature difference between the hot end and the cold end. The combination of different metals induces certain signals that allow effective temperature measurement. Thermocouple probes vary depending on the requirements, in our case we opted for a thermocouple with a 10 mm probe for the tank and a thermocouple with a 100 mm probe for the cell.*[42]*

2.3.7 *Rotameters.* The rotameter or flowmeter is an industrial flow meter used to measure the flow of liquids and gases. The rotameter consists of a tube and a float. The float's response to flow changes is linear, and a flow range of 10 to 1 is standard*[43]*. Conclusively the ZEAST rotameter shown in Figure **7**7 is chosen, this one with a measurement range of 3 LPM (liters per minute) being among the desired measurement ranges for small-scale prototyping

Figure 7. Rotameters

2.4 Mechanical system

2.4.1 *H2 cell.* The cell, shown in Figure **8**8 consists of two positives exposed, three negative and sixteen neutral ones separated from each other by neoprene packaging and a separating membrane, two connections for liquid inlet and two for gas outlet, two 16 x 16 x 0.8 cm acrylic plates and two reinforcement frames.

Figure 8. H2 Cell

2.4.2 *Water tank.* The tank shall be connected to the cell and the bubblers by means of a hose circuit. In addition, it will have a temperature sensor and a level sensor that will send a signal to the microcontroller when the water level is very low.

2.4.3 *Deposit bubblers.* The bubbling devices are the devices responsible for removing excess water that comes out of the cell, as well as protecting it from possible gas combustion, which would cause a dangerous explosion. The bubblers were manually made from PVC pipe as it is a corrosion-resistant and easy-to-machine material.

2.4.4 *Power supply.* This 300w power supply will be responsible for sending electrical energy to the cells, so that they can perform the electrochemical electrolysis process.

2.4.5 *Instrument panel.* The instrument panel is where the flow meters, LCD display, LED indicators, control buttons and power switch will go. The material of this is satin stainless steel, for greater durability.

Figure 9. Instrument panel

2.4.6 Structure. The frame or chassis where the elements that make up the system are housed in general consists mainly of a 3/4-inch square tube (See Figure 10).

Figure 10. Structure

2.5 Flow diagram of electronic control system

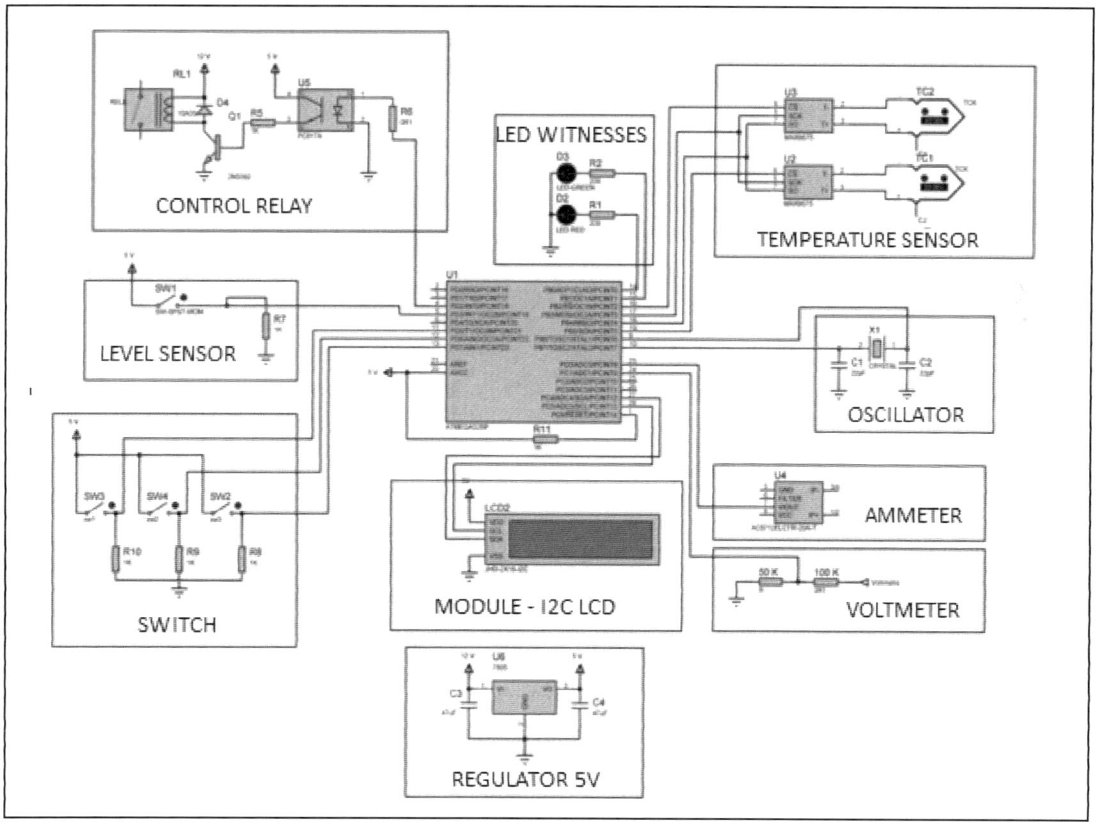

Figure 11. Flow diagram of electronic control system

The connection diagram shown in **Figure *11*.** , is made taking into account the requirements for the operation of the electronic system in general, this circuit is proposed that will connect the temperature sensors, level sensor, LCD display and other elements, to the microcontroller atmega 328P, this for the time of the assembly have a correct idea of what are the connection pins between the different devices

2.6 Module connection diagram

After CAD modeling, the identification of the elements of the module, the development of the electronic control system and its schematization, it is important to highlight the physical connection diagram of the module presented in **Figure *12***, which allows the assembly to be known in detail. made at the time of building the module.

It is important to highlight in the connection diagram, that in the H2 cell is where all the electrochemical process happens. Which is powered by a water tank that integrates a level sensor. The hydrogen and oxygen produced are deposited in a special container called "bubbling tank", being essential to know the outflow of gases generated by the H2 cell through a pair of flowmeters or rotameters, connected to the bubbling out.

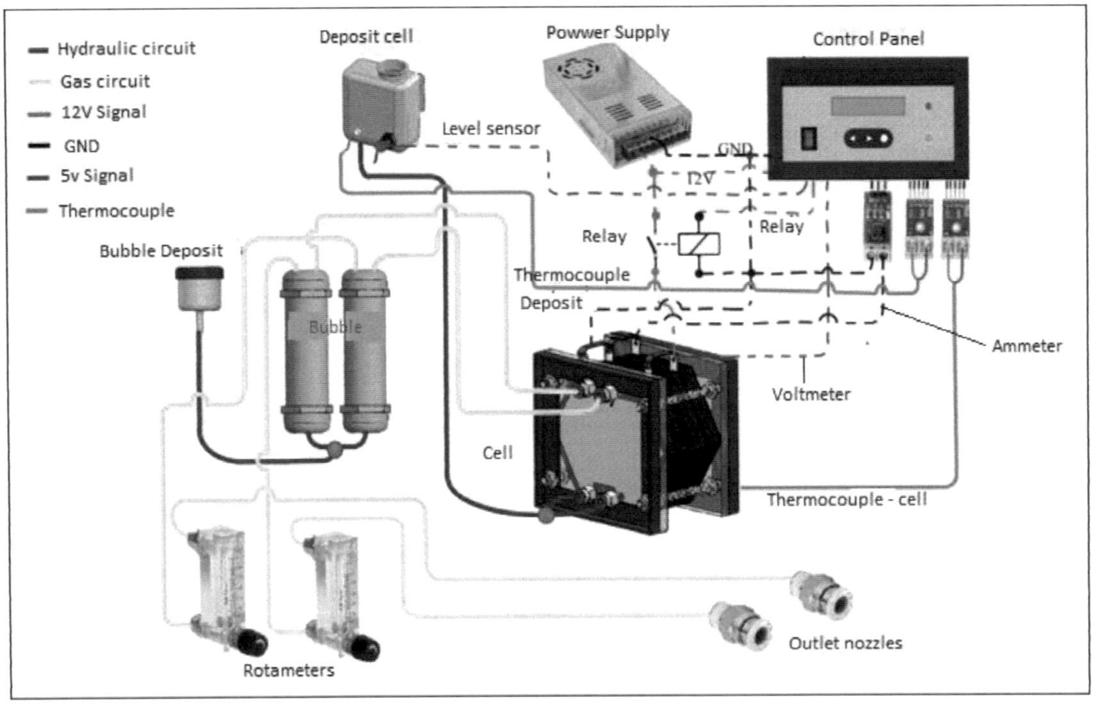

Figure 12. Model connection diagram

2.7 Experimentation

The experimentation process involves the functional checking of the different sensors implemented in the system, in order to guarantee that the measurements indicated on the module or LCD screen are correct. Starting, with the level sensor, in which manual tests are carried out, checking its proper operation. Subsequently, the tank and cell temperature sensors are verified, both must have the same or very close values when the cell is empty; the sensors present similar measurements as evidenced in figure x, corroborating their correct operation. (See **Figure *13***)

Figure 13. Deposit temperature and cells respectively

On the other hand, a crucial factor when carrying out the tests with the electrolysis module are the electrical variables of voltage and current, since these will allow projecting the consumption of the system and, based on this value, being able to determine the generation capacity of the photovoltaic solar system for future implementation. To measure the values, a voltage and current divider can be applied, or simply using Ohm's Law depending on the proposed electrical circuit. **Figure *14*** ... shows the voltage and current values calculated by the developed algorithm, contrasted with the values evidenced by a measuring instrument or multimeter.

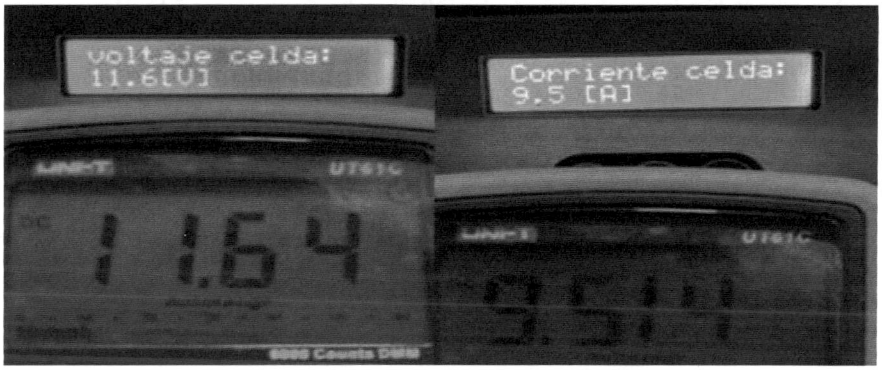

Figure 14. Comparison of voltages and cell current

Finally, with the parameters of the sensors adjusted and the verification of the voltage and current meters, a series of tests were performed on the module where the flow values were verified with respect to the hydrogen and oxygen gases, with which determined the production of the developed system.

3. Results and analysis

Figure *15* shows the final result of the dimensioning process that included, modeling, component selection, programming and module assembly. The final product is a scale model of an electrolysis system based on the use of water, which is versatile and portable, and also has direct current power source feeding in order to generate a sustainable system in the future.

Figure 15. Electrolysis module proposed

Once the module finishes its configuration and the measurement instruments are calibrated and / or verified, we highlight the indication through the LCD display of real data. The module is put into operation and a series of experimental tests are carried out that show production of hydrogen and oxygen, measured through two flow indicators or rotameters as presented in **Figure *16***, with a hydrogen flow of 2 liters per minute and an oxygen flow of 1 liter per minute. Hydrogen was produced through the connection of the plates to the positive terminal of the source and oxygen in the plates connected to the positive terminal.

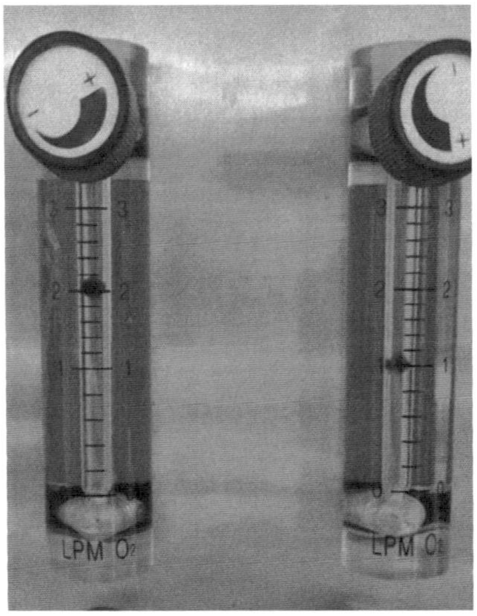

Figure 16. Rotameters

The module produced 1 litter of oxygen per minute and 2 litters of hydrogen per minute with a current draw of 16 amps at a constant voltage of 11.6 Volts. **Figure *17*** presents the values of hydrogen and oxygen flow with current variation, where it was shown that the flow of gases varies with the flow of current consumed by the cell, provided that the voltage remains constant. This increase in the flow of gases may be the product of the electrolytic solution of the water, that is, of the amount of sodium hydroxide that varies with respect to the electrical resistance of the water, generating the current change or the temperature change, product of water heating for the production of hydrogen despite not being a system designed to produce it in the vapor phase, which is usually more efficient.

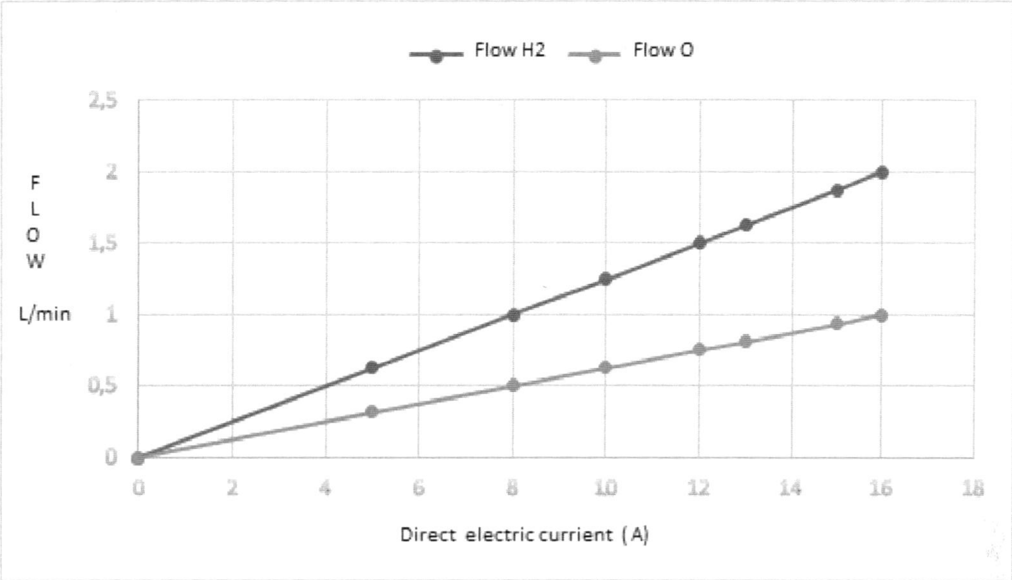

Figure 17. Flow (H2 ; O2) relative to the current

Additionally, the temperature reached by the cell with a current consumption of 16 amps is 65 ° C, with respect to the tank temperature which is 26 ° C, with a temperature increase of 39 ° C. This incurs a 150% increase compared to initial system temperature. Finally, the figure 18 shows the relationship between the flows of hydrogen and oxygen gases with respect to temperature, where a significant increase in temperature is evident as the gas is produced. If we compare it with the previous figure, the current increase is linear and the temperature increase is not.

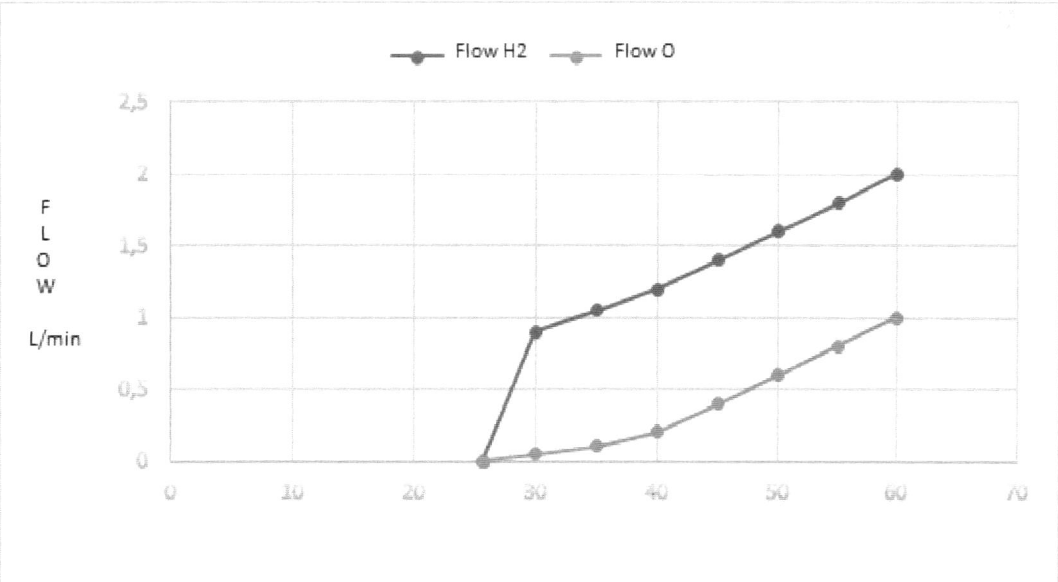

Figure 18. Flow (H2; O2) from temperature

4. Conclusions

The flow of hydrogen and oxygen produced is related to the cell current h2, and this current depends on the amount of dilute sodium hydroxide in the water, since it increases or decreases the electrical conductivity of the water, obtaining a maximum current consumed from 16 amps for our case study with a production of 2 liters per minute of hydrogen and one liter per minute of oxygen. Two aspects are important to highlight; The first is that in the present study the composition of the water was not analyzed, so the amount of sodium hydroxide present is unknown, and the second is that, depending on the region where the equipment is used, production may be affected or benefited by the composition of the water.

The temperature in the cell is another factor that directly influences the amount of production of hydrogen and oxygen by the electrolysis module. In the system presented in this Paper, it was observed that with the increase in the flow of gases, in the same way, the increase in temperature up to 60 ° C. It is important to highlight that the system benefits from the increase in temperature, it cannot be compared with a Large-scale water vapor electrolysis system that handles temperatures up to 700 ° C.

The current system presents losses due to convection, conduction and radiation; this is because the increase in the temperature of the cell causes the water to evaporate, generating humidity present in the exhaust gases. For the current model, saline gases were not measured, so their composition is unknown.

The current system is a functional module, capable of producing hydrogen and oxygen from the use of a direct current source. The system was designed in such a way that the module can supply electrical consumption through the power source for a photovoltaic solar system that generates the demand for the prototype. The use of a 12v charge regulator and a system between two and three panels connected in parallel is recommended. This recommendation is made from a manual estimation product of the ampere hour method, for this reason it is important to validate the chosen system through a specialized simulation.

References

[1] K. J. Laidler, «Historia de la electrólisis», *Educ. Quím.*, vol. 1, n.º 3, Art. n.º 3, ago. 2018, doi: http://dx.doi.org/10.22201/fq.18708404e.1990.3.66993.

[2] «Significado de Electrólisis», *Significados*. https://www.significados.com/electrolisis/ (accedido ago. 02, 2020).

[3] «INGENIERÍA Y COMPETITIVIDAD». https://revistaingenieria.univalle.edu.co/ (accedido ago. 02, 2020).

[4] R. Mostaza Mangas, «Reacciones de oxidación-reducción y aplicaciones», 2018, Accedido: ago. 02, 2020. [En línea]. Disponible en: http://uvadoc.uva.es/handle/10324/31281.

[5] «Significado de Conductividad eléctrica», *Significados*. https://www.significados.com/conductividad-electrica/ (accedido ago. 02, 2020).

[6] «La Electrólisis». https://www.quimicas.net/2015/08/la-electrolisis.html (accedido ago. 02, 2020).

[7] J. Tapias, C. Sandoval, y J. Sánchez, «ANÁLISIS DE PROSPECTIVA DEL SECTOR ENERGÉTICO DE COLOMBIA, PARA LA INTEGRACIÓN DE FUENTES FOTOVOLTAICAS EN LOS SISTEMAS DE DISTRIBUCIÓN DE ENERGÍA ELÉCTRICA APLICANDO UNA REVISIÓN EN BASES DE DATOS CIENTÍFICAS», *Rev. Colomb. Tecnol. Av. RCTA*, vol. 2, nov. 2018, doi: 10.24054/16927257.v32.n32.2018.3034.

[8] A. Pinto, «Procesos de electrolisis», Accedido: ago. 02, 2020. [En línea]. Disponible en: https://www.academia.edu/30317238/Procesos_de_electrolisis.

[9] «Procesos electrolíticos y leyes de Faraday», *Cienciaonthecrest*, ago. 13, 2015. https://cienciaonthecrest.com/2015/08/13/procesos-electroliticos/ (accedido ago. 02, 2020).

[10] «Electroquimica.pdf». Accedido: ago. 02, 2020. [En línea]. Disponible en: https://fisquiweb.es/Apuntes/Apuntes2Qui/Electroquimica3.pdf.

[11] «que son los electrones internos? - Brainly.lat». https://brainly.lat/tarea/4419186 (accedido ago. 02, 2020).

[12] S. M. Martínez Amaya y V. E. Ramírez Dilas, «Producción de Energía Eléctrica y Gas de Síntesis a Partir de la Electrólisis del Agua con Energía Solar y uso del Co2», *Univ. Nac. Pedro Ruiz Gallo*, abr. 2018, Accedido: ago. 02, 2020. [En línea]. Disponible en: http://repositorio.unprg.edu.pe/handle/UNPRG/1794.

[13] B. E. Tarazona-Romero, Á. Campos-Celador, Y. A. Muñoz-Maldonado, C. L. Sandoval-Rodríguez, y J. G. Ascanio-Villabona, «Prototype of lineal solar collector Fresnel»:, *Visión Electrónica*, vol. 14, n.º 1, Art. n.º 1, ene. 2020, doi: 10.14483/22484728.16013.

[14] B. E. Tarazona-Romero, A. Campos-Celador, Y. A. Muñoz-Maldonado, J. G. Ascanio-Villabona, M. A. Duran-Sarmiento, y A. D. Rincón-Quintero, «Development of a Fresnel Artisanal System for the Production of Hot Water or Steam», en *Recent Advances in Electrical Engineering, Electronics and Energy*, oct. 2020, pp. 196-209, doi: 10.1007/978-3-030-72212-8_15.

[15] «Energias Renovables». Accedido: ago. 02, 2020. [En línea]. Disponible en: https://www.energias-renovables.com/ficheroenergias/productos/pdf/ConPago/Revista571.pdf.

[16] J. G. Ascanio-Villabona, C. L. Sandoval-Rodriguez, A. D. Rincón-Quintero, B. E. Tarazona-Romero, y R. E. Paez-Castro, «Building a prototype for functional analysis of the energy potential of the water flow in pipe ½ "using microturbines applied to Unidades Tecnológicas de Santander», *IOP Conf. Ser. Mater. Sci. Eng.*, vol. 844, n.º 1, p. 012056, may 2020, doi: 10.1088/1757-899X/844/1/012056.

[17] «Agotamiento de los combustibles fósiles y emisiones de CO2...» http://www.usc.es/econo/RGE/Vol19_1/castelan/resu1c.htm (accedido ago. 02, 2020).

[18] A. D. Rincón-Quintero *et al.*, «Manufacture of hybrid pieces using recycled R-PET, polypropylene PP and cocoa pod husks ash CPHA, by pneumatic injection controlled with LabVIEW Software and Arduino Hardware», *IOP Conf. Ser. Mater. Sci. Eng.*, vol. 844, p. 012054, jun. 2020, doi: 10.1088/1757-899X/844/1/012054.

[19] «INFORMACIÓN TECNICA SOBRE GASES DE EFECTO INVERNADERO Y EL CAMBIO CLIMÁTICO». Accedido: ago. 02, 2020. [En línea]. Disponible en: http://www.ideam.gov.co/documents/21021/21138/Gases+de+Efecto+Invernadero+y+el+Cambio+Climatico.pdf/7fabbbd2-9300-4280-befe-c11cf15f06dd.

[20] J. G. Ascanio-Villabona, L. A. D. Portillo-Valdés, O. Lengerke-Pérez, B. E. T. Romero, A. D. Rincón-Quintero, y M. A. Durán-Sarmiento, «Analysis of the Energy Potential of a Tangential Microturbine for Application in a Passivhaus Environment», en *Recent Advances in Electrical Engineering, Electronics and Energy*, oct. 2020, pp. 181-195, doi: 10.1007/978-3-030-72212-8_14.

[21] D. G. P. Darquea, «Estudio de emisiones contaminantes utilizando combustibles locales», *INNOVA Res. J.*, vol. 3, n.º 3, pp. 23-34, 2018.

[22] C. L. Sandoval-Rodriguez, B. E. Tarazona-Romero, O. Lengerke-Perez, C. G. Cárdenas-Arias, D. C. D. Diaz, y O. A. A. Cárdenas, «Descriptive Study of a Rotary Machine Affected by Misalignment and Imbalance Applying the Wavelet Transform», en *Recent Advances in Electrical Engineering, Electronics and Energy*, oct. 2020, pp. 226-242, doi: 10.1007/978-3-030-72212-8_17.

[23] Y. G. Besa y F. J. P. Lucena, «PRODUCCIÓN DE HIDRÓGENO A PARTIR DE ENERGÍA SOLAR FOTOVOLTAICA. ANÁLISIS DE SENSIBILIDAD DE PARÁMETROS DE DISEÑO», p. 136.

[24] B. M. Roman Padilla y B. C. Arevalo Cordovilla, «Análisis de la producción de Hidrógeno a partir de energía solar fotovoltaica», *Repos. Univ. Estatal Milagro*, may 2018, Accedido: ago. 02, 2020. [En línea]. Disponible en: http://repositorio.unemi.edu.ec//handle/123456789/3894.

[25] «La importancia de las energías renovables | ACCIONA». https://www.acciona.com/es/energias-renovables/ (accedido ago. 02, 2020).

[26] A. D. Rincón-Quintero, L. A. D. Portillo-Valdés, A. Meneses-Jácome, J. G. Ascanio-Villabona, B. E. Tarazona-Romero, y M. A. Durán-Sarmiento, «Performance Evaluation and Effectiveness of a Solar-Biomass Hybrid Dryer for Drying Homogeneous of Cocoa Beans Using LabView Software

and Arduino Hardware», en *Recent Advances in Electrical Engineering, Electronics and Energy*, oct. 2020, pp. 238-252, doi: 10.1007/978-3-030-72208-1_18.
[27] R. Chaubey, S. Sahu, O. O. James, y S. Maity, «A review on development of industrial processes and emerging techniques for production of hydrogen from renewable and sustainable sources», *Renew. Sustain. Energy Rev.*, vol. 23, pp. 443-462, jul. 2013, doi: 10.1016/j.rser.2013.02.019.
[28] «Evaluación de un sistema doméstico DC utilizando una fuente solar fotovoltaica». https://repository.usta.edu.co/handle/11634/17666 (accedido ago. 02, 2020).
[29] B. E. Tarazona, C. L. S. R, C. G. C. Arias, J. G. A. V, y J. J. V. N, «Detection of structural alterations in metal bodies: An approximation using Fourier transform and principal component analysis (PCA)», *Sci. Tech.*, vol. 25, n.º 2, Art. n.º 2, jun. 2020, doi: 10.22517/23447214.23501.
[30] «La efectividad de un modelo metodológico mixto para la enseñanza-aprendizaje de español como lengua extranjera». https://scielo.conicyt.cl/scielo.php?script=sci_arttext&pid=S0718-93032009000100004 (accedido ago. 02, 2020).
[31] Colegio Oficial de Ingenieros de Telecomunicación y Grupo de Nuevas Actividades Profesionales, *Energía solar fotovoltaica*. Madrid: Colegio Oficial Ingenieros de Telecomunicación, 2007.
[32] «DESARROLLO-METODOLOGICO-PARA-EL-ANALISIS-DE-LA-VIABILIDAD-DE-UN-PROYECTO-DE-MICROGENERACION.pdf». Accedido: ago. 02, 2020. [En línea]. Disponible en: https://www.researchgate.net/profile/I_Vera-Romero/publication/315642430_DESARROLLO_METODOLOGICO_PARA_EL_ANALISIS_DE_LA_VIABILIDAD_DE_UN_PROYECTO_DE_MICROGENERACION/links/58d6ae2345851533788c9a12/DESARROLLO-METODOLOGICO-PARA-EL-ANALISIS-DE-LA-VIABILIDAD-DE-UN-PROYECTO-DE-MICROGENERACION.pdf.
[33] L. A. Sánchez, «DISEÑO, CONSTRUCCIÓN Y EVALUACIÓN DE UN MINI- SISTEMA ELECTROLÍTICO PARA USO DIDÁCTICO.», p. 52.
[34] P. Erice y D. Alejandro, «Evaluación Técnico Económica de una Planta de Producción de Hidrógeno Mediante Electrólisis de Agua Utilizando Energía Eléctrica Producida con Celdas Fotovoltaicas de Alta Eficiencia», 2009, Accedido: ago. 02, 2020. [En línea]. Disponible en: http://repositorio.uchile.cl/handle/2250/103384.
[35] «Desarrollo de una metodología para la elaboración de tarjetas PCB de forma óptima'». Accedido: ago. 02, 2020. [En línea]. Disponible en: http://ribuni.uni.edu.ni/2265/1/92097.pdf.
[36] B. E. Romero-Tarazona, C. L. Rodriguez-Sandoval, J. G. Villabonai-Ascanio, y A. D. Rincón-Quintero, «Development of an artificial vision system that allows non-destructive testing on flat concrete slabs for surface crack detection by processing of digital images in MATLAB», *IOP Conf. Ser. Mater. Sci. Eng.*, vol. 844, p. 012058, jun. 2020, doi: 10.1088/1757-899X/844/1/012058.
[37] «ATMega328P Microcontroller Pinout, Pin Configuration, Features & Datasheet». https://components101.com/microcontrollers/atmega328p-pinout-features-datasheet (accedido ago. 02, 2020).
[38] «Arduino UNO R3 CH340», *Ingenieria Colectiva*, mar. 20, 2020. https://siticed.com.mx/2020/03/20/arduino-uno-r3/ (accedido ago. 02, 2020).
[39] «Módulo sensor de corriente ACS712 5 A», *Electronilab*. https://electronilab.co/tienda/modulo-sensor-de-corriente-acs712-5/ (accedido ago. 02, 2020).
[40] «ACS712 Current Sensor: Features, How it works, Arduino Guide», *Latest open tech from seeed studio*, feb. 15, 2020. /blog/2020/02/15/acs712-current-sensor-features-how-it-works-arduino-guide/ (accedido ago. 02, 2020).
[41] «MAX6675 Módulo interfaz para termopar tipo K», *Geek Factory*. http://www.geekfactory.mx/tienda/modulos-para-desarrollo/max6675-modulo-interfaz-termopar/ (accedido ago. 02, 2020).
[42] «Termopar tipo K - WIKA España». https://www.wika.es/landingpage_thermocouple_es_es.WIKA (accedido ago. 02, 2020).
[43] «Rotámetro». https://es.omega.com/prodinfo/rotametros.html (accedido ago. 02, 2020).

Electromechanical Hand Prototype for the Simulation of the Opening and Closing Movement.

C L Sandoval-Rodriguez[1-2], E Y Veslin- Díaz[1], B E Tarazona-Romero[1-3], J G Ascanio- Villabona [1-3], C G Cárdenas-Arias[1-4], C A Angulo-Julio[1] , .

[1] Energy Systems, Automation and Control Research Group (GISEAC), Design and materials research group (DIMAT), Electromechanical Engineering, Faculty of Natural Sciences and Engineering, Unidades Tecnologicas de Santander (UTS), Road No 9-82 students, 680,005, Bucaramanga, Colombia.
[2] Doctoral Program in Electronic and Communications, Department of Communications Engineering, University of the Basque Country, UPV/EHU, 48013 Bilbao, Spain.
[3] Doctoral Program in Energy Efficiency and Sustainability in Engineering and Architecture, Department of Machines and Thermal Engines, University of the Basque Country (UPV / EHU), Engineer Torres Quevedo Square, 1, 48013, Bilbao, BI, Spain.
[4] Doctoral Program in Engineering of Materials and Sustainable Processes, University of the Basque Country, UPV/EHU, 48013 Bilbao, Spain.

csandoval@correo.uts.edu.co

Abstract. This work aims to describe the construction of a low-cost functional prototype of an electromechanical hand with opening and closing capacity. The device, built on an ABS filament extruder, is actuated by a MG995 high torque servo motor, controlled by an Arduino-UNO R3 board, from the processing of electromyographic signals carried out by: MyoWare Muscle Sensor. The realization and functional evaluation of the device is developed in six phases: The first is to select the thermo-deformable materials, the second is to select the electronic components; the third phase is the design of the electromechanical hand prosthesis, the fourth phase was the construction of the mechanical parts; the fifth phase is the development of the control program on the electronic card; and the sixth phase of the project consists of validating the developed prototype by observing its opening and closing through the oscilloscope. Due to the low cost of development and ease of reproducibility (because it is built under the open-source philosophy), this work is expected to make a contribution to the disabled community, especially the one with limited resources.

Keywords: **Hand prototype, Electromiogrphic Signals, 3d print**

1. Introduction.

THIS paper describes the construction of a functional electromechanical hand model for opening and closing functions. Therefore, the information on the materials, control, methods and methodology is described in it.
The device has a muscle sensor (EMG) brand Myoware, used in the development of biomechanical projects [1] [2] y [3] , and an Arduino UNO electronic board is used as controller.

This work can generate a contribution to the disabled community, especially those with limited resources, by providing a low-cost device which can be improved and replicated, since this system was built with the philosophy of open access (open-source). There are currently 2,632,255 people with disabilities, that is, 6.4 percent of the general population [4]. Of this important percentage, 20.8% have a permanent alteration of the body, hands, arms and legs. On the other hand, according to data from the same registry, 49.3% of all disabled people belong to the subsidized regime, while 29.7% do not have any type of affiliation to a general health system. Therefore, the realization of low-cost prostheses responds to an important need for the lives of a large number of low-income Colombians without the economic capacity to use commercial prostheses.

Rapid prototyping developments, the integration of manufacturing with CAD programs, open access to biological information through repositories, and even the making available of components for modification and construction, has been shown to significantly reduce prosthetic design and construction costs [5]. For this reason, for the execution of this prosthesis, the UTS make available a 3D printing machine with autonomous electrical and mechanical positioning, while the hand is designed on the Repetier free access platform. In turn, a low-cost controller is implemented for the processing of the EMG signals and the actuation of the device. For the development of this project, a descriptive methodology was used with a qualitative and experimental approach divided into five stages: 1) bibliographic review of the state of the art, 2) consultation with experts, 3) definition and selection of components, 4) final tests prototype validation and 5) system results.

2. Prototype development: Manufacture of the external structure of the prototype

For the design and manufacture of the structure of the hand, CAD and Repetier software were used, in order to create the virtual model that will later be printed and assembled. Repetier is a free software for 3D printing with several applications for robotics developments [6] which also allows remote printing and storage of developed parts [7].

1) *Mounting files to Reptier-Host software:* Then the flexy-hand file is loaded [2] representing the structure of the hand in CAD, loaded in the Repetier – host software see (Fig. 1). The design was prepared and limited for the development of 3D printing, and the construction parameters such as the type of material to be printed, internal structure of the pieces, speed and quality of the printing were adjusted.

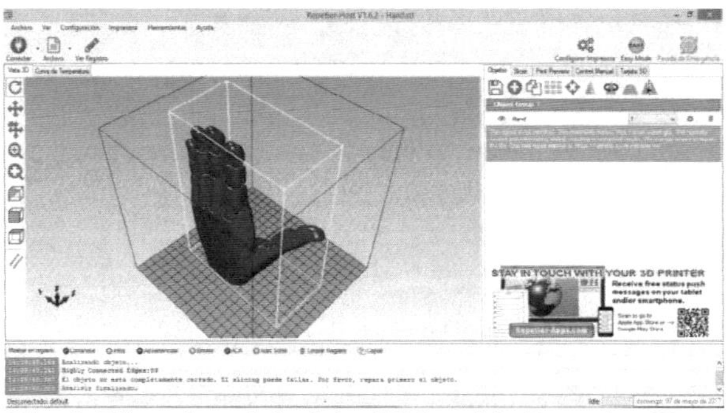

Fig. 1. Virtual model of the 3D hand in Repetier-Host.

2) *Printed Components:* The components are grouped in the program template and subsequently the print order is executed. How material was ABS selected. Printing took around 10 hours in total for printing only part of the palm and fingers. Additionally, it took approximately 4 hours to print the rest of the components. The pieces were built using a honeycomb-type internal filling, which is the most recommended internal structure for biomedical applications given its structural

capabilities in which low weight, high rigidity and porosity stand out [8]. In Fig. 2 the printing result is shown.

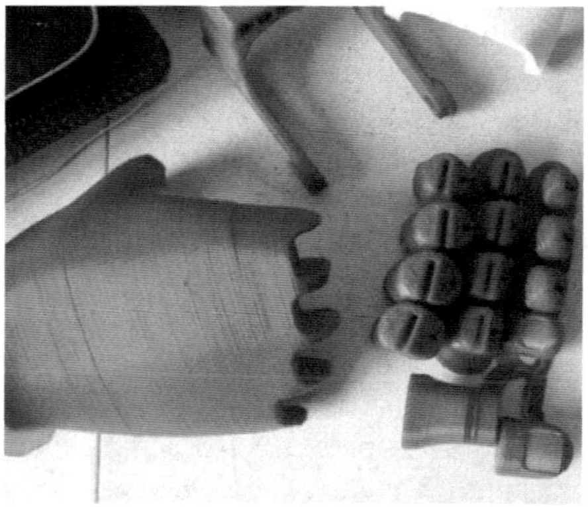

Fig. 2. Resulting of printed components

3) *Assembling the components:* A type of joint made of silicone was designed to join the fingers of the palms (Fig. 3, left) These joints were made from 3D printed molds and injected silicone. This operation allowed creating a set of 14 joints to unite all the fingers. The silicone joints were inserted into the fingers through horizontal grooves in such a way that they remain locked. This adjustment allows the movement of the pieces without disengagement (Fig. 3, right).

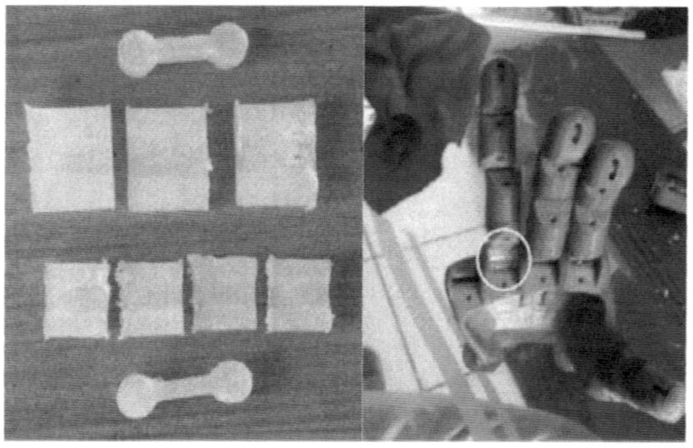

Fig. 3. Silicone finger joints, left. Using the feet to develop the ensemble, right.

4) *Installation of the finger tension cables:* Nylon threads are inserted through each of the holes in the hand designed for this purpose, the process begins from each upper end of the fingers, through a hook in which a tie is made to hold it, and later the tensioner is inserted through each of the ducts until reaching the outlet located in the palm of the hand (Fig. 4).

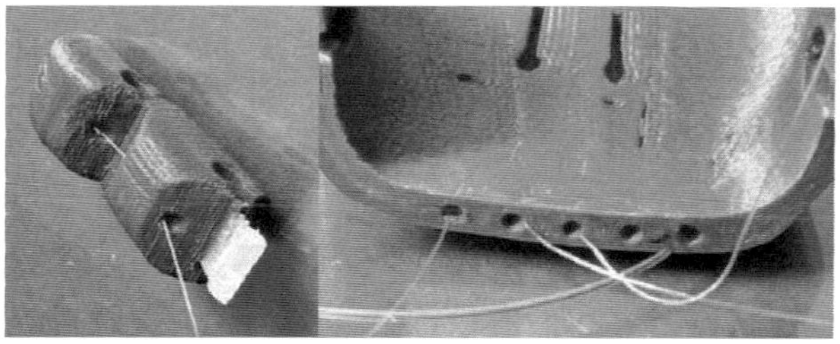

Fig. 4. Installation of the tensioning cables, left, assemble through one finger, right, exit of the tensioner by the palm of the hand.

5) *Assembly of all components:* The assembled pieces make up the final structure of the hand, to which the electronic and mechanical elements that will give it programmed mobility will be adapted. In Fig. 5 we can see the fingers already attached, a piece was designed, which is attached to the wrist of the hand, this piece is in the form of a tablet, and its purpose is to contain an Arduino uno R3 board, which will be used as control system for the mechanism.

3. Prototype development: Linking the electronic components to the control board

The connection of all electrical and electronic components was carried out in accordance with the requirements for which it was desired to develop the project. A muscle sensor was used to read the biceps activation of the arm, electrical conductors and electrode pads complement the installation of the sensor, meanwhile to generate the movement a high torque servo motor was implemented.

1) *Arduino R3 Control Board:* The board used to control the movements is the Arduino uno R3 which has an open code of programming. This card was selected for its versatility, low cost, easy installation, as well as having a complete hardware with multiple power inputs, an Atmel AVR microcontroller and USB port. Arduino UNO has demonstrated its ability to process multiple biological signals with biomedical applications, such as the control of heart beat [9]. electromyography [10], and control of elbow movement [11], among others.

2) *Muscle Sensor:* To read the muscle signals, a Myoware electromyographic (EMG) sensor was implemented. This device has the peculiarity of having both sensor and transducer and control software incorporated in a single device, which works with Arduino hardware and software. The electronic card of the sensor, works to measure the electrical signal of the muscle fiber cells, performing preprocessing of this, through a band-pass filter and a rectification process, the signal is converted into an analog signal from 0 to Vs (source voltage, set to 6 V) that is easily read by the microcontroller. The magnitude of the signal is directly proportional to the activity in the muscle. Its operation is simple, since the electrodes have the function of capturing the amplitude of the muscle associated with the opening movement of the hand and it is configured to interpret the output voltage and possible presence of the activity to assist [12].

3) *MG995 servo motor:* The servomotor used is the Tower Pro MG995, which provides up to 11Kg-cm of torque and metal gears and high robustness, designed to work with high loads. This motor was selected for its applicability in biomedical projects [13], [14] and [15] in addition to being compatible with the control card.

4) *Connection and assembly of all electrical and electronic elements:* The electronic card and the servomotor were mounted to the red piece, which functions as a base for the servo and the electrical components, the servo was located on its bases and successively mounted on top of the servo the tablet that carries the electronic card was located. In Fig. 5 you can see the profile of the assembly of these two elements. Both the servo and the card tablet are attached by means of screws.

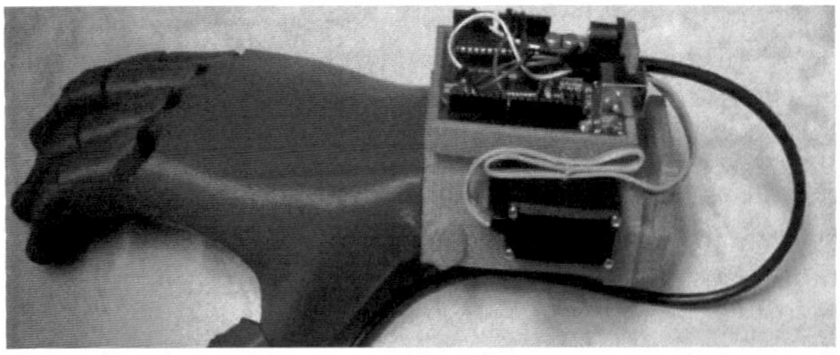

Fig. 5. Visualization of the assembly of the servo and electronic card.

The electrical elements were linked according to the traditional servo motor connection system, and the manufacturer's recommendations to the Arduino board, occupying a PWM output to control the servo motor [16], an analog input from the Arduino was also used for the signal EMG sensor. Both devices were powered with 5V available from an external source so as not to limit the servo motor current. Fig. 6 shows the final assembly of the prototype with the sensor attached and its implantation in the arm to perform the functional tests.

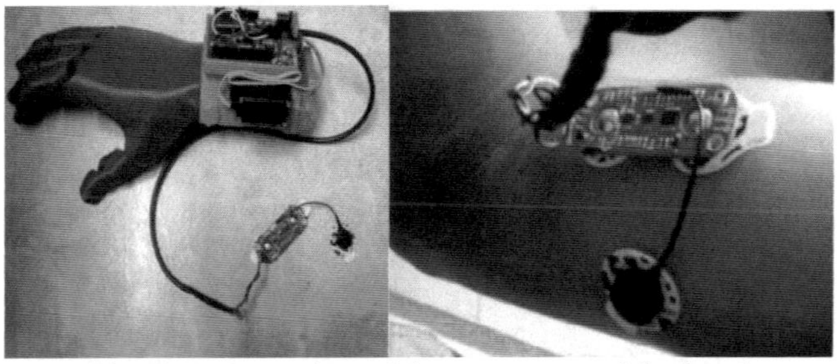

Fig. 6. Presentation of the prototype of the hand.

4. Prototype development: Control code programmed in Arduino IDE

The electronic control board uses Arduino software. This is an open source program. For this work, the ARDUINO 1.8.1 version was used to generate the lines of programming code. The servo.h library was necessary for managing the servomotor to associate the servo with the algorithm. Unlike the sensor, which does not require any Library. A 10-line control code was developed which can be seen in Fig. 7. The programming code is designed so that when the muscle signal is activated and by means of a preset value, the <If - else> operation determines if it should close the hand, otherwise if the muscle signal is attenuated the hand should be opened. To keep the hand closed, the muscle contraction must be constant. The general flow diagram of the program is presented in Fig. 8.

sensor_de_musculo_simple Arduino 1.8.1
Archivo Editar Programa Herramientas Ayuda
sensor_de_musculo_simple

```
#include <Servo.h>

Servo myservo;

const int threshValue = 250;

void setup() {

myservo.attach(3);
}

void loop() {
int value = analogRead(A3);

 if(value < threshValue)
 {
  myservo.writeMicroseconds (800);
 }
 else
 {
  myservo.writeMicroseconds (2250);
 }
 }
```

Fig. 7. Programming Code in the Arduino IDE.

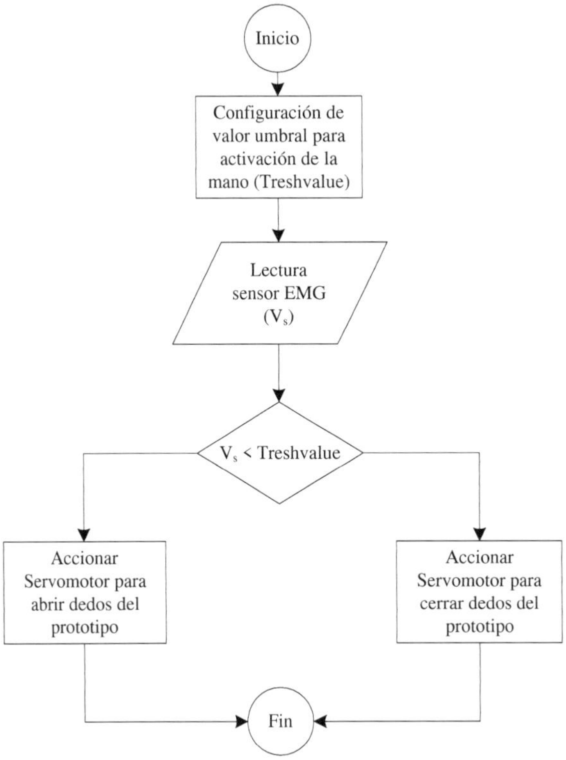

Fig. 8. Flow diagram of the Programming logic implemented in the controller

In Fig. 9, you can see all together the sensor signal captured by the oscilloscope, the muscle contraction of the forearm and the hand printed in 3D with its fingers closed. This answer meets the objectives set.

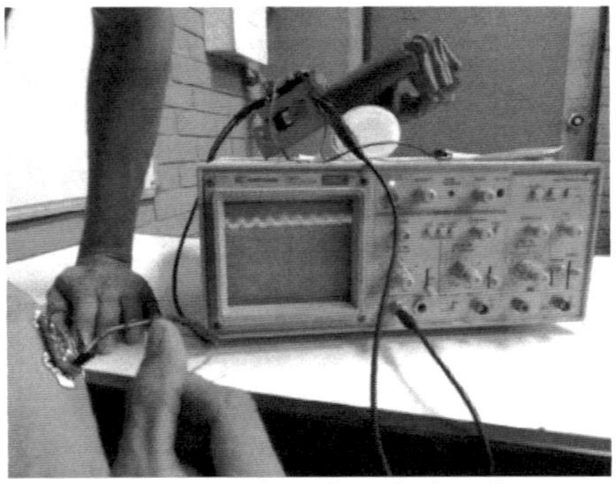

Fig. 9. Prototype response to a biceps contraction signal, the closure of the mechanism by action of the muscular response processed by the Arduino controller is observed

5. Discussion

The expected results of the development of this work was to obtain a 3D printed hand that had a good appearance that is capable of responding to muscle contractions, more specifically, that replicates the opening and closing movement of the hand.

The device was controlled by an Arduino microcontroller, which processed the information from a muscle sensor. The evidential type tests carried out on the electromechanical hand prototype yielded as results the visualization of the output signal of the muscle sensor and the verification of the opening and closing movement according to the muscular contractions performed by the volunteer. To guarantee the repeatability of the controller, 8 different tests were carried out, in which the behavior of the hand was replicated.

It was observed that it is necessary to locate the EMG sensor on an area of the arm where there is a large muscle such as the biceps, placing the floating electrode parallel to the muscle sensor, it was also found that the electrodes should not be reused for different practices , since contact is lost between the tests and the deposit of body sweat and environmental dust that increased the impedance of the electrode, making it difficult to read the signal.

On the other hand, this handheld prototype is very easy to operate, since the card does not require any configuration, so it only requires an electrical supply and adhere the electrodes to the volunteer's skin to perform the reading, and control this device at via Arduino.

Although the results are satisfactory, we consider that it is necessary to link the device to other muscles to observe their response. Also, evaluate the force and speed with which the movements occur. Even carry out tests on people with an amputated limb to observe what threshold values must be considered to activate the hand.

6. Conclusions

The manufacture of the pieces obtained from the virtual models was carried out correctly, obtaining pieces with the specified dimensions, which allowed a coupling with adequate tolerances between each piece, which, given the selected material and the type of printing presented an acceptable physical appearance and considerable stamina.

The connection of all the electronic components was carried out using Dupont type electrical conductors, following the electrical scheme suggested by the manufacturer, and the technical information from the collected literature.

Using the IDE Arduino software, the open programming code for the prototype was developed, in which the description of the function of each line of code was written in each of the lines of programming code to facilitate its interpretation and to be able to develop future modifications. .

Functional tests were carried out for the validation of the prototype, developing 8 validation tests, yielding satisfactory operating results, in which there were no operating failures, taking into account that it is essential to always have electrode pads in good condition and not have been reused, additionally the power supply must be higher than 5V to guarantee the operation of the servomotor and the sensor. Performing tests with the oscilloscope to have a second way of verifying the operation of the hand, the output signal of the sensor was obtained and compared with the operation of the hand during muscle contraction, observing the replication of the movement made by the volunteer .

Finally, future experiments are desired, observing the operation of the hand with other muscles and integrating a more robust control system that allows controlling the opening and closing of the hand in different positions, to obtain a behavior more similar to that of the human hand.

References

[1] R. Madhu Mohan, C. Girisha, K. Vijayalakshmi, C. Karthik, G. Harish, and M. Piyush, "Behavioral Studies of Myoware Sensor During Automated Prosthetic Implementation," *Res. Rev. A J. Life Sci.*, vol. 10, no. 1, pp. 10–15, Jan. 2020, doi:

10.37591/RRJOLS.V10I1.2563tatistics and sampling. 13th Edition. ECOE Editions, pp. 303-310. 2012
[2] L. Crawford and D. Vavra, "Typing with EMG using MyoWare," Prosthet. Orthot. Int., pp. 1–5, Jun. 2016.
[3] R. Rangwani and H. Park, "Vibration Induced Proprioceptive Modulation in Surface-EMG Based Control of a Robotic Arm," in International IEEE/EMBS Conference on Neural Engineering, NER, 2019, vol. 2019-March, pp. 1105–1108, doi: 10.1109/NER.2019.8717117.
[4] Fundación Saldarriaga Concha, "Caracterización sobre Discapacidad a Nivel Nacional," Bogotá, 2008
[5] J. M. Zuniga, A. M. Carson, J. M. Peck, T. Kalina, R. Srivastava, and K. Peck, "The development of a low-cost three-dimensional printed shoulder, arm, and hand prostheses for children."
[6] R. Celi, A. Sempertegui, D. Morocho, D. Loza, D. Alulema, and M. Proano, "Study, design and construction of a 3D printer implemented through a delta robot," in CHILECON 2015 - 2015 IEEE Chilean Conference on Electrical, Electronics Engineering, Information and Communication Technologies, Proceedings of IEEE Chilecon 2015, 2016, pp. 717–722, doi: 10.1109/Chilecon.2015.7404650.
[7] R. K. Megalingam et al., "Cost-Effective, Custom-Made, 3D Printer Design and Fabrication for Educational Purposes," in Proceedings of the 2nd International Conference on Trends in Electronics and Informatics, ICOEI 2018, 2018, pp. 10–15, doi: 10.1109/ICOEI.2018.8553893.
[8] R. Hedayati, M. Sadighi, M. Mohammadi-Aghdam, and A. A. Zadpoor, "Mechanical properties of additively manufactured octagonal honeycombs," Mater. Sci. Eng. C, vol. 69, pp. 1307–1317, Dec. 2016, doi: 10.1016/j.msec.2016.08.020.
[9] N. J. Farin, S. M. A. Sharif, and I. Mobin, "An Intelligent Sensor Based System for Real Time Heart Rate Monitoring (HRM)," Intell. Control Autom., vol. 7, pp. 55–62, 2016, doi: 10.4236/ica.2016.72006.
[10] A. H. A. Majid, M. A. A. Rahman, S. A. Mazlan, and H. Zamzuri, "Biosignals based intelligent control interface for current-induced physiological devices," in 2015 10th Asian Control Conference: Emerging Control Techniques for a Sustainable World, ASCC 2015, 2015, doi: 10.1109/ASCC.2015.7244514.
[11] H. Yang, C. Xiang, H. Han, and L. Hao, "Inverse kinematics modeling and motion control of PAM bionic elbow joint," in 2015 IEEE International Conference on Robotics and Biomimetics, IEEE-ROBIO 2015, 2015, pp. 1347–1352, doi: 10.1109/ROBIO.2015.7418958.
[12] A. Z. bin Tahir and B. of E. T. in Electronics, "EMG sensor for robotic applications," Jun. 2016.
[13] S. Thanakodi, S. M. H. Azhar, and A. Miskon, "Prosthetic hand controlled by wireless flex sensor on EOD robot," in IFMBE Proceedings, 2016, vol. 56, pp. 163–167, doi: 10.1007/978-981-10-0266-3_33.
[14] J. Tomás Mora, "Diseño y montaje de brazo robótico móvil: Diseño electrónico para el control de motores basado en Arduino," May 2015.
[15] K. E. Kulenthiiran, M. K. Hettihewa, R. M. C. H. Jayasinghe, H. M. K. P. Kulathunga, R. S. Dahanayake, and R. M. T. P. Rajakaruna, "CONTROLING MOVEMENTS OF A ROBOTIC ARM THROUGH CAPTURING AND REPLICATING MOTION OF HUMAN ARM," SAITM-RSEA, 2014.
[16] M. H. M. Sidik and S. A. C. Ghani, "Volume Measuring System Using Arduino for Automatic Liquid Filling Machine," 2017.

IOP Conf. Series: Materials Science and Engineering 1154 (2021) 012036 doi:10.1088/1757-899X/1154/1/012036

Development of A Station for The Measurement of Environmental Conditions in The Unidades Tecnologicas De Santander (UTS), Using Telemetry and Database Management with Raspberry Pi.

A Quintero Ruiz[1-2], C L Sandoval-Rodriguez[1-2], A D Rincon-Quintero[1-3], B E Tarazona-Romero[1-3], C G Cárdenas-Arias[1-4], J G Ascanio- Villabona [1-3].

[1] Energy Systems, Automation and Control Research Group (GISEAC), Design and materials research group (DIMAT), Electromechanical Engineering, Faculty of Natural Sciences and Engineering, Unidades Tecnologicas de Santander (UTS), Road No 9-82 students, 680,005, Bucaramanga, Colombia.

[2] Doctoral Program in Electronic and Communications, Department of Communications Engineering, University of the Basque Country, UPV/EHU, 48013 Bilbao, Spain.

[3] Doctoral Program in Energy Efficiency and Sustainability in Engineering and Architecture, Department of Machines and Thermal Engines, University of the Basque Country (UPV / EHU), Engineer Torres Quevedo Square, 1, 48013, Bilbao, BI, Spain.

[4] Doctoral Program in Engineering of Materials and Sustainable Processes, University of the Basque Country, UPV/EHU, 48013 Bilbao, Spain

Abstract. This article focuses on the development of an electronic remote data acquisition system which will have the following sensors adapted: thermometer, hygrometer and barometer. From the measurements of these will be obtained data of temperature, relative humidity, atmospheric pressure, which will be stored in the form of records in a data recorder using Raspberry Pi and through wireless communication the data will be transmitted when the software requires it to a module Central of visualization, which according to some parameters will automatically update the histories according to the defined times, or at the request of the user. It will be implemented following a process of research, planning, execution and validation. It implies the necessary information for the implementation, in the planning process the different systems to be used will be divided into 3 groups, measurement and registration system, communication system and a control system composed of the software, and then the acquisition will proceed of the equipment and accessories necessary to execute the assembly. In the execution process, the physical implementation of all the systems will be carried out and the programming of the data acquisition, communication and data storage systems will be developed.
Keywords: Meteorological Station, Telemetry, Raspberry PI.

1. Introduction

In order to carry out measurements of environmental conditions in places that are difficult to access and to view these data remotely in order to study and analyse them, it is necessary to have a data acquisition system, storage and remote data transmission to a viewing station, This system called SCADA by its acronym in English Supervisory Control and Data Acquisition. (Rodríguez, 2007) SCADA systems are diverse using devices based on Arduino, Raspberry Pi or RTU designed with analogue or digital inputs to connect all kinds of sensors and equipped with communication devices either by radio frequency, GPRS, or even with connection. to internet through WIFI. However, RTU-based systems are too powerful and expensive and are generally used in the industry for robust SCADA systems with a large number of sensors and control devices.

2. Materials and Method

An investigation was carried out that according to promote the use of renewable energies in the consumption of electrical energy in the institution, a meteorological monitoring system is necessary to determine the behavior of the climatological variables to optimize the viability and design for a project self-generation through solar panels.

2.1 Sensor Selection

Temperature Sensor Selection

For the selection of the temperature sensor, the accuracy of the sensor, the practicality to install and its electrical output were taken into account. For this, the DHT 22 and LM35 sensors were taken into account

Table 1. Temperature Sensor Comparison

Sensor	**LM 35**	**DHT 22**	**RTD PT100 (3493A)**
Rank	-55° C to 150° C	-40° C to 80° C	-25 ° C to + 105 ° C
Accuracy	± 0.5° C	± 0.5° C	± 0.25° C
Departure	10 mV / °C	V	Ohm
Response time	40 ms	2 s	1 s

The RTD PT100 is selected since according to its characteristics it allows us a greater range, greater accuracy, if necessary a better resolution can be configured, the RTDs have a linearity in a fairly wide range and their architecture is much more resistant for outdoor environments Unlike the other sensors compared, since it is not an electronic circuit, it is not affected by humidity nor does it require the use of bakelites for its connections and the connection cable is already protected from the elements. But this, having a resistance output, it is necessary to use an analog to digital input converter since the inputs of the Raspberry Pi card where the sensors are connected are digital.

Atmospheric Pressure Sensor Selection

For the atmospheric pressure sensor, as in the temperature sensor, the range, the accuracy and its electrical output were taken into account to be able to be connected to the Raspberry Pi. Among the options you have the BMP180 BMP280 and BME280.

Table 2. Atmospheric Pressure Sensors Comparison

Sensor	**BMP 180**	**BMP 280**	**BME 280**
Rank	300 to 1100 hPa	300 to 1100 hPa	300 to 1100 hPa
Resolution	0.01 hPa	0.0016 hPa	0.0018 hPa
Accuracy	+2.0 hPa	± 1 hPa	± 1 hPA
Departure	I^2C and SPI	I^2C and SPI	I^2C and SPI
Sampling rate	120 Hz	182 Hz	182 Hz

According to the comparison, the highest accuracy and shortest response time are obtained with the BME 280 sensor, they all handle the same output interface and the BME280 also includes a relative humidity sensor. In the BMP280 sensor the resolution is better, the disadvantage is that it does not include the relative humidity sensor, because of this the BME280 sensor is selected, which allows the space of the sensors to be as compact.

IOP Conf. Series: Materials Science and Engineering 1154 (2021) 012036 doi:10.1088/1757-899X/1154/1/012036

Relative Humidity Sensor Selection

According to the specifications in the table below, the DHT22 sensor was selected since its specifications are better in terms of response time and accuracy compared to the other sensors, in the same way the cost of the sensor is very low.

Table 3. Relative Humidity Sensor Specifications

Humidity Sensor	AM2302 DHT22	HS1101	DHT11	Hih4000001
Rank	0% to 100% RH	0% to 100% RH	20% to 90% RH	0% to 100% RH
Resolution	0.1% RH	0.1% RH	1% RH	0.1% RH
Accuracy	± 2% RH	± 2% RH	± 4% RH	± 3.5% RH
Response time	2 s	5 s	10 s	5 s

Source: Bosch Technical Data Sheet

2.2 Controller Selection

Raspberry Module Selection

When selecting the controller for data acquisition and storage, the characteristics of each of the Raspberry Pi models on the market are taken into account.

Table 4. Raspberrys Model Comparison

Modelos	Raspberry Pi 3 Model B	Raspberry Pi 2 Model B	Model B+	Model A+	Model A	CMDK
Chip Procesador	Broadcom BCM2837 64Bit ARMv7	Broadcom BCM2836 32bit ARMv7	Broadcom BCM2835 32bit ARMv6 SoC full HD	Broadcom BCM2835 32bit ARMv6 SoC full HD	Broadcom BCM2835 32bit ARMv6 SoC full HD	Broadcom BCM2835 32bit ARMv6 SoC full HD
GPU	Videocore IV	Videocore IV	Videocore IV	Videocore IV	Videocore IV	Videocore IV
Velocidad del procesador	QUAD Core @1250 MHz	QUAD Core @900 MHz	Single Core @700 MHz	Single Core @700 MHz	Single Core @700 MHz	Single Core @700 MHz
RAM	1GB SDRAM @ 400 MHz	1GB SDRAM @ 400 MHz	512 MB SDRAM @ 400 MHz	256 MB SDRAM @ 400 MHz	256 MB SDRAM @ 400 MHz	512 MB SDRAM @ 400 MHz
Almacenamiento	MicroSD	MicroSD	MicroSD	MicroSD	SDCard	4GB eMMC
USB 2.0	4 puertos USB	4 puertos USB	4 puertos USB	1 puerto USB	1 puerto USB	1 puerto USB
Power Draw / voltage	2.5A @ 5V	1.8A @ 5V	1.8A @ 5V	1.8A @ 5V	1.2A @ 5V	1.8A @ 5V
GPIO	40 pin	40 pin	40 pin	40 pin	26 pin	120 pin
Puerto Ethernet	Si	Si	Si	No	No	No
Wi-Fi	Built in	No	No	No	No	No
Bluetooth LE	Built in	No	No	No	No	No

A comparison is made of the Raspberry Pi 3 B and the Raspberry Pi 2 B models, as they are the last ones released by the manufacturer, they present the best hardware specifications, and thanks to the comparison, the following is easily observed:

- QUAD Core Broadcom BCB2837 64 bit ARMv7 processor.
- Processor speed has increased from 900MHz on Pi 2 to 1.25 Ghz on RPi 3 B.
- Up to 2.5 amps upgraded switching power supply, meaning you can now power even more powerful devices via USB ports.

The main differences are the 64-bit quad-core CPU, Wi-Fi, and Bluetooth. However, the improved power management should mean where the Broadcom's Pi3 provides better support. In terms of size it is identical, all the connectors and mounting holes are in the same place. Also, the Pi3's performance is roughly 50-60% faster than the Pi2. (Foundation RP, 2017)

Knowing this data, the Raspberry Pi 3 B is chosen for its performance and provides a better operation of the devices to be used.

3. Results And Discusion

In the selection of the necessary teams for the station of measurement with telemetry and management of database selected two Raspberrys Pi 3, one of these with internal memory of 16 gb, connected like system of measurement composed of a controller TTGO Arduino One SX1278 which receives the signals of measurement of a sensor of temperature RTD, sensor of atmospheric pressure and relative humidity selected according to the specifications of these, the system had of this form in union of a Raspberry Pi with Arduino, being able to this last make work only with the sensors and system of storage, but wanted that the project in the station of measurement had a system of control more powerful as it is it his operating system Rasbian for the work with files, between other advantages explained in his technical index card. The system of measurement has a solar signpost like feeding with battery to give the autonomy in hours of no solar generation. The station of visualizations has the Raspberry Pi 3 with internal memory of 32 gb connected to a screen led of 7 in with source of feeding of 110v and regulation to 5v.

For the autonomy of feeding obtained a length of 6 hours without solar light, using a solar signpost of 20 W for proofs with which explained , but the results of autonomy did not allow an operation of 24 hours because of the way in programmed the communication between both Raspberry since these keep lit with the communication established still in the moments in that it does not send information, to reduce the consumption because of this needed of modifications in the operation of the station of measurement as in the solar signpost that installed for the Raspberry PI 3, by this was necessary deactivate true services of the system to consume less energy and attain with a solar signpost of 40 W an autonomy 24/7 while the battery find in good conditions. With more time of improvement of the programming in a future project where add more sensors could improve the communication to be able to sleep the Raspberry in the moments in that it do not send information.

It determined diminish the active functions of the Raspberry in the station of measurement of form to consume less improves having the following resulted before and after the changes made in east, show the initial data and finals because of the big quantity of data taken.

Table 5. Initial Data Proofs of Autonomy Station of Measurement

Date	Temperature (°C)	Relative humidity (%)	Atmospheric pressure (psi)
16/06/2019 18:54	27.3	59.3	8821
16/06/2019 18:55	27.2	59.2	8821
16/06/2019 18:56	27.1	59.2	8821
16/06/2019 18:57	27.1	59.2	8821
16/06/2019 18:58	27.2	59.6	8823
16/06/2019 18:59	27.2	59.6	8822

IOP Conf. Series: Materials Science and Engineering 1154 (2021) 012036 doi:10.1088/1757-899X/1154/1/012036

16/06/2019 19:00	27.2	59.9	8824
16/06/2019 19:01	27.2	59.9	8823

Table 6. Final Data Proof of Autonomy Station of Measurement

Date	Temperature (°C)	Relative humidity (%)	Atmospheric pressure (psi)
16/06/2019 23:09	24.4	68.2	8865
16/06/2019 23:10	24.3	67.9	8864
16/06/2019 23:11	24.3	68	8864
16/06/2019 23:12	24.3	67.9	8864
16/06/2019 23:13	24.3	67.9	8864
16/06/2019 23:17	24.1	67.7	8865

The following services of the system to consume less energy and attain with a solar signpost of 40 W an autonomy 24/7 while the battery find in good conditions.

Service of LAN/WIFI
Service of VNC
Service of SSH.
Service of audio

The only services that always have to be activated in the Raspberry Pi of the station of measurement are I2C, SPI and serial.

3.1 Proofs of Distance Modules of Communication Arduino Lora Sx1278

Figure 1. Space of Proofs for Measurements

The proofs of measurement of distance of the modules of communication make in the sector of Girón in the neighborhood Banks of the Rio taking like reference the spaces of proofs in inner (indoor) and in outsides on line of sight between the two modules (outdoor).

The proofs look for to find the maximum scope with which can obtain the best results of communication so much on line of sight as in inner spaces. All the proofs make in the same geographic space taking into account the characteristics of the terrain regarding elevation

Table 7. Combination of the main parameters configurables

Ancho de banda BW (kHz)		Factor de alcance SF		Tasa de codificación de errores. CR	
Hardware	Software RADIO.BW	Hardware	Software RADIO.sf	Hardware	Software RADIO.cr
500	BW500	6 → 64 chips / symbol (modo utilizado para FSK)	SF_6	4/5	CR_4_5
250	BW250	7 → 128 chips / symbol	SF_7	4/6	CR_4_6
125	BW125	8 → 256 chips / symbol	SF_8	4/7	CR_4_7
62.5*	BW62_5*	9 → 512 chips / symbol	SF_9	4/8	CR_4_8
41.7*	BW41_7*	10 → 1024 chips / symbol	SF_10		
31.25*	BW31_25*	11 → 2048 chips / symbol	SF_11		
20.8*	BW20_8*	12 → 4096 chips / symbol	SF_12		
15.6*	BW15_6*				
10.4*	BW10_4*				
7.8*	BW7_8*				

In the table 16 observe the parameters of configuration by software that take part in the modification of the chip of communication and that has to take into account to the hour to make the proofs.

Said parameters take part directly in the maximum distance of communication of the modules and that have to configure of such form that fulfil the requirements of design and the possible maximum distance that can obtain .

Table 8 Experimental Data made with the modulate LoRa SX1278

# Conf	Parámetros configurables por software			Consumo de corriente mA					Tiempos segundos		Consumo energético mili-Vatios s		Comentarios
	SF	BW	CR	I_{TXmin}	I_{TX}	I_{TXmax}	I_{RX}	I_{Sleep}	t_{tx}	t_{rx}	W_{tx}	W_{rx}	
1	9	500	4_5	45	46	47	11,5	1,7	0,025	0,03	3,80	1,14	Estable
2	12	500	4_5	25	52,5	80	11,5	1,7	0,21	0,23	36,38	8,73	Estable
3	9	125	4_5	34	47	60	11,7	1,7	0,105	0,13	16,29	5,02	Estable
4	12	125	4_5	37	58,5	80	11,7	1,7	0,8	0,9	154,44	34,75	Estable
5	9	61,5	4_5	20	40	60	11,7	1,7	0,205	0,26	27,06	10,04	Estable
6	12	61,5	4_5	80	85	90	11,2	1,7	1,7	1,8	476,85	66,53	Estable
7	9	41,7	4_5	40	60	80	11	1,7	0,3	0,4	59,40	14,52	Estable
8	**12**	**41,7**	**4_5**	80	**85**	90	**11**	**1,7**	**2,4**	**2,6**	**673,20**	**94,38**	**Estable**
9	9	31,5	4_5	50	65	80	11,5	1,7	0,4	0,52	85,80	19,73	Estable
10	**12**	**32,5**	**4_5**	80	**80**	80	**11**	**1,7**	**3,5**	**4**	**924,00**	**145,2**	**Estable**
11	9	20,8	4_5	60	75	90	11	1,7	X	X	X	X	[illegible]
12	7	20,8	4_5	30	45	60	11	1,7	X	X	X	X	[illegible]

According to the data observed in the table 16 establish the data configurables by software the following:

SF (Factor of spectrum widened or of scope) = 12

BW (Bandwidth) = 41.7

CR (Tax of coding of errors) = 4/5

The configuration selected ensures the maximum effectiveness in the transmission of the data, but involves that it spends more time in the process of coding of the same, which does not affect in the process due to the fact that the time of reading of the sensors is of 60 seconds.

It selects this configuration because it is the one who presents the best values of current and power of transmission and reception without that they have lost of packages of data sent to comparison of the configuration of the line 10 (Table 16) that presents the best values of power of transmission but the stability of the signal is not very good since they lose packages of data in the transmission and reception of the same. By such reason selects the configuration posed to the start.

3.2 Proofs with obstacles (indoor)

The proofs make carrying the module of visualizations connected to a battery and in state of operation displaces by the streets expecting obtain the greater distance of communication between the two modules.

For this proof connects the modulate to the PC to verify the tax of sent of data between the modules and verify the maximum distance to which receive said data.

Figure 2. Proofs Indoor Lora SX1278

To obtain the maximum distance indoor takes the measure that launches the application maps and decompose the planes x-and with the measure of the maximum height to which finds the station of measurement. The maximum distance between the modules obtain it using the theorem of Pitágoras (1). In it formulates it we enter the values of the height of the building and the maximum distance of plant that obtained in the measurement.

Figure 3.Diagram to find maximum distance of communication between devices in inner spaces

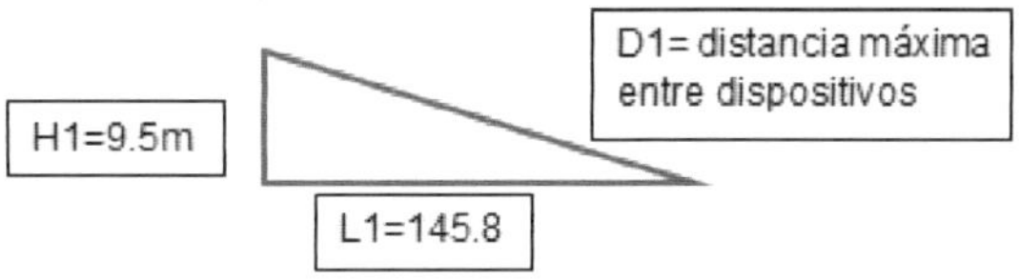

$$H1\ \ 2 + L1\ \ 2 = D1 \quad \textbf{2 (1)}$$
$$D1 = \sqrt{(H1\ \ 2 + L1\ \ 2)} \quad \textbf{(2)}$$
$$D1 = \sqrt{(9.5^2 + 145.8^2)} \quad \textbf{(3)}$$
$$D1 = 146.11m \quad \textbf{(4)}$$

The maximum distance obtained for applications of communication (indoor) with the modules Arduino LoRa SX1278 TTGO is of 146.11 meters. After this distance the packages of data begin to lose or take much more time of the expected what indicates us that this is the recommended distance to use said modules of communication.

3.3 Proofs with line of sight (outdoor)

Figure 4. Proofs outdoor Lora SX1278

IOP Conf. Series: Materials Science and Engineering 1154 (2021) 012036 doi:10.1088/1757-899X/1154/1/012036

The proofs of maximum scope on line of sight limit to the distance that observes in the figure since it was not possible to situate a zone cleared to greater distance and of easy access, due to the fact that the majority of the terrains neighboring are of private property and in the directions in where it can observe the main roads of access are hindering the principal condition of the proof that is to line of sight between the devices.

The data that obtain to this distance fulfil in the whole with the parameters established by software for the configuration of each one of the modules and is possible that can reach a bigger distance with the same parameters without that there is loss of data, which leaves posed for future essays and proofs of laboratory.

Figure 5. Diagram to find maximum distance of communication between devices in External spaces

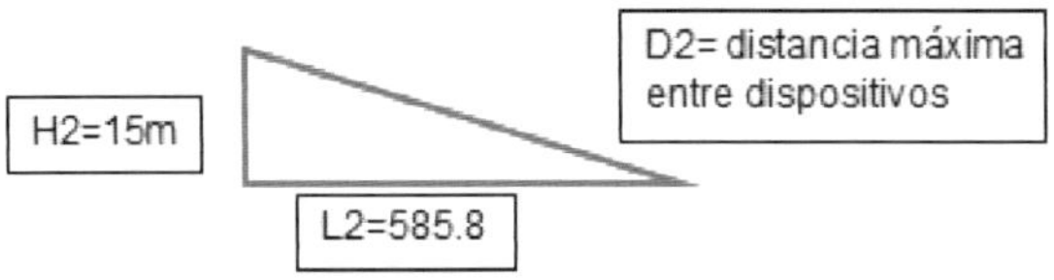

$$H2\ \ 2 + L2\ \ 2 = D^2 \quad (1)$$
$$D2 = \sqrt{}(H2\ \ 2 + L2\ \ 2) \quad (2)$$
$$D2 = \sqrt{}(15\ \ 2 + 585.8^2) \quad (3)$$
$$D2 = 586m \quad (4)$$

3.4 Proofs of Measurement With Regard To Team Of Reference

It treated to obtain the data of the environmental stations surrounding of entities of the state to the places of the proofs, but was not possible. The environmental stations that have the public information did not find near to the places of the proofs, even to heights very distant what did that the error of comparison was very big and was not a conclusive proof. Because of this used a team certified by laboratory accredited by the National Organism of Accreditation of Colombia (ONAC) like reference for the proofs of comparison what ensures a traceability to national and international patterns in each variable. Although it does not treat of a calibration since for this requires of some environmental conditions uniforms for the proof if it obtained an error of the proofs what allowed an adjust in the measurement of the team to approach us to the measurement that obtained of the team of reference.

The team used is a mini environmental station mark Kestrel Instruments reference 5000 serial 2355316 of the which did proofs in temperature, relative humidity and atmospheric pressure, the 3 variables find calibrated and certificates by the laboratory accredited Conamet with the following respective reports CLT 151120, CLH40820, CLP 67620. The respective accreditation of the laboratory can be consulted in the page of the ONAC in the section of laboratories of calibration accredited in the following link: http://onac.org.co/certificados/09-lac-008.pdf

IOP Conf. Series: Materials Science and Engineering 1154 (2021) 012036 doi:10.1088/1757-899X/1154/1/012036

Figure 6 Test Measurement In Installations UTS

Figure 7 Serial Instrument Kestrel

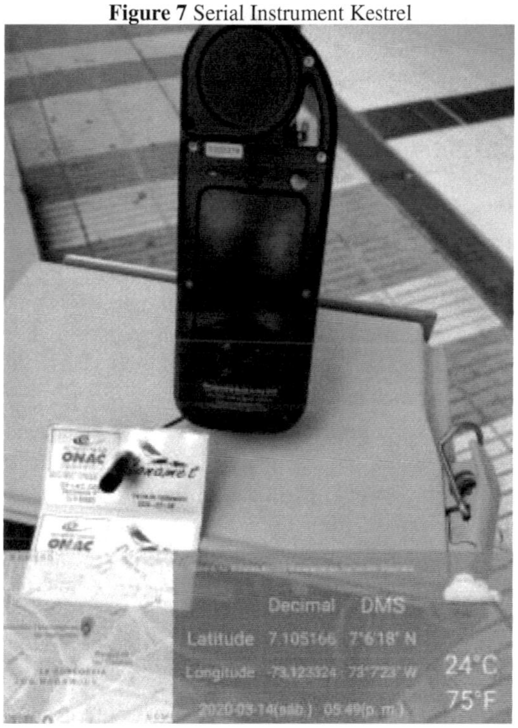

3.5 Resulted Errors Comparison Measurement Of the measurements

Obtained calculated a minute-by-minute error between each variable measured by the team Kestrel 5000 and the environmental station, and these 61 errors calculations for each variable took out an average, to obtain the results of measurement.

Figure 8. Errors Calculated of the comparison of measurement between the Kestrel 5000 and the environmental station.

1	Fecha	Error Temperatura (°C)	Error Humedad Relativa(%)	Error Presión Atmosferica (hPa)
2	14/03/2020 18:38	0.7	-0.1	-0.5
3	14/03/2020 18:39	0.6	0.4	-0.5
4	14/03/2020 18:40	0.5	-0.2	-0.5
5	14/03/2020 18:41	0.5	0	-0.5
56	14/03/2020 19:32	0.5	2.2	0.4
57	14/03/2020 19:33	0.2	2.6	0.4
58	14/03/2020 19:34	0.3	3	0.4
59	14/03/2020 19:35	0.4	3.1	0.5
60	14/03/2020 19:36	0.3	3.5	0.5
61	14/03/2020 19:37	0.4	3.6	0.5
62	14/03/2020 19:38	0.4	3.5	0.5
63	Promedio	0.449	1.267	0.002

Source: Authors in Excel

As it observes in the figure obtain an error in temperature of 0.5°C, 1.267 % in Relative Humidity and 0.002 hPa in Atmospheric Pressure, which are data quite near to the station of reference, by what the station of Measurement guarantees some correct measurements, these data even can be improved with an adjust finer that for the request of the station is not necessary.

4. Conclusion

It developed a system of measurement of 3 different sensors for the measurement of environmental conditions, temperature, relative humidity and atmospheric pressure. Composed by electronic systems of control Raspberry Pi and Arduino adapted to the system and in a setting of tripod in which they had all the systems, with an autonomous feeding by means of a solar signpost and a battery of backrest, this establishes communication by radiofrequencies by means of modules Lora that allowed us a communication of 586 m stable without obstacles with the station of visualization in which they can be observed the minute-by-minute measurements by means of a table and charts in an included screen in the system, all made in language Python.

With the setting made of the solar signpost of 40 W according to the calculations of feeding obtained an autonomy 24/7 that requires for the operation and communication of the sensors during all the time of operation as long as the battery find in good conditions so that it allow the complete load of this.

The communication remained stable in a distance of 146 m with obstacles with a low percentage of stray of data, this because of the big quantity of buildings and possible interferences of signal what hampers to follow increasing the distance without having effects in the stable communication. Even so, the quantity of stray data are very low with regard to the big quantity of data that transmits since this transmits each min registered and is the sufficient functional mind for a station of measurement since for these allows a register of each 5 min.

The programming transmits each data measured taken of a database stored in the station of visualization and handled like independent variables what allows to keep the value of each measurement without affectation or changes and without limit of characters of the values measured, something that in a principle presented inconvenient of limit of quantity of characters when transmitting all the measurements of an hour specifies stored what forced to make operations of divisions affecting the values of the measurements transmitted to reduce characters.

The costs were a bit higher of the initially budgeted as it had to consider purchase the modules lora with his systems of antennas and include a regulator of solar signpost of equal form increased the power of the solar signpost by what the value of east increased, however, in spite of this the cost of the complete team keeps on being much more economic that a commercial environmental station.

5. References

ANDROIDPC. (14 March 2016). *ANDROIDPC.* Obtained of ANDROIDPC: https://androidpc.es/review-raspberry-pi-3/2/

Autosolar. (09 of 06 of 2017). *Autosolar.* Obtained of https://autosolar.es/blog/aspectos-tecnicos/diferencias-entre-silicio-monocristalino-y-multicristalino-o-policristalino

Ten, L. G. (24 of 04 of 2008). *Tiempo.com.* Obtained of https://www.tiempo.com/ram/1696/manualidades-meteorolgicas-una-garita-de-platos-que-funciona/

Foundation, R. P. (2017). Datasheet Rasberry PI 3 model b+. Recovered the 04 of 2018, of https://static.raspberrypi.org/files/product-briefs/raspberry-pi-model-bplus-product-brief.pdf

Foundation, R. P. (s.f.). *Raspberrypi.org.* Recovered the 04 of 2018, of https://www.raspberrypi.org/documentation/setup/

Full Mecanica. (2014). *Full Mecanica.* Obtained of http://www.fullmecanica.com/definiciones/f/1214-fuerza-del-viento-presion-del-viento

IDEAM. (s.f.). *IDEAM.* Obtained of Institute of Hydrology, Meteorology and Environmental Studies.

Méndez, Or. (05 of 04 of 2016). *330 ohms.* Obtained of 330 ohms: https://330ohms.com/blogs/blog/114183172-arduino-mini-micro-y-nano-que-tan-diferentes-son

Ministry of Environment, House and Territorial Development. (October of 2010). *MANUAL OF DESIGN OF SYSTEMS OF SURVEILLANCE OF THE QUALITY OF THE AIR.* Obtained of Ministry of Environment and Territorial Development: http://www.minambiente.gov.co/images/asuntosambientalesysectorialyurbana/pdf/contaminacion_atmosferica/protocolo_calidad_del_aire_-_manual_dise%C3%B1or.Pdf

National Instruments. (2019). *National Instruments.* Recovered the 1 of septiembe of 2019, of http://www.ni.com/es-co.html

OMM. (2014). *Guide of Instruments and Methods of Meteorological Observation.* Ginebra: Organisation Meterológica World-wide.

Orozco, To., & Rodríguez, C. (2015). *Design and Implementation of a Prototype of Meteorological Station.* San Juan of Pasture: University of Nariño.

Rodríguez, To. (2007). *Systems SCADA* (2 ed.). Barcelona: Marcombo.

Román, J. (2015). *Telemetry of Meteorological Station.* Guayaquil: Polytechnical Upper school of the Seaboard.

Santos, H. (2017). *Design and implementation of a system domótico based in Raspberry Pi.* Madrid: University Carlos III of Madrid.

Silva, R. (s.f.). *GIT HUB.* Obtained of GIT HUB: https://github.com/jieter/python-lora

Spark, W. (2019). *Weather Spark.* Obtained of Weather Spark: https://es.weatherspark.com/y/24381/clima-promedio-en-bucaramanga-colombia-durante-todo-el-a%C3%B1or#Sections-Sources

SSDielect. (10 of 03 of 2019). *SSDielect.* Obtained of SSDielect: https://ssdielect.com/es/sistemas-de-desarrollo-esp8266-wifi/1198-ttgo-433mhz.html

Tobajas, To. (2016). *Design and Implementation of a Meteorological Station with Raspberry PI.* Barcelona: Open University of Catalonia.

Unit of Planeación Mining Energetic. (October of 2017). *Balance of Natural Gas.* Recovered the 04 of 04 of 2018, of UPME: http://www1.upme.gov.co/hidrocarburos/publicaciones/balance%20GAS%20NATURAL%202017-2026%20VERSION%20FINAL.Pdf

XM. (2017). *Integral report of Sustainable Management.* Recovered the 03 of 04 of 2018, of XM: http://www.xm.com.co/documents/pdf/informe_integral_xm.pdf

IOP Conf. Series: Materials Science and Engineering 1154 (2021) 012037 doi:10.1088/1757-899X/1154/1/012037

Risk indicators of mass removal phenomena according to the Mora - Vahrson method, applied in Pitalito and Campoalegre municipalities

L Pedreros[1], **S Rivera**[1]

[1] Facultad de Ingeniería Civil, División de Ingenierías, Universidad Santo Tomás, Ak 9 N 51-11, Bogotá D.C, Colombia

Email: laura.pedreros@usantotomas.edu.co

Abstract. The massive removal phenomena have been one of the most frequent natural disasters in the world, causing thousands of deaths, victims, damage to homes and diseases. In Pitalito and Campoalegre, municipalities of Huila department in Colombia have occurred various events such as heavy rains, earthquakes, which have caused landslides, floods, among others, to affect the economy, the community and transportation. For this reason, a study was carried out on the area's most prone to suffer these phenomena to take preventive measures in favor of the protection of the population, the resources of management and the planning of civil works. For the proposed object, the Mora-Varshon method was used, which allows classifying the degree of susceptibility to landslides in which the areas are found. Also, various factors or parameters was evaluated such as the soil moisture, lithology, slope, seismicity and rain, each of these indicators were obtained using information from IDEAM, Servicio Geologico Colombiano (SGC) and using geographic information for geoprocessing in the ArcGIS software to realize a mapping to indicate the susceptibility to landslides, classifying the areas of the municipalities such as very low, low, medium, moderate, high or very high.

1. Introduction

The mass removal phenomena are events that have been generated over the years, many of these phenomena are stronger than others, causing landslides of great magnitude directly affecting the economy, mobility, the community, the environment and also causing human losses. These are caused by various variables that are linked to natural and anthropic factors, which must be thoroughly studied to prevent any disaster that may be generated, and thus, protect the community, informal settlements and for the possible development of civil works; these is because a study is carried out on the susceptibility of landslides that can be generated in the municipalities of Pitalito and Campoalegre, belonging to the department of Huila.

The article is focused on determining the level of mass removal of Pitalito and Campoalegre municipalities, for this will be made the creation of a geodatabase that will allow a mapping in ArcGIS in order to evaluate which is the indicator of seismicity in these areas, taking into account factors such as precipitation, lithology, soil moisture, slope and seismicity, which will be obtained through data from IDEAM, the Servicio Geologico Colombiano and through the use of geographic information; It is important to emphasize that the Mora-Vahrson method is developed through the aforementioned factors, which originated in Costa Rica as a consequence of an earthquake that occurred in that place and seeks to be able to make decisions regarding the planning of land use, exploitation of natural resources, the planning of infrastructures, urbanism and roads. In this way, the methodology shown in figure (1) is proposed and implemented, which is composed of three stages, in the first instance, the study of the mora - Vahrson method, in the second instance the elaboration of the different maps to identify the triggers in the ArcGIS software, finally, the assessment of the area's most prone to landslide susceptibility.

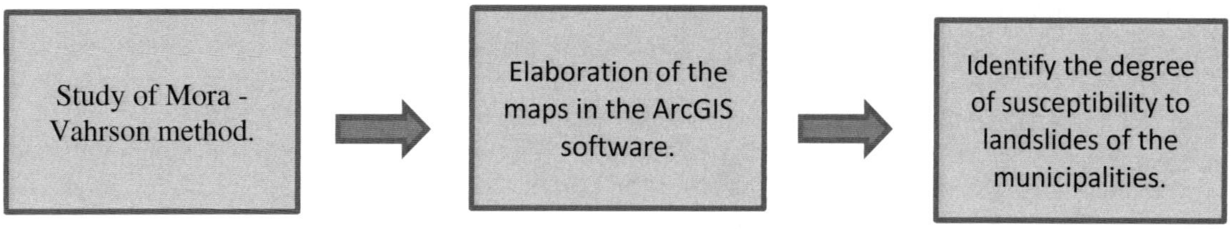

Figure 1. Methodology Used.

2. Location of the municipalities

2.1. Pitalito

Pitalito is a municipality that belongs to the department of Huila, which has a territorial extension of 653 km2, located at 1 ° 51 '16 "North and 76 ° 3' 6" West and is approximately 196 km from the capital of Huila It is also located south of the department at 1,318 meters above sea level, with an average temperature of 20 ° C. Pitalito limits to the north with the municipalities of Saladoblanco and Elías, to the south with the Municipality of Palestina, to the east with the Municipalities of Acevedo and to the west with the municipalities Isnos and San Agustín. Pitalito is part of the highest basin of the Magdalena River and as water currents are the Osoguico, Narajos, Balseros, Granadillos, and clear rivers, among others [1]. In terms of soils, the municipality is characterized by having six different types, among which are: mountain soils in a cold and humid climate, mountain soils with a medium humid climate, additionally it has the presence of piedmont soils with a climate medium and humid, lowland soils with a medium and humid climate, highland soils with a medium and humid climate and finally with valley soils with a medium and humid climate [2].

The territory is formed from a system of geological faults of probable activity, the most likely faults are the Chillurco faults, heading east - west along the southern foothills of the Chillurco ridge, the Guachicos fault heading northeast - southwest through the valley of the same name and the Suaza fault to the east of the municipality through the Suaza - Acevedo valley [2].

2.2. Campoalegre

The phenomena of mass removal are events that have been generated over the years, many of these phenomena are stronger than others, causing landslides of great magnitude directly affecting the economy, mobility, the community, the environment and also causing human losses. These are caused by various variables that are linked to natural and anthropic factors, which must be thoroughly studied to prevent any disaster that may be generated, and thus, protect the community, informal settlements and for the possible development of civil works; these is because a study is carried out on the susceptibility of landslides that can be generated in the municipalities of Pitalito and Campoalegre, belonging to the department of Huila.

The article is focused on determining the level of seismic threat of the municipalities of Pitalito and Campoalegre, for this will be made the creation of a geodatabase that will allow a mapping in ArcGIS in order to evaluate which is the indicator of seismicity in these areas, taking into account factors such as precipitation, lithology, soil moisture, slope and seismicity, which will be obtained through data from IDEAM, the Servicio Geologico Colombiano and through the use of geographic information; It is important to emphasize that the Mora-Vahrson method is developed through the aforementioned factors, which originated in Costa Rica as a consequence of an earthquake that occurred in that place and seeks to be able to make decisions regarding the planning of land use, exploitation of natural resources, the planning of infrastructures, urbanism and roads. In this way, the methodology shown in figure (1) is proposed and implemented, which is composed of three stages, in the first instance, the study of the mora - Vahrson method, in the second instance the elaboration of the different maps to identify the triggers in the ArcGIS software, finally, the assessment of the area's most prone to landslide susceptibility.

3. Mass removal phenomena

The mass removal phenomena or also known as slope failures, landslides, among others, are displacements from a higher to a lower level of rocks, fine material, rubble and earth [5], either quickly or slowly, which are presented essentially as a consequence of the force of gravity and with the help of other factors such as precipitation, seismicity, slope, the impact of human action, among others.

Mass removal phenomena are classified into several types, which according to their magnitude can generate disasters in communities, directly affecting homes, people's lives, communication routes and the economy; To avoid these consequences, it is necessary to monitor the disasters that have occurred in the area in the past, in order to prevent possible disasters in the future, however, their prediction turns out to be highly complex, since it is not possible to determine exactly the time, day, place and magnitude [6]. Table (1) shows the classification of the types of mass removal phenomena according to Cruden & Varnes [7][23].

Table 1. Classification of mass removal phenomena Varnes 1978 [7][23].

Type of movement	Type of material
Fall	Rock – Soil
Toppling	Rock – Soil
Rotational Sliding	Rock – Soil
Translational Sliding	Rock – Coarse Grain – Fine Grain
Flows	Rock – Coarse Grain – Fine Grain
Lateral Extensions	Rock – Coarse Grain – Fine Grain

In the first instance, according to the previous classification, falls are defined as the detachment of material of various sizes from a slope in a rapid manner, presenting rock falls in free fall, also generating jumps and bearings; These characteristics depend on the slope in which the slope is, in the second instance the overturns (topples) are identified as the rotation of soft rocks or detrital materials towards the outside of the slope and according to the slope they depends on whether they are fast or slow, these movements are propagated by the presence of water, ice in the joints, by expansion and by seismicity [8].

Additionally, in the third instance of the aforementioned classification, it can be evidenced that the landslides that are rapid movements are generated by the low cohesion of the soil, predominantly blocks, sands and gravels, where the material moves along the fault surface and reaches the bottom of the slope without any form, these movements contain mostly solid material and what remains is water [3], these landslides can be classified as rotational where the fault surface is curved or concave and transnational that Unlike the rotational ones, the fault surface is flat [9]. On the other hand, with regard to flow, it is characterized by the sliding of rock, soil or small-sized earth along the slope. This flow can be wet or dry and in many occasions this phenomenon is generated after landslides, which causes water to be absorbed very easily in the faults. Finally, with regard to the lateral extensions, these are characterized by presenting themselves in two types of movement, where one of them completely alters the material traditionally in rocky material while the other occurs only in cohesive soils when previously liquefaction has already occurred [10].

4. Mora – Vahrson Method

The Mora-Vahrson method was born in Costa Rica, created by Sergio Mora and Wilhelm Vahrson, exactly in 1991. This method has been widely used in Central America [11] given its accuracy in predicting the threats of phenomena in addition, mass removal is used to obtain the degree of susceptibility to landslides of a place, in order to identify the magnitude to which the area is prone. This method considers several factors provided by the observation and measurement of morpho-dynamic parameters [12], where it proposes that some parameters help in particular to predict the instability of the slope, which are divided into two groups that can be found in the literature with different names such as susceptibility factors or passive elements (P) and trigger factors (D) or active elements, equation (1) illustrates how to determine the susceptibility to landslides (S) according to Mora - Vahrson.

$$S = P \times D \quad (1)$$

According to the previous equation, the passive elements (P) are constituted in the first place by lithology (Sl), which is the study of rocks, where characteristics such as the size of the particles and the grain are taken into account, the physical-chemical characteristics, the type of transport and the texture; as a second constitutive parameter is the soil moisture (Sh), which depends on the air temperature, the rains that have occurred, if it is near the sea, if there are plants, among others [13], It is also known as the relationship between the weight of water contained in it and the weight of its solid phase [15], and the last factor is the slope (Sp) that covers the dynamic morphology of the relief, that is, it is that parameter that symbolizes the natural roughness of a zone [22],which is characterized according to its geometry, so the more slope there is, the frequency of landslides also increases, so it is pertinent to take into account the five categories of slope that can be found: flat from 0 ° to 1 °, slightly smooth from 1 ° at 3 °, mild from 3 ° to 5 °, moderate from 5 ° to 15 ° and strong greater than 15 ° [13], in such a way that taking into account these three parameters (P) is determined, as shown below in equation (2).

$$P = Sl \times Sh \times Sp \quad (2)$$

On the other hand, within the category of tripping factors (D) it should be mentioned that it is constituted by the seismicity parameter (Ds) of the area in the first place, which refers to the capacity that a certain area has to experience earthquakes and is closely related to the collision of tectonic plates, and secondly by the data of the intensity of Rain or precipitation (Dp), which refers to the event that occurs when the atmosphere becomes saturated with water vapor, condensing and last rushing. Next, the way to determine (D) is illustrated in equation (3).

$$D = Ds + Dp \quad (3)$$

5. Characterization of passive and triggering factors

The elaboration of the corresponding maps in the ArcGIS software of each factor, identifies the value in which each one of these is found. In order to obtain a correct result, each data obtained by mapping must be classified according to the values provided by Mora -Vahrson in the proposed method.

5.1. Lithology (Sl)

The preparation of the lithological map in the ArcGIS software is carried out taking into account the information from the 2015 geological atlas map of Colombia provided by the Colombian Geological Service (SGC), said map shows geological and lithological information found in the study municipalities as illustrated in table (2), according to this information, the correct classification is made with the characterization that Mora - Vahrson expresses in his study as evidenced in table (3).

Table 2. Geology y lithology of the study municipalities.

	Geology Nomenclature	Lithology
Campoalegre	J – Pi	Plutonic igneous rock of intermediate composition
	Q – al	Alluvial deposit
	MP3NP1 – Mag2	High-grade metamorphic rock
	Q – Ca	Fan deposit
	e6e9 – Sc	Sedimentary rock of continental environment
	J – VCc	Volcanoclastic rock of continental environment
	n4n6 – Sc	Sedimentary rock of continental environment
	N2Q1 - VCc	Volcanoclastic rock of continental environment
Pitalito	J – Pi	Plutonic igneous rock of intermediate composition
	Q – t	Terrace deposit
	Q – al	Alluvial deposit
	C – Sctm	Continental sedimentary rock – transitional - maritime
	Q2 – l	Palludal deposito
	N2 - VCc	Volcanoclastic rock of continental environment
	Q – Ca	Fan deposit
	B6k6 – Stm	Continental sedimentary rock – transitional - maritime
	J – VCc	Volcanoclastic rock of continental environment

Table 3. Classification of the lithology parameter according to Mora – Vahrson [14]

Lithology	**Physical – mechanical Characteristics**	**Degree of Susceptibility**	**Factor Value SI**
Alluviums: thick, permeable, compact, with low water table. Limestones: hard, permeable Intrusive: slightly cracked, low water table. Basalts, andesite, ignimbrites and similar, healthy, permeable and not very fissured. Healthy metamorphic rocks, little cracking, low water table	Healthy with little or no weathering, high shear strength, healthy cracks without fillers.	Low	1
Slightly altered sedimentary rocks, solid stratification, little fissures, low water table. Intrusive rocks, hard limestone, lava, ignimbrites or metamorphic medium cracked or altered, water table at intermediate depths.	Medium to high shear strength, sharable fractures.	Moderated	2
Sedimentary rocks, intrusive rocks, hard limestone, lava, ignimbrites, poorly welded or moderately altered metamorphic tuffs. Colluviums, lahars, sands, slightly compacted regolithic soils, poor drainage, relatively high-water tables.	Moderate to medium shear strength, significant fracturing.	Medium	3
Fluvial lacustrine alluviums, pyroclastic soils with little compaction, sectors of hydrothermal alteration, strongly altered and fractured rocks with stratification and foliations in favour of the slope, with clayey fillings, shallow phreatic levels.	Moderate to low cut resistance.	High	4
Alluvial, colluvial and regolithic materials of very low mechanical quality with an advanced state of alteration, poor drainage, categories 3 and 4 are included with very shallow phreatic levels, subject to very high hydrodynamic gradients.	Very low cutting resistance, soft materials with many fines.	Very high	5

Once the classification is made, we proceed in Arcgis to add a column in the attribute table of the map obtained by the Colombian geological service, where the respective value for each lithological part of the municipalities will be added, then the shapefiles are transformed. of the polygon to raster according to the given value, thus obtaining the lithological map as illustrated in figures (2) and (3) with values of Sl 1, 3, 4 and 5, that is, a degree of low, medium, high and very high susceptibility, while in Campoalegre it has a low, medium and high degree of susceptibility.

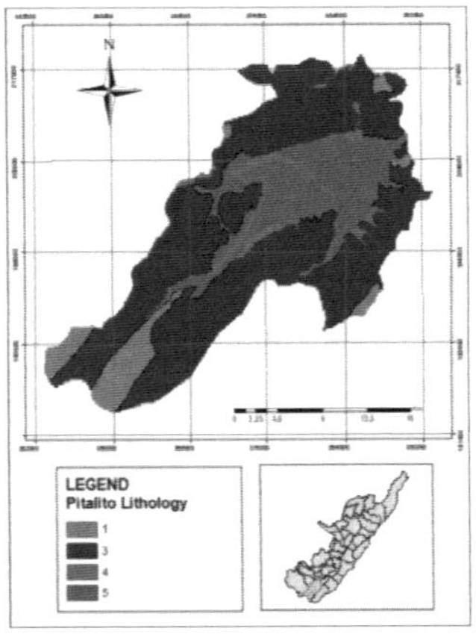

Figure 2. Lithological map of Pitalito municipality.

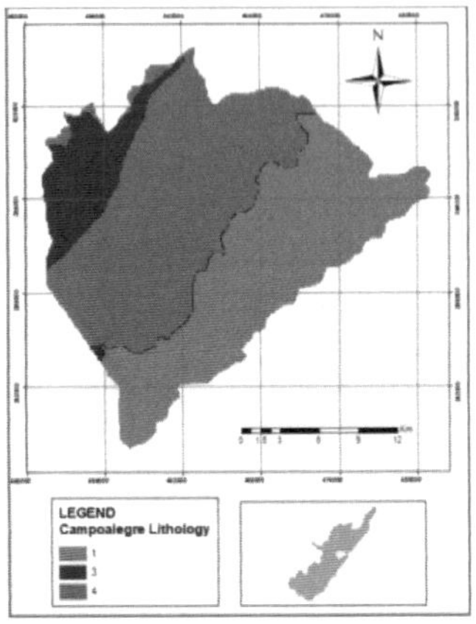

Figure 3. Lithological map of Campoalegre municipality.

5.2. Soil moisture (Sh)

Mora. vahrson sets out some criteria to be able to classify the range in which the soil moisture is found, taking into account the mean monthly precipitation values of the meteorological stations in the area for the years of existence of measurements [16], for the present case They were obtained according to information provided by IDEAM. Table (4) shows the classification of precipitation values according to the monthly average, that is, each monthly data obtained is assigned a value of 0, 1 or 2 depending on the range in which it is found, from In the same way in table (5) the true value of soil moisture is obtained, this is achieved by adding the value assigned at 12 months according to table (4) and according to the result it is classified according to the ranges given by Mora Vahrson.

Table 4. Classification of average monthly precipitation values Mora – Vahrson [14].

Average monthly rainfall (mm/month)	Value
< 125	0
125 - 250	1
250 <	2

Table 5. Soil moisture parameter (Sh) Mora – Vahrson [14].

Accumulated value of precipitation rate	Classification	Sh Factor
0 – 4	Very low	1
5 - 9	Low	2
10 - 14	Medium	3
15 - 29	High	4
20 - 24	Very high	5

In order to obtain greater accuracy against this parameter, a weighted average is carried out to complement the missing data from each of the stations, for this the height of precipitation recorded in the auxiliary stations (Hpa, Hpb, Hpc), the precipitation height of the missing station (Hpx), the average annual precipitation of the auxiliary stations (Pa, Pb, Pc) and finally the average annual precipitation of the study station, as evidenced in the expression (4).

$$hpx = \frac{1}{3}[\frac{px}{pa} * hpa + \frac{px}{pb} * hpb + \frac{px}{pc} * hpc] \quad (4)$$

Once the explained the procedure, in the ArcGIS software a table with the monthly precipitation values must be inserted, thus generating twelve maps of each month, these twelve maps are used to obtain the IDW tool since through this, they are made isohyets in order to evaluate the different soil moisture parameters in each municipality, which must be reclassified according to table (5), after that they are added using the Raster calculator tool to obtain a single general map, thus obtaining a soil moisture value in the municipality of Pitalito of 1 and 2, classified as very low and low as illustrated in figure (4) and in the municipality of Campoalegre of 2, classified as low, as illustrated in figure (5)

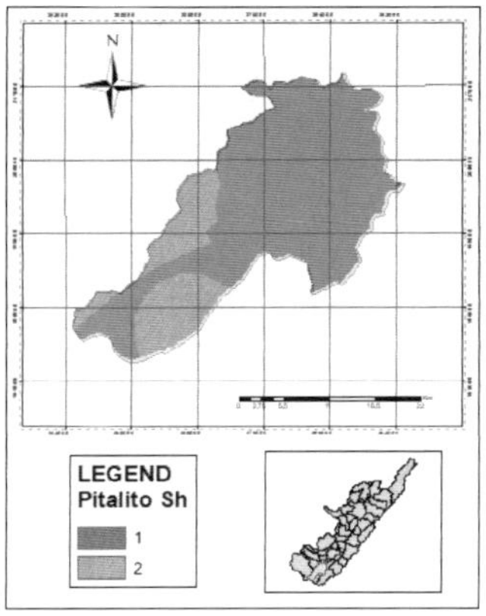

Figure 4. Soil moisture map in the Pitalito municipality.

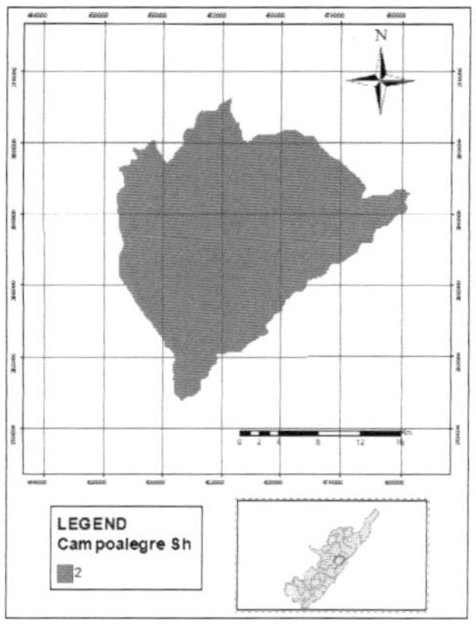

Figure 5. Soil moisture map in the Campoalegre municipality.

5.3. Slope (Sp)

The classification of the slope according to Mora Vahrson is illustrated in table (6), which is used in the elaboration of the maps of the study municipalities, so it is necessary to load a satellite image in the software, in order to of obtaining a DEM, which consists of a digital elevation model and provides content of slopes and elevations of an area. Subsequently, the flows of water present in the municipalities are established taking into account the coordinate (WGS 1984 zone 18N), in order to delimit the basin and extract the morphometric characteristics, where the behaviour of the water currents can be analysed, as well their physical characteristics.

Table 6. Slope Parameter (Sp) Mora - Vahrson [14].

Equivalence in degrees	Classification	Factor
0 – 4,29	Very low	0
4,3 – 9,93	Low	1
9,94 – 16,7	Moderated	2
16,71 – 26,57	Medium	3
23,58 – 38,66	High	4
>38,66	Very high	5

To continue with the elaboration of the map, now the attention be focused on the construction of the contour lines in the two municipalities, which generate the triangulation to define the mountainous part of the study places, in order to obtain the final value corresponding to the slopes, illustrated in figures (6) and (7), by using the Slopes option which provides information on the range of the slope and the Reclassify tool to classify the slope in the ranges defined in the method Mora-Vahrson.

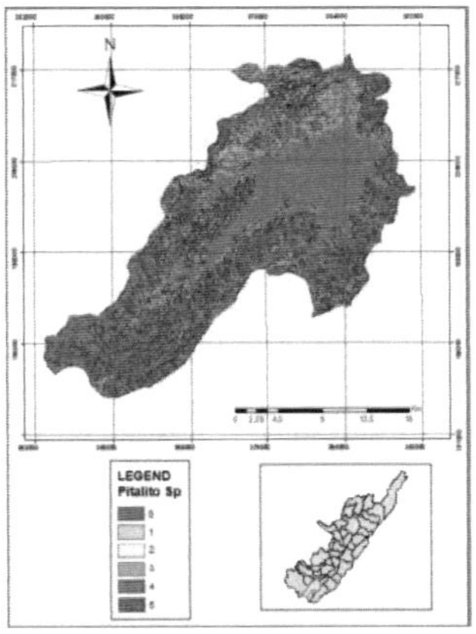

Figure 6. Slope map of Pitalito municipality.

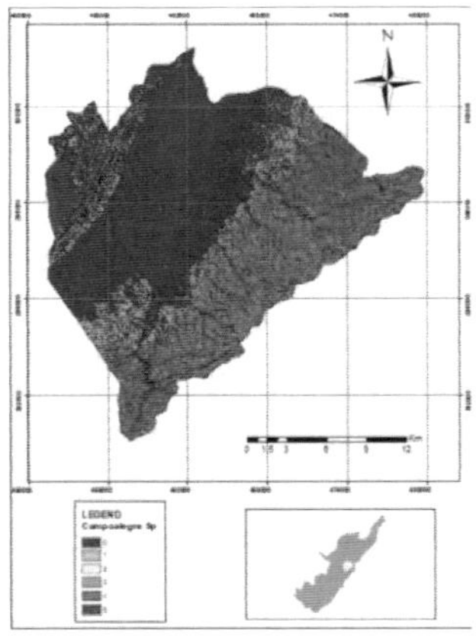

Figure 7. Slope map of Campoalegre municipality.

5.4. Seismicity (Ds)

Mora - Vahrson classifies the seismic activity of a place through the establishment of parameters that are illustrated in table (7), where the intensity on the Mercalli scale is a parameter whose values refer to the qualitative exposure of the consequence of the earthquakes, on the other hand, there are the peak acceleration values (% g) and the PGA acceleration value (m / s2), where the peak acceleration is a percentage of gravity, so that in Colombian regulations, it corresponds to Aa value defined in NSR-10 (Colombian earthquake resistant construction regulations); that is, by multiplying each value (Aa) in percentage with the value of gravity (9.81m / s2), the value of the acceleration PGA is obtained.

Table 7. Seismicity parameter (Ds) Mora – Vahrson [14].

Intensity	Peak Acceleration (%g)	Acceleration PGA (m/s2)	Classification	Ds Value
III	1 – 12	0,098 – 1,226	Mild	1
IV	13 – 20	1,227 – 2,011	Very Low	2
V	21 – 29	2,012 – 2,894	Low	3
VI	30 – 37	2,895 – 3,679	Moderated	4
VII	38 – 44	3,680 – 4,365	Medium	5
VIII	45 – 55	4,366 – 5,445	High	6
IX	56 – 65	5,446 – 6,426	Strong	7
X	66 – 73	6,427 – 7,210	Quite Strong	8
XI	74 – 85	7,211 – 8,388	Very Strong	9
XII	>85	>8,389	Extremely Strong	10

To obtain the Ds value, the threat map of the Colombian Geological Institute (NTC) of the Colombian Geological Institute was identified in order to obtain the value of the effective peak acceleration (Aa) and the value of the threat. From the above, it was determined that the municipalities of Campoalegre and Pitalito are in a high seismic hazard zone of "8" and have an effective peak acceleration of 0.3.

The seismicity factor is classified according to a moderate intensity, identified with the number 4, since when multiply the value of the effective peak acceleration (Aa = 0.3) with gravity (g = 9.81) as illustrated in equation (5). It should be clarified that in order to carry out a more in-depth study of the value of 0.3 in the respective municipalities, an interpolation was made to each map with the Aa values of neighboring municipalities, in table 7.1 the chosen municipalities with their respective effective peak acceleration are shown.

$$Ds = Aa * g(\frac{m}{s2}) \tag{5}$$

Table 7.1 Effective peak acceleration Aa.

Study municipality	Neighboring municipality	Aa value
PITALITO	San Agustín	0.25
	Isnos	0,25
	Solado blanco	0,25
	Elías	0,3
	Timaná	0,3
	Acevedo	0,3
	Palestina	0,3
CAMPOALEGRE	Yaguará	0,25
	Hobo	0,3
	Algeciras	0,3
	Rivera	0,3
	Palermo	0,25

Using the IDW tool in the ArcGIS software, the data is interpolated, thus obtaining two values in each map, these values range from 0.27 to 0.3, which when approximating the closest value for each municipality is 0, 3, obtaining, as well as a Ds value of 4 as mentioned above, with a moderate classification as illustrated in figures (8) and (9).

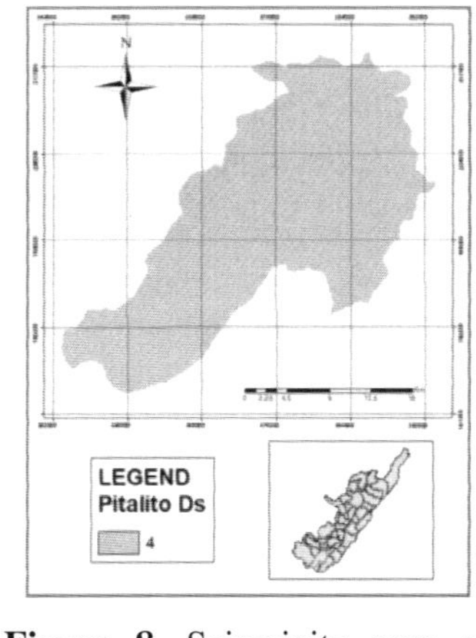

Figure 8. Seismicity map of Pitalito municipality.

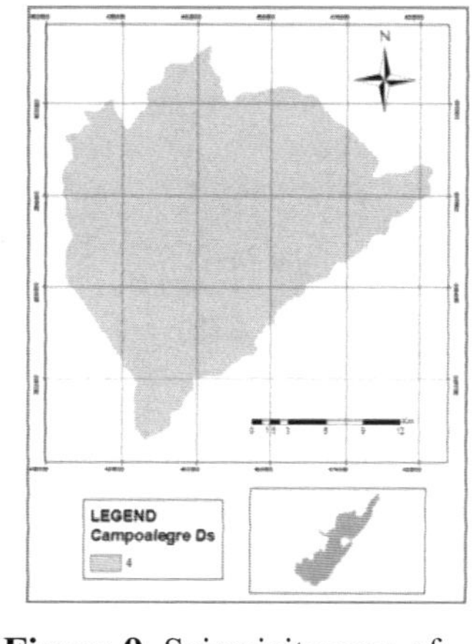

Figure 9. Seismicity map of Campoalegre municipality.

5.5. Precipitation (Dp)

To determine this factor within the municipalities under study, it should be taken into account the precipitation in 24 hours of the stations of the same, however, to determine the value of this parameter, a return period of 100 years is estimated through the final value with the Gumbel distribution as stated by Mora - Vahrson, which allows to determine extreme values for flows or rainfall [17], for which the statistical distributions of hydrology are used, since these mathematical functions are related with the magnitude of an event with its probability of happening [18]. To do this, the factor (F (x)) is taken into account, which indicates the probability that a value lower than (x), which is the return period that will occur in a precipitation or flow, in the same way (Sn) which corresponds to the deviation used to evaluate the dispersion of data as seen in equation (6) and which in turn depends on the sums of the deviations of each point (x) from the previously mean calculated and divided by the number of data.

$$Sn = \sqrt{\frac{\Sigma(x-\bar{x})^2}{n}} \tag{6}$$

In the same way, to complement the Gunbel distribution, the mean is required, which is determined by the average of the maximum values per year, of the values of μy and σy, which are obtained from table (8), where the values of According to the number of data to be taken, however, in case the number of data is not found in the table, the values must be interpolated in order to find the exact value; On the other hand, α and u are taken into account, which are parameters used to determine the value of the precipitation of each station in a return period of 100 years, these values in turn are established by means of equations (7) and (8), where α depends on σy and the standard deviation, while u depends on the mean, μy and α. Finally, the most important value is determined, which is x, which corresponds to the precipitation of each station with a return period of 100 years, for which it is determined by equation (9).

Table 8. Values of the variables μy y σy [19].

Data number	μy	σy
10	0,4952	0,9496
15	0,5128	1,0206
20	0,5236	1,0628
25	0,5309	1,0914
30	0,5362	1,1124
35	0,5403	1,1285
40	0,5436	1,1413
45	0,5463	1,1518
50	0,5485	1,1607
55	0,5504	1,1682
60	0,5521	1,1747
65	0,5535	1,1803
70	0,5548	1,1854
75	0,5559	1,1898
80	0,5569	1,1938
85	0,5578	1,1974
90	0,5586	1,2007
95	0,5593	1,2037
100	0,5600	1,2065

$$\alpha = \frac{\sigma y}{Sx} \quad (7)$$

$$u = \bar{x} - \frac{uy}{\alpha} \quad (8)$$

$$X = -\ln(-\ln(F(x))) * \alpha + u \quad (9)$$

Finally, once the value of X has been obtained in each station, this value is classified in table (9) established by Mora-Vahrson; to generate the map in the ArcGIS software mentioned above; For this, a table is exported with the precipitation values, that is, with the X values that refer to the maximum precipitation in 24 of each study station, in order to generate an isoyetal map with the help of the IDW tool. , since the isohyets are those that link all the points that have the same precipitation information in a period of time.

Table 9. Precipitation parameter Mora -Vahrson [14].

Accumulated value of precipitation index	Classification	Facto Sh
< 100	Very Low	1
101 - 200	Low	2
201 - 300	Medium	3
301 – 400	High	4
> 400	Very Low	5

According to the isohyets obtained, a classification is carried out using the Reclassify tool to be able to link the data obtained with the classification provided by the aforementioned authors, as illustrated in figures (10), where Pitalito has a precipitation factor of 2 and 3, classified as low and medium, while in figure (11) a Campoalegre precipitation factor of 1 and 2 is obtained, classified as very low and low.

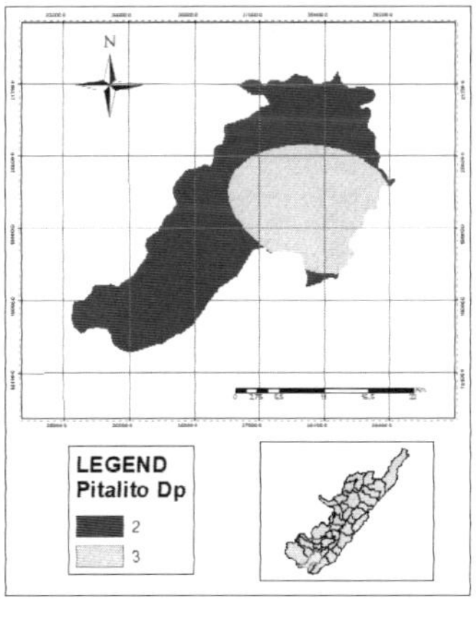

Figure 10. Precipitation map of Pitalito municipality.

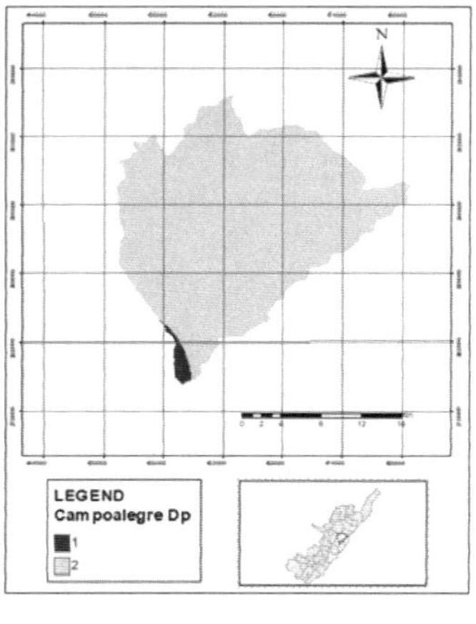

Figure 11. Precipitation map of Campoalegre municipality.

6. Susceptibility to Landslides (S)

Mora - Vahrson classifies the susceptibility to landslides of a place using criteria illustrated in table (10). To identify the municipalities in which landslide potential they are, the respective calculation illustrated in equation (1) must be made. To do this, the respective multiplication is done in the ArcGIS software using the Raster Calculator tool, so that the map obtained is reclassified according to the values provided by the aforementioned authors.

Table 10. Susceptibility to landslides Mora - Vahrson [14].

Classification	Sliding Potential	Sh Factor
I	Very low	Stable sectors, no corrective measures are required, the influence of surrounding sectors with moderate to very high susceptibility must be considered. Sectors suitable for high-density urban use and location of essential buildings such as hospitals, education centers, police stations, fire stations and others.
II	Low	Stable sectors that require minor corrective measures, only in the case of large infrastructure works, the influence of neighbouring sectors with moderate very high susceptibility must be considered. Sectors suitable for high-density urban use and location of essential buildings such as hospitals, education centers, police stations, fire stations, etc. Sectors with poorly compacted fillings are of special care
III	Moderated	No infrastructure should be allowed to be built without geotechnical studies and improvement of the site's earthworks, groundwater management, land bio-stabilization, etc. Sectors with poorly compacted landfills are of special concern. Recommended for low-density urban uses. condition.
IV	High	Landslides (<50%) in case of High probability major earthquakes and high intensity rainfall in order to use them, detailed stability studies must be carried out and corrective measures implemented to ensure the stability of the sector, otherwise they must be maintained as protection areas.
V	Very High	Very high probability of landslides (>50%) in case of major earthquakes and high intensity rainfall Forbidden to be used for urban purposes, it is recommended to use them as protection areas.

According to the previous table, it can be seen that the study municipalities show a susceptibility to landslide that ranges from very low to high, with the potential for moderate landslides prevailing as illustrated in figures (12) and (13), therefore that it is necessary to have prevention when developing civil works either due to the conditions of the land or when changing its composition when carrying out construction of buildings, bridges and roads, that is to say that the way in which the land is located is evaluated before, having in the same way the urban planning and vital lines [24], since if man comes to modify the components of the surface of land and slopes when cutting, filling, deforestation, among others, can generate as a consequence a destabilization due to the alteration of stresses.

To avoid the aforementioned alterations, it is necessary that infrastructure works respect protected areas such as streams, rivers, among others, mainly in that area with a steep slope, in addition to implementing adequate water management, in the same way, techniques of engineering using biology and technology in soils in order to reduce impacts and stabilize.

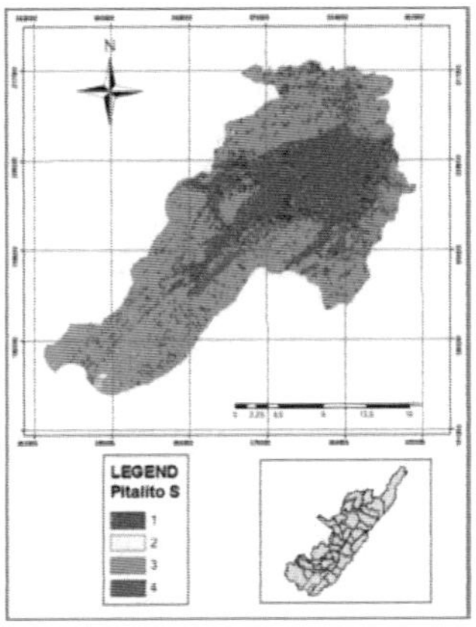

Figure 12. Landslide susceptibility map of Pitalito municipality.

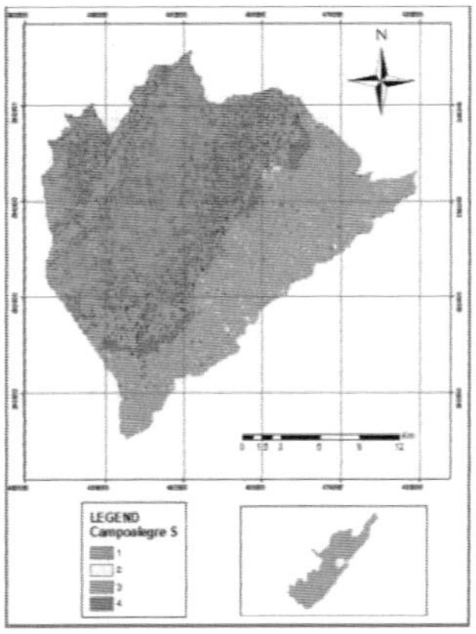

Figure 13. Landslide susceptibility map of Campoalegre municipality.

7. Conclusion

The application of the Mora - Vahrson methodology presents a final map that addresses the factors that impact and influence the development of mass removal phenomena, providing information that leads to an analysis of the presence of morpho dynamic processes in an area; Factors such as lithology and slope are permanent values throughout the study area, so that precipitation, soil moisture and earthquake, which, since they are not constant, are the ones that produce the greatest change at the time of the survey. zoning, that is, they are the variables that have the most relevant influence on the phenomena of mass removal that may occur, however, this method does not address erosive processes, which makes it an approximate but not totally true method. that could be presented.

Taking into account the final maps of susceptibility to landslide in the municipalities of Campoalegre and Pitalito, it can be observed that Mora -Vahrson present a simple way of being able to identify the places with the highest risk of landslides through the use of geographic information, thus finding that in the south and east of the municipality of Campoalegre, exactly in the paths of Otas, Guayabo, Chia, Vilaco bajo, Vilaco alto, Buenavista, El Roble, San Milguel, Palmar Alto, El esmero, Los planes, El Peñon are the areas The most vulnerable areas, while in the municipality of Pitalito the most vulnerable areas are those at the ends of the municipality, by which exactly the south, part of the east and west of the municipality are affected, thus impacting villages such as El Cedro, Montecristo, La Cristalina, El Alto de la Cruz, El Meson, Las Brisas, El Diamante, Puerto Lleras, Costa Rica, Resinas, Barranquilla, Buenos Aires, Monserrate, among others.

One of the events that can occur the most and have occurred previously is floods, since these occur more frequently when there are rains, so it can cause overflows in rivers and streams in Campoalegre such as Neiva River, Otas Ravines, Rios Frio Ravines, Magdalena River, among others, thus affecting homes, roads, bridges, aqueducts, crops, among others; However, in the location of these rivers and streams there are few houses that could be affected. On the other hand, Pitalito is among the two municipalities, which has presented the most events since it is more prone to mass removal phenomena, in addition, due to its location it is exposed to flooding by the Yaguilga Ravines, Magdalena River, among others, thus affecting roads and homes.

For the aforementioned, it is necessary to carry out technical studies in order to establish actions that allow correcting and preventing the risk and if it is necessary to carry out a rehabilitation of the areas referring to the forest variety, that is an ecological restoration, in the same way the study is made to be used as a focus when making decisions, whether for planning the places, to carry out in-depth studies, determine the correct use of the land, and finally so that when carrying out civil works it is taken into account a correct methodology when building, in order to reduce the risks of a landslide.

8. References

[1] Universidad Nacional De Colombia 1999 *Plan de ordenamiento territorial* (Pitalito - Colombia) p 10-30
[2] Rico M 2016 *Plan de desarrollo municipio de Pitalito* (Pitalito - Colombia) P 49.
[3] Sercoin 2014 *Amenaza por movimientos en masa, plan de ordenamiento territorial* (Campoalegre -Huila) p 2-22
[4] Pgirs *Diagnóstico integral* (Campoalegre – Colombia) p 1-25
[5] Indiger 2020 *Caracterización General del Escenario de Riesgo*
[6] Ojeda G, Lacreu H and Sosa G 2007 *Atlas de recursos geoambiéntales* (Potrero de los Funes) p 43-48
[7] Lara M and Sepúlveda S 2008 *Remociones en masa* (Chile: Universidad de chile) p 2-4
[8] Lario J and Bardají T 2016 *Introducción a los riegos geológicos* (Madrid: UNED. Universidad Nacional de Educación a Distancia) p 52-68
[9] Ingeominas 2002 *Catálogo nacional de movimientos en masa* (Bogotá) p 6-36
[10] Jaramillo J and Pasato J 2016 *Aplicación del método mora vahrson para la clasificación de la susceptibilidad a los deslizamientos de la vía macas - Riobamba en la parroquia zuñac* (Ecuador: Escuela Superior Politécnica de Chimborazo Extensión Morona Santiago) p 21-25.
[11] Herrera L, Mena Y and Martínez R *Aplicación de métodos indirectos para el análisis de susceptibilidad de deslizamiento en la subcuenca del río gatuncillo, panama* (Panama) p 4-8.
[12] Barrantes G, Barretes O and Núñez O 2011 *Effectiveness of the modified Mora-Vahrson methodology in the case of landslides caused by the cinchona earthquake* (Costa Rica: Universidad Nacional Heredia) p 142-149.
[13] Orozco O 2010 *Atlas regional de impacto derivados de las actividades petroleras en coatzacoalcos, Veracruz.* p 26
[14] Solorzano J and Paz D 2018 *Elaboración de un mapa de suceptibilidad a deslizamientos en el cerro San Eduardo de la ciudad de Guayaquil usando el software Arcgis (Ecuador: Escuela Superior Politécnica del Litoral) p 4*
[15] Badillo J Rodríguez R 2005 *Mecánica de suelos Tomo 1. Fundamentos de lamecánica de suelos (*México, D.F) p. 61
[16] Rodríguez J Quintana C and Rivera H 2013 *Landslide hazard zoning in urban areas by the mora & vahrson analysis method: Case of study* (Pamplona: Universidad de Pamplona) p 15 – 20
[17] Díaz F, Álzate B, González, An and Ruiz G *Estudio comparativo de metodologías de zonificación de amenaza por movimientos en masa aplicado al sector rural de Villavicencio* (Bogotá D.C: Universidad Nacional de Colombia) p 1-3.
[18] Mujica S and Pacheco H 2013 *Methodology for the generation of a model map for ubication of threats of landslides in the river camurí grande, in the state vargas, Venezuela* (Revista de investigación vol 37) (Venezuela: Universidad Simón Bolivar) p 215-235.
[19] Sánchez J 2013 *Cálculos estadísticos en hidrología* (España: Universidad de Salamanca) p 1-11.
[22] Mora S and Vahrson, W (1994*). Macrozonation methodology for landslide hazard determination.* (Costa Rica), P 49-58.
[23] Cruden D and Varnes D. (1996). *Landslides: investigation and mitigation. chapter 3 - landslide types and processes.* (Washington) P 40-65.
[24] Mora R, Vahrson W and Mora S. (1992). *Mapa de Amenaza de Deslizamientos, Valle Central, Costa Rica.* (Costa Rica) P 16.

Analysis of Envelopes Applied To Bioclimatic Architectures.

J G Ascanio- Villabona[1 3], C L Sandoval-Rodriguez[1 2], A D Rincon-Quintero[1 3], B E Tarazona- Romero[1 3], C G Cárdenas-Arias[1 4], A Quintero Ruiz[1 2], K T Jaimes-Quintero[1]

[1] Energy Systems, Automation and Control Research Group (GISEAC), Design and materials research group (DIMAT), Electromechanical Engineering, Faculty of Natural Sciences and Engineering, Unidades Tecnológicas de Santander (UTS), Road No 9-82 students, 680,005, Bucaramanga, Colombia.
[2] Doctoral Program in Electronic and Communications, Department of Communications Engineering, University of the Basque Country, UPV/EHU, 48013 Bilbao, Spain.
[3] Doctoral Program in Energy Efficiency and Sustainability in Engineering and Architecture, Department of Machines and Thermal Engines, University of the Basque Country (UPV / EHU), Engineer Torres Quevedo Square, 1, 48013, Bilbao, BI, Spain.
[4] Doctoral Program in Engineering of Materials and Sustainable Processes, University of the Basque Country, UPV/EHU, 48013 Bilbao, Spain

Abstract. In the last decade, retrofitting strategies have been reviewed to improve energy efficiency and reduce the environmental impact of existing buildings [1]. A modernization strategy consists of an architectural design based on bioclimatic characteristics. The exploitation of climatic indicators and the use of passive strategies are among the central principles of this design approach [2]. The main objective of this study is to provide an extensive review of the different building envelopes with the adoption of high-performance materials or systems to improve human thermal comfort.
Keywords: Energy saving, Human thermal comfort, Bioclimatic design, Energy Efficiency, Building envelope.

1 Introduction

In 2010, the construction sector consumed 32% of energy, and was responsible for a third of global CO2 emissions [3]. Pérez Lombard et al. [4] identified the increase in energy consumption of the residential and commercial sectors that reached 20-40% of the total world energy use [5], where half of the energy consumption of the building is dedicated to heating systems , ventilation and air conditioning (HVAC), since, both in hot and cold climates, active systems must heat or cool the current building in conditions that are comfortable for the inhabitants [6].

Currently worldwide, the greatest concern for designers and researchers is the environmental issue, especially the excessive consumption of fossil fuels and the consequent danger of climate change. Among the many solutions suggested by global institutions, the use of renewable energy is considered the most suitable factor. These energy approaches can be generalized under the title of sustainable architecture [7], ecological architecture, zero energy [8], passive solar design and bioclimatic design.

Bioclimatic design [9], refers to the architectural design approach that uses both solar energy and related environmental resources to provide thermal comfort indoors and outdoors. According to ASHRAE Standard 55 [10], hermal comfort is defined as: "The condition of the mind that expresses satisfaction with the thermal environment and is assessed by subjective assessment."

The thermal performance of buildings is related to three main categories: microclimate of the building environment, building physics, and required internal thermal comfort [11]. Among these categories, construction physics specifies the building envelope such as roofs, walls, windows, doors and floors, among its parameters, significantly affecting energy consumption [12], due to the fact that buildings release or obtain heat through of their wraps [13], [14].

Approximately 50% of the total energy consumption of the building is affected by the elements of the building envelope [15]. Therefore, applying enclosures with highly insulated materials [16], [17] or energy efficient equipment such as heat recovery systems [18], can contribute to reducing the energy demand in the home [19], [20].

However, the decision to adopt new envelope systems is difficult because it is a multi-criteria problem that encompasses ecological, economic, social and other dimensions [21]. The downside with modernization also includes limitations such as influences from the existing context [22].

This document provides a review of various envelope alternatives currently applied, such as innovative concepts to replace or retrofit bioclimatic building envelopes. Therefore, this research incorporates as many envelopes as possible as upgrade options into the decision process to optimize the synthesized performance of potential envelopes without subjective material selections.

2 Background of the investigation.

This article reviews the previous studies on the envelopes applied to bioclimatic buildings, determining parameters, as variables to minimize the energy consumption of the house. The studies reviewed include indexed journal articles, conference proceedings, books, dissertations, and reports.

However, during this review, other relevant research is discussed if necessary, to illustrate the idea. The reviewed articles studied the implementation of optimization in the design of building envelopes. In addition, commercial and residential cases are included. Other parameters, such as HVAC systems, controls, renewables, etc., are also included in the reviewed studies as long as the building envelope and energy parameters are primarily considered use / demand.

3 Literature review

To compile the documents, keywords such as 'construction envelope', 'energy consumption' were used. Searches are also carried out with the combination of the main keywords and the different optimization methods. Later in Table 1, the references of the relevant documents that meet the selection criteria are included.

Table 1. Bibliographic review of related studies

Author	Year	Country	Summary	Doi
Piggot, J, et al. [23]	2019	Chile	This research is to analyze the thermal performance of vertical envelope solutions for wooden partition walls used in Canadian establishments and to evaluate how these technologies contribute to reducing the energy demand of educational establishments in Chile.	http://dx.doi.org/10.7764/rdlc.18.1.201
Cassandro, R. [24]	2018	Colombia	A composite panel prototype, type SIP (Structural Insulated Panel), is proposed as a structural thermal panel, as a housing envelope wall made with renewable materials of natural origin, specifically recycled guadua and cardboard.	http://dx.doi.org/10.14718/revarq.2018.20.2.2116
Mazzoco, M. et al. [25]	2018	Argentina	In this research the thermal-energy behavior of a social housing in the central region of Argentina is studied. The objective is to energetically evaluate the house for a historical period between 1960 and 2011 and obtain the thermo-physical model through simulation with ECOTECT and SIMEDIF to propose its rehabilitation.	http://dx.doi.org/10.1590/s1678-86212018000400302

Author	Year	Country	Summary	Doi
Osuna, I. et al. [26]	2017	Colombia	The thermal behavior of an experimental planted roof prototype is determined, which can be installed on inclined roof systems in corrugated sheets of fiber cement, widely used in our environment.	http://dx.doi.org/10.14718/revarq.2017.19.1.1109
Molina, M y Fernández, P. [27]	2013	España	Comparative analysis of the thermal envelope and energy demand between two types of housing, the palloza and the traditional house, in the Os Ancares mountain area, in northwestern Spain.	http://dx.doi.org/10.4067/S0718-915X2013000200008
Luciani, S. et al. [28]	2018	Colombia	With the aim of contributing to the reduction of impacts in the construction of buildings, various ventilated and conventional façade systems were designed, involving opaque facades, plant elements and air chambers.	http://dx.doi.org/10.14718/revarq.2018.20.2.1726
Mercado, M., et al. [29]	2010	Argentina	The present work evaluates: (a) the thermal-energy quality of a social housing, by means of in-situ measurements under conditions of real use; (b) the necessary energy requirement through a balance; (c) the simulation of the home in the SIMEDIF program, where the model was adjusted and easy and inexpensive improvements were tested; and (d) a qualitative survey of the thermal sensations.	http://dx.doi.org/10.1590/S1678-86212010000200006
Pérez, N., et al. [30]	2018	Brasil	The objective of this work is to demonstrate the benefits that building systems with green envelopes can offer to internal environments when temperatures are lower.	http://dx.doi.org/10.4067/S0718-50732018000100015
Pérez, H. et al. [31]	2013	Chile	In this study, the analysis on the design and bioclimatic architecture is carried out, taking as a premise the sustainability between user, environment and built space.	http://dx.doi.org/10.4067/S0718-915X2013000200010
Ayelen, M. et al. [32]	2018	Argentina	The present work proposes as a general objective to carry out a review of the state of the opto-thermal evaluation of materials and components of the urban-building envelope, in relation to the transfer context and the possibilities of regional technological development for the estimation of energy properties and indicators of the urban-building envelope.	https://doi.org/10.22320/07190700.2018.08.02.05
Ramos, A. [33]	2017	Chile	In the study presented in this article, a mathematical model is developed for the real-time simulation of a winemaking envelope representative of the national winemaking technology - reinforced concrete-, and thermal monitoring of both the wine stored in this interior as in a stainless steel one, with the use of HOBO U-12 temperature sensors.	https://doi.org/10.22320/07190700.2017.07.02.04
Carvalho, C., et al. [34]	2017	Brasil	The objective of this work is to evaluate the thermal comfort of a multi-family housing building using the Mahoney limit values, comparing the results with those obtained from the normative values for each of the bioclimatic zones.	http://dx.doi.org/10.1590/s1678-86212017000100131

Author	Year	Country	Summary	Doi
Balter, J. et al. [35]	2016	Argentina	The objective of the work is to evaluate the energy efficiency and the habitability conditions of housing units in high-rise buildings. Four cases were selected in the City-Oasis of Mendoza, Argentina: two located below and two above the tree canopy. Enclosures of the mass and light type were evaluated. Hygro-thermal measurements were carried out for two years in situ, used to validate models using the Energy Plus simulation program.	http://dx.doi.org/10.1590/s1678-86212016000100059
Matheus, C. et al [36]	2016	Brasil	This work presents three different studies of thermal behavior with vegetal coatings: 1) a case study with vegetal cover, 2) an experimental study with a living wall and 3) a case study with a green wall, all with comparative and simultaneous measurements between Surfaces. exposed and protected by green.	http://dx.doi.org/10.1590/s1678-86212016000100061
Melisa, G. et al. [37]	2016	Chile	This work proposes the search for alternative materials for thermal insulation of houses, reusing resources. These materials could contribute to both the improvement of the existing home (recycled) and the new one.	http://dx.doi.org/10.4067/S0718-83582016000100004
Schmitt, C. [38]	2013	Chile	Considerations of environmental responsibility, the search for improvements in the indicators of thermal transmission and infiltrations for glass facades and the existence of new technologies for modeling pieces have stimulated the incorporation of wood in curtain wall structures.	http://dx.doi.org/10.4067/S0717-69962013000200017
Bengochea, M. et al. [39]	2011	España	The ventilated façade is a building envelope system based on the combination of ceramic panels with the air chamber in order to promote energy savings under appropriate conditions, especially in the face of cooling needs.	10.3989/cyv.142011
Williamson, J. et al. [40]	2016	España	Alternative compounds in the envelope sector are obtained by stretch extrusion of narrow interlocking panel sections and profiles.	10.20868/ade.2016.3195
Varela, S. et al. [41]	2018	España	In the present work, the thermal behavior of the facade of a typical building from the 1950s that has been rehabilitated with an external thermal insulation system (ETICS) is analyzed.	10.20868 / ade.2018.3851
Ottelé M, et al. [42]	2011	Italia	This article discusses a comparative life cycle analysis (LCA) located in the Netherlands for: a conventionally built European brick facade, a directly green facade, an indirectly green facade (supported by a steel mesh), a facade covered with a living wall system based on planters and a facade covered with a living wall system based on layers of felt.	https://doi.org/10.1016/j.enbuild.2011.09.010
Katia Perini & Paolo Rosasco [43]	2013	Italia	This document presents a cost-benefit analysis of different vertical greening systems - green facades and living wall systems - considering the personal and social costs and benefits during their life cycle.	https://doi.org/10.1016/j.buildenv.2013.08.012

Source: Authors.

The building envelope parameters that cause energy consumption in a building include architectural features and construction materials. The relationship between the design of the building envelope and the energy consumption data can be identified by iteration changing the variables [44], [45]. Based on the studies, the influencing variables are listed below:

- Orientation of the building [46], [47], [48], [49], [50].
- Building geometry: aspect ratio and compactness ratio (SA / V) [51], [52], [53], [54].
- Construction materials for building envelopes [55], [56], [57].
- Geometric position and density of fenestrations [wall-window relationship (WWR)] [58], [59], [60].
- Sun protection devices [61], [62], [63], [64], .
- Type, size and schedule of the heating, ventilation and air conditioning system (HVAC) [65], [66], [67], [68], [69].
- Management of energy storage. (thermal storage with phase change materials) [64], [70], [71].
- Lighting schedules and indoor lighting power density (LPD) [63], [72], [73], [74].
- Location and weather [75], [76], [77].

4 Results

The environmental impact and energy consumption of conventional materials for construction, according to Diulio, et al, [78], such as cement, concrete, clay and PVC, represents one of the great global concerns, which makes it necessary to rethink the architecture of the building going back to its origins in the construction sector, using local materials with low energy costs and minimal environmental impact. The research made it possible to evaluate through the impact of the envelope on the demand for heating energy, how the thermal behavior of recycled-type envelope materials make a significant saving (18.89%) that generates a reduction of the thermostat temperature by one degree from 21 ° C to 20 ° C. It shows that without thermal improvement in the envelope that covers the masonry, this is not possible since in previous years the user had to set the thermostat at 24 ° C to obtain poor thermal comfort with high specific energy consumption. This caveat is important since if the recommendation is proposed to society to lower the thermostat for buildings that have central air conditioning by one degree, it will only be effective in enclosures that have a sufficient level of thermal insulation in order to increase the interior surface temperature in enclosures, opaque and transparent, thus helping to subtract from large amounts of thermal contamination.

Given these provisions for the minimization of environmental impacts product of conventional architectures, Escorcia et al, [79] in their research to validate the thermal reconditioning of homes for Chile, recognizes effective actions to reduce consumption through passive strategies and the development of clean energy and renewable policies that also coincide with the thermal improvement of envelopes (walls, windows, roofing and floors. Estimates of energy consumption for heating using the classic methodology, annual consumption equation, which contains the degrees- annual days of the respective climatic zone and the characteristics of the home, that is, its volumetric transmittance and the volume of the home.

However, in a proposal for thermal reconditioning of homes in Chile, Pavez and Meza [80] concluded that, although different factors could explain this phenomenon of temperature increase; however, higher thermal resistance is the most likely alternative and so is the use of the thermal inertia of the masonry as the envelope material for insulation. Although it could also be the case of concrete as a coating, probably due to its second highest thermal resistance; but the authors' choice that the thermal inertia of concrete is taken advantage of, apart because economically it has the characteristic that it is achieved as a disused material.

In Colombia, social housing presents a problem associated with the selection and use of a materiality consistent with climate change and the conditions of thermal and light comfort. Therefore, Medina and Escobar [81] present a case study with 114 thermal and 18 light simulations to have a support in their recommendations on which ones establish the envelopes as the barriers between the interior and exterior of the houses that, therefore, Therefore, they have a great implication in terms of comfort and, as a consequence, play a key role in the harmony between the climate of the place and the comfort conditions in the spaces. They are efficient envelopes with respect to environmental conditions. The authors affirm that, with the aggravation of ignorance or the scarce application of passive strategies of adaptation to climate in architecture, and the absence of notions of control of thermal and light comfort, which exacerbates the problem of comfort and low ventilation, because the envelopes are a much more complex element, which goes beyond a functional value, since they are the components with the largest area and incidence in the thermal and light control of the building.

They summarize everything in their analysis on envelopes, categorizing the materials in such a way that they were applied to simulations divided into the following three groups:

a) Traditional materials, which correspond to those commonly used in construction in Colombia, such as: concrete, brick, wood, plaster, concrete blocks, aluminum, 3 mm glass, clay tile, PVC tile, zinc tile, cement, rubber, asphalt.
b) Avant-garde materials, which correspond to materials that have evolved their usual use such as: double glazing or glass with an air chamber, lightweight aggregate concrete, plasterboard, fiberglass polyester, fiber-cement, polycarbonate, glass wool and galvanized steel.
c) Innovative materials that offer quite significant positive impacts, based on chemistry and state-of-the-art technology to obtain them: expanded polystyrene, porous aluminum, polyethylene airgel, ETFE, glasses with argon gas and glasses with solar panels.

The materials that are usually used in the construction of Colombia do not present an unfavorable thermal and light behavior in cold conditions, but if it is to comply with the conditions in cold and hot climates and apart from the economic conditions in the use of materials non-conventional, Cassandro [24] in his analysis of the thermal envelope material, points out that the envelope of a building has as its main function to maintain a constant thermal insulation and includes all the elements that divide or separate the exterior from the interior; Its main purpose is to maintain the average internal comfort climate between 18 and 20 ° C. Then, it shows that in the country there are materials such as guadua waste chips and recycled cardboard; where its results of resistance to compression, thermal and acoustic insulation prove to be a viable and competitive product compared to those currently marketed and show that it is a sustainable alternative within the principles of passive energy.

Based on the choice of efficient envelopes, after carrying out thermal and light simulations [28] the simulations to recognize the performance of PVC material compared to conventional materials such as masonry, it was recorded that PVC reduced the temperature, on July 21, which is a day of extreme temperature, compared to the masonry that reduced the internal temperature of 2.2 °C, that is, the model composed of PVC presented better performance. If the base of the envelope is in PVC, the discussion should also focus on the cost-performance ratio, considering that, although the ventilated façade system can decrease the interior temperature in the building.

Analyzing the situation of envelopes in Argentina according to Ayelen [32], infrared thermography is used as a method to estimate the surface temperature of the components of the envelopes of a building from data taken indoors. Through in situ measurements of temperatures and heat fluxes in the envelopes of houses and buildings.

For high-rise buildings, Balter, et al, [35] responds with respect to materiality to the conception of a building with a concrete structure and glass skin, and although the structure is significant due to the seismic nature of the region, it predominates a transparent aesthetic. It has 76% of transparent materials in its envelope and the average thermal resistance of the total envelope is 0.26 m2 ° C / W. Then, as the improvements in the materials were studied with respect to the known variables, it is suggested that you select for these large buildings an opaque 0.25m solid brick envelope with 0.05m expanded polyurethane insulation (K = 0.78 W / m2 ° C), and in the light case, work continues with an opaque reinforced concrete envelope of 0.40 m (K = 2.54W / m2 ° C). In both buildings, the transparent envelope selected is hermetic double glazing (DVH) of 6 mm with a low emissivity interior face (K = 1.80 W / m2 ° C).

A study that evaluates alternative thermal insulation with cheap materials is that of Melisa [37] who recommends plastic bottles, recycled expanded polystyrene, corrugated cardboard, wheat straw and recycled polyethylene, as the materials that comprise a feasible alternative for replication by family farmers, since the parameters analyzed in the feasibility study determine that they are easy to implement, very low cost and with an optimal response in terms of thermal insulation. Although the analysis of the life cycle of these materials must be taken into account, although it is a task to be considered in the future, it would allow to specify the feasibility of their use from an integral point of view.

According to Ramos [33], energy efficiency in the envelopes used in Argentine industries is characterized by few specific solutions. The definition of efficient enclosure prototypes from the rational use of energy is a task under development in numerous published works.

Taking buildings with natural conditions in climates of high temperature and high humidity in Brazil, Carvalho, et al [34] exposes that the heat exchange occurs in fluctuating conditions, since the components of the envelope are periodically heated and cooled due to variations in external temperature and incident solar radiation. Under such conditions, the thermal capacity becomes a property with a decisive effect in determining the internal thermal conditions.

Matheus, et al [36] with vegetal coatings and evaluate their ability to mitigate undesirable heat gains in the indoor environment. It is expected, through the results, to indicate the potential of integrating vegetation in buildings and developing projects that use this resource as a passive conditioning strategy. The case study showed that green roof compared to ceramic tiles causes thermal lag, resulting in a more stable temperature for the built environment. According to the temperature data of the surface of the green roof, in relation to the ceramic tile roof, there is an average thermal damping of 9ºC at the hottest times of the day, and an average delay of 2 h.

In recent years, several studies in Spain have addressed this issue, based primarily on obtaining the energy consumption and CO2 emissions of the existing building stock and its possibilities for improvement. Kurtz, et al [82] study the thermal envelopes of old Spanish houses, then the author resolves that his architecture must reform the envelopes in the blind part of facades with solid brick, which is the most used solution or solid concrete block It has a high insulating capacity suitable for all or almost all levels of Ld. Hollow brick facades are not always sufficient, depending on their thickness and Ld of the site.

Pérez [83] states that when the conditions of Spanish buildings are subjected to moderate or high temperatures, and the interior conditions of comfort close to 20ºC, the materials that make up the enclosures are usually found at temperatures above 10ºC. , taking into account the relative humidity of 55%, which is within the range of interior spaces without high humidity production and the even higher values of environmental relative humidity that Spain handles, make it not realistic to consider that the enclosures that are going to handle have a 50% resistance percentage. The thermal calculation carried out in this way can cause inadequate and insufficient designs of the thermal envelope, energy losses higher than those initially foreseen and a higher energy consumption associated with the air conditioning installations.

Finally, Williamson, et al [40] implemented two varieties; an expanded polystyrene bead insulation with a bonding agent, installing them in Edinburgh properties, using a white bead with a thermal conductivity value of 0.040 W / (m · K).

5 Conclusions

This work has carried out considerable research to investigate different building envelopes. Therefore, it provides an overview of the research developments by various authors in different countries, reviewing related literature available from 2001 to 2020, to highlight the requirement of the document selection criteria, where it was found that:

The studies reviewed confirm that building envelope measurements have the potential to significantly reduce the energy footprint of the built environment. Heat transmission through opaque building enclosures has been shown to be a significant energy loss mechanism, particularly in residential buildings, as they exhibit lower sell-to-wall ratios than glass-dominated office skyscrapers. This is why adding insulation to walls that are not otherwise insulated is very effective; All the studies reviewed indicate improvements by adding insulation or increasing the insulation thickness, albeit with decreasing returns as the insulation thickness increases.

In order to reduce the heating of external surfaces, two primary strategies have been reported: the use of external cladding systems that reduce solar absorption (reduction up to 1.9% in building cooling) as well as the use of vegetation both as shadow generators as detached from the building to green the roof and walls.

Windows and glazing are a relevant link to achieve better energy efficiency in building closings. Therefore, changes from single panel glazing to dual panel solar control can have a significant effect (more than 12% in the case of an average residential building). On the other hand, external shading in the form of blinds and / or vegetation has proven to be highly effective (between 28-34% in cooling energy); in addition to improving lighting and solar glare. In addition to the above, the proper orientation of the window reduces energy consumption by up to 55%. Natural ventilation is also a factor that represents a great challenge, however, the use of natural ventilation strategies, present savings of up to 30% in cooling energy savings.

The previous conclusions indicate that very significant energy savings can be achieved in buildings by addressing the building envelope and implementing low-energy ventilation strategies. However, to achieve these savings effectively, it is not only enough to implement these strategies in new construction but also to create awareness and develop modernization measures that can be applied to the various homes built in times when regulations on energy efficiency were scarce or null, being rehabilitated so that they can comply with the current standard parameters in energy efficiency.

References

[1] S. Changa, D. Castro-Lacouture y Y. Yamagata, «Decision support for retrofitting building envelopes using multi-objective optimization under uncertainties,» *Journal of Building Engineering,* vol. 32, 2020.

[2] A. B. Daemei, S. R. Eghbali y E. M. Khotbehsara, «Bioclimatic design strategies: A guideline to enhance human thermal comfort in Cfa climate zones,» *Journal of Building Engineering,* vol. 25, 2019.

[3] D. Ü.-V. L. F, C. Susana, S. Camila y B. K. P. Heating, «Heating and cooling energy trends and drivers in buildings,» *Renewable and Sustainable Energy Reviews,* vol. 41, pp. 85-98, 2015.

[4] LuisPérez-Lombard, JoséOrtiz y ChristinePout, «A review on buildings energy consumption information,» *Energy and Buildings,* vol. 40, pp. 394-398, 2008.

[5] J. G. Ascanio-Villabona, C. L. Sandoval-Rodriguez, A. D. Rincón-Quintero, B. E. Tarazona-Romero y R. E. Paez-Castro, «Building a prototype for functional analysis of the energy potential of the water flow in pipe ½ "using microturbines applied to Unidades Tecnológicas de Santander,» *IOP Conference Series: Materials Science and Engineering,* vol. 844, 2019.

[6] S. Gyamfi, SusanKrumdieck y T. Urmee, «Residential peak electricity demand response—Highlights of some behavioural issues,» *Renewable and Sustainable Energy Reviews,* vol. 25, pp. 71-77, 2013.

[7] G. Farmer y S. Guy, «Reinterpreting Sustainable Architecture: The Place of Technology,» *Journal of Architectural Education,* vol. 54, nº 3, pp. 140 - 148, 2001.

[8] A. Marszal, P. Heiselberg, J. Bourrelle, E. Musall, K. Voss, I. Sartori y A. Napolitano, «Zero Energy Building – A review of definitions and calculation methodologies,» *Energy and Buildings,* vol. 43, pp. 971-979, 2011.

[9] D. L. Jones y J. Hudson, Architecture and the environment : bioclimatic building design, United kingdom: L. King, London, 1998.

[10] ASHRAE Standard 55, « Thermal Environmental Conditions for Human Occupancy,» ASHRAE Inc, Atlanta, 2013..

[11] J. S.Gero, NevilleD'Cruz y A. D.Radford, «Energy in context: A multicriteria model for building design,» *Building and Environment,* vol. 18, pp. 99-107, 1983.

[12] International Energy Agency- IEA, «Technology Roadmap Energy Efficient Building Envelopes,» 2013. [En línea]. Available: https://www.iea.org/publications/freepublications/publication/TechnologyRoadmapEnergyEfficientBuildingEnvelopes.pdf. [Último acceso: 2 08 2020].

[13] S. Jige, Q. Q. Li, G. Augenbroe, J. Brown y P. P.-J. Yang, «A GIS-based Energy Balance Modeling System for Urban Solar Buildings,» *Energy Procedia,* vol. 75, pp. 2946-2952, 2015.

[14] S.-H. Lee, «Intermittent Heating and Cooling Load Calculation Method -Comparing with ISO 13790,» *Architectural research,* vol. 14, pp. 11-18, 2012.

[15] L. Mavromatidis, A. Bykalyuk y H. Lequay, «Development of polynomial regression models for composite dynamic envelopes' thermal performance forecasting,» *Applied Energy,* vol. 104, pp. 379-391, 2013.

[16] F. P. Chantrelle, H. Lahmidi, W. Keilholz, M. E. Mankibi y P. Michel, «Development of a multicriteria tool for optimizing the renovation of buildings,» *Applied Energy,* vol. 88, pp. 1386-1394, 2011.

[17] J. G. Ascanio-Villabona, A. D. Rincón-Quintero, C. G. Cardenas-Arias y C. L. Sandoval-Rodriguez, «Incidence of corrosion in low voltage electrical conductor,» *IOP Conference Series: Materials Science and Engineering,* vol. 844, 2019.

[18] S.E.Chidiac, E.J.C.Catania, E.Morofsky y S.Foo, «Effectiveness of single and multiple energy retrofit measures on the energy consumption of office buildings,» *Energy,* vol. 36, pp. 5037-5052, 2011.

[19] S. Chang, D. Castro-Lacouture, F. Dutt y P.-J. Yang, «Framework for evaluating and optimizing algae façades using closed-loop simulation analysis integrated with BIM,» *Energy Procedia,* vol. 143, pp. 237-244, 2017.

[20] T. F. H. Manz, «Thermal simulation of buildings with double-skin façades,» *Energy and Buildings,* vol. 37, pp. 1114-1121, 2005.

[21] E. Asadi, M. G. d. Silva, C. H. Antunes, L. Dias y L. Glicksmanf, «Multi-objective optimization for building retrofit: A model using genetic algorithm and artificial neural network and an application,» *Energy and Buildings,* vol. 81, pp. 444-456, 2014.

[22] A. Nutkiewicz, R. K. Jain y R. Bardhan, «Energy modeling of urban informal settlement redevelopment: Exploring design parameters for optimal thermal comfort in Dharavi, Mumbai, India,» *Applied Energy,* vol. 231, pp. 433-445, 2018.

[23] J. Piggot, M. B. Piderit y P. Blanchet, «Energy assessment of wood-frame vertical envelope solutions applied in educational establishments of southern Chile,» *Revista de la Construcción,* vol. 18, nº 1, pp. 201-213, 2019.

[24] R. Cassandro-Cajiao, «Muro panel térmico estructural compuesto en guadua y cartón : modelo experimental aplicado al clima de la zona cafetera.,» *Revista de Arquitectura (Bogotá),* vol. 20, nº 2, pp. 90-109, 2018.

[25] M. P. Mazzocco, C. Filippín, H. Sulaiman y S. F. Larsen, «Energy performance of a social dwelling in Argentina and its retrofitting based on thermal simulation,» *Ambiente Construído,* vol. 18, nº 4, 2018.

[26] I. Osuna-Motta, C. Herrera-Cáceres y O. López-Bernal, «Techo plantado como dispositivo de climatización pasiva en el trópico.,» *Tecnología, medio ambiente y sostenibilidad. ,* vol. 19, nº 1, 2017.

[27] M. Molina Huelva y P. Fernández Ans, «Evolution of Thermal Behavior in Stone and Thatched Roof Traditional Houses. Enhancement of a Sustainable Model in Northwest Spain,» *Revista de la Construcción,* vol. 12, nº 2, 2013.

[28] S. Luciani-Mejía, R. Velasco-Gómez y R. Hudson, «Ecoenvolventes : análisis del uso de fachadas ventiladas en clima cálido-húmedo,» *Tecnología, medioambiente y sostenibilidad,* vol. 20, nº 2, 2018.

[29] M. V. Mercado, A. Esteves y C. Filippín, «Thermal-energy efficiency of a social house in the city of Mendoza, Argentina,» *Ambiente Construído,* vol. 10, nº 2, 2010.

[30] N. P. Gallardo, A. Rogério, G. Z. F. Neves, F. A. Vecchia y V. F. Roriz, «Respuesta Térmica al frío en edificios con cubiertas verdes para clima tropical. Fachadas y cubiertas verdes,» *Revista ingeniería de construcción,* vol. 33, nº 1, 2018.

[31] H. Pérez Castro, J. Flores y A. López, «Analisys of Ventilation induced in a Livable Space through a hydro-solar Chimney System,» *Revista de la construcción,* vol. 12, nº 2, 2013.

[32] A. M. Villalba, N. L. Alchapar, E. N. Corre, A. E. Pattini y L. Santoni, «Métodos de evaluación opto-térmica de materiales y componentes de la envolvente edilicia. Situación en Argentina,» *Hábitat Sustentable,* vol. 8, nº 2, pp. 64-79, 2018.

[33] A. I. Ramos-Sanz, «Potencial de demanda energética cero en envolventes industriales: El caso de la vinificación en zonas de gran amplitud térmica.,» *Hábitat Sustentable,* vol. 7, nº 2, pp. 40-49, 2017.

[34] C. C. Ferreira, H. A. d. Souza y E. S. d. Assis, «Discussion of the limits of the thermal properties of building envelopes according to Brazilian thermal performance standards,» *Ambiente Construído,* vol. 17, nº 1, 2017.

[35] J. Balter, C. Ganem y C. Discoli, «Heavy and light high-rise buildings in oasis-cities: thermal and energy assessment of apartments above and below the tree canopy in Mendoza, Argentina,» *Ambiente Construído,* vol. 16, nº 1, 2016.

[36] C. Matheus, F. D. N. Caetano, D. D. d. O. Morelli y L. C. Labaki, «Thermal performance of green envelopes in buildings in the Brazilian southeast,» *Ambiente Construído,* vol. 16, nº 1, 2016.

[37] G. M. Viegas, C. Walsh y M. V. Barros, «Qualitative and quantitative evaluation of alternative thermal insulation. The case of family-based farming,» *Revista INVI,* vol. 31, nº 86, 2016.

[38] C. Schmitt, «Fachadas vidriadas estructuradas en madera,» *ARQ,* nº 84, 2013.

[39] M. M. Pérez, G. L. Patiño, M. A. B. Escribano y P. A. L. Jiménez, «Cuantificación de la eficiencia de la fachada cerámica ventilada mediante técnicas de la mecánica de fluidos computacional,» *Boletín de la Sociedad Española de Cerámica y Vidrio,* vol. 50, nº 2, pp. 99-108, 2011.

[40] J. B. Williamson, J. Stinson, A. Reid y J. Currie, «Improving The Thermal Performance of "Hard-to-Treat" Historic Buildings in Edinburgh's New Town,» *Anales de edificación* , vol. 12, nº 1, 2016.

[41] S. Varela, C. Viñas, A. Rodríguez y P. Aguilera, «Analysis of the thermal behaviour of the ETICS system: Rehabilitated building in Madrid,» *Anales de edificación,* vol. 4, nº 4, 2018.

[42] M. Ottelé, K. Perini, A. Fraaij, E. Haas y R. Raiteri, «Comparative life cycle analysis for green façades and living wall systems,» *Energy and Buildings,* vol. 43, nº 12, pp. 3419-3429, 2011.

[43] K. Perini y P. Rosasco, «Cost–benefit analysis for green façades and living wall systems,» *Building and Environment,* vol. 70, pp. 110-121, 2013.

[44] S. L. B. &. C. S. Chungloo, «Parametric Analysis of Energy Efficient Building Envelope of Thailand,» Asian J. Energy Environ, 2001.

[45] C.K.Cheung, R.J.Fuller y M.B.Luther, «Energy-efficient envelope design for high-rise apartments,» *Energy and Buildings,* vol. 37, nº 1, pp. 37-48, 2005.

[46] A. R. G. Carbonari y P. Romagnoni, «Optimal orientation and automatic control of Optimal orientation and automatic control of,» de *PLEA 2001- The 18th Conference on Passive and Low: Energy Architecture*, Florianópolis, BRAZIL, 2001.

[47] «Cost effectiveness in hvac by building envelope optimization.,» 2012. [En línea]. Available: http://www.redalyc.org/articulo.oa?id=281724144004. [Último acceso: 3 Agosto 2020].

[48] F. Goia, M. Haase y M. Perino, «Optimizing the configuration of a façade module for office buildings by means of integrated thermal and lighting simulations in a total energy perspective,» *Applied Energy,* vol. 108, pp. 515-527, 2013.

[49] F. Favoino, Q. Jin y M. Overend, «Towards an ideal adaptive glazed façade for office buildings,» de *6th International Conference on Sustainability in Energy and Buildings 2014, SEB-14*, 2014.

[50] B. Romero-Tarazona, C. Rodriguez-Sandoval, J. Villabonai-Ascanio y A. D. Rincón-Quintero, «Development of an artificial vision system that allows non-destructive testing on flat concrete slabs for surface crack detection by processing of digital images in MATLAB,» *IOP Conference Series: Materials Science and Engineering,* vol. 844, 2019.

[51] P. McKeen y A. S. Fung, «The Effect of Building Aspect Ratio on Energy Efficiency: A Case Study for Multi-Unit Residential Buildings in Canada.,» *Buildings,* vol. 4, pp. 336-354, 2014.

[52] A. Perera y M. Sirimanna, «A Novel Simulation Based Evolutionary Algorithm to Optimize Building Envelope for Energy Efficient Building.,» de *7th International Conference*, 2014.

[53] M. Košir, T. Gostiša y Ž. Kristl, «Search For An Optimised building envelope configuration during early design phase with regard to the heating and cooling energy consumption.,» In Presented at the CESB16-Central Europe towards Sustainable Building 2016: : Innovations for Sustainable Future, Prague, Czech Republic., 2016.

[54] S. Kang, S.-g. Yong, J. J. H. Kim, H. Cho y J. & Koo, «Automated processes of estimating the heating and cooling load for building envelope design optimization.,» *Building Simulation,* vol. 11, nº 2, pp. 219-233, 2017.

[55] S. Manu, J. Wong, R. Rawal, P. C. Thomas, S. Kumar y A. Deshmukh, «An initial parametric evaluation of the impact of the energy conservation building code of india on commercial building sector,» International Building Performance Simulation Association, Sydney, 2011.

[56] R. S. Srinivasan, W. W. Braham, D. E. Campbell y D. C. Curcija, «Building envelope optimization using emergy analysis.,» de *Proceedings of Building Simulation 2011: 12th Conference of International Building Performance Simulation Association*, Sydney, 2011.

[57] O. T. Karaguzel y K. P. Lam, «Simulation based parametric analysis Part-I: One Factor at-a-Time (OAT) Evaluation of Enclosure Measures for building 661,» Centre for buiding Performance and Diagonostics, Carnegia Mellon University, Pittsburgh, Pennsylvania, 2012.

[58] S. M. Saridar, «The impact of façade design on daylighting performance in office buildings the case of Beirut, Lebano. the case of Beirut, Lebano.,» University of Newcastle, 2004.

[59] N. Ibrahim y A. Zain-Ahmed, «Energy saving due to daylighting: a simplified prediction tool for wall envelope design of air-conditioned office buildings in Malaysia.,» *Built Environment ournal,* vol. 3, nº 1, pp. 63-75, 2006.

[60] C. Velan, «Evaluation of system design parameters for optimum energy performance of a green IT building. Doctoral thesis,» Dr. M.G.R. educational and research institute, Chennai, 2013.

[61] C.-M. Lai y Y.-H. Wang, «Energy-saving potential of building envelope designs in residential houses in Taiwan,» *Energies,* vol. 4, pp. 2061-2076, 2011.

[62] M. Ferrara, M. S. E. Filippia y V. Cravinoa, «A simulation-based optimization method for the integrative design.,» *Energy Procedia,* vol. 78, p. 2608 – 2613, 2015.

[63] M. Pesenti, G. Masera y F. Fiorito, «Shaping an Origami shading device through visual and thermal simulations,» *Energy Procedia,* vol. 78, pp. 346-351, 2015.

[64] F. Ascione, R. F. De Masi, F. R. S. de Rossi y G. P. Vanoli, «Optimization of building envelope design for nZEBs in Mediterranean climate: Performance analysis of residential case study,» *Applied Energy,* vol. 183, p. 938–957, 2016.

[65] Y. Y. R. Pan y Z. Huang, «Energy modeling of two office buildings with data center for green building design,» *Energy and Buildings,* vol. 40, p. 1145–1152, 2008.

[66] W. Xu, K. P. Lam, A. Chong y O. T. Karaguzel, «Multi-Objective Optimization of Building Envelope, Lighting and HVAC Systems Designs,» ASHRAE and IBPSA-USA., Salt Lake City, UT, 2016.

[67] A. D. Rincón-Quintero, L. A. D. Portillo-Valdés, A. Meneses-Jacomé, C. G. Cárdenas-Arias, C. L. Sandoval-Rodriguez, J. G. Ascanio-Villabona, B. E. Tarazona-Romero, J. A. Sandoval-Baldión1, J. C. Ortiz-Ríos1 y R. A. Mantilla-Olejua, «Manufacture of hybrid pieces using recycled R-PET, polypropylene PP and cocoa pod husks ash CPHA, by pneumatic injection controlled with LabVIEW Software and Arduino Hardware,» *IOP Conference Series: Materials Science and Engineering,* vol. 844, 2019.

[68] CIBSE, «Environmental Design: CIBSE Guide A,» seventh ed, Bros, Great Britain,, 2006.

[69] C. L. Sandoval-Rodriguez, J. G. A. Villabona, C. G. Cárdenas-Arias, A. D. Rincon-Quintero y B. E. Tarazona-Romero, «Characterization of the mechanical vibration signals associated with unbalance and misalignment in rotating machines, using the cepstrum transformation and the principal component analysis,» *IOP Conference Series: Materials Science and Engineering,* vol. 844, 2019.

[70] A. A. Chernousov y B. Y. Chan, «Numerical simulation of thermal mass enhanced envelopes for office buildings in subtropical climate zones.,» *Energy and Buildings,* vol. 118, p. 214–225, 2016.

[71] C. L. Sandoval-Rodriguez, J. G. A. Villabona, C. G. Cárdenas-Arias, A. D. Rincon-Quintero y B. E. Tarazona-Romero, «Characterization of the mechanical vibration signals associated with unbalance and misalignment in rotating machines, using the cepstrum transformation and the principal component analysis,» *IOP Conference Series: Materials Science and Engineering,* vol. 844, 2019.

[72] M. Mayhoub y R. Labib, «Towards a solution for the inevitable use of glazed facades in the arid regions via a parametric design approach,» The Proceedings of CIE, Manchester, UK.

[73] A. Kirimtat, B. K. Koyunbaba, I. Chatzikonstanti y S. Sariyildiz, «Review of simulation modeling for shading devices in buildings,» *Renewableand Sustainable Energy Reviews,* vol. 53, p. 23–49, 2016.

[74] C. G. Cárdenas-Arias, A. D. Rincón-Quintero, A. Santos-Jaimes, C. L. Sandoval-Rodriguez, D. F. Rojas-Gomez y S. J. Ardila-Galvis, «Elasticity modulus variation of the AISI SAE 1045 steel subjected to corrosion process by chloride using tension test destructive,» *IOP Conference Series: Materials Science and Engineering,* vol. 844, 2019.

[75] I. Da-Silva y E. B. Ssekulima, «In International conference, towards sustainable energy solutions for the developing world,» de *Domestic use of energy conference. Cape Peninsula University of Technology*, Cape Town, South Africa, 2011.

[76] T. M. Echenagucia, A. Capozzoli, Y. Cascone y M. Sassone, «The early design stage of a building envelope: Multi-objective search through heating, cooling and lighting energy performance analysis,» *Applied Energy,* vol. 154, pp. 577-591, 2015.

[77] F. D. Luca, H. Voll y M. Thalfeldt, «Horizontal or Vertical? Windows' layout selection for shading devices optimization,» *Management of Environmental Quality: An International Journal,* vol. 27, nº 6, pp. 623-63, 2016.

[78] M. d. l. P. Diulio, G. R. Netto y R. a. C. J. D. Berardi, «Impacto de la envolvente en la demanda de energía en calefacción residencial de la región metropolitana de La Plata, tomando como

caso testigo el reciclado energético de una vivienda,» *Ambiente Construído,* vol. 16, nº 1, pp. 55-70, 2016.

[79] O. ESCORCIA OYOLA, «Validación del reacondicionamiento térmico de viviendas para la reconstrucción pos-terremoto 2010: Dichato, Chile.,» *Revista de la construcción,* vol. 12, nº 2, pp. 54-71, 2013.

[80] D. PAVEZ-GALLEGOS y L. MEZA-MARIN, «Aplicación de un subsidio de reacondicionamiento térmico de viviendas en la Región Metropolitana de Chile.,» *Revista de la Construcción,* vol. 17, nº 3, pp. 401-411, 2018.

[81] M.-P. N y E.-S. J, «Envolventes eficientes : relación entre condiciones ambientales espacios confortables y simulaciones digitales,» *Revista de Arquitectura,* vol. 21, nº 1, pp. 90-109, 2019.

[82] F. Kurtz, M. Monzón y B. López-Mesa, «Obsolescencia de la envolvente térmica y acústica de la vivienda social de la postguerra española en áreas urbanas vulnerables. El caso de Zaragoza,» *Informes De La Construcción,* vol. 67, nº 1, 2015.

[83] J. L. P. Galaso, «Simbiosis entre clima, lugar y arquitectura. Desarrollo de estrategías bioclimáticas aplicadas en la Costa del Sol Occidental. Tesis doctoral,» Universidad de Málaga, Málaga. España, 2015.

IOP Conf. Series: Materials Science and Engineering 1154 (2021) 012039 doi:10.1088/1757-899X/1154/1/012039

Matrix design for guided bend tests under the codes AWS D1.1, AWS D1.5, ASME Section IX and standard NTC 523 for specimens with thicknesses equal to or less than 3/8 inches

Manuel Pinzón Candelario[1,2], Germán Torres Bermúdez[2], Yhon F Soto Serna[2] and J N Zuleta[1]

[1]Servicio Nacional de Aprendizaje SENA , Regional Risaralda, Pereira, Colombia
[2]Universidad Tecnológica de Pereira, Facultad de Ingeniería, Pereira, Colombia

Abstract. This article presents the design of a guided bending system, which allows integrating the AWS D1.1, AWS D1.5, ASME Section IX and NTC 523 codes. The study meets the requirements of these standards and reduces use of different matrices for testing specimens with thicknesses equal to or less than 3/8 of an inch.

1. Introduction

The industry needs to obtain certifications for its processes and products that guarantee high competitiveness in current markets, it implies complying with international standards related to production, administration, quality policies, the environment, among others. In this context, the metalworking, hydrocarbon and construction industries require quality evaluation of welded joints for their different processes. The guided bend test is a method that is required to qualify the suitability of personnel and welding procedures before carrying out the work [1].

The codes AWS D1.1, AWS D1.5, ASME Section IX and the NTC 523 standard, establish requirements in the shape and dimensions of the matrix to perform this test [2-5]; Each type of specimen requires a different matrix to evaluate the same property, increasing the costs of acquisition, storage and handling of these tools.

It is proposed to design a system that meets the mechanical conditions that the guided bending test requires. The components must be manufactured with the necessary dimensions to comply with the regulations that govern it and be adaptable to any destructive testing machine.

1.1. Guided bend. The test consists of applying a centered load by means of a plunger that acts on the specimen (where the weld is located), supported by two shoulders of a matrix (Figure 1) until it takes the shape of a "U". The result allows to examine cracks or gaps in the welded joint [6].

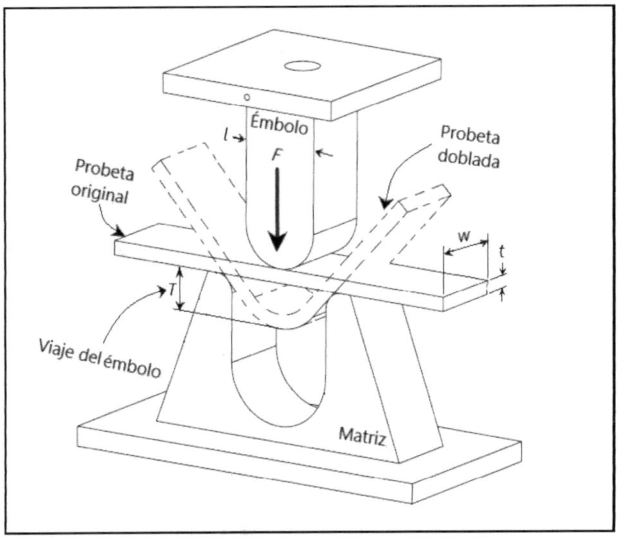

Figure 1. Guided bend test.

Where *l* is the diameter of the plunger, *T* is the displacement, *w* and *t* are the width and thickness of the specimen respectively, and *F* is the force applied on the specimen.

2. Matrix design

The design of the system that integrates the mentioned codes is shown in figure 2. The description of each component is presented in Table 1.

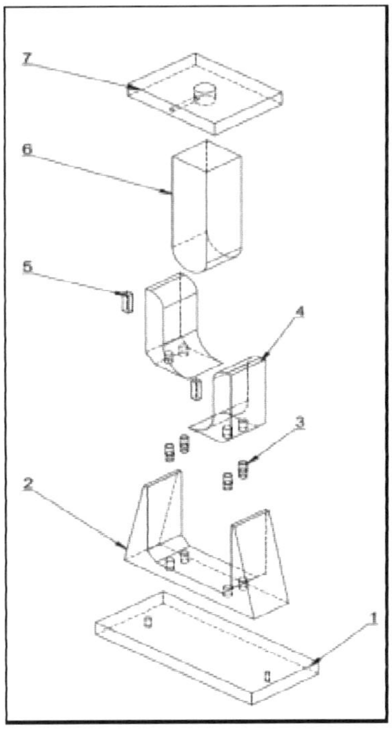

Figure 2. Device components.

Table 1. Device components for guided bend tests.

Item	Components	Description
1	Matrix base	Rectangular section plate as matrix support.
2	Frame	Structure that serves as shoulder support
3	Pins	Fixing elements for the shoulders that exchanged according to the code.
4	Shoulders	Dispositivos que guían el doblez de la probeta. Varían según la norma.
5	shoulder stop	Devices that guide the bending of the specimen. It varies by code.
6	Plunger	System component that applies force to the specimen. It varies by code.
7	Plunger base	Rectangular section plate that is assembled to the plunger.

In the proposed design it can be seen that the only components that must be changed, depending on the code to be applied, are the shoulders and the plunger; the other components remain constant. Figure 3 shows the dimensions of the matrix frame.

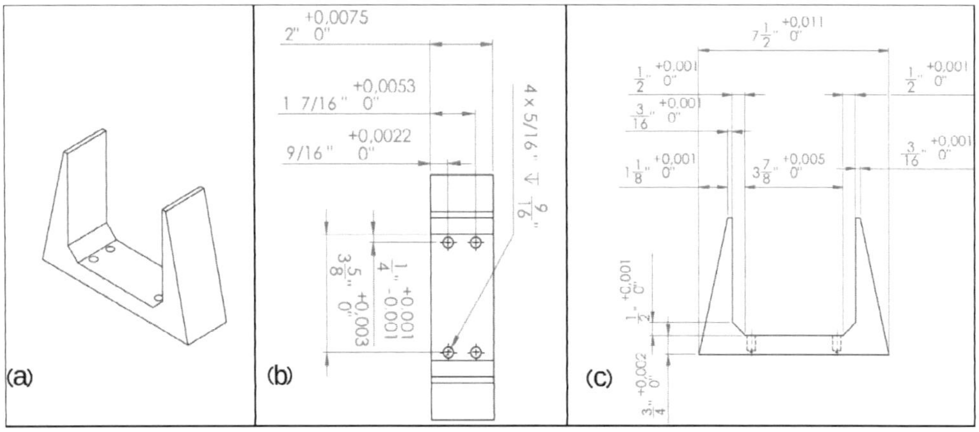

Figure 3. Matrix frame. (a) General view. (b) Top view. (c) Side view.

The AWS code establishes requirements according to three yield strength (*YS*) ranges and they are the same for D1.1 and D1.5. Figure 4 shows the dimensions of the shoulders according to the AWS code for the intervals: (a, b, c) *YS* less than 50ksi, (d, e, f) *YS* between 50ksi and 90ksi, (g, h, i) *YS* greater than 90ksi, respectively. Shoulder dimensions according to NTC 523 and ASME Code Section IX are shown in figures 5 and 6.

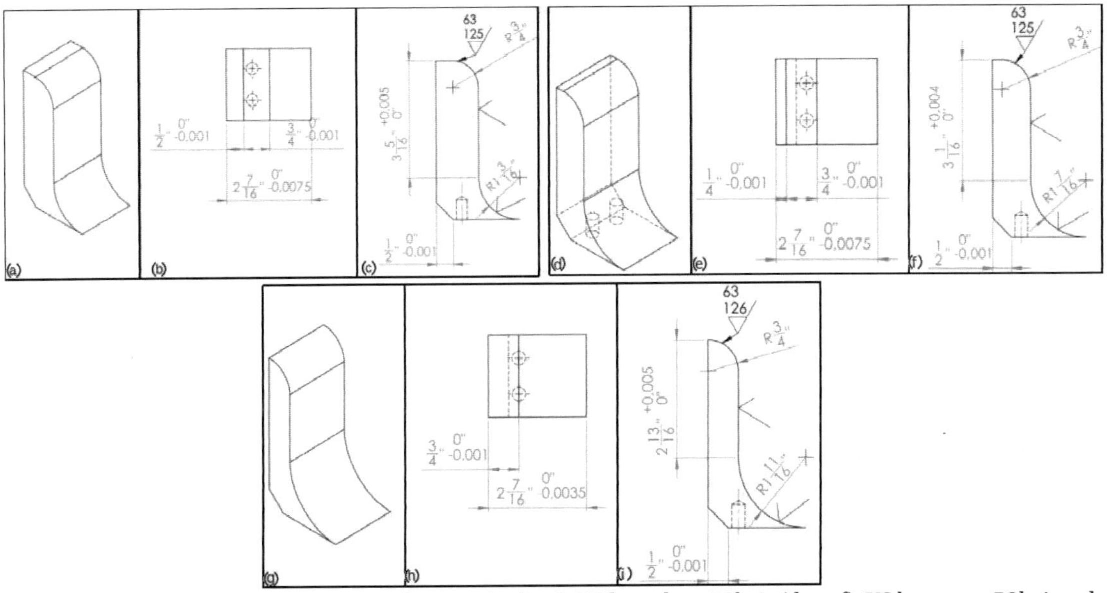

Figure 4. Shoulders. AWS D1.1 and D1.5. (a, b, c) *YS* less than 50ksi, (d, e, f) *YS* between 50ksi and 90ksi, (g, h, i) *YS* greater than 90ksi. (a, d, g) General view. (b, e, h) Top view. (c, f, g) Side view.

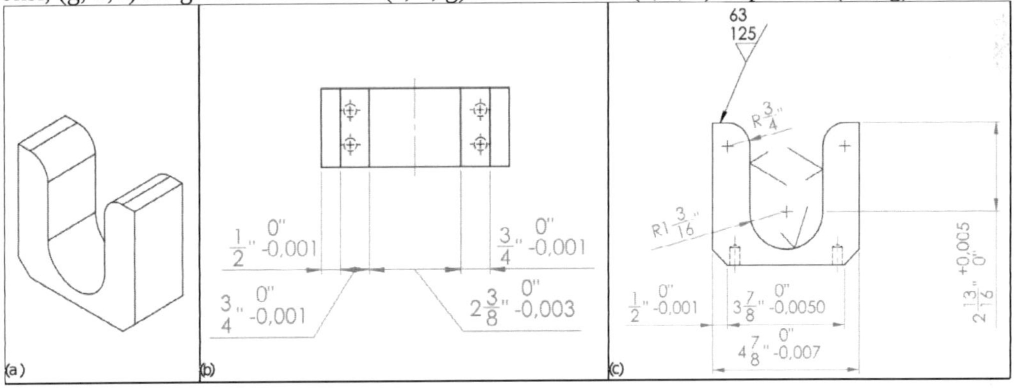

Figure 5. Shoulders NTC 523 standard (a) General view. (b) Top view. (c) Side view.

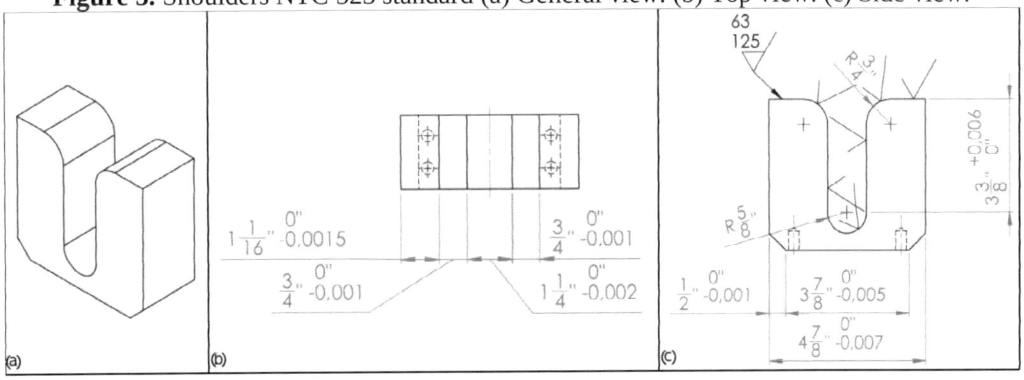

Figure 6. Shoulders ASME SECTION IX code (a) General view. (b) Top view. (c) Side view.

Figure 7 shows plunger designs and dimensions according to requirements. The NTC 523 standard plunger maintains the same dimensions required by the AWS code for *YS* less than 50 ksi [2].

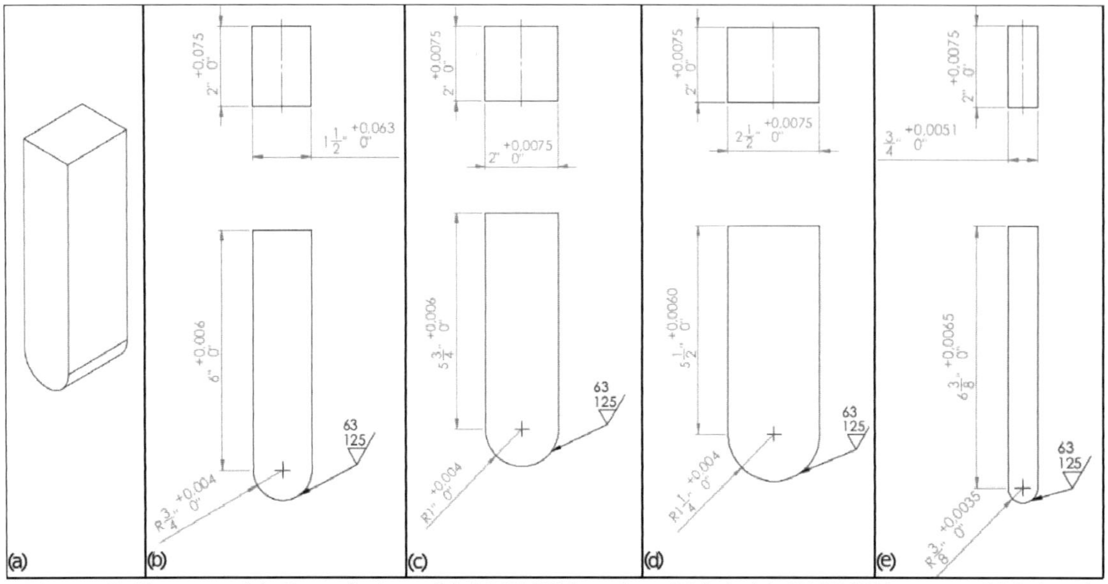

Figure 7. Plungers. (a) General view. (b) AWS *YS* less than 50 ksi. (c) AWS *YS* between 50 ksi and 90 ksi. (d) AWS *YS* greater than 90 ksi and NTC 523. (e) ASME Section IX.

3. Simulation results

3.1. Tensile strength. The material selected for the design calculations was quenched and tempered AISI SAE 4140 steel, from the family of tool steels, whose combined performance of mechanical strength, resistance to fatigue, toughness and limited corrosion, make it a suitable material for the manufacture of the tooling, for being easy to acquire and low cost [7].

The tensile strength (*S*) in external fibers of a beam of thickness 2*c* and moment of inertia *I*, subject to a moment *M* is given by [8],

$$S=\frac{M\cdot c}{I} \tag{1}$$

For the case of a beam with a rectangular cross section (figure 1) it follows that,

$$M=\frac{F\cdot l}{4};\quad c=\frac{t}{2};\quad I=\frac{w\cdot t^3}{12}$$

Therefore, the bending force F necessary to bring the specimen to the *YS* can be estimated by means of Eq. (2) [9],

$$F=\frac{0{,}67\cdot S\cdot w\cdot t^2}{l} \tag{2}$$

For the specimen to exceed the *YS*, Eq. (2) is multiplied by two, obtaining the following,

$$F=\frac{1{,}33\cdot S\cdot w\cdot t^2}{l} \tag{3}$$

The values of the forces according to the parameters of each code are shown in Table 2.

Table 2. *YS* Forces.

Code or standard	Specimen width *w* (mm)	Specimen thickness *t* (mm)	Plunger diameter *l* (mm)	Tensile strength *S* (MPa)	Force *F* (N)
AWS (<50ksi)	38,1	9,52	98,43	420	19596,2
AWS (50-90ksi)	38,1	9,52	111,13	723,8	29912,7
AWS (>90ksi)	38,1	9,52	123,83	1100	40795,9
NTC 523	38,1	9,52	98,43	1100	51323,4
ASME Section IX	38,1	4,76	69,85	1100	18245,1

3.2. Simulation results. The "Simulation" tool from SolidWorks design CAD was used. The Von Mises maximum stress criterion recommended for ductile materials was applied, and the safety factor *FDS* was calculated for each force value (Table 2). The FDS less than 1 (one) indicates that the material exceeded the *YS* and it was brought to plastic zone. *FDS* greater than 1 indicates that the material remained in the elastic zone [8, 10]. Figures 8 and 9 show the *FDS* of the specimen according to the parameters of Table 2.

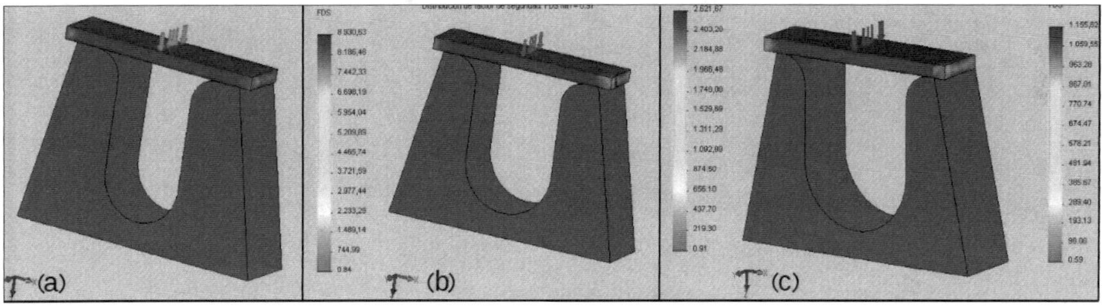

Figure 8. *FDS* Specimen under AWS code. (a) *YS* less than 50 ksi, (b) *YS* between 50ksi - 90ksi and (c) *YS* greater than 90ksi.

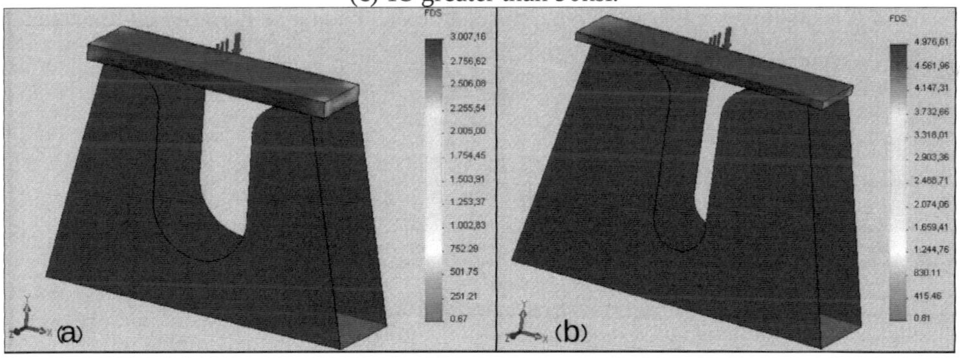

Figure 9. *FDS* Specimen under (a) NTC 523 standard. (b) ASME Section IX code

For the *FDS* simulation of the matrix frame and shoulders, the AWS code (*YS* greater than 90ksi) and the NTC 523 standard were applied, which require a greater bending force compared to the other codes. The force used in AWS (*YS* greater than 90ksi) is 40,759.9 N and in NTC 523 it is 51,323.4 N (Table 2). Figure 10 shows the simulation of the matrix frame and figure 11 of the shoulders, respectively.

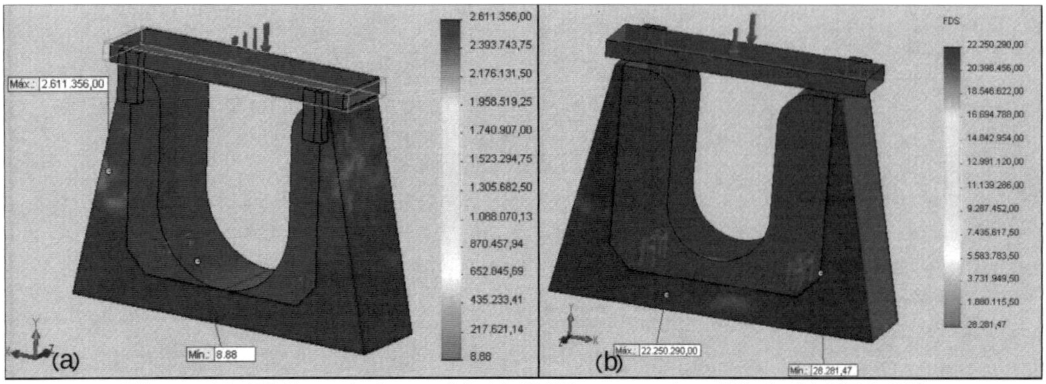

Figure 10. *FDS* matrix frame (a) AWS *YS* greater than 90ksi. (b) NTC 523

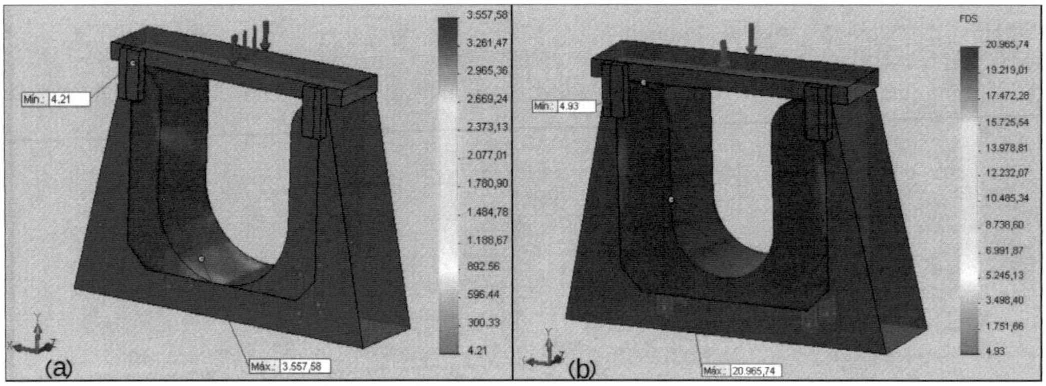

Figure 11. *FDS* shoulders (a) AWS *YS* greater than 90ksi. (b) NTC 523.

The lowest *FDS* values of all the simulations are presented in Table 3. In it it is observed that the *FDS* values for the matrix frame and shoulders components are above 1, this indicates that the material remains in its elastic zone and does not reach the *YS*. On the other hand, in the case of the specimens, it is confirmed that the forces applied to them exceed their *YS*, taking them to the plastic zone.

Table 3. Minimum *FDS* safety factor.

Component	Code or standard	*FDS* (min)
Specimens	AWS (*YS* less than 50ksi)	0,84
	AWS (*YS* between 50ksi and 90ksi)	0,91
	AWS (*YS* greater than 90ksi)	0,59
	NTC 523	0,67
	ASME Section IX	0,81
Matrix frame	AWS (*YS* greater than 90ksi)	8,88
	NTC 523	28,47
Shoulders	AWS (*YS* greater than 90ksi)	4,21
	NTC 523	4,93

4. Conclusions
It was possible to design a set of interchangeable elements that constitute a matrix to carry out guided bending tests of welded specimens under the AWS D1.1, AWS D1.5, ASME Section IX codes and NTC 523 standard. A design model was established of matrices for bending tests, in which the use of materials and manufacturing processes can be reduced, by not requiring the construction of individual matrices for each one.

The FDS values for the matrix frame and shoulders components do not reach the yield strength and the material remains in the elastic zone. In addition, the specimens exceed the yield strength, and are taken to the plastic zone.

5. References
[1] Shinn, G. C. 1973. Lets Quantify Our Welding Instruction. Agricultural Education Magazine. **46** 6125-6.
[2] Instituto Colombiano de Normas Técnicas y Certificación. 1997. Ensayo guiado de doblamiento para determinar la ductilidad de las soldaduras. NTC 523. *Primera actualización. Bogotá D.C. El instituto*. 1-6 p.
[3] American Welding Society. 2006. Structural Welding Code-Steel.AWS D1.1/D1.1M:2006. 20 ed.Miami,USA:AWS. 1-2-3-96-120-122 p.
[4] American Welding Society. Bridge Welding Code. 2002. AASHTO/AWS D1.5 M/D1.5:2002. Miami, USA:AWS. 1-2-3-133-174-175-177 p.
[5] The American Society of Mechanical Engineers. 2010. ASME Boiler and Pressure Vessel Code. IX Qualification Standard for Welding and Brazing Procedures. Welders Brazers, and Welding and Brazing Operators. New York,USA. 1-2 p.
[6] García, A., Salas, R., Centeno, L., & Velázquez del Rosario, A. 2011. Metalurgia de uniones soldadas de aceros disímiles (ASTM A240–A537) y comportamiento mecánico ante cargas monotónica y cíclica. *Revista latinoamericana de metalurgia y materiales.* **32** 1136-48.
[7] Moreno, F. R., Salazar, W. J., & Sáenz, L. A. 2005. Evaluación del comportamiento mecánico a la fatiga en aceros AISI 4340 y AISI 4140 tratados térmicamente con recocidos y normalizados. *Revista ingeniería UC,* **12**(3), 40-45.
[8] Rilley, W., Sturges, L., Morris, D. 2001. Mechanics of Materials. *Ed. Limusa-Wiley*. 350-352
[9] Kalpakjian, S., Schmid, S. 2002. Manufactura, Ingeniería y Tecnología Vol. I. 5Ed. *Pearson*. 1294.
[10] Ríos, L., Roncancio, E. 2007 "Análisis y desarrollo de un programa de selección rápida de factores de seguridad, para diseño de elementos mecánicos". *Scientia et Technica Año XIII.* **35** 254.

Sizing of a sustainable condensation system for the production of sweet water through the use of photovoltaic panels in La Mesa de los Santos, Santander, Colombia.

B E Tarazona-Romero[1 2], A-J Rodriguez Nieves [1] C L Sandoval-Rodriguez[1 3], J G Ascanio-Villabona[1 2], M-A Duran-Sarmiento[1 2] and A-D Rincon-Quintero[1 2]

[1] Automation and Control Energy Systems Research Group (GISEAC), Faculty of Natural Sciences and Engineering, Electromechanical engineering, Unidades Tecnologicas de Santander (UTS), Student Street No 9-82, 680005, Bucaramanga, Colombia.
[2] Energy in Building research group (ENEDI), Doctoral Program in Energy Efficiency and Sustainability in Engineering and Architecture, Department of Machines and Thermal Engines, University of the Basque Country (UPV / EHU), Engineer Torres Quevedo Square, 1, 48013, Bilbao, BI, Spain.
[3] Doctoral Program in Electronic and Communications, Department of Communications Engineering, University of the Basque Country, UPV/EHU, 48013 Bilbao, Spain.

Abstract. The objective of the project is to carry out the dimensioning of a photovoltaic system that allows its integration with a condensation system for the production of fresh water in isolated areas (Rural), in such a way that a sustainable condensation application capable of producing water is generated. sweet to supply small groups of inhabitants in areas where the basic supply of liquid is deficient. The study area is the Department of Santander, Colombia, specifically in the Municipality of los Santos and the area known as Mesa de los Santos, in which difficult geographical conditions are presented such as: altitude of 1650 meters above sea level and deficiency in aquifers. The methodological development proposes the analysis of the meteorological conditions of the area, the identification, selection and dimensioning of the condensing system and the photovoltaic system through the amp-hour method and the software system advisor model (SAM). Additionally, a CAD model is developed to obtain the complete schematization of the model and carry out a simulation process to analyze the behavior of the integration of the aforementioned systems. Finally, a techno-economic analysis of the proposed system is carried out.

Keywords: Electrolysis of water, prototype, Hydrogen production, photovoltaic system and CAD modeling.

1 Introduction

Water is the axis of the sustainable development of the regions and a key element in the socioeconomic development, the production of food, energy, ecosystems and for the survival of the communities [1] [2]. In addition to this, water plays a key role in adapting to climate change, generating a direct connection between the environment and society. [3]. Additionally, water is a right [4], which has been affected for various reasons such as: global warming [5], population growth [6], exploitation of fossil fuels, deforestation, natural phenomena, among others [7].

The scarcity of water in the world, currently generates 10% of the deaths of children in the world, it is estimated, on the other hand, that 8 out of 10 people do not have access to drinking water and live in rural or isolated areas, around 842,000 people in the world die from stomach affections as a result of the consumption of undrinkable water and that at least 1.8 billion people worldwide have access only to contaminated water sources [8]. In the 2019 World Water Development Report, it was noted that improving water management in the world is essential to eradicate poverty [9]. According to the World Resources Institute (WRI), the Middle East and North Africa are the most affected areas in the world by lack of water [10] [11]. In Latin America and the Caribbean, the growing population in the suburbs around large cities has caused a supply impact, with rural areas being the most affected for the supply of drinking water. [12].

In Colombia, access to water is considered a fundamental right [4]. For this reason, the use of all sectors of the country must be guaranteed. The National Water Study (ENA) in 2018 presented by the Institute of Hydrology, Meteorology and Environmental Studies (IDEAM) reveals that in capital cities such as Bogotá, Cali, Cartagena, Cúcuta, Bucaramanga, Montería and Tunja present a moderate water deficit While the Orinoquia, Pacific and Amazonia regions present surpluses, I feel the rural area in all cases the most affected [13] [14].

In Santander, Colombia, specifically in La mesa de los santos, there is currently a water shortage problem. Los santos is a municipality located 1,650 meters above sea level. This municipality is made up of several rural areas from which primary economic activities such as agriculture, poultry, mining and secondary economic activities such as tourism are derived, economic activities that are affected by the little supply of fresh water since it does not have many water sources, the closest river is the Chicamocha river and it is located in the lower part of the plateau at a height difference of 1200 meters and the low rainfall currently makes it even more difficult for communities to develop in rural areas. Eventual water pumping solutions are limited by electrical energy, either by high consumption required to drive the water from the Chicamocha river or by the absence of electrical power networks that allow energy to be transported to the required points. [15] [16].

Currently, renewable energies are widely applied in different world sectors, somewhat remedying the problem of water shortages [17] [18]. Desalination systems are a viable alternative for the drinking water production process, but their high operating cost makes it difficult to implement them in isolated areas with low economic power [19] [20]. This is why solar photovoltaic [21] and thermo solar [22] [23], have been integrated into these processes but continue to be costly. Thus, various small-scale solutions have been developed such as: desalination by conventional and solar distillation [24] [25], pumped supply by photovoltaic panels [26] [27], condensation systems assisted by Solar electric energy[28], among others, the latter, focuses on the use of a conventional cooling system under modifications to produce fresh water[29] [30].

The main objective of this paper is to present the dimensioning of a condensation system integrated into a photovoltaic electric power generation system, to produce fresh water, taking as an application the municipality of Mesa de los Santos, Santander, Colombia, generating the calculation of viable investment costs. The methodological development of the paper is presented in section 2 as well as the materials selected for the simulation process. Later in section 3 the sizing of the condensation system is presented, as well as the sizing of the solar system through the comparison between a static or direct method called ampere-hours and a dynamic methodology through simulation in the System Advisor software. Model (SAM) of free access, allowing to determine an autonomous system, capable of feeding the condensing system load. In the same section, the modeling in SolidWorks Software and its simulation are described to determine its behavior, as well as the techno-economic study of the proposed final prototype or model. Finally, section 4 describes the most relevant conclusions of the project. It is important to highlight that the project system is a sustainable system with easily accessible elements in the region or metropolitan area of implementation, positively impacting communities with basic needs in rural areas.

2 Methods and Materials

2.1 Methodology

For the development of the sustainable condensation system, Figure 1 presents the methodological work flow diagram, starting with the basic analysis of the meteorological conditions of the area through open access databases, then the module sizing is carried out, which consists of two subsystems: the condensation subsystem and the photovoltaic electrical energy production subsystem. The first system must satisfy the needs of the project and produce water, from the condensation process through the use of a refrigeration equipment. The second system must have the capacity to supply energy to the consumption of the condensing subsystem, applying the amp-hour calculation methodology contrasted with the Software system advisor model (SAM), finally determining the size of the photovoltaic subsystem required.

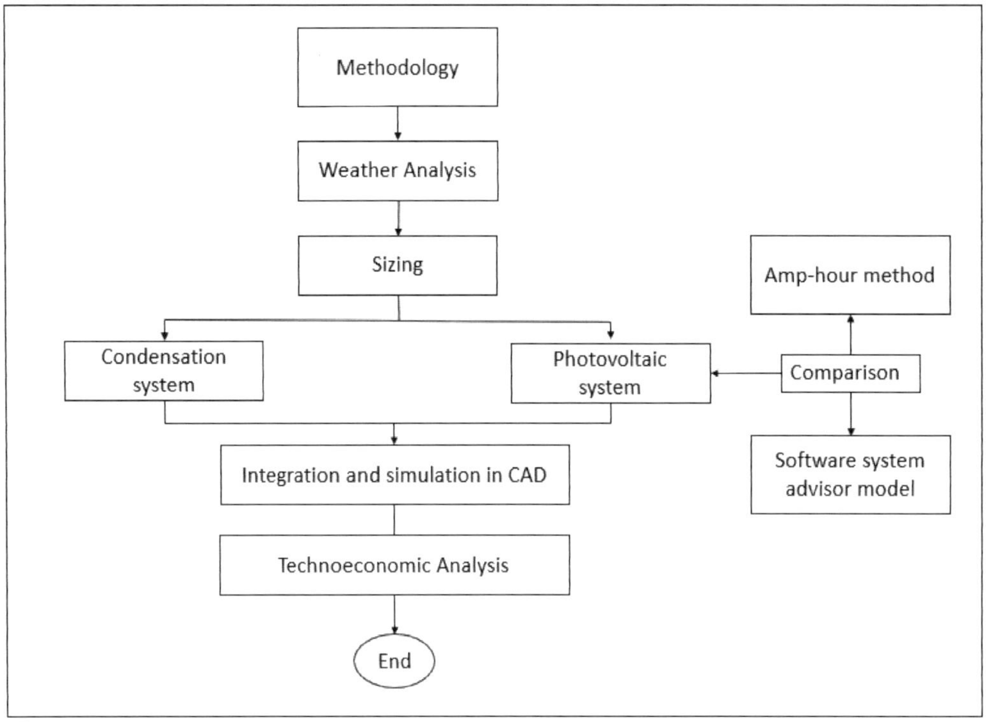

Figure 1. Working diagram

Subsequently, the condensation system is modeled by using Solid Work, in order to have the equipment in a schematic way and perform a simulation of the behavior of the final sustainable condensation system. Finally, a techno-economic study is developed in order to know the future implementation costs of the project.

2.2 *Software's.*

The programs or software's used in the development of the project are mentioned below.

2.2.1 SolidWorks. SolidWorks *[31]* is a CAD software that allows mechanical modeling in 2D and 3D, which allows real-scale designs of systems and their simulation, based on established parameters. Which will be used to model the proposed condensation system model and its simulation, to analyse its behaviour.

2.2.2 Genetron Properties V.1.4. Genetron proprieties v1.4 *[32]* is an open access software that allows simulating the refrigeration cycle with different types of refrigerants and exports the results in exportable tables to Excel, as well as Pressure-Enthalpy and Temperature-Enthalpy diagrams.

2.2.3 Systems Advisor Model (SAM). System Advisor Model (SAM) *[33]* is a free techno-economic software for modeling various types of renewable energy systems where the possibility of developing photovoltaic systems from small to large scale is highlighted.

2.3 Condensation System. The equipment necessary for the development of the condensation system for the simulation is described below:

2.3.1 Evaporator. The evaporator is the element within the refrigeration equipment which the refrigerant reaches it at a low temperature, it runs through the framework of pipes that allows it to absorb heat from the surrounding environment through the aluminum fins, in a few words its function will be to reduce the temperature of the control volume until the air reaches condensation temperature. The evaporator to be used for the simulation is the one produced by the company Thermocoil de Colombia, with reference E 2x2x1000 4T.

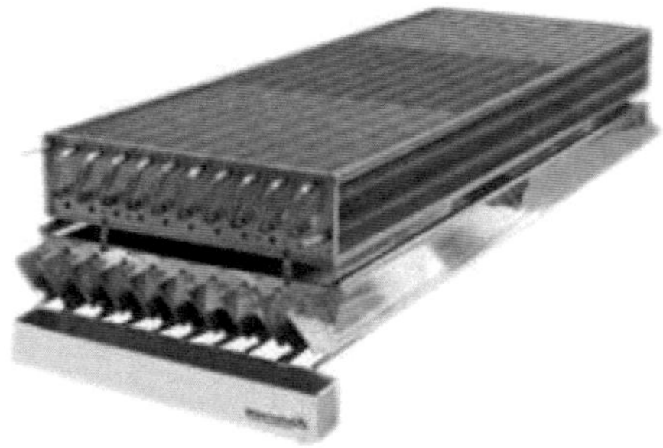

Figure 2. Evaporator *[34]*

2.3.2 Condenser. The condenser function will be to release hot air to the environment as a result of the understanding of the refrigerant, its manufacture is very similar to the evaporator since its very task as a heat exchanger, the difference is that the condenser yields heat to the environment and the evaporator receives heat from the environment. The condenser to be used has a performance of 544 Watt and is from the thermocoil company of Colombia.

Figure 3. Condenser. *[35]*

2.3.3 Compressor. The compressor generates a force compressing the gas that comes from the evaporator in a gaseous state. This pressure increases the temperature of the gas that returns to its liquid state and heats up, the power required to carry out this work is 145 W, Therefore, the commercial

Danfoss BD250GH compressor would be used, with an operating voltage of 48V, maximum power of 250 W.

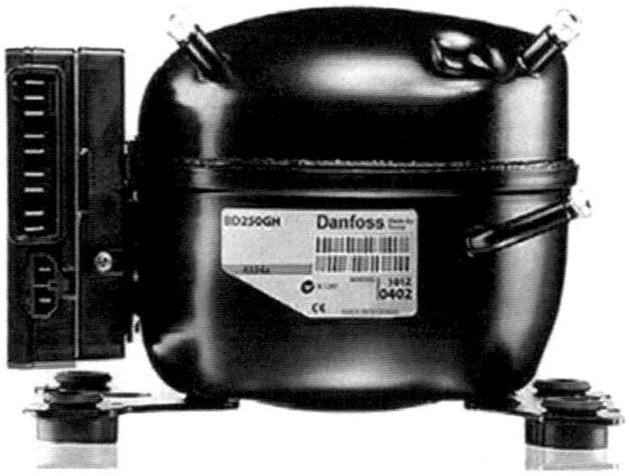

Figure 4. Compressor. *[36]*

2.3.4 *Expansion Valve.* Element that throttles the flow of the refrigerant liquid to produce a sudden pressure drop forcing the liquid to evaporate, as a second option it is also possible to place a capillary tube but it could not be electronically controlled to open or close according to the pressure need, this in order to increase or decrease the temperature as the humidity and temperature of the environment are measured.

2.3.5 *Fan.* The LWFD200-EO model direct current fan of the Longwell brand will be used as a design parameter due to its favorable technical characteristics, commercial value and air flow. Commercially, direct current fans do not have a variety of manufacturers, however, fans used as heat extractors for cooling processors and / or computers can be presented as another possible solution, in which their capacity would be compensated with a quantity of them.

2.3.6 *Control.* To turn it on and off, the variable to be censored will be the temperature and humidity variable, whose data will be obtained by the DTH22 censor and managed by a development board of the Arduino platform that in turn will give the order according to the need, to be used a relay board.

2.4 Photovoltaic panel

For the selection of the panel, the respective comparison of different brands recognized by the Organization of consumers and users OCU in terms of power, quality and commercial value is made, these characteristics are reflected in **Table *1***in order to select the JKM355M-66H solar panel of the manufacturer Jinko solar, since it appears to comply with established requirements, it has representation and marketing in Colombia.

Table 1. Solar Panel Selection

PANEL BRANS	MODEL	POWER	QUALITY	COST
SunPower	SPR-X21-345	345	97/100	441,65 €
REC	REC 280TP	280	94/100	379,00 €
Panasonic	VBHN3252J47	320	94/100	275,88 €
JINKO	JKM355M-66H	355	92/100	125,72 €
AXITEC	AXI265P	265	90/100	210,00 €
TRINA	TSM-270PD05	270	87/100	284,35 €
ALEXO	X59	320	86/100	350,90 €
LG	LG295S1C-A5	295	85/100	490,00 €
SOLARWORLD	SW280	280	84/100	266,20 €
ATERSA A265-P	A265-P	265	84/100	192,39 €

3. Results and analysis

3.1 Site

The municipality of Los Santos is located in the central region of the department of Santander, in the northeast of the country, on the western slope of the Eastern Cordillera, with an area of 284.74 km2. It limits to the north with the municipalities of Girón and Piedecuesta, to the south with Jordán and Villanueva, to the east with the municipality of Aratoca and to the west with the municipality of Zapatoca., Hydrographically limits to the east and south with the Chicamocha river that separates it from Aratoca, Jordán and Villanueva; to the west with the Sogamoso river that separates it from Zapatoca [15]. The municipal head is located according to its geographical coordinates at latitude 6.756643, and longitude -73.103154 at a height of 1.650 meters above sea level. This location is taken as a geographical reference for the Project Development since it has an average annual radiation based on the IDEAM of 5.13 kWh/m^2

The data of the meteorological conditions of the area is presented in the **Table** *2*, which was taken from the website "Nasa Prediction of World wine Energy Recourses". This location is taken as a geographical reference for the Project Development since it has an average annual radiation based on the IDEAM of 5.13 kWh/m^2.

Table 2. Metereological data, Mesa de los Santos, Santander

Month	Relative Humidity %	g water /kg dry	Dew Point °C	Dry buld Temperature °C	Humidity buld temperature °C	Solar irradiation kWh/m^2
January	73,84	14,02	16,3	21,13	15,94	5,19
February	69,56	14,10	16,25	22,01	15,85	5,08
March	75,53	14,98	17,15	21,63	16,86	4,83
April	82,1	15.47	17,62	20,78	17,41	5,15
May	83,08	15.51	17,64	20,63	17,42	4,98
Jun	81,22	15.16	17,29	20,6	17	5,03
July	75,79	14.09	16,23	20,6	15,82	5,52
August	74,54	14.28	16,42	21,08	16,04	5,66
September	74,84	14.42	16,5	21,17	16,13	5,28
October	83,74	14.97	17,11	19,94	16,84	5,1
November	86,65	15.56	17,68	19,99	17,47	5,01
December	82	15.31	17,47	20,64	17,18	4,74
Anual Average	78,57	14.858	16,97	20,85	16,66	5,13

3.2 Operating parameters of the condensing system

The water condensation system uses a conventional refrigeration cycle and the elements that make up its system, where the air is cooled and, in turn, is sent through aluminium channels within an estimated control volume and / or study equivalent to $6m^3$. This volume is taken into consideration according to the refrigeration capacity for the temperatures to work. The basic operating parameters of the refrigeration system are based on the data taking into account the temperatures of the dry bulb, wet bulb and dew point set out in **Table** *2*, the resulting sizing provides information on the power for the compressor or the unit of refrigeration, this element for its electrical consumption is an important part of the total load to be considered in the photovoltaic system

After the identification of the meteorological parameters, equation 1 is applied, which allows the calculation of the cooling capacity of the condensation system[37]:

$$C = 230\, x\, V + (\#\, PyE\, x\, 476) \tag{1}$$

Where:
C = cooling capacity in BTU
230 = Factor calculated for Latin America "Maximum temperature of 40 ° C" (given in BTU / hm³)
V = Volume of the area where the equipment will be installed, Length x Height x Width in m³
# PyE = # of people + Appliances installed in the area
476 = Profit and loss factors contributed by each person and / or appliance
Being, # PyE = 1, so:

$$C = 230\, x\, 6 + (1x\, 476)$$

$$C = 230\, x\, V + (\#\, PyE\, x\, 476)$$

$$C = 1856\, Btu$$

$$C = 1856\, BTU * \frac{0.29307107\ \mathrm{W}}{1\, BTU}$$

$$C = 543.93W$$

In addition to the value of cooling capacity, the evaporator temperature must be taken into consideration since it is an element of the cooling system in charge of reducing the temperature of the surrounding air, therefore, this is estimated at 0 ° C since its temperature differential together with the ambient temperature, they affect the dew temperature for the average condensation for the study place.With these parameters, the data obtained for a basic refrigeration cycle is integrated into the free-use software Genetron properties V. 1.4 from the company Honeywell (See Figure 5).

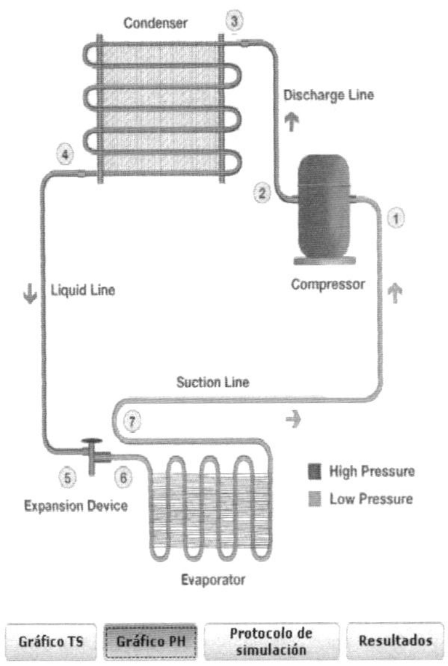

Figure 5. Basic Refrigeration Cycle

3.3 Photovoltaic System

For the sizing of the photovoltaic system, it is essential to use the electrical energy consumption of the refrigeration system, which will be in operation for 8 hours per day, the table presents the estimate of the load that is 1580 W / day.

Table 3. Load estimate

TYPE OF LOAD C.D.	RATED POWER (W)	OPERATING TIME (H)	DAILY POWER (W/DIA)
Compressor BD250GH.2	145	8	1160
Evaporator fan	25	8	200
Condenser fan	25	8	200
Control electronics	2,5	8	20
Total			1580

3.3.1 Amp-Hour Methods. Once the fixed installed power has been defined, the real energy required is determined taking as a reference the standard efficiency of the batteries of 90%, being a factor that directly affects the power consumption reducing it to 1755 W. Subsequently, the coefficient is determined system performance taking into account the data in the Table 4 and applying equation 2.

Table 4. Approximate energy losses per component in a PV installation [38]

CONCEPTO	PÉRDIDAS EN %
Temperature del panel	0,45
Dirt and dust	4
Self-discharge of batteries	0,5
Performance of the loading and unloading process	7
Regulators	7
Wiring from panels to regulator	1,5
Wiring the rest of the installation	4

Taking the above information, equation 2 can be applied and the performance ratio (RT) determined

$$PR = \left(1 - \frac{P1}{100}\right) X \left(1 - \frac{P1}{100}\right) X..X \left(1 - \frac{Pn}{100}\right) \tag{2}$$

Replacing the values of the table in equation 2 we obtain:

$$PR = \left(1 - \frac{0,45}{100}\right) x \left(1 - \frac{4}{100}\right) x \left(1 - \frac{0,45}{100}\right) x \left(1 - \frac{0,5}{100}\right) x \left(1 - \frac{7}{100}\right) x \left(1 - \frac{7}{100}\right) x \left(1 - \frac{1,5}{100}\right) x \left(1 - \frac{4}{100}\right)$$

$$PR = 0.78$$

Then, we proceed to determine the number of panels required by the system to feed the capacitor charge, this calculation is developed with the application of equation 3.

$$\#de\ paneles = \frac{L}{WpXHSPXFsXPR} \tag{3}$$

Where:

L = actual daily energy required.
Wp = peak watts of the photovoltaic module used in the installation.
HSP = peak sun hours incident on the plane of the panels.
FS = shade factor or shade loss coefficient
PR is the global factor of losses or performance ratio.

The selected panel is described in the materials section and with it the number of units required is estimated. The connection of the panels will be in parallel as this will allow to maintain the voltage between the panels and add the maximum current delivered by each one.

$$\#de\ paneles = \frac{1755\ W}{355WX4.75X0.95X0.78}$$

$$\#de\ paneles = 1.4045$$

$$\#de\ paneles \cong 2.$$

Additionally, the auxiliary equipment of the photovoltaic system is determined where a battery with a capacity of 183 Amp-hours, a charge regulator of 100 Amp-hours and an electrical conductor of $1.32mm^2$ is required. Finally, the One-Line Diagram of the dimensioned system is made, which is presented in the Figure 6.

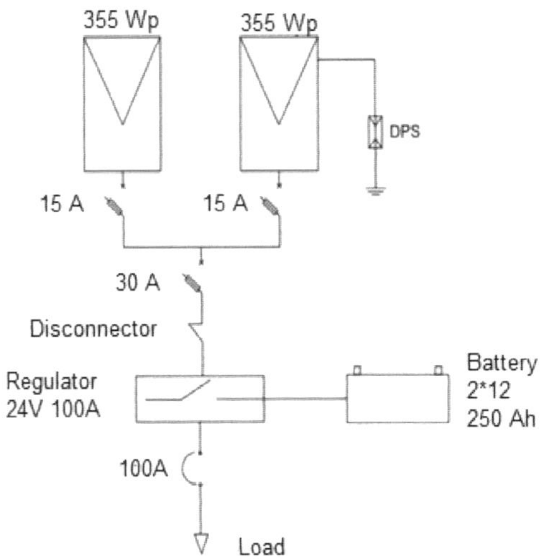

Figure 6. One-line Diagram

3.3.2 *Sizing with System Advisor Model.* The SAM simulation used the same meteorological data as the amp-hour calculation, as well as the same electrical energy consumption and the photovoltaic panel, differs from the previous estimate in the use of the SMA America reference inverter: SB2500V, the size of the system of project to 4 kWdc, being more than double the installed power, in order to cover the losses due to shadows and inefficiency of the solar panels. The simulation determines that 7 photovoltaic panels are required to cover consumption and Figure 7 shows the average annual production of the system where it is guaranteed that the power required by the system for its operation is produced from 10 am to 2 pm. As a plus point, a surplus will be generated that can be used to supply some equipment in the area (For this, an additional study must be carried out)

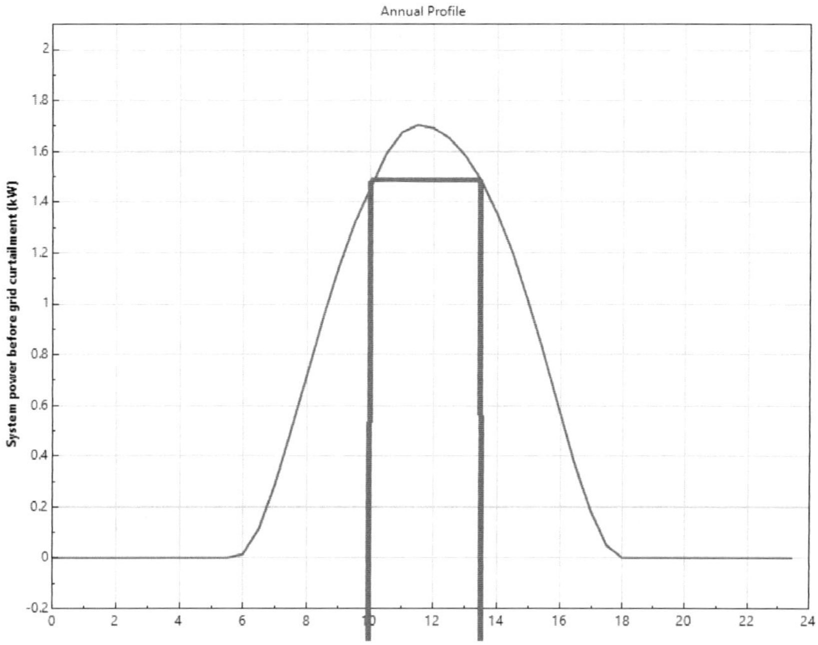

Figure 7. Annual Production -SAM

Finally, the one-line connection diagram of the system simulated by SAM is made (See **Figure *8***).

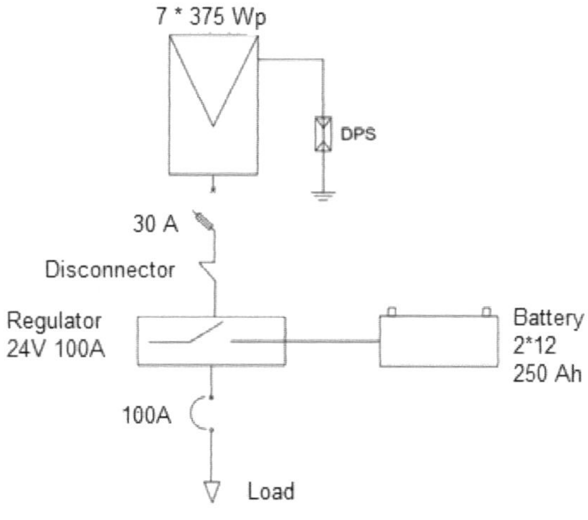

Figure 8. one-line Diagram – SAM

3.4 Comparison of sizing methods

Table 5 presents the data of the Ampere-hour method and the simulation in SAM, where there is evidence of a difference in the results of each developed, this is because SAM in its dynamic simulation model takes into consideration the interaction of the radiation of the place , the photovoltaic generator and the inverter, obtained as a result that the necessary power must be dimensioned from 3 Kw, so that when generating it delivers the power of 1.5kw, therefore, SAM allows the user to select for the expected generation of seven panels of the reference JKM375HC66M, in the same way the result is evidenced by Figure 7, which, as already mentioned, the power required by the system on average annually is obtained from 10 am to 2 pm.

Table 5. Methods comparison

Analysis method	Selected panel	Number of panels	Installed power
Amp-hours	JKM355M-66H	2	710 w
SAM	JKM375HC66M	7	2625 w

For the purposes of the application, the data obtained by SAM regarding the photovoltaic generator and other equipment such as the regulator and batteries will be taken into account and those previously estimated within the calculations of the amp-hour method. These elements are the ones that will make up the team under study and economic evaluation.

3.5 Simulation in CAD – Real's system

The modeling and simulation were carried out in the SolidWorks Software on a real scale, integrating the condensation system (refrigeration cycle) with the photovoltaic power system, the final result is presented in the **Figure 9**.

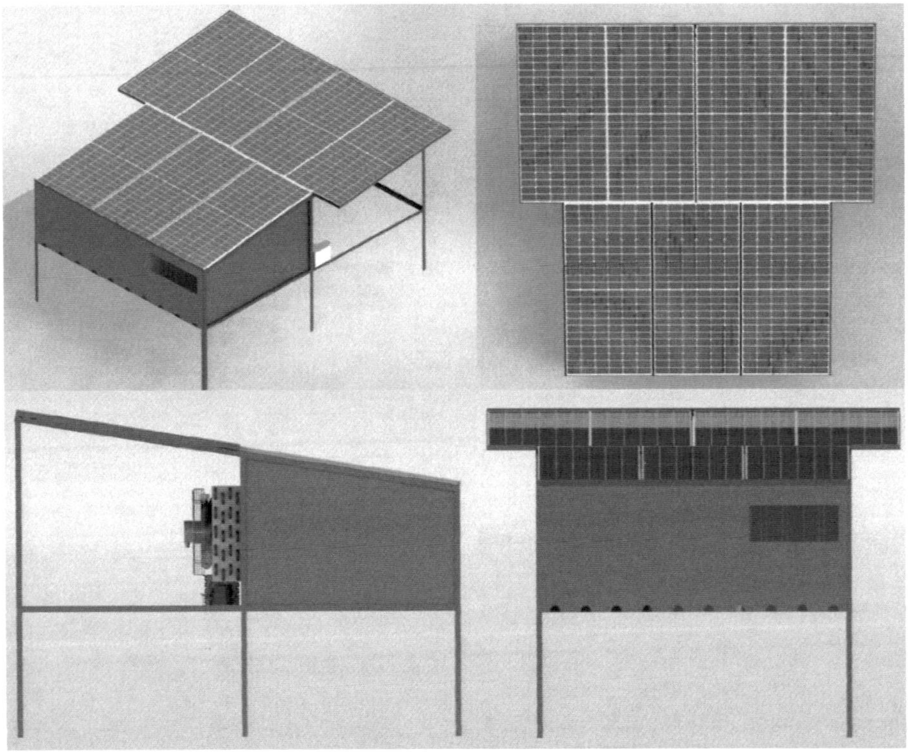

Figure 9. CAD Modeling

The simulation process was carried out in the following way: it starts with the air circulation in a control volume of 6m3 at a rate of 950m3 / h or 15.83m3 / s this value is due to the air flow delivered by the fan, in the same way in this way, they enter data for air temperature at the inlet of -5 ° C so that once it is inside the control volume, it brings the air to dew point temperature. Of the 415 interactions, the software simulated the behavior of the air within the control volume as shown in the **Figure *10***, with constant temperature and fluid circulation throughout the control volume.

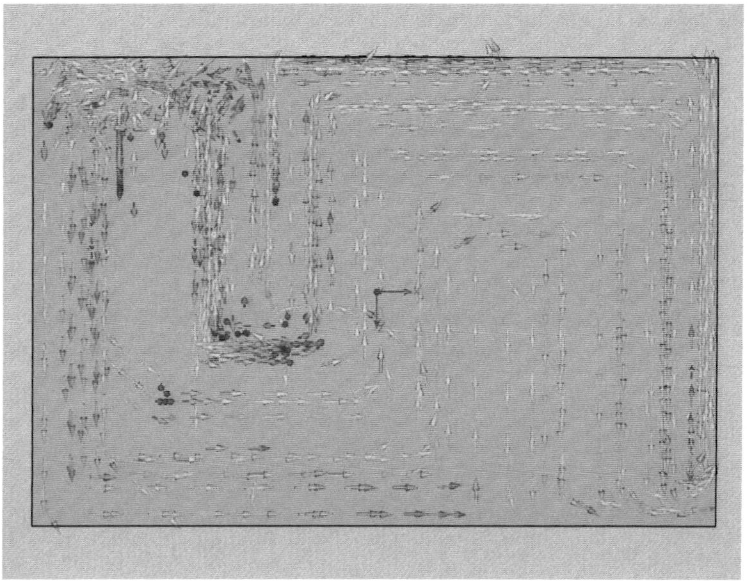

Figure 10. Simulation

Finally, the software tabulates the results of the simulation represented in Table 6. Simulation's result , where interesting data for the operational feasibility of the equipment are observed.

Table 6. Simulation's result

Local Parameter	Minimum	Maximum	Average	Bulk Average	Surface Area [m^2]
Pressure [Pa]	101325,00	917548,63	345563,85	366717,82	0,04
Density (Fluid) [kg/m^3]	2,01	3,14	2,53	2,51	0,04
Velocity [m/s]	201,40	781,49	671,40	688,50	0,04
Velocity (X) [m/s]	-366,39	474,86	47,35	39,96	0,04
Velocity (Y) [m/s]	-436,94	-24,58	-154,71	-151,69	0,04
Velocity (Z) [m/s]	37,54	765,83	608,62	635,19	0,04
Mach Number [ ]	0,29	1,13	0,97	1,00	0,04
Mass Fraction of Water [ ]	0,01	0,01	0,01	0,01	0,04
Mass Fraction of Air [ ]	0,99	0,99	0,99	0,99	0,04
Temperature (Fluid) [°C]	-2,00	2,00	0,00	0,00	0,04
Temperature (Solid) [°C]	-4181,91	2403,78	-810,06		0,04
Specific Humidity [kg/kg]	1,00	1,00	1,00	1,00	0,04
Absolute Humidity [kg/m^3]	0,01	0,02	0,02	0,02	0,04

3.6 Technoeconomic study

Table *7* shows the value of each of the elements that will make up both the cooling system and the photovoltaic generation system, taking into account manufacturing labor, consumable materials such as cables, connectors, hardware, among others. and the utility of the whole system.

Table 7. Cost system

Item	Material	Quantity	Unit Value	Total cost
1	Photovoltaic panel - Jinko solar JKM375HC66M	7	172,00 €	1.204,00 €
2	Regulator JN-MPPT-C 100A	1	330,00 €	330,00 €
3	Battery 12V 200A	2	239,00 €	478,00 €
4	Compressor BD250GH.2	1	152,00 €	152,00 €
5	Fan LWFD200-EO	2	24,00 €	48,00 €
6	Evaporator coil	1	50,00 €	50,00 €
7	Condenser coil	1	50,00 €	50,00 €
8	Expansion Valve Danfoss 068N2012	2	27,00 €	54,00 €
9	Electronic control kit	1	18,00 €	18,00 €
10	Metallic structure	1	500,00 €	500,00 €
11	Electrical protection kit	1	60,00 €	60,00 €
			Total material value	2.944,00 €
			Expendable 4%	117,76 €
			Workforce	250,00 €
			Utility 15%	496,76 €
			Total	3.808,52 €

3.6.1 Economic profitability. For profitability, it is taken into account that in Colombia a 20-liter container (bottle) has a commercial value of 2.2 Euros, if a daily expenditure of these 20 liters in a year is considered they are equivalent to 7,200 liters or 792 euros, it is also It has an average of 3.8% of inflation to obtain the evolution of the price of this product over the years.

Table 8. Profitability

VAN	9722
TIR	21%
Payback (annual)	6

The initial investment begins to generate profits from the sixth year, on the other hand the net present value (NPV) with a discount rate of 2.5% at twenty years generates € 9,722 in profits with an internal rate of return (IRR) 21% as evidenced in **Table *8***.

4. Conclusions

The state of the art presents evidence of technological developments and research, in which an attempt has been made to carry out this type of system looking for the condensation of the water in the air, but its main problem lies in the application of thermodynamics, which is why this Research sought to bring the air to dew temperature and thus improve the thermal efficiency of the equipment and therefore energy efficiency.

The refrigeration equipment was selected in continuous current in order to reduce costs when integrating the two systems (refrigeration-photovoltaic system) to propose a sustainable condensation system, which was reflected in the selection of the inverter without ignoring that at the time if working in alternating current, there would be gains in cooling power since a conventional compressor could be used.

The amp-hour method gives you an ideal equipment sizing and selection methodology, neglecting the losses existing in the photovoltaic system. Therefore, it is recommended to take it as an initial method, for the subsequent expansion of a dynamic system that allows a real evaluation of the operation of the system.

The increase in the number of panels from one dimensioning method to another did not significantly affect the budget and the economic profitability of the system. The flexibility of SAM to perform the dynamism of the system for one year is highlighted, allowing to have an adequate projection of the behavior of the photovoltaic system, guaranteeing the supply of electrical energy to the condensing system, which in turn, will remain active for around 8 hours a day producing fresh water.

Finally, a sustainable water condensation system is obtained, through the integration of a conventional cooling system and a photovoltaic system, having a reasonable investment cost and a payback time or return on investment, starting from 6 year. Taking into account that the useful life of the photovoltaic system is approximately 20 years, the initial investment would be repaid for 14 years.

References

[1] L. Li *et al.*, «Global trends in water and sediment fluxes of the world's large rivers», *Sci. Bull.*, vol. 65, n.º 1, pp. 62-69, ene. 2020, doi: 10.1016/j.scib.2019.09.012.

[2] G. Deng, F. Lu, L. Wu, y C. Xu, «Social network analysis of virtual water trade among major countries in the world», *Sci. Total Environ.*, p. 142043, ago. 2020, doi: 10.1016/j.scitotenv.2020.142043.

[3] R. Mukhopadhyay, S. M. Karisiddaiah, y J. Mukhopadhyay, «Chapter 2 - Scientific Assessment», en *Climate Change*, R. Mukhopadhyay, S. M. Karisiddaiah, y J. Mukhopadhyay, Eds. Elsevier, 2018, pp. 15-53.

[4] J. Echeverría Molina y S. Anaya Morales, «El derecho humano al agua potable en Colombia: Decisiones del Estado y de los particulares», *Vniversitas*, vol. 67, n.º 136, pp. 1-14, ene. 2018, doi: 10.11144/Javeriana.vj136.dhap.

[5] S. Ramírez y E. Ramón, «El calentamiento global y la degradación de la ozonosfera (Un análisis científico desde la diversidad de criterios)», *Rev. Integra Educ.*, vol. 8, n.º 3, pp. 43-54, dic. 2015.

[6] J. Muñoz-García y A. P. Acosta, «Población y Medio Ambiente: Consecuencias ambientales del crecimiento demográfico», *El Campo Las Cienc. Las Artes*, n.º 139, pp. 27-56, 2002.

[7] United Nations, «El papel de los combustibles fósiles en un sistema energético sostenible | Naciones Unidas», *United Nations*, 2020. https://www.un.org/es/chronicle/article/el-papel-de-los-combustibles-fosiles-en-un-sistema-energetico-sostenible (accedido ago. 28, 2020).

[8] Ayuda en Accion, «Escasez de agua, un problema mundial», *Ayuda en Acción*, dic. 19, 2017. https://ayudaenaccion.org/ong/blog/sostenibilidad/escasez-de-agua-problema-mundial/ (accedido ago. 28, 2020).

[9] UNESCO, «Informe del 2019 – No dejar a nadie atrás», *UNESCO*, mar. 07, 2019. https://es.unesco.org/water-security/wwap/wwdr/2019 (accedido ago. 28, 2020).

[10] World Resources Institute, «Water», *World Resources Institute*, 2019. https://www.wri.org/our-work/topics/water (accedido ago. 28, 2020).

[11] J. G. Ascanio-Villabona, C. L. Sandoval-Rodriguez, A. D. Rincón-Quintero, B. E. Tarazona-Romero, y R. E. Paez-Castro, «Building a prototype for functional analysis of the energy potential of the water flow in pipe ½ "using microturbines applied to Unidades Tecnológicas de Santander», *IOP Conf. Ser. Mater. Sci. Eng.*, vol. 844, p. 012056, jun. 2020, doi: 10.1088/1757-899X/844/1/012056.

[12] Ayuda en Accion, «La guerra del agua: países con mayor escasez de agua», *Ayuda en Acción*, feb. 11, 2020. https://ayudaenaccion.org/ong/blog/sostenibilidad/guerra-agua-paises-escasez/ (accedido ago. 28, 2020).

[13] A. J. Paz Cardona, «Calidad, abastecimiento y saneamiento: los grandes retos del agua en Colombia», *Noticias ambientales*, mar. 23, 2019. https://es.mongabay.com/2019/03/colombia-estudio-nacional-agua-ideam/ (accedido ago. 28, 2020).

[14] J. Tapias, C. Sandoval, y J. Sánchez, «ANÁLISIS DE PROSPECTIVA DEL SECTOR ENERGÉTICO DE COLOMBIA, PARA LA INTEGRACIÓN DE FUENTES FOTOVOLTAICAS EN LOS SISTEMAS DE DISTRIBUCIÓN DE ENERGÍA ELÉCTRICA APLICANDO UNA REVISIÓN EN BASES DE DATOS CIENTÍFICAS», *Rev. Colomb. Tecnol. Av. RCTA*, vol. 2, nov. 2018, doi: 10.24054/16927257.v32.n32.2018.3034.

[15] L. J. F. Duarte, «OPTIMIZACIÓN DE LA COBERTURA Y ABASTECIMIENTO DEL SERVICIO PÚBLICO DE AGUA POTABLE EN EL MUNICIPIO DE LOS SANTOS, SANTANDER», Posgrado, UDES, Santander, 2019.

[16] J. G. Ascanio-Villabona, L. A. D. Portillo-Valdés, O. Lengerke-Pérez, B. E. T. Romero, A. D. Rincón-Quintero, y M. A. Durán-Sarmiento, «Analysis of the Energy Potential of a Tangential Microturbine for Application in a Passivhaus Environment», en *Recent Advances in Electrical Engineering, Electronics and Energy*, oct. 2020, pp. 181-195, doi: 10.1007/978-3-030-72212-8_14.

[17] E. Ahmadi, B. McLellan, y T. Tezuka, «The economic synergies of modelling the renewable energy-water nexus towards sustainability», *Renew. Energy*, ago. 2020, doi: 10.1016/j.renene.2020.08.059.

[18] A. D. Rincón-Quintero *et al.*, «Manufacture of hybrid pieces using recycled R-PET, polypropylene PP and cocoa pod husks ash CPHA, by pneumatic injection controlled with LabVIEW Software and Arduino Hardware», *IOP Conf. Ser. Mater. Sci. Eng.*, vol. 844, p. 012054, jun. 2020, doi: 10.1088/1757-899X/844/1/012054.

[19] A. G. Olabi, K. Elsaid, M. K. H. Rabaia, A. A. Askalany, y M. A. Abdelkareem, «Waste heat-driven desalination systems: Perspective», *Energy*, vol. 209, p. 118373, oct. 2020, doi: 10.1016/j.energy.2020.118373.

[20] K. Elsaid, E. T. Sayed, M. A. Abdelkareem, A. Baroutaji, y A. G. Olabi, «Environmental impact of desalination processes: Mitigation and control strategies», *Sci. Total Environ.*, vol. 740, p. 140125, oct. 2020, doi: 10.1016/j.scitotenv.2020.140125.

[21] G. M. Vargas Gil, R. B. Aguiar Cunha, S. G. Di Santo, R. M. Monaro, F. F. Costa, y A. J. Sguarezi Filho, «Photovoltaic energy in South America: Current state and grid regulation for large-scale and distributed photovoltaic systems», *Renew. Energy*, ago. 2020, doi: 10.1016/j.renene.2020.08.022.

[22] B. E. Tarazona-Romero, Á. Campos-Celador, Y. A. Muñoz-Maldonado, C. L. Sandoval-Rodríguez, y J. G. Ascanio-Villabona, «Prototype of lineal solar collector Fresnel»:, *Visión Electrónica*, vol. 14, n.º 1, Art. n.º 1, ene. 2020, doi: 10.14483/22484728.16013.

[23] B. E. Tarazona-Romero, A. Campos-Celador, Y. A. Muñoz-Maldonado, J. G. Ascanio-Villabona, M. A. Duran-Sarmiento, y A. D. Rincón-Quintero, «Development of a Fresnel Artisanal System for the Production of Hot Water or Steam», en *Recent Advances in Electrical Engineering, Electronics and Energy*, oct. 2020, pp. 196-209, doi: 10.1007/978-3-030-72212-8_15.

[24] Q. Chen *et al.*, «A self-sustainable solar desalination system using direct spray technology», *Energy*, vol. 205, p. 118037, ago. 2020, doi: 10.1016/j.energy.2020.118037.

[25] B. E. Tarazona, C. L. S. R, C. G. C. Arias, J. G. A. V, y J. J. V. N, «Detection of structural alterations in metal bodies: An approximation using Fourier transform and principal component analysis (PCA)», *Sci. Tech.*, vol. 25, n.º 2, Art. n.º 2, jun. 2020, doi: 10.22517/23447214.23501.

[26] S. Statkic, B. Jovanovic, A. Micic, N. Arsic, y S. Jović, «Adaptive neuro fuzzy selection of the most important factors for photovoltaic pumping system performance prediction», *J. Build. Eng.*, vol. 30, p. 101242, jul. 2020, doi: 10.1016/j.jobe.2020.101242.

[27] A. D. Rincón-Quintero, L. A. D. Portillo-Valdés, A. Meneses-Jácome, J. G. Ascanio-Villabona, B. E. Tarazona-Romero, y M. A. Durán-Sarmiento, «Performance Evaluation and Effectiveness of a Solar-Biomass Hybrid Dryer for Drying Homogeneous of Cocoa Beans Using LabView Software and Arduino Hardware», en *Recent Advances in Electrical Engineering, Electronics and Energy*, oct. 2020, pp. 238-252, doi: 10.1007/978-3-030-72208-1_18.

[28] C. L. Sandoval-Rodriguez, B. E. Tarazona-Romero, O. Lengerke-Perez, C. G. Cárdenas-Arias, D. C. D. Diaz, y O. A. A. Cárdenas, «Descriptive Study of a Rotary Machine Affected by Misalignment and Imbalance Applying the Wavelet Transform», en *Recent Advances in Electrical Engineering, Electronics and Energy*, oct. 2020, pp. 226-242, doi: 10.1007/978-3-030-72212-8_17.
[29] P. Saji Raveendran y S. Joseph Sekhar, «Experimental studies on the performance improvement of household refrigerator connected to domestic water system with a water-cooled condenser in tropical regions», *Appl. Therm. Eng.*, vol. 179, p. 115684, oct. 2020, doi: 10.1016/j.applthermaleng.2020.115684.
[30] H. Hassan, M. S. Ahmed, M. Fathy, y M. S. Yousef, «Impact of salty water medium and condenser on the performance of single acting solar still incorporated with parabolic trough collector», *Desalination*, vol. 480, p. 114324, abr. 2020, doi: 10.1016/j.desal.2020.114324.
[31] SolidWorks Corporation, «SOLIDWORKS», *Dassault systemes-SolidWorks*. https://www.solidworks.com/es (accedido ago. 20, 2020).
[32] Honeywell, «Genetron Properties Suite | European Refrigerants», *Refrigerants*. https://www.honeywell-refrigerants.com/europe/genetron-properties-suite/ (accedido ago. 20, 2020).
[33] NREL, «Home - System Advisor Model (SAM)», *NATIONAL RENEWABLE ENERGY LABORATORY*, sep. 09, 2018. https://sam.nrel.gov/ (accedido ago. 20, 2020).
[34] Frigoristas Wordpress, «EVAPORADORES», *Instalaciones Frigoricicas Comerciales e Industriales*, may 16, 2016. https://frigoristas.wordpress.com/evaporadores/ (accedido ago. 20, 2020).
[35] Frigoristas Wordpress, «CONDESADORES», *Instalaciones Frigoricicas Comerciales e Industriales*, jun. 05, 2016. https://frigoristas.wordpress.com/condensadores/ (accedido ago. 20, 2020).
[36] Danffoss, «Compresores de CC para telecomunicaciones», *Danfoss*. https://www.danfoss.com/es-es/products/compressors/dcs/compressors-for-refrigeration/telecom-dc-compressors/ (accedido ago. 20, 2020).
[37] N. Quadri, *Manual de Aire Acondicionado y Calefacción: Calculo y Diseño*, Alsani. Alsina, 2013.
[38] A. Ruiz Gomez, «Energia Solar Fotovoltaica. Autoxconsumo. Energia solar termica», 2.ª ed., España: UDIMA, 2019.

IOP Conf. Series: Materials Science and Engineering 1154 (2021) 012041 doi:10.1088/1757-899X/1154/1/012041

Synthesis of starch powder from different organic wastes: A green approach to a valuable material

D P Sanabria[1], H I Farfán[1], D N Quintana[1], I V Abril[1], H F Castro[1], R A Paredes[1], K L Roa[1, 2]

[1] Universidad Pedagógica y Tecnológica de Colombia, Sogamoso, Colombia

[2] Centro de Investigación en Ciencia Aplicada y Tecnología Avanzada, Instituto Politécnico Nacional, Ciudad de México, México

E-mail: karol.roa@uptc.edu.co

Abstract. Plastic pollution has become one of the main causes of irreversible damages to the environment. Despite the well-known adverse effects, the lack of culture in recycling is causing overcrowded landfills with tons of this disposable material. Developing new products is necessary to replace plastics with eco-friendly materials, in this way, biopolymers are a suitable alternative. Biopolymer synthesis requires a natural polymer known as starch, which is a common compound in roots, tubers, fruits, cereals and aquatic plants. In the present work, we study different solid organic wastes such as cassava, potatoes, plantain, corn and two types of algae *Elodea Canadensis* and *Myriophyllum Quitense*, in order to discover its feasibility as starch sources. For this purpose, we established an extraction method taking into account the main structure of each organic waste. By means of Fourier-transform infrared spectroscopy (FTIR) and iodine test, we obtained the chemical structures and principal characteristics of each starch. Synthesized powders exhibited characteristics of thermoplastic starches making them available for its potential use in biopolymers.

1. Introduction

Plastics derived from petroleum have become necessary in several activities of human population thanks to its benefits; however, its massive use is causing negative impacts to the earth. This environmental problem has stimulated research to discover and propose new materials that can replace plastics [1]. Accumulation of plastics rises up to 4900 millions of tons (almost 60% of total production) in the atmosphere, according to current trends and dynamics of global waste management, by 2050 almost 9000 millions of tons will be recycled, 12000 millions of tons incinerated and 12000 millions of tons left in landfills or in natural environments [2].

To develop eco-friendly alternatives to replace plastics is necessary to find a cheap, abundant and renewable raw material; in this way, starch currently represents a widely used source for plastic production with biodegradable properties. One natural source of starch is food; its wide consumption is constantly creating high amounts of organic wastes that can potentially cause an environmental impact. These organic wastes has an important content of starch that can be used [3]. Taking into account that in Colombia, roots, fruits and tubers are involved within the food basket, a high consumption of these could generate high amounts of organic husks. Nowadays the lack of recycling culture involve an inadequate management of the organic wastes becoming sources of contamination [4].

Another source of available starch that is causing significant environmental impacts is aquatic plants; the principal cause is due to its endemic growth, which reduces the extension of lakes and lagoons. This non-controlled growth cause decomposition, water quality alterations, poisoning of species and ecological imbalance. Due to fertilizers used in the peripheral of the lacustrine body, algaes modified its chemical composition causing a negative impact on other species [5].

One of the alternatives to mitigate these environmental problems are biopolymers, a class of polymers made from renewable raw materials or derived from environment. These natural sources makes them a sustainable option in terms of production. To develop natural polymers or biopolymers is necessary a starch source. This polysaccharide is stored as energy reserve in plants; common starch sources are roots (cassava), tubers (potatoes), aquatic plants (*Elodea Canadensis* and *Myriophyllum Quitense*), fruits (plantain) and cereals (corn) [6].

Several works previously reported different extraction methods of starch, which include the wet or traditional method. This method consisted of a solution of lactic acid and sulfur dioxide, which would generate a change in the starch due to nixtamalization; a variation of the wet method is carried out with the use of distilled water in other cases 25% sodium hydroxide [8] or buffer solutions [9] are used as solvents. To accelerate the drying process and make an efficient separation, the centrifuge is usually used [10], another procedure consist of decant for a long time suspending the material in water until the sediment is obtained. Another existing method is through the dry route, in this procedure is necessary to carry out repeated grinding, thus modifying the sugar content of the starch [11]. Currently, there are methods whose purpose is to increase efficiency of extraction process using ultrasound-assisted equipment. The method uses frozen roots, which has shown that 10 minutes of sonication promotes an increase yield of starch. Additionally, it was shown that does not modify the structure of the starch granule [12] making this method viable for the extraction of starch without changing the structure and optimizing the yield.

In the present work, we investigate five different wastes as potential starch sources these include: potato peels, cassava, plantain, corn (leaves and teas) and aquatic plants. To our knowledge, there are reduced studies linked to starch extraction from the proposed wastes; therefore, findings of this work are necessary and opening for further quantitative studies. Synthesis of starch carried out three phases; in the first stage, we designed the extraction process, in the second we performed an iodine test for each starch and last, we studied by means of Fourier-transform infrared spectroscopy (FTIR) each sample in order to discover principal bands of chemical bonds formed. Finally, according to the extraction efficiency we propose which organic wastes are potential starch sources to obtain biopolymers with molecular balanced structures.

2. Materials and methods

2.1. Extraction process from organic wastes

Collected organic wastes used in this study such as corn husks, potato, plantain and cassava peels were from different food companies located in Sogamoso, Colombia. The extraction process of starch was designed based on the wet method [7] and is depicted in Figure 1. The process consisted of seven phases, each one completed as follows:

- Selection: In this phase, we removed damaged and irregular peels from collected organic wastes.
- Washing: Organic wastes were subjected to three washes to remove impurities; first, raw materials were cleaned carefully in a container with deionized water. Then, peels were disinfected using 0.5% of hypochlorite mixed with 1000 mL of distilled water; solution was used per 1000 g of peels.
- Grinding: Raw materials were grinded using a disc mill to obtain a homogenous mass with less polysaccharides.

- Filtration: 1000 g of mass was mixed with 1000 mL of deionized water was added and flocculated for 2 hours to allow separation of starch granules from mass. Then, the mixture was passed through a gauze filter to separate raw fiber from liquid. This process is repeated twice with the dry matter remaining.
- Sedimentation: Filtered liquid was left at room temperature for 12 hours to allow accumulation of starch particles. The sediment was washed using 200 mL of water per gram of starch, until a clean material was obtained and finally decanted.
- Drying: The starch was dried at 40°C for 24 hours in an electric stove.
- Milling and sieving: Dry starch was passed through No. 40 sieve (400 microns).

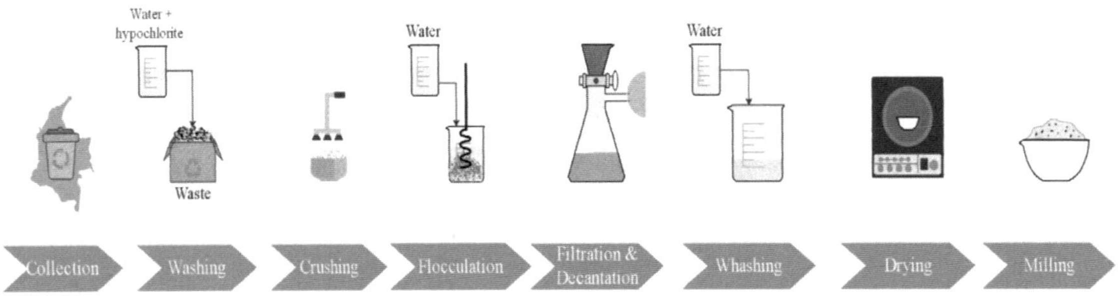

Figure 1. Scheme of extraction process from wastes: corn husks, potato, plantain and cassava peels.

2.2. Extraction process from algaes

Two types of algaes were used *Elodea Canadensis* and *Myriophyllum Quitense,* both collected from Lake Tota located in Boyacá, Colombia. As is showed in Figure 2 treatment of algaes and extraction process of starch involved eight phases, described as follows:

- Washing: Aquatic plants were carefully cleaned with distilled water and disinfected with 0.5% hypochlorite. Then, algaes were washed using 1000 mL of water per 1000 g of aquatic plants.
- Drying: Cleaned algaes were left on a surface and dried for 1 week at room temperature, avoiding sun exposure.
- Crushing: Using an electric disk mill algaes were reduced to 400 microns.
- Filtration: 50 g of grinded aquatic plants were mixed with 700 mL of distilled water; the mixture was flocculated for 2 hours to allow the release of starch particles. Then, the fiber was filtered from liquid. This process was repeated twice with the remaining dry matter.
- Agitation: The mixture was stirred on a hotplate at 30 °C for 30 minutes and then filtered once.
- Sedimentation: The mixture was left for 24 hours to allow decantation of starch particles.
- Drying: Sediment was dried using a stove at 40 °C for 24 hours.
- Milling and sieving: Dried mass was milled to 400 microns using a mesh No. 40.

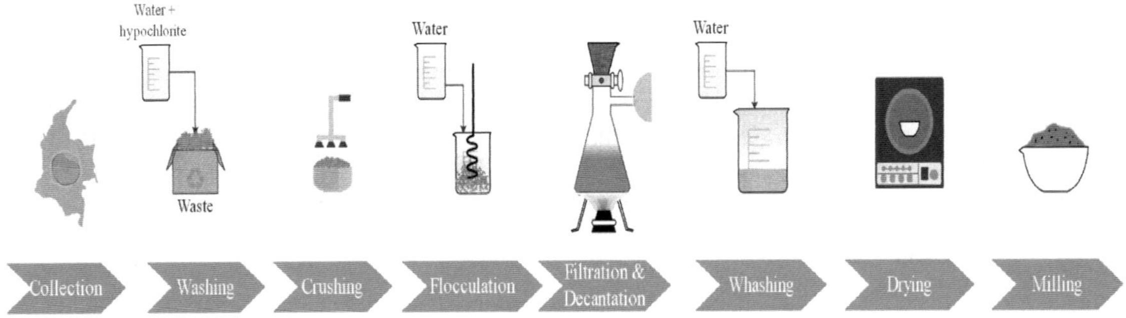

Figure 2. Scheme of extraction process from algaes *Elodea Canadensis* and *Myriophylum Quitense*.

The starch obtained of each sample was characterized through Fourier-transform infrared spectroscopy (FTIR), for which the dry sample was mixed with KBr and pressed to obtain a tablet, which was introduced in a Nicolet ™ iS50 FTIR spectrometer. For each of the samples, 32 scans were recorded in the region of 400 to 4000 cm^{-1} [16].

3. Results & Discussion

3.1. Extraction efficiency

The amount of starch obtained in weight is shown in table 1 extraction efficiency obtained from potato was 7.3%. Compared with an industrialized one with a yield of 15.56 to 17.76% [5] of starch from potato residues, in this case, efficiency variation was affected by the extraction method used. The method was carried out based on Gonzales L, Gómez S and Abad works, which were proposed as an effective industrialized development. Based on works of Villalobos M, López P, Rodríguez and Prado M, the extraction process is similar to the proposed in the present work and presented a starch content of approximately 7 to 8 % [13]. The increase of efficiency in extraction of starch from potato peels, an industrialized process is more convenient. In comparison with Bilgin work, starch extraction of plantain peels has a higher starch efficiency reaching 11.3%. Works from Giraldo J, Cuarán J, García L, and Pardo L, showed an efficiency of 12.3% [14]. The differences could be associated with quality of plantain peels or differences in extraction processes. It should be noted that grain size could affected the extraction efficiency of starch, since two solid phases remained once suspended water was removed. Characteristic gray colour was observed as well, which could be characteristic of traces of proteins, lipids and small starch granules. The last due to its low density which could not separate from water remaining in white colour on surface [15]. The efficiency calculation was determined by equation (1), determining the percentage efficiency (E) of each starch extracted from organic wastes and algaes:

$$E = \frac{weight\ from\ starch\ extracted}{weight\ sample\ processed} * 100 \tag{1}$$

Table 1. Extraction efficiencies of potential starches as a percentage.

Type of starch	% efficiency of obtaining
Potato	7,3
Plantain	6,7
Elodea Canadensis	4,15
Myriophyllum Quitense	3,82

3.2. Analysis of Fourier-transform infrared spectroscopy (FTIR)

The spectrum depicted in Figure 3 shows C-C elongation signals at 764 cm^{-1}, the signal 860 cm^{-1} allows displays a deformation of CH2; the range of 900 to 1250 cm^{-1} corresponds to the mostly asymmetric stretching of the C-O bond vibrations, which are associated with the elongations of the polysaccharide particles [17]. Besides, a decrease in the signals attributed to bending vibrations is observed in 1243 cm^{-1} y 1650 cm^{-1}. Also by elongation (3250-3900 cm^{-1}) of the O-H groups; this implies that the samples examined are hydrophilic. It is also noted that at 1344 cm^{-1} there is a deformation of CH2.

Concerning the peak of the 1462 cm^{-1} which is why its deformation is evident. The signal at 2920 cm^{-1} corresponds to the C-H stretches of the glucose unit of the starch molecule; additionally, it was observed that the 1740 cm^{-1} in each of the samples corresponds to the voltage vibration of C=O. The results obtained showed considerable similarities with the typical band of a thermoplastic starch [18]. The links identified are shown in Table 2.

Table 2. Bonds resulting from Fourier-transform infrared spectroscopy.

Wavenumber (cm^{-1})	Type of bond	Description
3250-3900	O-H elongation	Stretches related to intra and intermolecular hydrogen bonds.
2920	Elongation CH	
1740	Elongation C=O	Hydroxyl groups confer hydrophilic properties.
1650	Bending (O-H)(Water)	
1462	CH deformation or CH2 vibration	
1344	C-O-H bending, CH2 vibration	Characteristic bonds of crystalline starch/ amorphous starch region
1243	O-H bending	
1010, 1080, 1150	C-O and C-C Stretch	
900	Elongation C-O	Anhydroglucose units that are stretched in the starch.
860	Deformation CH2	
764	C-C elongation	

According to the infrared transmission spectroscopy (Figure 4), the combinations of starches that could be compatible for the manufacture of bioplastics are those that have flexions, elongations and deformations (near wave numbers) in similar fragments, that is, they present chemical bonds related; those mixtures were the potato-plantain, and corn-cassava [20].

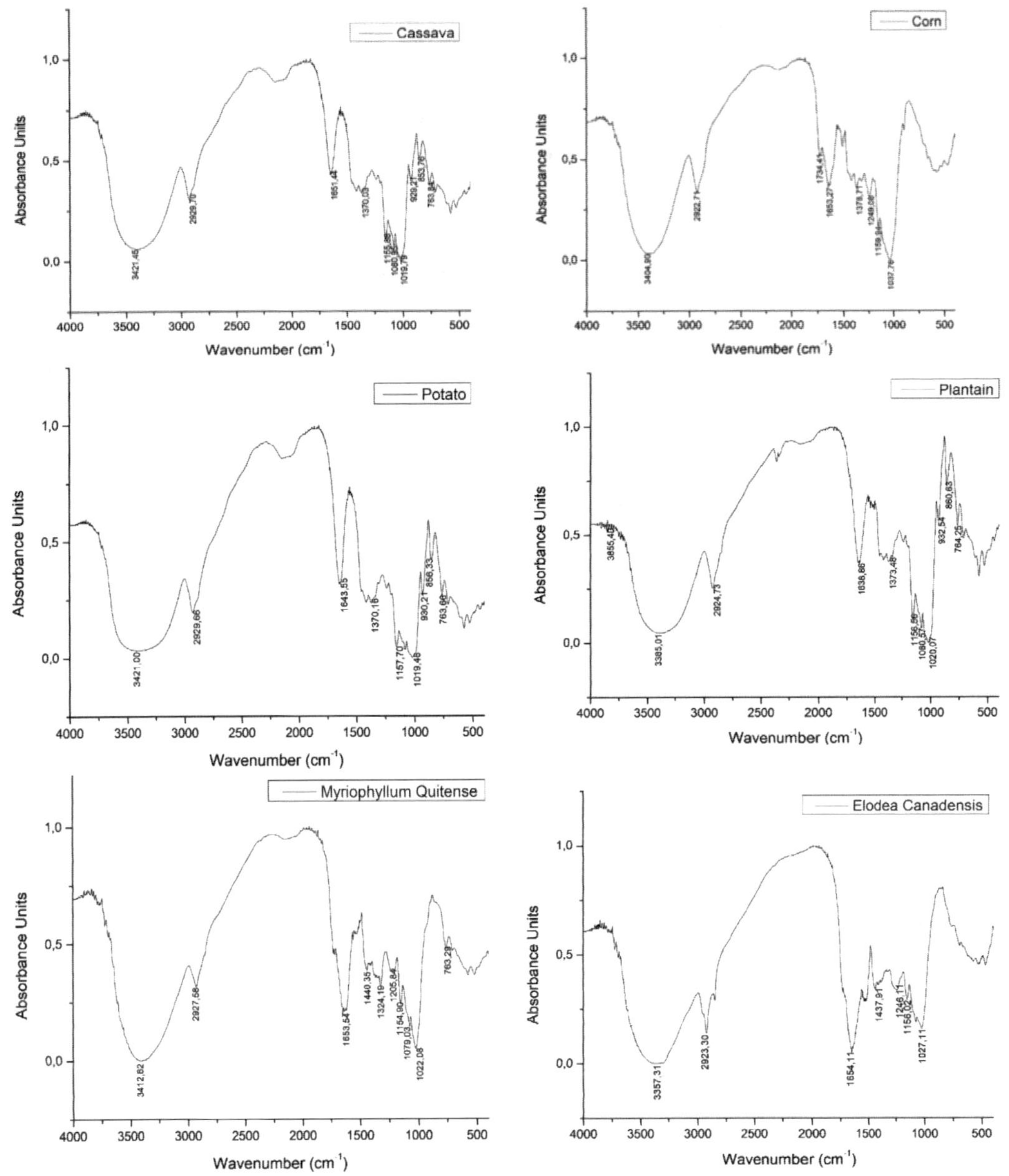

Figure 3. FTIR of the starches studied (cassava, corn, potato, plantain, *Myriophyllum Quitense* and *Elodea Canadensis*)

3.3. Iodine test

To verify starch structures in the organic material and algaes, the iodine test was performed [21]. The procedure consist of adding 0.25 mL of iodine to a solution of 10 mL of distilled water with 1 g of material to be tested. The reaction occurs when amylose is present, which has a non-branched helical structure that allows the iodine to penetrate the helix of the starch, thus modifying the properties of light absorption until obtain a blue color, although high concentrations displayed black color. As shown in figure 4, after test was performed black color was displayed showing the presence of starch.

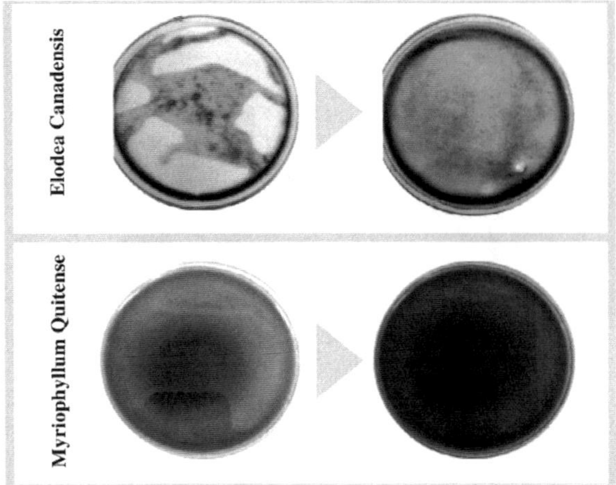

Figure 4. Iodine test performed on *Elodea Canadensis* and *Myriophyllum Quitense* algaes.

4. Conclusions

Extraction efficiencies obtained showed that algaes used presented approximately 50% less starch than organic peels, although the extraction process was carried out by the same method, and algaes were subjected to an additional flocculation procedure using temperature.

The behavior found in starches extracted from potato, cassava and corn residues presented IR bands that corresponded to typical structures of starch granules. Also, aquatic plants *Elodea Canadensis* and *Myriophyllum Quitense*, possess the characteristics of the starch molecule, for this reason, it is possible, in a next phase to give it a sustainable use as a biopolymer.

According to the FTIR analysis, starches have chemical functions similar to each other, because they present the deformations, flexions and peaks on the same wavenumbers, therefore, the carbohydrates as more similar bases for the elaboration of the bioplastics, were potato-plantain, and corn-cassava, varying the quantities of the aquatic plants for the biopolymer.

The results of the iodine test applied to the material obtained from the aquatic plants showed a positive result for the presence of starch, the most consistent coloration was in the Myriophyllum Quitense concerning the Elodea Canadensis; which corroborates the existence of bonds associated with the polysaccharide molecules detected by means of the analysis indicated by the infrared transmission spectroscopy.

5. Future work

Based on the results obtained we recommend to characterized the fiber obtained as co-products from the starch extraction process in order to take advantage of them for further applications.

References

[1] Kwak H. W., Lee H., Lee M. E., and Jin, H. J. 2018. Facile and green fabrication of silk sericin films reinforced with bamboo-derived cellulose nanofibrils. *Journal of Cleaner Production*, **200**, 1034-42.

[2] Geyer, R., Jambeck, J. R., & Law, K. L. 2017. Production, use, and fate of all plastics ever made. *Science Advances*, **3**(7).

del Rey Castillo, E., Almesfer, N., Saggi, O., & Ingham, J. M. 2020. Light-weight concrete with artificial aggregate manufactured from plastic waste. *Constr build mater*, **265**, 120199.

[3] Gutiérrez, A. S., Eras, J. J. C., Hens, L., & Vandecasteele, C. 2020. The energy potential of agriculture,

agroindustrial, livestock, and slaughterhouse biomass wastes through direct combustion and anaerobic digestion. The case of Colombia. *Journal of Cleaner Production*, 122317.

[4] Gonzalez, L. V. P., Gómez, S. P. M., & Abad, P. A. G. 2017. Aprovechamiento de residuos agroindustriales en Colombia. *RIAA*, **8**(2), 141-150.

[5] Acuña-Alonso, C., Lorenzo, O., Álvarez, X., Cancela, Á., Valero, E., & Sánchez, Á. 2020. Influence of microcystis sp. and freshwater algae on pH: Changes in their growth associated with sediment. *Environmental Pollution*, 114435.

[6] Chakraborty, I., Pallen, S., Shetty, Y., Roy, N., & Mazumder, N. 2020. Advanced microscopy techniques for revealing molecular structure of starch granules. *Biophysical Reviews*, **12**(1), 105-122.

[7] Fonseca-López, D., Monroy-Fonseca, L., & Rodríguez-Molano, C. 2019. Effect of the method of extraction of canna indica L. starch on its physicochemical properties. *Revista Lasallista De Investigación*, **16**(2), 44-52.

[8] Medina, C., Paredes, A., Rodríguez, M. E., Moreno, M., Camacho, D. B., García, D., & Ojeda, C. 2010. Evaluación de dos métodos de extracción de almidón a partir de cotiledones de mango. *Bioagro*, **22**(1), 67-74.

[9] Viña, S. Z., & García, M. A. 2016. Starch extraction process coupled to protein recovery from leguminous tuberous roots (Pachyrhizus ahipa). *Carbohydrate polymers*, **152**, 231-240.

[10] Flores-Gorosquera, E., García-Suárez, F. J., Flores-Huicochea, E., Núñez-Santiago, M. C., González-Soto, R. A., & Bello-Pérez, L. A. 2004. Rendimiento del proceso de extracción de almidón a partir de frutos de plátano (Musa paradisiaca). Estudio en planta piloto. *Acta Científica Venezolana*, **55**(1), 86-90.

[11] Cobana, M., & Antezana, R. 2007. Proceso de extracción de almidón de yuca por vía seca. *Revista Boliviana de Química*, **24**(1), 78-84.

[12] González-Lemus, L. B., Calderón-Domínguez, G., de la Paz Salgado-Cruz, M., Díaz-Ramírez, M., Ramírez-Miranda, M., Chanona-Pérez, J. J., . . . Farrera-Rebollo, R. R. 2018. Ultrasound-assisted extraction of starch from frozen jicama (P. erosus) roots: Effect on yield, structural characteristics and thermal properties. *CYTA - Journal of Food*, **16**(1), 738-746.

[13] Villalobos, M., López, M., Rodríguez, P., & Prado, M. 2014. Obtención de almidón a partir de los residuos de papa del mercado Abastos. *Ciencias de la Ingeniería y Tecnología Handbook,* **6**, 266-271.

[14] Giraldo, J. L., Cuarán, J. C. C., García, L. V. A., & Pardo, L. M. F. 2014. Usos potenciales de la cáscara de banano: elaboración de un bioplástico. *Revista colombiana de investigaciones agroindustriales*, **1**(1), 7-21.

[15] Ziegler, V., da Silva Timm, N., Ferreira, C. D., Goebel, J. T., Pohndorf, R. S., & de Oliveira, M. 2020. Effects of drying temperature of red popcorn grains on the morphology, technological, and digestibility properties of starch. *International Journal of Biological Macromolecules*, **145**, 568-574.

[16] Sánchez-Rivera, M. M., Flores-Ramírez, I., Zamudio-Flores, P. B., González-Soto, R. A., Rodríguez-Ambríz, S. L., & Bello-Pérez, L. A. 2010. Acetylation of banana (Musa paradisiaca L.) and maize (Zea mays L.) starches using a microwave heating procedure and iodine as catalyst: Partial characterization. *Starch‐Stärke*, **62**(3-4), 155-164.

[17] Pourfarzad, A., Habibi Najafi, M. B., Haddad Khodaparast, M. H., & Khayyat, M. H. 2015. Serish inulin and wheat biopolymers interactions in model systems as a basis for understanding the impact of inulin on bread properties: A FTIR investigation. *Journal of Food Science and Technology*, **52**(12), 7964-7973.

[18] Kizil, R., Irudayaraj, J., & Seetharaman, K. 2002. Characterization of irradiated starches by using FT-Raman and FTIR spectroscopy. *Journal of agricultural and food chemistry*, **50**(14), 3912-3918.

Mina, J., Valadez-Gonzalez, A., Herrera-Franco, P., Zuluaga, F., & Delvasto, S. 2011. Physicochemical characterization of natural and acetylated thermoplastic cassava starch. *Dyna*, **78**(166), 174-182.

[19] Xiong J., Li Q., Shi Z., and Ye, J. 2017. Interactions between wheat starch and cellulose derivatives in short-term retrogradation: Rheology and FTIR study. *Food Research International,* **100,** 858-863.

Ramesh, S., Liew, C., & Arof, A. K. 2011. Ion conducting corn starch biopolymer electrolytes doped with ionic liquid 1-butyl-3-methylimidazolium hexafluorophosphate. *Journal of Non-Crystalline Solids*, **357**(21), 3654-3660.

[20] Gonzalez, J. A., Vallejo, J., Martínez, H. N., Cerecero, R., & López, L. 2019. Effect of chemical modification of titanium dioxide particles via silanization under properties of chitosan/potato-starch films. *Revista Mexicana De Ingeniera Quimica*, **18**(3), 913-927.

[21] Martín-Sánchez, M., Martín-Sánchez, M. T., & Pinto, G. 2013. Lugol reactive: History of discovery and teaching applications. *Educacion Quimica*, **24**(1), 31-36.

SEPIC - MPPT controller with photovoltaic energy for groundwater extraction

C L Corzo[1], R A Núñez[1], J I Flórez[2], P A Rodríguez[1] and K J Plata[1]

[1]Electronic Engineering Program, Unidades Tecnológicas de Santander -UTS-, Calle de los estudiantes # 9-82 Ciudadela Real de Minas, Bucaramanga, Colombia
[2]Master in Industrial Controls, Universidad de Pamplona, Calle 5 # 3-93, Pamplona, Colombia

ccorzo@correo.uts.edu.co

Abstract. The model-based design methodology is presented for the development of an experimental SEPIC-MPPT controller powered by a photovoltaic module, designed to extract water from a deep well using a DC submersible motor pump. To specify the proposed methodology, a photovoltaic panel was characterized under various conditions of solar radiation and the modeling of its Voltage-Current relationship was carried out. The best tilt angle for the installation was determined to be nine degrees (9°). Similarly, a SEPIC controller was developed from model-based design. The results obtained showed an average adjustment error of less than 4%, on the performance parameters contrasted in the modeled / experimental system. The efficiency achieved by the SEPIC-MPPT controller was 85.26%. It is concluded that the methodology implemented for the design is coherent and valid for the development of SEPIC-MPPT controllers with application in underground water extraction systems, in areas with photovoltaic potential.

Keywords: *Maximum power monitoring algorithm, Digital control, SEPIC controller, Submersible motor pump, Photovoltaic system.*

1. Introduction
The incorporation of renewable and clean energy sources is a priority generated by the depletion of conventional sources due to the overexploitation of natural resources [1], which is emerging as a viable alternative to solve the problem of electricity supply in non-interconnected areas existing in the Colombian rural sector; in which the data show that the average solar irradiation is around 4.3 KW / m2 [2] [3]. To take advantage of the energy potential, the scientific community has improved the efficiency of photovoltaic systems with technologies such as controllers with algorithms for monitoring the maximum power point [4] [5]. Therefore, this research proposed the development of an experimental prototype using the technique of model-based design, the integration of a solar panel as an energy source, an electronic power controller to improve the efficiency of the system and an algorithm control for monitoring the maximum power point. The methodology aims to develop a controller that applies in the efficient operation of a submersible motor pump to extract available underground water in a deep well, with solar energy.

IOP Conf. Series: Materials Science and Engineering 1154 (2021) 012042 doi:10.1088/1757-899X/1154/1/012042

2. Methodology

As a methodological strategy for the development of the research, the model-based design with the SIMULINK® tool was used, as well as the design of experiments for the development of a functional prototype of the SEPIC-MPPT controller, powered by photovoltaic energy, which used the Perturb & Observe technique for tracking the maximum power point. The experimental modeling and validation of the panel was carried out, with emphasis on the evaluation of the angle of inclination in the installation. In addition, the modeling and implementation of the SEPIC-MPPT controller was carried out with management of the Perturb & Observe algorithm, which was validated in the functional hardware prototype, developed on a programmable platform with ARM-CORTEX-M3 architecture. Static load modeling was also implemented (DC powered submersible motor pump) in which the characteristic curve that relates power in operation with flow and head was identified. The results obtained established the adjustment error between the modeled controller and the experimental controller and the efficiency of the system.

3. Materials and methods

For the modeling and development of the experimental prototype of the proposed controller, the characterization of a solar panel, the design of a DC-DC converter, an algorithm for monitoring the maximum power point and the characterization of water extraction systems were carried out. underground based on submersible motor pumps. These developments are described below.

3.1. Solar panel modeling

It started from the analysis to establish the parallel resistance and series [6], which is based on equations (1), (2) and (3) presented below.

$$I = I_{pv} - I_o \left[exp\left(\frac{V+R_sI}{\alpha V_t Ns}\right) - 1\right] - \frac{V+R_sI}{R_p} \tag{1}$$

$$I_o = \frac{I_{sc}+K_I\Delta_T}{exp\left(\frac{V_{oc}+K_V\Delta_T}{\alpha V_t}\right)-1} \tag{2}$$

$$I_{pv} = (I_{pvn} + K_I\Delta_T)\frac{G}{K_n} \tag{3}$$

Where I_{pv} is the photovoltaic current, I_o the nominal saturation current of the diode, V the voltage across the cell, α the ideality factor of the diode, V_t thermal voltage of the diode, Ns the number of cells in series, I output current through the cell, R_s series resistance, R_p parallel resistance, I_{sc} short circuit current, V_{oc} open circuit voltage, Δ_T difference between current and nominal temperature, G current solar irradiation, K_n nominal solar irradiation, K_I current coefficient and K_V voltage coefficient. The coefficients were assumed according to photovoltaic panels with similar characteristics. The expression corresponding to the parallel resistance was derived, which is presented in equation (4).

$$R_{p/min} = \frac{V_{mp}}{I_{scn}-I_{mp}} - \frac{V_{ocn}-V_{mp}}{I_{mp}} \tag{4}$$

Through the iteration of the expressions with the Newton-Raphson methodology, the values of the missing parameters were obtained, $R_p = 89.786278$, $R_s = 0.350000$, $a = 1.200000$, $R_{pv} = 6.024692$ y $R_{Io} = 4.763799e^{-13}$. With the data found, the panel was modeled (figure 1).

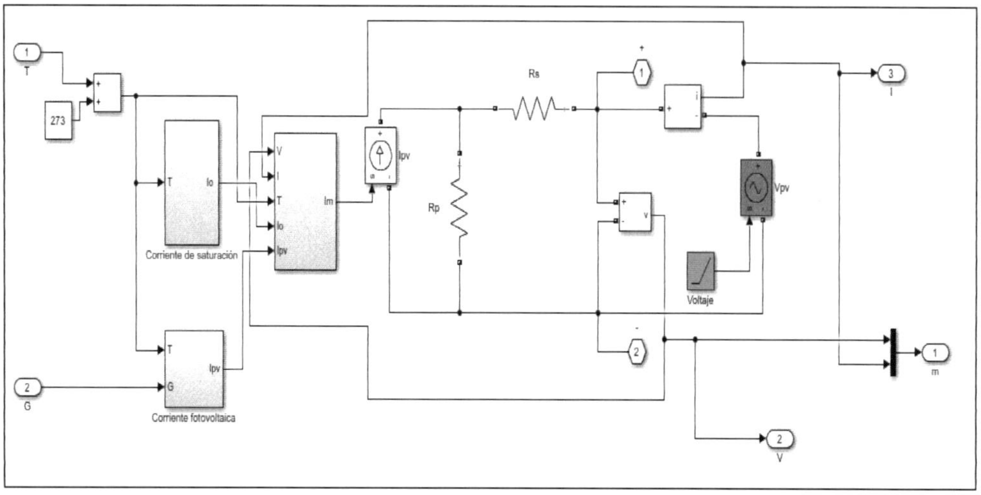

Figure 1. Complete model of the available monocrystalline solar panel.

To integrate a variable resistive load that would allow deriving the power characteristic of the solar panel, a SIMULINK® model was implemented, which connected eight 27 ohm and 20W resistors sequentially and in parallel, with the purpose of establishing the behavior of the panel model with variable load. The model made is presented in (figure 2).

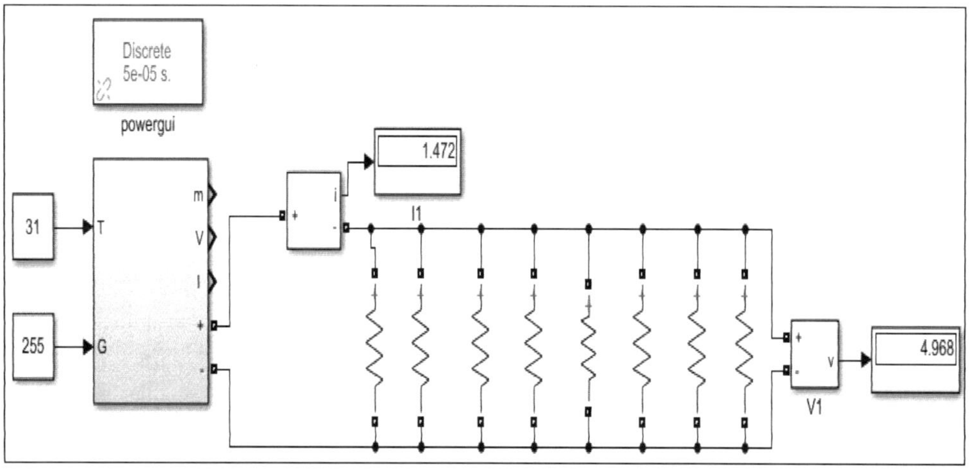

Figure 2. Model of eight resistors in parallel connected to the panel for load simulation

The procedure began with a resistive load connected in parallel with the output of the solar panel subjected to a radiation of 712 W / m^2, an inclination of nine degrees, an ambient temperature of 27 ºC and a panel temperature of 36 ºC. The process presented in [7] [8] was adapted to carry out the load change experimentally, through the sequential addition of resistors in parallel until reaching the minimum resistance value when connecting eight resistors. In each of the steps, the current and voltage values were taken, to contrast them with those generated in the experimental characterization of the same system.

3.2. Characterization of the photovoltaic panel available

This process started from the on-board specifications available in the panel used for the investigation, listed in (table 1). Similarly, a calibrated Middleton SK01-D pyranometer with a resolution of 1mV per W / m2 and a Datalogger connected to the CoolTerm serial acquisition program was used for radiation measurement.

Table 1. On-board specifications for monocrystalline solar panel available.

Specification	Value
Nominal Power (Pmax)	100 W
Output Tolerance (Pmax)	3 %
Max Voltage (Vmp)	18.0 V
Max Current (Imp)	5.56 A
Open circuit voltage	22.32 V
Short circuit current	6.0 A

The proper tilt of the panel with respect to the horizontal global solar radiation was verified by the response of the panel at three different tilt angles (0º, 9º and 15º) based on the latitude of the panel location, which was 7º 06 '19, 05 ". Experimentally, the power curves obtained in the three test angles were compared with simulated loads using eight resistors of 27 ohms and 20W and a variation step of 3.375 ohms.

3.3. SEPIC controller modeling

The SEPIC controller modeling considered contributions from the comparison carried out in [9] and the analyzes presented in [10] [11] as a starting point. The design requirements considered the parameters Minimum input voltage = 15 V, Maximum input voltage = 22 V, Diode conduction voltage = 0.7 V, Output voltage = 12.6 V and Output current = 1.1 A. With these considerations performed the calculations described below.

3.3.1. Duty cycle selection. The operation of the controller in continuous conduction was chosen, the maximum and minimum critical duty cycles were considered and calculated, through equations (5) and (6).

$$Dmax = \left(\frac{(vout)+(vD)}{(vinMin)+(vout)+(vD)}\right) = \left(\frac{(12.6)+(0.70)}{(15)+(12.6)+(0.70)}\right) = 0.47 \quad (5)$$

$$Dmin = \left(\frac{(vout)+(vD)}{(vinMax)+(vout)+(vD)}\right)\left(\frac{(12.6)+(0.70)}{(22)+(12.6)+(0.70)}\right) = 0.3768 \quad (6)$$

3.3.2. Inductance selection. For the calculation of inductances, a peak-to-peak current oscillation of 40% was considered in relation to the maximum input current at minimum voltage. Equations (7), (8), (9) and (10) were used for the calculation.

$$DeltaIL = \left((Iout) * \left(\frac{vout}{vinMin}\right) * (0.4)\right) = \left((1.1) * \left(\frac{12.6}{15}\right) * (0.4)\right) = 0.3696A \quad (7)$$

$$L1 = L2 = L = \left(\frac{(vinMin)*(Dmax)}{(DeltaIL)*(fsw)}\right) = \left(\frac{(15)*(0.47)}{(0.3696)*(90000)}\right) = 211.92 * 10^{-6}H \quad (8)$$

$$iL1pico = \frac{(Iout)*(vout+vD)*(1.2)}{(vinMin)} = \frac{(1.1)*(12.6+0.70)*(1.2)}{(15)} = 1.1704A \quad (9)$$

$$iL2pico = (Iout) * \left(1 + \frac{(0.4)}{(2)}\right) = (1.1) * \left(1 + \frac{(0.4)}{(2)}\right) = 1.3200A \quad (10)$$

3.3.3. Mosfet transistor selection. Minimum threshold voltage Vth (min), ignition resistance RDS (ON), gate drain load QGD and maximum drain voltage to source, VDS (max) were used as base parameters in the selection. The RFZ44N transistor was chosen, which has Rds_{on} = 0.032 ohms, Qgd = 25 * 10-9, Vds = 5V, Ig = 1.47A. Based on these values, the current and power requirements of the switch were calculated using equations (11), (12) and (13).

$$IQ1pico = iL1pico + iL2pico = 1.1704 + 1.32 = 2.4904A \tag{11}$$

$$IQ1rms = \left((Iout) * \sqrt{\left(\frac{(vout+vinMin+vD)*(vout+vD)}{(vinMin)^2}\right)}\right) = 1.4227\ A \tag{12}$$

$$pQ1 = \left(\begin{matrix}(IQ1rms)^2 * (Rds_{on}) * (Dmax) + \\ (vinMin + vout) * (IQ1pico) * \left(\frac{(Qgd)*(fsw)}{Ig}\right)\end{matrix}\right) = 0.1356\ W \tag{13}$$

3.3.4. Output diode selection. In the SEPIC, the current that flows through the diode is the same that passes through the MOSFET, which is the maximum current of the switch. The minimum peak voltage across the diode was calculated using equation (14).

$$vRD1 = vinMax + vout = 22 + 12.6 = 34.6V \tag{14}$$

The Schottky diode reference SR350 was chosen, which exceeds the calculated parameter.

3.3.5. Coupling capacitor selection. The selection of the capacitor (Cs) depends on the RMS current, and the ripple voltage (DeltaCs). Equations (15), (16) and (17) were used to calculate this component.

$$Icsrms = Iout * \sqrt{\left(\frac{vout+vD}{vinMin}\right)} = 1.1 * \sqrt{\left(\frac{12.6+0.70}{15}\right)} = 1.0358A \tag{15}$$

$$DeltaCS = \left(\frac{(Iout)*(Dmax)}{(CS)*(fsw)}\right) \tag{16}$$

$$CS = \left(\frac{(Iout)*(Dmax)}{(DeltaCS)*(fsw)}\right) = \left(\frac{(1.1)*(0.47)}{(1.16*10\text{^}-3)*(90000)}\right) = 5000uF \tag{17}$$

3.3.6. Output capacitor selection. The output capacitor must be able to handle the RMS current calculated by equation (17) and flowing through the MOSFET switch. The calculation of the capacitor was carried out using equations (18) and (19).

$$ESR = \left(\frac{(vripple*12.6)*(0.5)}{iL1pico+iL2pico}\right) = \left(\frac{(0.02*12.6)*(0.5)}{1.1704+1.32}\right) = 0.0506 \tag{18}$$

$$Cout = \left(\frac{(Iout)*(Dmax)}{(vripple)*(12.6)*(0.5)*(fsw)}\right) = \left(\frac{(1.1)*(0.47)}{(0.02)*(12.6)*(0.5)*(90000)}\right) = 45.58uF \tag{19}$$

3.3.7. Input capacitor selection. The waveform of the input current is triangular so the inductor ensures that the ripple seen by the capacitor is low. The RMS current in the input capacitor is defined by equation (20) and a minimum value of 10uF is established for this component as it is not a critical component.

$$ICinrms = \left(\frac{DeltaIL}{\sqrt{(12)}}\right) = \left(\frac{0.3696}{\sqrt{(12)}}\right) = 0.1067A \tag{20}$$

With the calculations carried out, the SEPIC controller model was generated in which an input for the useful cycle signal generated by a Perturb & Observe algorithm, and an output multiplexed with voltage and current signals for the respective measurements, was provided. The SIMULINK® model implemented for the controller is presented in (figure 3).

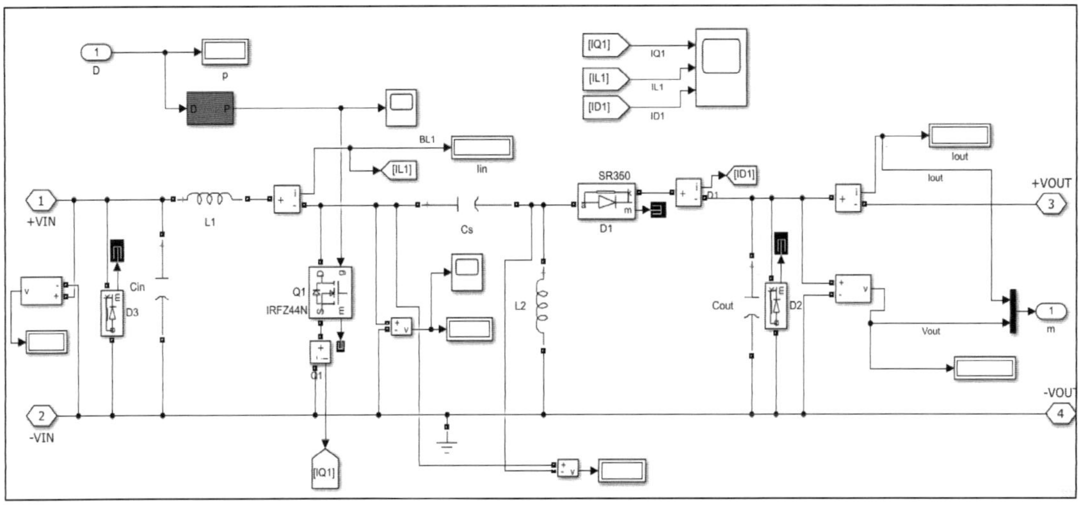

Figura 3. Modelo del controlador SEPIC.

3.4. SEPIC-MPPT controller implementation

The controller was implemented on the SAM3X8E processor with ARM Cortex-M3 architecture, available on an Arduino-DUE development board, which has 512 KBytes of program memory, 96 KBytes of data memory and multiple input / output modules. This processor has a 32-bit data bus, and operates at a frequency of 84 MHz, the circuit components calculated for modeling the controller were also integrated on a PCB, as well as the voltage and current sensor modules for the corresponding validations. The experimental prototype is shown in (figure 4).

Figura 4. *Prototipo experimental SEPIC-MPPT implementado*

3.5. Static modeling of the motor pump for groundwater extraction

The hydraulic behavior of a motor pump relates the flow with parameters such as required operating power, total manometric head, operating voltage and current, among others [12] [13]. In the absence of these values, an experimental characterization was necessary to determine the dynamics of the device, in which the power was established as an independent variable and as dependent variables the flow, the height at which the water extraction is carried out, the performance, operating voltage and current. Based on the stated method and based on the technical specifications available in the data sheet of a DC powered submersible motor pump (Table 2), an experiment was designed to derive the required characteristics.

Table 2. Technical specifications of DC powered submersible motor pump, available.

Specification	Value
Material	ABS-Steel
Size	(80x48x63) mm
Input diameter	12 mm
Output diameter	6.9 mm
Voltage	6-12 V
Nominal Current	1.2 A
Power	18.8 W
Max Flow	700 L/h
Water Head Max.	5 m
Water temperature	60 ºC

In the implemented experiment, two six-liter tanks were located for water storage. One fixed and the other with the possibility of changing the height. The motor pump was energized with 12.68 Volts to fill the tank with variable height and ten repetitions were performed for each height. The fill time, flow rate, current and power absorbed by the motor pump were measured and averaged. The measured values are presented in (table 3).

Table 3. Experimental characterization of the submersible motor pump

Power (W)	Current (A)	*Flow (L/S)*	Height (m)	Time (S)
10,66	0,841	0,012265508	4,00	489,2
10,79	0,851	0,018525310	3,67	323,9
11,72	0,924	0,041097441	3,09	146,0
13,33	1,051	0,078027236	2,46	76,90
13,95	1,100	0,098699101	1,82	60,80
14,07	1,110	0,106203008	1,22	56,50
14,38	1,134	0,123076923	0,60	48,75

With the data of the experimental characterization and with the use of the MATLAB® CFTOOL tool, the characteristic curves of the submersible motor pump were generated, which are presented in the results section.

3.6. Implementation of the Perturb & Observe algorithm

The P&O algorithm is initialized with the product of the current voltage of the photovoltaic panel, and the duty cycle that changes in constant increments [14-16]. The algorithm establishes that, if the difference between the current power and the previous power is positive and the difference of the current voltage with the previous voltage is also positive, the duty cycle is decreased to increase the output voltage so that the MPPT. If the difference in powers was positive and the difference in voltages was negative, the duty cycle is increased until the MPPT is reached. If both differences are negative, the control action to take is to decrease the duty cycle to increase the output voltage of the solar panel until it reaches the MPPT. Finally, if the power difference is negative and the voltage difference was positive, the duty cycle is increased until the maximum power point is reached [17]. The flow diagram of the Perturb & Observe algorithm integrated into the controller model is presented in (figure 5).

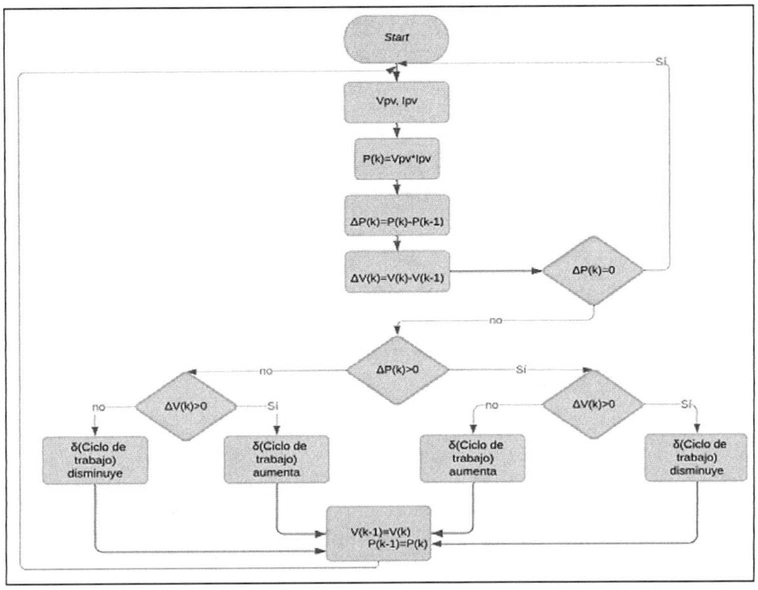

Figure 5. Perturb and Observe algorithm flow chart built into controller

4. Results

After validating the proposed implementations for both the modeling and the experimental prototype of the different components that make up the SEPIC-MPPT controller, the results achieved are presented. The modeling of the solar panel and simulated resistive loads were compared with the actual connected resistive loads. Similarly, the modeled controller was contrasted with the experimental prototype and its operation was validated without the P&O - MPPT algorithm and with the integration of the algorithm. The results obtained for each component are presented below.

4.1. Solar panel modeling

The solar panel model developed in SIMULINK® based on the Standard Test Conditions -STC-, generated the power curves presented in (figure 6). The solar radiation curve corresponding to 1000 W / m^2 (upper curve) was taken as a reference to contrast with the results of the characterization of the solar panel.

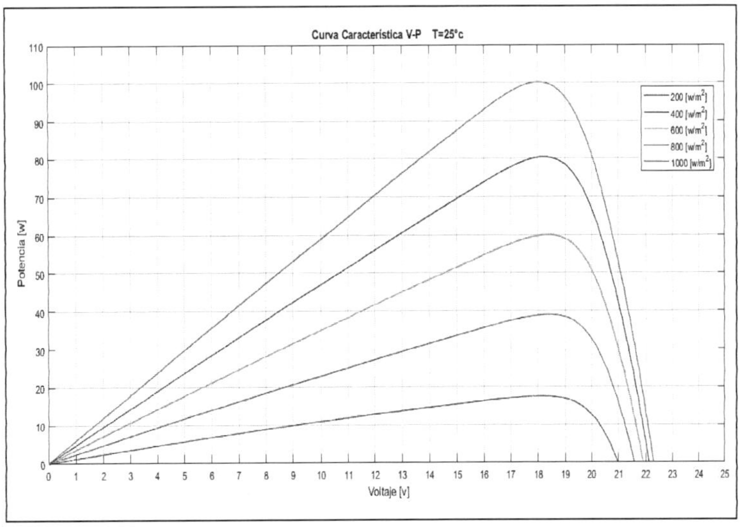

Figure 6. Solar panel power curves modeled by SIMULINK ®

4.2. Experimental characterization of the solar panel

The curves obtained for the inclination angles used in the validation with solar radiation of 1000 W / m^2 were derived from measurements made with the sun at the zenith. In (figure 7) the response for the three values of inclination tested are presented.

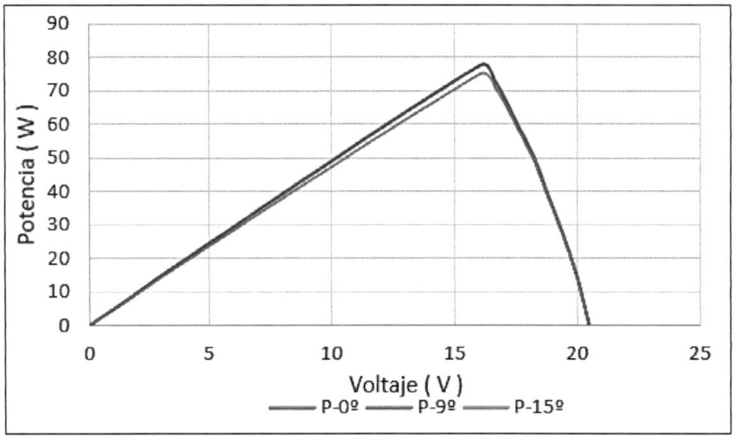

Figure 7. Result of the experimental characterization of the solar panel.

When comparing the results obtained in the modeling with the power values measured in the solar panel installed with an inclination of nine degrees and with the sun at the zenith (1000 W / m^2), a small difference is appreciated between the power generated and the characteristic delivered by the simulation. This difference is tolerable and valid to consider the model developed in the implementation of the SEPIC controller.

4.3. Characteristic Voltage - Current with resistive loads

The curves obtained with the simulated and real resistances connected in parallel with a radiation of 712 W / m^2 were compared to establish the corresponding adjustment error. The results obtained for the Voltage - Current characteristic with the resistive load connected in parallel to the panel are presented in (figure 8).

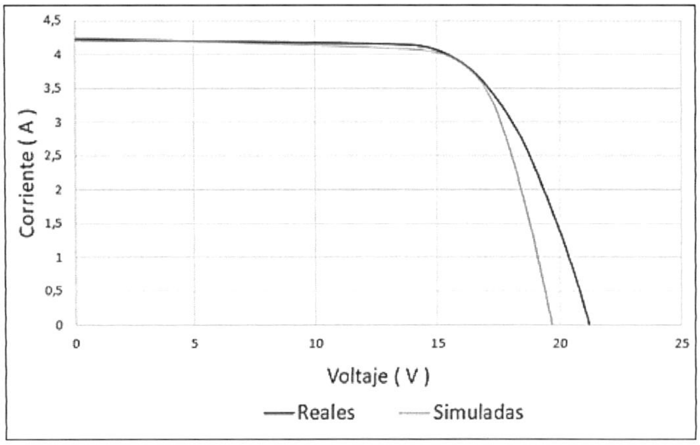

Figure 8. Characteristic Voltage - Current with resistive loads

The average adjustment error found in the comparison between the simulated model and the experimental model with the resistors connected was 2.59% for voltage and 2.70% for current.

4.4. Static modeling of the submersible motor pump

From the experimental characterization of the motor pump, the flow/head (figure 9) and flow/power/head (figure 10) curves for groundwater extraction were generated using the Cftool and Fitting Tool of MATLAB®.

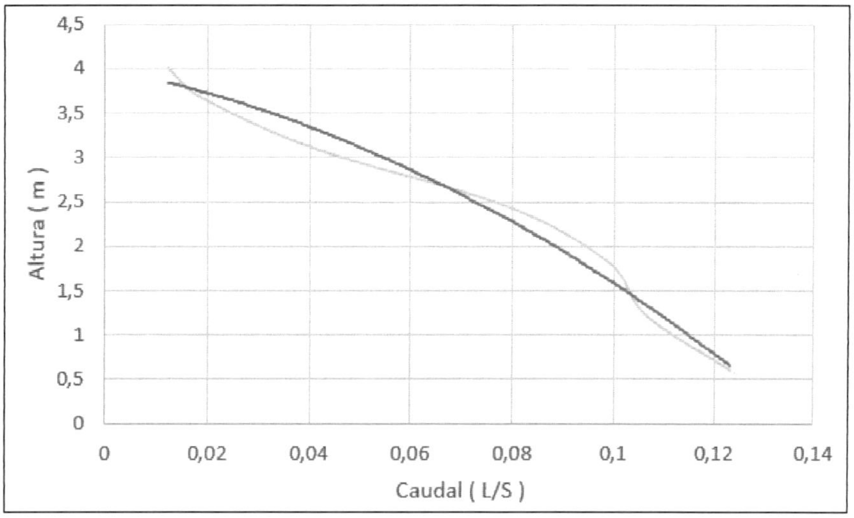

Figure 9. Characteristic Flow/Height of the submersible motor pump

Based on the recorded data, the function $h(c) = -129.4c^2 - 11.21c + 3.995$ with determination coefficient $R^2 = 0.984$ was generated, which describes the dynamics of the submersible motor pump to integrate it into the system model. It is inferred from (figure 9) the decrease in flow rate as the extraction head increases.

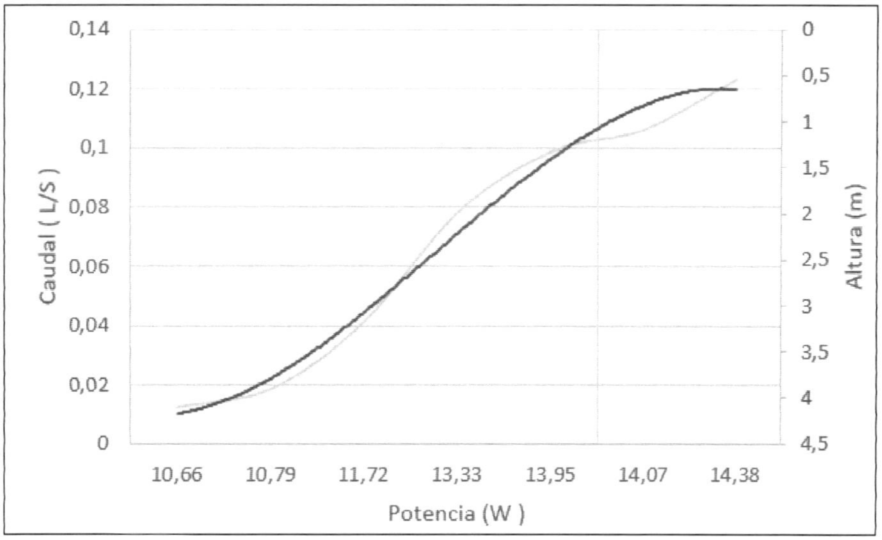

Figure 10. Characteristic Power/Flow/Height of the submersible motor pump

The function $c(p) = -0.001p^3 + 0.0108p^2 + 0.0138p + 0.0141$ was established with coefficient of determination $R^2 = 0.9862$ that describes the behavior of the flow as a function of the power, which is modulated by the height extraction.

4.5. SEPIC-MPPT controller modeling

The SEPIC-MPPT controller (figure 11) was implemented with the calculated components and the P&O algorithm with a frequency of 82 KHz.

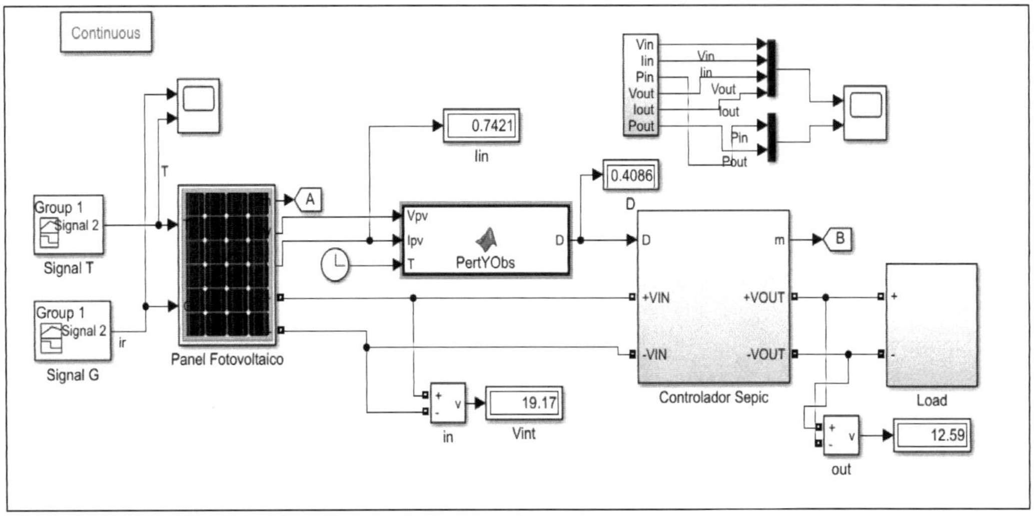

Figure 11. Complete SEPIC-MPPT experimental model in SIMULINK ®

The voltage, current and power values at the input and output of the controller, as well as the radiation and temperature were recorded using a Datalogger and averaged for analysis. The calculated averages are presented in (table 4).

Table 4. Performance parameters of the modeled / experimental system

Parameter	Modeling		Experimental		Error %		
	No-MPPT	MPPT	No-MPPT	MPPT	No-MPPT	MPPT	Prototype
Vin (V)	20,17	20,25	19,38	20,06	-4,08	-0,95	-3,30
Iin (A)	0,58	0,71	0,60	0,72	3,33	1,39	-1,40
Pin (W)	11,63	14,35	11,70	14,62	0,60	1,85	0,68
Vout (V)	11,24	12,53	11,21	12,55	-0,27	0,16	2,68
Iout (A)	0,93	1,04	0,96	1,02	3,12	-1,96	2,59
Pout (W)	10,49	13,09	10,75	12,84	2,42	-1,95	2,24
Rad. W/m^2	980,86	998,23	980,86	998,23	0,00	0,00	0,00
Temp. °C	24,53	24,56	24,53	24,56	0,00	0,00	0,00

The evaluation of the performance parameters of the modeled / experimental controller was carried out in reference to the Standard Test Conditions -STC-. The results obtained indicate an oscillation in the adjustment error of less than ± 5% for the system without MPPT algorithm. For the system with the MPPT algorithm, the oscillation in the adjustment error was less than ± 2%. In order to obtain absolute indicators, the results of the test with MPPT and without MPPT were averaged for the experimental prototype and a variation of less than 4% was determined.

To validate the efficiency of the experimental controller, three measurements were made with different solar radiation and operating temperature, in which both the input power and the output power were measured. The measurements were made with the application of the MPPT algorithm and without the application of the MPPT algorithm. In (table 5) the results obtained are presented, which show an efficiency of 65.61% for the controller without MPPT and 85.26% for the controller with MPPT.

Table 5. Efficiency of the SEPIC-MPPT experimental system.

Parameter	No-MPPT			MPPT		
Rad. W/m^2	320,00	735,00	830,00	330,00	730,00	763,00
Temp. ºC	36,50	27,75	30,50	29,75	29,00	29,25
Pin (W)	7,72	10,45	10,51	14,20	14,07	14,33
Pout (W)	4,90	6,94	7,04	12,65	11,50	12,18
η %	63,47	66,36	66,99	89,07	81,73	84,99

5. 5. Conclusions

Based on the results obtained in reference to the methodology implemented to develop an experimental SEPIC-MPPT controller powered by photovoltaic energy with application in the extraction of groundwater, the following conclusions were derived.

5.1. Solar panel tilt angle

It is inferred from the inspection on the curves presented in (Figures 6 and 7), that the highest power level was reached with the installation of the panel at nine degrees (9º) based on the latitude of the geographical location, which was 7 06'19.05 ".

5.2. Load resistor bank

It is concluded based on the results with respect to the simulated / experimental resistance bank, that the adjustment error of the Voltage - Current characteristic generated (figure 8) is less than 3%, which indicates coherence between the simulation and the characterization, that validates the design specifications integrated into the model in reference to the behavior of the solar panel.

5.3. Static model of the motor pump

The model derived from the experimental evaluation of the submersible motorized pump presented a coefficient of determination R^2 of 98.4% (figure 9), in the adjustment to the dynamics of the motorized pump, which validates the result for its integration into the experimental controller model. There was also evidence of a restriction in the extraction height (figure 10), which imposes a minimum height for the depth of groundwater extraction from 0.5 meters, which is equivalent to 12.5% of the total depth evaluated.

5.4. SEPIC-MPPT controller

From the results obtained, the complete model of the SEPIC-MPPT controller (figure 11) and the experimental prototype (figure 4) were developed on which the performance parameters were evaluated (table-4). It was evidenced from the results that the MPPT algorithm decreased the percentage of error in the performance parameters (± 2%) in contrast to the performance of the controller without MPPT (± 5%). By averaging the results, it is concluded that the average adjustment error was less than ± 4%, so that the developed solution is coherent and valid for extrapolation to higher operating conditions.

5.5. SEPIC-MPPT Controller Efficiency

In relation to the efficiency of the experimental controller and based on the results obtained by subjecting the system to various radiation and temperature conditions (table 5), the efficiency achieved by the controller without MPPT was 65.61%, while for the controller with MPPT, the efficiency achieved was 85.26%. It is concluded from these results that the SEPIC-MPPT controller is feasible to be implemented in underground water extraction systems since its loss is less than 15%, which represents a high level of use of the photovoltaic potential available in the study region.

References

[1] Rodríguez D J, Hernández J y Jaramillo A 2017 Estimation of Mean Monthly Global Solar Radiation Using Model Based on Sunshine Hours for Colombia *2017 IEEE 44th Photovoltaic Specialist Conference (PVSC), Washington, DC* pp 1143-1146

[2] Duque E, Isaza A, Ortiz P, Chica S, Lujan A and Molina J 2017 Urban sets innovation: Design of a solar tree PV system for charging mobile devices in Medellin — Colombia *2017 IEEE 6th International Conference on Renewable Energy Research and Applications (ICRERA), San Diego, CA* pp. 495-498

[3] Diaz R R, Alarcón A L y and Moreno R J 2013 Monitoring system for global solar radiation, temperature, current and power for a photovoltaic system interconnected with the electricity distribution network in Bogota *2013 IEEE 56th International Midwest Symposium on Circuits and Systems (MWSCAS), Columbus, OH* pp 485-488

[4] Azeem I, Baig M M and Uddin M H 2018 A Strategy to Evaluate MPPT Techniques *2018 Asian Conference on Energy, Power and Transportation Electrification (ACEPT), Singapore, 2018* pp. 1-4

[5] Singh A and Shimi S L 2017 MATLAB / SIMULINK Simulation of PV System Based on MPPT in Variable Irradiance with EV Battery as Load *2017 IEEE International Conference on Computational Intelligence and Computing Research (ICCIC), Coimbatore* pp. 1-4

[6] Fares M A, Atik L, Bachir G and Aillerie M 2017 Photovoltaic panels characterization and experimental testing *Energy Procedia* **119** 945–52

[7] Chacón J J, Ortiz R A, Archila J E, Mantilla M A, Botero M A and Petit J F 2017 Prototype for the characterization of photovoltaic panels based on a SEPIC converter *2017 IEEE Workshop on Power Electronics and Power Quality Applications (PEPQA)* pp 1–6

[8] Brito E M d. S, Antônio A d. S, Cupertino A F and Pereira H A 2014 Characterization of solar panel using capacitive load *2014 11th IEEE/IAS International Conference on Industry Applications* pp 1–7

[9] Soedibyo, Amri B and Ashari M 2015 The comparative study of Buck-boost, Cuk, Sepic and Zeta converters for maximum power point tracking photovoltaic using P&O method *2015 2nd International Conference on Information Technology, Computer, and Electrical Engineering (ICITACEE)* pp 327–32

[10] Deepak, Pachauri R K and Chauhan Y K 2016 Modeling and simulation analysis of PV fed Cuk, Sepic, Zeta and Luo DC-DC converter *2016 IEEE 1st International Conference on Power Electronics, Intelligent Control and Energy Systems (ICPEICES)* pp 1–6

[11] Gali V and Amrutha P B 2016 Fast dynamic response of SEPIC converter based photovoltaic DC motor drive for water pumping system *2016 International Conference on Circuit, Power and Computing Technologies (ICCPCT)* (IEEE) pp 1–5

[12] Yadav K, Kumar A, Sastry O S and Wandhare R 2019 Solar photovoltaics pumps operating head selection for the optimum efficiency *Renew. Energy* **134** 169–77

[13] Li G, Jin Y, Akram M W and Chen X 2017 Research and current status of the solar photovoltaic water pumping system – A review *Renew. Sustain. Energy Rev.* **79** 440–58

[14] Saleh A, Azmi K S F, Hardianto T and Hadi W 2018 Comparison of MPPT Fuzzy Logic Controller Based on Perturb and Observe (P&O) and Incremental Conductance (InC) Algorithm On Buck-Boost Converter *2018 2nd International Conference on Electrical Engineering and Informatics (ICon EEI)* pp 154–8

[15] Majstorović M, Mršević D, Đurić B, Milešević M, Stević Z and Despotović Ž V 2020 Implementation of MPPT Methods with SEPIC Converter *2020 19th International Symposium INFOTEH-JAHORINA (INFOTEH)* pp 1–6

[16] Kokare A, Patil S and Bacchav L 2017 Implementation of a highly efficient MPPT technique for a PV system using sepic converter *2017 International Conference on Information, Communication, Instrumentation and Control (ICICIC)* pp 1–5

[17] Nigam A and Gupta A K 2016 Performance and simulation between conventional and improved perturb & observe MPPT algorithm for solar PVcell using MATLAB/Simulink *2016 International Conference on Control, Computing, Communication and Materials* pp 1–4

IOP Conf. Series: Materials Science and Engineering 1154 (2021) 012043 doi:10.1088/1757-899X/1154/1/012043

Preparation and morphological and physicochemical characterization of pillared clays from bentonite

A C Padilla[1], J Castro-Suarez[2] and F Colpas-Castillo[1]

[1] Grupo de investigación Carboquimica, Programa de Química, Facultad de Ciencias Exactas y Naturales, Universidad de Cartagena, Colombia.

[2] Area de Basicas Exactas, Universodad del Sinú, Seccional Cartagena, Colombia

Email: fredycolpas@hotmail.com

Abstract. Mineral grade and commercial grade bentonite pillared clays were prepared with SnO_2. This is to create adsorbents for removal of contaminants from water. The initial and modified clays were characterized by X-ray fluorescence (XRF), infrared spectroscopy (FT-IR), scanning electron microscopy (SEM) and nitrogen sortometry. The results of the chemical analysis (XRF) clearly confirm the incorporation of tin in the modified materials. The analysis by SEM and nitrogen sortometry verified the successful modification of the clays by incorporation of tin dioxide species, and the appearance of mesoporous structures with an increase in the values of surface area and porosity.

1. Introduction

In recent years, the so-called emerging pollutants (or micropollutants) have attracted considerable interest. They are compounds of diverse origin and chemical nature, whose presence and consequences in the environment have gone unnoticed. They are present in water in low concentrations of ngL^{-1} to μgL^{-1} and are considered very harmful to human health and the environment. They can cause various effects in organisms, such as chronic toxicity, endocrine disruption and bioaccumulation [1].

One of the main problems of this type of pollutant is that current wastewater treatment plants are unable to eliminate them. These are designed to remove organic matter and nutrients in higher concentrations (gL^{-1}). Therefore, emerging pollutants are present in surface waters, groundwater and for human consumption [2].

The methods used to treat emerging pollutants can be classified into three families: physical-chemical treatments, biological treatments, hybrid processes and advanced oxidation processes (AOPs). One of the most studied physical-chemical treatments for water treatment are adsorption processes, especially for the elimination of organic pollutants. That is why they have also been widely studied for their application in the elimination of emerging pollutants [8].

The most commonly used adsorbents are activated carbons: granular activated carbon (GAC) or powdered activated carbon (PAC). their main advantage is that they do not generate toxic products and have a high adsorption capacity [3-8]. The adsorption efficiency is a function of the properties of the adsorbent used, such as surface area, morphology and surface chemistry [4].

There are also other studies using other types of adsorbents, such as zeolites, clays or carbon nanotubes. Among the clay minerals, there are smectites or montmorillonites, which are characterized by having a moderate interlaminar load with good expansion capacity, which allows them to "swell" when they are introduced in an aqueous medium and to be chemically modified by intercalation-pillarization with bulky molecular species [5].

Pillarization consists of introducing, in the interlaminar space of a smectite, a very voluminous polycation that, after calcination, gives rise to a stable oxide that determines a fixed and permanent porosity of controlled size (molecular sieves). The fields of application of these materials (pillared clays) are very wide due to the possibility of intercalating different types of pillars, making them suitable for specific adsorption, molecular separation, or heterogeneous catalysis processes in a wide range of reactions [6].

Due to the above, it is very important to find alternative solutions for the degradation of emerging pollutants present in bodies of water. The objective of this work is to obtain from the preparation of new materials such as pillared clays, to use them in the degradation of emerging pollutants.

2. Materials and methods

2.1. Materials

The mineral bentonite (Bt) used in this study was supplied by Mining and Commercial Projects of Colombia S A S (Puerto Colombia, Atlántico) and the commercial Na-K-Bentonite was acquired from Sigma-Aldrich. Mineral bentonite and commercial bentonite were pretreated with a Na_2CO_3 solution (0.5M) for 24h and washed thoroughly with copious amounts of deionized water. The pillarizing agent was prepared by the dropwise addition of 50 mL chloride of tin (II), SnCl2 · $2H_2O$ (0.4M) to 50 mL of KOH (0.4M) with continuous stirring at 70 °C for 24h. 5.0g of Bt were mixed with 400 mL of deionized water and then 100 mL of the pillarizing agent was added to the clay suspension, with continuous stirring at 60 °C for 24 hours. The result of the suspension was washed with abundant deionized water by means of centrifugation around 6 replicates. The collected material was dried at 80 °C for 24h and calcined at 500 °C for 3h. This process was performed for both mineral grade bentonite and commercial grade bentonite.

2.2. Methods

Nitrogen isotherms were measured at -196 °C using a Gemini 2375 instrument (Micromeritics). Before each measurement the samples were degassed under vacuum at 160 °C for 4h. The surface area (BET) was calculated from the isotherms. Spectroscopy (FT-IR) was performed in the Nicolet iS10 spectrophotometer, by means of the preparation method of pellet. The KBr pellet with the sample was subjected to heating at 100 °C for 1h.

Scanning electron microscopy (SEM) was performed in an FEI microscope (ESEM) Quanta 200, with a resolution of 10 μm at 25 kV. The analyzes were carried out with the samples previously subjected to heating at 100 °C for 1h. A thin layer of gold was coated, as it was a poorly conductive sample. The elemental chemical analysis was carried out on an X-ray fluorescence spectrometer, MagixPro PW-2440 Philips (WDXRF) equipped with a Rhodium tube, with a maximum power of 4KW. The semi-quantitative analysis was carried out with the SemiQ 5 software, making 11 scanning, in order to detect all the elements present in the sample.

3. Results and discussion

3.1. Samples' characterization

The changes in porosity when pillared are reflected in nitrogen absorption, see Figure 1 and 2. The isotherms present higher nitrogen adsorption volumes in the pillared bentonites than in the starting bentonites, highlighting the increase in the porosity of bentonites as a consequence of tin modification. The shape of all isotherms allows them to be classified as type II, according to the IUPAC classification [7]. This is typical of solids with macro and mesopores. Furthermore, the desorption branch of the isotherms forms hysteresis type H3, which is attributed to pores in the form of slits with non-uniform sizes [7, 9, 10].

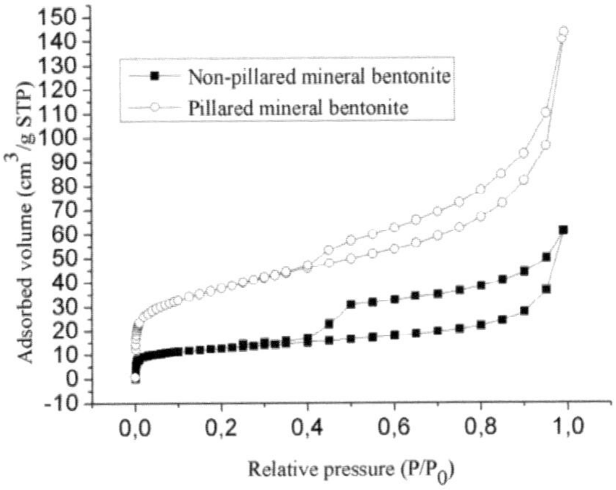

Figure 1. Comparison of the adsorption isotherms of the mineral bentonite.

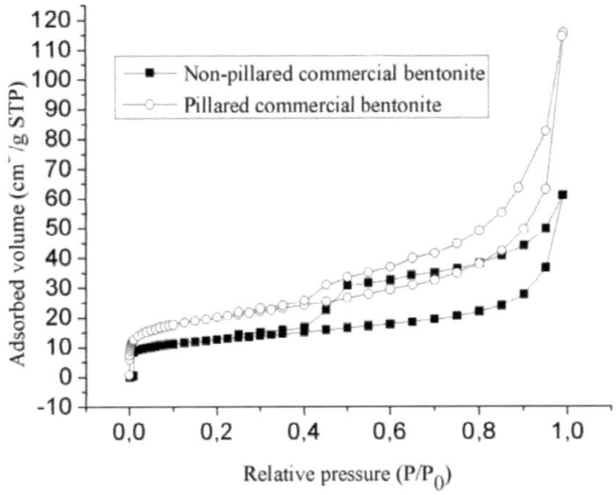

Figure 2. Comparison of adsorption isotherms of commercial bentonite.

The parameters of the porous structure calculated with the StarDriver V2.03 software are shown in Table 1. As can be seen, the surface area and the volume of the micropore increased almost two times after pillarization. The increase in surface area and porosity must be attributed not only to the pillarization process, but also to the deposition of tin oxide on the outer surface of the bentonite flakes and the aggregation of these flakes. The micropore volume value of unpillared commercial bentonite is due to the drying process that typically adds soda ash (anhydrous sodium carbonate). This serves to improve the expansion properties of bentonite, which is then processed through coal or gas rotary dryers, to which, if too much heat is applied, many of the properties of bentonite can be destroyed; in this case, the micropore volume was affected by this process.

Table 1. Comparison of the textural properties of mineral and commercial bentonite.

	BET surface area (m^2g^{-1})	Micropore area (m^2g^{-1})	External surface area (m^2g^{-1})	Pore size (nm)	Micropore volume (cm^3g^{-1})
Bt-min not pillared	44	13	31	4.20	0.005
Bt-min pillared	131	21	109	3.30	0.010
Bt-com not pillared	4	0,62	4	24.02	0.000
Bt-com pillared	70	16	53	4.99	0.007

The spectra (FT-IR) are shown in Figure 3, where you can see the characteristic bands at 3620 cm^{-1}, this correspond to stretching vibrations of structural OH groups of aluminum in octahedral coordination; the bands observed at 1635 cm^{-1} are the deformations of the O-H bond of the adsorbed water and the bands observed at 1019 and 980 cm^{-1} correspond to stretching vibrations of Si-O in the structure of the clay mineral [9, 10]. It can be seen that there are small displacements in the bands at the time of pillarizing the clays, this is due to the stretching vibrations of Sn-O observed at 548 and 552 cm^{-1}, which gives us indications that there was a modification in the clay mineral structure.

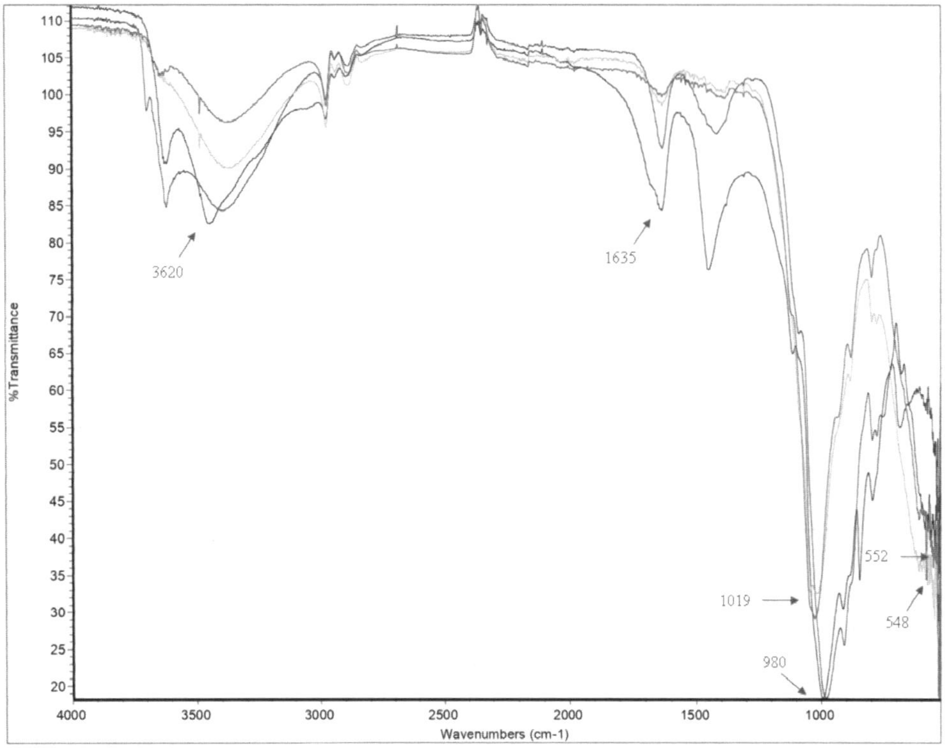

Figure 3. Infrared spectra of mineral bentonite and commercial bentonite.

The SEM micrographs of the starting mineral bentonite and the pillared mineral bentonite are shown in Figures 4 and 5. These images reveal the variations in morphologies. The surface of the starting bentonite is smooth with respect to the pillared bentonite, which shows a less smooth surface, this is due to the formation of grains with highly eroded surfaces and can be associated with the delamination of the mineral [9,10].

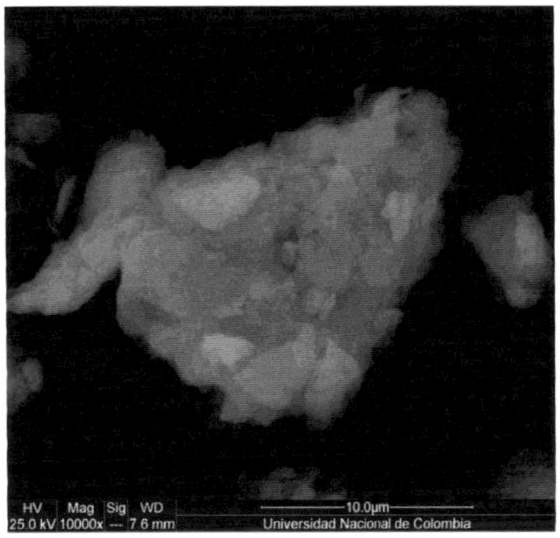

Figure 4. Micrograph of mineral bentonite without pillars.

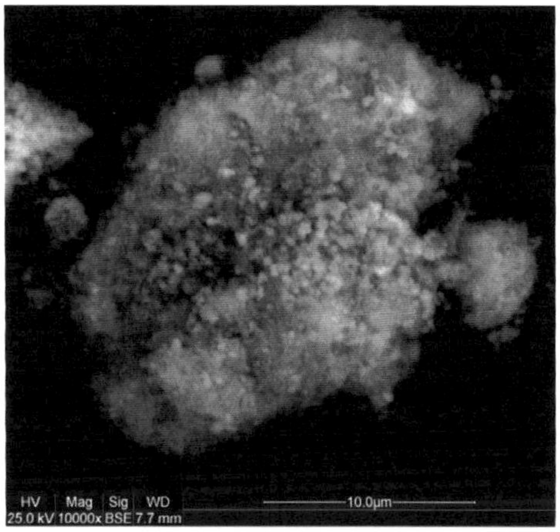

Figure 5. Micrograph of mineral bentonite pillared.

X-ray fluorescence analysis was performed to identify and quantify the elements that were present in both the starting bentonite and the pillared bentonite. Table 2 shows the starting bentonite that has a high content of silicon (52.44%) and aluminum (20.01%), this being the typical composition of a montmorillonite, in this case bentonite. A high content of sodium (10.86%) can also be seen, which is characteristic of sodium bentonites. From the analysis of the pillared bentonite, table 2 shows an increase in potassium (1.01%) and a notable decrease in sodium (0.46%) with respect to the starting bentonite. This is due to the cation exchange that took place. A high content of tin (28.79%) is also seen, evidencing the effective incorporation of SnO_2 species in the pillarization process. In addition, table 2 shows a slight decrease in silicon (41.1%) and aluminum (13.91%) and this is attributed to the incorporation of the pillar.

Table 2. Chemical analysis of the starting bentonite and the pillared bentonite.

Elemento y/o Compuesto	Bt-min $_{\text{without pillars}}$ (%)	Bt- min $_{\text{pillared}}$ (%)
Sn	-	28,79
Si	52,44	41,1
Al	20,01	13,91
Na	10,86	0,46
Fe	8,25	10,35
Mg	5,90	1,13
Ca	1,80	0,92
K	0,74	1,01
Ti	-	1,45
Cl	-	0,14

4. Conclusions

According to the results obtained in this work, it is concluded that the bentonites modified by the incorporation of tin oxide through the pillarization process showed an improvement in the textural parameters, increasing the volume of the micropores, the surface area and generating controlled porosity in solids. This indicates that the SnO_2 particles not only intercalated within the space of the interlayer of clay, but also precipitated on the outer surface of the clay flakes.

Acknowledgements

Thanks to the instrumental analysis laboratories of the University of Cartagena. Fuel and Energy Research Laboratory (LICE) of the National University of Colombia. Thanks to Juan Diego Usme, Carlos Oviedo and Dr. Yazmin Agámez for their welcome in the research group (LICE) and for their help in sortometry, XRF, SEM and FT-IR measurements.

References

[1] Virkutyte, J., Varma, R.S. & Jegatheesan, V. (2010). Treatment of Micropollutants in Water and Wastewater, IWA Publishing, London.

[2] Giger, W., Brunner, P.H. & Schaffner, C. (1984). 4-Nonylphenol in sewage sludge: accumulation of toxic metabolites from non-ionic surfactants. Science, 225, 623–625.

[3] Estevinho, B.N., Martins, I., Ratola, N., Alves, A. & Santos, L. (2007). Removal of 2,4. dichlorophenol and pentachlorophenol from waters by sorption using coal fly ash from a Portuguese thermal power plant. J. Hazard. Mater., 143 (1-2), 533-540.

[4] Rivera-Utrilla, J. Sánchez-Polo, M., Ferro-García, M.A., Prados-Joya, G. & Ocampo-Pérez, R. (2013). Pharmaceuticals as emerging contaminants and their removal from water. A review. Chemosphere, 93 (7), 1268-1287.

[5] Carriazo, J., Molina, M., y Saavedra, M. (2009). Arcillas pilarizadas: un proyecto de síntesis inorgánica en el laboratorio, Revista educación en ingeniería Asociación Colombiana de Facultades de Ingeniería -ACOFI- www.acofi.edu.co, 8, 26-31.

[6] Centi, G., Perathoner, S. (2008). Catalysis by layered materials: A review. *Microporous and Mesoporous Mater*. 107, 3-15.

[7] Leofanti, G.; Padovan, M.; Tozzola, G.; Venturelli, B. Surface área and pore texture of catalysts. *Catal. Today*. 1998. 41 (1-3): 207-219.

[8] Cabrera-Lafaurie, W. A.; Román, F. R.; Hernández-Maldonado, A. J. Transition metal modified and partially calcined inorganic-organic pillared clays for the adsorption of salicylic acid, clofibric acid, carbamazepine, and caffeine from water. *J. Colloid Interface Sci.* 2012, *386*(1), 381-391.

[9] Vidal, C. dos Santos, A. do Nascimento, R. Bandosz, J. (2014). Reactive adsorption of pharmaceuticals on tin oxide pillared montmorillonite: Effect of visible light exposure. Chemical Engineering Journal, 259, 865-875.

[10] Torres, J., Carriazo, J., Sanabria, N. Síntesis y caracterización de bentonita modificada con especies de TiO_2 y Fe-TiO_2 obtenidas del mineral ilmenita. Rev. Colomb. Quim. 2013, 42 (1), xx-xx.

IOP Conf. Series: Materials Science and Engineering 1154 (2021) 012044 doi:10.1088/1757-899X/1154/1/012044

Extraction of Humic Acids from Different Origins: A Review

H Quiroz-Urzola[1], J R Castro-Suarez[2] and F Colpas-Castillo[1]

[1] Grupo de Carboquimica, Programa de Química, Faculta de Ciencias Exactas y Naturales, Universidad de Cartagena, Cartagena-Colombia
[2] Universidad del Sinú, Cartagena-Colombia

e-mail : fcolpasc1@unicartagena.edu.co

Abstract. The synthesis of molecules such as humic acids is a good strategy for the remediation of eroded soils. Since the synthesis and extraction of these carbon molecules do not represent a process of energy liberation of the carbon, it is also a low-cost process. Bad practices in soils, such as the indiscriminate use of pesticides and pesticides, cause damage to the soil that can be difficult to reverse, depending on the type of exposure and resilience they have. For example, the Atlantic coast is the most affected by salinization, according to Ideam a phenomenon related to incorrect agricultural activities, devastating tourism, and the impact of the dry climate. Soil erosion is an effect that has a negative influence on activities such as agriculture since the erosion of the soils prevents or causes the agricultural activity to be low or almost nil. Humic acids are molecules that can retain cations in their macroporous structure. Therefore, it is important to study what is the affinity that these macromolecules have with the cations of Zn, Ca, K, Mg and therefore, to know with which of them the affinity is greater depending on the type of carbon studied. The objective of this monographic study is to review the conceptual part of humic acids, their different forms of extraction, and their applications reported by different authors working in this field.

1. Humic Substances

Humic substances are long chains of numerous stable carbons that have physical, chemical, and biological effects on the soil, related to agricultural activity. They are one of the main sources of N, P, and S for plants, they influence the mobility of non-ionic organic compounds such as pesticides and pollutants, removing them from aqueous solutions, they retain nutrients due to their cation exchange properties [1].

According to Huelva (2013) [2], Humic Substances (HS) are very complex macromolecules. These have different functional groups that allow them to act as polyelectrolytes of weak acids and be sites of reaction with different chemical agents; It is accepted that this consideration is valid for each of its fractions where Humic Acids (HA) are also included as one of its components. Most of these reaction sites are attributed to the presence of oxygen in the OH functional groups of alcohols, phenols, carboxyl, and others, being those of carboxyl and phenolic groups the ones with the greatest interaction and quantity.

For Seguel et al., 2020 [1] and Janoš, 2003 [3], Humic substances are made up of three parts. These are Humins, Humic Acids, and Fulvic Acids, the last two can be separated in different pH media, Fulvic Acids, for example, are solubilized in pH acidic medium, while Humic Acids are soluble in alkaline mediums and humins are insoluble at all pH values.

2. Humic acids

These are macromolecules that are formed from organic, fossil, and plant residues through microbial decomposition carried out mainly by fungi, they can be found in peat, leonardite, and compost [4,5].

According to Noboa (2019) [6], he defines humic and fulvic acids as complex organic molecules formed by the decomposition of organic matter. These directly influence the fertility of the soil, while contributing significantly to its stability, affecting the absorption of nutrients and as a direct consequence, optimal growth, and development of the plant.

However, Mosquera (2007), states that humic acids are complex aromatic macromolecules with amino-acids, amino-sugars, peptides, and aliphatic compounds that are involved in the bonds between aromatic groups. The hypothetical structure for humic acids contains free and bonded phenolic groups, quinone structures, nitrogen and oxygen as bridges, and carboxylic acid groups at various locations on the aromatic rings (3); In addition, they are molecules in which the presence of aromatic rings of the di and trihydroxybenzene type is observed and nitrogen as a structural component that indicates the presence of carbohydrate and protein residues.

Other authors argue that AH are mixtures of carbohydrates, proteins, and lipids from both plants and microbial origin, together with partial degradations of lignin and tannins, and with microbial materials such as melanins, or are pseudo structures defined as hypothetical molecular constitutions with elements, structures and functional groups similar and consistent with some or all of the observed properties of a given mixture, which are not molecular structures, nor are they average structures or an average structural model [7-9].

Finally, Pantoja et al. [10], identifies humic acids as "A fraction of humic substances, made up of aromatic and polyaromatic nuclei, linked through aliphatic chains. These have a diversity of functional groups, which allow them to exercise various functions in the soil-plant relationship ".

3. Importance of humic acids

humic acids such as fulvic acids are of great importance in crops since they prevent the soil from compacting; They help to transfer nutrients from the soil to the plant, increase the water retention capacity, increase the speed of seed germination and stimulate the proliferation of the microflora present in the soil, increasing the soil's moisture retention capacity. In general, it is estimated that humic acids can retain water in a proportion of twenty times its weight. They increase the cationic and anionic exchange capacity of the soil and thus its capacity to provide nutrients to the crop, as well as its storage in the soil [11-12].

HA promote development and plant growth in crops of agronomic interest [13-15].

It has been shown that there is a relationship between the chemical properties and functions of humic acids and the changes in biological activity that are manifested in the growth of the plant that occurs after the chemical modification of humic acids [2,16].

According to Solórzano [4], 2019; Rodríguez [5], The water absorption capacity of humic acids is due to their large branched structure by aliphatic side chains of amino acids, peptides, and aliphatic acids, where this corresponds to around 65%, the remainder corresponds to structures with an aromatic character; These facilitate the rotation of cations in the medium due to their cation exchange capacity and ease of chelation.

4. Methodology for Humic acid extraction

For the extraction of humic acids, separation is carried out in an alkaline medium where the humic acids are precipitated. For this extraction, many times some type of oxidation is carried out that increases the original amount of HA found in the carbonaceous matrix, this can be carried out through various types of oxidizing agents, even starting from different raw materials. Next, we will look at some of the many ways that can be used to extract humic acids.

Vasquez et al. (2019) [17], uses a factorial type experimental design, with a 2^2x3 arrangement, testing two levels of time (3h and 6h), three levels of concentration, 30, 40, and 50% of oxidizing agent (H_2O_2) and two levels of particle size, 0.25mm (60 mesh) and 0.125mm (120 mesh). For sample preparation, the coal from the Cerrejón mine, located in the department of La Guajira (Colombia), was crushed in a ball mill (Pulvex brand) and sieved to two particle sizes, using sieves with mesh number 60 and 120. Demineralization and solvent removal was carried out according to the procedure suggested by Anillo-Correa et al. [18], Aqueous Oxidation was performed according to [18]. In the characterization of the materials, the carbon and humic acid samples were characterized by infrared spectrometry (FTIR). Proximal studies, in accordance with ASTM (2012) standards, to determine the percentage of moisture, volatile matter (% MV), ash, percentage of fixed carbon (% CF) and the% CF / MV ratio.

Santana [19], shows different methods for the extraction of humic substances, one of them is through the determination of humic substances in sediments by focused ultrasound and ultraviolet-visible spectroscopy. The development of a new and rapid methodology to determine the amount of humic substances (HSs) present in sediment samples is deeply optimized with this method. Emphasis is placed on alkaline extraction because traditional methods required high volumes and long extraction times. The extraction of liquids-solids by focused ultrasound (FUSLE) is carried out, followed by the separation of SHs in acidic media. The determination of it using visible ultraviolet spectroscopy turned out to be a very suitable analytical procedure to accelerate the extraction of HSs, specifically humic acid (HA) and fulvic acid (AF). All the variables that affect the extraction process (sodium hydroxide concentration and volume, ultrasound power, cycles, and time) were selected using the analysis of variance and the multivariate design. The optimal values of these variables are the power of continuous application of the (FUSLE), 95% (10 cycles), and 15 mL of NaOH 1 mol/ L^{-1}. p.12 [20]

Santana refers to Susic (2016) [21], to classify humic acid extraction methodologies and achieves to classify them according to the type of extractant and material to be extracted into three types, mild, medium, and strong extractants, see table 1.

The methodology proposed by Anillo et al (2013) [18] was the following: For oxidation with air, 50g of sample and an airflow of 100 mL/min were used in a porcelain reactor with a fluidized bed. This process was carried out at temperatures of 180, 200, and 220 °C. The oxidation in an aqueous medium was carried out with hydrogen peroxide (Merck) and with nitric acid (Merck). For oxidation with hydrogen peroxide, 5 g of air-oxidized carbon, 30 mL of a 30% (v / v) solution of H_2O_2, and 15 mL of concentrated acetic acid (Merck 99-100%) were added to an Erlenmeyer flask. This system was heated to 60 °C and kept under continuous stirring for 6 and 12 h. oxidation with Nitric acid was carried out by mixing 5 g of oxidized carbon with air with 30 mL of HNO_3 at 30% (m / v). This system was heated to 60 °C and kept under continuous stirring for 6 and 12 h. For the extraction of humic acids from oxidized coals, 100 mL of NaOH (0.1 M) (Sigma-Aldrich) were added to 5 g of these materials, maintaining this system for 1 h under continuous stirring. Subsequently, the solution was filtered under vacuum and 100 mL of HCl (0.1 M) (Merck) were added to the filtrate to precipitate humic acids. This system was kept at rest for 24 h and was subsequently centrifuged at 3600 rpm for 10 min. The colloid obtained after centrifugation was washed with portions of ethanol and finally dried at 100 °C for 1 h to be subjected to different analyzes. The oxidation process in aqueous solution was carried out in triplicate in the case of the two oxidizing agents, not being observed standard deviations greater than 1.3% in any of the cases.

According to the International Society of Humic Substances (2008), the methodology used for the extraction of humic acid is based on two methods, which are based on two fundamental stages, first the extraction of humic substances using NaOH as extractants, which is recommended by IHSS, (2008) and Santi, (2000) [22] or KOH (Rocha et al., [23]) and then the separation of humic and fulvic acids from the acidification of humic substances with diluted HCl or H_2SO_4. The best extraction percentages are obtained with NaOH and HCl.

Table 1. Types of extractants used to extract humic acids, taken from reference [19]

Extractant		Fundaments	References
strong	NaOH	It is the most used procedure for the extraction of humic substances of commercial type, due to its high percentage of extraction (80%).	al., 2016; Scaglia et al., 2016; Lukyanov et al., 2016; Fuentes et al.,
	KOH		Saito et al., 2014; Lukyanov et al., 2016; Fuentes et al.,
medium	$Na_4P_2O_7$, EDTA	These extractants produce minimal alterations in the matter organic. In some cases they are used together with alkaline products to increase their effectiveness.	Fuentes et al., 2016; Asing et al., 2009
soft	EDTA, acetylacetone, cupferron and various types of organic solvents such as tetrahydrofuran	One way to increase effectiveness is to mix these reagents with other products, such as urea at high concentrations. Obviously, although the alterations that occur in the extracted organic matter are minor, the effectiveness shown by these products is much lower.	Hemati et al., 2012
Formic Acid (HCOOH)		It is obtained from 55% extraction of organic matter in mineral soils, up to 80% in compost. Formic acid extraction is the most efficient for where most organic substances are partially humified.	Ramos, 2000

4.1. Methodology used in our research

The Carbon sample from the Cerrejón mine (Guajira - Colombia) was provided by Professor Yazmin Agamez Pertuz from the National University of Colombia. This was crushed and sieved to a particle size between 1mm and 150μm corresponding to mesh size # 18 and # 100 respectively.

4.1.1. Oxidation. For the oxidation in an aqueous medium, two different oxidizing agents were used, hydrogen peroxide (H_2O_2) at 30% (v / v) and nitric acid (HNO_3) at 30% (v / v). The oxidation time was 10 hours.

For oxidation with hydrogen peroxide, 5 g of carbon (C10), 15 mL of glacial acetic acid, and 30 mL of hydrogen peroxide at 30% (v / v) were added in a 100mL flat-bottom balloon in a proportion of 1:2, this system was immersed in a water bath on a heating plate with magnetic stirring. The system was heated to 60 °C and kept under constant stirring for 10 h.

For oxidation with nitric acid, 5 g of carbon (C10), 15 mL of glacial acetic acid, and 30 mL of nitric acid at 30% (v / v) were added in a proportion of 1:2, this system was immersed in a water bath on a heating plate with magnetic stirring. The system was heated to 60 °C and kept under constant stirring for 10 h.

4.1.2. Extraction of humic acid. For the extraction of humic acids, sodium hydroxide (NaOH) was used.

For the extraction of the oxidized humic acids with hydrogen peroxide, 100 mL of NaOH (0.1 M) were added, maintaining this system for 1 hour under continuous stirring and heating at 60ºC. Subsequently, the solution was filtered under vacuum and 100 mL of HCl (0.1 M) were added to the filtrate to precipitate humic acids. This system was kept at rest for 24 hours well covered to avoid contact with air, then it was filtered under vacuum again and the precipitate obtained was washed with portions of distilled water and 2 portions of 5 mL of ethanol and the humic acid obtained was dried at a temperature of 100 °C for 1 h. finally, it was subjected to different analyzes for its respective characterization.

For the extraction of the oxidized humic acids with nitric acid, 100 mL of NaOH (0.1 M) were added, maintaining this system for 1 hour under continuous stirring and heating at 60ºC. Subsequently, the solution was filtered under vacuum and 100 mL of HCl (0.1 M) were added to the filtrate to precipitate humic acids. This system was kept at rest for 24 hours well covered to avoid contact with air, then it was filtered under vacuum again and the precipitate obtained was washed with portions of distilled water and 2 portions of 5 mL of ethanol and the humic acid obtained was dried at a temperature of 100 °C for 1 h, finally it was subjected to different analyzes for its respective characterization.

5. Applications

5.1. *Bioactivity of humic acids.*

5.1.1. First application. Huelva and Martínez and other members of the Agrarian University of Havana worked with humic acids obtained through vermicompost of bovine manure. The chemical modifications in the structure of the HA were analyzed to determine the structural changes through Diffuse Reflectance Fourier Transform Infrared Spectroscopy (DRIFT) techniques and, through their spectra, the hydrophobicity (HI) and condensation (IC) indices of molecules were calculated [2].
Table 2. shows the content of C obtained by humic acids after the structural modification and the optical coefficient E4 / E6, where the observed values for C are values that are within the range of the values reported in the scientific literature and compared with other humic acids extracted with other methods and from other origins [24].

Table 2. Percent of C and coefficient E4 / E6 of vermicompost humic acids and their derivatives, taken from reference [2].

humic substances	percent of C	E4 / E6
AH	56.53 b	5.25 b
AH-Ac	60.75 a	4.93 c
AH-Met	66.70 a	3.78 d
AH-NH_3	55. 65 b	6. 54

The percentage of C was lower in the AH-NH_3 treated with Thionyl due to the formation of amides because the entering group is -NH_2, since a blockage is formed between the Thionyl chloride and the anhydride which creates a steric impediment for the molecule [2].,

5.1.2. Second *a*pplicatio*n.* In this case, the Federal Rural University of Rio de Janeiro in Brazil carried out studies on rice crops. This to measure the protection capacity of rice plants that were treated with humic acids against water stress. For this study, two different humic acids identified as AHC and AHB from Cuba and Brazil respectively were used, where according to the partial characterization of these materials, the only differences significant structural between the two different humic acids were evidenced in the percentage of Carbon C% and oxygen O%, according to this, the rice plants with humic acids were subjected to water stress for 96 hours, where it was shown that those samples of rice plants that interacted with humic acids preserved the permeability of the membrane. This study was carried out because humic acids provide plants with the ability to adapt to abiotic stresses such as salinity, the presence of heavy metals at toxic levels, and water deficiencies [16].

5.1.3. Third application. In a study carried out at the University of Chile, positive and significant results were obtained in Haplohumult soil samples that were quite fragile due to constant and intensive agricultural use. Humic acid was used to test the water retention capacity, increase in porosity, increase of up to 24% in wheat yield, and among other factors. This study was carried out in the Araucanía Region of southern Chile, where it was determined that the use of humic acid contributed to a better wheat crop yield to more grains per ears and higher biomass production. In this study, 30 grams of humic acid were added in 20 liters of water, spreading them evenly all over the soil study, there were different types of soils, T0, T1, and T2, with T0 being an area of soil without humic acid, T1 the area of soil with humic acid and T2 a soil native to the study area, two soil samples were taken at different depths, the first from 0-10cm deep and the second from 30-40cm deep. Wilson and Valenzuela state in their study that the wheat root system is considerably formed in the first 10 cm and decreases considerably according to the depth of the soil [1, 25]

5.1.4. Fourth application. The Research Group of Organic Matter and Biostimulants of the Department of Chemistry of the Agrarian University of Havana, obtained an aqueous extract of HS from vermicompost of bovine manure, the biostimulation of different doses of the extract of vermicompost has been verified in crops of agronomic interest such as corn, lettuce, tomato and beans. The foliar application of these extracts in tomato plants promoted the biological development of the plants, as well as the agricultural productivity in indicators such as the mass of the fruit and the yield for two consecutive years. The physical-chemical characterization of these extracts has shown the presence of humic substances such as humic and fulvic acids, phytohormones, beneficial microorganisms, amino acids, and essential elements that could contribute to their biostimulant action [24, 26-29].

6. Conclusions.
The extraction of humic acids is a good alternative for the remediation of eroded soils. In addition to being a low-cost practice, it also represents an alternative that does not leave polluting residues of the material where it is extracted. HA are capable of capturing contaminating materials in their matrix due to their great multi-substituted structure with -OH and -COOH groups, due to their great ionic affinity, which makes it a technique for improving soil quality as it contributes with ease nutrients to the soil and plants.

References
[1] Seguel, O., Parra, C., Homer, I., Kremer, C., & Beyá-Marshall, V. (2020). Efecto del ácido húmico sobre las propiedades físicas de un Haplohumult cultivado con trigo. Agro Sur, 47(3), 27-38.

[2] Huelva López, R., Martínez Balmori, D., Calderín García, A., Hernández González, O. L., & Guridi Izquierdo, F. (2013). Propiedades químicas y química-físicas de derivados estructurales de ácidos húmicos obtenidos de vermicompost. Actividad biológica. *Revista Ciencias Técnicas Agropecuarias*, *22*(2), 56-60.

[3] Janoš, P. (2003). Separation methods in the chemistry of humic substances. *Journal of Chromatography A*, *983*(1-2), 1-18.

[4] Solórzano Cedeño, A. E. (2019). *Efecto de quitosano, hongos micorrízicos y ácidos húmicos sobre el crecimiento y desarrollo en variedades de pimiento (Capcicum annuum L) bajo condiciones protegidas* (Bachelor's thesis, Quevedo-UTEQ).

[5] Rodríguez, F. (2017). Sustancias húmicas: Origen, caracterización y uso en la

agricultura. *Obtenido de https://www. intagri. com/articulos/nutricion-vegetal/acidos-humicosfulvicos-nutricion-vegetal.*

[6] Noboa Tobar, F. J. (2019). Efecto de la aplicación de tres productos a base de ácidos húmicos y fúlvicos sobre el crecimiento y desarrollo de plántulas de cacao (Theobroma cacao L.) en la zona de Valencia, provincia de Los Ríos (Bachelor's thesis, Quevedo-UTEQ).

[7] Mosquera, C., Bravo, I., & Hansen, E. (2007). Comportamiento estructural de los ácidos húmicos obtenidos de un suelo Andisol del Departamento del Cauca. *Revista Colombiana de Química*, *36*(1), 31-41.

[8] Burdon, J. (2001). Are the traditional concepts of the structures of humic substances realistic?. *Soil science*, *166*(11), 752-769.

[9] MacCarthy, P. (2001). The principles of humic substances. *Soil Science*, *166*(11), 738-751

[10] Pantoja Guerra, M., Almanza Pérez, Y., & Valero Valero, N. (2016). Evaluación del efecto auxin-like de ácidos húmicos en maíz mediante análisis digital de imágenes. *Revista UDCA Actualidad & Divulgación Científica*, *19*(2), 361-369.

[11] Guerrero, A. (2006). Efecto de tres bioestimulantes comerciales en el crecimiento de los tallos de proteas, Leucadendron sp. Cv. Safari Sunset. Universidad Técnica del Norte. Ibarra-Ecuador.

[12] Melo López, L. (2006). Análisis y caracterización de ácidos fúlvicos y su interacción con algunos metales pesados.

[13] Baldotto, L. E. B., Baldotto, M. A., Giro, V. B., Canellas, L. P., Olivares, F. L., & Bressan-Smith, R. (2009). Desempenho do abacaxizeiro'Vitória'em resposta à aplicação de ácidos húmicos durante a aclimatação. *Revista Brasileira de Ciência do Solo*, *33*(4), 979-990.

[14] Zandonadi, D. B., Canellas, L. P., & Façanha, A. R. (2007). Indolacetic and humic acids induce lateral root development through a concerted plasmalemma and tonoplast H+ pumps activation. *Planta*, *225*(6), 1583-

[15] Rodda, M. R. C., Canellas, L. P., Façanha, A. R., Zandonadi, D. B., Guerra, J. G. M., Almeida, D. L. D., & Santos, G. D. A. (2006). Estímulo no crescimento e na hidrólise de ATP em raízes de alface tratadas com humatos de vermicomposto: II-Efeito da fonte de vermicomposto. *Revista Brasileira de Ciência do Solo*, *30*(4), 657-664.

[16] Guridi-Izquierdo, Fernando, Calderín-García, Andrés, Louro-Berbara, Ricardo L., Martínez-Balmori, Dariellys, & Rosquete-Bassó, Mayelín. (2017). Los ácidos húmicos de vermicompost protegen a plantas de arroz (Oryza sativa L.) contra un estrés hídrico posterior. *Cultivos Tropicales*, *38*(2), 53-60

[17] Vásquez Osorio, M., Colpas Castillo, F., & Tarón Dunoyer, A. (2019). Influencia del tamaño de partícula en la extracción de ácidos húmicos de carbón bituminoso. *Revista UDCA Actualidad & Divulgación Científica*, *22*(1).

[18] Anillo-Correa, R., Colpas-Castillo, F., & Meza-Fuentes, E. (2013). Aumento del contenido de ácidos húmicos en un carbón de bajo rango a través de la oxidación con aire y con peróxido de hidrogeno o ácido nítrico. *Química Nova*, *36*(3), 387-392

[19] Santana, E. S. *TÍTULO: Optimización de extracción de sustancias húm estiércol de ganado vacuno* (Doctoral dissertation, Universidad de Matanzas).

[20] Raposo, J. C., Villanueva, U., Olivares, M., & Madariaga, J. M. (2016). Determination of humic substances in sediments by focused ultrasound extraction and ultraviolet visible spectroscopy. *Microchemical Journal*, *128*, 26-33.

[21] Susic, M. (2016). Replenishing humic acids in agricultural soils. *Agronomy*, *6*(4), 45

[22] SANTI, L. P., Goenadi, D. H., Widiastuti, H., & Mardiana, N. (2000). Extraction and characterization of humic acid from plantation's solid organic waste composts. *Menara Perkebunan*, *68*(2), 29-36.

[23] Rocha, J. C., Rosa, A. H., & Furlan, M. (1998). An alternative methodology for the extraction of humic substances from organic soils. *Journal of the Brazilian Chemical Society*, *9*(1), 51-56.

[24] Calderín García, A., Pimentel Quintero, J., Martínez Balmori, D., Huelva López, R., & Guridi Izquierdo, F. (2016). Efeitos no cultivo do milho de um extrato líquido humificado residual,

obtido a partir de vermicomposto. *Revista Ciencias Técnicas Agropecuarias*, *25*(1), 38-43.

[25] Wilson, M. G., & Valenzuela, O. R. (1998). EVALUACIÓN DEL SISTEMA RADICAL DEL CULTIVO DE TRIGO MEDIANTE DOS MÉTODOS DE MEDICIÓN[1].

[26] Hernandez, O. L., Calderín, A., Huelva, R., Martínez-Balmori, D., Guridi, F.,

[27] Hernández del Valle, G., Hernández González, O., Guridi Izquierdo, F., & Arbelo Fortes, N. (2012). Influencia de la siembra directa y las aplicaciones foliares de extracto líquido de Vermicompost en el crecimiento y rendimiento del frijol (Phaseolus vulgaris L.) cv. cc-25-9. *Revista Ciencias Técnicas Agropecuarias*, *21*(2), 86-90.

[28] Hernández, O. (2010). *Modificaciones al proceso de extracción de sustancias húmicas. Efectos biológicos* (Doctoral dissertation, Tesis de Maestría].[Mayabeque, Cuba]: Universidad Agraria de la Habana).

[29] Arteaga, M., Garcés, N., Guridi, F., Pino, J. A., López, A., Menéndez, J. L., & Cartaya, O. (2006). Evaluación de las aplicaciones foliares de humus líquido en el cultivo del tomate (Lycopersicon esculentum Mill) var. Amalia en condiciones de producción. *Cultivos tropicales*, *27*(3), 95-101.

IOP Conf. Series: Materials Science and Engineering 1154 (2021) 012045 doi:10.1088/1757-899X/1154/1/012045

Portable Device for Continuous Measurement (DIPORMEC)

H. A. Rodriguez-Arias[1]

[1] Research Group in Robotics, Automation, Telecommunications, Electricity, Electronics and Computation. (GIRATEC), Faculty of Engineering Sciences, Antonio de Arevalo University Foundation, Cartagena, Colombia, e-mail: harold.rodriguez@unitecnar.edu.co

Abstract. Nowadays, the conditions of the workplace are having a very important role in different areas. Therefore, a device was created to measure variables, for the time being, of quantity, of light, Lumens and Lux, temperature and relative humidity. Besides, this device continuously saves these variables in a file on an SD card to be analysed later.
Another feature of this device is that it is portable which means it works with built-in batteries or with a universal connector, so that it can work easily in a workplace (offices).
For this work it is shown how what we call cloning of the luminosity sensor was performed, for this particular case, with an LDR the amount of Lux is determined, taking as reference the light sensor of the Samsung A7 cell phone, and thus generating the equation that governs the sensor. In the connection diagram, the possibility of connecting other sensors that do not require adjustment work such as the one of brightness is left open, taking full advantage of the Arduino card's capacity, but for the moment it is not part of the study of this work.

1. Introduction

Measurements with DIPORMEC can be taken without requiring a PC or other processing and storage device, as it was provided with its own processor (Arduino UNO microcontroller card) and a storage SD card. Therefore, it is expected that its measurements were not greater to that of a lotion box, which makes it truly portable. In addition, it can be monitored through Bluetooth or RF (X-Bee) in case of being required, this in order to see the samplings in real time. However, it should be noted that these samplings are recorded in the SD memory to be able to be analyze in Excel or another program.

The particularity of DIPORMEC is that it is able to make multiple measurements at the time and for long periods. The most important part is to validate the sensors. For this reason, a technique called sensor cloning was used in this part of the project. This technique consists of making simultaneous measurements with a calibrated device and the economic sensor, In addition, it help to find a function that leads the sensor used in DIPORMEC to approach the measurement made with the calibrated equipment, having at least an R2 ratio higher than 0.98. Thus, costs low and multi-sensing equipment is built.

2. Materials and methods

2.1. Problem statement

Jobs must meet special characteristics so that workers have specific and optimal conditions for a good performance and to avoid illnesses due to non-compliance with these conditions. The main idea of this project is to create a prototype device that takes information continuously during the day of the amount of light or lumens in a workplace. This information is saved on an SD card to be downloaded and then the data is interpreted on a PC in a program such as Excel. The device can save the information of a week taking data up to every minute or other time period as required. It is also possible to attach other variables such as the amount of noise and even temperature and specific humidity. Therefore, at this point the question arises: How can measurements be acquired continuously of different variables such as lumens or temperature with a low-cost portable device?

The following methodology is followed in order to build DIPORMEC at low cost as an initiative to develop equipment and support the provision of UNITECNAR's laboratories with useful tools for teaching - learning and generation of own technology (see figure 1).

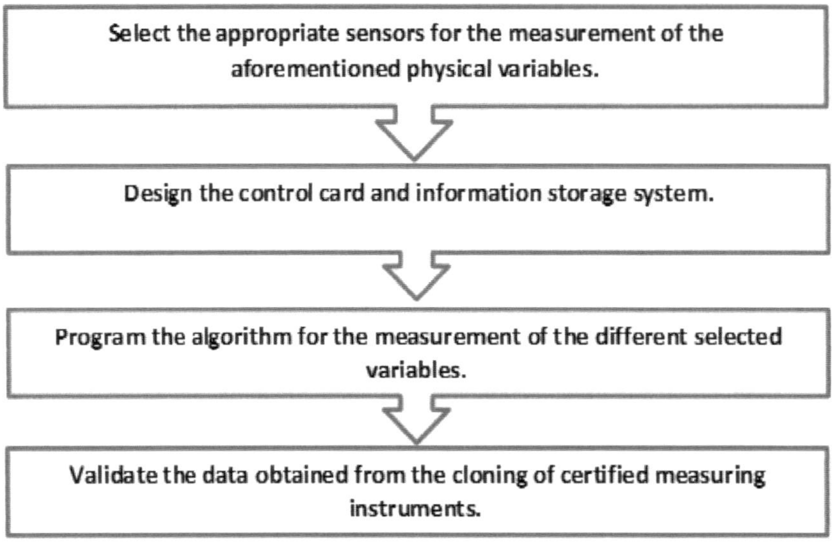

Figure 1. Project methodology

3. Phases implemented to the construction of DIPORMEC

To be able to assemble this project, it is necessary to take into account the configuration steps that the different cards and sensors bring. However, the development of the application is important in order to efficiently and accurately capture the information or data obtained from the sensors.

The main elements used in DIPORMEC are described below, as well as some of the sensors that are integrated into the device, with their respective characteristics.

3.1. Arduino Nano Card

The Arduino Nano (see figure 2) is a small, complete and an easy board to use. It is based on the ATmega328P (Arduino Nano 3.x). It has about the same functionality of the Arduino Duemilanove, [3] but in a different package. It lacks just a DC power connector. Finally, it works with a USB Mini-B cable instead of a standard one. [4]

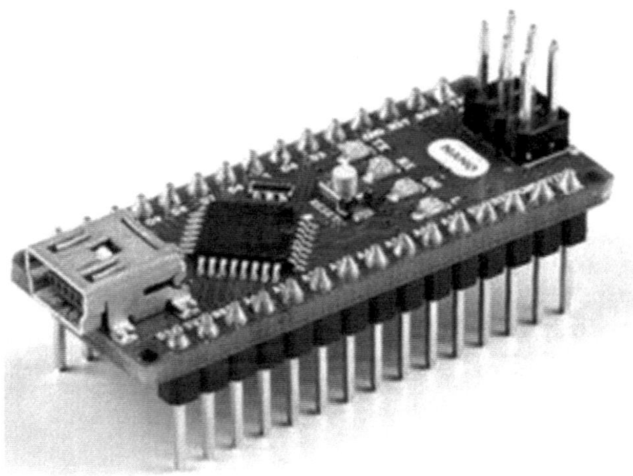

Figure 2. Arduino NANO [4]

Chracteristics:

- ATmega328 microcontroller
- AVR architecture
- Operating voltage 5 V
- Flash memory 32 KB of which 2 KB used by the boot manager
- SRAM 2 KB
- Clock speed 16 MHz
- IN 8 analog plugs
- EEPROM 1 KB
- Current flows through the connection bold of E/S 40 mA (Pins of E / S)
- Input voltage 7-12 V
- Pins of E / S digitals 22 (6 which are PWM)
- Output PWM 6
- The power consumption 19 mA
- PCB size 18 x 45 mm
- Weight 7 g

3.2. Micro SD module for Arduino

The SD and micro SD cards have become a standard which have displaced other data storage media due to their large capacity and small size. For this reason they have been integrated into a large number of devices, being currently frequent components in computers, tablets and smartphones, among others.

Within the world of Arduino, it is possible to find low cost readers for both SD and micro SD cards (see figure 3). The first to appear was the SD readers and later the micro SD. Therefore, in general, the modules with micro SD are more modern models than those of SD.

In both types of readers, reading can be done through the SPI bus. Although they may have other interfaces, such as I2C bus or UART, it is usually preferable to use SPI because of its high transfer rate.

Regarding the cards used, SD cards or SDSC (Standard Capacity) or SDHC (High Capacity) can be used, but not SDXC (Extended Capacity). It must be formatted in file system FAT16 or FAT32 [5].

Figure 3. Micro SD writing and reading module for Arduino [5]

3.3. Real Time Clock or RTC

A real time clock (RTC) is an electronic device that allows obtaining time measurements in the temporary units that we use daily.

The term RTC was created to differentiate this type of clocks from the usual electronic clocks, which simply measure time by counting pulses of a signal, without having a direct relationship with temporal units.

On the contrary, the RTCs are more similar to the clocks and calendars that we usually use which work with seconds, minutes, hours, days, weeks, months and years.

RTCs are usually formed by a crystal resonator integrated with the necessary electronics to correctly post the passage of time. The RTC electronics take into account the peculiarities of our way of measuring time, such as the sexagesimal system, the months with different days, or the leap years [5].

The RTCs offer the advantage of reducing energy consumption, providing greater precision and freeing Arduino from having to record the time. In addition, RTCs often incorporate some type of battery that allows maintaining the value of time in case of loss of power.

In the world of homemade electronics and Arduino, there are two common RTCs which are the DS1307 and the DS3231, both manufactured by Maxim (formerly Dallas Semiconductor). The DS3231 has a much higher accuracy and can be considered as a substitute for the DS1307.

The temperature variations in the model DS1307, which affect the time measurement of the resonator crystals, translate into errors in an accumulated phase shift. Because of this, the DS1307 suffers a time lag, which can be 1 or 2 minutes a day [5].

To solve this, the DS3231 (see figure 4) incorporates temperature measurement and compensation guaranteeing an accuracy of at least 2ppm, which is equivalent to a maximum delay of 172ms / day or one second every 6 days. In the real world, they usually get higher accuracies equivalent to phase shifts of 1-2 seconds per month.

The communication in both models is done through the I2C bus, so it is easy to obtain the measured data. The supply voltage is 4.5 to 5.5 for the DS1307 and 2.3 to 5.5V for the DS3231 [5].

Figure 4. RTC DS3231 [5]

Characteristics

- RTC High accuracy which handles all functions for the maintenance of date/time
- Accuracy of ± 2ppm operating at a temperature of 0 ° C to + 40°C
- Module which has DS3231 clock and I2C EEPROM memory
- It has a backup battery
- Record of seconds, minutes, hours, day of the week, date, month and year with compensation of leap years up to 2100
- The DS3231 includes a temperature sensor with accuracy of +/- 3 degrees centigrade
- 2 programmable alarms per hour/date
- Programmable square signal output

IOP Conf. Series: Materials Science and Engineering 1154 (2021) 012045 doi:10.1088/1757-899X/1154/1/012045

3.4. Bluetooth module HC-06

This module allows you to add wireless connectivity through a serial TTL interface between Microcontrollers (PIC, Arduino) and other devices such as PCs, laptops or smartphones. The Bluetooth module HC-06 (See figure 5) is configured at the factory to work as a slave, that is, ready to listen for connection requests [6].

Figure 5. Bluetooth module HC-06 [6]

Characteristics

- Operation Voltage: 3.3V / 5V.
- Operating Current: <40 mA
- Current sleep mode: <1mA
- Chip: BC417143
- Reach 10 meters
- Transmission speed: 1200bps up to 1.3Mbps
- Baudrate by default: 9600,8,1, n.
- Bluetooth: V2.0+EDR
- Cable length: 21.5cm
- Frequency: 2.4 GHz ISM band
- Modulation: GFSK (Gaussian Frequency Shift Keying)
- Emission power: 4 dBm, class 2
- Sensivity: -84dBm a 0.1% VER
- Asynchronous speed: 2.1Mbps (max.) / 160 kbps
- Synchronous speed: 1Mbps / 1Mbps
- Security: Authentication and encryption
- Interface: Bluetooth - Serial port UART TTL

3.5. Temperature and humidity sensor DTH11 for Arduino

The DHT11 (see figure 6) is a low cost digital temperature and humidity sensor. It uses a capacitive humidity sensor and a thermistor to measure the surrounding air, and displays the data via a digital signal on the data pin (no analog input pins). It is quite simple to use, but requires careful synchronization to take data. The only drawback of this sensor, it is that you can only obtain new data once every 2 seconds, so the readings that can be performed are minimum every 2 seconds [7].

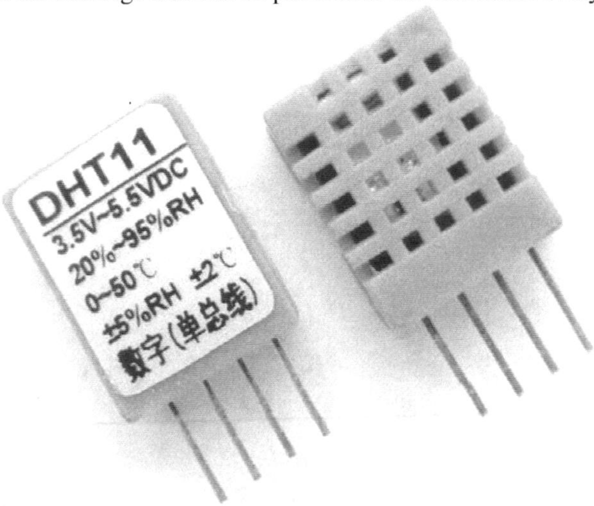

Figure 6. Sensor DHT11 [7]

Characteristics

- Power: 3Vdc ≤ Vcc ≤ 5Vdc
- Temperature measurement range: 0 to 50 ° C
- Accuracy of temperature measurement: ± 2.0°C
- Resolution Temperature: 0.1 ° C
- Humidity measurement range: 20% to 90% RH.
- Accuracy of humidity measurement: 4% RH.
- Moisture Resolution: 1% RH
- Sensing time: 1 sec.

3.6. Light intensity sensor module LDR

This module is made up of LDR (see figure 7) or photoresistor, which is sensitive to ambient light intensity exposure, in order to determine the brightness and light intensity of the medium. This module establishes the brightness threshold through a digital output, providing a high or low voltage level, depending on the configuration of the potentiometer. Real information can be taken through the analog output of the module to a microcontroller or Arduino, obtaining more exact values of luminosity [8].

Characteristics

- To detect light intensity from the environment
- Adjustable sensitivity by potentiometer
- Use the comparator LM393 for greater stability
- Installation hole for easy use
- Power indicator (red LED)
- Digital output indicator D0 (green LED)
- 4-wire connection
- Dimensions 30 x 15mm

Figure 7. Light intensity sensor module LDR [8]

As previously mentioned, more sensors can be added, such as sound, air quality, among others. Nevertheless the prototype is built with those mentioned here for the moment, on a test basis, making clear its expansion in its next stage.

4. Results

Among the most relevant results so far, such us, the selection of the sensors and other electronic elements, as well as programming strategies, the authors want to highlight the work done to take the values of certain variables, in this case the method used to measure the intensity of light.

As one of the final objectives of DIPORMEC is to make a low-cost but reliable device when taking the measurements, the sensor cloning methodology was worked. This latter is nothing more than trying to deliver the value of lux of the light intensity measurement which it was not taken directly with a photometer, but with another more economic sensor, for this project the LDR module.

The method consists of simultaneously taking values with the two sensors and collecting data for the same measure. In this instance, a controlled environment was built consisting of an internal black box, 40cm on each side which it is the ceiling. A Led lamp was placed which was connected to a power source of 12V, and a rheostat to be able to vary the amount of light delivered. Besides, in the lower part equally distanced, the LDR module was located and the light sensor was used in this exercise and it was incorporated in a Samsung Galaxy A7, which delivers the measurements in lux directly as shown in Figure 8. Some of the improvements that will be made to the prototype, it is to do this same exercise but with a certified photometer, to reach a true approximation of the data.

Then, it is proceeded to start completely dark and the data is taken from the Smartphone in Lux and the LDR module in volts. The light intensity was increased, turning the rheostat until delivering the maximum delivered by the Led lamp. Each time the intensity was left a while, the measures were stabilized and the data was taken which table 1 arose.

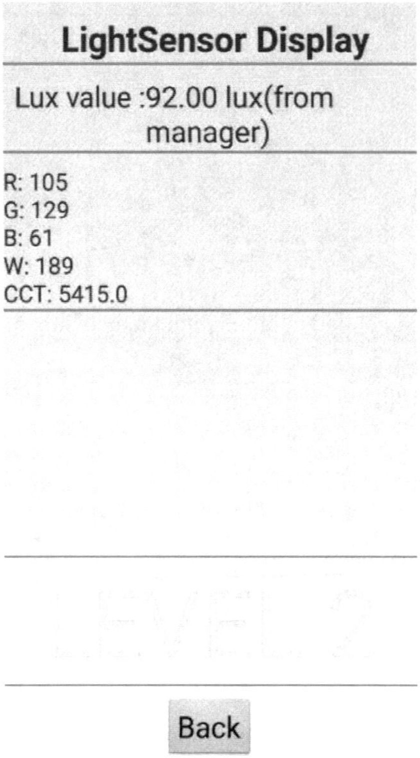

Figure 8. Samsung Galaxy A7 Smartphone brightness sensor

Table 1. Data from LDR module and the Smartphone

LDR module Volts	Smartphone LUX
0	0
1,12	151
1,57	285
1,96	465
2,03	510
2,17	605
2,31	714
2,41	811
2,5	905

With this information a graph of volts Vs Lux was made (see figure 9) and the respective curve adjustment was ready. It was tested with linear, parabolic adjustment but the one that exceeded R2 greater than 0.98 was the third degree polynomial, with a R2 = 0.9998.

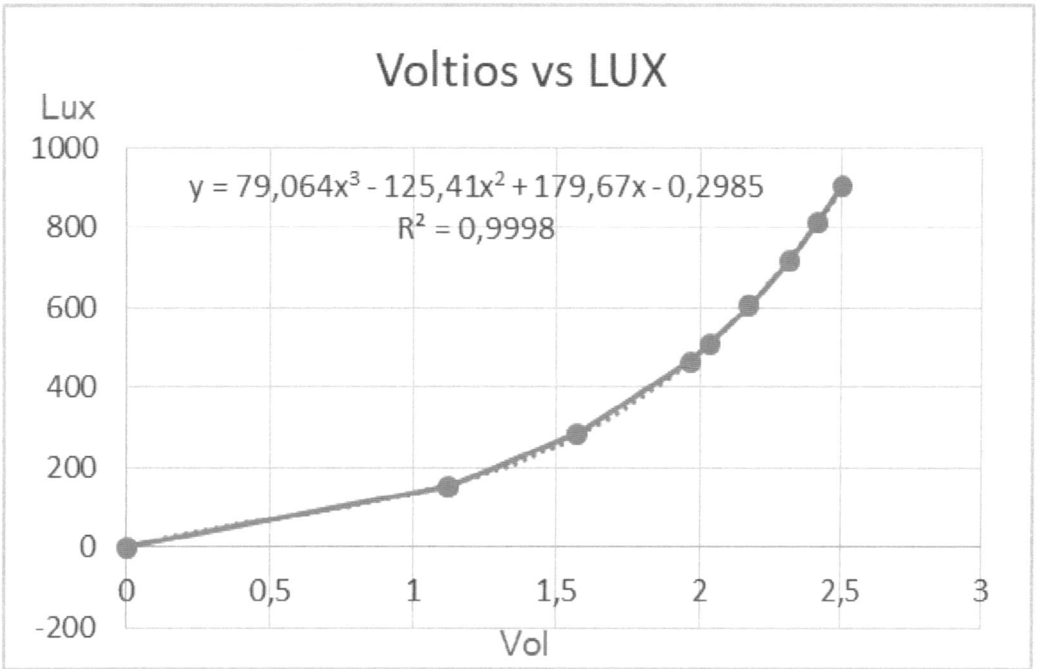

Figure 9. Relation between LDR module volts and LUX of the Smartphone

This means that equation (1) is the one that represents the value in LUX measured from the LDR module is:

$$LUX = 79{,}064v^3 - 125{,}41v^2 + 179{,}67v - 0{,}2985 \qquad (1)$$

Where:

LUX: is the value of the light intensity measured with the LDR module.
v: is the value in volts taken from the LDR module with one of the analog inputs of the Arduino card.

The behavior of the curve was expected to be linear, but the value of R2 was 0.8677 and the relative error (2), sometimes exceeded 12%, so it was decided to move to a quadratic curve adjustment, which was not bad, without However, the maximum relative error exceeded 6%, and what was wanted was a relative error below 5%, to have an acceptable cloned sensor accuracy.

The relative error was calculated as follows:

$$RE= ((Vr-Vm)/Vr)*100\% \qquad (2)$$

Where:

RE: It is the relative error and is given in percentage, %.
Vr: It is the real value.
Vm: It is the measured value.

For this reason, it is tested with an adjustment of order 3 and a maximum relative error of 2.4% is achieved, well below the desired 5%, and it was observed that the higher the measurement, the relative error decreases, even in Some cases were 0% or 0.1%, and it really is not that it adds more computational cost, for that reason it was decided to take equation (1), and its R2 was 0.9998 very close to 1 which is what you want.

The Arduino card was programmed and the exercise was repeated, this time the value was not written in volts delivered by the LDR, but in LUX, doing the treatment in the Arduino card. The data were recorded in table 2. Besides, the error was also calculated relative in percentage, finding as maximum error of measurement of 2.45%, and as the intensity of light is higher, the error decreases.

Table 2. Data of the ldr module and the smartphone in lux

LDR module LUX	Smartphone LUX	Relative Errors %
154,70	151,00	2,45
278,63	285,00	2,24
465,39	465,00	0,08
509,03	510,00	0,19
606,94	605,00	0,32
720,11	714,00	0,86
811,01	811,00	0,00
900,44	905,00	0,50

Then, it was proceeded to compare the two curves, and the result at sight is almost identical (see Figure 10). This was done in order to verify that the measurement with the LDR module is a clone or simulates the measurement of the Smartphone's light sensor.

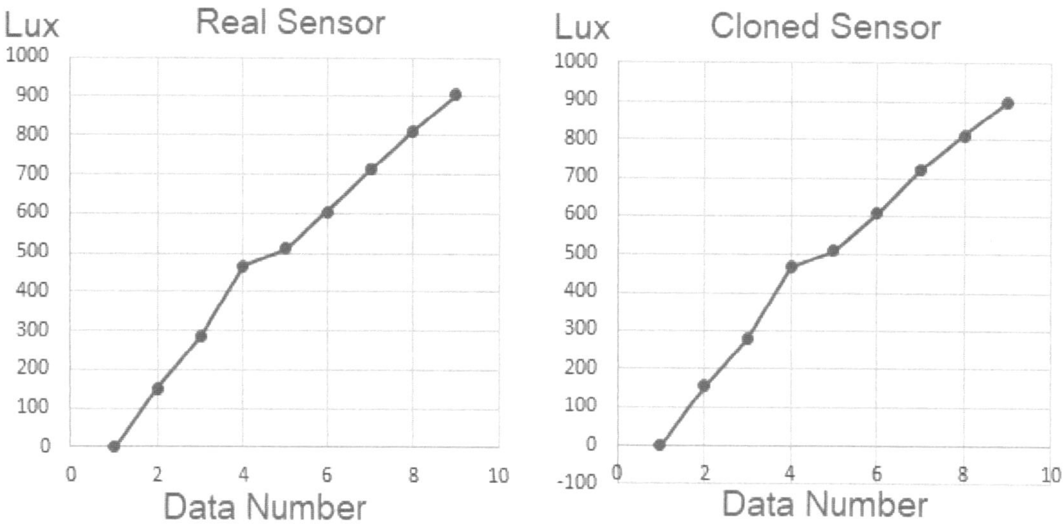

Figure 10. Comparison between the real sensor and the clone sensor

In this way a low cost sensor is obtained, but it responds very closely to a specialized sensor and with a much higher cost. It is thought to do the same with the noise sensor.

The connection of the system can be observed in detail figure 11 in order to readers to have a better idea about the assembly of DIPORMEC, making it clear that sensors 3, 4 and 5 are optional, there could even be many more. However, the card is built and programmed according to the needs.

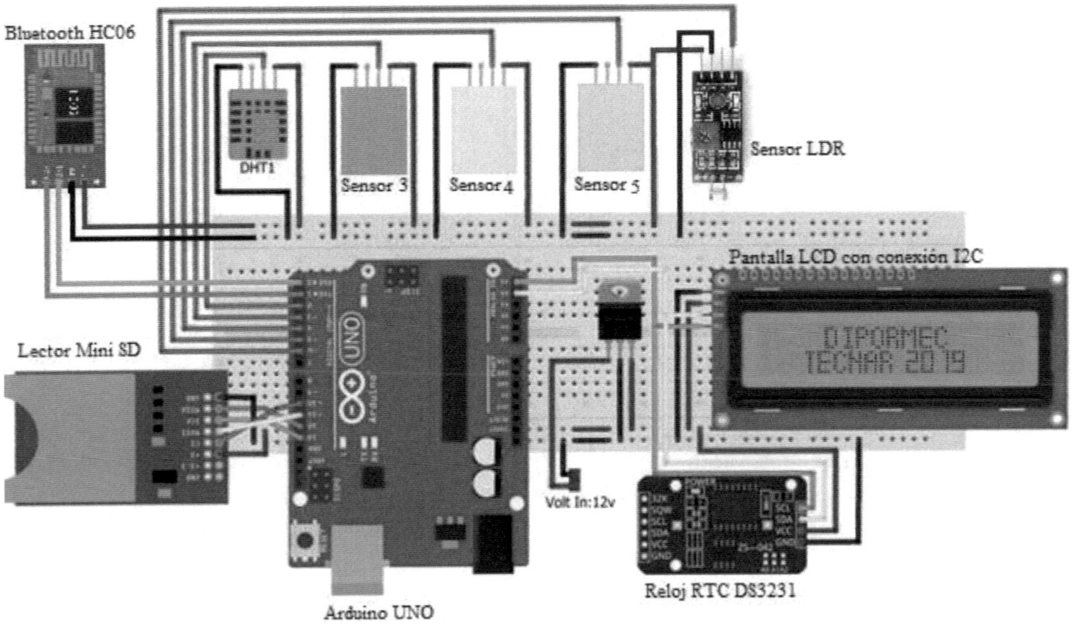

Figure 11. Connection's diagram among the DIPORMEC components

5. Conclusions

As a first conclusion, the construction of a portable device for the continuous acquisition of data (DIPORMEC) facilitates some tasks. This first prototype confirms this because it measures comfort variables in the workplace, and not only with the typical measurement with specialized instruments, whose measurements are made at a specific time, but also it can be monitored daily, weekly or monthly if necessary, so that preventive actions can be taken as required.

DIPORMEC is a device that at cost price, value of materials one by one, does not exceed 20 USD. Besides, it provides the possibility of having a photometer, a thermometer and a relative humidity meter at the same time, and with a few dollars more, noise measurement can be added and even air quality metering.

In the market some equipment is obtained for up to $ 14, but these only give one variable, and it does not have the possibility of expanding to more variables, DIPORMEC is an ideal solution for measuring multi-simple variables and that allows to expand to new sensors since even There are free analogs and digital input pins, and it will allow you to save to an SD card, and simultaneously transmit via Bluetooth and WiFi. A device of this type exceeds $ 80, and is not expandable to new variables.

The cloning sensor technique turns out to be very relevant in terms of lowering costs. However, these sensors cannot be compared in durability and other characteristics of the sensors of recognized brands.

Acknowledgments

Our gratitude to the Fundación Tecnológica Antonio de Arévalo - TECNAR - for supporting and financing the project. Moreover, it helped potentiating the technological development of the institution. Likewise, thanks are expressed to the students of the SIRATEC Research Seedbed, ROBOTEC TEAM and SIMEC for their dedication and performance to carry out the activities and contribute in one way or another to this project.

References

[1] Meetthings, «meetthings.com,» 2018. [En línea]. Available: http://meetthings.com/blog/index.php/17-11-2011/que-es-un-lumen/. [Último acceso: 1 08 2018].
[2] gsmspain, «gsmspain.com,» 2018. [En línea]. Available: http://www.gsmspain.com/glosario/?palabra=SD. [Último acceso: 15 09 2018].
[3] Tienda.bricogeek, «www.tienda.bricogeek.com,» 2018. [En línea]. Available: https://tienda.bricogeek.com/descatalogado/10-arduino-duemilanove-usb.html. [Último acceso: 15 08 2018].
[4] Arduino, «Arduino,» 2018. [En línea]. Available: https://store.arduino.cc/usa/arduino-nano. [Último acceso: 7 7 2018].
[5] L. Llamas, «Ingeniería, Informática y diseño.,» 2018. [En línea]. Available: https://www.luisllamas.es/tarjeta-micro-sd-arduino/. [Último acceso: 20 7 2018].
[6] TodoMicro, «TodoMicro La tecnología en tus manos,» 2018. [En línea]. Available: https://www.todomicro.com.ar/arduino/24-modulo-bluetooth-hc-06.html. [Último acceso: 2 8 2018].
[7] Electronilab, «www.electronilab.co,» 2018. [En línea]. Available: https://electronilab.co/tienda/sensor-de-temperatura-y-humedad-dht11/. [Último acceso: 06 09 2018].
[8] Ardobot, «Ardobot,» 2018. [En línea]. Available: www.ardobot.com/modulo-sensor-de-intensidad-de-luz-ldr.html. [Último acceso: 10 09 2018].

IOP Conf. Series: Materials Science and Engineering 1154 (2021) 012046 doi:10.1088/1757-899X/1154/1/012046

Radial distribution Systems solving in GAMS - Practice Implementation

L A Rojas-Martinez[1], OD Montoya[2], J A Martinez-Trespalacios[3], C. Rojas[4], C. Pacheco[4], john R. Castro-Suarez[5]

[1]Fundación Universitaria Tecnologico Comfenalco, Cartagena de Indias - Colombia
[2]Universidad Tecnologica de bolivar, Cartagena de Indias - Colombia
[3]Institución Tecnológica Colegio Mayor de Bolívar, Cartagena de Indias - Colombia
[4]Universidad de Sucre, Sincelejo - Colombia
[5]Universidad del Sinu, UNISINU, Cartagena - Colombia

E-mail: [1]luis.rojas@tecnocomfenalco.edu.co

Abstract. In the following work, the practical implementation of a radial distribution system will be carried out. This work is done in the GAMS software.In the present work the analysis will be carried out to three radial systems of 10, 33, 69 nodes. In which it is expected to analyze the best location of distributed generation systems (GD) and batteries (B), in order to obtain an objective function (Z) that represent the best performance of each system. The systems have a single source of generation and will be assigned Distributed Generators and Batteries. In the same way taking into account the variation of the systems, it is expected to analyze the active power losses, the reactive power losses, the power of the battery and the charge state of the battery in a period of 24 hours. The analysis was done with the GAMS software.

1. Introduction

Distributed Generation (DG) can be defined as electricity generation by small-scale technologies to produce electricity near the loads fed. Over the last decade, energy and environment have become pivotal concerns to all countries in the world. Most of them have been encouraged through regulations such as the Paris agreement (2016) to face climate change issues [1]. Electricity coming from renewable energy sources accounted for almost 25% of the world's capacity in 2018 [2] and it is forecasted to be expanded by 50% between 2019 and 2024, led by solar photovoltaic (PV) [3]. Widespread penetration of these mature renewable energy and other distributed generation technologies at home and abroad have allowed to incorporate them into the distribution network [4, 5]. As a consequence, there is a global trend of deregulation and decentralization of the electricity market, and subsequently, there is an imminent need for more flexible electric systems. Advances in research have demonstrated that introducing DG in power distribution can bring several benefits including voltage stability, power losses reduction, grid reinforcement, and cost reduction during on-peak periods among others [4, 6]. However, the unappropriated location of DGs can cause adverse effects including increased power losses, as DG can affect power flow distribution in distribution networks. Therefore, it is necessary to find the optimal locations and sizing of DGs to increase their benefits. To date, there have been many approaches reported to solve this problem. These approaches can be grouped into three categories: Classical optimization [7, 8, 9] Meta-heuristic [10, 11, 12] and analytical approaches [13].

Classical methods includes linear programming (LP) and optimal power flow (OPF). The first, is probably the easiest to implement, but not all the models allows reducing them into a set of linear equations [14]. The second, can optimize very complex problems with several variables, finding the best operating levels for electric power plants to satisfy demands given while minimizing operating cost [9].

Regarding analytical approaches, Naik et al. proposed analytical expressions for optimal siting and sizing of DG in radial distributions, based on changes in active and reactive components of branch currents cause by the DG placement [15]. As well, Elsaiah pointed to optimal placement and sizing of distributed generation on power distribution systems for loss reduction by proposing non-iterative and direct equations able to avoid convergence issues [16].

However, when problems becomes non-linear or solutions gets high dimensions, both above mentioned method may fall (CITE). Meta-heuristic techniques are the most widely used to overcome these problems. These methods are mostly based on nature behavior. Currently, several studies have used metaheuristic methods to address these problems. For instance, in Junjie's et al. study, authors analyzed the problem of location of Distributed Generation (DG) based on an immune algorithm. They solved the problem of the energy planning of the distribution networks in smart grids and dynamic planning as a final optimal solution considering costs associated to environmental compensation, traditional DG capacity and maintenance costs [17]. Shena and Provas, developed a Krill Herd (KH) based algorithm to solve optimal placement of distributed generator problem; reducing power losses and energy cost in distribution radial networks [18]. Zhiyun et al. proposed a novel methodology to solve the smart distribution network with the goal of minimizing power losses and enhance voltage stability using an algorithm based on the electromagnetism-like mechanism (ELM) [19]. Grisales and Gonzalez solved the problem based on Population-Based Incremental Learning (PBIL) and Particle Swarm Optimization PSO techniques [20]. Likewise, Setyawan and Soeprijanto used a artificial intelligence technique Symbiotic Organism Search (SOS) for sizing distributed generation and Loss Reduction Sensitivity Factor (LRSF) for optimal the placement of DG [21]. Ali et. al use the Ant-Lion (AL) algorithm for optimal location and sizing of renewable DG [22]. Sanjay et al. used the hybrid gray wolf optimizer (HGWO) to solve the allocation of DG [23]

The main drawback meta-heuristic methods is finding the optimal hyper-parameters. The solution provided by them is near optimal solution for a problem with numerous variables. However, this solution may change when changing algorithm start point or hyper-parameters, as a consequence, the effectiveness could be decreased. In addition, parameters selection task can be more complicated when using ensemble models.

Although the high efficiency of DG to supply energy sources, there are still challenges to face abrupt increases in electricity demand. The incorporation of energy storage (ES) devices represents a potential solution not only to satisfy the 100% of electricity demand [24], but also for use the stored energy to be supplied during on-peak periods for cost-reductions [25]. Lithium-based batteries the most used in the industry today, are experiencing its lowest cost in history, more than 85% cheaper than in 2010, bringing electricity cost from this source down to almost four-fifths during the period 2010-2019 and is forecasted that by 2023, battery average prices will fall up to 90% compared with 2010 prices [26, 27, 28]. These reductions are thanks to increasing order size and market uptake in battery electric vehicles sales. However, unappropriated placement and sizing of ES devices and DG altogether may implies technical (transient stability) and economical (cost) impacts [29]. Currently, several approaches have pointed to include ES and DG altogether to maximize benefits on the grid side. For instance, Erdinc et al. addressed the sizing of distributed generation and energy storage for smart households under a demand response strategy over a long-term analysis period [30]. Conducted an approach pointing to

IOP Conf. Series: Materials Science and Engineering 1154 (2021) 012046 doi:10.1088/1757-899X/1154/1/012046

coordinated operation of a neighborhood of smart households considering the vehicle to grid (V2G) concepts, as well as other ES units and DG [31]. In Cao's et al. study, authors propose a dynamic coordination optimization for active distribution network considering ES devices [32] on a 33-node test feeder.

In this paper, we addressed the problem by using the General Algebraic Modelling System (GAMS) to determine the optimal location and sizing of DG and ES units in radial distribution systems by considering active and reactive power losses, and battery operational parameters such as state of charge during a 24-hours period. Early stages of this approach are presented in [33], were a MATLAB-based GUI was developed to perform analysis of power losses in radial distribution networks considering DG units. In this work, GAMS was selected as nonlinear optimizing package to solve this problem using the large-scale nonlinear optimization CONOPT solver. Finally, we conducted three test systems with 10, 33 and 69 nodes respectively, used for simulation purposes in order to confirm both the effectiveness and robustness of the nonlinear model, and the proposed GAMS solution methodology.

2. Objective function

The objective function, represented by Equation 1, targets the minimization of the operative cost of each of the systems.

$$z = min\left(\sum_{i=1}^{g}\sum_{t=1}^{24} C\left(i,t\right) P_{Gconv}\left(i,t\right) \Delta t\right) \quad (1)$$

Where $\cdot C\left(i,t\right)$, is the purchase cost of energy, P_{Gconv}, is the Power Generation of a Conventional Generator connected to node i for each period t, Δt is delta of the time period. $P_{Gdistrib}$, is the Power Generation of a Distributed Generator. P_{Gbat}, is the Power Generation of the Batteries. Finally, P_{Dispa} in the Dispatched Power.

2.0.1. Constraints Eq. (2) represents the active power balance at each node in the network .

$$\begin{aligned} P_{Gconv}\left(i,t\right) + P_{Gdistrib}\left(i,t\right) + P_{Gbat}\left(i,t\right) - \\ P_{Disp}\left(i,t\right) = V\left(i,t\right)\sum_{j=1}^{n} Y\left(i,j\right) V\left(i,j\right) \\ cos\left(\theta\left(i,t\right) - \theta\left(i,t\right) - \delta\left(i,j\right)\right) \\ \left\{\forall i \in \Omega_N\right\} \end{aligned} \quad (2)$$

Eq. (3) represents the reactive power balance at each node in the network .

$$\begin{aligned} Q_{Gconv}\left(i,t\right) + Q_{Gdistrib}\left(i,t\right) + Q_{Gbat}\left(i,t\right) - \\ Q_{Disp}\left(i,t\right) = V\left(i,t\right)\sum_{j=1}^{n} Y\left(i,j\right) V\left(i,j\right) \\ sin\left(\theta\left(i,t\right) - \theta\left(i,t\right) - \delta\left(i,j\right)\right) \\ \left\{\forall i \in \Omega_N\right\} \end{aligned} \quad (3)$$

Eq. 4 represent capacity limits of voltage at each node of the network.

$$V^{min} \leq V\left(i,t\right) \leq V^{max} \forall i \in \Omega_N \quad (4)$$

Eq. 5 represents the capacity limits for Active power coming from CG.

$$P_{Gconv}^{min} \leq P_{Gconv}(i,t) \leq P_{Gconv}^{max} \forall i \in \Omega_{Gconv} \tag{5}$$

Eq. 6 represents the capacity limits for active power coming from DG.

$$P_{Gdistrib}^{min} \leq P_{Gdistrib}(i,t) \leq P_{Gdistrib}^{max} \forall i \in \Omega_{Gdistrib} \tag{6}$$

Eq. 7 represents the capacity limits for reactive power coming from CG.

$$Q_{Gconv}^{min} \leq Q_{Gconv}(i,t) \leq Q_{Gconv}^{max} \forall i \in \Omega_{Gconv} \tag{7}$$

Eq. 8 represents the capacity limits for reactive power coming from DG.

$$Q_{Gdistrib}^{min} \leq Q_{Gdistrib}(i,t) \leq Q_{Gdistrib}^{max} \forall i \in \Omega_{Gdistrib} \tag{8}$$

ES units were constrained specifically on the State-of-the-charge (SoC). Eqs 11 and 10 models the charge and discharge processes. Eqs 13 and 12 bounds the time of charge. Finally, Eq. 13 bounds the SoC.

$$SoC_{bat}(i,t) = SoC_{bat}(i,t-1) + \gamma_{bat} P_{Gbat(i,t)} \tag{9}$$

$$SoC_{bat}(i,t) = SoC_{bat}(i,t-1) - \gamma_{bat} P_{Gbat(i,t)} \tag{10}$$

$$SoC_{bat}(i,t=1) = SoC_{initial}(i,t) \tag{11}$$

$$SoC_{bat}(i,t=24) = SoC_{final}(i,t) \tag{12}$$

$$0 \leq SoC(i,t) \leq 1, \forall i \in \Omega_N \tag{13}$$

Where γ_{bat}, is the charging efficiency of the batteries, $SoC_{initial}$ is the initial state of charge of the batteries. SoC_{final} is the final state of charge of the batteries.

3. General modeling system: GAMS implementation

4. Test systems and simulation cases

This section introduces the electrical configuration and the test system information, for the radial distribution systems used in this work for validating the proposed formulation in the GAMS package. Three test system were used: a 10-node and a 33-node test systems, and a 69-node test feeder. Detailed information of these test systems is presented below.

5. Optimal flow cases

In this section we will be giving the specifications of each of the radial systems, radial systems of 10 nodes, radial system of 33 nodes and radial system of 69 nodes. In each one of them will be given the index of generators, the index of nodes, the index of times, the index of distributed generators, the index of batteries to be used, the association of generators and nodes; by means of which the point where the main generator will be established, the association of distributed generators and nodes; through which it will be established which is the best place to locate the system of distributed generators and the association of batteries and nodes; By means of the latter it is expected to improve the performance of the system, improving the losses and obtain a better performance of the system, since there is no interruption with the main generation of the system. This process will be presented step by step explaining the process mounted GAMS software screen to screen.

IOP Conf. Series: Materials Science and Engineering 1154 (2021) 012046 doi:10.1088/1757-899X/1154/1/012046

```
****************** OPTIMUM FLOW PROBLEM ******************************
****************** Master's Course in Engineering ********************
****************** Technological University of Bolivar ***************

SETS
G              Generator Index                             /G1/
N              Nodes Index                                 /N1*N10/
T              Time Index                                  /T0*T23/
GD             Distributed Generators Index                /GD1*GD2/
B              Battery Index                               /B1/
MAP(G,N)       Association Index of Generators and Nodes  /G1.N1/
MAPGD(GD,N)    Index of Distributed Generators and Nodes  /GD1.N5,GD2.N10/
MAPB(B,N)      Association Index of Battery and Node      /B1.N10/
;
```

Figure 1. SETS

```
TABLE GDATA(G,*) Input Data of Generators

     PMIN PMAX QMIN QMAX COST
*    (W)  (W)  (VAr) (VAr) (E/Wh)
G1   0.0  3.0  -1.0  2.0   6;
```

Figure 2. Input data of generators

```
TABLE GDATA(G,*) Input Data of Generators

     PMIN PMAX QMIN QMAX COST
*    (W)  (W)  (VAr) (VAr) (E/Wh)
G1   0.0  3.0  -1.0  2.0   6;
```

Figure 3. PMIN - PMAX - QMIN - QMAX - COST

5.1. Optimal flow cases tf 10 nodes

In this introductory part of the GAMS code, the respective variations will be made in order to establish the best location of the two systems of Distributed Generation and the best location of the storage system. The radial system to take into account to perform the analysis is a radial system of 10 nodes and nine lines, See Figure. 16. The YBUS information is in the Table 1.

Table 1. YBUS Data System TF10

Node i	Node j	R [ ]	X [ ]	P[kW]	Q[kVAr]	V[p.u]
1	2	0.1233	0.4127	1840	460	0.9929
2	3	0.2467	0.6051	980	340	0.9823
3	4	0.7469	12.050	1790	446	0.9581
4	5	0.6984	0.6084	1598	1840	0.9427
5	6	19.837	17.276	1610	600	0.9116
6	7	0.9057	0.7886	780	110	0.9015
7	8	20.552	11.640	1150	60	0.8832
8	9	47.953	27.160	980	130	0.8527
9	10	53.434	30.264	1640	200	0.8313

To establish the analysis, the variation of the Distributed Generators will be presented as follows:

Index of Distributed Generators and Nodes

```
TABLE LINE(N,N,*) Line Input Data

                     Y                            PHI

 N1.N1               1228.16130920897             -1.28047364644308
 N2.N1               1228.16130920897              1.86111900714671
 N1.N2               1228.16130920897              1.86111900714671
 N2.N2               2035.41542707382             -1.24202231373257
 N3.N2               809.539838140677              1.95792399450852
 N2.N3               809.539838140677              1.95792399450852
 N3.N3               1179.08776879658             -1.13080644929435
 N4.N3               373.13854150419              2.12567215566408
 N3.N4               373.13854150419              2.12567215566408
 N4.N4               934.181061368825             -0.834676197190369
 N5.N4               571.128742969098              2.42495643472672
 N4.N5               571.128742969098              2.42495643472672
 N5.N5               772.229199939385             -0.716601242958232
 N6.N5               201.100458311739              2.42509074279997
 N5.N6               201.100458311739              2.42509074279997
 N6.N6               641.600108727414             -0.716427691100239
 N7.N6               440.499651222427              2.42519884586163
 N6.N7               440.499651222427              2.42519884586163
 N7.N7               661.470353209691             -0.648718978787816
 N8.N7               223.968784486594              2.62626956823296
 N7.N8               223.968784486594              2.62626956823296
 N8.N8               319.957932266175             -0.51532755661009
 N9.N8               95.9891477870433              2.62625466432998
 N8.N9               95.9891477870433              2.62625466432998
 N9.N9               182.13245266444              -0.515335494917402
 N10.N9              86.1433048780279              2.62625993810839
 N9.N10              86.1433048780279              2.62625993810839
 N10.N10             86.1433048780279             -0.515332715481401;
```

Figure 4. Line Input data

```
--
TABLE BUS(N,*) Input Data YBUS
          VMIN     VMAX     PL       QL
*         (V)      (V)      (W)      (VAr)
N1        0.8      1.1      0.0      0.0
N2        0.8      1.1      1.84     0.46
N3        0.8      1.1      0.98     0.34
N4        0.8      1.1      1.79     0.446
N5        0.8      1.1      1.598    1.84
N6        0.8      1.1      1.61     0.6
N7        0.8      1.1      0.78     0.11
N8        0.8      1.1      1.15     0.06
N9        0.8      1.1      0.98     0.13
N10       0.8      1.1      1.64     0.2;
```

Figure 5. Input data YBUS

[(1)]GD1.N4,GD2,N10 GD1.N5,GD2,N10 GD1.N6,GD2.N10 GD1.N7,GD2,N10 GD1.N8,GD2,N10 GD1.N9,GD2,N10

To complement the analysis, the Association of the Battery and Nodes will be presented as follows:

Association Index of Battery and Node

[(1)]B1.N10

```
TABLE TIEMPO(T,*)  Input Data Time
        DEMAND
*        ($)
T0        1
T1        1
T2     0.3584
T3     0.3847
T4     0.4996
T5     0.7908
T6     0.8372
T7     0.9577
T8     0.8196
T9     0.8030
T10    0.8342
T11    0.6861
T12    0.7547
T13    0.6890
T14    0.6633
T15    0.6540
T16    0.6548
T17    0.7272
T18    0.9418
T19    1.0000
T20    0.9253
T21    0.9031
T22    0.7458
T23    0.5478;
```

Figure 6. Input data time

```
TABLE POT(GD,*)  Input Data Power Distributed Generators
     Cap
GD1 2.0
GD2 2.5;
- - -
```

Figure 7. Input data power distributed generators

```
TABLE BATTS(B,*)  Input Data SoC Batteries
     SoCmin  SoCmax  SoC0  SoCf  phib  pbmin  pbmax
B1  0.5     1       0.5   0.5   0.5   -0.5   0.5
- - -
```

Figure 8. Input data SoC batteries

5.2. Optimal flow cases tf 33 nodes

To present the analysis process done in GAMS software as the first instance, it will be presented the respective variations will be made in order to establish the best location of the two systems of Distributed Generation and the best location of the storage system. The radial system to

```
SCALAR
PI /3.1416/;
```

Figure 9. Input scalar

```
** The optimization variables are declared **

VARIABLES
z            value of the objective function
p(G,T)       active power generated by the generator G
pgd(GD,T)    power generated by GD
q(G,T)       Reactive power generated by the generator G
v(N,T)       modulus of tension in node N
d(N,T)       angle of tension in node N
pb(B,T)      Battery power B
SoC(B,T)     Battery state of Charge B
Ploss(T)     Losses of active power
Qloss(T)     Losses of recactive power;
```

Figure 10. Optimization Variables declared

```
** Assignment of limits of the variables **
**p.lo(G)=GDATA(G,'PMIN'); p.up(G)=GDATA(G,'PMAX');
**q.lo(G)=GDATA(G,'QMIN'); q.up(G)=GDATA(G,'QMAX');
v.lo(N,T)=BUS(N,'VMIN'); v.up(N,T)=BUS(N,'VMAX');
d.lo(N,T)=-PI; d.up(N,T)=PI;
pgd.lo(GD,T) =0; pgd.up(GD,T) = GDS(T,GD);
pb.lo(B,T) = BATTS(B,'pbmin'); pb.up(B,T) = BATTS(B,'pbmax');
SoC.lo(B,T) = BATTS(B,'SoCmin'); SoC.up(B,T) = BATTS(B,'SoCmax');
** El nudo 3 se toma como nudo de referencia, por
** eso se fija su angulo a cero.
d.fx('N1',T)=0;
v.fx('N1',T)=1;
```

Figure 11. Assignment of limits of the variables

```
ALIAS(N,NP);
** Las matrices de datos Z y PHI se definen simetricas mediante
** la condicion $(ORD(N) GT ORD(NP)) sobre los conjuntos N y NP.
*LINE(N,NP,'Y')$(ORD(N) GT ORD(NP))=LINE(NP,N,'Y');
*LINE(N,NP,'PHI')$(ORD(N) GT ORD(NP))=LINE(NP,N,'PHI');
** Se declaran las restricciones.
```

Figure 12. input data variables

take into account to perform the analysis is a radial system of 33 nodes and 32 lines, See Figure. 17. The YBUS information is in the Table 2.

To establish the analysis, the variation of the Distributed Generators will be presented as follows:

Index of Distributed Generators and Nodes

[(1)]GD1.N7,GD2.N20 GD1.N8,GD2.N21 GD1.N9,GD2,N22 GD1.N10,GD2,N23 GD1.N11.GD2,N24 GD1.N12.GD2,N25

```
EQUATIONS
COST         objective function
PBAL(N,T)    restrictions of active power balance
QBAL(N,T)    restrictions of reactive power balance
Qper(T)      Equation of reactive power losses
Pper(T)      Equation of losses active power
SoCB(B,T)    Battery state of charge equation
;
```

Figure 13. Equations

```
COST.. z =e= SUM(T,ploss(T));

PBAL(N,T).. SUM(G$MAP(G,N),p(G,T))+ SUM(B$MAPB(B,N),pb(B,T)) +
SUM(GD$MAPGD(GD,N),pgd(GD,T)*POT(GD,'Cap')) -
BUS(N,'PL')*TIEMPO(T,'DEMANDA')=e=v(N,T)*SUM(NP,LINE(N,NP,'Y')*v(NP,T)*
COS(d(N,T)-d(NP,T) - LINE(N,NP,'PHI')));

QBAL(N,T).. SUM(G$MAP(G,N),q(G,T))-BUS(N,'QL')*TIEMPO(T,'DEMANDA')=e= v(N,T)*
SUM(NP,LINE(N,NP,'Y')*v(NP,T)*SIN(d(N,T)-d(NP,T)-LINE(N,NP,'PHI')));

pper(T).. Ploss(T) =e= SUM(N,v(N,T)*SUM(NP,LINE(N,NP,'Y')*v(NP,T)*COS(d(N,T)-
d(NP,T) -LINE(N,NP,'PHI'))));

qper(T).. qloss(T) =e= SUM(N,v(N,T)*SUM(NP,LINE(N,NP,'Y')*v(NP,T)*SIN(d(N,T)-
d(NP,T)-LINE(N,NP,'PHI'))));

SoCB(B,T)$(ord(T) gt 1).. SoC(B,T) =e= SoC(B,T-1) - BATTS(B,'phib')* pb(B,T);

d.fx('N1',T)=0;

v.fx('N1',T)=1.0;

SoC.fx(B,'T0')= BATTS(B,'SoC0');

SoC.fx(B,'T23')= BATTS(B,'SoCf');

```

Figure 14. Equations

```
MODEL foc /ALL/;
SOLVE foc USING nlp MINIMIZING z;
option decimals =5;
display v.l,ploss.l,qloss.l,z.l,pb.l,SoC.l
```

Figure 15. Equations

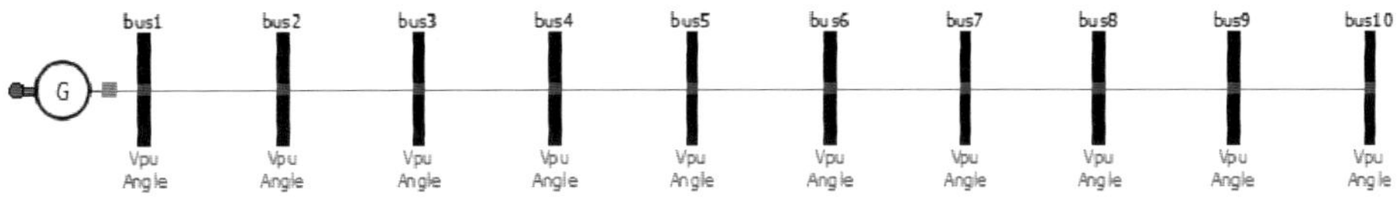

Figure 16. Radial system 10 nodes

To complement the analysis, the Association of the Battery and Nodes will be presented as follows:

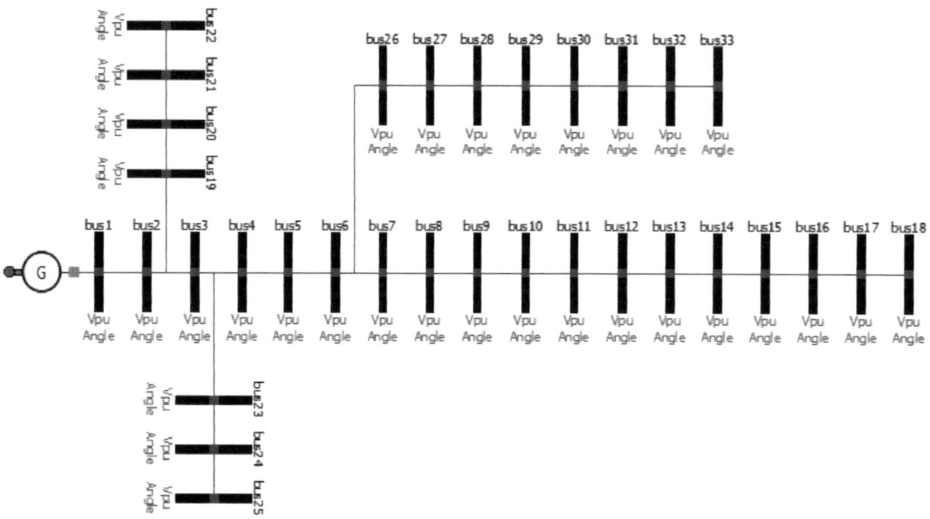

Figure 17. Radial system 33 nodes

Table 2. YBUS Data System TF33

Node i	Node j	R [ ]	X [ ]	P[kW]	Q[kVAr]	V[p.u]
1	2	0.0922	0.0477	100	60	0.9970
2	3	0.4930	0.2511	90	40	0.9829
3	4	0.3660	0.1864	120	80	0.9754
4	5	0.3811	0.1941	60	30	0.9679
5	6	0.8190	0.7070	60	20	0.9495
6	7	0.1872	0.6188	200	100	0.9459
7	8	1.7114	1.2351	200	100	0.9323
8	9	1.0300	0.7400	60	20	0.9260
9	10	1.0400	0.7400	60	20	0.9201
10	11	0.1966	0.0650	45	30	0.9192
11	12	0.3744	0.1238	60	35	0.9177
12	13	1.4680	1.1550	60	35	0.9115
13	14	0.5416	0.7129	120	80	0.9092
14	15	0.5910	0.5260	60	10	0.9078
15	16	0.7463	0.5450	60	20	0.9064
16	17	1.2890	1.7210	60	20	0.9044
17	18	0.7320	0.5740	90	40	0.9038
2	19	0.1640	0.1565	90	40	0.9965
19	20	1.5042	1.3554	90	40	0.9929
20	21	0.4095	0.4784	90	40	0.9922
21	22	0.7089	0.9373	90	40	0.9916
3	23	0.4512	0.3083	90	50	0.9793
23	24	0.8980	0.7091	420	200	0.9726
24	25	0.8960	0.7011	420	200	0.9693
6	26	0.2030	0.1034	60	25	0.9475
26	27	0.2842	0.1447	60	25	0.9450
27	28	1.0590	0.9337	60	20	0.9335
28	29	0.8042	0.7006	120	70	0.9253
29	30	0.5075	0.2585	200	600	0.9218
30	31	0.9744	0.9630	150	70	0.9176
31	32	0.3105	0.3619	210	100	0.9167
32	33	0.3410	0.5302	60	40	0.9164

Association Index of Battery and Node
$[(1)]_{B1.N15}$

5.3. Optimal flow cases tf 69 nodes

To present the analysis process done in GAMS software as the first instance, it will be presented the respective variations will be made in order to establish the best location of the two systems of Distributed Generation and the best location of the storage system. The radial system to take into account to perform the analysis is a radial system of 69 nodes and 68 lines, See Figure. 18. The YBUS information is in the Table 3.

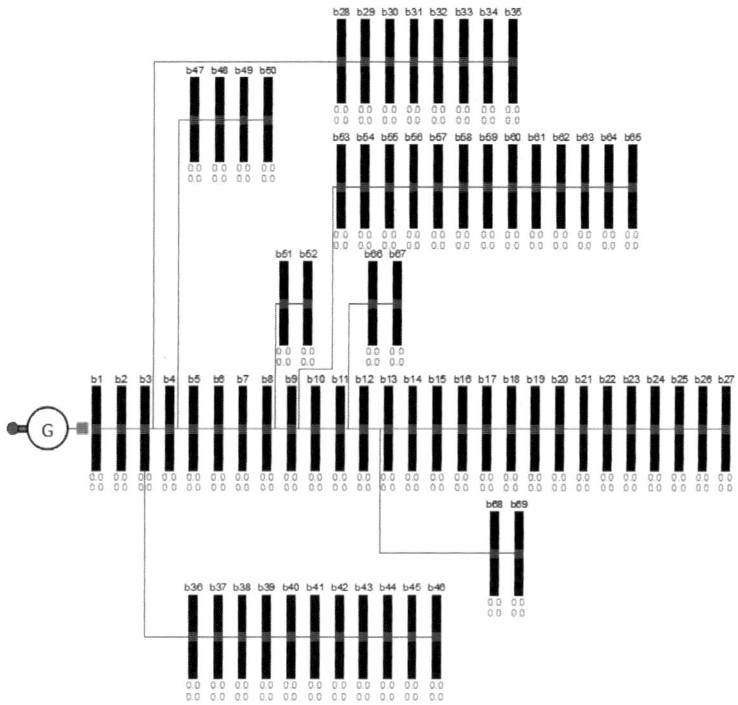

Figure 18. Radial system 69 nodes

To establish the analysis, the variation of the Distributed Generators will be presented as follows:

Index of Distributed Generators and Nodes

[(1)]GD1.N17,GD2,N49 GD1.N88,GD2,N50 GD1.N19,GD2,N51 GD1.N20,GD2,N52 GD1.N21,GD2,N53 GD1.N22,GD2,N54

To complement the analysis, the Association of the Battery and Nodes will be presented as follows:

Association Index of Battery and Node

[(1)]B1.N68

6. Results

6.1. Analysis Result GAMS tf10 nodes

Taking into account the variations with the Distributed Generator G1, the Distributed Generator G2, and the storage system connected to the node N10. The following objective function is obtained. See Table 4.

In the first test configuration it can be seen that the objective function z reaches up to 4.04893, in the second configuration it can be seen that the value increases to 4.28028, in the third configuration it shows a significant decrease with respect to the previous two with a value of z equal 3.62215, in the fourth configuration it is observed that the value decreases and the objective function z shows a value of 3.5376, in the fifth configuration when the GD 1 is connected to node 8 the function decreases further showing a z equal to 3.46973. Taking into account this progress it was to be expected that when testing with the GD1 configuration connected to node

Table 3. Radial system 69 nodes

Node i	Node j	R [ ]	X [ ]	P[kW]	Q[kVAr]	Node i	Node j	R [ ]	X [ ]	P[kW]	Q[kVAr]
1	2	0.0005	0.0012	0	0	3	36	0.0044	0.0108	26	18.55
2	3	0.0005	0.0012	0	0	36	37	0.0640	0.1565	26	18.55
3	4	0.0015	0.0036	0	0	37	38	0.1053	0.1230	0	0
4	5	0.0251	0.0294	0	0	38	39	0.0304	0.0355	24	17
5	6	0.366	0.1864	0	0	39	40	0.0018	0.0021	24	17
6	7	0.3811	0.1941	2.6	2.2	40	41	0.7283	0.8509	102	1
7	8	0.0922	0.047	40.4	30	41	42	0.3100	0.3623	0	0
8	9	0.0493	0.0251	75	54	42	43	0.0410	0.0478	6	4.3
9	10	0.8190	0.2707	30	22	43	44	0.0092	0.0116	0	0
10	11	0.1872	0.0619	28	19	44	45	0.1089	0.1373	39.22	26.3
11	12	0.7114	0.2351	145	104	45	46	0.0009	0.0012	39.22	26.3
12	13	10.300	0.3400	145	104	4	47	0.0034	0.0084	0	0
13	14	10.440	0.3450	8	5	47	48	0.0851	0.2083	79	56.4
14	15	10.580	0.3496	8	5	48	49	0.2898	0.7091	384.7	274.5
15	16	0.1966	0.0650	0	0	49	50	0.0822	0.2011	384.7	274.5
16	17	0.3744	0.1238	45	30	8	51	0.0928	0.0473	40.5	28.3
17	18	0.0047	0.0016	60	35	51	52	0.3319	0.1140	3.6	2.7
18	19	0.3276	0.1083	60	35	9	53	0.1740	0.0886	4.35	3.5
19	20	0.2106	0.0690	0	0	53	54	0.2030	0.1034	26.4	19
20	21	0.3416	0.1129	1	0.6	54	55	0.2842	0.1447	24	17.2
21	22	0.0140	0.0046	114	81	55	56	0.2813	0.1433	0	0
22	23	0.1591	0.0526	5	3.5	56	57	15.900	0.5337	0	0
23	24	0.3463	0.1145	0	0	57	58	0.7837	0.2630	0	0
24	25	0.7488	0.2475	28	20	58	59	0.3042	0.1006	100	72
25	26	0.3089	0.1021	0	0	59	60	0.3861	0.1172	0	0
26	27	0.1732	0.0572	14	10	60	61	0.5075	0.2585	1244	888
3	28	0.0044	0.0108	14	10	61	62	0.0974	0.0496	32	33
28	29	0.0640	0.1565	26	18.6	62	63	0.1450	0.0738	0	0
29	30	0.3978	0.1315	26	18.6	63	64	0.7105	0.3619	227	162
30	31	0.0702	0.0232	0	0	64	65	10.410	0.5302	59	42
31	32	0.3510	0.1160	0	0	11	66	0.2012	0.0611	18	33
32	33	0.8390	0.2816	0	0	66	67	0.0047	0.0014	18	33
33	34	17.080	0.5646	10	10	12	68	0.7394	0.2444	28	20
34	35	14.740	0.4873	14	14	68	69	0.0047	0.0016	28	20

Table 4. Objective Function Analysis tf10 nodes

GD	N	GD	N	B	N	z
1	5	2	10	1	10	4.04893
1	4	2	10	1	10	4.28028
1	6	2	10	1	10	3.62215
1	7	2	10	1	10	3.5376
1	8	2	10	1	10	3.46973
1	9	2	10	1	10	3.68020
1	10	2	10	1	10	4.10496

9, GD2 connected to node 10 and Battery connected to node 10; but the response of the system shows a high z of 3.68020. A similar behavior is observed in the following combination when GD1 is connected to node 10, GD2 to node 10 and Battery to node 10. In this case z showed a value of 4.10496. With which it can be concluded that in this radial system of 10 nodes to move away from the main conventional generation can be observed better efficiencies of the system but only to a certain extent.

Taking into account the best performance of the system tf10 nodes locating the distributed generators in nodes 8, 10 and the storage system in node 10. It is presented in Table 5. The active power losses. It is presented in Table 6. Losses of reactive power. It is presented in Table 7. The state of charge of the SoC battery.

Table 5. Losses of active power tf10 nodes

Time	Ploss.L	Time	Ploss.L	Time	Ploss.L	Time	Ploss.L	Time	Ploss.L
T0	0.31434	T1	0.36495	T2	0.03082	T3	0.03560	T4	0.06075
T5	0.16016	T6	0.16674	T7	0.17258	T8	0.10379	T9	0.09418
T10	0.10109	T11	0.06100	T12	0.07638	T13	0.06115	T14	0.05602
T15	0.05617	T16	0.05870	T17	0.09483	T18	0.24630	T19	0.29422
T20	0.28169	T21	0.30704	T22	0.19411	T23	0.07711		

6.2. Analysis Result GAMS tf33 nodes

Taking into account the variations with the Distributed Generator G1, the Distributed Generator G2, and the storage system connected to the different nodes. The following objective function

Table 6. Losses of reactive power tf10 nodes

Time	Qloss.L	Time	Qloss.L	Time	Qloss.L	Time	Qloss.L	Time	Qloss.L
T0	0.47762	T1	0.54222	T2	0.04397	T3	0.05077	T4	0.08649
T5	0.23812	T6	0.25220	T7	0.27562	T8	0.16886	T9	0.15460
T10	0.16594	T11	0.10102	T12	0.12636	T13	0.10118	T14	0.09263
T15	0.09423	T16	0.09916	T17	0.15701	T18	0.37527	T19	0.45129
T20	0.42353	T21	0.45262	T22	0.28967	T23	0.11690		

Table 7. Battery State of Charge B tf10 nodes

Time	SoC.L	Time	SoC.L	Time	SoC.L	Time	SoC.L	Time	SoC.L
T0	0.50000	T1	0.50000	T2	0.50019	T3	0.75018	T4	1.00000
T5	0.88994	T6	0.71795	T7	0.50000	T8	0.53822	T9	0.65698
T10	0.66286	T11	0.86475	T12	0.87232	T13	0.96436	T14	1.00000
T15	1.00000	T16	1.00000	T17	1.00000	T18	0.99734	T19	0.77221
T20	0.63287	T21	0.50000	T22	0.50000	T23	0.50000		

is obtained. See Table 8.

In the first test configuration it can be seen that the objective function z reaches up to 2.07447, in the second configuration it can be seen that the value decreases to 2.06209, in the third configuration it shows a significant increase with respect to the previous two with a value of z equal 2.075470, in the fourth configuration it is observed that the value decreases and the objective function z shows a value of 1.754244, in the fifth configuration when the GD 1 is connected to node 11 the z function increases a little big again with respect to the previous showing a z equal to 1.849709. Taking into account this progress it was to be expected that when testing with the GD1 configuration connected to node 12, GD2 connected to node 25 and Battery connected to node 20; but the response of the system shows a high z of 1.899920. With which it can be concluded that in this radial system of 33 nodes for proven cases the best performance of the system tf33 nodes locating the distributed generators GD1 in nodes 10, GD2 in nodes 23 and the storage system in node 18. It is presented in Table 9. The active power losses. It is presented in Table 10. Losses of reactive power. It is presented in Table 11. The state of charge of the SoC battery.

Table 8. Objective Function Analysis tf33 nodes

GD	N	GD	N	B	N	z
1	7	2	20	1	15	2.07447
1	8	2	21	1	16	2.06209
1	9	2	22	1	17	2.075470
1	10	2	23	1	18	1.754244
1	11	2	24	1	19	1.849709
1	12	2	25	1	20	1.899920

Table 9. Losses of active power tf33 nodes

Time	Ploss.L	Time	Ploss.L	Time	Ploss.L	Time	Ploss.L	Time	Ploss.L
T0	0.12550	T1	0.17115	T2	0.02783	T3	0.03005	T4	0.04174
T5	0.08653	T6	0.08921	T7	0.09638	T8	0.07119	T9	0.06840
T10	0.07370	T11	0.05055	T12	0.06065	T13	0.05095	T14	0.04743
T15	0.04619	T16	0.04629	T17	0.05647	T18	0.09444	T19	0.10916
T20	0.09489	T21	0.10369	T22	0.07113	T23	0.04072		

Table 10. Losses of recactive power tf33 nodes

Time	Qloss.L	Time	Qloss.L	Time	Qloss.L	Time	Qloss.L	Time	Qloss.L
T0	0.08874	T1	0.12230	T2	0.02060	T3	0.02211	T4	0.03012
T5	0.06136	T6	0.06280	T7	0.06755	T8	0.05014	T9	0.04820
T10	0.05188	T11	0.03579	T12	0.04280	T13	0.03607	T14	0.03362
T15	0.03276	T16	0.03283	T17	0.03990	T18	0.06614	T19	0.07554
T20	0.06550	T21	0.07278	T22	0.05022	T23	0.02888		

Table 11. Battery State of Charge B tf33 nodes

Time	SoC.L	Time	SoC.L	Time	SoC.L	Time	SoC.L	Time	SoC.L
T0	0.50000	T1	0.50000	T2	0.57206	T3	0.63400	T4	0.65158
T5	0.55553	T6	0.50000	T7	0.50000	T8	0.53372	T9	0.57016
T10	0.60149	T11	0.65708	T12	0.70144	T13	0.75656	T14	0.81589
T15	0.87676	T16	0.93748	T17	0.98634	T18	1.00000	T19	0.98947
T20	0.91523	T21	0.70338	T22	0.56218	T23	0.50000		

6.3. Analysis Result GAMS tf69 nodes

Taking into account the variations with the Distributed Generator G1, the Distributed Generator G2, and the storage system connected to the different nodes. The following objective function is obtained. See Table 12.

In the first test configuration it can be seen that the objective function z reaches up to 2.85431, in the second configuration it can be seen that the value is 2.854375, in the third configuration it shows a significant decrease with respect to the previous two with a value of z equal 2.450519, in the fourth configuration it is observed that the value increases and the objective function z shows a value of 2.556052, in the fifth configuration when the GD 1 is connected to node 21; the z function decreases a little big again with respect to the previous showing a z equal to 2.333838. Taking into account this progress it was to be expected that when testing with the GD1 configuration connected to node 22, GD2 connected to node 54 and Battery connected to node 68; we can see the response of the system shows a decrease of the objective function z with a value of 2.267984. With which it can be concluded that in this radial system of 69 nodes for proven cases the best performance of the system tf69 nodes locating the distributed generators GD1 in nodes 22, GD2 in nodes 54 and the storage system in node 68. It is presented in Table 13. The active power losses. It is presented in Table 14. Losses of reactive power. It is presented in Table 15. The state of charge of the SoC battery.

Table 12. Objective Function Analysis tf69

GD	N	GD	N	B	N	z
1	17	2	49	1	68	2.85431
1	18	2	50	1	68	2.854375
1	19	2	51	1	68	2.450519
1	20	2	52	1	68	2.556052
1	21	2	53	1	68	2.333838
1	22	2	54	1	68	2.267984

Table 13. Losses of active power tf69 nodes

c Time	Ploss.L	Time	Ploss.L	Time	Ploss.L	Time	Ploss.L	Time	Ploss.L
T0	0.15887	T1	0.16672	T2	0.02012	T3	0.02304	T4	0.03842
T5	0.09750	T6	0.10697	T7	0.14164	T8	0.10235	T9	0.09811
T10	0.10617	T11	0.07101	T12	0.08632	T13	0.07162	T14	0.06628
T15	0.06425	T16	0.06440	T17	0.08033	T18	0.14176	T19	0.16226
T20	0.13609	T21	0.13084	T22	0.08653	T23	0.04637		

Table 14. Losses of reactive power tf69 nodes

Time	Qloss.L	Time	Qloss.L	Time	Qloss.L	Time	Qloss.L	Time	Qloss.L
T0	0.06888	T1	0.07217	T2	0.00869	T3	0.00997	T4	0.01667
T5	0.04230	T6	0.04666	T7	0.06170	T8	0.04463	T9	0.04278
T10	0.04629	T11	0.03097	T12	0.03765	T13	0.03124	T14	0.02891
T15	0.02804	T16	0.02811	T17	0.03513	T18	0.06208	T19	0.07107
T20	0.05957	T21	0.05699	T22	0.03762	T23	0.02011		

7. Conclusions

Taking into account the best performance of the system tf10 nodes locating the distributed generators in nodes 8, 10 and the storage system in node 10. It is presented in Table 5. The

Table 15. Losses of recactive power tf33 nodes

Time	Soc.L	Time	Soc.L	Time	Soc.L	Time	Soc.L	Time	Soc.L
T0	0.50000	T1	0.50000	T2	0.54192	T3	0.57662	T4	0.57968
T5	0.50178	T6	0.50000	T7	0.52197	T8	0.56459	T9	0.60968
T10	0.65013	T11	0.71261	T12	0.76490	T13	0.82695	T14	0.89282
T15	0.94609	T16	0.99603	T17	1.00000	T18	0.93850	T19	0.84706
T20	0.79615	T21	0.59424	T22	0.50000	T23	0.50000		

active power losses. It is presented in Table 6. Losses of reactive power. It is presented in Table 7. The state of charge of the SoC battery.

Taking into account this progress it was to be expected that when testing with the GD1 configuration connected to node 12, GD2 connected to node 25 and Battery connected to node 20; but the response of the system shows a high z of 1.899920. With which it can be concluded that in this radial system of 33 nodes for proven cases the best performance of the system tf33 nodes locating the distributed generators GD1 in nodes 10, GD2 in nodes 23 and the storage system in node 18. It is presented in Table 9. The active power losses. It is presented in Table 10. Losses of reactive power. It is presented in Table 11. The state of charge of the SoC battery.

With which it can be concluded that in this radial system of 69 nodes for proven cases the best performance of the system tf69 nodes locating the distributed generators GD1 in nodes 22, GD2 in nodes 54 and the storage system in node 68. It is presented in Table 13. The active power losses. It is presented in Table 14. Losses of reactive power. It is presented in Table 15. The state of charge of the SoC battery.

References

[1] J. Rogelj, M. Den Elzen, N. Höhne, T. Fransen, H. Fekete, H. Winkler, R. Schaeffer, F. Sha, K. Riahi, M. Meinshausen, Paris agreement climate proposals need a boost to keep warming well below 2 c, Nature 534 (7609) (2016) 631–639.

[2] E. Global, Outlook 2018 (iea, 2018).

[3] I. IEA, W. UNSD, et al., Who. tracking sdg 7: The energy progress report 2019 (2019).

[4] G. Pepermans, J. Driesen, D. Haeseldonckx, R. Belmans, W. D'haeseleer, Distributed generation: definition, benefits and issues, Energy policy 33 (6) (2005) 787–798.

[5] D. Isle, M. Vaziri, M. Zarghami, S. Vadhva, Review of concepts to increase distributed generation into the distribution network, in: 2014 Sixth Annual IEEE Green Technologies Conference, IEEE, 2014, pp. 118–125.

[6] P. A. Daly, J. Morrison, Understanding the potential benefits of distributed generation on power delivery systems, in: 2001 Rural Electric Power Conference. Papers Presented at the 45th Annual Conference (Cat. No. 01CH37214), IEEE, 2001, pp. A2–1.

[7] Y. Wen, C. Guo, D. S. Kirschen, S. Dong, Enhanced security-constrained opf with distributed battery energy storage, IEEE Transactions on Power Systems 30 (1) (2014) 98–108.

[8] G. P. Harrison, A. Piccolo, P. Siano, A. R. Wallace, Hybrid ga and opf evaluation of network capacity for distributed generation connections, Electric Power Systems Research 78 (3) (2008) 392–398.

[9] C. J. Dent, L. F. Ochoa, G. P. Harrison, Network distributed generation capacity analysis using opf with voltage step constraints, IEEE Transactions on Power systems 25 (1) (2010) 296–304.

[10] S. Ganguly, D. Samajpati, Distributed generation allocation on radial distribution networks under uncertainties of load and generation using genetic algorithm, IEEE Transactions on Sustainable Energy 6 (3) (2015) 688–697.

[11] M. Sedghi, M. Aliakbar-Golkar, M.-R. Haghifam, Distribution network expansion considering distributed generation and storage units using modified pso algorithm, International Journal of Electrical Power & Energy Systems 52 (2013) 221–230.

[12] M. Kasaei, Optimal placement of distributed generation and capacitor in distribution networks by ant colony algorithm, International Journal on Technical and Physical Problems of Engineering (IJTPE) (20) (2014) 52–56.

[13] D. Q. Hung, N. Mithulananthan, R. Bansal, Analytical strategies for renewable distributed generation integration considering energy loss minimization, Applied Energy 105 (2013) 75–85.

[14] Ž. Popović, V. D. Kerleta, D. Popović, Hybrid simulated annealing and mixed integer linear programming

algorithm for optimal planning of radial distribution networks with distributed generation, Electric Power Systems Research 108 (2014) 211–222.
[15] S. N. G. Naik, D. K. Khatod, M. P. Sharma, Analytical approach for optimal siting and sizing of distributed generation in radial distribution networks, IET Generation, Transmission & Distribution 9 (3) (2014) 209–220.
[16] S. Elsaiah, M. Benidris, J. Mitra, Analytical approach for placement and sizing of distributed generation on distribution systems, IET Generation, Transmission & Distribution 8 (6) (2014) 1039–1049.
[17] M. Junjie, W. Yulong, L. Yang, Size and location of distributed generation in distribution system based on immune algorithm, Systems Engineering Procedia 4 (2012) 124 – 132, information Engineering and Complexity Science - Part II. doi:https://doi.org/10.1016/j.sepro.2011.11.057.
URL http://www.sciencedirect.com/science/article/pii/S2211381911002104
[18] S. Sultana, P. K. Roy, Krill herd algorithm for optimal location of distributed generator in radial distribution system, Applied Soft Computing 40 (2016) 391 – 404. doi:https://doi.org/10.1016/j.asoc.2015.11.036.
URL http://www.sciencedirect.com/science/article/pii/S1568494615007541
[19] Z. Gong, Q. Chen, K. Sun, Novel methodology solving distribution network reconfiguration with dg placement, The Journal of Engineering 2019 (16) (2019) 1668–1674. doi:10.1049/joe.2018.8521.
[20] L. F. Grisales-Noreña, D. Gonzalez Montoya, C. A. Ramos-Paja, Optimal sizing and location of distributed generators based on pbil and pso techniques, Energies 11 (4) (2018). doi:10.3390/en11041018.
URL https://www.mdpi.com/1996-1073/11/4/1018
[21] G. Setyawan, Umar, A. Soeprijanto, O. Penangsang, Symbiotic organism search for sizing and optimal location of distributed generation using novel sensitivity factor, IOP Conference Series: Materials Science and Engineering 588 (2019) 012021. doi:10.1088/1757-899x/588/1/012021.
URL https://doi.org/10.1088%2F1757-899x%2F588%2F1%2F012021
[22] E. Ali, S. A. Elazim, A. Abdelaziz, Ant lion optimization algorithm for optimal location and sizing of renewable distributed generations, Renewable Energy 101 (2017) 1311 – 1324. doi:https://doi.org/10.1016/j.renene.2016.09.023.
URL http://www.sciencedirect.com/science/article/pii/S0960148116308126
[23] R. Sanjay, T. Jayabarathi, T. Raghunathan, V. Ramesh, N. Mithulananthan, Optimal allocation of distributed generation using hybrid grey wolf optimizer, IEEE Access 5 (2017) 14807–14818. doi:10.1109/ACCESS.2017.2726586.
[24] X. Zhang, H. Chen, Y. Xu, W. Li, F. He, H. Guo, Y. Huang, Distributed generation with energy storage systems: A case study, Applied Energy 204 (2017) 1251–1263.
[25] E. García, Ì. Isaac, Demand response systems for integrating energy storage batteries for residential users, in: 2016 IEEE Ecuador Technical Chapters Meeting (ETCM), IEEE, 2016, pp. 1–6.
[26] M. A. Pellow, H. Ambrose, D. Mulvaney, R. Betita, S. Shaw, Research gaps in environmental life cycle assessments of lithium ion batteries for grid-scale stationary energy storage systems: End-of-life options and other issues, Sustainable Materials and Technologies (2019) e00120.
[27] A. Chaianong, A. Bangviwat, C. Menke, B. Breitschopf, W. Eichhammer, Customer economics of residential pv–battery systems in thailand, Renewable Energy 146 (2020) 297–308.
[28] A. Gailani, M. Al-Greer, M. Short, T. Crosbie, Degradation cost analysis of li-ion batteries in the capacity market with different degradation models, Electronics 9 (1) (2020) 90.
[29] A. K. Srivastava, A. A. Kumar, N. N. Schulz, Impact of distributed generations with energy storage devices on the electric grid, IEEE Systems Journal 6 (1) (2012) 110–117.
[30] O. Erdinc, N. G. Paterakis, I. N. Pappi, A. G. Bakirtzis, J. P. Catalão, A new perspective for sizing of distributed generation and energy storage for smart households under demand response, Applied Energy 143 (2015) 26–37.
[31] N. G. Paterakis, O. Erdinç, I. N. Pappi, A. G. Bakirtzis, J. P. Catalão, Coordinated operation of a neighborhood of smart households comprising electric vehicles, energy storage and distributed generation, IEEE Transactions on smart grid 7 (6) (2016) 2736–2747.
[32] M. Cao, H. Wang, K. Li, Z. Wu, L. Sun, Dynamic coordination optimization for active distribution network considering energy storage system, in: Proceedings of PURPLE MOUNTAIN FORUM 2019-International Forum on Smart Grid Protection and Control, Springer, 2020, pp. 675–688.
[33] L. A. Rojas, O. D. Montoya, J. Campillo, Analysis of power losses in electric distribution system using a matlab-based graphical user interface (gui), in: J. C. Figueroa-García, M. Duarte-González, S. Jaramillo-Isaza, A. D. Orjuela-Cañon, Y. Díaz-Gutierrez (Eds.), Applied Computer Sciences in Engineering, Springer International Publishing, Cham, 2019, pp. 565–576.

Design and development of a PWA - Progressive Web Application, to consult the diary and programming of a technological event.

C J Gómez-Sierra[1]

[1] Comsis: Research group in computing, information systems and software development - Antonio de Arevalo University Foundation, Cartagena - Colombia,
E-mail : cesar.gomez@unitecnar.edu.co

Abstract. Expotecnología is an event aimed at students, teachers and professionals in the areas of systems, computing, petrochemicals, process control and industrial instrumentation, electronics and telecommunications, electricity, electromechanics and civil works, with the purpose of addressing not only the challenges that accompany the various activities in which related professionals perform, but also the problems, projects and programs of the academic units in which they are trained. In addition, it is a conducive space to enhance their role in the profession, research and academy. Currently, there is a website for the Expotecnología event, which can be seen here: https://expotecnologia.unitecnar.edu.co. But all the scheduling and schedule management is consulted manually by means of flyers, billboards or email. In this sense, it is necessary to develop a progressive web application, where all the optimized processes can be carried out, in addition the participants to the event would not have any inconvenience to know the day, time and place where they will be presented, because it can also work without internet connection (offline).

1. Introduction

Understanding the concept in a very simple way, PWA (Progressive Web Application) are applications that work in the same way as native or traditional ones, because they are developed in an environment and interface of similar navigation and interaction with users. The main and true difference between these types of technologies is that, in the PWA, it is not necessary to download from any application store such as Google Play or the App Store, because these are executed directly in the browser of the device where it is displayed. Progressive Web Applications are considered as the evolution of applications developed in HTML5 with the help of Service Workers, which allows running all services in browsers, but in the background. Therefore, these PWAs can be charged almost immediately, even without internet connectivity. The applications called PWA are mobile web applications that take advantage of the new possibilities and APIs offered by the new Web technologies, such as Service Workers (a script that runs in the background and allows the implementation of functionalities that do not require a web page or user interaction. user) and Web App Manifest (JSON file that allows you to specify application metadata such as name, color and icon that distinguishes it), among others. This allows a web application to incorporate some of the historically unique features of native applications, such as offline operation, receiving push notifications and having an access icon in the application launcher [1].

Hybrid applications and, more recently, PWA (Progressive Web Application) applications are emerging as the trend in the way of accessing and serving content. PWAs allow emulating the concept of a native application, supported by the technologies provided by web browsers [2]. During this project the benefits of using various technologies to serve pages in a friendly and simple way will be explained, such as Cache Storage, Service Workers, Manifest (Json), JavaScript, Firebase, HTML5 and CSS3.

Currently the logistics of the agenda and programming of the Expotecnología event at Unitecnar, is carried out manually. Seeking to optimize this process and to be able to visualize the complete schedule from the day it starts, the schedules, the name of the exhibitor and the spaces where it will be presented, it is necessary to venture into new technologies such as PWA (Progressive Web Application), which provide us various benefits and comforts.

A PWA is a much lighter application than a native application, its download and installation is much easier, updates are made directly on the website and can work without an Internet connection. The updates are reflected directly in the application that is on the mobile device and can work on Android and IOS platforms.

2. Materials and methods

2.1. Problem statement

The development of this project is within the framework of quantitative research, through which all the related variables can be measured, an analysis of the results obtained and all the conclusions can be expressed. Quantitative research is a process based on testing a theory composed of variables, measured with numbers and analyzed using statistical techniques. The goal of quantitative methods is to determine whether the predictive generalization of a theory is true [3].

The main objective of this project is that all the people who participate in it as speakers, have the possibility of having a tool that avoids confusion when looking for the place and time of their presentation. Basically, by means of a query through your identity card, you can observe through a responsive graphical interface, all the information stored in a database (name of the speaker, title of the presentation, day, time and place) using Google's Firebase, through the application's graphical interface developed with JavaScript, HTML5, CSS3 and Service Worker implementation. Under this premise, the question arose: How to develop a low-cost application with free software, which shows an electronic agenda through a query, of a technological event? Through this project, the production and development of applications is encouraged within the group of students who participate in the research seedbeds at Unitecnar (see figure 1).

Figure 1. Responsive Graphical Interface

3. Stages in the development of the PWA

Initially, for the development of this application, it was necessary to investigate and know what the Firebase technology was and how it works (platform for web and mobile development, developed by Google) and then make the implementation with all the technologies described below.

3.1. Google Firebase

The Firebase Realtime Database is a NoSQL cloud-based database that syncs data across all clients in real-time, and provides offline functionality. Data is stored in the Realtime database as JSON, and all connected clients share one instance, automatically receiving updates with the newest data [4].

With a prior registration to the Application and free of charge, the database was created and all the information of the speakers registered and accepted to the technological event called Expotecnología was stored. In this way, and having the database connected through the JavaScript code embedded in the HTML5, it was possible to generate the desired queries (see figure 2).

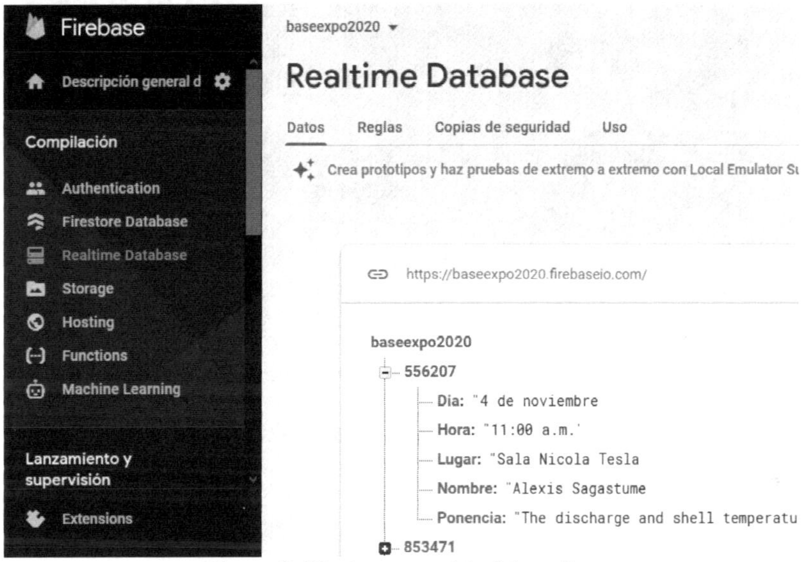

Figure 2. Firebase graphical interface

3.2. HTML5 language

Through a simple, friendly Code, using AJAX requests, JSON requests and the most important HTML5 classes, shaping through CCS3, it is possible to incorporate these advanced technologies to implement modern and pertinent solutions in information systems (see figure 3).

HTML has been used as the standard method for developing web pages since it was first introduced in the early1990s. Over the past twenty years, HTML, with the help of JavaScript and CSS, has enabled the web to evolve from a simple document-sharing device into a complex and dynamic platform for application development. The introduction of JavaScript allowed web developers to construct intricate web applications with functionality rivaling that of modern desktop applications. However, most web applications require functionality beyond what JavaScript alone can provide; this forces developer to supplement these needs with server-side technologies such as PHP, Perl, or Ruby.

IOP Conf. Series: Materials Science and Engineering 1154 (2021) 012047 doi:10.1088/1757-899X/1154/1/012047

This process is often tedious, as the programmer must develop different portions of the application in different languages. As hardware inevitably evolves to meet consumers requirements, web applications must grow more complex to compete with their desktop counterparts. To meet this growing need, HTML5 introduces methods for storing data that may allow future web applications to see vast improvements to performance, security, and privacy, without unnecessary supplemental languages [5].

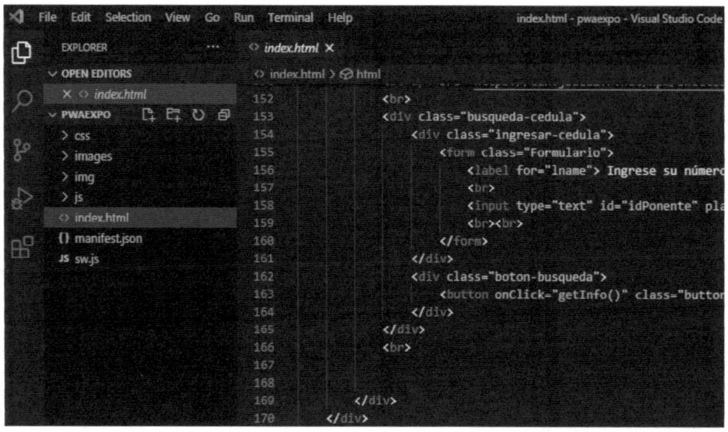

Figure 3. Code implementation in HTML5 Language.

3.3. Using the Sweet Alert 2 library and JavaScript

Through the incorporation of the JavaScript library called Sweet Alert 2, dynamic and incredible interfaces are achieved to implement in web projects. A fairly simple and clean Code was used in the development of this progressive web application (see figure 4).

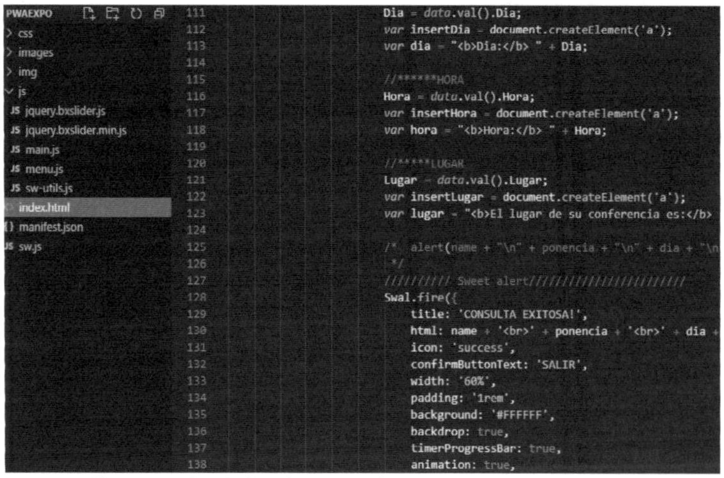

Figure 4. Code implementation in JavaScript Language.

JavaScript is not the only scripting language; there are others such as VBScript and Perl. So why choose JavaScript over the others, the main reason for choosing JavaScript is its widespread use and availability. Both of the most commonly used browsers, IE and Firefox, support JavaScript, as do almost all of the less commonly used browsers. So, you can assume that most people browsing your web site will have a version of JavaScript installed, though it is possible to use a browser's options to disable it. Of the other scripting languages already mentioned, VBScript, which can be used for the same pur-poses as JavaScript, is supported only by Internet Explorer running on the Windows operating system, and Perl is not used at all in web browsers. JavaScript is also very versatile and not just limited to use within a web page. For example, it can be used in Windows to automate computer-administration tasks and inside Adobe Acrobat PDF fi les to control the display of the page just as in web pages, although Acrobat uses a more limited version of JavaScript. However, the question of which scripting language is more powerful and useful has no real answer. Pretty much everything that can be done in JavaScript can be done in VBScript, and vice versa [6].

3.4. Design Responsive and CCS3 language

The term responsive web design was popularized by an article that web designer and developer Ethan Marcotte wrote in 2010. The goal of responsive web design is to make a web page look equally good regardless of the screen size of a device. Before the introduction of responsive web design, web designers and developers created most websites by following the principles of pixel perfect web design. Pixel perfect web design treats a web page like a page from a magazine. In this approach, the mock-up of a web page is first created in Photoshop, and then a developer recreates that design to fit a web browser. The goal of pixel-perfect web design is to make a web page resemble the original mock-up as much as possible.

The greatest advantage of responsive web design is that a library does not have to maintain and update more than one set of content. Another advantage is that there is no need for additional promotion of the library's mobile website, since whenever a library patron accesses the library website on a mobile device, the web-site automatically adjusts its layout to be mobile-friendly. This is a great plus considering that libraries often experience difficulty in promoting their mobile offerings such as down-loadable e-books, tablets for borrowing, and a mobile website. From the library patron's perspective, a library website with responsive web design presents all the information found in the full website. This content-parity enables patrons using a mobile device to find and do almost everything that they can do on the full desktop site as long as their mobile devices support the features in the full desktop site. (Flash on the mobile Safari web browser would be the example of an unsupported feature. [7]. (see figure 5-6).

Figure 5. iPhone cell phone

Figure 6. Android cell phone

3.5. Use of Service Worker and manifest Json

A service worker is a fairly robust tool that creates a sequence of commands, which are run by the browser in the background. It really is a JavaScript file that continues to run even though the website is closed. The Json manifest causes the Web Application to tell the browser what elements to display for each Web page, such as color, splash screen, etc. In addition, all these technologies cause an icon to be installed on the cell phone without the need to install anything additional, or download any application.

4. Results

As a result of this research and development work of a progressive, completely responsive and clean Web Application, a very friendly interface was obtained, which is currently working mounted on a private server for its implementation (http: // www. cesargoweb.com/pwaexpo/). It should be noted that it is fully scalable and upgradeable and can be adapted to any technological event or institution of higher education. The application is deployed without problems and runs at an optimal speed, which leads us to be able to navigate smoothly from any electronic device, even without having an internet connection, due to its local storage in browsers, another important advantage of working. with Service Worker, Json Manifest and JavaScript (see figure 7).

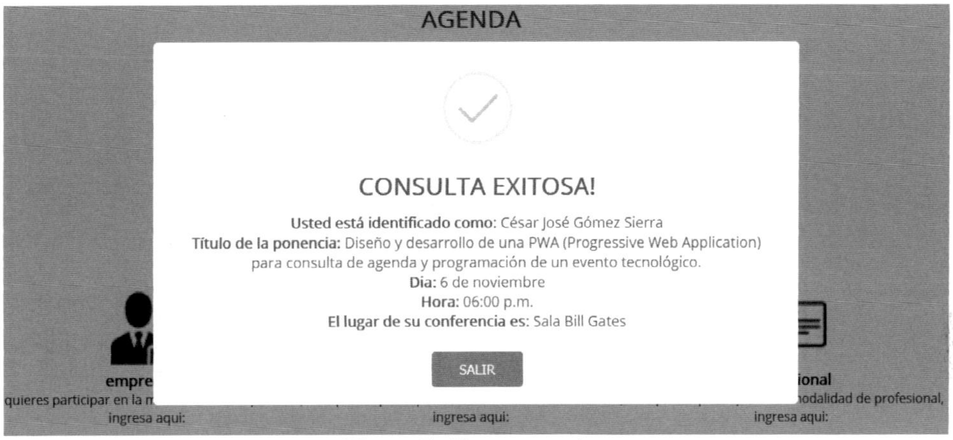

Figure 7. Computer view

5. Conclusions

Progressive web applications or PWA, have become the new trend in application development that make the most of all the benefits offered by information and communication technologies in the web field. Unlike native or conventional applications, we can run PWAs in any browser or operating system, presenting many advantages to the user, such as push notifications, increased functionality depending on the device on which we deploy it and also its performance. That is, they behave like a native Application but using web technology and consuming less system resources as such.

Among the most important features of progressive web applications, we have:

• They can be run directly on the desktop, without having to download it from the Play Store or App Store and consuming fewer resources
• Progressive web applications work the same for each user, regardless of the web browser they are using.
• Its Responsive & Clean design makes it work on any device, computer, Laptop, Tablet or Smart phone.
• They are generally developed to be deployed under the HTTPS protocol, with its SSL certificate.
• All progressive web applications are kept up to date with the latest versions
• They are easily identifiable, more like an Application than a website

• They can be shared only with their URL, without the need to install.

The most important of these PWAs are the Service workers, which are scripts that are executed between the browser and the server to form the fundamental basis of development and using the web cache to show the results immediately. These Service workers are so important that they can use the cache immediately above to view the application when the internet connection is unstable or slow. As successful cases of progressive web application implementation we have: AliExpress, Twitter, Lancôme, Nikkei, among others.

Acknowledgments
A special thanks to the Antonio de Arevalo University Foundation - Unitecnar, for the support provided. Likewise, all the students of the SEDMOW research hotbed are thanked for the great effort they put into all the activities proposed to achieve this project.

References

[1] Aguirre, V., Ortu, A., Delía, L. N., Thomas, P. J., Corbalán, L. C., Cáseres, G., & Pesado, P. M. (2019). PWA para unificar el desarrollo Desktop, Web y Mobile. In XXV Congreso Argentino de Ciencias de la Computación (CACIC) (Universidad Nacional de Río Cuarto, Córdoba, 14 al 18 de octubre de 2019).
[2] Lanza Ortega, E. (2019). Aplicación Web Progresiva (PWA) para la gestión de pagos de estacionamiento en superficie.
[3] Abawi, K. (2008). Qualitative and quantitative research. World health, 2(1), 1-10.
[4] Moroney, L. (2017). The firebase realtime database. In The Definitive Guide to Firebase (pp. 51-71). Apress, Berkeley, CA.
[5] West, W., & Pulimood, S. M. (2012). Analysis of privacy and security in HTML5 web storage. Journal of Computing Sciences in Colleges, 27(3), 80-87.
[6] Wilton, P. (2004). Beginning JavaScript. John Wiley & Sons

IOP Conf. Series: Materials Science and Engineering 1154 (2021) 012048 doi:10.1088/1757-899X/1154/1/012048

Approximation of Cross-tier interference in HETNET using Stochastic Geometry

Y Tarriba-Lezama[1] , LValdez -Cervantes
[1,2]Antonio de Arévalo University Foundation
[1] University of Cauca
[1] yesid.tarriba@unitecnar.edu.co,yesidtarriba@unicauca.edu.co
[2] decano.fadi@unitecnar.edu.co

Abstract. This work describes an approximation model to interpret in a heterogeneous two-tier network using stochastic geometry, the behavior of inter-tier interference or cross tier. and compares in terms of distributions the existing relationship between the intensity variables at each tier and the transmission powers of BS.The relationship between intensity functions of the distributions of each BS level and between the transmission powers will be considered

1. Introduction

The performance of any wireless network is conditioned by many evaluations metrics, in the point of view the physical layer the limits of performance the interference show the important factor to describe how the network is working ,the interference have two origins in a HETNET , cross-tier interference and co-tier interference so called inter-tier interference and intra-tier interference respectly[1] . The stochastic geometry is a mathematical tool used to puntual process for model the how is the behaviour of variable in the network [2] .in this work we will use this tool for an estimate the cross-tier whith some considerate characteristic that simplify the model but is a viable approximation generalized. In an two tier network we will used the PPP (Puntual Process of Poisson) for model of distribution of *BSM*(Macro Base Station) and *BSF* (Femto Base station) y we will a compare the behavior the PPP and Power of transmition in each tier

2. Model and Considerations

Below is described a multi-tier heterogeneous network where we define RF c*haracteristics as frequency and* distribution power segmented by each tier.We define this difference as the tie*r at which we assume conditions* that favor to describe an approximation of interference between tier. We determine at the topo*logy level the* distribution the Base stations (BS) at each level are modelled by PPP with La*mbda k in such a way that their* condition fulfills that: $\lambda_F > \lambda_M$ where λ_M, λ_F .Is the intensity function for the distribution of the BSM and BSF. respectively. Our model is a two-Tier network distributed both under the h*omogeneous PPP and which has as* level variables ,$Teir_k = (\lambda_k > Pt_k)$. The stochastic geometry can, depending *on the distance between BS and* coverage, establish the interference. This principle starts from the integral geometry *as indicated in [3]. within the* considerations to have for our heterogeneous network model, it is posed below:

- *The noise power PN = 0 is postulated.*
- *The effect on the download link is considered.*
- *We neglect the effects of multipath fading and shadowing.*
- *The possible additional tier will be considered iid.*
- *The Antenna patter are circular and considered isotropic the propagation*

The metric that we are going to consider is the interference between BS of each level called cross-tier interference. This is aggravated by the fact that ***BSF*** are usually installed and Manipulate by their end users[4]. The term of cross-tier interference defines way on stochastic geometry. The download link for the base stations will present areas where the areas for ***BSM*** and ***BSF*** converge. This effect is debasing to the performance of the network, considering that the use of frequencies in the base stations coincide or share radioelectric resources.

$$SIR_K = \frac{Pt_M}{Pt_F} \quad (1)$$

The equation (1) show the general shows the general description of the signal-to-interference ratio. Due to the distribution of the base stations PPP distribution generates SIR as a stochastic process. In Figure 1 show this model for description of HETNET of two-tier whit fixed power and BS distribuided under PPP

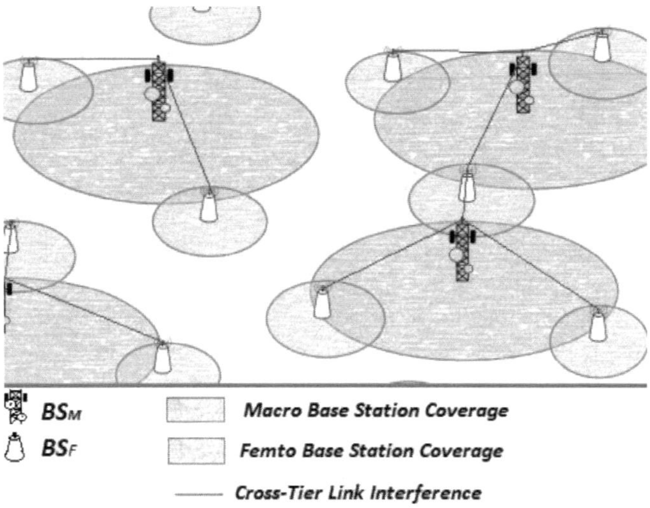

Figure. 1. Model of Tow-Tier HETNET considered under PPP

3. Interference Description

Interference as a performance metric establishes the relationship between the transmission power of a node at a level as a result of the sum of the powers of the remaining signals of the network in which it is considered that they share radio resources and have interference in the area of coverage [5], In equation (1) the generalization of the SIR is described, for the case of a multilevel network we would establish the relationship in terms of the sum of the signal power. In the description of the interference in HETNET, there are scenarios that increase, become more complex, the greater the number of tiers in the network. For the two level case study we could discover:

- Interference Co-Tier between BSM.
- Interference Co-Tier between BSF.
- Interference Cross- Tier between BSM and BSF

The SIR (Signal Interference Ratio) seen from BSm is described below in which by means of the equation.

$$SIR = \frac{Pt_M}{\sum_{k=1}^{K} I_k} \tag{2}$$

Where ***Ik*** is the total interference of a given tier. We propose the deployment in R^2 of the ***BSM*** and ***BSF*** in such a way that the strategy to mitigate the interference would be the FFR efficiently [6] In this work, the estimation of cross tier interference was described, bearing in mind the previously announced considerations in works such as [7] establish the estimate of cross tier interference's considering the low coverage model in coverage regions that have a dependence on the tier distributed function.

$$I_{CT} = \frac{G_K L_K P_{CT}}{\ell R_K} \tag{3}$$

The equations (3) describes the model of Cross tier interference's considering the fading at each tier and the circular coverage model of the BS. It is possible to consider that within the exceptions listed in this work it is possible to consider a good approximation of the results.

4. Simulations and Approximation

Next we will describe the procedure for calculating the interference areas between the coverage regions of each tier. Initially, it is to be considered that the interference in the HETNET network under the PPP distribution is in itself a stochastic process in which several dependency approaches are planned and among these we will consider that the distances between BS of each tier are random [8]. we propose the equation of the signal to interference relationship for the approximation.

$$SIR_K = \frac{N_M Pt_M}{N_F Pt_F} \tag{4}$$

N_M *and* N_F *are number of* BSM ***and*** BSF *respectly*

$$N_M = PoissonRND(\lambda_M S) \tag{5}$$

$$N_F = PoissonRND(\lambda_F S) \tag{6}$$

S *is the area of net wok in* Km^2 *According to our model, let's set the distribution in a plane* $\mathbb{R}^2$

$$D = \sqrt{(X_M - X_F)^2 + (Y - Y_F)^2} \tag{7}$$

In equation.

$$D_{MF} = ((R_M + R_F)^2 - D^2) * (D^2 - (R_F + R_M)^2) \tag{8}$$

In equation 7 X_M;Y_M and X_F;Y_F are the Cartesian position of any BS in the Area and equations 8 D_{MF} is the distance between BS_M and BS_F and R_M, R_F are the Radius of ***BSM*** and ***BSF*** respectly.

$$I_{Cross} = R_M^2 arcos\left(\frac{R_M^2 - R_F^2 + D^2}{2DR_M}\right) + R_F^2 arcos\left(\frac{R_F^2 - R_M^2 + D^2}{2DR_F}\right) - \frac{1}{2}\sqrt{D_{MF}} \tag{9}$$

In relation to the equation 4 the formal planet of the SIR is established together with the equation 9,we define:

$$SIR = \frac{N_M PT_M}{\sum I_{Cross} + \sum I_{Overlap}} \tag{10}$$

In the Figure 2 we can see the possibilities to extract the interceptions in the interference areas that will determine the total interference of the network

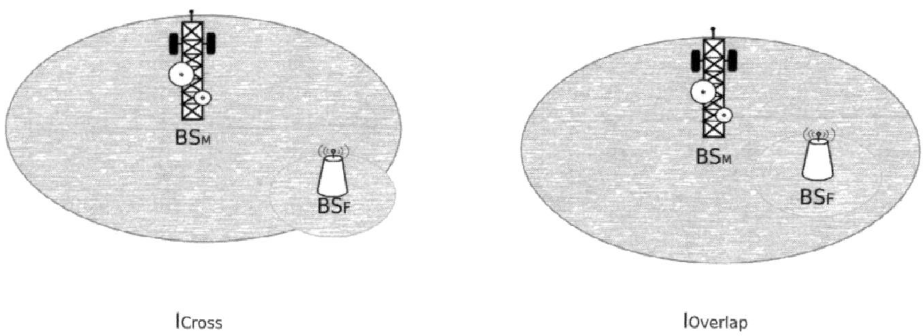

Figure. 2. ICross and IOverlap for SIR.

4.1 Simulations Parameters

In this section we will describe the simulation process and the discussion of the results of the network simulation to describe the cross-tier interference shown in table 1. In which two BS distributions under the PPP distribution with different intensities and different powers generate the distribution of coverage as shown in Figure 3 that uses equation 8 to establish in $\mathbb{R}^2$ the distances between the nodes between Tier for and with Equation 9 calculates the intercepts, having an input variable in this process the distance that is random The realizations of the graph generate the interference matrix where the options shown in Figure 2 are considered and equation 10 is applied from here we obtain that the representation of the SIR contaminating the interference between tier is possible to approximate it by means of a Gamma distribution function. The gamma distribution models the relationship of SIR VS the probability where in stages, compliance requires a minimum of up to 5 realizations to exercise the approximation. For our case we carry out 85 realizations and once the results of the realizations are obtained we proceed to adjust the distribution to which best is adjustment to the result this procedure is carried out for the estimation of greater likelihood and is verified with the goodness adjustment tests In the Figure 3 show one realizations of distribution at HETNET two-tier. The sign ***Lc*** represent the downlink whit interference between tier. This ***Lc*** is considered by the nature of an signal, random and it is who is considered of study to determine in the equations7 The results based on table 1. shows us the approximation in Figure 4 where in the discrete distribution or histrogram it is performed in adjustment to the gamma distribution function as mentioned before using the Kolmogorov-Smirnov test as a goodness test, that with the data shown in table 2 it is shown in that the Cross-tier SIR is considered a good approximation through the gamma distribution. The CDF of the results are shown in Figure 5, up to here they described them with established initial values that match the characteristics of a HETNET, next we are going to carry out the variation of the HETNET parameters that describe the domain characteristics. Among which the intensity functions and the transmission power are noted

Table 1. Parameters of simulation for HETNET

Index	Parameter	Value	Variable
1	Lambda $_M$	1	λ_M
2	Lambda $_F$	4	λ_F
3	Number of Tier	2	k
4	Distribution	Homogeneous PPP	
5	$Power_M$	25W	Pt_M
6	$Power_F$	8W	Pt_M
7	Area of Network	$1000Km^2$	S

We will describe the distribution fit through the results. The modifications are proposed where the behavior of the interference is described as the relationships between the variables vary. Where the claim between increasing and decreasing lams is shown in Figure 6, and the relationship between the transmission powers also increasing and decreasing. This analysis starts with describing in general terms.

Table 2. Result Distribution Fitting

Approximation Distribution	**Shape**	**Scale**
Gamma	12,8772	11,8239

Table 3. Goodness-of-Fit Tests

	Gamma
Dplus	0,0870596
Dminus	0,0739218
DN	0,0870596
P-Value	0,557139

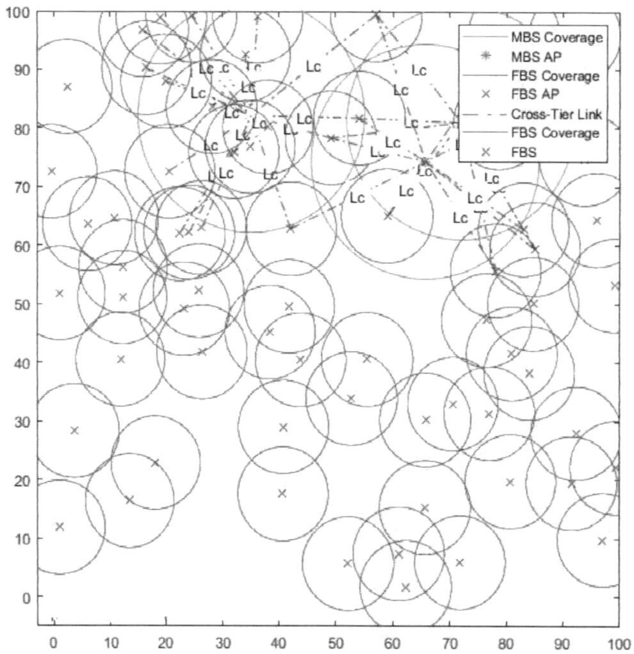

Figure. 3. Interference Cross for HETNET in $\mathbb{R}^2$.

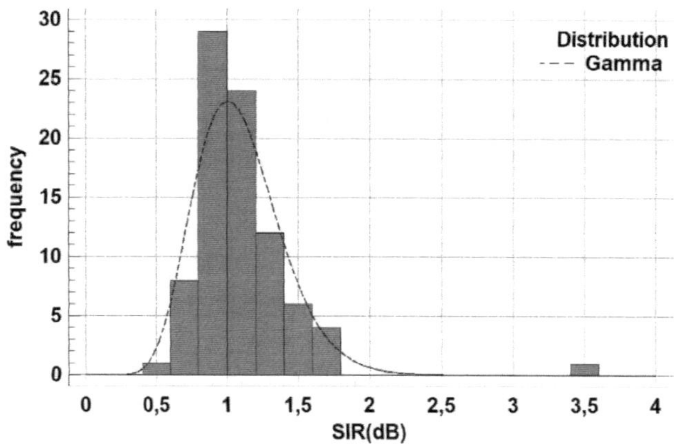

Figure. 4. PDF SIR Approximation whit parameter of Table 1

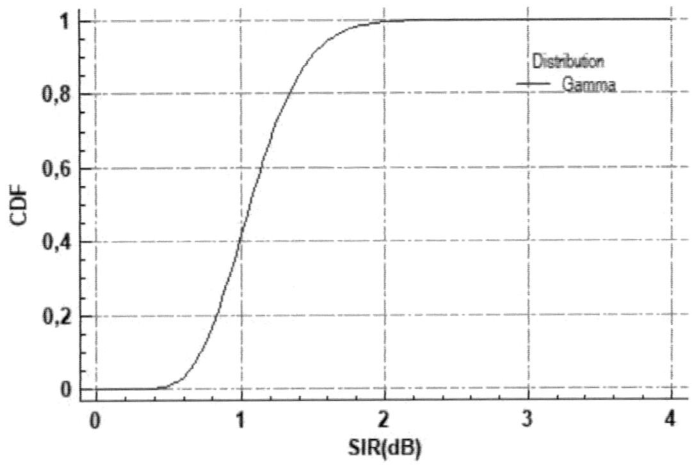

Figure. 5. CDF SIR whit parameter of Table 1

how the PDF and CDF share of cross-tint interference is as the variables are modified

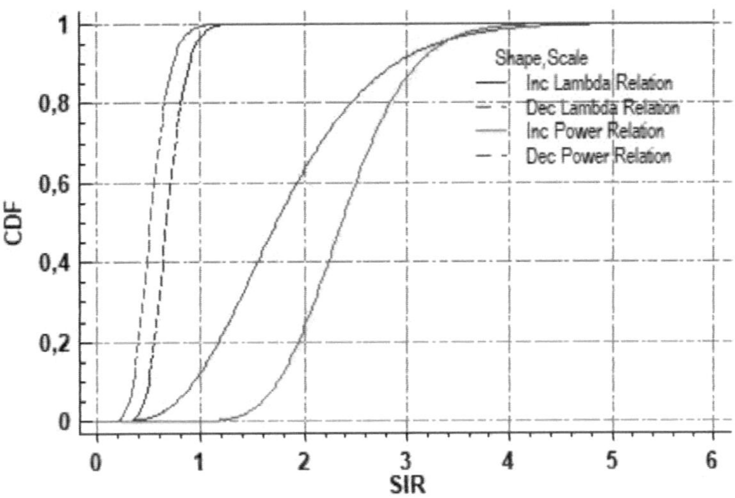

Figure. 6. CDF SIR whit parameter of Table 1

5. Conclusions.

From the results obtained, it is valid to note that the gamma function is considered a good approximation to model cross tier inference in a HETNET network. Considering the use of stochastic geometry to model the distribution of nodes, this function is present when describing the SIR of the network.
and as it is shown in Figure 6, the CDF tends to coincide when the relationships between transmission power and intensity functions are similar. This indicates that the cross tier interference is a function dependent on the intensity functions and the power ratio

IOP Conf. Series: Materials Science and Engineering 1154 (2021) 012048 doi:10.1088/1757-899X/1154/1/012048

References

[1] H. ElSawy, E. Hossain, M. Haenggi, "Stochastic geometry for modeling, analysis, and design of multi-tier and cognitive cellular wireless networks: A survey," *IEEE Communications Surveys & Tutorials*, vol. 15, no. 3, pp. 996–1019 (2013).

[2] M. Haenggi, J. G. Andrews, F. Baccelli, O. Dousse, M. Franceschetti, "Stochastic geometry and random graphs for the analysis and design of wireless networks," *IEEE journal on selected areas in communications*, vol. 27, no. 7, pp. 1029–1046 (2009).

[3] V. Petrov, D. Moltchanov, P. Kustarev, J. M. Jornet, Y. Koucheryavy, "On the use of integral geometry for interference modeling and analysis in wireless networks," *IEEE Communications Letters*, vol. 20, no. 12, pp. 2530–2533 (2016).

[4] W. C. Cheung, T. Q. Quek, M. Kountouris, "Stochastic analysis of two-tier networks: Effect of spectrum allocation," presented at the *2011 IEEE International Conference on Acoustics, Speech and Signal Processing (ICASSP)*, pp. 2964–2967 (2011).

[5] M. Peng, C. Wang, J. Li, H. Xiang, V. Lau, "Recent advances in underlay heterogeneous networks: Interference control, resource allocation, and self-organization," *IEEE Communications Surveys & Tutorials*, vol. 17, no. 2, pp. 700–729 (2015).

[6] M. M. Hussain, M. A. Ali, A. A. Memon, "Stochastic geometry based interference alignment in two-tier heterogenous cellular networks," presented at the *2018 International Conference on Computing, Mathematics and Engineering Technologies (iCoMET)*, pp. 1–5 (2018).

[7] R. W. Heath, M. Kountouris, T. Bai, "Modeling heterogeneous network interference using Poisson point pro cesses," *IEEE Transactions on Signal Processing*, vol. 61, no. 16, pp. 4114–4126 (2013).

[8] H. P. Keeler, B. Błaszczyszyn, M. K. Karray, "SINR based coverage probability in cellular networks under multiple connections," presented at the *IEEE Intl. Symposium on Inf. Theory (ISIT)*, pp. 1167–1171 (2013).

Teaching equipment of newton's second law

Hugo Armando Gallego Becerra
Teaching Physics Department, Technological University of Pereira.
Sebastián Martínez García
Teaching Physical Engineering Program, Technological University of Pereira.
Luis Guillermo Henao Melo
Teaching Physics Department, Technological University of Pereira.

ugo@utp.edu.co

Abstract. Mechanic is the branch of physics that is responsible for the study of the movement and balance of bodies in a system, within this can also be considered the bodies at rest. For the first case the scientist Isaac Newton considered the study of the movement of bodies, where he defines that any physical system which is subjected to different actions, as long as such actions are not nullified, will result in a change in the state of the body. The force is the magnitude that quantitatively measures the intensity, direction and direction of that interactions, having said that, if the summation of all forces of a system is different from zero, it can be concluded that the system will be in motion and therefore have an associated acceleration. Based on the above definition, was executed by the research group Design and Construction of Prototypes for Demonstration Experiments (DICOPED) in the design and construction of a prototype that allows to verify the Newton's Second Law, looking to develop an autonomous, robust and low-cost prototype, that allows the use of the technology and tools present in the means, encourage and facilitate the teaching of basic concepts but no less important of physics to students of basic and higher education.

1. Introduction

Developing equipment for physics education, is a little-explored field by academic world, but it is a valuable tool to achieve entrenching basic physics concepts. DICOPED research group has worked on the development of teaching equipment that allows teachers in the classroom or laboratories to show experimental forming of physical phenomena. Likewise, the use of the human and technological resources present in the means, allow to develop autonomous prototypes that satisfy the present needs in educational environments at the local, national level and why not in the international future.

2. Methodological Design

Four phases were proposed for the development of the autonomous prototype by the experimental demonstration of Newton's Second Law:

- The first is based on the understanding of the theoretical concepts of Newton's Second Law.
- The second consists of the design and construction of a mechanical model that allows experimental verification of Newton's Second Law.

- The third is based on electronic development and coding of electronic system akin to the requirements of mechanical assembly.
- The fourth phase consists of the development of a guide which allows to experimentally verify the results obtained with the completed prototype.

The detailed development will be shown below.

3. Theoretical Framework

3.1 Newton's Second Law

Newton's Second Law or Law of Force is also known as the Fundamental Law of Dynamics, it determines a proportional relationship between force and variation of the amount of movement or linear moment of a body [1].

This law declares that when an object experiment a force, that force causes an acceleration in the object. The previous acceleration is therefore proportional to the intensity of the force, going in direction of the applied force and it is inversely proportional to the mass of the object.

Therefore, if a resulting non-zero force acts on a moving body, this force will modify the state of movement of the body by changing the direction or magnitude of its speed.

3.2 Amount of Movement

The amount of movement, linear moment or momentum is a vector magnitude, which in classical mechanics is defined as the product of the body mass and its speed at a given moment [2] (see Figure 1).

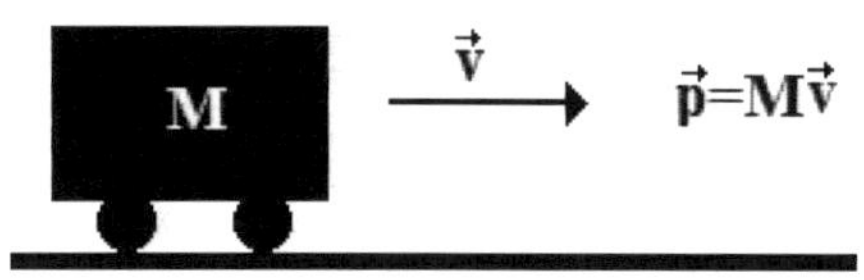

Figure 1, Linear momentum diagram.

From the above it can be concluded that a body in motion has a non-zero linear momentum, otherwise the body is at rest. The units of linear momentum in the international system are .

The amount of movement is a vector magnitude, that is, to describe it, you must know its magnitude, direction, which has for direction the same direction of the velocity of the body. For this reason, it is sometimes confused with speed. This is not correct since the moment relates the mass of the object which considerably influences the amount of movement of the body.

It is important to keep in mind that the original form of Newton's Second Law directly relates the concept of linear momentum and the resulting net force acting on it, for inertial frames of reference and systems of constant mass, since:

$$\vec{a} = \frac{d\vec{v}}{dt} \qquad (1)$$

$$\sum \vec{F} = m\frac{d\vec{v}}{dt} = \frac{d}{dt}\left(m\vec{v}\right) \quad (2)$$

Therefore, knowing that mv is p, then the previous expression is understood as the change in time of the linear momentum of a particle is equal to the resulting force acting on it, the above can be written as:

$$\sum \vec{F} = \frac{d\vec{p}}{dt} \qquad (3)$$

It can be deduced that the resulting force is zero, the linear moment of the particle must be constant. [3]

IOP Conf. Series: Materials Science and Engineering 1154 (2021) 012049 doi:10.1088/1757-899X/1154/1/012049

4. Design and Construction of Prototype

Next, the process of the design and construction of the prototype will be detailed, for which a preliminary analysis was carried out for the adequate selection of materials and available technology to build a robust and low-cost prototype.

4.1 Mechanical design of the prototype

The mechanical design of the prototype consists of a plane with an approximate length of 60 cm, the surface of the plane has a rail on which a variable-mass cart will move (inserting masses on the vertical axis that the cart has in its upper part), for this there is a system of copper wheels which, when moving on a low friction surface, allow to avoid energy loss due to friction in the movement of the system, in Figure. 2 the described assembly is shown.

Figure 2, Base with rail and mobile cart

Likewise, the system has a 4 cm diameter pulley which is shown in Figure. 3, which rotates around a low friction axis, said pulley will be complemented with an adhesive that will allow to determine the angular acceleration of the movement of the system once the device is docked.

Figure 3, Marked pulley.

To complete the mechanical system, there is a universal mass support, which allows varying the mass that is suspended in the air, which can be seen in Figure. 4.

Figure 4, Mounting with universal mass support.

In the previous figure you can see the equipment, the CNY70 sensor is located at the back of the pulley, its power supply and output are at the other end of the rail, which facilitates and avoids the interference of cables or connectors in movement, this can be seen in Figure. 5.

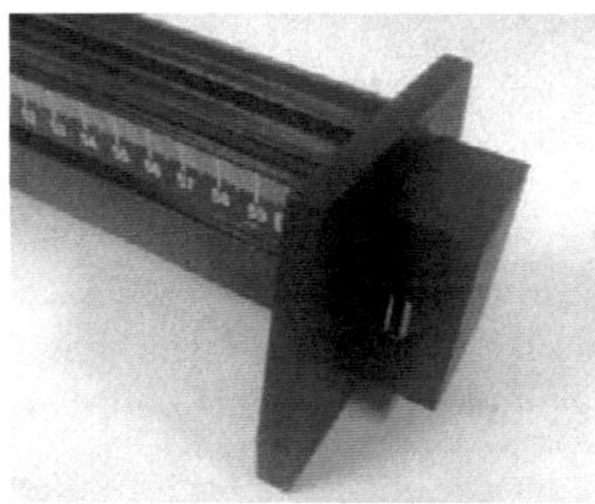

Figure 5, CNY70 sensor connectors.

Likewise, it was necessary to design a box that would be coupled to the electronic part of the prototype, for this acrylic was used which is a resistant and easy-to-use material, in this box the Arduino was housed in conjunction with the Shield Key-Pad that allows user interaction with the system.

Based on the CorelDraw design application, which offers powerful drawing, text handling, vectorization, color separation, and special effects features, the prototype parts were designed.

The designed parts are described below (see Figure. 6, Figure. 7, Figure. 8, Figure. 9).

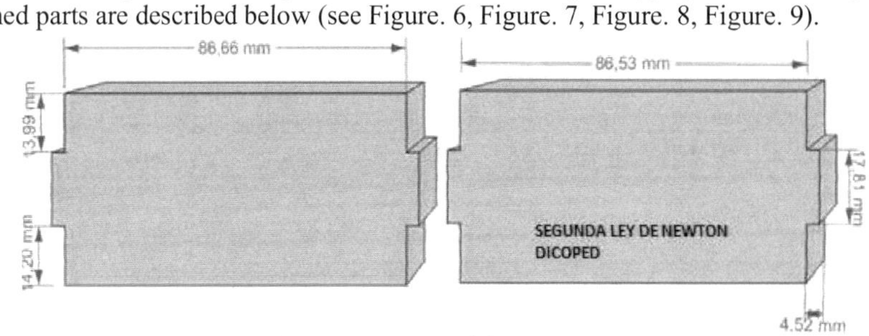

Figure 6, Rear and front covers.

IOP Conf. Series: Materials Science and Engineering 1154 (2021) 012049 doi:10.1088/1757-899X/1154/1/012049

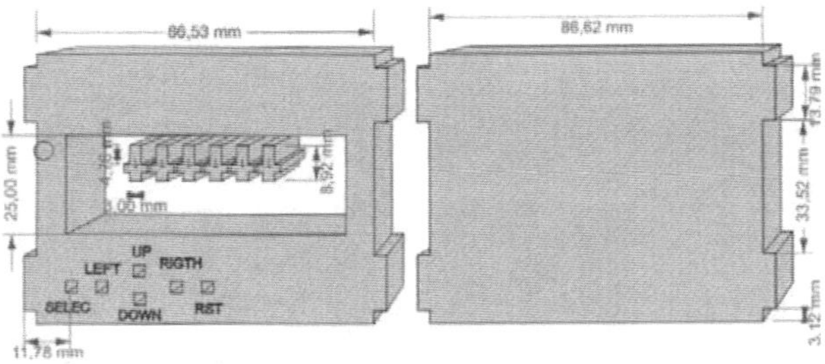

Figure 7, Top and bottom caps.

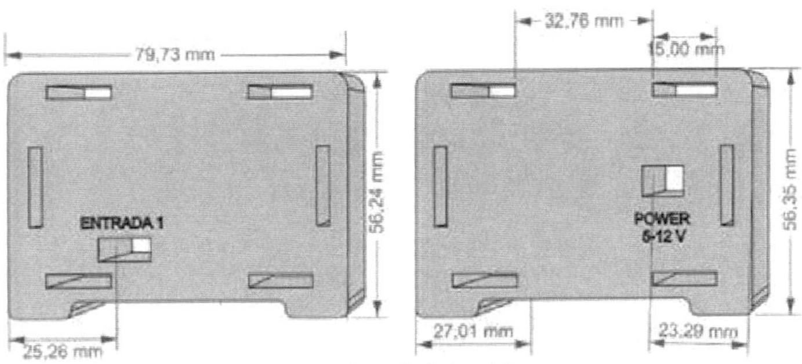

Figure 8, Left and right side covers.

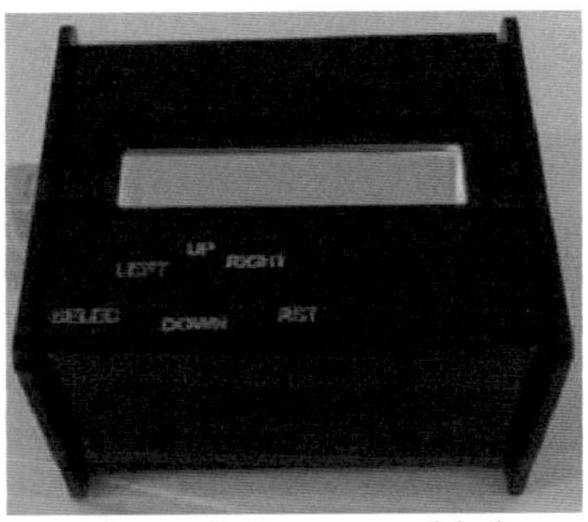

Figure 9, Box assembled.

4.2 Electronic prototype design

The prototype developed consists of a control platform based on Arduino, which is linked to a Shield containing the keyboard and the LCD screen, this device allows the user to interact with the Arduino and define the application execution flow (see Figure. 10).

Figure 10, Arduino – Shield LCD Keyboard.

The electronic design implemented consists of the aforementioned element and a CNY70 sensor that allows to measure the time it takes for the rotation of the pulley present at one of the ends of the prototype.

The signal delivered by the CNY70 sensor, which is an IR emitter and receiver type sensor, is shown in the following image (see Figure. 11).

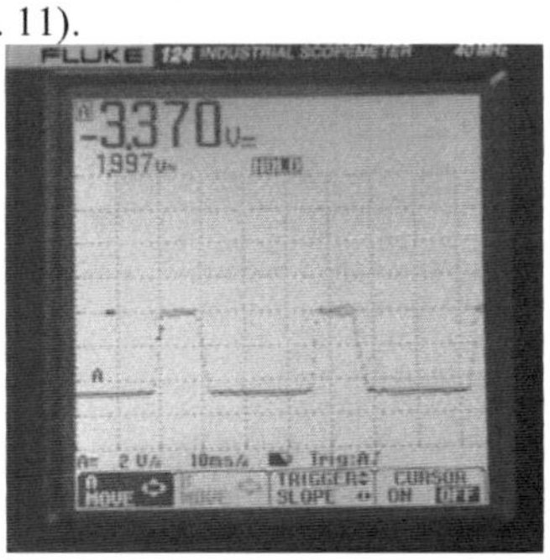

Figure 11, Output signal.

Since the above signal is not a clean signal, processing was performed to improve the precision in taking experimental data. For this, a Smith Trigger was used, which acts as an ideal conditioner since it transforms the sensor output signal to pure logic levels.

For this application, a positive feedback op amp was used. Reference levels are adjusted with resistance.

Normally NOT gates with Schmitt trigger with TTL technology are used. The symbol inside the triangle in Figure. 12 denotes that the logic gate is Schmitt-trigger [4].

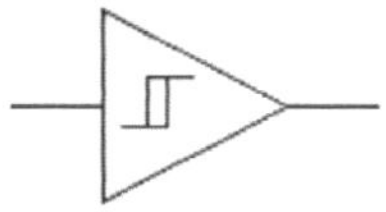

Figure 12, Smith Trigger symbol.

The arrangement of the following figure provides a good response of the output signal since the transfer curve shows a behavior like the one shown in the figure (see Figure. 13).

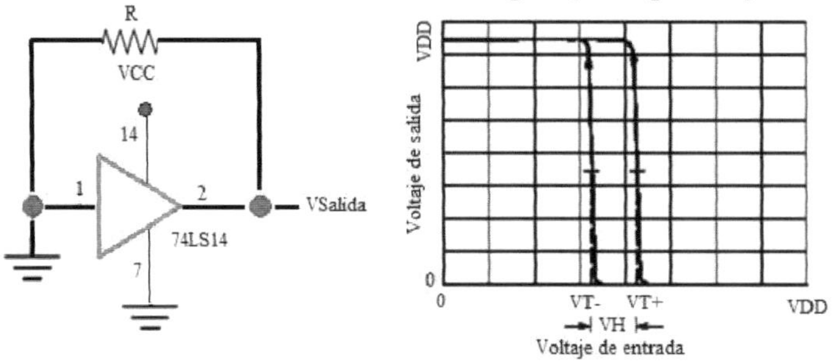

Figure 13, Circuit to eliminate Schmitt-trigger noise - Transfer curve.

Finally, after performing the described processing, an output signal is obtained in accordance with the desired requirements for the application (see Figure. 14).

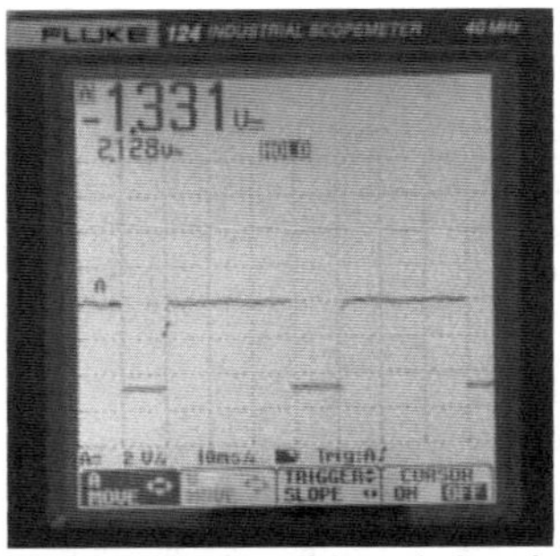

Figure 14, Conditioned signal.

ARES, a Proteus tool suitable for the design of printed circuit boards (PCBs), was used to design the printed circuit board. It is fully integrated with the ISIS tool.

Once the electronic schema has been designed in ISIS (see Figure. 15), the list of networks (NETLIST) is automatically generated. A network is a group of pins interconnected to each other and the network list is a list of all the networks that make up the layout. ARES receives this list of networks to design the printed circuit. In this way, it is ensured that the board will have the pins connected to each other in an identical way as defined in the electronic diagram.

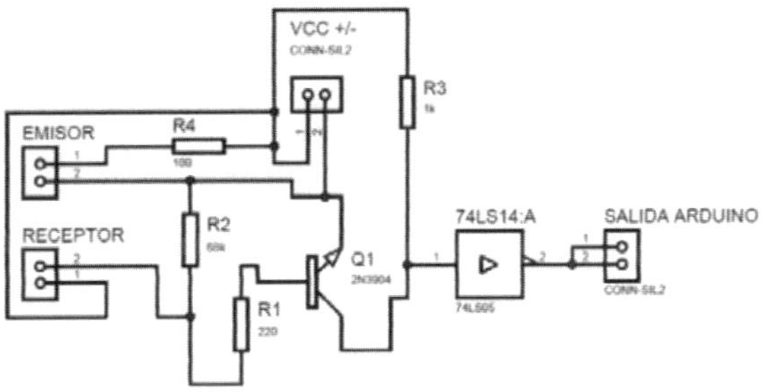

Figure 15, Design of the electronic schema in ISIS.

Below is the result obtained in the ARES of the circuit designed in PROTEUS (see Figure. 16).

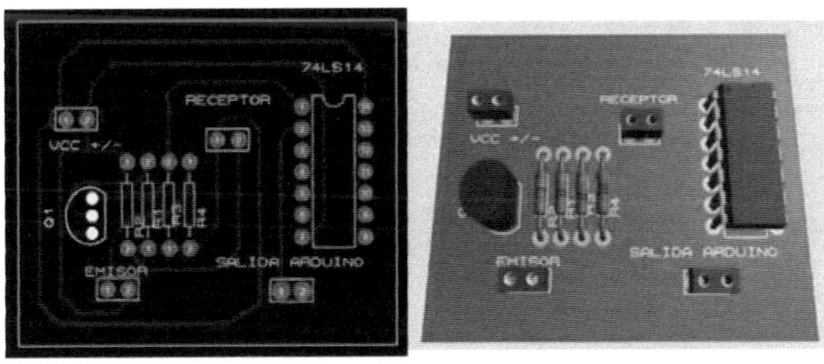

Figure 16, PCB result in ARES - 3D PCB in ARES.

Finally, when printing the circuit, the tracks that must be burned on the virgin copper card are obtained to carry out the final assembly of the circuit (see Figure. 17).

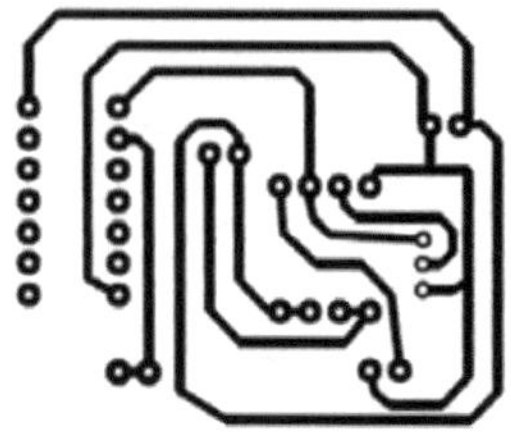

Figure 17, Printing on photo paper

For the control of the device, the Arduino was encoded, in such a way that it carried out the previously established and thus bring to a successful conclusion the experimental stage for the verification of Newton's Second Law.

5. Results

Finally, the result is a prototype that allows experimental verification of Newton's Second Law, for this, in addition to the design and construction of the prototype, a guide for experimental practice was developed, which is shown below.

Newton's Second Law Experimental Practice Guide

5.1 Introduction

In every physical system, a body will remain in a state of rest or movement at constant speed unless an action occurs that alters its state. From the above, it can be concluded that a force is the measure of a physical interaction that affects the state of movement of the bodies present in a system, these changes in movement are associated with a change in speed which leads to the definition acceleration.

Based on the above, it can be affirmed that there must be some relationship between the concepts of force and acceleration, as well as the study of the behavior of this phenomenon that has an answer in Newton's second law.

5.2 Objectives

- **i.** Experimentally determine the acceleration of a body, when the system presents a sum of forces other than 0.
- **ii.** Construct a graph of the angular position based on time.
- **iii.** Analyze and describe the behavior of a system from a graph.
- **iv.** Give experimental support to the mathematical relationship between force, mass and acceleration.

6. Conceptual Framework

6.1. Newton's Second Law

Newton's Second Law states that the acceleration of an object is proportional to the net force applied to it, which can be expressed mathematically as:

$$\vec{F} = M\vec{a} \tag{4}$$

Where:

- $\vec{F} \rightarrow$ Net force on the system.

- $M \rightarrow$ System mass.
- $\vec{a} \rightarrow$ Acceleration experienced by the system.

This acceleration is a vector quantity and its direction is given by the direction of the resulting net force, that is, the force resulting from the sum of the forces present in the system [5].

In this practice, the previous statement will be verified experimentally, for this we will take a mass object m_2 and it will accelerate generating on it a known force. This will be done by tying an inextensible thread of negligible mass at one end of the mass, and completing the system with a pulley that will allow suspending a mass m_1 vertically, which will result in a tension T on the thread, which will finally be the force on the mass m_2, the system described in the Figure. 1.

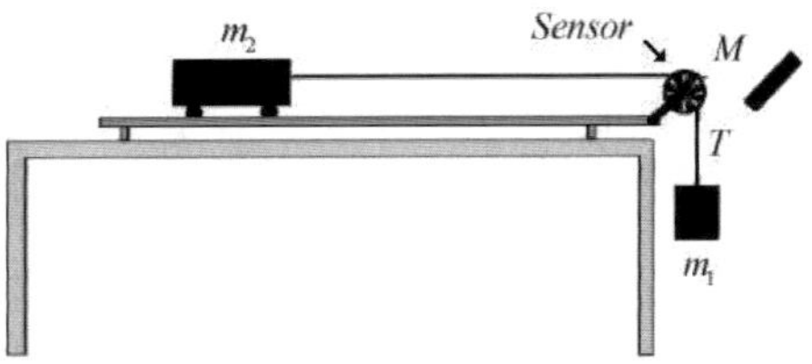

Figure 18, Equipment schematic.

The system is made up of two bodies, which are initially at rest, once they are released, they accelerate due to the existing forces in the system. If the free-body diagram for the system is made, the following analysis is obtained.

6.1.1. Theoretical analysis of movement

For the analysis of the system, the mass of the pulley and the negligible thread must be assumed as the first measurement. Once this has been stated, the force analysis is carried out according to Newton's laws.
Sum of forces for mass m_2:

$$T_1 - F_R = m_2 a \tag{5}$$

Sum of forces for the mass m_1:

$$m_1 g - T_2 = m_1 a \tag{6}$$

Where is considered F_R as the friction force between the carriage and the rail surface, m_2 is the mass of the car, T_1 is the tension of the thread on the carriage, m_1 corresponds to the suspended mass on the universal bracket, the tension T_2 is directly related to the hanging mass and a is the acceleration of the system. From the previous analysis, when operating the equations, we obtain:

$$a = \frac{m_1 g - F_R}{m_1 + m_2} \tag{7}$$

Once the analysis has been carried out on the bodies, the analysis is carried out on the pulley, assuming that the pulley has a non-negligible mass, therefore, when applying Newton's laws, we arrive at:

$$T_1 - F_R = m_2 a \tag{8}$$

$$m_1 g - T_2 = m_1 a \tag{9}$$

$$T_1 - T_2 = \frac{I\alpha}{R} = \frac{Ia}{R^2} \tag{10}$$

In such a way that in the previous expression R it is the radius of the pulley, α is the angular acceleration of the pulley and I is the moment of inertia of the pulley $\left(I = \frac{1}{2}Mr^2\right)$. When performing the respective analysis and solving the acceleration of the previous equations, we obtain:

$$a = \frac{m_1 g - F_R}{m_1 + m_2 + \frac{I}{R^2}} \tag{11}$$

Finally, the experimental study of the system is carried out as a function of mass, working with the expression:

$$\left(m_1 + m_2 + \frac{I}{R^2}\right)a = m_1 g - F_R \tag{12}$$

Plotting the left side of the equality as a function of yields a curve that resembles a mathematical expression similar to the one shown below.

$$f(x) = gx - F_R \tag{13}$$

Which corresponds to the equation of a line, where the friction force would be the cut of the line with the vertical axis and its slope would be directly related to the value of gravity. For greater reliability of the data obtained, it is necessary to determine the correlation coefficient to determine the correspondence and trend of the measurements. It is important to keep in mind that the suspended mass must be large enough compared to the mass of the rope to obtain good experimental results.

6.2. Procedure

6.2.1. Assembly of Equipment

- **i.** With the rail and the control box identified, proceed to place the equipment on a horizontal table, verify that it is level (with the device located on the rail).
- **ii.** Identify output 1 of the control box, connect the USB cable from the control box to the sensor connector located on the rail pulley.
- **iii.** Take the power source to energize the control box through the 5 V input and connect it to 110 V.
- **iv.** Place the low-friction cart at the end opposite the pulley on the track and tie a string to one end. The other end must be free for mass support.
- **v.** Identify the mass support and measure its initial mass, keep the data.
- **vi.** Put a mass of 250 grams in the cart and measure the total mass of the cart, keep the value taken.
- **vii.** Pass the string through the pulley at the end of the track and hang the mass support from that end.

6.2.2. Experimental procedure

- **i.** The radius of the pulley used is $r =$ ___________ cm.
- **ii.** The mass of the pulley used is $M =$ ___________ g.

iii. Measure the mass of the mass support, this mass will be called m_1.

iv. Measure the mass of the cart with the washers, this mass will be called m_2.

v. For each of the data collection determine the total mass of the system, consider the masses m_1, m_2 y $M/2$.

vi. Position the cart on the opposite end of the pulley, make sure that the mass support is suspended in the air and has a considerable distance to the ground.

vii. To carry out the data collection, take the control box and follow the following steps:

- Verify that the title "Newton's Second Law" appears on the screen.
- Press the "SELECT" key to continue with the experiment, "Release the carriage" will appear on the screen and prevent the carriage from colliding with the data pulley. Take the control box and follow the steps below:
- Once the movement is finished, "t1 = xxxx" will appear on the screen, use the "UP" and "DOWN" keys to scroll through the 16 different times associated with the interaction of the pulley with the sensor.

viii. Press the "RESET" key to repeat the procedure.

ix. Repeat the above procedure for at least 4 different masses.

x. Determine the angular variation for each of the times obtained, for this, bear in mind that: $\theta_i = \frac{2\pi}{10} i$ where $i = 1, 2, 3, ..., 16$.

xi. With the data obtained in the previous numerals, make a table containing the angles (in radians) related to each of the time measurements (in seconds).

6.2.3. Analysis

i. Make a graph of the angle in radians as a function of time for each of the assemblies. Describe what the behavior of the graph is. Indicate that you can conclude from this behavior.

If in the previous numeral the graph obtained was not a straight line, perform a polynomial regression. What can you conclude from the results obtained? What physical meaning does this behavior have?

7. Conclusions

Finally, it is worth highlighting the importance of the development of prototypes for the experimental demonstration of fundamental physical laws, such as in this case Newton's Second Law, the developed prototype allows the hand of its guide, to experimentally verify the value of the acceleration of fall free. [6]

Likewise, it is possible to show that a large budget is not necessary for the development of projects such as the one shown in the article, where with a basic budget and the knowledge acquired in the academy, it is possible to contribute to the teaching of physics, which of one way or another impacts the interest of the current generation and encourages a taste for knowledge of basic sciences.

It is important to highlight that the DICOPED research group works continuously in the development of demonstration prototypes, strengthening the study of physics every day. Not only in the field of classical mechanics, if not in fields such as electricity and slightly more complex physical phenomena such as waves and light.

Recognitions

The authors wish to thank God, as well as the entire work team of the DICOPED research group, professors, graduates and students who in one way or another contributed to the design and construction of the prototype for the experimental demonstration of Newton's Second Law, in the same way to each of the people outside the university who collaborated with their knowledge and tools in the development of the necessary elements to carry out the project development.

References

[1] ANDALUSIAN SOCIETY OF MATHEMATICAL EDUCATION. Newton's Laws [online]. Available on the Web: http://thales.cica.es/rd/Recursos/rd98/Fisica/02/leyes.html

[2] LANDAU & LIFSHITZ. *Mechanics*. Ed. Reverté, Barcelona, 1991. ISBN 84-291-4081-6

[3] BOLIVARIAN REPUBLIC OF VENEZUELA, MINISTRY OF POPULAR POWER FOR HIGHER EDUCATION. Newton's Second Law of Rigid Body Dynamics [online]. Available on the Web:
http://nestormorenodr.blogspot.com.co/p/13-segunda-ley-de-newton-de-la-dinamica.html

[4] ESPINOSA CARD, Rubén Darío. *ANALOG ELECTRONIC DESIGN*. Manizales, 2009. 226 pp.

[5] ANDALUSIAN SOCIETY OF MATHEMATICAL EDUCATION. Newton's Laws [online]. Available on the Web: http://thales.cica.es/rd/Recursos/rd98/Fisica/02/leyes.html

[6] Design and Construction of Prototypes for Physics Experiments I. Between Science and Engineering. ISSN 1909-8367. Year 5 No. 9. 210 Pages. From page 186 to 199.

AUTHOR INDEX